ENCYCLOPAEDIA
OF
ALGAE

ENCYCLOPAEDIA
OF
ALGAE

Vol. 1

Edited by

A.M. Bagulia

ANMOL PUBLICATIONS PVT. LTD.

NEW DELHI - 110 002 (INDIA)

ANMOL PUBLICATIONS PVT. LTD.
H.O.: 4374/4B, Ansari Road, Darya Ganj,
New Delhi-110 002 (India)
Ph.: 23278000, 23261597
B.O.: No. 1015, Ist Main Road, BSK IIIrd Stage
IIIrd Phase, IIIrd Block,
Bangalore - 560 085 (India)
Visit us at: www.anmolpublications.com

Encyclopaedia of Algae

First Published, 2008

ISBN 978-81-261-3528-8 (Set)

PRINTED IN INDIA

Printed at Mehra Offset Press, Delhi.

Contents

Preface

The first plants on earth were algae and these still thrive in a range of aquatic habitats. The land plants have evolved from the algae, more specifically green algae. Some 400 million years ago freshwater, green, filamentous algae invaded the earth. Algae are used as fertilizer and energy-source (biodiesel), etc.

Algae encompass several groups of living aquatic organism that get light-energy through photosynthesis, using it for transforming inorganic substances into organic substances. Algae make the photosynthetic organism found in most habitats. These are of single-celled to multicellular complex species, such as 'Giant Kelps' that grow to 65 metres in length. Conventionally, these have been regarded as simple plants, in the study of Botany and referred to as Eukaryota. Chromophyta have bacterial characteristics and some authorities consider them to be Protists.

Protists are traditionally considered more animal-like. The prokaryotic forms, referred to as blue-green algae are only half-algae with a mixture of bacterial characteristics. But they are quite different from the bacteria and are referred to as Cynochloronta. All other forms are known as true Eukaryota algae, in the study of Botany. They have a nucleus enclosed within a membrane. The protoctists are defined as eukaryotic microorganism with the exception of animals and plants but including fungi and algae.

All algae have photosynthetic machinery derived from cyanobacteria, and produce oxygen as a by-product of photosynthesis, unlike non-cyanobacterial photosynthetic bacteria. It is estimated that algae produce about 73 to 87 per cent of the net global production of oxygen, available for respiration.

Most of the simpler algae are unicellular flagellates or amoeboids, but colonial and non-motile forms have developed independently among several groups. Some more organisational levels, more than one of which may occur in the life cycle of a specie, are: Colonial, Capsoid, Coccoid, Palmelloid, Filamentous, and Parenchymatous.

This eminent work presents perfect and realistic features of the algae in an interesting manner. Expectedly, this study would generate a separate discipline in the study of algae.

Editor

1

Introduction

Concept of Algae

There is no easy definition of an algae. Algae are generally microscopic organisms, are generally thought of as simple aquatic plants which do not have roots, stems or leaves and have primitive methods of reproduction. However, some algae display primitive animal features such as motility, while blue-green algae differ markedly from plants and all other algae, in that they have a cellular structure and function that is more common to bacteria than to the plant kingdom.

Algae live in a wide range of aquatic environments and are a natural component of most aquatic ecosystems. Additionally, a great many are also terrestrial, living in soil, snow, or in association with other organisms, especially fungi (as lichens), and animals. Aquatic algae are found in both fresh and marine waters. They range in size from large kelps (metres in length) to those visible only under a microscope.

Symptoms of Tracing Algae

Algae vary considerably in size, shape, and growth form. They can be:

- Single-celled

- Many celled - either colonially or as filaments of cells; or
- Elaborate plant bodies with differentiated cell types

Main habitat preferences are:

- Free floating in the water column (planktonic). These comprise the microscopic unicellular algae and colonial and filamentous algae, known as "phytoplankton".
- Growing as a film on rocks on the bottom (benthic) or on plants growing in the water (epiphytic). These may be single-celled or small colonial and filamentous species.
- Growing out into the water column but attached to a substrate at one point. These comprise the larger filamentous algae, and macroalgae (e.g. seaweeds).

Types of Algae

The main groups of algae found in Australian freshwater are:

- Green algae (Chlorophyceae)
- Diatoms (Bacillariophyceae)
- Euglenoids (Euglenineae)
- Blue-green algae (Cyanophyceae)

However, there are also other less common groups of algae that do occur in freshwater...

Green Algae

Green algae range in size from microscopic to large plants, and can be single-celled, colonial, or filamentous. Some of the single-celled and colonial green algae have small tails or "flagella" attached to each cell, which they use to swim with. However, many green algae are non-motile. Green algae may be either planktonic or attached.

They show the greatest diversity of shapes, sizes and species of any group of freshwater algae. Green chloroplasts are frequently observable within the cells of green algae when looked at under a microscope.

Blue-Green Algae

Blue-green algae or *Cyanobacteria* are microscopic cells that grow naturally in Australian fresh and salt waters. They are a type of bacteria, but in some ways act like plants by using sunlight to manufacture carbohydrates from carbon dioxide and water, a process know as photosynthesis. In doing do, they release oxygen. They grow in dams, rivers, creeks, reservoirs, lakes and even hot springs.

Blue-green Algae Produced Highly Potent Toxins

The main cause of concern about blue-green algae is the ability of some to produce highly potent toxins. There are four different forms of toxins that can be produced:

- *Hepatotoxins:* These attack the liver and other internal organs of the poisoned victim. Some have also been identified as cancer promoting substances.
- *Neurotoxins:* These act as neuromuscular blocking agents, leading to respiratory arrest.
- *Endotoxins:* These are contact irritants, and can cause severe dermatitis and conjunctivitis in people coming into contact with the algae through swimming or showering. They may also cause stomach cramps, nausea, fever and headaches if consumed. Their presence in airborne droplets can cause asthma. Some are also thought to be possible tumour promoters, although this has yet to be shown.
- *Non-specific Toxins:* These are relatively slow acting general toxins which progressively damage most organs, including the liver.

Stock Deaths

In Australia, no recorded human deaths have been attributed to blue-green algal toxins. However, many stock deaths have been documented. The toxins produced can persist in water for weeks. The toxins can also be concentrated by shellfish, which poses a potential health risk if they are consumed.

Toxic Blue-Green Algae

The four main toxic blue-green algae in Australia are:

- *Anabaena,*
- *Microcystis,*
- *Cylindrospermopsis, and*
- *Nodularia.*

Anabaena, and *Microcystis* are the two main bloom-forming genera in NSW waters. *Anabaena* forms long chains of cells, called a trichome, which sometimes grows in a spiral, depending on the species.

Microsystis aeruginosa is most common in lakes and reservoirs. It forms irregularly shaped colonies of cells up to 1 to 2 mm wide that can be visible to the naked eye. *Microsystis* blooms can be quite persistent lasting for months, or even years in some cases.

Nodularia often forms thinner scums than those of *Anabaena* and *Microcystis* blooms. *Nodularia,* like *Anabaena,* forms chains of cells or trichomes. Although it occurs in fresh waters, it is more common in brackish waters.

Cylindrospermopsis is commonly thought of as subtropical blue-green algae, but it also occurs in more temperate regions during the summer, including parts of New South Wales. It has very tiny cells that form chains or trichomes. It is a freshwater species, and causes problems in town water supply systems due to its highly potent toxins.

Irritant Blue-Green Algae

Not all blue-green algal species are toxic, and even different strains of the same species may differ, with some being highly toxic and others non-toxic.

All blue-green algae however, contain lipopolysaccharides, which act as contact irritants, Even if the other more potent blue-green algal toxins are not present, the presence of these contact irritants may make the water unsuitable for body contact or recreation if the blue-green algae are present in bloom proportions.

A number of other blue-green algae have been shown to be toxic overseas, but not yet so in Australia. Therefore, these too should be treated with caution when present in bloom proportions.

Causes of Blooms

High Nutrient Load: Blue-green algal blooms are natural phenomena and while it is not exactly clear what triggers a bloom, excess human sources of nutrients such as fertilizers and sewage certainly can increase the intensity of blooms (i.e. greater number of algae).

Thermal Stratification

One of the most important factors triggering blue-green algal blooms appears to be a lack of mixing of surface and deeper water layers in a river or reservoir. In lakes and reservoirs, mixing is mainly controlled by wind and temperature. Through the summer months the surface waters heat up resulting in a warmer top layer and cooler bottom layer which do not mix. In rivers, mixing is mainly caused by flow. Flows from headwaters can decrease or stop during drought conditions allowing thermal stratification to develop. Weirs and extraction of water for irrigation and stock watering also reduce flow in rivers.

Algae Floating to Surface

Some blue-green algae can float to the surface under these conditions having access to all the light in the top waters (photic region) and nutrients in the top and bottom waters. This allows the algae to flourish and bloom. Some other algae are motile and can swim to the photic region under these conditions.

Many other factors play a role in the formation of blue-green algal blooms including temperature, salinity, zooplankton grazing, pH and turbidity.

Development of Algae

The circumstances conducive for the growth of Algae are as follows:

- Ornamental ponds and Golf courses
- Stations & Mining operations
- Sewage treatment works, power
- Household water supplies
- Farm dams & Stock troughs
- Rice crops
- Aquaculture
- Swimming pools.

2

Significance of Algae

Algae encompass several groups of relatively simple living aquatic organisms that capture light energy through photosynthesis, using it to convert inorganic substances into organic matter.

Algae are photosynthetic organisms that occur in most habitats. Algae vary from small, single-celled species to complex multicellular species, such as the giant kelps that grow to 65 metres in length.

Although algae have conventionally been regarded as simple plants, they actually span more than one domain, including both Eukaryote and Bacteria, as well as more than one kingdom, including plants and protists, the latter being traditionally considered more animal-like. Thus, algae do not represent a single evolutionary direction or line but a level of organisation that may have developed several times in the early history of life on Earth.

Algae range from single-cell organisms to multicellular organisms, some with fairly complex differentiated form and (if marine) called seaweeds. All lack leaves, roots, flowers, seeds and other organ structures that characterise higher plants (vascular plants). They are distinguished from other protozoa in that they are photoautotrophic although this is not a hard and fast distinction as some groups contain members that are mixotrophic, deriving energy both from photosynthesis and uptake of organic carbon either by osmotrophy, myzotrophy, or phagotrophy. Some unicellular species rely entirely on external energy sources and have reduced or lost their photosynthetic apparatus.

All algae have photosynthetic machinery ultimately derived from the cyanobacteria, and so produce oxygen as a by-product of photosynthesis, unlike non-cyanobacterial photosynthetic bacteria. It is estimated that algae produce about 73 to 87 per cent of the net global production of oxygen, which is available to humans and other terrestrial animals for respiration.

Algae are usually found in damp places or bodies of water and thus are common in terrestrial as well as aquatic environments. However, terrestrial algae are usually rather inconspicuous and far more common in moist, tropical regions than dry ones, because algae lack vascular tissues and other adaptations to live on land. Algae can, however, endure dryness and other conditions in symbiosis with a fungus as lichen.

The various sorts of algae play significant roles in aquatic ecology. Microscopic forms that live suspended in the water column — called phytoplankton — provide the food base for most marine food chains. In very high densities (so-called algal blooms) these algae may discolour the water and outcompete or poison other life forms. Seaweeds grow mostly in shallow marine waters. Some are used as human food or harvested for useful substances such as agar or fertilizer.

The study of marine and freshwater algae is called phycology or algology.

The US Algal collection is represented by almost 300,000 accessioned and inventoried herbarium specimens.

Anyone who has walked on the slippery rocks of the seashore will be aware of organisms, which are recognisable as belonging to a group known as algae. A very different form of algae will be familiar to swimming pool owners. The seashore algae are large, slippery, and firmly attached to the rocks, while the swimming pool algae are small, often unicellular, and either float freely in the water colouring it green, or coat the sides of the pool with a green or brownish film. What common features do these organisms share that make them all algae, yet separate them from other living organisms? What types of organisms are recognised as algae? In what environments do algae occur?

Algae are all of the things cited in the previous slide, and grow in all of these places, but as a group they are notoriously difficult to define. Biologically, the name "algae" is given to a group of organisms of mixed affinity. The word itself has no taxonomic significance whatsoever. The algae do, however, share a number of features, which allow them to be treated as a group, although as a group with a very mixed evolutionary history. Algae are constructed fairly simply. They generally do not have vascular tissue, and they do not show the high level of organ differentiation of the familiar, more complex plants. It might be argued however that this distinction doesn't apply to the more advanced brown algae, which do have a certain degree of organ differentiation, and which even have a type of vascular tissue.

Algae have naked reproductive structures. This means that there are no protective layers of cells surrounding reproductive structures. None of the algae have reached even the level of organisation in reproductive structures shown by the archegoniate plants, such as the mosses and liverworts.

Most algae are photoautotrophic, which means that they can make their own food materials through photosynthesis by using sunlight, water and carbon dioxide. A few algae are not photoautotrophic, but they belong to groups which are usually autotrophs.

Incidentally, the word algae is plural so we say, for example, *Chlorella* and *Spirogyra* are algae. The singular form is algae so we say, for example, *Chlorella* is an algae. When used as an adjective, we say algal, as for example algal cells. In common with all plants, most algae contain chlorophyll-a, as well as various other photosynthetic pigments. This is true even though, strictly speaking, not all algae belong to the plant kingdom. Algae show a broad range of complexity, as might be expected in a group of organisms with such a mixed background. They range in complexity from tiny, microscopic forms, to very complex forms such as the kelps.

Algae include both prokaryotic and eukaryotic organisms. The Cyanobacteria (sometimes called Cyanophyta or blue-green algae) and a relatively recently discovered algal division called the Prochlorophyta are both prokaryotic divisions, while all other algae are eukaryotes.

The prokaryotic algae range only from unicells and colonies to the simplest of branched filaments.

Algae occur in virtually any habitat on earth as long as long as water is found there at some time, even if it is just moisture which might be present for a very short time.

Algae may be found as free-floating phytoplankton, which form the base of food webs in large water bodies.

There are also algae, which live, attached to rocks and other substrata at the bottom of bodies of water such as the sea. Algae may occur as epiphytes on higher plants, or on other algae.

All major bodies of water have algae in abundance, including lakes, small streams, large rivers, and even waterfalls. Algae occur in fresh water, to the saline water of the sea, and even in salt pans. There are also algae that thrive in the heated water of hot springs.

In the sea they may occur below the range of tidal exposure — in the subtidal zone.

Growing in the intertidal zone, algae are subjected to a number of stresses and disturbances. At low tide, they may bake in the sweltering sun or even get rained on by fresh water.

In some parts of the world, intertidal algae are even scoured by sea ice, yet they persist in living in this environment.

Those algae, which live attached to the bottom of a water body are called benthic algae, and the ecosystems of which they are part are referred to as benthos. Small, microscopic algae, which drift about in bodies of water, such as lakes and oceans, are called phytoplankton. Phytoplankton are important in freshwater and marine food webs, and are probably responsible for producing much of the oxygen that we breathe. Some forms of algae are even able to grow in Arctic and Antarctic sea ice, where they can be quite productive and support a whole associated food web. Some algae can even grow on the seabed, beneath a thick blanket of Arctic or Antarctic sea ice, even though they are in total darkness for a considerable part of the year. Algae are found in snow too. In some parts of the world, blooms of snow algae may paint the snow beds red in spring.

You may be surprised to find that algae even occur in the driest deserts. In some areas of the Namib Desert in Namibia, and the Richtersveld in South Africa, one often finds many quartz stones scattered about on the ground. Since quartz is quite translucent, the stones permit a considerable amount of light to pass through, so there is sufficient light for photosynthesis to take place underneath the stones.

A small amount of moisture may be retained in the soil under the quartz stones, so unicellular algae are able to grow underneath them.

There is even a unicellular green algae called *Prototheca* which causes disease in humans, although like this specimen, you have to be very ill already to get it. It produces skin lesions, mainly in patients whose immune systems have been damaged by other serious diseases. And of course there are the algae that enter into symbiosis with other organisms, for example, the symbiotic organisms that we call lichens.

The stony corals, which construct coral reefs in warm tropical seas are only able to build up these massive and beautiful structures because they have formed a symbiotic partnership with tiny single-celled algae called zooxanthellae.

The zooxanthellae which live in the tissues of the coral, share with it the organic products of their photosynthesis, as well as helping the coral with the construction of its limestone skeleton.

Even the chloroplast of land plants had its origin as a blue-green algae that lived within the cells of the ancestral organism. Such a special symbiotic relationship, where one organism lives inside the cells of another, is called endosymbiosis.

This fascinating group of organisms forms the basis for the science of phycology — the study of algae. Historically, algae were treated as belonging to the plant kingdom under a two kingdom system in which all living things were considered either plants or animals. More recently, it has been recognised that algae fall into several kingdoms. Some, such as the green algae, are plants. Others fall under the kingdom Protista. Prokaryotic algae belong to the kingdom Monera together with the bacteria, but this kingdom should probably be divided into more than one. It has even been suggested that the red algae should have a kingdom of their own. Thus it is easy to see that the 'algae' constitute an artificial grouping of organisms. They are dealt with together because of historical beliefs and for the sake of convenience.

Many phycologists, however, still treat the algae as though they were part of the plant kingdom, although recognising that the prokaryotic algae (the blue-green algae) belong with the bacteria. This is mostly so because university structures do not yet reflect the new systems of classification, and courses dealing with algae are usually offered as "Botany" courses.

Although algae may be treated as though they belonged to a single related group in spite of their differences, it must be recognised that they are polyphyletic. That is, there are many lines of evolution leading up to those organisms that are today called 'algae' and studied by phycologists.

The algae are polyphyletic because of the separate prokaryotic (blue-green algae and prochlorophytes) and eukaryotic (all other algae) lines of evolution. Within the eukaryotic algae there may also be many lines of evolution. For example, the red algae probably do not share a common eukaryotic ancestor with any of the other algae. That is why it has been suggested that they should have their own kingdom.

Thus it is seen that the algae are a diverse assemblage of unrelated organisms that are characterised by:

- Simple construction;
- Naked reproductive structures;
- *Photoautotrophic* nutrition;
- Chlorophyll-a.

Algae are a diverse group of organisms that cause mayhem even when professional systematists and biologists try to define them in relation to other organisms. There are sufficient morphological, physiological and ecological similarities between them to make the term 'algae' an acceptable one, yet it seems that the term 'algae' means different things to different people.

Through the years, algal classification has been very unstable. Placements at the kingdom and lower levels, are currently still dynamic. It is interesting to trace the changes in their position in the hierarchy of life.

2-Kingdom System

The plant and animal kingdoms have enjoyed a long tradition in formal taxonomy. Since the beginning of biology, organisms have been classified as either plants or animals because humans have developed the bias that only these two kingdoms exist in life. We live in a terrestrial, macroscopic world where we rarely encounter organisms that fail to fit neatly in the plant-animal dichotomy. Therefore observational and subjective characters were used for clear-cut classification. Important microscopic traits were overlooked, and as a result, weak criteria were used to justify placements of organisms as either plants or animals.

Algae, like the bacteria and fungi, were often assigned to the plant kingdom. Properties for including the algae into the plant kingdom were their ability to make their own food by photosynthesis, their structural similarity to land plants and the fact that the larger forms were observed to be sedentary. Eukaryotic unicellular organisms with chloroplasts were also called plants. Bacteria fell under the plant banner because of their rigid cell walls that behaved similarly to cellulose-type cell walls of land plants. Although fungi are not photosynthetic and have little in common structurally with plants, their inability to move about supported their classification as plants.

Unicellular creatures that move and ingest food protozoa — were called animals. Microbes such as *Euglena* that move but are photosynthetic were claimed by both botanists and zoologists, and showed up in the taxonomies of both the plant and animal kingdoms.

Clearly, the 2-Kingdom System was riddled with conceptual and operational problems. Technological advances, such as refined microscopy further proved that the system was oversimplified and therefore not widely acceptable.

5-Kingdom System

In 1969, Robert H Whittaker suggested a scheme with additional kingdoms as an alternative to the 2-Kingdom System. The 5 kingdoms were Monera, Protista, Plants, Animals and Fungi. The kingdoms fundamentally split the prokaryotes from the eukaryotes by placing the prokaryotes on their own as Kingdom Monera. It is still considered to be the most natural and unambiguous kingdom within this system, its boundary set by distinct prokaryotic organisation of the nucleus. The prokaryotes are bacteria, which include the Cyanophyta, the blue-green algae. Despite the prokaryotic organisation of the Cyanophyta, they possess unique photosynthetic pigments not present in the rest of the group, and they release free oxygen as a by-product of photosynthesis. Organisms of the other four kingdoms all consist of cells organised on the eukaryotic plan.

Kingdom Protista, otherwise called Proctotista, is the least homogenous group and also the hardest to define. It is something of a grab-bag containing all eukaryotes that do not fit the definitions of plants, animals and fungi. The plant, animal and fungi kingdoms are multicellular eukaryotes, each of these defined mainly by their nutritional modes. Some workers have severely criticised the use of these characters to form these groups, claiming that they defined adaptive traits rather than taxa, and were thus entirely unsuitable in revealing phylogenetic relationships.

Most algae are traditionally considered as a plant subkingdom within the 5-Kingdom classification. The diagnostic characters of the algal group as a whole were ill-defined, but nevertheless vastly different from the well-defined traits of the other two plant subkingdoms, namely the bryophytes and vascular land plants. Other biologists who were convinced that not all algae are plants revised the classification, preferring algae to be placed in Kingdom Protista, with only some multicellular phyla, particularly the Chlorophyta, Rhodophyta and Phaeophyta, remaining as plants. Then there were other biologists who regarded some of these multicellular forms to be placed in Kingdom Protista.

The general classification of groups within the algae is based partly upon the nature of the chlorophylls and accessory pigments present in the photosynthetic membranes. Additional clues to their taxonomic affinities include chloroplast structure, the chemistry of cell walls, number and position of flagellae, and the form of food reserves in cells. These properties are used to categorise algae and assess relationships amongst them. However, their phylogenetic relationships and evolutionary lineages are still not reflected adequately by the 5-kingdom system.

Major revisions were made to the 5-Kingdom System to

form the 8-Kingdom System. The final product was the culmination of the work by many biologists over a period of time.

8-Kingdom System

Not only is this scheme the most recently proposed, but also the most controversial. Its construction stresses that the term 'algae' wrongly tried to combine organisms with such diversity and differential evolutionary histories into a single group. Under the 8-Kingdom System, algal members are dispersed in vastly different kingdoms that are distinct from each other. Based on advanced genetic and ultrastructural studies, this system is generally supported by a handful of biologists who advocate the need to either elevate various algal members to a higher taxonomic status or to qualify them more distinctively. This has led to the reorganisation of many groups in such a way that their diagnostic characters suggest their evolutionary relationships to other groups more accurately.

The 8-Kingdom System was born from the phylogenetic hypothesis that the Archaebacteria diverged from the main eubacterial lineage at an early stage at the dawn of life. The Eubacteria which contains the Cyanophyta, then evolved linearly and only much later formed the oldest member of the Eukaryota, the Superkingdom Archeozoa. This superkingdom in turn evolved in a 1-way fashion to give rise to Superkingdom Metakaryota, the most basal member being Kingdom Protozoa. Also contained within the Metakaryota, are the four highest eukaryotic kingdoms derived as four distinct individual lineages from the Protozoa, namely Kingdoms Plantae, Animalia, Fungi and Chromista.

Archaezoa comprise only three phyla (Archaemoebae, Metamonada, Microsporidia), which differ from all other eukaryotes in lacking mitochondria, peroxosomes, Golgi dictyosomes and cisternae, and probably also in having 70S rather than 80S ribosomes. All Metakaryota have either mitochondria or hydrogenosomes, but never both. The ancestral metakaryote probably evolved by the simultaneous conversion of three different endosymbionts into chloroplasts, mitochondria and peroxisomes. The ancestral chromist probably evolved during the later stages of this symbiotic process by the symbiotic merging into a single cell of two different lineages of these early metakaryotes. Thus, chromists and metakaryotes both had a symbiotic origin, but eukaryotes did not.

Algal Groups

Cyanobacteria are aquatic and photosynthetic, that is, they live in the water, and can manufacture their own food. Because they are bacteria, they are quite small and usually unicellular, though they often grow in colonies large enough to see. They have the distinction of being the oldest known fossils, more than 3.5 billion years old, in fact! They are one of the largest and most important groups of bacteria on earth.

Many Proterozoic oil deposits are attributed to the activity of cyanobacteria. They are also important providers of nitrogen fertilizer in the cultivation of rice and beans. The cyanobacteria have also been tremendously important in shaping the course of evolution and ecological change throughout earth's history. The oxygen atmosphere that we depend on was generated by numerous cyanobacteria during the Archaean and Proterozoic Eras. Before that time, the atmosphere had a very different chemistry.

The other great contribution of the cyanobacteria is the origin of plants. The chloroplast with which plants make food for themselves is actually a cyanobacterium living within the plant's cells. Sometime in the late Proterozoic, or in the early Cambrian, cyanobacteria began to take up residence within certain eukaryote cells, making food for the eukaryote host in return for a home. This event is known as endosymbiosis, and is also the origin of the eukaryotic mitochondrion.

Because they are photosynthetic and aquatic, cyanobacteria are often called "blue-green algae". This name is convenient for talking about organisms in the water that make their own food, but does not reflect any relationship between the cyanobacteria and other organisms called algae. Cyanobacteria are relatives of the bacteria, not eukaryotes, and it is only the *chloroplast* in eukaryotic algae to which the cyanobacteria are related.

The cyanobacteria have an extensive fossil record. The oldest known fossils, in fact, are cyanobacteria from Archaean rocks of western Australia, dated 3.5 billion years old. This may be somewhat surprising, since the oldest rocks are only a little older: 3.8 billion years old!

Cyanobacteria are among the easiest microfossils to recognise. Morphologies in the group have remained much the same for billions of years, and they may leave chemical fossils behind as well, in the form of breakdown products from pigments. Small fossilised cyanobacteria have been extracted from Pre-cambrian rock, and studied through the use of SEM and TEM (scanning and transmission electron microscopy).

Many Proterozoic oil deposits are attributed to the activity of cyanobacteria, such as *Gloeocapsomorpha*. Small concentrically layered structures called pisolites are also the result of fossilised bacteria. Cyanobacteria are otherwise rarely preserved in rocks other than chert, though some possible blue-green bacteria have been recovered from shale.

At right is a layered stromatolite, produced by the activity of ancient cyanobacteria. The layers were produced as calcium carbonate precipitated over the growing mat of bacterial filaments; photosynthesis in the bacteria depleted carbon dioxide in the surrounding water, initiating the precipitation. The minerals, along with grains of sediment precipitating from the water, were then trapped within the sticky layer of mucilage that surrounds the bacterial colonies, which then continued to grow upwards through the sediment to form a new layer. As this process occurred over and over again, the layers of sediment were created. This process still occurs today; Shark Bay in western Australia is well known for the stromatolite "turfs" rising along its beaches.

In some cases, the stromatolites were infiltrated with a mineral-rich solution which fossilised the bacteria along with the layers, but more often only the layers are preserved.

The oldest stromatolites date to the Early Archaean, and they became abundant by the end of the Archaean. In the Proterozoic, stromatolites were widespread on earth, and were ecologically important as the first reefs. By the close of the Proterozoic, the abundance of stromatolites decreased markedly, though cyanobacteria continued to leave a fossil record, such as *Langiella* and *Kidstoniella* known from the Lower Devonian Rhynie chert.

The cyanobacteria have also been tremendously important in shaping the course of evolution and ecological change throughout earth's history. The oxygen atmosphere that we depend on was generated by numerous cyanobacteria photosynthesising during the Archaean and Proterozoic Era. Before that time, the atmosphere had a very different chemistry, unsuitable for life as we know it today.

Cyanobacteria, Photosynthetic

Though cyanobacteria do not have a great diversity of form, and though they are microscopic, they are rich in chemical diversity. Cyanobacteria get their name from the bluish pigment phycocyanin, which they use to capture light for photosynthesis. They also contain chlorophyll a, the same photosynthetic pigment that plants use. In fact the chloroplast in plants is a symbiotic cyanobacterium, taken up by a green algal ancestor of the plants sometime in the Pre-cambrian. However, not all "blue-green" bacteria are blue; some common forms are red or pink from the pigment phycoerythrin. These bacteria are often found growing on greenhouse glass, or around sinks and drains. The Red Sea gets its name from occasional blooms of a reddish species of *Oscillatoria*, and African flamingos get their pink colour from eating *Spirulina*.

Whatever their colour, cyanobacteria are photosynthetic, and so can manufacture their own food. This has caused them to be dubbed "blue-green algae", though they have no relationship to any of the various eukaryotic algae. The term "algae" merely refers to any aquatic organisms capable of photosynthesis, and so applies to several groups.

Cyanobacteria, Important in the Nitrogen Cycle

Cyanobacteria are very important organisms for the health and growth of many plants. They are one of very few groups of organisms that can convert inert atmospheric nitrogen into an organic form, such as nitrate or ammonia. It is these "fixed" forms of nitrogen which plants need for their growth, and must obtain from the soil. Fertilizers work the way they do in part because they contain additional fixed nitrogen which plants can then absorb through their roots.

Nitrification cannot occur in the presence of oxygen, so nitrogen is fixed in specialised cells called heterocysts. These cells have an especially thickened wall that contains an anaerobic environment.

Many plants, especially legumes, have formed symbiotic relations with nitrifying bacteria, providing specialised tissues in their roots or stems to house the bacteria, in return for organic nitrogen. This has been used to great advantage in the cultivation of rice, where the floating fern *Azolla* is actively distributed among the rice paddies. The fern houses colonies of the cyanobacterium *Anabaena* in its leaves, where it fixes nitrogen. The ferns then provide an inexpensive natural fertilizer and nitrogen source for the rice plants when they die at the end of the season.

Cyanobacteria also form symbiotic relationships with many fungi, forming complex symbiotic "organisms" known as lichens.

Nutritious or Poisonous?

The cyanobacterium *Spirulina* has long been valued as a food source; it is high in protein, and can be cultivated in ponds quite easily. In tropical countries, it may be a very important part of the diet, and was eaten regularly by the Aztecs; it is also served in several oriental dishes. In the US, the popularity of *Spirulina* is primarily as a "health food", being sold in stores as a dried powder or in tablet form.

Many other species of cyanobacteria produce populations that are toxic to humans and animals. Blue-green pond scums have been linked to the poisoning of cattle and dogs, and occasionally people. It is therefore not recommended that wild populations be gathered and eaten without some knowledge of the organisms involved.

Cyanobacteria may cause other problems as well; a species of *Lyngbya* is responsible for one of the skin irritations commonly known as "swimmer's itch."

Cyanobacteria are among the easiest microfossils to recognise. They are larger than other bacteria, and morphologies in the group have remained much the same for billions of years.

Like other bacteria, cyanobacteria have no nucleus or internal membrane systems. In many species, however, the external membrane has been folded to increase total surface area. The "packets" created in the cell membrane by these foldings are the surfaces across which the chemical reactions of photosynthesis take place.

All cyanobacteria are unicellular, though many grow in colonies or filaments, often surrounded by a gelatinous or mucilaginous sheath. *Nostoc*, pictured below at left, may produce spherical colonies as much as three or four centimetres in diameter. These are large enough to be mistaken for amphibian eggs. Fossilised cyanobacterial mounds are called stromatolites.

Despite the fact that they lack flagella, many cyanobacteria are capable of moving about. *Oscillatoria*, pictured above at right, gets its name from the "swaying" motion of its filaments when observed under the microscope. No one has yet determined how these filaments are able to do this.

Red Algae

Red algae are red because of the presence of the pigment phycoerythrin; this pigment reflects red light and absorbs blue light. Because blue light penetrates water to a greater depth than light of longer wavelengths, these pigments allow red algae to photosynthesise and live at somewhat greater depths than most other "algae". Some rhodophytes have very little phycoerythrin, and may appear green or bluish from the chlorophyll and other pigments present in them.

In Asia, rhodophytes are important sources of food, such as nori. The high vitamin and protein content of this food makes it attractive, as does the relative simplicity of cultivation, which began in Japan more than 300 years ago.

Some rhodophytes are also important in the formation of tropical reefs, an activity with which they have been involved for millions of years; in some Pacific atolls, red algae have contributed far more to reef structure than other organisms, even more than corals. These reef-building rhodophytes are called coralline algae, because they secrete a hard shell of carbonate around themselves, in much the same way that corals do.

Coralline algae are important carbonate producers in present-day and fossil environments and are supposed to be valuable facies indicators.

However, corallines are a highly complicated group. The identification of fossil coralline algae bears several problems and the intraspecific variability is very high. Their distribution is very wide spread: Corallines occur from the tropics to polar regions, and from the intertidal down to more than 200 m water depth.

Encrusting coralline algae (Families Sporolithaceae and Corallinaceae, belonging to the Rhodophyta) are characterised by calcified cell walls forming a rigid skeleton. Living coralline algae have a red thallus surface ("thallus" = plant body), which is caused by the red pigments (therefore, they belong to the Red Algae).

Coralline algae are one of the most abundant carbonate producers. Encrusting corallines are expected to be important binders and encrusters in present-day and fossil reefs.

- In addition to their binding function in coral reefs, encrusting coralline algae may also form "algal reefs".
- However, not all of the taxa need a stable substrate. Coralline algae may also form unattached nodules (= rhodoliths) and their branches may form thick sediment accumulations (= maerl).
- Erect forms, composed of flexible branches (geniculate coralline algae).
- Encrusting forms (non-geniculate, encrusting coralline algae).

3

Classification of Algae

Cyanobacteria

Cyanobacteria (blue + bacterium) also known as Cyanophyta is a phylum (or "division") of Bacteria that obtain their energy through photosynthesis. They are often still referred to as blue-green algae, although they are in fact prokaryotes. The description is primarily used to reflect their appearance and ecological role rather than their evolutionary lineage. The name Cyanobacteria comes from the colour of the bacteria, Cyan, the bacteria do not use or produce Cyanide whose chemical prefix is cyano. Putative fossil traces of cyanobacteria have been found from around 3.8 billion years ago (b.y.a.) They are a significant component of the marine nitrogen cycle and an important primary producer in many areas of the ocean. Their ability to perform oxygenic (plant-like) photosynthesis is thought to have converted the early reducing atmosphere into an oxidising one, which dramatically changed the life forms on Earth and provoked an explosion of biodiversity.

Forms

Cyanobacteria are found in almost every conceivable habitat, from oceans to fresh water to bare rock to soil. Most are found in fresh water, while others are marine, occur in damp soil, or even temporarily moistened rocks in deserts. A few are endosymbionts in lichens, plants, various protists, or sponges and provide energy for the host. Some live in the fur of sloths, providing a form of camouflage.

Cyanobacteria include unicellular and colonial species. Colonies may form filaments, sheets or even hollow balls. Some filamentous colonies show the ability to differentiate into several different cell types: vegetative cells, the normal, photosynthetic cells that are formed under favourable growing conditions; akinetes, the climate-resistant spores that may form when environmental conditions become harsh; and thick-walled heterocysts, which contain the enzyme nitrogenase, vital for nitrogen fixation.

Heterocysts may also form under the appropriate environmental conditions (anoxic) wherever nitrogen is necessary. Heterocyst-forming species are specialised for nitrogen fixation and are able to fix nitrogen gas, which cannot be used by plants, into ammonia (NH_3), nitrites (NO_2^-) or nitrates (NO_3^-), which can be absorbed by plants and converted to protein and nucleic acids. The rice paddies of Asia, which produce about 75 per cent of the world's rice, could not do so were it not for healthy populations of nitrogen-fixing cyanobacteria in the rice paddy fertilizer.

Many cyanobacteria also form motile filaments, called hormogonia, that travel away from the main biomass to bud and form new colonies elsewhere. The cells in a hormogonium are often thinner than in the vegetative state, and the cells on either end of the motile chain may be tapered. In order to break away from the parent colony, a hormogonium often must tear apart a weaker cell in a filament, called a necridium.

Each individual cell of a cyanobacterium typically has a thick, gelatinous cell wall, which stains gram-negative. They lack flagella, but hormogonia and some unicellular species may move about by gliding along surfaces. In water columns some cyanobacteria float by forming gas vesicles, like in archaea.

Photosynthesis

Cyanobacteria have an elaborate and highly organised system of internal membranes, which function in photosynthesis. Photosynthesis in cyanobacteria generally uses water as an electron donor and produces oxygen as a by-product, though some may also use hydrogen sulphide as occurs among other photosynthetic bacteria. Carbon dioxide is reduced to form carbohydrates via the Calvin cycle. In most forms the photosynthetic machinery is embedded into folds of the cell membrane, called thylakoids.

The large amounts of oxygen in the atmosphere are considered to have been first created by the activities of ancient cyanobacteria. Due to their ability to fix nitrogen in aerobic conditions they are often found as symbionts with a number of other groups of organisms such as fungi (lichens), corals, pteridophytes (Azolla), angiosperms (Gunnera) etc.

Cyanobacteria are the only group of organisms that are able to reduce nitrogen and carbon in aerobic conditions, a fact that may be responsible for their evolutionary and ecological success. The water-oxidising photosynthesis is accomplished by coupling the activity of photosystem (PS) II and I (Z-scheme). In anaerobic conditions, they are also able to use only PS I — cyclic photophosphorylation — with electron donors other than water (hydrogen sulphide, thiosulphate, or even molecular hydrogen) just like purple photosynthetic bacteria.

Furthermore, they share an archaebacterial property, which is the ability to reduce elemental sulphur by anaerobic respiration in the dark. Perhaps the most intriguing thing about these organisms is that their photosynthetic electron transport shares the same compartment as the components of respiratory electron transport. Actually, their plasma membrane contains only components of the respiratory chain, while the thylakoid membrane hosts both respiratory and photosynthetic electron transport.

Attached to thylakoid membrane, phycobilisomes act as light harvesting antennae for the photosystems. The phycobilisome components (phycobiliproteins) are responsible for the blue-green pigmentation of most cyanobacteria. The variations to this theme is mainly due to carotenoids and phycoerythrins, which give the cells the red-brownish colouration. In some cyanobacteria, the colour of light influences the composition of phycobilisomes. In green light, the cells accumulate more phycoerythrin, whereas in red light they produce more phycocyanin. Thus the bacteria appear green in red light and red in green light. This process is known as complementary chromatic adaptation and is a way for the cells to maximise the use of available light for photosynthesis.

A few genera, however, lack phycobilisomes and have chlorophyll *b* instead (Prochloron, Prochlorococcus, Prochlorothrix). These were originally grouped together as the prochlorophytes or chloroxybacteria, but appear to have developed in several different lines of cyanobacteria. For this reason they are now considered as part of cyanobacterial group.

Relationship to Chloroplasts

Chloroplasts found in eukaryotes (algae and higher plants) likely evolved from an endosymbiotic relation with cyanobacteria. This endosymbiotic theory is supported by various structural and genetic similarities. Primary chloroplasts are found among the green plants, where they contain chlorophyll *b*, and among the red algae and glaucophytes, where they contain phycobilins. It now appears that these chloroplasts probably had a single origin, in an ancestor of the clade called Primoplantae. Other algae likely took their chloroplasts from these forms by secondary endosymbiosis or ingestion.

It was once thought that the mitochondria in eukaryotes also developed from an endosymbiotic relationship with cyanobacteria. Mitochondria are believed to have originated not from cyanobacteria but from an ancestor of Rickettsia.

Classification

The cyanobacteria were traditionally classified by morphology into five sections, referred to by the numerals I-V. The first three - Chroococcales, Pleurocapsales, and Oscillatoriales - are not supported by phylogenetic studies. However, the latter two - Nostocales and Stigonematales - are monophyletic, and make up the heterocystous cyanobacteria. The members of Chroococales are unicellular and usually aggregated in colonies. The classic taxonomic criterion has been the cell morphology and the plane of cell division. In Pleurocapsales, the cells have the ability to form internal spores (baeocytes). The rest of the sections include filamentous species. In Oscillatoriales, the cells are uniseriately arranged and do not form specialised cells (akinets and heterocysts). In Nostocalles and Stigonematalles the cells have the ability to develop heterocysts in certain conditions. Stigonematales, unlike Nostocalles include species with truly branched trichome. Most taxa included in the phylum or division Cyanobacteria have not yet been validly published under the Bacteriological Code. Except:

- The classes Chroobacteria, Hormogoneae and Gloeobacteria.
- The orders Chroococcales, Gloeobacterales, Nostocales, Oscillatoriales, Pleurocapsales and Stigonematales.
- The families Prochloraceae and Prochlorotrichaceae.
- The genera Halospirulina, Planktothricoides, Prochlorococcus, Prochloron, Prochlorothrix.

Biotechnology and Applications

Certain cyanobacteria produce cyanotoxins like Anatoxin-a, Anatoxin-as, Aplysiatoxin, Cylindrospermopsin, Domoic acid, Microcystin LR, Nodularin R (from Nodularia), or Saxitoxin. Sometimes a mass-reproduction of cyanobacteria results in algal blooms.

The unicellular cyanobacterium *Synechocystis* sp. PCC 6803 was the first photosynthetic organism whose genome was completely sequenced (in 1996, by the Kazusa Research Institute, Japan). It continues to be an important model organism.

At least one secondary metabolite, cyanovirin, has shown to possess anti-HIV activity.

Some cyanobacteria are sold as food, notably *Aphanizomenon flosaquae* and *Arthrospira platensis* (Spirulina). It has been suggested that they could be a much more substantial part of human food supplies, as a kind of superfood.

Along with algae, some hydrogen producing cyanobacteria are being considered as an alternative energy source, notably at Oregon State University, in research supported by the US Department of Energy, Princeton University, Colorado School of Mines as well as at Uppsala University, Sweden.

Health Risks

Some species of cyanobacteria produce neurotoxins, hepatotoxins, cytotoxins, and endotoxins, making them dangerous to animals and humans. Several cases of human poisoning have been documented but a lack of knowledge prevents an accurate assessment of the risks.

Prokaryotic Algae

Traditionally the Cyanobacteria have been included among the algae, referred to as the *cyanophytes* or *blue-green algae*, (the term "algae" refers to any aquatic organisms capable of photosynthesis) though some recent treatises on algae specifically exclude them. Cyanobacteria are some of the oldest organisms to appear in the fossil record dating back to the Precambrian, possibly as far as about 3.5 billion years. Ancient cyanobacteria likely produced much of the oxygen in the Earth's atmosphere.

Cyanobacteria can be unicellular, colonial, or filamentous. They have a prokaryotic cell structure typical of bacteria and conduct photosynthesis on specialised cytoplasmic membranes called thylakoid membranes, rather than in organelles. Some filamentous blue-green algae have specialised cells, termed heterocysts, in which nitrogen fixation occurs.

Eukaryotic Algae

All other algae are eukaryotes and conduct photosynthesis within membrane-bound structures (organelles) called chloroplasts. Chloroplasts contain DNA and are similar in structure to cyanobacteria, presumably representing reduced cyanobacterial endosymbionts. The exact nature of the chloroplasts is different among the different lines of algae, reflecting different endosymbiotic events. Note many of these groups contain some members that are no longer photosynthetic. Some retain plastids, but not chloroplasts, while others have lost them entirely.

Green Algae

Green algae are microscopic protists; found in all aquatic environments, including marine, freshwater and brackish water.

The green algae (singular: *green algae*) are the large group of algae from which the embryophytes (higher plants) emerged. As such, they form a paraphyletic group, variously included among the Plantae or with the Protista. The green algae include unicellular and colonial flagellates, usually but not always with two flagella per cell, as well as various colonial, coccoid, and filamentous forms. In the Charales, the closest relatives of higher plants, full differentiaıion of tissues occurs. There are about 1000 — 2000 species of green algae. Many species live most of their lives as single-cells, other species form colonies or long filaments.

A few other organisms rely on green algae to conduct photosynthesis for them. The chloroplasts in euglenids and chlororachniophytes were presumably acquired from ingested green algae, and in the latter retain a vestigial nucleus (nucleomorph). Some species of green algae, particularly of genera Trebouxia or Pseudo-trebouxia (Trebouxiophyceae), can be found in symbiotic associations with fungi to form lichens. In general the fungal species that partner in lichens cannot live on their own, while the algal species is often found living in nature without the fungus.

Cellular Structure

Almost all forms have chloroplasts. These contain chlorophylls *a* and *b*, giving them a bright green colour (as well as the accessory pigments beta carotene and xanthophylls), and have stacked thylakoids.

All green algae have mitochondria with flat cristae. When present, flagella are typically anchored by a cross-shaped system of microtubules, but these are absent among the higher plants and charophytes. Flagella are used to move the organism. Green algae usually have cell walls containing cellulose, and undergo open mitosis without centrioles.

Origins

The chloroplasts of green algae are bound by a double membrane, so presumably they were acquired by direct endosymbiosis of cyanobacteria. A number of cyanobacteria show similar pigmentation, but this appears to have arisen more than once, and the chloroplasts of green algae are no longer considered closely related to such forms. Instead, the green algae probably share a common origin with the red algae.

Classification

Green algae are often classified with their embryophyte descendants in the green plant clade Viridiplantae (or Chlorobionta). Viridiplantae, together with red algae and glaucophyte algae, form the supergroup Primoplantae, also known as Archaeplastida or Plantae *sensu lato*.

- Chlorophyta:
  - — Chlorophyceae,
  - — Ulvophyceae, and
  - — Trebouxiophyceae.

- Chlorokybales,
- Klebsormidiales,
- Zygnematales,
- Desmidiales,
- Coleochaetales, and
- Charales (stoneworts).

The orders outside the Chlorophyta are often grouped as the division Charophyta, which is paraphyletic to higher plants, together comprising the Streptophyta. Sometimes the Charophyta is restricted to the Charales, and a division Gamophyta is introduced for the Zygnematales and Desmidiales. In older systems the Chlorophyta may be taken to include all the green algae, but taken as above they appear to form a monophyletic group.

Reproduction

Green algae are eukaryotic organisms. Sexual reproduction varies from fusion of identical cells (isogamy) to fertilization of a large non-motile cell by a smaller motile one (oogamy). However, these traits show some variation, most notably among the basal green algae, called prasinophytes.

Haploid algae cells (containing only one copy of their DNA) can fuse with other haploid cells to form diploid zygotes. When filamentous algae do this, they form bridges between cells, and leave empty cell walls behind that can be easily distinguished under the light microscope. This process is called *conjugation*.

The species of *Ulva* are reproductively isomorphic, the diploid vegetative phase is the site of meiosis and releases haploid zoospores which germinate and grow producing a haploid phase alternating with the vegetative phase.

Green Algae and the Origin of Land Plants

Twenty years ago, a relatively slim volume with chapters by leading chlorophycologists celebrated the systematics of green algae (Irvine and John, 1984), a field that was undergoing rapid and fascinating changes, both in content and theory. "The present period may be termed the 'Age of Ultrastructure' in green algal systematics," wrote Frank Round in the introductory chapter, which summarised the history and state of the art. Round (1984) argued that light microscopy had laid the foundation in the preceding two centuries, but that the foundation was largely descriptive—alpha taxonomy in the most restricted sense. Ultrastructure, he asserted, had enlarged and presumably would continue to expand our horizons to unify systematics of green algae and overcome the fragmented alpha taxonomy that had dominated the field. Little did Round know that this golden age of green algal systematics was about to go platinum. Molecular systematics, in concert with a rigorous theoretical approach to data analysis and hypothesis testing (Theriot, 1992; Swofford et al., 1996), would at first complement and then transform the age of ultrastructure and usher in the "Age of Molecules."

Deconstructing Hypotheses of Relationships of the Green Algae and Land Plants

A link between green algae and land plants has been clear to biologists for centuries, since before Darwin and the advent of evolutionary thinking and phylogenetics. Recent new data on morphology, genes, and genomes, as well as new ways of analysing and synthesising information, are only the most recent in a long history of change in our understanding of these so-called "primitive" plants. This review focuses primarily on research that has led to both some radical restructuring of the classification of algae and some satisfying confirmations of the careful observations of earlier workers.

First and foremost, green algae, the division Chlorophyta of Smith (1950), are undoubtedly monophyletic with embryophyte green plants, although the Chlorophyta in this sense is paraphyletic (Mattox and Stewart, 1984; Mishler and Churchill, 1985; McCourt, 1995). Embryophytes (land plants; bryophytes and vascular plants) are clearly descended from green algal-like ancestors, but the sister of the embryophytes includes only a few green algae. The remainder of Chlorophyta constitutes a monophyletic group. This major bifurcation in green plant evolution implies a single common ancestor to the two lineages, but, given the diversity of unicellular green algae and our growing understanding of them, there may be additional lineages outside this major bifurcation.

Green Algae?

The term *algae* is not phylogenetically meaningful without qualifiers. Algae in general and green algae in particular are difficult to define to the exclusion of other phylogenetically related organisms that are not algae. This difficulty is a reflection of recent data on algae as well as the way phylogenetic thinking has permeated classification. Green algae are photosynthetic eukaryotes bearing double membrane-bound plastids containing chlorophyll *a* and *b*, accessory pigments found in embryophytes (beta carotene and xanthophylls), and a unique stellate structure linking nine pairs of microtubules in the flagellar base. Starch is stored inside the plastid and cell walls when present are usually composed of cellulose.

The plastids of green algae are descended from a common prokaryotic ancestor, for which descendants are endosymbiotic in the host cells of a number of other eukaryotic lineages. These plastids are termed primary, i.e., derived directly from a free-living prokaryotic ancestor, although a secondary origin has been proposed. Plastid-bearing lineages permeate all of the other major clades of algae.

Green Algal Diversity

Mostly microscopic and rarely more than a metre in greatest dimension, the green algae make up for their lack in size with diversity of growth habit and fine details of their cellular architecture. Body (thallus) size and habit ranges from microscopic swimming or non-motile forms (e.g., nanoplankton, benthos, or lichen phycobionts) to macroscopic (benthic attached forms). Thallus structure runs the gamut of complexity, from swimming and non-motile unicells, to filaments, colonies, and various levels of tissue organisation (pseudo-parenchymatous, parenchymatous, or thalloid) and branching morphologies. Unicells are spherical to elongate, with or without flagella, scales, and wall layers or other coverings (e.g., loricas). Filaments generally exhibit cylindrical cells arranged end-to-end, although chains of irregularly shaped cells are known. Unbranched (*Oedogonium*) and branching (*Draparnaldia*) forms are known, and many branching forms have attenuated terminal filament tips (*Chaetophora*). Colonies of various sizes occur, from pairs of cells (*Euastropsis*) to thousands (*Hydrodiction*). Cells in colonies may be joined by gelatinous strands or share a common parental wall. Colonies range in form from small sarcinoid packets (nonlinear clusters of cells; *Chlorokybus*) to aggregates of thousands of swimming cells (*Volvox*). Branching forms may be simple bifurcating or reticulating networks of filaments, but a few achieve a complexity that can be called tissuelike (*Nitella*). Cells may be uninucleate or coenocytic, in which many nuclei are dispersed throughout the cytoplasm of so-called giant cells (*Caulerpa*).

The systematics of green algae, and algae in general, has been driven by observational tools due to the apparent simplicity of the organisms to the naked eye and due to the small size and cryptic features of the organisms. Linnaeus recognised only five artificially inclusive genera (*Tremella, Fucus, Ulva, Conferva, Corallina*), plus *Chara* and *Volvox*, the last as an animal-like plant (Prescott, 1951). Light microscopy opened the windows on the microstructures of algae and provided the bulk of observational data for a long time. As described earlier, electron microscopy and molecular data have revolutionised our understanding of green algal phylogeny, which is reflected in modern classification.

Previous green algal taxonomy had grouped organisms based on growth habit, and several lineages were inferred by arranging taxa according to evolutionary "tendencies". Motile ancestral green algal unicells were hypothesised to have given rise to distinct lines of increasing size and complexity, with each exhibiting a variation on a theme. One line resulted in motile colonies, another in non-motile branching thalli or colonies (with motile reproductive cells retained in the life cycle), and a third in large non-motile coenocytic forms.

The view that the cellular features involved in the vital processes of cell division and swimming (of gametes or asexual zoospores) would be highly conserved evolutionarily led to numerous comparative studies targeting the mitotic, cytokinetic, and swimming apparatus of the cell. The flagellar apparatus, with its flagellar basal bodies and axonemes and rootlets of microtubules, has been painstakingly compared across a large number of green algae. To a lesser extent, plastid structure has been important for diagnosis of some groups. This focus came about through research by early workers that revealed evolutionarily conservative characters that cut across misleading convergence in vegetative morphology. In the age of molecular systematics, we are evaluating hypotheses formulated from comparative ultrastructure studies over the last 30 years, and adding new hypotheses as well.

Ultrastructural work showed that filamentousness, coloniality, coccoid habit (nonmotile, lacking flagella), and many other vegetative features evolved numerous times and were generally unreliable as characters marking monophyletic groups. The simple sequencing of forms, flagellate coccoid unbranched filament branched filament tissuelike, may have occurred repeatedly in many lineages.

Close on the heels of this first wave of new data and new methods of analysis employing a phylogenetic approach came a molecular revolution. By and large, the refutation of the classical tendencies as organising principals in algal evolution and classification was confirmed, and a new dogma arose, which has been refined and enlarged but retains the major scaffolding erected by ultrastructure and biochemistry.

Molecular studies in green algae have been largely driven by slightly earlier studies in embryophytes and, to a lesser extent, cyanobacteria. This is due largely to the development of primers for many genes that were first studied in embryophytes. The green algae lineage (eukaryotes containing primary green plastids) originated as much as 1500 million years ago, and the divergence of land plants occurred perhaps 700 Mya or more likely 425–490 Mya. Despite the antiquity of a shared common ancestor of land plants and their nearest relatives, ribosomal DNA and many plastid genes are recognisable as homologs in green plants and algae. As a result, molecular phylogenetics of green algae expanded rapidly as methodologies and approaches were transferred to algal taxa.

The complete nuclear genome for *Chlamydomonas reinhardtii* was released in early 2004, and annotation is an ongoing process (E. Harris, Duke University, personal communication). Organellar genomes to date include five plastid and 10 mitochondrial genomes, and several other green algal plastid or mitochondrial genomes should soon be published (C. Lemieux, Lavall University, personal communication). Several complete genome sequencing projects are also underway (A. Grossman, Stanford University; B. Palenik, Scripps Institution of Oceanography, personal communications).

With the advent of molecular data has come the possibility of finding characters that transcend rampant morphological homoplasy revealed by ultrastructure. However, a recurrent theme of many molecular studies has been a pattern of relatively well-resolved distal branches of a phylogenetic tree with weak or poor support for the internal or deeper divergences. Chapman et al. (1998) blamed this result on the limitations of sequence data and the antiquity of green algae: a stochastically changing molecule cannot be expected to retain enough signal to resolve events that happened nearly simultaneously in the ancient past. We suspect that a major reason is that most studies have employed only one or two genes and that more data can resolve these major ambiguities.

The green algae have evolved in two major lineages. One, which we refer to as the chlorophyte clade, includes the majority of what have been traditionally called green algae—flagellate green unicells and colonies (e.g., *Chlamydomonas, Volvox*), filamentous branched (e.g., *Chaetomorpha, Cladophora*) and unbranched forms (e.g., *Oedogonium*), green seaweeds (e.g., *Ulva, Codium*), many soil algae (e.g., *Chlorella*), terrestrial epiphytes (e.g., *Trentopohlia*), and many phycobionts (e.g., *Trebouxia*). The other lineage, the charophyte clade, contains a smaller number of green algal taxa, although some (e.g., *Spirogyra, Chara*) are widespread and as familiar as any algae. Charophyte algal thalli include swimming unicells, filaments (branched and unbranched), ornate unicells, and fairly complex forms that have been called parenchymatous. Charophyte algae are found in fresh water, with a few ranging into brackish habitats, and several groups live in soils, crusts, and other aerial settings. A third group of taxa, called prasinophytes, consists of "primitive" — appearing unicells of uncertain affinity—that is these taxa are probably members of one of the two main clades or more likely are representatives of other early-diverging clades.

Within the chlorophyte clade are three well-supported groups: chlorophytes, trebouxiophytes, and ulvophytes. The charophyte clade comprises at least five small but distinct groups of green algae leading to a highly diverse clade of land plants.

Classification of Green Algae

For the purposes of this review, we will use the classification, which gives division, class, and order names of major groups with informal names in parentheses. This is not intended to be a definitive taxonomic revision of green algal classification, but we anticipate that such a revision will incorporate the basic scheme used here (C. F. Delwiche, University of Maryland, personal communication). The use of some terms or prefixes (e.g., charo-) is inevitably confusing because of the historical claim that such terms have on us. In other cases, paraphyly of traditional classes, orders, families, genera, or even species, makes classification difficult.

Chlorophyte Clade

This clade contains three major groups and the majority of described species of green algae: chlorophytes, trebouxiophytes, and ulvophytes. Mattox and Stewart (1984) assigned the groups class-level rank, i.e., Chlorophyceae, Pleurastrophyceae (now at least in part Trebouxiophyceae), and Ulvophyceae. All members of this clade have swimming cells with two or four anterior flagella. Within the cell, the flagellar basal bodies are associated with four microtubular rootlets. All three groups share the character of cruciately arranged rootlets that alternate between two and higher numbers of microtubules (X-2-X-2). From group to group, differences in the offset from this cruciate pattern are seen, and arrangements are categorised according to relative positioning. When viewed from above the cell, the basal bodies and rootlets can have a perfect cruciate pattern (i.e., with basal bodies directly opposed, DO) or they are offset in a counterclockwise (CCW) or clockwise (CW) position. The ancestral condition has been inferred to be a CCW offset, with CW and DO the derived states (chlorophytes).

Molecular work, primarily on the small subunit of ribosomal DNA (18S rDNA) has strongly supported the monophyly of this triad of green algal groups and shows the ulvophytes as sister to a clade containing chlorophytes and trebouxiophytes. A combined analysis of morphology and 18S rDNA data strongly supported the monophyly of the three groups. Later studies provided more evidence for this topology, often with high support, while greatly increasing taxon sampling.

Charophyte Clade

This group contains a number of green algae plus a large number of what are considered to be the mostly highly derived green autotrophs, the land plants. Nomenclaturally, the group has led a confused life. Mattox and Stewart (1984) placed the algae in this group in Charophyceae, although the exclusion of the land plants made this taxonomic arrangement paraphyletic. Graham and Wilcox (2000a) acknowledged the paraphyly of the group and used the term "charophyceans" to refer to them. Bremer et al. (1987) assigned the division name Streptophyta to the green algae plus land plants, although Jeffrey (1982) had used this name more restrictedly, including only stoneworts (Charales) and embryophytes (archegoniate land plants). We will refer to the green algal groups of the charophyte clade as charophyte algae. The clade (charophyte + embryophytes) is characterised by biflagellate cells (when motile cells are present), with asymmetrically inserted flagella and two dissimilar flagellar roots (including a multilayered structure, or MLS, and a smaller root), persistent mitotic spindles, open mitosis, and several enzyme systems not found in other green algae.

Prasinophytes

Prasinophyte algae have received attention recently because they are some of the important bloom-forming marine planktonic algaethat are known mixotrophs and can account for a significant component of biomass in marine planktonic systems. Prasinophyceae also occupy a critical position at the base of the green algal tree of life and are viewed as the form of cell most closely representing the first green algae, or "ancestral green flagellate" (AGF). One of the first green algae to be examined using transmission electron microscopy was a member of class Prasinophyceae. Prasinophyceae was proposed to segregate flagellate unicellular green algae with organic body scales from the other green flagellates.

Members of Prasinophyceae have diverse morphologies, ranging from bean- to star-shaped cells with one to eight flagella often inserted in a flagellar pit and up to seven distinct types of organic scales. Recently, coccoid forms have been added to this group (Fawley et al., 2000). They occur in marine and brackish water, although some members live in freshwater habitats (e.g., *Pedinomonas*, *Mesostigma*). Prasinophyte algae are among the smallest of the eukaryotic planktonic marine flagellates (Zignone et al., 2002). Sexual reproduction has only been demonstrated in *Nephroselmis olivacea* (Suda et al., 1989).

Scale morphology has been used to differentiate the major groups of prasinophytes (Norris, 1980; Melkonian, 1984; Moestrup, 1984). In all but a few genera, organic scales are produced in the Golgi apparatus and coat the body and flagella. Some taxa possess up to seven distinct scale types, but others have a single type of scale. Flagellar behaviour and morphology also reflect a tremendous diversity in this group. Some prasinophyte taxa have cruciately arranged rootlets, but others have asymmetrical rootlets. Some cells push with undulating flagella, and others swim with flagella forward.

The number of rootlets varies between two and four and can include associated system I (SMAC) and system II (rhizoplast) fibres. *Pyramimonas octopus* has eight flagella and at least 60 flagellar structures connecting the basal bodies (Moestrup and Hori, 1989). Multilayered structures (MLSs) are found in flagellate sperm cells of embryophytes and occur in *Mesostigma*, a putative charophyte algae, and Pyramimonadales but are absent from all other groups of prasinophytes and all other taxa in the chlorophyte clade (assuming that the MLS-like structures of Trentopohliales are not homologous; Chapman, 1984). MLSs have also been described in dinoflagellates and jakobids (Wilcox, 1989; O'Kelly, 1992), and if these are homologous to those in green plants, then the presence of an MLS can be interpreted as the ancestral condition in the green algae. The structure of the flagellar transition region has been used to distinguish among some of the orders and provides evidence for the relationship between *Mesostigma* and charophytes (Melkonian, 1984).

Characteristic flagellar hairs can also be used to distinguish among the main groups (Marin and Melkonian, 1994). Lateral flagellar hairs occur in all studied members, but tip hairs do not occur on flagella of Chlorodendrales and *Nephroselmis*. Ultrastructural differences in the manner of mitosis and cytokinesis are also apparent. Most members have persistent telophase spindles, the exception being members of Chlorodendrales.

The diversity in cell shape, number of flagella, flagellar apparatus organisation, organic scales covering the cell surface and flagella, and differences in cell division led to the conclusion that Prasinophyceae as proposed are not a monophyletic group. Various taxonomic treatments based on morphological and ultrastructural data have been proposed. Moestrup (1991) segregated the uniflagellate, non-scaled taxa *Pedinomonas, Marsupiomonas, Resultor*, and *Scourfieldia* sp. M561 into the class Pedinophyceae. Moestrup and Throndsen (1988) reclassified Prasinophyceae (excluding *Micromonas*).

Such dramatic morphological variation in unicells led to striking differences in opinion regarding the AGF. Some authors hypothesised that smaller taxa such as *Pedinomonas*, with their simpler flagellar apparatus and scales, were the best candidates (Moestrup, 1991). Other authors proposed that the AGF was a larger, more complex, and multiflagellate taxon. O'Kelly (1992) suggested that the complex flagellar apparatuses found in the larger, multiflagellate taxa were adaptations for prey capture in these algae or in their recent ancestors. Food particle ingestion and digestion has been shown in members of the Pyramimonadales (Bell and Laybourn-Parry, 2003) and has also been suggested (Delwiche, 1999) as a means of obtaining a cyanobacterial symbiont that ultimately became permanently incorporated as the plastid.

The advent of molecular systematics permitted evaluation of many morphologically and ultrastructurally based hypotheses regarding diversity and evolution of scaly, green flagellates, most of which are detailed in three comprehensive reviews (Melkonian, 1984; O'Kelly, 1992; Sym and Pienaar, 1993). Molecular analysis of 18S rDNA data have echoed the morphological diversity seen in prasinophytes, identifying at minimum seven separate lineages that form a grade at the base of the green tree of life (Kantz et al., 1990; Marin and Melkonian, 1994; Steinkotter et al., 1994; Nakayama et al., 1998; Fawley et al., 2000; Zignone et al., 2002). It is evident that most (perhaps all) of the remaining orders require further attention and should be reclassified.

Pyramimonadales

This monophyletic order is usually considered as sister to the rest of Prasinophytes and includes the quadriflagellate taxa *Halosphaera, Cymbomonas, Pyramimonas,* and *Pterosperma*. Some taxa have complicated scales. Moestrup et al. (2003) detailed the ultrastructural evidence for a close relationship of *Cymbomonas tetramitiformis* with *Halosphaera*. However, these authors also concluded that *Cymbomonas* possesses scales similar to *Mamiella*. It may be that the presence/absence of scales or a theca is a phylogenetically informative character, whereas scale morphology is not phylogenetically useful. Some members of this order, such as *Halosphaera, Pterosperma,* and *Cymbomonas* (Moestrup et al., 2003), are known to produce a cyst or phycoma stage, which at least in *Cymbomonas* contains two chloroplasts and thus is likely the result of sexual reproduction. The resistant walls of phycomata appear in the fossil record in the early Cambrian and perhaps earlier (Tappan, 1980).

Mamiellales

This order includes some of the smallest eukaryotes known, e.g., *Crustomastix* (3–5 μm long). Cells have one or two laterally inserted flagella and a total of two flagellar roots per cell. Some members lack scales, but most have a single form of scale over the body and flagella (Nakayama et al., 2000). Molecular data indicate that this group also includes coccoid taxa. Presumably all taxa contain the pigment prasinoxanthin, but Zignone et al. (2002) identified siphonoxanthin in *Crustomastix*. Mamiellales are usually reconstructed as sister to the rest after Pyramimonadales.

Mesostigmatophyceae

Mesostigma viride is the only member of this class. This asymmetrical cell with two laterally inserted flagella was recently placed in a separate class with the charophyte

Chaetosphaeridium based on molecular and cellular evidence (e.g., presence of similar maple-leaf-shaped scales on flagella). The flagellar rootlet system of *Mesostigma* has an MLS, unlike most other prasinophytes (except for Pyramimonadales). Although a separate class is warranted, the inclusion of *Chaetosphaeridium* has been refuted.

Pedinophyceae

This class includes *Pedinomonas* and *Resultor*, tiny (<3 μm) uniflagellate cells without scales. The flagellar apparatus is unusual among prasinophytes, consisting of one emergent flagellum and an additional basal body. The rootlets have an X-2-X-2 pattern, and the two basal bodies are offset in a counterclockwise orientation (Moestrup, 1991). Dividing cells have a persistent telophase spindle. Given this combination of characteristics, the phylogenetic position of Pedinophyceae still remains a puzzle. Contrasting hypotheses are that *Pedinomonas* represents either a reduced member of Mamiellales or a reduced ulvophycean taxon (Melkonian, 1990b). The only molecular study to include *Pedinomonas* placed it as sister of the green algae (Kantz et al., 1990), but because this analysis lacked an outgroup, there is no way to interpret its phylogenetic position.

Pseudoscourfieldiales

Members of this order (*Pseudoscourfieldia marina*, *Nephroselmis pyriformis*, and *N. olivacea*) have two flagella of unequal length, two body scale layers, and two flagellar scale layers. The flagellar apparatus is complex, having three flagellar roots. Molecular data resolve this group into either one or two clades of flagellate and coccoid taxa, both of which are usually placed as sister to Mamiellales.

Chlorodendrales

There is strong support for the monophyly of this order of flagellates and evidence that they share characters with the UTC clade, rather than with the other Prasinophyceae. The members of this group are distinct from other prasinophytes in many morphological and ultrastructural features: they possess a metacentric spindle that collapses at telophase; paired flagella that beat in a breast stroke pattern; and an X-2-X-2 configuration of the flagellar rootlets. These algae are also distinct in possessing a theca, which is formed by fusion of the outer layer of scales. The striking similarity between some members of the UTC clade and Chlorodendrales led Mattox and Stewart (1984) to reclassify *Tetraselmis* into a new class Pleurastrophyceae, along with *Pleurastrum, Trebouxia,* and *Pseudotrebouxia* (the latter genera are now placed in Trebouxiophyceae, Friedl, 1995, 1996). Analysis of molecular data always place *Tetraselmis* as sister to the morphologically more complex UTC clade.

Coccoids

In addition to the aforementioned lineages, Fawley et al. (2000) identified at least two well-supported, yet unnamed, clades of coccoid taxa from existing culture collection isolates that had not previously been sampled. These taxa were suspected members of the Prasinophyceae because they possess prasinophyte-specific pigments. It is evident that molecular data have corroborated and even enhanced our understanding of the diversity and non-monophyly of prasinophyte algae evident from ultrastructural data. Molecular data also have been particularly important in demonstrating that the earliest prasinophyte lineages (closest relatives of the AGF) were large, complex, scaly, and multiflagellate rather than small and naked. Molecular analysis have also confirmed a close phylogenetic relationship of *Tetraselmis* and the UTC clade. As more taxa and genes are sampled, we are gaining greater detail about the evolution of this grade of green algae; however, important questions remain about the influence of data analysis, data set composition, and taxon sampling on the phylogenetic placement of these individual lineages and taxa. Taxonomic revisions to classify the phylogenetic lineages now treated as orders within the Prasinophyceae should be forthcoming as greater resolution is achieved and analysis based on data from more than a single gene are published.

Ulvophytes

This group was named the class Ulvophyceae by Mattox and Stewart (1984), but because of the uncertain status of the class as a clade, we will refer to the group collectively as ulvophytes. Ulvophytes are diverse morphologically and ecologically and comprise some of the more strikingly beautiful green algae, extant or otherwise (Berger and Kaever, 1992). The group is predominantly marine and includes some of the best-known green seaweeds, such as the sea lettuce *Ulva* (Hayden and Waaland, 2002), the weedy *Codium* (Goff et al., 1992), *Caulerpa* (Meinesz, 1999), and the model organism *Acetabularia* (Mandoli, 1998). Other filamentous genera dominate localised freshwater habitats (*Cladophora, Rhizoclonium,* and *Pithophora*), sometimes to the detriment of human use (Lembi and Waaland, 1988).

Thallus forms include non-motile unicells, branched and unbranched filaments, filmy membranes (mono- or distromatic), and cushiony forms of compacted tubes. Many ulvophytes have multinucleate thalli and a siphonous construction, i.e., with few or no cross walls, which makes the thallus one giant, multinucleate cell. The surfaces of thalli of many marine ulvophytes are lightly to heavily calcified, and some species are important contributors to coral reef structure and common in the fossil record (Butterfield et al., 1988; Berger and Kaever, 1992). At the ultrastructural level, bi- and quadriflagellate motile cells have a cruciate (X-2-X-2) flagellar root system with CCW offset and overlapped basal bodies, with scales and rhizoplasts. Cytokinesis occurs by furrowing, with a closed persistent spindle (Mattox and Stewart, 1984; O'Kelly and Floyd, 1984; Sluiman, 1989).

The diplobiontic life cycle (free-living gametophyte and sporophyte phases, which may be iso- or heteromorphic; Bold and Wynne, 1985) is found in four of the five groups of ulvophytes. This type of life cycle, though neither uniform nor universal in the group, contrasts to that of other green algae, which generally have a predominant haploid vegetative phase and a single-celled, often dormant, zygote as the diploid stage (haplobiontic, Bold and Wynne, 1985). The diplobiontic life cycle is thus largely absent from green algae in freshwater habitats.

Because of the lack of clear synapomorphies for the ulvophytes, monophyly has been an open question since the group was established (O'Kelly and Floyd, 1984; Bremer, 1985; Mishler and Churchill, 1985). Analysis using molecular data are lacking. Preliminary studies of 18S and 26S rRNA (Zechman et al., 1990) indicated that Ulvophyceae of Mattox and Stewart (1984) are not monophyletic, but a verdict awaits addition of further taxa and genes. Antiquity of the group and possible large numbers of extinctions may make recovery of the relationships among the major clades within the ulvophytes difficult with a single gene such as 18S rDNA.

Within the ulvophyte lineage, five well-demarcated groups are recognised (O'Kelly and Floyd, 1984), which are usually ranked as orders, although their elevation to class is favoured by some authors (e.g., Van den Hoek et al., 1995).

Ulotrichales

This group includes non-motile unicells (*Codiolum*), branched (*Acrosiphonia*) and unbranched filaments (*Ulothrix*), and bladelike forms (*Monostroma*). The order is fairly common in fresh and salt water and on solid substrates or other algae. The simple unbranched filamentous thallus of *Hormidium* led to its being classified in Ulotrichales, whereas cytokinesis and zoospore morphology resulted in its renaming (*Klebsormidium*) and removal to an entirely different major group (charophytes; Silva et al., 1972; Mattox and Stewart, 1984). Assignment to *Klebsormidium* was confirmed by molecular data (Karol et al., 2001).

Ulvales

This order contains several seaweed genera familiar worldwide, including *Ulva* and *Enteromorpha*, with membranous or tubular thalli, respectively. Recent work indicates that these genera are paraphyletic and *Enteromorpha* has been synonymised with *Ulva* (Hayden and Waaland, 2002; Hayden et al., 2003). Morphogenetic switching between the tubular and membranous form may have evolved multiple times in the group (Tan et al., 1999).

Cladophorales (Including Siphonocladales)

Cladophorales contain branched and unbranched siphonous (multinucleate) green algae. This group includes one of the most commonly encountered freshwater and marine genera, the branching filamentous *Cladophora*. The genus has been monographed by Van den Hoek and colleagues (1963, 1982, 1984; van den Hoek and Chihara, 2000), who placed it in a separate order, Cladophorales. Several recent studies of 18S rDNA sequences (Bakker et al., 1994; Hanyuda et al., 2002) indicate that *Cladophora* is paraphyletic or polyphyletic to other genera in the group (*Chaetomorpha, Rhizoclonium, Microdictyon,* and *Pithophora*), although additional data are needed to resolve relationships of this group. Hanyuda et al. (2002) provided clear evidence that marine *Cladophora* has given rise to freshwater forms several times.

Caulerpales and Dasycladales

Caulerpales and Dasycladales are two marine groups of distinctive and often beautiful, siphonous seaweeds. The highly invasive *Caulerpa taxifolia* is a member of Caulerpales. Molecular studies of *Caulerpa* using the *tufA* gene have indicated that many morphological species are not monophyletic and that later diverging lineages in the genus are diversifying faster than ancient ones (Fama et al., 2002).

Dasycladales comprise two families, Dasycladaceae and Polyphysaceae (formerly Acetabulariaceae, Silva et al., 1996), that are estimated to have diverged some 400 Mya based on fossil evidence (Berger and Kaever, 1992); molecular data indicate a more recent split approximately 265 Mya (Olsen et al., 1994). The group includes the model organism, mermaid's wine glass (*Acetabularia* spp.). Zechman et al. (1990) studied rDNA sequences derived from RNA transcripts, and later Olsen et al. (1994) studied 18S rDNA sequences in Dasycladaceae and Polyphysaceae. Both analysis found support for monophyly of the former family. All dasyclads possess a stemloop deletion unique among green algae (Olsen et al., 1994). Olsen et al. (1994) used distance and parsimony methods and supported monophyly of the two families, but only when the phylogenetic tree was unrooted; the authors suggested that rooting the tree with ulvophycean outgroups removed signal and resolution because the outgroups were so distantly related to the ingroup families. Olsen et al. (1994) also found that several genera, including *Acetabularia*, were paraphyletic.

Berger et al. (2003) sampled 17 of 19 species from Polyphysaceae (Acetabulariaceae) in a study of 18S rDNA sequences and corroborated the hypothesis of monophyly of Polyphysaceae and paraphyly of Dasycladaceae. Zechman (2003) sampled a different gene, the plastid *rbcL*, in taxa from the two families and performed parsimony, likelihood, and Bayesian analysis. In contrast to the earlier 18S rDNA studies, Zechman (2003) concluded that both families were paraphyletic, not just Dasycladaceae. He also found a similar problem of paraphyly of genera. Berger et al. (2003) mapped features of cap formation onto the 18S rDNA tree and proposed a series of name changes in genera that would reconcile monophyletic groups on the tree and names of genera. Clearly, there is much to be done in this group, and additional multigene studies to resolve the phylogeny of the two families are underway (F. Zechman, California State University at Fresno, personal communication).

Trentopohliales

With an interesting mixture of ultrastructural characters, this group is sometimes omitted from the ulvophytes or included with several disparate green algal lineages (Chapman, 1984). Although entirely terrestrial, the group has marine relatives. A sea-to-land transition or vice versa would be unique among green, red, and brown algae (Graham and Wilcox, 2000).

Trentopohlians have phragmoplast-like cytokinesis (Chapman and Henk, 1986; Chapman et al., 2001), an MLS-like flagellar structure and unusual zoospores (Graham, 1984), and "primitive"-type plasmodesmata, three characters reminiscent of charophyte algae (Chapman, 1984; Chapman et al., 1998). However, 18S rDNA sequence data place these algae firmly in the chlorophyte clade, most likely in the ulvophytes. Clearly, this is one group that bears further investigation of both morphology and DNA sequences.

Chlorophyceae

Chlorophyceae are a monophyletic group that includes some of the most familiar of the microscopic green algae, including many model organisms. The unicellular flagellate *Chlamydomonas* has been used to study flagellar motion and swimming (Mitchell, 2000), photosynthesis mutations (Niyogi, 1999), and plastid genome modifications in secondarily non-photosynthetic taxa (Vernon et al., 2001). Colonial green algae such as *Volvox* have been models for the evolution of multicellularity, cell differentiation, and colony motility (Hoops, 1997; Kirk, 2003). Chlorophycean algae *Scenedesmus* and *Pediastrum* are also important palaeoecological or limnological indicators (Nielsen and Sorensen, 1992; Komarek and Jankovska, 2001).

Green algae in this class have a great range of vegetative morphology, from coccoid to swimming unicells, colonies, and simple flattened thalli to unbranched and branched filaments. All have a haplobiontic life cycle (zygotic meiosis). Orders previously recognised were defined on their vegetative morphology, form of sexual reproduction (either isogamy, anisogamy, or oogamy), and mode of asexual reproduction (zoosporic or autosporic). During cell division, mitosis is closed and cytokinesis involves a phycoplast system of microtubules, sometimes combined with furrowing. Swimming cells are vegetative cells, zoospores (asexual), or gametes, with two, four, or hundreds of flagella. Cells with two or four flagella have cruciate (X-2-X-2) rootlets and flagella that are displaced in a "clockwise" (CW, 1–7 o'clock) direction or are "directly opposed" (DO, 12–6 o'clock). In some swimming colonial forms such as *Volvox*, the flagellar apparatus undergoes developmental modification and the flagella are reoriented for colony swimming (Hoops, 1997). Other variants include taxa with two unequal flagella (*Heterochlamydomonas*) or flagella emerging from a pit (*Hafniomonas*).

A flood of primarily 18S rDNA data collected from chlorophycean green algae in the last two decades has given rise to dramatic modifications at every level of classification. A number of the traditional orders (Chlorellales, Chlorococcales, Chlorosarcinales), originally circumscribed using vegetative morphology, are now known to contain phylogenetically unrelated taxa. This is especially true for the groups that are morphologically separate and exhibit convergent evolution towards reduced morphology or cases in which *absence* of a trait was used as a main distinguishing feature of an order. Numerous evolutionary losses of motile cells are found across the chlorophycean green algae. Members of the Chlorellales (now mostly in the Trebouxiophyceae) completely lack motile cells, and thus the phylogenetically useful features that come from flagellar ultrastructure cannot be scored for these algae (Huss and Sogin, 1990).

Not only are these traditional orders polyphyletic, many of the species within genera are also being reclassified into different taxonomic groups (even classes), largely supported by molecular data. In nearly all cases that have been (or can be) examined, ultrastructural and other cellular features are congruent with the molecular reclassifications, even though the ultrastructural data alone are often insufficient to support the reclassifications. Based on molecular data and corroborated by a great deal of biochemical evidence, *Chlorella* was split into numerous groups spread across Chlorophyceae and Trebouxiophyceae (Huss and Sogin, 1990; Huss et al., 1999). Buchheim et al. (1990) showed that *Chlamydomonas* is not monophyletic. Since then, many additional genera have been shown to be highly polyphyletic. Congeners of *Neochloris* and *Characium* are now in three classes (Watanabe and Floyd, 1989; Lewis et al., 1992; Kouwets, 1995; Watanabe et al., 2000). Friedl and O'Kelly (2002) used ultrastructural and molecular data to distribute four species of the chlorosarcinoid genus *Planophila* into Ulvophyceae (two species into two distinct clades), one into Trebouxiophyceae, and a fourth species into Chaetopeltidales (Chlorophyceae). Species of *Chlorococcum* were separated into Chlamydmonadales and Ulvophyceae (Watanabe et al., 2001; Krienitz et al., 2003).

Thus, molecular data have provided evidence of convergent morphological evolution in many of the characters used to distinguish unicellular genera of chlorophycean green algae. The results of these studies only emphasise the need for additional molecular investigations, especially those that explore morphological character evolution in Chlorophyceae. Two overarching themes are evident in Chlorophyceae.

First, superficial similarities in body plans can be misleading, especially for simpler forms, and second, a range of morphology is represented within a single clade, with the coccoid form being nearly ubiquitous. To paraphrase J. B. S. Haldane (cited in Evans et al., 2000), nature must have had an inordinate fondness for coccoids.

CW Clade (Chlamydomonadales)

This clade includes a large number of taxa in Chlamydomonadales, plus taxa formerly placed in Volvocales, Dunaliellales, Chlorococcales, Tetrasporales, Chlorosarcinales, and Chaetophorales (Lewis et al., 1992; Wilcox et al., 1992; Buchheim et al., 1994; Nakayama et al., 1996a, b; Watanabe et al., 1996; Booton et al., 1998a, b; Krienitz et al., 2003). Vegetative morphology includes biflagellate unicells and colonies, quadraflagellate unicells, coccoid unicells, weakly branching filaments, or packets that are sometimes in a gelatinous matrix. All of the taxa possess the CW orientation of flagella and rootlets.

Buchheim et al. (1996, 1997), among others, demonstrated that the large and well-recognised genus *Chlamydomonas* represents at least five lineages within the CW clade. Some of the lineages share organismal characters, such as the pigment loroxanthin (Fawley and Buchheim, 1995) and autolysin groups (Buchheim et al., 1990). Wall-less unicellular taxa in the related order Dunaliellales (sensu Ettl) were also shown to be non-monophyletic, but instead interspersed in the CW clade. The unusual *Hafniomonas* is interpreted as having a modified CCW flagellar apparatus orientation or as a relative of Chaetopeltidales, is also a member of the CW clade (Nakayama et al., 1996a).

The colonial flagellates previously placed in Volvocales are not monophyletic based on molecular data (Buchheim and Chapman, 1991; Buchheim et al., 1994) and do not represent a tidy progression of evolution by building ever larger colonies from *Chlamydomonas*-like ancestors with developmentally modified flagellar apparatus ultrastructure. Some colonial genera (*Gonium*) have been shown to be monophyletic, although most colonial forms with larger cell numbers are not (Nozaki et al., 2000).

Sphaeropleales (DO clade)

As emended by Deason et al. (1991), this order includes vegetatively non-motile unicellular or colonial taxa that have biflagellate zoospores with the DO flagellar apparatus arrangement: *Sphaeroplea, Atractomorpha, Neochloris, Hydrodictyon,* and *Pediastrum.* All of these taxa possess basal body core connections (Wilcox and Floyd, 1988). With an increase in the number of taxa for which sequence data are available, there is evidence of an expanded DO clade that includes additional zoosporic (*Bracteacoccus, Schroederia*; Buchheim et al., 2001; Lewis, 1997) and some strictly autosporic genera such as *Ankistrodesmus, Scenedesmus, Selanastrum, Monoraphidium,* and *Pectodictyon* (Krienitz et al., 2001, 2003). Monophyly of the DO clade is generally weakly supported by phylogenetic analysis of molecular data, even those that have used data from two genes (Buchheim et al., 2001; Wolf et al., 2002; Shoup and Lewis, 2003).

Oedogoniales

This order includes three genera, *Oedogonium*, *Oedocladium*, and *Bulbochaete*, and approximately 600 species, all of which grow attached to submerged surfaces in freshwater habitats. All members of this order form simple or branched filaments. The familiar *Oedogonium* is often used as an example of oogamous sexual reproduction in green algae. All genera produce unusual motile cells (either asexual zoospores or male gametes) with an anterior ring of flagella (stephanokont). The flagellar apparatus of these cells has been studied extensively (Pickett-Heaps, 1975), and is clearly unlike the swimming cells in other groups of green algae. Given the strikingly unusual ultrastructure of the swimming cells, homology assessment with flagellar characters found in the other groups is difficult. Pickett-Heaps (1975) hypothesised that this group represents a "basal" lineage that gave rise to other filamentous forms. Analysis of molecular data (18S and 26S rDNA) have indicated that the order is clearly monophyletic (Booton et al., 1998b; Buchheim et al., 2001) and is often placed sister to the rest of Chlorophyceae. However, this placement varies, often with differing resolutions that include the Chaetopeltidales and Chaetophorales. Because all three orders represent "long branches" in the tree, a robust placement is not often obtained or depends on method of analysis.

Chaetopeltidales

This order of four genera was proposed by O'Kelly (1994) to include taxa that produce quadriflagellate motile cells with a perfectly cruciate (DO) flagellar apparatus orientation. The vegetative morphologies found in this order include flattened thalli (*Chaetopeltis*) and a sarcinoid genus *Floydiella* (formerly *Planophila*; Friedl and O'Kelly, 2002). Analysis of 18S rDNA data place this group, albeit weakly, with the biflagellate DO taxa (Booton et al., 1998a; Krienitz et al., 2003). Analysis of combined 18S and 26S rDNA data were also inconclusive, but they often resulted in topologies with this order outside the two clades of biflagellate taxa, closer to the base of Chlorophyceae (Buchheim et al., 2001; Shoup and Lewis, 2003).

Chaetophorales

Members of the order have unbranched or branched filamentous vegetative bodies and produce quadriflagellate motile cells with upper and lower pairs of basal bodies in a CW + CW arrangement. Phylogenetic analysis of 18S rDNA data by Nakayama et al. (1996a) placed *Chaetophora incrassata* as sister to the rest of Chlorophyceae. With expanded sampling, Booton et al. (1998b) placed Chaetophorales nearest the biflagellate taxa with the CW orientation, although this placement was very weakly supported. As with Oedogoniales and Chaetopeltidales, the monophyly of the order is supported, but its relative placement remains unresolved.

Incertae Sedis

Topologies resulting from the analysis of two genes (18S and 26S rDNA) have indicated a distinctive chlorophycean clade, adjacent to Sphaeropleales, which includes the filamentous genus *Cylindrocapsa* and the unicellular *Trochiscia* and *Treubaria*. Previous hypotheses indicated an alliance of *Cylindrocapsa* with Sphaeropleales based on morphology of the pyrenoid (region within plastid containing a high concentration of the enzyme ribulose-1, 5-bisophosphate carboxylase [rubisco] and associated with starch synthesis), but this relationship is not supported with molecular data (Buchheim et al., 2001). Another unnamed group of freshwater picoplanktonic taxa, the *Mychonastes* clade (Krienitz et al., 2003), forms a sister group to Oedogoniales at the base of Chlorophyceae. A new order will need to be established eventually to accommodate this clade. In addition, several flagellate taxa, including species of *Carteria* and *Hafniomonas*, form a grade at the base of the CW clade and do not correspond to a named group (Nakayama et al., 1996a; Hoham et al., 2002).

Trebouxiophytes

In their 1984 classification, Mattox and Stewart proposed the class Pleurastrophyceae to accommodate freshwater algae with the CCW flagellar apparatus orientation, a metacentric spindle, and phycoplast-mediated cytokinesis. This class included the non-motile unicells *Pleurastrum* and *Trebouxia*, the unicellular flagellate *Tetraselmis*, and *Microthamnion*, a filamentous algae with short branches. Conversely, Melkonian (1990a) chose to group these same taxa (excluding the flagellate *Tetraselmis*) into a separate order Microthamniales. Additional studies expanded membership of this group to include the terrestrial algae *Leptosira*, which produces unbranched uniseriate filaments (Lokhorst and Rongen, 1994). An early molecular analysis based on just a few taxa by Kantz et al. (1990) provided evidence for the non-monophyly of Pleurastrophyceae.

Friedl and Zeltner (1994) and Friedl (1995), using 18S rDNA data, demonstrated the designated type of the class Pleurastrophyceae, *Pleurastrum insigne*, was actually a member of Chlorophyceae. In addition, Friedl provided evidence for a distinct clade (class Trebouxiophyceae) of many of the taxa that were formerly in Pleurastrophyceae (such as *Microthamnion*), but not related to Ulvophyceae or Chlorophyceae. In agreement with Melkonian (1990a), Friedl (1995) excluded the flagellate *Tetraselmis* from the group. Subsequently, through the use of 18S rDNA data, an increasing number of taxa have been added to this order, including some *Chlorellas* and *Oocystis*, autosporic taxa for which collection of data on motile cell ultrastructure is impossible. Although many members of Trebouxiophyceae (*Trebouxia*) participate in lichen symbioses and several workers defined the group primarily on lichen phycobionts, this class now includes a growing number of free-living planktonic or terrestrial species. *Prototheca*, a secondarily non-photosynthetic coccoid algae, and picoplanktonic coccoids such as *Nannochloris* are members of this class.

Members of Trebouxiophyceae reproduce asexually by autospores or zoospores. Sexually reproductive stages have not been observed directly in any of the trebouxiophyte algae; however, one recent application of phylogenetic data collected for lichen photobionts sheds light on this topic. Kroken and Taylor (2000) used fine-grained sampling of isolates of *Trebouxia jamesii*, a photobiont of the fungal genus *Letharia*, to provide evidence of a recombining population structure.

The trebouxiophytes were shown to be monophyletic (Friedl, 1995), but because some molecular studies do not recover monophyly or recovered weak support for monophyly (Krienitz et al., 2003), this question will need further attention. Again, designation of this group was based on molecular data and a suite of morphological characters, many or most of which have the plesiomorphic character state. For example, the CCW flagellar orientation is shared with Ulvophyceae. At least five distinct lineages are recovered with 18S rDNA data (e.g., Krienitz et al., 2003), four of which correspond to named groups.

Trebouxiales

This order represents the "lichen algae group" and includes zoosporic taxa, like *Trebouxia*, which are lichen phycobionts (Ahmadjian, 1993). Note that not all lichenised algae are in this group; other photobionts include cyanobacteria, *Trentepohlia* (Ulvophyceae), and *Heterococcus* (tribophytes). Several genus-level taxonomic changes have been made within this order. For example, *Friedmannia* was shown to be nested within the genus *Myrmecia* and was renamed (Friedl, 1995).

Microthamniales

The highly branched filamentous *Microthamnion* and the unicellular *Fusochloris* (former member of *Neochloris*, Chlorophyceae) are supported as a clade, related to but distinct from Trebouxiales. Both taxa are zoosporic and have the characteristic biflagellate swimming cells of Trebouxiophyceae. A close phylogenetic relationship between such morphologically different taxa once again illustrates the rapid evolution of vegetative morphology.

Prasiolales

This group includes zoospore-forming taxa, which are unicellular, short filaments, or small sheetlike thalli. These algae are commonly found in a great range of environmental conditions, including freshwater, marine, and terrestrial habitats such as on moist concrete walls and rocks, and tree bark (Rindi et al., 1999; Handa et al., 2003; Rindi and Guiry, 2003). Members of this group are also often encountered in cold deserts (e.g., *Pabia signiensis* from Antarctica; Friedl and O'Kelly, 2002). *Prasiola* was formerly in Ulvophyceae, but analysis of molecular data support its membership in Trebouxiophyceae (Friedl and O'Kelly, 2002). Handa et al. (2003) also resolved a close phylogenetic relationship of a corticolous species of *Stichococcus* with *Prasiola*. Naw and Hara (2002) identified isolates of *Prasiola* with a vegetative body that resembles *Enteromorpha* but are distant from Ulvophyceae and can be distinguished on the basis of the stellate plastids.

Chlorellales

Molecular data have had a great impact first on placing members of Chlorellales in Trebouxiophyceae rather than Chlorophyceae and then on clarifying relationships within Chlorellales. This order includes two well-supported families of autosporic species, Chlorellaceae and Oocystaceae. *Chlorella* (previously classified in Chlorellales, Chlorophyceae) was shown to be paraphyletic with many (or most) members in Trebouxiophyceae (Huss and Sogin, 1990; Huss et al., 1999). Yamamoto et al. (2003) investigated the genus *Nannochloris* using 18S rDNA and actin genes and determined non-monophyly of these small planktonic, coccoid forms. Their molecular phylogenetic lineages corresponded to cell division patterns. The non-photosynthetic *Prototheca* (Huss and Sogin, 1990; Ueno et al., 2003), the Oocystaceae (Hepperle et al., 2000), and the spiny *Lagerheimia* (Krienitz et al., 2003) are now additional members of this order. Clearly, major rearrangements are still possible for algae referred to with some frustration as LRGTs, or "little round green things" (Graham and Wilcox, 2000a).

Charophyte Algae and Land Plants

Charophyte algae are relatively poor in species diversity but are paraphyletic to a hugely diverse group, the land plants (~500 000 species). Smith (1950) recognised Charophyceae as one of two classes of green algae (the other was Chlorophyceae), but he included only Charales (stoneworts) in the group. Mattox and Stewart (1984) revised the composition of the group greatly by including four other orders. Charophyceae sensu Mattox and Stewart shared features of glycolate metabolism, cytokinesis (phragmoplast or similar structures in more-derived forms), and motile cell structure (asymmetrical, with an MLS associated with flagella) that were recognised as both distinct from that in other green algae and shared by embryophytes. Early on, Stewart and Mattox (1975) and colleagues (Pickett-Heaps and Marchant, 1972; Pickett-Heaps, 1975) recognised this assemblage of algae as members of one of two divergent green algal lineages. These early workers recognised the implication of the findings: an assemblage of green algae living today are direct descendants of an ancestor shared with land plants (Pickett-Heaps, 1969, 1972). Phycologists still tended to separate the charophycean green algae from land plants, but other botanists pointed out the monophyly of the group (Mishler and Churchill, 1984, 1985; Bremer, 1985; Bremer et al., 1987).

Using data on biochemistry (Stewart and Mattox, 1972), flagellar structure (Melkonian, 1984), and cytokinesis, Mattox and Stewart (1984) codified Charophyceae to comprise five orders. In general, molecular studies have supported the monophyly of individual orders, with some modifications. In addition, one or more prasinophycean taxa may be members of the charophyte lineage (i.e., *Mesostigma*, Melkonian, 1989; Karol et al., 2001; Turmel et al., 2002c).

Mesostigmatophyceae

This asymmetrical unicell is singular in more ways than one. Its flagellar structure first indicated that it was a member of the charophyte clade (Melkonian, 1989). Subsequent molecular analysis of 18S rDNA led to the hypothesis that *Mesostigma* is sister to *Chaetosphaeridium*, and these two genera were placed in a new class, Mesostigmatophyceae (Marin and Melkonian, 1999). Later molecular studies indicated that this grouping was artifactual and that *Chaetosphaeridium* was a member of Coleochaetales, in which the genus had long been classified (Delwiche et al., 2002). This left *Mesostigma* as an unusual algae for which phylogenetic placement is still a subject of contention. Delwiche et al. (2002) using *rbcL* sequences and Karol et al. (2001) using a four-gene analysis (*rbcL, atpB, nad5,* and 18S rDNA) found *Mesostigma* to be sister to other charophyte algae and embryophytes. In contrast, other studies have found *Mesostigma* to be sister to the rest of the Chlorobionta, i.e., sister to Chlorophyta and Charophyta. These studies, based on chloroplast small and large subunit rDNA sequences (Turmel et al., 2002a) and multiprotein sequences from genome sequences of the mitochondrion (Turmel et al., 2002c) and plastid (Lemieux et al., 2000), are consistent with pigment studies. Yoshii et al. (2003) identified 9'-cis neoxanthin in *Mesostigma* although all other green plants have trans-neoxanthin. The authors interpreted this result as an indication of the early divergence of *Mesostigma*, prior to the evolution of the cis-isomerases necessary to convert trans to cis (in all other green plants). In contrast, Martin et al. (2002) analysed 274 protein-coding genes in analysis of 16 plastid genomes, including *Mesostigma*, two other non-charophyte green algae, and six embryophytes, and found a topology congruent with nesting of *Mesostigma* within the charophyte algae-embryophyte clade as in Karol et al. (2001). Clearly, genomic sampling is just beginning and should clarify these positions.

Chlorokybales

This order is monotypic, with thalli of sarcinoid packets of cells that grow subaerially (Rogers et al., 1980). Biflagellate scaly zoospores are produced, with hairy flagella that possess an MLS in the root. The four-gene analysis of Karol et al. (2001) and Delwiche et al. (2002) placed this species in the charophyte algae, near the base of the lineage.

Klebsormidiales

These algae are simple unbranched filaments with parietal laminate or lobed chloroplasts. Zoospores have an MLS-type root and two asymmetrical flagella. *Klebsormidium* was proposed by Silva et al. (1972) to remove confusion attending the name *Hormidium* when the algae was considered a member of Chlorophyceae. *Klebsormidium* contains approximately 24 species, and Lokhorst (1996) monographed European taxa. Two genera (*Stichococcus* and *Raphidonema*) thought to be members of the order are not, but *Entransia fimbriata*, originally placed by Hughes (1948) in Zygnematales, is now allied with *Klebsormidium* (McCourt et al., 2000; Karol et al., 2001; Turmel et al., 2002a). Cook (2004) studied morphology of this species, and although she did not find zoospores, she found empty zoosporangia, which distinguish it from Zygnematales.

Zygnematales and Desmidiales

Mattox and Stewart (1984) combined these two orders in one, Zygnematales, but wall ornamentation and structure has led to the demarcation of placoderm desmids in Desmidiales (Gerrath, 2003). Thalli are unicellular or filamentous, with one colonial genus. Two morphological synapomorphies unite the group: (1) flagella are entirely lacking at any stage in the life cycle, and (2) sexual reproduction involves conjugation, the fusion of non-flagellate gametes that move actively or passively through a tube or within a gelatinous envelope. Gametes are usually isomorphic but sometimes exhibit differences in motility (i.e., one amoeboid male gamete moves; Hoshaw et al., 1990). These algae represent the most species-rich group of charophyte algae, with approximately 4000 species (Gerrath, 2003). Family classification based on ultrastructure of the cell wall (Mix, 1972; Brook, 1981) has been supported by molecular studies. The derived condition of highly ornamented walls with elaborate pores is found in four families of placoderm desmids. Desmidiales and placoderm desmid families appear to be monophyletic (McCourt et al., 2000; Gontcharov et al., 2003). However, two families with a plesiomorphic cell wall condition (smooth, unornamented) are either paraphyletic in their original definition or members of the sister of placoderm desmids (McCourt et al., 2000; Gontcharov et al., 2003). These molecular analysis indicate that more complex forms evolved from simple filaments and morphological switching from unicells to filaments occurred several times.

Coleochaetales

This group contains two genera (*Coleochaete* and *Chaetosphaeridium*) and about 20 species, which have a morphology and life history that makes them important as model systems for understanding the evolution of embryophytes (Graham, 1993, 1996; Graham and Wilcox, 2000b). *Coleochaete* thalli are mostly tightly arranged discs of cells, although some are made of more loosely branching filaments. *Chaetosphaeridium* thalli are filamentous. Characteristic of the order are distinctive sheathed hairs that are extensions of the cell wall containing a small amount of cytoplasm. Zygotes of *Coleochaete* are retained on the maternal gametophyte in some species, and placental transfer cells may transfer nutrients to the zygote (Graham and Wilcox, 1983; Graham, 1985, 1996).

Studies of 18S rDNA sequences cast doubt on the monophyly of Coleochaetales and indicated that *Chaetosphaeridium* was sister to the rest of the charophyte lineage (Sluiman and Guihall, 1999), a conclusion that held up despite the discovery of a fungal artifact in the sequence (Cimino et al., 2000; Sluiman, 2000). An 18S rDNA analysis placed *Chaetosphaeridium* as sister to the flagellate *Mesostigma* (Marin and Melkonian, 1999), but subsequent analysis of *rbcL* (Delwiche et al., 2002), chloroplast small and large subunit rDNA (Turmel et al., 2002a), and a four-gene analysis of plastid, nuclear, and mitochondrial genes (Karol et al., 2001) presented convincing evidence that *Chaetosphaeridium* is sister to *Coleochaete*.

Relationships of *Coleochaete* have been studied using *rbcL* (Delwiche et al., 2002), including endophytic species that live within the cell wall of *Nitella* (Cimino and Delwiche, 2002).

Charales

This group contains six extant genera with several hundred species, collectively called stoneworts or charophytes in the palaeontological literature (Grambast, 1974; Feist et al., in press). Thalli are attached by rhizoids to sandy or silty substrates in quiet freshwater habitats and range from a few centimetres to several decimetres in height. The thallus consists of a central axis of large multinucleate internodal cells, with whorls of branchlets radiating from nodes of uninucleate cells; a single meristematic cell initiates growth at the apex of the axis and branchlets. This whorled branching habit is convergent with some aquatic angiosperms such as *Ceratophyllum* and *Myriophyllum*. Calcium carbonate accumulates on the surfaces of many species (hence the name stoneworts), which partly accounts for the rich fossil record of more than 80 genera and 10 families stretching back to the upper Silurian (Feist and Grambast-Fessard, 1991; Gensel and Edwards, 1993). Reproduction is oogamous, and sperm morphology is complex (Garbary et al., 1993).

Molecular studies have supported monophyly of the sole extant family, Characeae (McCourt et al., 1996, 1999; Meiers et al., 1997, 1999). This family is distinguished morphologically from extinct taxa but clearly sister to them (Feist et al., in press). *Chara* species appear to be monophyletic, although the results indicate that conventional taxonomy (Wood and Imahori, 1965) of tribes, sections, and subsections within *Chara* are in need of revision (Meiers et al., 1997, 1999; McCourt et al., 1999). Species of *Nitella* have been studied using *rbcL* (Sakayama et al., 2002) and *atpB* sequences (Sakayama et al., 2004), and results indicate that taxonomy within this genus is also in need of revision. Oospore surface morphology appears to hold promise for such revision (Sakayama et al., 2002, 2004).

Origin of Land Plants

A Clade of their Own: Standard textbooks on botany and phycology acknowledge a world view in which the green algae as a whole are no longer monophyletic to the exclusion of land plants (embryophytes; Raven et al., 1999; Graham and Wilcox, 2000a; Graham et al., 2002; Judd et al., 2002). The molecular data on this point support the already strong case from ultrastructural data and are overwhelming, coming from the nuclear ribosomal repeat unit, mainly the small (18S) subunit, but including 5S and large subunit (26S) rDNA sequences as well (reviews in Chapman and Buchheim, 1991; McCourt, 1995; Melkonian and Surek, 1995; Chapman et al., 1998). Several plastid genes sampled broadly enough to address this question yield similar results: *rbcL* (Manhart, 1994; McCourt et al., 1996, 2000; Delwiche et al., 2002) and small and large subunit rDNA (Turmel et al., 2002a). Other molecular synapomorphies supporting the charophyte clade (though not uniformly present or sampled in all groups) are genomic features not found in the chlorophyte clade: transfer of the *tufA* gene from the plastid to the nucleus in the common ancestor of some charophyte algae (Zygnematales, Coleochaetales, Charales) and embryophytes (Baldauf and Palmer, 1990), patterns of insertion of introns in two plastid tRNA genes (Manhart and Palmer, 1990), an ORF for *matK* in a group II intron of the *trnK* exon of Charales and possibly some other charophyte algae (Sanders et al., 2003), and insertion of an intron into the VATPase A gene in Coleochaetales (Starke and Gogarten, 1993).

Despite the convincing evidence for monophyly of charophyte algae and embryophytes, the topology of the charophyte tree has shifted dramatically. Specifically, the exact identity of the group sister to embryophytes has proven to be elusive. Various groups or combinations of groups received support from different sources of data. For example, several analysis of 18S rDNA sequences yielded a topology in which Charales was placed with strong support as sister to the other four orders plus land plants (Friedl, 1997; Huss and Kranz,

1997). In contrast, *rbcL* data and some 18S rDNA data indicated that Charales plus Coleochaetales (McCourt et al., 1996, 2000) or Charales alone (Melkonian and Surek, 1995; Delwiche et al., 2002) were sister to land plants. Such conflicts usually indicate a dearth of data in terms of taxa or characters or both. Another recent study sampled small and large subunit plastid rDNA genes broadly across all major charophyte and chlorophyte groups and five bryophytes (Turmel et al., 2002a) and found strong support for monophyly of the major groups, with the exception of *Mesostigma*, which, as mentioned, was positioned sister to both chlorophyte and charophyte clades. However, the analysis could not resolve with confidence the sister taxon of land plants.

Karol et al. (2001) sampled one mitochondrial (*nad5*), one nuclear (18S rDNA), and two plastid (*rbcL, atpB*) genes (total of 5147 nucleotides) across the major charophyte groups (26 species), plus eight land plants and six chlorophyte outgroups. Bayesian, maximum parsimony, and maximum likelihood analysis all supported monophyly of Charales and land plants. Turmel et al. (2003) sequenced the entire mitochondrial genome of *Chara vulgaris* L., and their analysis were congruent with the conclusions of Karol et al. (2001). Gene loss from the mitochondrion, and group I and II intron distributions in the *Chara* mtDNA clearly allied this algae with that of land plant mtDNA. Maximum likelihood analysis of 23 protein sequences also strongly supported monophyly of *Chara* with land plants, although the number of taxa in the analysis (six) adds a note of caution to the result. Clearly, more genes and more taxa are needed to evaluate this result (Rokas et al., 2003).

Extreme Greens and the Multiple Origins of Terrestrial Green Plants

One of the exciting insights of the last few years has been the discovery of diverse green algae that can inhabit "extreme" habitats. More intensive sampling, coupled with culture-based methods and "environmental sampling," have expanded our understanding of green algal diversity and of the conditions necessary for survival and growth of green algae. For example, although we have known about the occurrence of green algae in saline environments for a long time, recent studies provide an expanded view of the extent to which green algae can exist in extreme environments. The chlorophycean algae *Dunaliella salina*, one of the major species used in carotenoid production, can exist at supersaline conditions (in waters that are >10 per cent salt, Padwahl and Singh, 2003). *Dunaliella acidophila*, another extremophilic species, grows at extremely low pH (<2). The ability to do so depends on active proton elimination (Gokhmann et al., 2000). Zettler et al. (2002) reported both charophyte and chlorophyte green algae, among other eukaryotes, living in acidic (pH 2) waters with high levels of heavy metals. Edgomb et al. (2002) used environmental sampling to uncover deep-sea hydrothermal vent eukaryotes related to the prasinophyte, *Mantoniella*.

Besides being found in highly saline and acidic waters, green algae are also common on land. Colonisation of terrestrial environments by green plants is one of the most important events in the history of the Earth. However, the embryophyte transition to land is just one of numerous such transitions in the green plastid-bearing lineage. Mapping the distribution of known terrestrial lineages on a generalised green plant cladogram illustrates that the evolution to terrestrial environments is neither uncommon nor phylogenetically limited. Terrestrial green algae are found in at least six major clades (classes or orders) exclusive of the embryophytes. Terrestrial lineages seem to be absent in the prasinophyte algae but are found in both chlorophyte and charophyte clades. Multiple terrestrial lineages exist within Chlorophyceae, Trebouxiophyceae, and charophytes. All of the known terrestrial algae are inferred to have aquatic ancestors, and, in fact, some are closely related to aquatic taxa (Hoham et al., 2002; Lewis and Flechtner, 2002).

Terrestrial algae are morphologically similar to aquatic taxa and include unicellular coccoids to filaments. Many spend a majority of the vegetative life out of water, although their reproductive stages cannot be completed without water. Members of Trentepohliales (Ulvophyceae) are known from aerial habitats, often as pathogens in and on the leaves of citrus plants. They are also common on bark and rocks, developing a characteristic brilliant orange colour from secondary pigments. Within Trebouxiophyceae, many members are known lichen symbionts, and others are known from moist soils. Terrestrial members of Charophyceae include *Klebsormidium*, *Chlorokybus*, *Mesotaenium*, and *Cylindrocystis*; terrestrial Chlorophyceae include the filamentous genus *Fritschiella* and coccoid genera *Bracteacoccus*, *Chloromonas*, *Chlamydomonas*, and *Scenedesmus*.

Terrestrial algae grow in some of the most difficult habitats on earth, such as desert soils. Here, algae experience lengthy periods of desiccation and extremes in temperature and light levels. The earliest studies enumerating the green algae from deserts were based on vegetative morphology and recovered only a few genera (e.g., Cameron, 1960). However, morphological studies involving examination of alternate life cycle stages or studies using molecular analysis of green algae isolated from deserts of western North America have indicated a much larger number of genera than found previously and multiple transitions to arid habitats from aquatic ancestors within Chlorophyceae, Trebouxiophyceae, and charophytes such as *Klebsormidium* and *Cylindrocystis* (Flechtner et al., 1998; Lewis and Flechtner, 2002).

Psychrophilic algae are defined as those taxa having optimal growth below 10°C. Cold-loving taxa are found in Chlorophyceae and charophytes, including *Chloromonas*, *Chlamydomonas*, *Chlorosarcina*, *Mesotaenium*, and *Cylindrocystis*. Many cold-tolerant algae occur in numbers large enough to form red, pink, or green zones easily visible in snow. The unicellular biflagellate genus *Chloromonas* is one of the most common of the snow algae (Ling, 2001, 2002). *Chloromonas* is distinguished from *Chlamydomonas* by the absence of a pyrenoid, and the lack of pyrenoids and cold tolerance was thought to be correlated. However, molecular phylogenetic studies (Buchheim et al., 1990, 1997a, b; Hoham et al., 2002) concluded that neither *Chlamydomonas* nor *Chloromonas* is monophyletic, but rather that *Chloromonas* species were derived numerous independent times from *Chlamydomonas*. Hoham et al. (2002) showed that cold tolerance developed at least three times in the *Chlamydomonas* clade and that it was not correlated with the presence of a pyrenoid (Borkhsenious et al., 1998). At the deepest taxonomic levels, presence of a pyrenoid seems to be phylogenetically meaningful; all embryophytes (except for hornworts) lack pyrenoids. The utility of pyrenoid presence as a taxonomic character among closely related algae is now questionable because pyrenoids were shown to be under the regulation of a single gene in *Chlamydomonas* (Fukuzawa et al., 2001), and the transition between a pyrenoid-containing and a pyrenoidless state might not be complex.

Documented cases of independent transitions to terrestrial habitats provide important contrasts to what is known about physiological adaptations in embryophytes. Examples of the transition from aquatic to a variety of terrestrial habitats within the green algae are also exciting because they provide detailed information about the phylogenetic relationships of extreme greens to their relatives in more moderate habitats, thus offering an opportunity to study the timing and mechanisms of the transitions to these specific environments. Such studies were made possible only recently because of molecular phylogenetic analysis and the ability to separate taxa with confusingly similar vegetative morphology. These investigations will become increasingly feasible and interesting as more taxon-dense sampling covers a wider range of habitats.

The classification is, of course, not without faults, and some are serious. Nevertheless, we think it is as good as can presently be devised (Mattox and Stewart, 1984).

In 1984, Mattox, Stewart, and others working at that time no doubt felt that the accumulating body of new data had produced significant advances in green algal systematics and dramatically modified classification from a strictly vegetative basis to a focus on "evolutionarily conserved" ultrastructure. With this, we must agree. Progress in the molecular systematics of green algae rests on hundreds of ultrastructural and biochemical investigations that paved the way for evaluation of these explicit hypotheses. Many broad-scaled ultrastructure-based hypotheses have been corroborated by molecular data, yet many relationships now revealed by molecular data were never predicted or predictable based on ultrastructural data.

Molecular data have provided greater resolution than classifications based on ultrastructure. For example, the paraphyly of the charophytes was already indicated from cladistic analysis of non-molecular data, but molecular data further clarified the relationships by providing additional evidence for the inclusion of *Mesostigma* in the charophyte lineage (or placing it as the sister taxon to all green algae and plants) and for resolving Charales as sister (among extant lineages) to embryophytes. Sequence data have also provided valuable insight into the phylogenetic placement of taxa that were difficult to place based on morphology because they contain characters representative of more than one group (e.g., Trentepohliales, *Entransia*).

There are many striking differences between molecular and ultrastructurally based classifications. Only two of the five classes established from ultrastructural data (Chlorophyceae and Ulvophyceae) are recovered with molecular data, although the ulvophytes are often not resolved as monophyletic with molecular data. Many of the taxa currently included in Trebouxiophyceae would not have been placed in this class on the basis of morphology alone. Two other classes based upon ultrastructure (Prasinophyceae, Charophyceae) are grades that should be reclassified. Another important impact of molecular data has been resolution of relationships among the problematic coccoid green algae (especially of the strictly autosporic taxa without swimming cells). No doubt there will be many more profound changes to come ir the classification of these groups.

Molecular data are also expanding our view of diversity, both in familiar taxa and in newly revealed, previously unsampled taxa. The distinct lineages of paraphyletic *Chlamydomonas* were not distinguished because they possess nearly identical flagellar structure. Culture-based or "environmental sampling" studies that include molecular data have documented novel lineages (Phillips and Fawley, 2000; Lewis and Flechtner, 2003) that are now being characterised. If these new taxa correspond to critical or early diverging lineages, their inclusion in future analysis could have a profound impact on our understanding of relationships among the major groups of green algae.

Despite these advances, molecular data thus far have failed to resolve firmly some of the deeper branches of the green plant tree of life, including the monophyly of the ulvophyte algae, the relationships among lineages in the Chlorophyceae, and the topology of the prasinophyte algae. A contributing factor to this lack of resolution is a paucity of data compared to that from other plant groups. Only a small proportion of the known taxa have been included in molecular studies, and with few exceptions, only data from a single gene are available. A new focus on the collection of multiple genes or even genomes is underway (Qui and Lee, 2000; Chapman and Waters, 2002). The work of Karol et al. (2001), Lemieux et al. (2000), and Turmel et al. (2002a, b, c) have illustrated the power of resolving critical nodes with ever increasing amounts of data. In addition, C. J. O'Kelly (Bigelow Laboratory for Ocean Sciences, personal communication) and colleagues, funded through the Assembling the Tree of Life programme of the National Science Foundation (USA), are collecting plastid and mitochondrial genome-level data for 30 green algae, selected to represent some of the "difficult" regions of green plant phylogeny. There is optimism that, with enough data, the more difficult regions of the tree and incongruence among single-gene data sets will be resolved, although nodes representing rapid radiations may require data that are less clocklike (e.g., informative changes in morphology or genome structure that track divergence of short branches).

Morphological and ultrastructural studies will still have a role in the systematics of green algae in the future. Previous studies (e.g., Sluiman, 1985; Bremer et al., 1987; Garbary et al., 1993; Mishler et al., 1994) used explicit homology assessment and cladistic analysis of morphological, ultrastructural, and molecular data to uncover major early diverging lineages of green plants. Scotland et al. (2003) claim that morphological data are being superceded by molecular data because "much of the useful morphological diversity has already been scrutinised." This is clearly not the case for groups such as the green algae, for which new morphological data are still being obtained, albeit at a much slower rate than molecular data. Novel molecular lineages are prime targets for future ultrastructural studies, but molecular diversity is not always predictive of ultrastructural diversity (O'Kelly et al., 2004). If our goal is to circumscribe natural and recognisable groups, understand morphological evolution, and study adaptation, the collection of disembodied sequence data will be of little use without ultrastructural data. In the case of green algae, discovery and characterisation of unsampled diversity could have an immense impact in recovering the green plant tree of life.

4

Various Kinds of Algae

Most of the simpler algae are unicellular flagellates or amoeboids, but colonial and non-motile forms have developed independently among several of the groups. Some of the more common organisational levels, more than one of which may occur in the life cycle of a species, are:

- *Colonial* - small, regular groups of motile cells
- *Capsoid* - individual non-motile cells embedded in mucilage
- *Coccoid* - individual non-motile cells with cell walls
- *Palmelloid* - non-motile cells embedded in mucilage
- *Filamentous* - a string of non-motile cells connected together, sometimes branching
- *Parenchymatous* - cells forming a thallus with partial differentiation of tissues

In three — lines even higher levels of organisation have been reached, leading to organisms with full tissue differentiation. These are the brown algae — some of which may reach 50 m in length — the red algae, and the green algae. The most complex forms are found among the green algae, in a lineage that eventually led to the higher land plants. The point where these non-algal plants begin and algae stop is usually taken to be the presence of reproductive organs with protective cell layers, a characteristic not found in the other algae groups.

The first plants on earth were algae and these still thrive in a range of aquatic habitats today. The land plants evolved from the algae, more specifically green algae. Some 400 million years ago freshwater, green, filamentous algae invaded the land. These probably had an isomorphic alternation of generations and were probably heterotrichous. Fossils of isolated land plant spores suggest land plants may have been around as long as 475 million years ago.

Brown Algae

The Phaeophyceae (or brown algae, singular: brown algae) is a large group of mostly marine multicellular algae, including many seaweeds of colder Northern Hemisphere waters. They play an important role in marine environments both as food, and for the habitats they can form. For instance Macrocystis, a member of the Laminariales or kelps, may reach 60 m in length, and forms prominent underwater forests. Another example is Sargassum, which creates unique habitats in the tropical waters of the Sargasso Sea. This is one of the few areas where a large biomass of brown algae may be found in tropical waters. Many brown algae such as members of the order Fucales are commonly found along rocky seashores. Some members of the division are used as food for humans. Worldwide there are about 1500 — 2000 brown seaweed species.

Brown algae belong to a very large group, the Heterokontophyta, a eukaryotic group of organisms distinguished most prominently by having chloroplasts surrounded by four membranes, suggesting an origin from a symbiotic relationship between a basal eukaryote and another eukaryotic organism. Most brown algae contain the pigment fucoxanthin, which is responsible for the distinctive greenish-brown colour that gives them their name. Brown algae are unique among heterokonts in developing into multicellular forms with differentiated tissues, but they reproduce by means of flagellate spores, which closely resemble other heterokont cells. Genetic studies show their closest relatives to be the yellow-green algae.

Phaeophyta first appear in the fossil record in the Mesozoic, possibly as early as the Jurassic. Their occurrence as fossils is rare due to their generally soft-bodied habit, and scientists continue to debate the identification of some finds. Other algae groups, such as the red algae and green algae have a number of calcareous members, which are more likely to leave evidence in the fossil record than the soft bodies of the brown algae. Miocene fossils of a soft-bodied brown macro algae, Julescrania, have been found well-preserved in Monterey Formation diatomites, but few other certain fossils, particularly of older specimens are known in the fossil record.

Taxonomy

This is a list of the orders in the class Phaeophyceae:

- Ascoseirales Petrov
- Choristocarpales
- Cutleriales Oltmanns
- Desmarestiales Setchell & Gardner
- Dictyotales Kjellman
- Ectocarpales Setchell & Gardner
- Fucales Kylin
- Laminariales Migula
- Ralfsiales Nakamura
- Scytothamnales A. F. Peters & M. N. Clayton
- Sphacelariales Oltmanns
- Sporochnales Sauvageau
- Syringodermatales E. C. Henry
- Tilopteridale Bessey

A few species, such as Botrydium stoloniferum, are placed incertae sedis, or of uncertain position, as to order in this classification scheme.

Life Cycle: The life cycle shows great variability from one group to another. However, the life cycle of *Laminaria* consists of the diploid generation, that is the large plant well know to most people. It produces sporangia from specialised microscioic structures, these divide meiotically (meiosis) before they are released. As they are haploid, there are equal numbers of male and female spores. With the exception of the Fucales, all brown algae have a life cycle which consists of an alternation between morphologically haploid and diploid plants referred to as Monomorphic. A dimorphic life cycle consists of an alteration between dissimilar haploid and diploid plants.

Arame

Arame, or *Ecklonia bicyclis,* is a species of kelp best known for its use in Japanese cuisine. It is one of many species of seaweed used in Japanese dishes.

Arame kelp is also harvested as a source of alginate and iodine. It contains the storage polysaccharide laminarin and the tripeptide eisenin.

Ecklonia Radiata

Ecklonia Radiata is a species of kelp found in the Canary Islands, the Cape Verde Islands, Madagascar, Mauritania, Senegal, South Africa, Oman, southern Australia, Lord Howe Island, and New Zealand.

Ecklonia radiata is a common large kelp on reefs in the warm temperate parts of the southern hemisphere. It is often the dominant form of biogenic habitat on temperate reefs and can have strong effects on associated assemblages. *Ecklonia* is known to exhibit great morphological variation through geographic separation.

The kelp is attached to the substrate by means of a holdfast, which appears similar to the branching roots of a tree. Between the light brown branching processes (haptera) there is usually a wide range of invertebrate animals taking advantage of the protected micro-environment. Above the holdfast is a long, flexible stalk or stipe, much darker in colour. Finally, there is a much divided terminal blade, yellowish brown and often crinkled and frayed. The kelp is never completely exposed at low tide.

Kelp

Kelp are large seaweeds (algae), belonging to the brown algae and classified in the order Laminariales. Despite their appearance, they are not grouped with the normal aquatic or land plants (kingdom Plantae), but instead are included in either kingdom Protista or Chromista. There are about 30 different genera. Kelp grows in underwater forests (kelp forests) in clear, shallow oceans, requiring nutrient-rich water below about 20 °C. It offers protection to some sea creatures, or food for others. It is known for its high growth rate — the genus Macrocystis and Nereocystis *luetkeana* grow as fast as half a metre a day, ultimately reaching 30 to 50 m.

Through the 19th Century, the word "kelp" was closely associated with seaweeds that could be burned to obtain soda ash (primarily sodium carbonate). The seaweeds used included species from both the orders Laminariales and Fucales. The word "kelp" was also used directly to refer to these processed ashes.

Morphology

In most kelp, the thallus (or body), consists of flat or leaf-like structures known as blades. Blades originate from elongated stem-like structures, the stipes. The holdfast, a root-like structure anchors, the kelp to the substrate of the ocean. Gas-filled bladders (pneumatocysts) form at the base of blades of American species, such as *Nereocystis lueteana* and keep the kelp blades close to the surface.

Growth and Reproduction

Growth occurs at the base of the meristem, where the blades and stipe meet. Growth may be limited by grazing pressure, for example sea urchins can reduce entire areas to urchin barrens. The kelp life cycle involves a diploid sporophyte and haploid gametophyte stage. The haploid phase begins when the mature organism releases many spores, which then germinate to become male or female gametophytes. Sexual reproduction then results in the beginning of the diploid sporophyte stage which, if lucky, will develop into a mature plant.

Uses: Bongo kelp ash is rich in iodine and alkali. In great amount, kelp ash can be used in soap and glass production. Until the Leblanc process was commercialised in the early 1800s, burning of kelp in Scotland was one of the principal industrial sources of soda ash (predominantly sodium carbonate). Alginate, a kelp-derived carbohydrate, is used to thicken products such as ice cream, jelly, salad dressing, and toothpaste, as well as in manufactured goods. Giant kelp can be harvested fairly easily because of its surface canopy and growth habit of staying in deeper water.

Kelp is also used frequently in seaweed fertilizer, especially in the Channel Islands, where it is known as *vraic*.

Kombu (*Laminaria japonica* and others), several Pacific species of kelp, is a very important ingredient in Japanese cuisine. Kombu is used to flavour broths and stews (especially dashi), as a savory garnish (*tororo konbu*) for rice and other dishes, as a vegetable, and a primary ingredient in popular snacks (such as tsukudani). Transparent sheets of kelp (*oboro konbu*) are used as an edible decorative wrapping for rice and other foods.

Kombu can be used to soften beans during cooking, and to help convert indigestible sugars and thus reduce flatulence.

Kelp in History and Culture

During the Highland Clearances, many Scottish Highlanders were moved off their crofts, and went to industries such as fishing and kelping (producing soda ash from the ashes of kelp). At least until the 1820s, when there were steep falls in the price of kelp, landlords wanted to create pools of cheap or virtually free labour, supplied by families subsisting in new crofting townships. Kelp collection and processing was a very profitable way of using this labour, and landlords petitioned successfully for legislation designed to stop emigration. But the economic collapse of the kelp industry in northern Scotland led to further emigration, especially to North America.

Natives of the Falkland Islands are sometimes nicknamed "Kelpers", but this is not used much by themselves.

Prominent Species

- Bull-head kelp, Nereocystis luetkeana, a northwestern American species. Used by coastal indigenous peoples to create fishing nets.
- Giant kelp, Macrocystis pyrifera, the largest seaweed. Found in the Pacific coast of North America and South America.
- Kombu, Laminaria japonica and others, several edible species of kelp found in Japan.

Species of Laminaria in the British Isles

- Laminaria *digitata* (Hudson) J.V. Lamouroux (Oarweed; Tangle)
- Laminaria *hyperborea* (Gunnerus) Foslie (Curvie)
- Laminaria *ochroleuca* Bachelot de la Pylaie
- Laminaria *saccharina* (Linnaeus) J.V.Lamouroux (sea belt; sugar kelp; sugarwack)

Species of Laminaria Worldwide

A comprehensive listing of species in *Laminariales*:

- Laminaria *agardhii* (NE. America)
- Laminaria angustata (Japan)
- Laminaria *bongardina* Postels et Ruprecht (Bering Sea to California)

- Laminaria *cuneifolia* (NE. America)
- Laminaria *dentigera* Klellm. (California - America)
- Laminaria *digitata* (NE. America)
- Laminaria *ephemera* Setchell (Sitka, Alaska, to Monterey County, California - America)
- Laminaria *farlowii* Setchell (Santa Cruz, California, to Baja California - America)
- Laminaria *groenlandica* (NE. America)
- Laminaria japonica (Japan)
- Laminaria *longicruris* (NE. America)
- Laminaria *nigripes* (NE. America)
- Laminaria *ontermedia* (NE. America)
- Laminaria pallida Greville ex J.Agardh (South Africa)
- Laminaria *platymeris* (NE. America)
- Laminaria *saccharina* (Linnaeus) Lamouroux (Aleutian Islands, Alaska to southern California America)
- Laminaria *setchellii* Silva (Aleutian Islands, Alaska to Baja California America)
- Laminaria *sinclairii* (Harvey ex Hooker f. ex Harvey) Farlow, Anderson et Eaton (Hope Island, British Coluumbia to Los Angeles, California - America)
- Laminaria *solidungula* (NE. America)
- Laminaria *stenophylla* (NE. America)

Other Genera in the Laminariales which may be Considered as "Kelp"

- *Alaria marginata* Post. & Rupr. (Alaska and California - America
- *Costaria costata* (C.Ag.) Saunders Japan; Alaska, California - America)
- *Durvillaea potatorum* (Labillardière) Areschoug (Tasmania; Australia)
- *Ecklonia brevipes* J.Agardh (Australia; New Zealand)

- *Ecklonia maxima* (Osbeck) Papenfuss (South Africa)
- Ecklonia radiata (C.Agardh) J.Agardh (Australia; Tasmania; New Zealand; South Africa)
- *Eisena arborea* Aresch. (Vancouver Island, British Columbia, Montrey, Santa Catalina Island, California - America)
- *Egregia menziesii* (Turn.) Aresch.
- *Hedophyllum sessile (C.Ag.) Setch (Alaska,* California - *America)*
- *Macrocystis angustifolia* Bory (Australia; Tasmania and South Africa)
- *Pleurophycus gardneri* Setch. & Saund. (Alaska, California - America)
- *Pterygophora californica* Rupr. (Vancouver Island, British Columbia to Bahia del Ropsario, Baja Californis and California - America)

Interactions

Some animals are named after the kelp, either because they inhabit the same habitat as kelp or because they feed on kelp. These include:

- Kelp crab (Pugettia producta), the Pacific coast of North America.
- Kelpfish (blenny) (e.g., *Heterosticbus rostratus*, genus Gibbonsia), the Pacific coast of North America.
- Kelp Goose (kelp hen) (*Ocydromus fuscus*), South America and the Falkland Islands
- Kelp Pigeon (sheathbill) (*Chionis* sp), Antarctic

Ascoseirales

Ascoseirales is an order in the Phylum Phaeophyta or Brown Algae. It is a monotypic order of a single species *Ascoseira mirabilis* Skottsberg.

Bladder Wrack

Bladder Wrack (Fucus Vesiculosus L.) is a seaweed found on the coasts of the North Sea, the western Baltic Sea, and the

Atlantic and Pacific Oceans. It was the original source of iodine, discovered in 1811, and was used extensively to treat goitre, a swelling of the thyroid gland related to iodine deficiency. In the 1860s, it was claimed that bladder wrack, as a thyroid stimulant, could counter obesity by increasing the metabolic rate and, since then, it has been featured in numerous weight-loss remedies.

Bladder wrack is also known by the names *bladderwrack, black tang, rockweed, bladder Fucus, seawrack, sea oak, black tany, cut weed* and *rock.*

Description: *Fucus vesiculosus* is a very variable algae. It can grow to 100 cm or more and is easily recognised by the small gas filled vesicles which occur in pairs on either side of a central midrib running along the centre of the strap-like frond. Ascophyllum nodosum also has air vesicles, however, they are not in pairs but in series along a frond which is not flattened and without a midrib. Both are common species on most shores of the British Isles.

The large vegetative phase is diploid and gametophytic. Meiosis occurs during the formation of the gametes. The egg cells are formed on the female plants in conceptacles embedded in receptacles. The spermatozoids and egg cells are extruded from the conceptacles. Once fertilized the zygote settles and grows to form the new diploid gametophyte.

Distribution: The species is common especially on sheltered shores from the middle littoral to lower intertidal levels. It is rare on exposed shores where any specimens may be short, stunted and without the air vesicles.

Fucus vesiculosus is one of the most common algae on the shores of the British Isles. It is recorded from the Atlantic shores of Europe, the Baltic Sea, Greenland, Azores, Canary Islands and Madeira. On the Atlantic coast of North America from Ellesmere Island, Hudson Bay to North Carolina.

Consumption: A common food in Japan, it is used as an additive and flavouring in various food products in Europe.

Bladder wrack is commonly found as a component of kelp tablets or powders used as nutritional supplements. It is sometimes loosely called kelp, but that term technically refers to a different seaweed.

Primary chemical constituents of this plant include mucilage, algin, mannitol, beta-carotene, zeaxanthin, iodine, bromine, potassium, volatile oils, and many other minerals. The main use of bladder wrack (and other types of seaweed) in herbal medicine is as a source of iodine, an essential nutrient for the thyroid gland. Bladder wrack has proved most useful in the treatment of underactive thyroid glands (hypothyroidism) and goitre. Through the regulation of thyroid function, there is an improvement in all the associated symptoms. Where obesity is associated with thyroid trouble, this herb may be very helpful in reducing the excess weight. It has a reputation in helping the relief of rheumatism and rheumatoid arthritis, both used internally and as an external application upon inflamed joints.

A chemical constituent of bladder wrack called alginic acid swells upon contact with water; when taken orally, it forms a type of "seal" at the top of the stomach, and for this reason is used in several over-the-counter preparations for heartburn. The same constituent gives bladder wrack laxative properties as well. Other proposed uses of bladder wrack include treating atherosclerosis and strengthening immunity, although there is no scientific evidence at present that it works for these purposes.

Bladder wrack should not be used in cases of hyperthyroidism or cardiac problems, or during pregnancy and lactation. Excessive dosage (many times the recommended dosage) may lead to hyperthyroidism, tremor, increased pulse rate and elevated blood pressure.

Carpophyllum Maschalocarpum

Carpophyllum maschalocarpum, commonly known as flapjack, is a brown algae that grows in crevices of rocks at low tide levels. It is a common algae in New Zealand. They don't have branches, leaves or flowers like land plants. They are attached to rocks, not by roots, but by holdfasts. They have flexible stems and their tops are called fronds. Many look and feel like leather and usually have air-filled bladders or spongy cells inside the stems that help the water support them. They feed by using a process called photosynthesis.

Chordariales

Chordariales is an order in the Phylum Phaeophyta or Brown algae. Member of this order are may be filamentous, crustose with fused cells at the base or they may be terete and differentiated into a central medulla and an outer photosynthetic cortex. They have a sporphytic thallus usually aggregated to form a pseudo parenchyma.

As their general name suggests their pigmentation is brown.

Colpomenia Peregrina

Colpomenia peregrina is a seaweed (algae) not native to the British Isles but recorded in Ireland since 1934. It appears to have been introduced from the Pacific and was first noticed in Europe in 1906 on oyster beds. It has been recorded throughout the eastern north Atlantic from Norway and Sweden to Portugal. *Colpomenia peregrina* was first noticed in Europe in 1906, in Britain in 1908 and in Ireland in 1934.

Description: *Colpomenia peregrina* is a small brown algae (Phaeophyta), bladder-like, hollow and membranous, up to 9 cms across. The surface is thin and smooth but often collapsed or torn when older. Olive brown in colour and attached by rhizoidal filaments to rock at the base. There are two species in Europe: *C. sinuosa* and *C. peregrina*. *C. sinuosa* was present at least as far back as the 1840s in Spain and *C.peregrina* was introduced and first noticed by oyster fishermen in the Bay of Biscay in 1906. It was first noticed in Britain in 1907 in Cornwall and Dorset. The two species are superficially similar and in older texts, such as Knight and Parke (1931), *C. peregrina* is referred to as *C. sinuosa*.

Leathesia difformis (L.) Aresch is similar, it is yellow brown in colour, fleshy and mucilaginous in texture. It is globose and smooth when young becoming hollow and convoluted with age and growing to 5 cms in diameter. It is easily distinguished as it readily squashes when pressed under finger and thumb.

Ecology: In littoral rock pools, not exposed and in the sublittoral to 3m depth.

Specimens

There are specimens stored in the Ulster Museum Herbarium from counties in Northern Ireland, Co. Donegal and Wales under the catalogue numbers: F11254; F3136; F7675; F6154; F1682; F1693; F7491; F7674; F4254; F4254 and F1832.

Distribution

Irish Distribution: *Colpomenia peregrina* has been recorded in Ireland from Counties: Down, Donegal, Kerry, Galway, Clare and Cork. Apparently this algae was first recorded in Ireland by M.J.Lynn from Strangford Lough in March 1934 and from Lough Larne near Ballycarry and Magheramorne in 1935. It was also recorded from Portballintrae, on the north coast, and in the south at Lough Ine. In 1936, it was found at Rush (County Dublin) and at Killough (County Down). There are further records of from Portstewart (County Londonderry), cast ashore at Hood's Ferry, Islandmagee (opposite Larne), (County Antrim).

Britain: Generally around the British Isles. In Hardy and Guiry (2006) it is shown to be generally recorded around Ireland, south west England, Wales and the west coast of Scotland. Records from the east of Scotland are few and it not shown as present on the east or southeast coast of England. It is noted as a recent addition to the flora (as *C. sinuosa* in Knight and Parke (1931).

Cutleriales

Cutleriales is an order in the Phylum Phaeophyta or Brown algae. It is composed of 2-6 genera, based upon various circumscriptions. The circumscription in the taxobox (right) is the list of genera from AlgaeBase, although only 3 of these genera are listed as containing species, 11 species in *Cutleria*, and one each in *Microzonia*, and *Zanardinia*. Lee circumscribes the order with only the two genera, *Cutleria* and *Zanardinia*.

Cystoseira

Description: Cystoseira is characterised by highly differentiated basal and apical regions and presence of catenate pneumatocysts (air-vesicles). In Cystoseira old plants have an elongated main axis, and in time the primary laterals become proportionally elongated. Their lower parts are strongly flattened into 'foliar expansions' or basal leaves. Fertile regions which bear conceptacles are known as receptacles. These are normally found at the tips of the branches. Their basal and apical regions are highly differentiated. They posses catenate pnuematocysts (air vesicles). The aerocyst or air vesicles keep the organism erect, by causing it to float in strong currents.

Distribution: Cystoseira is one of the most widely distributed genera of the Fucales order and provides an essential habitat for many epiphytes, invertebrates, and fish. Cystoseira is found mostly in temperate regions of the Northern Hemisphere, such as the Mediterranean, Indian, and Pacific Oceans.

Dabberlocks

Alaria esculenta (Linnaeus) Greville.

Dabberlocks also called Winged kelp or Bladderlochs, is an edible seaweed. It is a traditional food along the coasts of the far north Atlantic Ocean.

There are about 12 species in the genus *Alaria* of these there is only one in the British Isles.

Description: *Alaria esculenta* grows to a maximum length of 2 m. The frond consists of a distinct midrib with wavy membranous lamina up to 7 cm wide on either side. The frond tapers towards the end. The base has a short stipe arising from a rhizoidal holdfast. The stipe may bear several sporophylls, club-shaped and up to 20 cm long and 5 cm broad. The whole frond is brown.

Ecology: Alaria esculents grows attached to rocks just below low water; it is common on rocky shores in exposed places.

Distribution

British Isles: Common with a distinct northwestern distribution.

World

Europe: British Isles,....North America: Northeast coast of North America.

Uses: *Alaria* is used either fresh or cooked in Greenland, Iceland, Scotland and Ireland.

Desmarestiales

Desmarestiales is an order in the Phylum Phaeophyta or Brown algae. Member of this order have terete or ligulate (flat) pinnately branched thalli attached by discoid holdfasts. They have a sporphytic thallus usually aggregated to form a pseudo parenchyma.

As their general name suggests their pigmentation is brown.

Dictyosiphonales

Dictyosiphonales is an order in the Phylum Phaeophyta or Brown Algae. Member of this order have a sporphytic thallus with true parenchyma formed by both longitudinal and transverse cell division with a thallus that is solid, tubular flat, saccate, branched or foliose.

The Dictyosiphonales include some common littoral seaweeds but the members of the order generally have a relatively undifferentiated structure without the conventional construction of a holdfast, stipe and lamina.

As their general name suggests their pigmentation is brown.

Dictyotales

Dictyotales is large order in the Phylum Phaeophyta or Brown Algae. Members of this order generally prefer warmer waters than other brown algae. One genus in this order is calcareous, *Padina*, the only calcareous member of this phylum.

Fucaceae

Fucaceae are a family of brown algae. There are 8 genera in the family, most notably Fucus, which includes several common seaweeds.

Fucales

Fucales is an order in the Phylum Phaeophyta or Brown algae. Members of this order are fucoids. The list of families in *Fucales*, as well as additional taxonomic information on algae, is publicly accessible at Algaebase.

The Class Phaeophyceae is included within the Division Heterokontophyta. This name comes from the Greek word *phaios* meaning "brown" and *phyton* meaning plant. They include some of the largest plants in the sea, some however, are small and fine in structure.

Classification: The Class Phaeophyceae is divided into the following orders.

- Ectocarpales
- Sphacelariales
- Syringodermatales
- Dictyotales
- Scytosiphonales
- Cutleriales
- Dictyosiphonales
- Chordariales
- Sporochnales
- Desmarestiales
- Laminariales

- Fucales
- Durvillaeales
- Ascoseirales

The Fucales include some of the more common littoral seaweeds and the members of the order have the typical seaweed construction of a holdfast, stipe and lamina. The lamina is often much branched and may include gas filled bladders. Growth is by division of the apical cells.

They are oogamous where there is fusion between the small male gamete and the large female gamete.

As their general name suggests their pigmentation is brown. All species are multicellular. Full details of the characteristics are complex and consist of details of the flagella, shape of the chloroplasts, structure of the cell walls and details of the life-cycle.

Fucus

Fucus is a genus of brown algae (seaweed) that lives in the intertidal zones of rocky shores. A common species found on the Atlantic coasts of Europe and North America is Fucus vesiculosus or bladder wrack. On the Pacific coast of North America, the most common species is Fucus gardneri.

Description: The thallus is perennial with an irregular or disc-shaped holdfast or with haptera. The erect portion of the thallus is dichotomous or subpinnately branched, flattened and with a distinct midrib. Gas-filled pneumatocysts (air-vesicles) are present in pairs in some species, one on either side of the midrib. The erect portion of the thallus bears cryptostomata and caecostomata (sterile surface cavities). The base of the thallus is stipe-like due to abrasion of the tissue lateral to the midrib and it is attached to the rock by a holdfast. The gametangia develop in conceptacles embedded in receptacles in the apices of the final branches. They may be monoecious or dioecious.

Ecology: In the British Isles, the larger brown algae occur on sheltered shores in fairly well defined zones along the shore from high water mark to below low water mark. On the more exposed shores not all of these species can be found and on very exposed shores few, if any, occur. Pelvetia canaliculata forms a zone at the top of the shore. Just below this *Fucus spiralis*, Fucus vesiculosus and Fucus serratus and Laminaria form clear zones, one below the other, along the shore down to low water mark. On sheltered shores Ascophyllum nodosum usually forms a broad and dominating zone along the shore at the mid-littoral. On the more expopsed shore little of it if any can be found. On very exposed shores few plants of these species occur while on very exposed shores none can be found. Other brown algae can be found at the low-littoral such as *Himanthalia, Laminaria saxatilis* and *Alaria esculenta*. Small green and red algae and animals occur under these large brown algae.

Uses: In Scotland and Norway, up until the mid 19th century, several seaweed species from *Fucus* and other genera were harvested, dried, burned to ash, and further processed to become "kelp," which was a type of soda ash that was less costly in Britain than the barilla imported from Spain (after consideration of the import duties). This kelp had an alkali content of about 2.5-5 per cent that was mainly sodium carbonate (Na_2CO_3); alkali is essential to soapmaking, glassmaking, and other industries. The seaweed was also used as fertilizer for crop land in the same areas in which it was harvested. The purest barilla had a sodium carbonate concentration of about 30 per cent.

It has been reported that to loosen dried earth, a machine which sticks probes a metre into the ground and loosens the earth by blasting air in under pressure. Dried seaweed (bladderwrack) can then be injected into the fissures to hold the drainage cracks open.

Medical: In 2005, it was announced that bacteria grown on *Fucus* have the ability to attack and kill the MRSA superbug.

Taxonomy

This list of species of *Fucus* excludes names of uncertain status.

- Fucus ceranoides L.
- Fucus chalonii Feldmann
- Fucus cottonii M.J. Wynne & Magne (=*Fucus cottonii* M.J.Wynne & Magne *nom. illeg.*)
- Fucus distichus L.
- Fucus evanescens C.Agardh
- Fucus gardneri P.C. Silva
- Fucus nereideus Lightfoot
- Fucus serratus L. — toothed wrack
- Fucus spermophorus L.
- Fucus spiralis L.
- Fucus tendo L.
- Fucus vesiculosus L.— bladder wrack
- Fucus virsoides J. Agardh

Life Cycle: These algae have a relatively simple life cycle and produce only one type of thallus which grows to a maximum size of 2 m. Fertile cavities, the conceptacles, containing the reproductive cells are immersed in the receptacles near the ends of the branches. After meiosis, oogonia and antheridia are produced and released, fertilization follows and the zygote develops directly into the diploid plant. It may be considered to be analogous to the life cycle of the flowering plant, however, in algae, the oogonia are released and fertilized in the sea while in flowering plants the ovules are fertilized while attached to the parent plant and then released as a fertilized seed.

Fucus Distichus

F. distichus is up to 10 cm long with a short stout cylindrical stipe, branching dichotomous, flat and with a midrib. *F. distichus* subsp. *edentatus* was first described from Shetland Islands by

Borgesen in 1903. Powell found *F. distichus* subsp. *anceps* on the north coast of Caithness. It had also been recorded from: Orkney Islands, Fair Isle, St Kilda and the Outer Hebrides in Scotland; in Ireland it had been recorded from Counties Clare, Donegal and Kerry. Two subspecies of *F. distichus* (subsp. *anceps* and subsp. *edentatus*) have been described from the British Isles.

Fucus distichus is the organism used as a model to study the development of cell polarity, since it forms an apolar zygote that can develop polarity given a varying number of gradients.

Fucus Serratus

F. serratus, toothed wrack, is the most distinctive of all the *Fucus* species. It clearly shows a distinctive serrated edge to the frond not shown by the other species of the genus.

Fucus Spiralis

F. spiralis is one of the three most common algae on the shores of the British Isles. It grows to about 40 cm long and does not show air bladders or a toothed edge as is found on *F.vesiculosus* and *F. serratus*. It forms a zone near the top of the shore above the zones of *F. vesiculosus* and *F. serratus*.

Fucus Vesiculosus

This is one of the most common species of *Fucus*, common on most shores in the mid-littoral. Readily identified by a distinct midrib and air vesicles in pairs on either side of the midrib.

Fucus Spiralis

Fucus Spiralis is a species of seaweed, a brown algae (Heterokontophyta, Phaeophyceae), living on the littoral shore of the Atlantic coasts of Europe and North America. It has the common name of spiral wrack. It is similar to Fucus vesiculosus and Fucus serratus the other common species of Fucus on the coasts of Europe. *Fucus spiralis, Fucus vesiculosus* and *Fucus serratus* along with Ascophyllum nodosum form the main and dominant seaweeds on rocky shores of the British Isles. These three species, along with two others Pelvetia canaliculata and Ascophyllum nodosum form a series of zones along the shore.

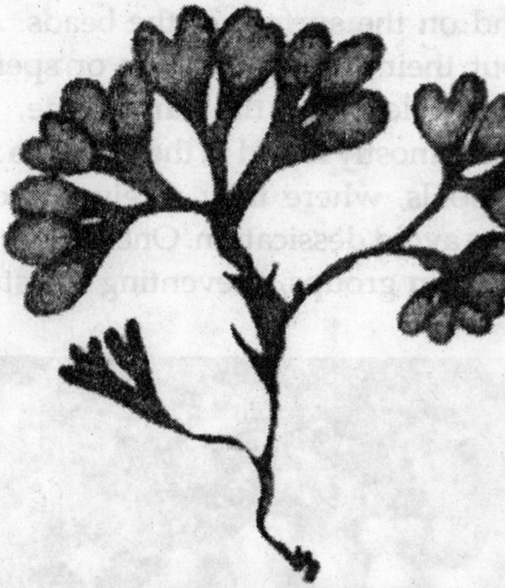

Distribution: *F. spiralis* is common on the western coasts of Europe, Canary Islands and Northeastern America.

Characteristics: *Fucus spiralis* has a prominent midrib and no gas bladders. The frond has a tendency to twist in a spiral, but this is not a reliable characteristic. The reproductive bodies form rounded swollen tips on their branches, which are usually in pairs. *F. spiralis* is olive brown and 150–200 mm long.

Hormosira Banksii

Neptune's Necklace (*Hormosira banksii*), also known as 'sea grapes' and 'bubbleweed', is a type of seaweed found in Australia and New Zealand. It is a common seaweed found in rock pools at mid-tide levels and is characterised by strings of olive-brown hollow beads. The beads have a slimy layer which keeps the moisture locked up inside. They also store gas, allowing them to stand up in water. This is allows them to obtain more sunlight and to move with the flow of the current. Each bead is a knobbly bladder filled with water that prevents dessication between tides. The unattached form of this seaweed (commonly found in mangroves) reproduce asexually, from broken fragments. The attached form of the seaweed reproduce sexually. Their sex organs are the pimple like features found on the surface of the beads. At high tide plant squeezes out their clusters of eggs or sperm in sticky masses. All plants release at the same time, maximising fertilization. They are mostly found in the mid-tide zone (littoral zone) or in rock pools, where they receive much light and enough sea water to avoid dessication. One of their behavioural adaptations is living in groups, preventing moisture loss and drying out.

The plants, especially when young, are edible

Knotted Wrack

Knotted Wrack or Egg Wrack (*Ascophyllum nodosum* (Linnaeus) Le Jolis) is a large, common, brown algae, in the Phaeophyceae, an edible seaweed of the northern Atlantic Ocean. It is a common species on the northwestern coast of Europe (from Spitzbergen to Portugal) including east Greenland, and the northeastern coast of North America.

Ascophyllum is very popular amongst the science community and has been claimed to be both the most active seaweed on the planet as well as the most researched by the academic community.

Description

Ascophyllum nodosum has long fronds with large egg-shaped air-bladders set in the fronds, not stalked, at regular intervals along the fronds which can reach up to 2 m in length. They are attached to rocks and boulders by a holdfast. The fronds are olive-brown in colour and somewhat compressed but without a midrib.

This seaweed grows quite slowly and can live for several decades; it may take approximately five years before becoming fertile.

Varieties and Forms

Several different varieties and forms of this species have been described.

Ascophyllum nodosum var. *minor* has been described from Larne Lough in Northern Ireland.

The species is found in a range of coastal habitats from sheltered estuaries to moderately exposed coasts, it often dominates the intertidal zone (although sub-tidal populations are known to exist in very clear waters).

There are free floating ecads of this species such as *Ascophyllum nodosum mackaii* Cotton, which is found at very sheltered locations, such as at the heads of Scottish sea lochs and Ireland.

It has been recorded as an accidental introduction to San Francisco, California, and eradicated as a potential invasive species there.

Ecology: *Ascophyllum nodosum* is found mostly on sheltered sites on shores in the mid-littoral where it can become dominant species in the littoral.

Distribution: Recorded in Europe: Faeröes (Faroe), Ireland, Britain and Isle of Man; Netherlands; North America: Bay of Fundy, Nova Scotia, Baffin Island, Hudson Strait, Labrador and Newfoundland.

Life History: Life history is of one diploid plant and gametes. The gametes are produced in conceptacles embedded in yellowish receptacles on short branches.

Notable Compounds

Ascophyllum nodosum is harvested for use in alginates, fertilizers and for the manufacture of seaweed meal for animal and human consumption. It has long been used as an organic and mainstream fertilizer for many varieties of crops due to its combination of both macronutrient, (e.g. N, P, K, Ca, Mg, S) and micronutrients (e.g. Mn, Cu, Fe, Zn, etc.) It also host to cytokinins, auxin-like, gibberellins, betaines, mannitol, organic acids, polysaccharides, amino acids, and proteins which are all very beneficial and widely used in agriculture.

Uses: *Ascophyllum nodosum* along with *Macrocystis pyfera*is harvested in Ireland, Scotland and Norway from which alginates are extracted. It is one of the world's principal alginate supply.

Kombu

Kombu or konbu are edible kelp from the genus Laminaria widely eaten in Northeast Asia.

Over 90 per cent of Japanese kombu is cultivated, and most is harvested, in Hokkaido. It is cultivated as far south as the Seto Inland Sea.

History: The earliest written record of kombu appeared in Shoku Nihongi in 797 as a gift and tax from the Tohoku Region. Its actual use is believed to be much earlier, most likely dating back to the Jomon period, but as it easily decomposes, no archaeological evidence can be found. During the Muromachi period, a newly developed drying technique allowed kombu to be stored for more than a few days and kombu became an important export from the Tohoku area. By the Edo period, as Hokkaido was colonised and shipment routes were organised, the use of kombu became widespread throughout Japan. Traditional Okinawan cuisine relies heavily on kombu as a part of the diet; this practice began in the Edo period. In Okinawa, the consumption of kombu per household is the highest of all prefectures. In the 20th century, a way to cultivate kombu was discovered and kombu became cheap and readily available everywhere.

Cooking: Kombu is used extensively in Japanese cuisines as one of the three main ingredients needed to make dashi, a soup stock. Kombu is usually sold dried or in a dried shred called "Oboro kombu". It may also be eaten fresh as sashimi. Making kombu dashi is simple though kombu dashi powder may also be used. A strip of dried kombu is often boiled from the very first step of making a dish and is commonly eaten after cooking.

It is also important in Chinese cuisine and Korean cuisine.

Kombu may be pickled with sweet and sour flavouring and are cut into small strips 5 or 6 centimetres long and 2 centimetres wide. These are often eaten as a snack with green tea.

Kombu is a rich source of glutamic acid, an amino acid responsible for umami, one of the five basic tastes.

It can be included when cooking beans to add nutrients and improve their digestibility.

Prominent Species

Japanese name followed by species:

- karafuto kombu *L. saccharina* contains mannitol and considered sweeter
- ma-kombu *L. Japonica*
- mitsuishi-kombu or dashi-kombu *L. angustata* commonly used in the making of dashi
- naga-kombu *L. longissima*
- Rishiri-kombu *L. ochotensis* commonly used for soup stock

Laminaria

Laminaria (a botanical name) is a genus of 31 species of brown algae (Phaeophyceae) all with the common name kelp.

This economically important genus is noted for being a long, leathery giant seaweed. It has a worldwide distribution.

Uses: *Laminaria* is used in the production of potassium chloride and iodine. It can be used medicinally to induce dilation of the cervix. It is also a nutritious food. The species *Laminaria japonica* is the main component of the Japanese food kombu.

Laminaria World-Wide

- *Laminaria angustata* Kjellman Japan
- *Laminaria dentigera* Kjellm. North America Pacific shores.
- *Laminaria ephemera* Setch. North America Pacific shores.
- *Laminaria farlowii* Setch. North America Pacific shores.
- *Laminaria sinclairii* (Harv.) Farl., Anders. & Eaton. North America Pacific shores.

Britis Isles

The following species have been recorded from the British Isles:-

- *Laminaria digitata* (Hudson) J.V. Lamouroux (Tangle)
- *Laminaria hyperborea* (Gunnerus) Foslie (Cuvie)
- *Laminaria ochroleuca* Bachelot de la Pylaie
- *Laminaria saccharina* (Linnaeus) J.V. Lamouroux (Sea belt)

- *Laminaria longricruris* Bachelot de la Pylaie (The occurrence of this species within the British Isles has not yet been confirmed).

Life History

The life history of *Laminaria* shows two stages. An alteration between a large diploid sporophyte and a very small, microscopic, haploid microthallus. The large phase is a diploid sporophyte, after meiosis it produces gametes with flagella. Half of these grow into male gametophytes and half into female gametophytes and form small microscopic branched filaments. The male gametophytes bear small unicellular antheridia which fertilizes the egg cell. The zygote then develops and grows to become the sporophyte.

Laminariaceae

Laminariaceae are a family of seaweeds, classified as brown algae. The family includes the largest known seaweeds: *Nereocystis* and *Macrocystis*.

Leathesia Difformis

Leathesia difformis, the sea potato, is a littoral brown algae in the phylum Phaeophyta and the order Chordariales which is commonly attached to other seaweeds and sometimes rocks. When young, the organism is solid but as it matures it become hollow and somewhat convoluted and has the appearance of a small leathery brown bag about the same size as a tennis ball. The texture is rubbery and the outer surface smooth.

Lessoniaceae

Lessoniaceae are a family of kelp. They are the largest of all algae, and are characterised by the single blade splitting into two. Some species have pneumatocysts at the base of their blades.

Taxonomy

The genera of this family can be hard to tell apart, but can be differentiated based on their habitat and the position of their pneumatocysts.

- Lessonia is relatively short, with multiple stipes, inhabiting intertidal or low tidal areas. This genus has no pneumatocysts.
- Macrocystis is the largest of the kelps, and has multiple stipes and multiple pneumatocysts on each stipe, with one blade each.
- Nereocystis has a single stipe, and a single pneumatocyst with multiple blades.
- Pelagophycus is much like *Nereocystis*, but has two separate stalks emerging from the pneumatocyst, from which sprout the blades, earning it its nickname of "elk kelp."
- Postelsia is intertidal, and for most of the time is exposed to the air. It has a single upright stipe, no pneumatocyst, and multiple hanging blades.

Other genera are:

- Dictyoneurum
- Lessoniopsis

Distribution

The distribution of Lessoniaceae includes the northeastern Pacific Ocean and southern Australian coasts.

Macrocystis

Macrocystis is a genus of kelp (Algae). This genus contains the largest of all the Phaeophyceae or brown algae. *Macrocystis* has pneumatocysts at the base of its blades. Sporophytes are perennial, and individual stipes may persist for many years. Common along the coast of the eastern Pacific Ocean, from central California to Baja California.

Description

Macrocystis is a genus of kelp, some species of which are so huge that the plants may grow to up to 60m. The stipes arise from a holdfast and branch three or four times from near the base. Blades develop at irregular intervals along the stipe. *M. pyrifera* grows to over 45 m long.

Life Cycle

The sporophyte has many sporangia located in its blades, which, through meiosis, releases haploid spores, which will grow into female and male gametophytes. These gametophytes grow through mitosis, and produce gametes, sperm and eggs. The gametophytes are microscopic. The female has larger and fewer cells then the male. The male releases its sperm, which finds its way to the female, following a pheremone released by the female, where it fertilizes the egg, to form the zygote, which, through mitosis, begins growth.

Growth

Juvenile giant kelp grow directly on the parent female gametophyte, extending one or two primary blades, and beginning a rudimentary holdfast, which will eventually cover the gametophyte completely. Growth occurs with lengthening of the stipe, and splitting of the blades. This occurs by means of small tears where the blade meets the stipe, which splits the stipe into two. Pneumatocysts grow after the first few blade splittings.

Species

There are four species of *Macrocystis* in the world. *Macrocystis angustifolia* Bory de Saint-Vincent; *M. intergrifolia* Bory de Saint-Vincent; *M. laevis* C.H.Hay; *M. pyrifera* C.Ag; and *M. pyrifera* var. *humboldtii* Bomplan.

Macrocystis pyrifera, giant kelp, is to be found in North America (Alaska to California), South America, South Africa and southern Australia. *Macrocystis integrifolia*, a smaller, intertidal species, is found on the Pacific coast of America (British Columbia to California) and South America. *M. pyrifera* grows to over 45 m long and can do so in one growing season, making it the organism with the fastest linear growth. *M. integrifolia* is much smaller, the sporangial thalli growing only to 6 m long; it is found on intertidal rocks or shallow subtidal rocks.

Nereocystis

Nereocystis (Greek for "mermaid's bladder") is a genus of kelp. It forms thick beds on rocks, and is an important part of kelp forests. There is only one species, *Nereocystis luetkeana*. It can grow to a maximum of 36 metres. *Nereocystis* has a holdfast of about 40 centimetres, and a single stipe, topped with a pneumatocyst containing carbon monoxide, from which sprout the numerous (about 30-64) blades. The blades may be up to 4 metres long, and up to 15 centimetres wide. It is usually annual, sometimes persisting up to 18 months. *Nereocystis* is the only kelp which will drop spore patches, so that the right concentration of spores lands near the parent's holdfast. It is common along the coast of the northeastern Pacific Ocean, from about Monterey, California to Aleutian Islands, Alaska.

Padina Pavonica

Padina pavonica, commonly known as Peacocks tail, is a brown algae found in Atlantic Ocean. Padina Pavonica is also the key primary ingredient to the highly acclaimed cosmetic product called "Pavonica" by Wagner Labs. This cosmetic skin moisturiser and anti-aging serum combines the wonderful properties of this pacific algae with other well known and very effective moisturising ingredients to create what many are referring to as "the ultimate skin luxury"

Pelvetia canaliculata

Canaliculata (channelled wrack) is a very common brown seaweed (Phaeophyta) found on the rocks of the upper shores of Europe. It grows to a maximum length of 15cms in dense tufts the fronds being deeply channelled on one side. They are irregularly dichotomously branched with terminal receptacles. The colour of *Pelvetia* is dark brown to almost black.

The channels help prevent the seaweed drying out, and a mucus layer helps reduce water loss. The reproductive bodies form swollen, irregularly shaped, deeply forked tips. It is distinguished from other large brown algae by the channels along the frond. It has no midrib, no air-vesicules and is the uppermost algae on the shore growing at or above high water mark. It grows on rocks, forming a zone along the upper shore at the upper littoral zone. *Pelvetia canaliculata* is common on the Atlantic shores of Europe from Iceland to Spain.

Pneumatocyst

Phytology: A Pneumatocyst is a large float containing gas found in brown algae. An organism may have more than one. They provide buoyancy to lift the blades towards the surface, allowing them to receive more sunlight for photosynthesis.

The proportion of gases in the pneumatocysts adjusts by equilibrating with the surrounding air or water. The pneumatocyst can hold O_2, CO_2 and CO.

Medicine: A pneumatocyst can also be an air bladder in the lungs which generally appears after a pulmonary trauma. These also go by the name of traumatic pneumatoceles.

Postelsia

Postelsia, or the Sea Palm (not to be confused with the southern sea palm), is a multicellular genus of kelp. There is only one species, *P. palmaeformis*. They are green in colour as juveniles, and change to a golden brown as they age, reaching a height of 50 to 75 centimetres. *Postelsia* is a partially protected species in California.

The sea palm was known by the natives of California by the name of Kakgunu-chale before any Europeans entered the region. *Postelsia* was first scientifically described by Franz Joseph Ruprecht (1814-1870) in 1852 from a specimen found near Bodega Bay in California. Ruprecht, an Austro-Hungarian who became curator of botany at the Academy of Sciences in St. Petersburg in 1839, studied seaweed specimens collected by botanist Ilya Vosnesensky, and published a paper describing one seagrass and five seaweeds, one of which was *Postelsia*. The sea palm has been used by several textbooks, such as the Campbell - Reece Biology textbook, as an example of multicellular protists, as well as an example of the class Phaeophyceae.

The binomial name of the sea palm, *Postelsia palmaeformis*, represents two things. The generic name honours Alexander Philipou Postels, an Estonian-born geologist and artist who worked with Ruprecht, while the specific name describes the alga's similarity to true palms.

Morphology

Postelsia has two distinct morphologies - one for its diploid sporophyte stage, which is the dominant portion of the life cycle, and one for its smaller, haploid gametophyte stage. Like all seaweeds, the sporophyte stage of *Postelsia* consists of a thallus, which is made up of a stem-like stipe topped with possibly over 100 leaf-like blades, and rests on a root-like holdfast. The holdfast anchors the organism to the rocks it lives on. The sea palm has no vascular system; the stipe is only for support of the organism and holds the fronds up over other organisms so they can receive more light. The stipe is merely a firm, hollow tube, able to withstand the open air of low tide conditions as well as the crashing waves of high tide. The blades are grooved, with the sporangia held within these grooves. The gametophyte stage is microscopic, consisting of only a few cells. The gametophytes produce sperm and eggs to create new sporophytes.

Life Cycle

The blades of the new sporophyte grow from an one or two initial blades by splitting. A tear forms in the middle of the blade at its base, which then continues along the entire length of the blade until it is split in two.

Like all phaeophytes, sea palms use the pigments chlorophyll *a*, chlorophyll *c*, fucoxanthin, and carotenes in photosynthesis. Their cell walls are composed of alginate. They use laminarin and mannitol for storage.

Juvenile Sea Palms

Like most brown algae, *Postelsia* goes through alternation of generations, and is an annual species. The diploid sporophyte produces, through meiosis, haploid spores, which drip down through the grooves in the blades onto the substrate, which may be mussels, barnacles, or bare rock. These spores develop, through mitosis, into small, multicellular haploid gametophytes, male and female. The male and female gametphytes create sperm and eggs, respectively. The sperm of the male reaches the female egg and fertilizes, resulting in a diploid zygote, which develops into a new sporophyte.

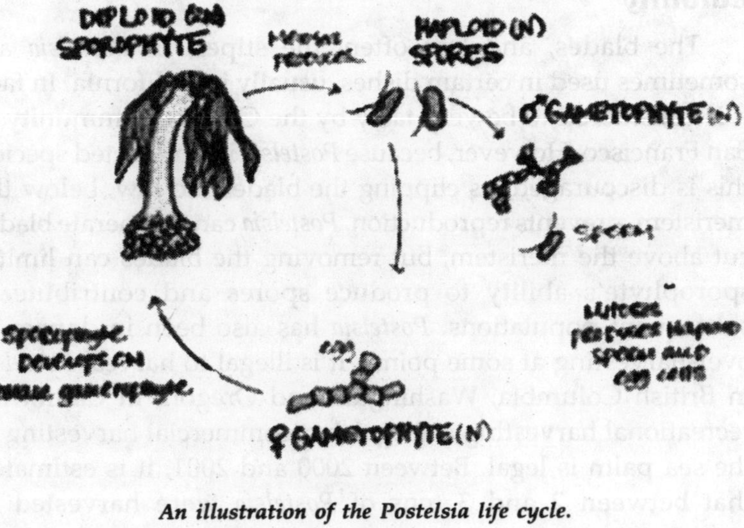

An illustration of the Postelsia life cycle.

Habitat

Sea palms are found on the rocky shores of western North America, from as far north as Vancouver Island, to the southern central coast of California. They live in the middle to upper intertidal zones in very wavy areas. High wave action may increase nutrient availability and moves the blades of the thallus, allowing more sunlight to reach the organism so that it can photosynthesise. In addition, the constant wave action removes competitors, such as mussels.

When *Postelsia* release their spores, they tend to fall within a few metres of the parent sporophyte, for two reasons. The first is that though spores are flagellated and can swim, they are often released at low tide and are deposited directly to the substrate below. Secondly, *Postelsia* gametophytes need to be close to each other in order for fertilization to occur. As such, sea palms tend to live very close to each other in large aggregations. Some juvenile sporophytes will grow on competing organisms, like mussels or barnacles, and rip them from the rocks when the waves come, gripping them with holdfasts of incredible strength.

Edibility

The blades, and less often, the stipes, of *Postelsia* are sometimes used in certain dishes, usually in California. In fact, it is considered to be very tasty by the Chinese community of San Francisco. However, because *Postelsia* is a protected species, this is discouraged, as clipping the blades too low, below the meristem, prevents reproduction. *Postelsia* can regenerate blades cut above the meristem, but removing the blades can limit a sporophyte's ability to produce spores and contribute to subsequent populations. *Postelsia* has also been in danger of over-harvesting at some points. It is illegal to harvest *Postelsia* in British Columbia, Washington and Oregon. In California, recreational harvesting is illegal, but commercial harvesting of the sea palm is legal. Between 2000 and 2001, it is estimated that between 2 and 3 tons of *Postelsia* were harvested in

California. The blades are eaten raw or are dried, and dried blades sell for up to US$45 per pound. Commercial harvesters of *Postelsia* must purchase a US$100 licence, pay a royalty to the State of California (US$24 per wet ton of algae harvested), and submit a monthly harvest log.

Pylaiella

Pylaiella (Mung) is a genus of seaweed (brown algae) that can be a nuisance due to its ability to coat people, ropes, animals, and more when it blooms close to the shore under particular conditions.

Sargassum

Sargassum is a genus of generally planktonic macroalgae (seaweed) in the order Fucales.

Description

The macroalgae, which may grow to a length of several metres, are generally brown or dark green in colour and are differentiated into holdfast, stipes, fronds and fruiting bodies. Some species have berrylike gas-filled bladders to keep the macroalgae afloat, thus promoting photosynthesis. Many macroalgae have rough sticky textures, which together with robust but flexible bodies, help to withstand strong currents.

The thick masses of *Sargassum* provide an environment for a distinctive and specialised group of marine forms, many of which are not found elsewhere.

Sargassum is commonly found in the beach drift near sargassum beds where they are also known as Gulfweed, and colloquially as *the weed of deceit*, a term also used to include all seaweed species washed up on shore.

Sargassum species are found throughout tropical areas of the world and are often the most obvious macrophyte in near-shore areas where sargassum beds often occur near coral reefs. The plants grow subtidally and attach to coral, rocks or shells in moderately exposed or sheltered rocky or pebble areas. In some cases (e.g. the Sargasso Sea), there are floating populations of sargassum.

Sargassum Muticum (Yendo) Fensholt

Description: *Sargassum muticum* is a large brown seaweed of the Division Phaeophyta. It grows attached to rocks by a perennial holdfast up to 5 cm in diameter. From this holdfast the main axis grows to a maximum of 5 cm high. The leaf-like laminae and primary lateral branches grow from this stipe. In warm waters it can grow to 12 m long, however, in British waters, it grows to no more than 4 m long. The holdfast gives rise to a single main axis with secondary and tertiary branches which are shed annually. Numerous small 2-6 mm stalked air vesicules provide buoyancy. The reproductive receptacles are also stalked and develop in the axils of leafy laminae. It is self fertile.

History: The Florida Keys and its smaller islands are well known for their high levels of *Sargassum* covering their shores. Gulfweed was observed by Columbus. Although it was formerly thought to cover the whole Sargasso Sea, making navigation impossible, it has since been found to occur only in drifts.

Sargassum mŭticum originally from Japan had by 15th March 1995 been found in Strangford Lough, County Down, Northern Ireland, this is an extension of the distribution of this invasive species. Herbarium specimens are now stored in the Ulster Museum (BEL catalogue numbers: F11241 - F11242; F11182 - F11185). The species was first found in the British Isles in the Isle of Wight in 1973. It is thought to have gained worldwide distribution through being transported with Japanese oysters. There is much concern about its impact on the coastal environment. It has become a great nuisance forming large detached mats, clogging marinas, recreational areas and other sports facilities. It can foul fishing lines, clog pipes of boats and trap debris.

The species is particularly tenacious with fast growth rates, high reproductive rates and an ability to spread vegetatively.

Sargassum is also cultivated and cleaned for use as an herbal remedy. Many Chinese herbalists prescribe powdered sargassum in paper packets of 0.5 gm, to be dissolved in warm water and drank as a tea. It is said to remove excess phlegm. When sold in this application, it is commonly referred to as Seaweed Sargassum Tea.

Distribution: *Sargassum muticum* has a range stretching from Nanaimo, British Columbia to Baja in California. In Europe, it now extends along the coasts of Great Britain, France, Scandinavia, Baltic Sea, Helgoland, Netherlands, Britain, Ireland, the Iberian Peninsula and into the Mediterranean from Italy and the Adriatic. It is recorded from Japan, China and Alaska.

Scytosiphon Lomentaria

Scytosiphon Lomentaria is a littoral brown seaweed with an irregularly lobed many filamentous form. It is a member of the Phaeophyta in the order Dictyosiphonales and grows attached to shells and stones in rock-pools and in near-shore waters. The attachment to the substrate is by a small disc shaped holdfast.

Seaweed

Seaweeds are any of a large number of marine benthic algae. They are macroscopic, multicellular, and macrothallic, in contrast with most other algae. Seaweed is often found in the seashore biome.

Taxonomy: Seaweeds are often confused with other photosynthetic organisms. Seaweeds are popularly described as plants, but biologists typically do not consider them true Plantae. They should not be confused either with plants, such as seagrasses (which are vascular plants). In addition, a few species of cyanobacteria bear a resemblance to seaweed algae. Many phycologists prefer the term "marine macroalgae" over "seaweeds".

History: In the early 19th century seaweeds were treated with disdain by some:

> There was a time when a student who, having collected some beautiful algae on the shore, showed the contents of his vasculum to the Professor of Botany, expressing a wish to get some information respecting them. The Professor looked at them, and putting on his spectacles, again looked at them, when, pushing them from him, he exclaimed: "Pooh! a parcel of Seaweeds, Sir; a parcel of Seaweeds!"

Structure

Seaweeds' appearance somewhat resembles non-arboreal terrestrial plants.

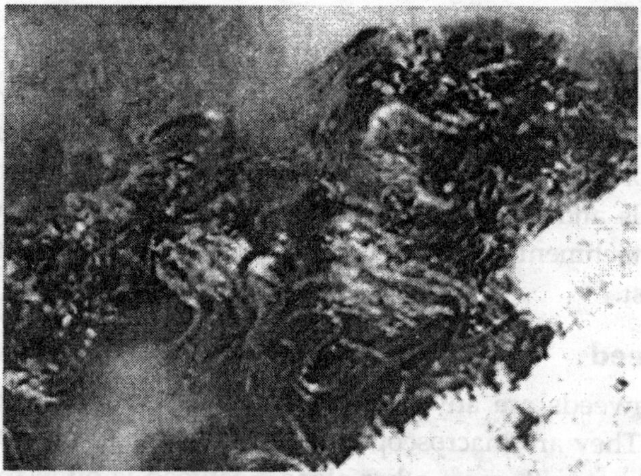

The top of a kelp forest in Otago, New Zealand

- thallus: the algal body
  - — lamina: a flattened structure that is somewhat leaf-like
    - (i) sorus: spore cluster
    - (ii) on Fucus, air bladders: float-assist organ (on blade)
    - (iii) on kelp, floats: float-assist organ (between lamina and stipe)
  - — stipe: a stem-like structure, may be absent

— holdfast: specialised basal structure providing attachment to a surface, often a rock or another algae.

The stipe and blade are collectively known as fronds.

Ecology

The ecology of seaweeds is dominated by two specific environmental requirements. These are the presence of seawater (or at least brackish water) and the presence of light sufficient to drive photosynthesis. A very common requirement is also to have a firm point of attachment. As a result, seaweeds are most commonly found in the littoral zone and within that zone more frequently on rocky shores than on sand or shingle. The ecological niches utilised by seaweeds are wide ranging. At the highest level are those that inhabit the zone that is only wetted by sea spray through top the deepest living that are attached to the sea-bed under several metres of water. In some parts of the world, the area colonised by littoral seaweeds can extend for several miles away from the shore. The limiting factor in such cases is the availability of sufficient sunlight to support photosynthesis. The deepest living seaweeds are the various kelps.

In addition to the familiar seashore seaweeds, a number of species have adapted to a fully planktonic niche and are free-floating, often with the assistance of gas filled sacs. Sargassum is one of the better know examples of this type of seaweed.

A number of species have adapted to the specialised environment of tidal rock pools. In this niche seaweeds are able to withstand rapidly changing temperature and salinity and even occasional drying.

Uses

Seaweeds are extensively used as food by coastal peoples, particularly in East Asia, e.g. China, Japan, Korea, Taiwan, and Vietnam, but also in Indonesia, Peru, the Canadian Maritimes, Scandinavia, Ireland, Wales, Philippines, and Scotland, among other places. For example, Porphyra is a red algae used in

Wales to make laverbread. In Asia, nori is a food composed of sheets of dried *Porphyra* and is used to wrap sushi or in soups. Chondrus crispus (commonly known as Irish moss or carrageen moss) is another red algae used in producing various food additives, along with Kappaphycus and various gigartinoid seaweeds.

Seaweeds are also harvested or cultivated for the extraction of alginate, agar and carrageenan, gelatinous substances collectively known as hydrocolloids or phycocolloids. Hydrocolloids have attained commercial significance, especially in food production as food additives. Agar is used in foods such as confectionery, meats and poultry products, desserts and beverages and moulded foods. Carrageenan is used in preparation of salad dressings and sauces, dietetic foods, and as a preservative in meat and fish products, dairy items and baked goods. Alginates enjoy many of the same uses as carrageenan, but are also used in production of industrial products such as paper coatings, adhesives, dyes, gels, explosives and in processes such as paper sizing, textile printing, hydro-mulching and drilling. The food industry exploits the gelling, water-retention, emulsifying and other physical properties of these hydrocolloids.

Medicine

In the biomedicine and pharmaceutical industries, alginates are used in wound dressings, and production of dental moulds and have a host of other applications. In microbiology research, agar is extensively used as culture medium. Carrageenans, alginates and agaroses (the latter are prepared from agar by purification), together with other lesser-known macroalgal polysaccharides, also have several important biological activities or applications in biomedicine.

A number of research studies have been conducted to investigate claims of seaweed's effects on human health. It has been asserted that seaweeds may have curative properties for tuberculosis, arthritis, colds and influenza, worm infestations and even tumours.

Other Uses

Other seaweeds may be used as seaweed fertilizer.

Sporochnales

Sporochnales is an order in the Phylum Phaeophyta or Brown algae. Member of this order are thread like algae growing by means of an intercalary row of dome shaped cells at the base of the hairs.

As their general name suggests their pigmentation is brown.

Toothed Wrack

Toothed wrack (*Fucus serratus* L.) is a seaweed of the north Atlantic Ocean. It is used in Ireland and France for the production of cosmetics and for thalassotherapy.

Description

Fucus serratus is readily distinguished from related taxa by the serrated edge of the fronds and by the lack of air vesicles, such as are found in F. vesiculosus.

The fronds are flat, about 2 cm wide, bifurcating, and up to 60 cm long including a short stalk. It is commonly found in high densities on rocky parts of the lower shore in the eastern north Atlantic.

Distribution

Fucus serratus is found along the Atlantic coast of Europe from Spitzbergen to Portugal and on the shores of northeast America. It introduced to Iceland and the Faroes by humans within the last 1000 years where it was first noted in a phycological survey in 1900. It grows very well on slow draining shores where it may occupy up to a third of the area of the entire seashore.

Wakame

Wakame, *Undaria pinnatifida,* is a type of edible kelp.

Health Effects

New studies conducted at Hokkaido University have found that a compound in wakame known as fucoxanthin can help burn fatty tissue. Studies in mice have shown that fucoxanthin induces expression of the fat-burning protein UCP1 that accumulates in fat tissue around the internal organs. Expression of UCP1 protein was significantly increased in mice fed fucoxanthin. Wakame is also used in topical beauty treatments.

Invasiveness

In New Zealand, wakame is a very serious weed, and was nominated one of the 100 worst invasive species in the world. It was first discovered in Wellington Harbour in 1987. It probably arrived accidentally in the late 1980s, via shipping from Asia, in ballast water. Native to cold temperate coastal areas of Japan, Korea and China, in recent decades, it has also established in France, Great Britain, Spain, Italy, Argentina and Australia.

Wakame is now found around much of southeastern New Zealand, and as far north as Auckland. It spreads in two ways: naturally, through the millions of microscopic spores released by each fertile organism, and through attachment to vessel hulls and marine farming equipment. It is a highly successful and fertile species, which makes it a serious invader. However, its impacts are not well understood and are likely to vary, depending on the location.

As Food

In Japan, wakame is distributed either dried or salted, and often used in soup, particularly miso soup, and salads. In China, it is called *qundaicai*. Chinese production is concentrated around Dalian. In Korea, it is called *miyeok* and used in salads or soup.

In Japan, Wakame is often found in Tofu salad, usually a simple side dish and consisting of Tofu, wakame and sometimes with a salad vegetable like cucumber. They are typically dressed with Japanese ingredients including soya sauce and vinegar/ rice vinegar.

Wakame is a rich source of EPA, an É-3 essential fatty acid. At over 400 mg/100 cal, it has one of the higher nutrient: calorie ratios, and among the very highest for a vegetarian source.

5

Golden and Brown Algae

The Phaeophyceae (or brown algae, singular: brown algae) is a large group of mostly marine multicellular algae, including many seaweeds of colder Northern Hemisphere waters. They play an important role in marine environments both as food, and for the habitats they can form. For instance Macrocystis, a member of the Laminariales or kelps, may reach 60 m in length, and forms prominent underwater forests. Another example is Sargassum, which creates unique habitats in the tropical waters of the Sargasso Sea. This is one of the few areas where a large biomass of brown algae may be found in tropical waters. Many brown algae such as members of the order Fucales are commonly found along rocky seashores. Some members of the division are used as food for humans.

Worldwide there are about 1500 — 2000 brown seaweed species

Ascoseirales

Ascoseirales is an order in the Phylum Phaeophyta or Brown algae. It is composed of 2-6 genera, based upon various circumscriptions.

- *Desmarestiales Setchell and Gardner:* Desmarestiales is an order in the Phylum Phaeophyta or Brown algae. Member of this order have terete or ligulate (flat) pinnately branched thalli attached by discoid holdfasts. They have a sporphytic thallus usually aggregated to form a pseudo parenchyma.

As their general name suggests their pigmentation is brown.

- *Dictyotales Kjellman:* Dictyotales is large order in the Phylum Phaeophyta or Brown Algae. Members of this order generally prefer warmer waters than other brown algae. One genus in this order is calcareous, *Padina*, the only calcareous member of this phylum.
- *Ectocarpales Setchell and Gardner:* Ectocarpales is a very large order in the Phylum Phaeophyta or Brown Algae. The order includes families with pseudoparenchymatous (Splachnidiaceae) or true parenchymatous (Scytosiphonaceae) tissue. Pseudoparenchymatous refers to a filamentous algae with cells packed very close together to give an appearance of parenchymatous tissue, the latter being composed of cells which can truly divide in three dimensions, unusual among the algae. Filamentous algae are composed of cells that divide along a single plane, allowing only elongation to form filaments of one or more rows of cells. Algae that can divide in two planes can form sheet-like thalli or bodies. Cells that can divide in a third plane potentially allow for the organism to develop a more complex body plan, and diversification of body plans into an erect thallus of some sort and a holdfast for attaching the upright portion to the substrate.

AlgaeBase uses the following families for this order:

- Acinetosporaceae Hamel ex J Feldmann
- Adenocystaceae F Rousseau, B de Reviers, M-C Leclerc, A Asensi, and R Delépine
- Chordariaceae Greville
- Chordariopsidaceae Kylin
- Ectocarpaceae C Agardh

- Ishigeaceae Okamura
- Scytosiphonaceae Farlow

AlgaeBase places Ralfsiaceae in its own order, the Ralfsiales Nakamura, and Splachnidiaceae in the Scytothamnales AF Peters and MN Clayton.

Golden Algae

The Golden Algae or chrysophytes are a large group of heterokont algae, found mostly in freshwater. Originally they were taken to include all such forms except the diatoms and multicellular brown algae, but since then they have been divided into several different groups based on pigmentation and cell structure. They are now usually restricted to a core group of closely related forms, distinguished primarily by the structure of the flagella in motile cells, also treated as an order Chromulinales. It is possible membership will be revised further as more species are studied in detail. They come in a variety of morphological types, originally treated as separate orders or families.

Most members are unicellular flagellates, with either two visible flagella, as in *Ochromonas*, or sometimes one, as in *Chromulina*. The Chromulinales as first defined by Pascher in 1910 included only the latter type, with the former treated as the order Ochromonadales. However, structural studies have revealed that short second flagellum or at least a second basal body is always present, so this is no longer considered a valid distinction. Most of these have no cell covering. Some have loricae or shells, such as *Dinobryon*, which is sessile and grows in branched colonies. Most forms with silicaceous scales are now considered a separate group, the synurids, but a few belong among the Chromulinales proper, such as *Paraphysomonas*.

Some members are generally amoeboid, with long branching cell extensions, though they pass through flagellate stages as well. *Chrysamoeba* and *Rhizochrysis* are typical of these. There is also one species, *Myxochrysis paradoxa*, which has a complex life cycle involving a multinucleate plasmodial stage, similar to those found in slime moulds. These were originally treated as the order Chrysamoebales. The superficially similar Rhizochromulina was once included here, but is now given its own order based on differences in the structure of the flagellate stage.

Other members are non-motile. Cells may be naked and embedded in mucilage, such as *Chrysosaccus*, or coccoid and surrounded by a cell wall, as in *Chrysosphaera*. A few are filamentous or even parenchymatous in organisation, such as *Phaeoplaca*. These were included in various older orders, most of the members of which are now included in separate groups. *Hydrurus* and its allies, freshwater genera which form branched gelatinous filaments, are often placed in the separate order Hydrurales but may belong here.

Green Algae

The "green algae" is the most diverse group of algae, with more than 7000 species growing in a variety of habitats. The "green algae" is a paraphyletic group because it excludes the Plantae. Like the plants, the green algae contain two forms of chlorophyll, which they use to capture light energy to fuel the manufacture of sugars, but unlike plants they are primarily aquatic. Because they are aquatic and manufacture their own food, these organisms are called "Algae," along with certain members of the Chromista, the Rhodophyta, and photosynthetic bacteria, even though they do not share a close relationship with any of these groups.

"Green Algae": Fossil Record

Though there are several Late Proterozoic fossils (both unicellular and multicellular) which superficially resemble green algae, fossils which can be assigned to the group with some confidence are not known earlier than the Cambrian. This is, perhaps in large part, because the group has traditionally been defined by pigment composition, and not with more easily identified morphological characters. Fossils were thus assigned to this group because of overall resemblance to modern taxa, Recent morphological work has added a number of characters which should aid in recognition of members of this group.

The Dasycladales has perhaps the best-known fossil record of any group of green algae. More than 120 fossil genera have been described, some dating back to Cambrian and Precambrian strata. This fossil diversity contrasts remarkably with the fact that today there are only eight extant genera, including the familiar "mermaid's cup" *Acetabularia*. The Dasycladales began to rapidly diversify in Middle Ordovician, and are common in all strata until the lower Cretaceous. Both living and extinct species are known primarily from warm marine waters,

Many members of the Dasycladales secrete lime (calcium carbonate) which increases their chances for preservation and later discovery as fossils. The group is easily recognised by their radial symmetry, with a central non-septate axis to which are attached whorls of lateral appendages which may or may not be branched. Fossil forms are cylindrical, club-shaped, or have a spherical appearance from the density of their branches.

Botryococcus is another "green algae" known from the Carboniferous onward. It is a colonial member of the Tetrasporales, and may be responsible for certain coal and petroleum deposits.

"Green Algae": Systematics, Part

It is now realised that the "green algae" should not be a formal taxonomic group, since they are paraphyletic with respect to the Plantae. In fact, one group of "green algae", the "charophytes" is more closely related to the plants than to the other "green algae". Like plants, the "green algae" have chlorophylls "a" and "b", and an alternation of generations. Unlike the plants, however, the alternation may be Isomorphic, so that the gametophyte and sporophyte are indistinguishable.

The major groups of "green algae" are distinguishable on

the basis of their Flagellar Insertion (the number and arrangement of the flagella that the cells have); their method of cell division; and their habitat.

Pleurastrophyceae

The Pleurastrophyceae comprises less than a dozen genera and a handful of species. The group is not as well-defined as the other groups of green algae, because it was first recognised as a group only in 1984. What makes the group distinctive is the way in which the cells divide during mitosis.

Pleurastrophyceans have a Metacentric Spindle, a trait rare among green algae and not known from other protist groups (Mattox & Stewart 1984). Unlike the usual pattern of cell division, the basal bodies or centrioles (depending on which is present) do not migrate to the poles of the cell before the spindle forms. It is believed that in most other groups, the basal bodies act as a centre from which the spindle formation and movement is coordinated, but this does not seem to be the case here.

A further oddity of pleurastophyte cell division is that the cleavage furrow that divides the cytoplasm forms from one side only, rather than pinching from both sides.

Like most members of the closely related Chlorophyceae, all Pleurastrophyceae are found in freshwater and soil, not in marine habitats. But, like the Ulvophyceae, they have an "11 o'clock-5 o'clock" flagellar arrangement. Unlike either group, all known reproduction is asexual.

Strictly speaking, the Pleurastrophyceae have no fossil record. However, the genera *Trebouxia*, *Pseudotrebouxia*, and *Myrmecia* are the most common green algae found associated with fungi to form lichens, which are known from fossils as early as the Devonian.

Chlorophyceae

Fresh-Water Green Algae: The Chlorophyceae are a large and important group of freshwater green algae. They include some of the most common species, as well as many members that are important both ecologically and scientifically.

There are approximately 350 genera and 2650 living species of chlorophyceans. They come in a wide variety of shapes and forms, including free-swimming unicellular species, colonies, non-flagellate unicells, filaments, and more. They also reproduce in a variety of ways, though all have a Haploid Life-Cycle, in which only the zygote cell is diploid. The zygote will often serve as a resting spore, able to lie dormant though potentially damaging environmental changes such as desiccation.

The Chlorophyceae includes three major groups distinguished primarily by basic differences in the arrangement of their flagellae:

- *Volvocales, Chaetophorales, & Chlorococcales:* Together make up more than half of all chlorophyceans. Members of these orders have an offset flagellar arrangement (1 o'clock-7 o'clock).
- *Chlorellales:* Members of this order have opposed flagellae (12 o'clock-6 o'clock), though some have only vestigial flagellae and so have not been definitively associated with this group. Similarities with members of the Chlorococcales make distinctions difficult.
- *Oedogoniales:* Members of this smallest group have a complex multiflagellate crown on their swimming spores. All are filamentous, oogamous, and have net-like chloroplasts.

Perhaps the most famous and important chlorophyceans are *Chlamydomonas* and *Volvox*. Both are important research organisms in biological laboratories. *Volvox*, pictured at the is a model organism for understanding cell death and cell interactions. *Chlamydomonas*, pictured above at left, though tiny and single-celled has been useful in countless studies, including freshwater ecology, cell division, physiology, genetic structure and function, and genomic evolution.

No less important are the Chlorellales, especially *Chlorella*. Though only about ten species are known in this genus, it plays an important role as Endosymbionts inside the tissues of other organisms. Sponges, polyps, ciliates, and forams all may house *Chlorella* internally, providing a home for the algae in exchange for its photosynthates.

More recently, this tiny algae has become popular as a "health food" for its concentrated proteins and chlorophyll, and is now raised commercially in large numbers, especially in Asia.

Many other species of Chlorophyceae are common. *Oedogonium* grows in freshwater worldwide, usually attached to other plants or algae; it has been a subject of intense study for its unusual cell division method. Species of Chlorococcales are frequently found in soil or mud, and some, such as *Apatococcus* may be found growing in bright green patches on walls and the sides of trees where moist conditions prevail.

The oldest fossils resembling Chlorophyceans come from the Proterozoic Bitter Springs Formation of central Australia. There is insufficient preservation to make a firm identification of these fossils, but other multicellular algae are known from this time (Butterfield et al. 1988), so the identification is not unreasonable.

More likely fossils of this group have been found in Middle Devonian rocks from New York State (Baschnagel 1966). Among these is *Paleooedogonium*, a fossil with striking resemblance to modern *Oedogonium*. The presence of this fossil and those of other freshwater green algae in the Devonian suggests that the group was quite diverse by this time.

Tertiary fossils of modern chlorophycean genera are also known, such as *Scenedesmus* in the Chlorellales (Fleming 1989), *Pediastrum* and *Botryococcus* (Gray 1960), and *Tetraedon* (Goth et al. 1988). The last of these is believed to be an important component in the formation of oil shales of the Tertiary, such as those in Messel, Germany.

Micromonadophyceae

The Micromonadophyceae are a "primitive" group of green flagellates that has only recently been recognised, and so is rélatively poorly known and little studied. The group is now attracting much attention, as it is believed to be representative of the earliest ancestors of the green algae and plants. However, the group is not based on any synapomorphies, and so may not be a natural, or Monophyletic, group.

Micromonads occur in both fresh and marine water, where they exist as free-swimming unicellular organisms. They will occasionally undergo encystment, losing their flagella, becoming dormant, and encasing themselves in a resistant wall. These cysts, such as the ones shown at left below, may be highly ornamented, and have many structural similarities to the spores of land plants. Such cysts are found as fossils as early as the Pre-cambrian, but their interpretation as micromonads is not entirely certain — there are other unicellular green algae which produce similar cysts.

Most micromonads have scales over the surface of their cell membrane and on their flagella. This is in contrast to the Cell Wall found in plants, and the Theca-like wall of the green algae. The scales arise within the Golgi Apparatus, and are then transferred by a vesicle to the apical end of the cell. The scales are often highly complex, and may be of several forms on a single cell, sometimes in several layers or arrangements on a single individual. Not all micromonads have these scales however; some have secondarily become naked through reduction of the cell and loss of scale production.

The micromonads have an unusual structure of flagellar attachment among green algae, and some species have a mix of flagellar attachments that are not normally found together in the other groups of the Chlorophyta. Instead of arising from a papilla, or bump, at the apex of the cell, the flagella extend out from an apical or lateral depression. Out of this depression may arise two, four, or eight flagellae. The wide variety of flagellar arrangements and numbers in this group of green algae suggest that it may be a holdover from an early diversification, and therefore, a basal group among green organisms.

The Micromonadophyceae was originally named the Prasinophyceae, and was created as a place to put a number of unicellular forms which were found to be very different from the Chlorophyceae, with which they had previously been classified. These taxa are now known to be different in having many primitive characters, and so the group is not based on derived characters as it ideally should be. A confounding problem came when the type genus of the Prasinophyceae, *Prasinocladus* was synonymised with *Tetraselmis*, a genus now placed in the Pleurastrophyceae. Since the name of the class relies on containing the defining genus, the old name Prasinophyceae can no longer be properly used.

The Micromonadophyceae was originally named the Prasinophyceae, and was created as a place to put a number of unicellular forms which were found to be very different from the Chlorophyceae, with which they had previously been classified. These taxa are now known to be different in having many primitive characters, and so the group is not based on derived characters as it ideally should be. A confounding problem came when the type genus of the Prasinophyceae, *Prasinocladus* was synonymised with *Tetraselmis*, a genus now placed in the Pleurastrophyceae. Since the name of the class relies on containing the defining genus, the old name Prasinophyceae can no longer be properly used.

Charophytes are well known to students of biology. Spirogyra, stoneworts, and desmids are all members of this freshwater group of "green algae". The charophytes have gained even more attention recently because they are now known to be the most closely related group of organisms to the Plantae. The charophytes are in fact, Paraphyletic with respect to the Plantae — some members of the group are more closely related to the plants than to the other charophytes, though there is still considerable debate as to the exact relationships among the various members.

Unlike the other green algae, the charophytes share with plants the Phragmoplast method of cell division, and the asymmetric arrangement of their sperm flagella. This latter character is only observable in the Charales and Coleochaetales, since members of the remaining orders do not produce flagellated gametes as part of their life cycle.

Zygnematales

The Zygnematales comprise 18 genera, with more than more than 600 species. *Spirogyra* and *Zygnema* are the best-known members of the order, and because of the beauty of these organisms when seen under a microscope, they are commonly used in biology classes as examples of the "green algae". However, the Zygnematales are not representative of the green algae in general, and have many unusual features which suggest a closer relationship to the plants than to the other green algae.

Almost all species of Zygnematales live in freshwater environments, where they are important as primary producers. They are commonly called "pondscum", and make up the filamentous Periphyton in ponds, growing on and around the larger aquatic plants.

Mature zygnemataleans are Filamentous and unbranched, growing as long chains of cells connected end-to-end. The filament grows when cells in the chain divide, developing into two new full-sized cells. Some species will produce Rhizoids at one end of the filament to help anchor them to objects in the water. Each cell is encased in a cell wall composed of three layers, two cellulosic inner layers and outer layer of mucus, which probably helps to keep the algae moist.

Like all organisms called "algae", zygnemataleans have Plastids which allow them to manufacture their own food. There is a great deal of variation in chloroplast morphology in the group. They are stellate in *Zygnema,* while in *Spirogyra* the chloroplasts are ribbonlike and are wound around the inside of the cell like a spiral, giving the genus its name. The photo at left shows this in *Spirogyra submaxima*. The "bumps" visible along the chloroplast are Pyrenoids used to store starch reserves.

Members of the order may reproduce in one of two ways, asexually by Fragmentation, or sexually by Conjugation. Conjugation is a very distinctive mode of reproduction which only occurs among algae in the Zygnematales and Desmidiales. The process begins with a parallel pairing of two filaments, aligning neighbouring cells. Then, a bridge-like Conjugation Tube is formed to connect neighbouring cells, and the amoeboid cells from one filament move over into the neighbouring filament. Alternatively, in some species, cells from both filaments will move into the cojugation tube. In either case, the cells meet and fuse, producing a diploid Zygote which forms a spore. The zygote will undergo meiosis directly to produce the next generation. Mature algae are therefore, haploid, though polyploidy is known in some taxa, resulting in size variation. Like plants, zygnemataleans have uniparental Inheritance of plastids; the plastids of only one gamete are passed on to the next generation.

Fossils of the group are mostly Cenozoic, though some Permian and Carboniferous fossils are also known. Zygospores are the most common fossils found; filaments appear to be much more fragile, and are only known from the Eocene. Fossil zygospores have been used to infer paleoecology of the regions in which they occur, suggesting that shallow freshwater pools or lakes were present in the past. Perhaps the best-known in the fossil record are the keeled spores of the genus *Debarya* (called *Peltacystia* when it occurs as fossil spores), though it is a rare algae today.

Desmidiales

The Desmids: The Desmidiales comprise about 40 genera of more than 10,000 species, making them a larger group than all the other charophytes combined. They are usually divided into two families based on the structure of the cell wall. The

Mesotaeniaceae, such as *Netrium* pictured above, have a homogeneous wall, while the Desmidiaceae, such as *Microasterias* pictured below, have a wall composed of two compartments connected by a narrow Isthmus, and straddled by a single plastid. It is this latter, larger family which gives the desmids their name, from the Greek word "desmos" meaning "bond".

Desmids are primarily found in freshwater habitats, such as ponds, rivers, and lakes. There, they may live as phytoplankton, on the bottom as benthic dwellers, or on the submerged portions of plants. They may also be found in saline waters, or in snow or ice. While most desmids are unicellular, many species grow as long filamentous colonies.

Like many other charophytes desmids have no flagella; they were lost at some point in the group's evolution. Reproduction therefore, cannot rely on motile gametes, but rather two cells will meet and join in a process called Conjugation. Once the two desmids have joined, their cytoplasm fuses into a single diploid cell, the Zygote, which encases itself in a thick wall. Later, the zygote will undergo meiosis to produce new haploid individuals. Growing desmid cells therefore have only a single set of chromosomes, and the zygote is the only diploid cell in the life cycle.

The thick-walled Zygospore which encases the zygote, is highly ornamented and very durable. As such, they have been found fossilised in Jurassic and Cretaceous deposits, and have been used to reconstruct paleoclimates.

Klebsormidiales

The Klebsormidiales is a small and poorly known group of charophyte algae. Members of this order grow as unbranched filaments, typically with a single Pyrenoid for storage of starch reserves, and one or two Chloroplasts against the inner side of each cell.

There are only three genera in the order, comprising approximately 40 species. The genera *Klebsormidium* and

Stichococcus are most frequently found growing in soil and on other moist substrates, but sometimes they may also be found growing in aquatic habitats. *Raphidonema*, the third genus of the order, is known primarily from montane snow fields at high elevation.

Unlike other charophytes, cell division in the Klebsormidiales occurs through formation of a cleavage furrow, but apparently neither a Phragmoplast nor a cell plate is formed. Still, in many ways, cytokinesis is similar to that in the Zygnematales, and for this reason a close relationship between the two orders has been suggested. The presence of a phragmoplast is the standard method of cell division both in the charophytes and in the plants, and the lack of this feature is one reason that the Klebsormidiales has only recently been reclassified into the Charophyceae.

Coleochaetales

The fifteen species of Coleochaetales have gained attention recently because they are implicated as the closest living relatives of the Plantae. Some species have Parenchymatous Growth, nurturing of the zygotes, and a fossil record of an appropriate age, all of which strengthen the implication. In addition, they are easily found and studied. We have found them, quite accidentally, attached to the leaves of freshly collected aquatic flowering plants while observing them under the microscope.

Plantae

The Plantae includes all land plants: mosses, ferns, conifers, flowering plants, and so on—an amazing range of diverse forms. With more than 250,000 species, they are second in size only to the arthropoda.

Plants have been around for a very long time. The plants first appeared in the Ordovician, but did not begin to resemble modern plants until the Late Silurian. By the close of the Devonian, about 360 million years ago, there were a wide variety of shapes and sizes of plants around, including tiny creeping plants and tall forest trees.

The most striking, and important, feature of plants is their green colour, the result of a pigment called chlorophyll. Plants use chlorophyll to capture light energy, which fuels the manufacture of food—sugar, starch, and other carbohydrates. Without these food sources, most life on earth would be impossible. There would still be mushrooms and algae, but there would be no fruits, vegetables, grains, or any animals (which ultimately rely on plants for their food too!)

Another important contribution of plants is their shaping of the environment. Think of a place without plants. The only such places on earth are the arctic wastelands, really arid deserts, and the deep ocean. Everywhere else, from the tundra to the rainforest to the desert, is populated by plants. In fact, when we think of a particular landscape, it is the plants which first come to mind. Try to picture a forest without trees, or a prairie without grasses. It is the plants which produce and maintain the terrestrial environment as we know it.

6

Green Algae

Acetabularia

Acetabularia are giant (0.5 to 10 cm long) unicellular and uninucleate marine green algae, with a characteristic umbrella-like shape. They are most famous from the experiments of Joachim Hammerling in the 1930's, where by transplanting parts of A. mediterranea and A. crenulata he showed that the genetic information of eukaryotes is contained in the nucleus. Species such as A. acetabulum are still used as model organisms for research on nucleus/cytoplasm relationships, cytoskeletal organisation and circadian rhythms.

Acetabularia push the limits of single-celled organisms, being some of the biggest examples found. They also have a large nucleus, being up to 100 micrometers big before meiosis.

There are about 30 species of Acetabularia, including:

- *Acetabularia acetabulum*
- *Acetabularia caliculus*
- *Acetabularia crenulata* (from Gulf of Mexico)
- *Acetabularia dentata*
- *Acetabularia farlowii*
- *Acetabularia kilneri*

- *Acetabularia major*
- *Acetabularia mediterranea* (mermaid's wine glass, from the Mediterranean Sea)
- *Acetabularia peniculus*
- *Acetabularia ryukyuensis*
- *Acetabularia schenkii*

Acetabularia Crenulata

Acetabularia crenulata, one of the many species known as mermaid's wineglass, is a form of green algae generally found in shallow tropical seas. It can be found growing in great abundance along stretches of the overseas highway to Key West, Florida bordering on Florida Bay. It has been used in some important research on nuclear and cytoplasmic relationships. During its life cycle it has one large nucleus.

Caulerpa Taxifolia

Caulerpa taxifolia is a species of seaweed (a type of algae) that has been commonly used as ornamentation in aquarium installations around the world. The algae has a stem which spreads horizontally just above the seafloor, and out of this stem there grow vertical fern-like pinnae, whose blades are flat like yew, hence the species name "taxifolia" (the genus of yew is "taxus"). The algae produces a large amount of a single chemical that is toxic to fish and other would-be predators. This is in contrast to other plants which produce a variety of toxins, but in reduced amounts. A specific strain of this algae was found to thrive in cold aquarium environments by the staff at the Wilhelmina Zoo in Stuttgart, Germany in 1980. In addition, the aquarium strain is less susceptible to the colder temperate waters of the Mediterranean than the naturally occurring tropical varieties.

Initial Infection

It appears that, in 1984, this seaweed was accidentally released into coastal waters of the Mediterranean Sea just below the Oceanographic Museum of Monaco. By 1989, the species had grown to cover a large area and was preventing other plants from growing, leading to the nickname "Killer Algae". The plant began to spread between ports along the Mediterranean coast. *C. taxifolia* has been discovered off the coasts of Australia and the United States, though none of those encroachments have been anywhere near the scale of what is happening in the Mediterranean.

For several years, the origin of the species was unclear. Marine biologist, Alexander Meinez first discovered the plant in the 1980s, and requested the help of the Monaco Oceanographic Museum, which sat right next to the first known *taxifolia* patch. However, the director of the museum argued that this invasion probably happened naturally, the result of ocean currents carrying a tropical species into the area. Bickering over whether the species was natural or invasive, and whether the museum had released it or not, contributed to a delay that allowed the plant to spread beyond control. The museum continued to deny releasing the plant, although former director Jacques-Yves Cousteau eventually expressed the belief that it was the only reasonable explanation.

C. taxifolia has no natural predators or competitors in the Mediterranean. It crowds out other fish and plants, and contains a strong toxin to most species. Regions that have been invaded by the plant now show that about half the expected number of fish have disappeared.

Reproduction Mechanism

The aquarium strain reproduces asexually, that is, vegetatively: the viscous, elastic white fluid inside the stem was found under the microscope to contain only male gametes. Rate of growth can be as fast as a centimetre per day. If any small part is severed from the rest of the algae, this small part will regrow into another algae. Anchors of ships and fishing nets can serve as carriers of Caulerpa. Thus the algae has been found to jump from the coast of one port city to the coast of another port city.

Other Infections

In 2000, the strain was found in the coast of California (USA.), near San Diego, and also in the coast of New South Wales, Australia. The California invasion was small enough to be considered controllable: it was covered with tarpaulin which was held down with sandbags at the edges of the infestation. Then chlorine was poured in through tubes which fed into certain openings in the tarpaulin: the interior of the tarpaulin fills up with chlorine and kills living organisms inside it, including the killer algae but also fish and plants. The killing of such other organisms was not desirable but was deemed preferable to letting the weed grow unchecked, which would, in the end, yield greater havoc on these same creatures.

The invasion in the Californian coast was most probably caused by an aquarium owner improperly dumping the contents, allowing *C. taxifolia* to flow through a storm sewer into the lagoon where the invasion was discovered. California has since passed a law forbidding the possession, sale or transport of *Caulerpa taxifolia* within the state. There is also a federal law under the Noxious Weed Act forbidding interstate sale and transport of the aquarium strain *Caulerpa*.

In July, 2006, the algae had been declared eradicated from the two Southern California locations (Hedionda Lagoon in Carlsbad and Seagate Lagoon in Huntington).

Possible Natural Control Method

Researchers at the University of Nice in France have been studying a tiny aquatic slug, similar to a snail but without a carapace on its back, which is known to be a natural predator of *C. taxifolia*. Called Elysia subornata, it was found off the coast of Florida, in waters warmer than those in the Mediterranean. This slug is believed to feed exclusively on Caulerpa taxifolia, by sticking its proboscis into the stem and sucking out the white viscous liquid inside the stem: this causes the algae to become limp, discoloured, and dead. As the slug does so, it absorbs the algae's poison. The slug has an enzyme which neutralises the noxious effect of the poison, and at the same time the poison protects the slug from being eaten by fish. This slug cannot survive in the cooler waters of the Mediterranean, however, and so is unable to check the growth of the killer weed.

Chaetosphaeridium Globosum

Chaetosphaeridium globosum is a one-celled algae which is thought to represent an ancient lineage of the green plants. It belongs to the class Mesostigmatophyceae (Hausmann) and subphylum Streptophyta. This organism exists in a filamentous form with one flagella per cell. It is a freshwater species. The flagellum is covered in scales in a 3 prong irregular shape called 'maple leafs'.

Researchers have found that the mitochondrial DNA of Chaetosphaeridium is markedly different than land plants indicating that land plants developed mitochondria separately from their protist progenitors. A very slight similarity exists between liverwort mtDNA and Chaetosphaeridium The chloroplast DNA is markedly similar though: indicating a close relationship had existed between the Viridiplantae and the clade that includes Chaetosphaeridium. This seems to argue that chloroplasts in green plants originated in pre-historic green algae; the family which includes Chaetosphaeridium globosum.

Chloroplasts are known to be captured (symbiotic) cyanobacteria with their own genetic code. This code is both transferred to the nucleus and retained in the chloroplast for the continuation of metabolic processes. This symbiosis, now proven by modern genomics, has shown us how Chaetosphaeridium globosum links ancient cyanobacteria with modern green plants like those in our garden.

Chara Fragilis

The Charales are an order of green algae, Chlorophyta, plants believed to be the closest relatives of the green land plants. They are branched, multicellular, chlorophyll-using plants that grow in fresh water. They may be called stoneworts, because the plants can become encrusted in lime (calcium carbonate) after some time. The "stem" is actually a central stalk consisting of giant, multinucleated cells. They are unique in having a whorl of small branchlets at each node in the stipe. In these whorls it is possible to see the phenomena cytoplasmic streaming very well. In fact, the streaming in *Chara* is the fastest recorded of any cell. Cytoplasmic streaming is caused by the microfilaments found inside the cell biology, as proven by the use of cytochalasin B to stop streaming.

There are about 400 taxa worldwide, with 33 species in Britain and Ireland.

Life History: The antheridia and oogonia are protected by a layer of sterile cells when mature; the oogonium is oblong in shape and consists of a single egg, while the spherical antheridium is packed with threadlike cells that produce spermatia. As a result, the Charales have the most complex structure of all green algae, if indeed they should be so labelled.

The possible ancestors of the land plants are also known as Brittleworts or Skunkweed. These curious labels arise from the fragility of their lime-encrusted stems, and from the foul odour these produce when stepped on.

Many botanists propose that the stoneworts and their relatives be placed in a phylum, division, sub-kingdom, or even kingdom by themselves, often named Charophyta. Their classification by taxonomists is currently undergoing much cladistic scrutiny.

Further DNA and RNA analysis may prove the Charophytes to be a crucial evolutionary link in the phylogenetic tree of life, the critical developmental step from the seaweeds towards the non-vascular and then vascular land plants.

Charophyta

The Charophyta are a division of green algae, including the closest relatives of the embryophyte plants. Because they exclude the embryophytes, they make a paraphyletic group (though they are occasionally restricted to simply the Charales or stoneworts, which are monophyletic). In some groups such as conjugating green algae, flagellate cells do not occur.

The latter group does engage in sexual reproduction, but motility does not involve flagella, which are totally lacking. Flagellate cells in the form of sperm are found in stoneworts (scientific name Charales) and in the group with the scientific name Coleochaetales (there is no common name).

Chlamydomonas

Chlamydomonas is a genus of green algae. They are unicellular flagellates. *Chlamydomonas* is used as a model organism for molecular biology, especially studies of flagellar motility and chloroplast dynamics, biogenesis, and genetics. One of the many striking features of Chlamydomonas is that it contains ion channels that are directly activated by light, such as channelrhodopsin.

Chlamydomonas Reinhardtii

Chlamydomonas reinhardtii is a motile single-celled green algae about 10 micrometers in diameter that swims with two flagella.

These algae are commonly found in soil and fresh water. They have a cell wall made of hydroxyproline-rich glycoproteins, a large cup-shaped chloroplast and an "eyespot" that senses light. Normal *Chlamydomonas* can grow on a simple medium of inorganic salts in the light, using photosynthesis to provide energy. They can also grow in total darkness if acetate is provided as a carbon source.

The *C. reinhardtii* wild type laboratory strain originates from an isolate made near Amherst, Massachusetts, in 1945.

C. Reinhardtii as a Model Organism

Chlamydomonas is used as a model organism for research on fundamental questions in cell and molecular biology such as:

- How do cells move?
- How do cells respond to light?
- How do cells recognise one another?
- How do cells regulate their proteome to control flagellar length?
- How do cells respond to changes in mineral nutrition? (nitrogen, sulphur etc.)

There are many known mutants of *C. reinhardtii*. These mutants are useful tools for studying a variety of biological processes, including flagellar motility, photosynthesis or protein synthesis.

Reproduction

Vegetative cells of the *reinhardtii* species are haploid with 17 small chromosomes. Under nitrogen starvation, haploid gametes develop. There are two mating types, identical in appearance and known as mt(+) and mt(-), which can fuse to form a diploid zygote. The zygote is not flagellated, and it serves as a dormant form of the species in the soil. In the light the zygote undergoes meiosis and releases four flagellated haploid cells that resume the vegetative life cycle.

Curious Fact: Under ideal growth conditions, cells may undergo two or three rounds of mitosis before the daughter cells are released from the old cell wall into the medium. Thus, a single growth step may result in 4 or 8 daughter cells per mother cell.

The cell cycle of this unicellular green algae can be synchronised by alternating periods of light and dark. The growth phase is dependent on light, whereas, after a point designated as the transition or commitment point, processes are light-independent.

Evolution

Chlamydomonas has been used to study different aspects of evolutionary biology and ecology. The fact that: it has a short generation time, it is both a heterotroph and a facultative autotroph, it can reproduce both sexually and asexually and that there is already a lot of genetic information available for the species has made it an organism of choice for many selection experiments.

Some examples (non exhaustive) of evolutionary work done with *Chlamydomonas* go from the evolution of sexual reproduction, the fitness effect of mutations, and the effect of adaptation to different levels of CO_2.

Genetics

The attractiveness of the algae as a model organism has recently increased with the release of several genomic resources to the public domain. The current draft of the *Chlamydomonas* nuclear genome sequence prepared by Joint Genome Institute of the US Dept. of Energy comprises 1557 scaffolds totalling 120 Mb. Roughly half of the genome is contained in 24 scaffolds all at least 1.6 Mb in length. The sequences of all three *C. reinhardtii* genomes are available.

Reinhardtii DNA Transformation Techniques

Gene transformation occurs mainly by homologous recombination in the chloroplast and heterologous recombination in the nucleus. The *C. reinhardtii* chloroplast genome can be transformed using microprojectile particle bombardment and the nuclear genome has been transformed with both glass bead agitation and electroporation. The biolistic procedure appears to be the most efficient way of introducing DNA into the chloroplast genome. This is probably because the chloroplast occupies over half of the volume of the cell providing the microprojectile with a large target. Electroporation has been shown to be the most efficient way of introducing DNA into the nuclear genome with maximum transformation frequencies two orders of magnitude higher than obtained using glass bead method.

Chlorella

Chlorella is a genus of single-celled green algae, belonging to the phylum Chlorophyta. It is spherical in shape, about 2 to 10 ¼m in diameter, and is without flagella. *Chlorella* contains the green photosynthetic pigments chlorophyll-a and -b in its chloroplast. Through photosynthesis, it multiplies rapidly requiring only carbon dioxide, water, sunlight, and a small amount of minerals to reproduce.

The name *Chlorella* is taken from the Greek word *chloros* meaning green and the Latin diminutive suffix *ella* meaning "small". The German biochemist Otto Heinrich Warburg received the Nobel Prize in Physiology or Medicine in 1931 for his study on photosynthesis in *Chlorella*. In 1961, Melvin Calvin of the University of California received the Nobel Prize in Chemistry for his research on the pathways of carbon dioxide assimilation in plants using *Chlorella*. In recent years, researchers have made less use of *Chlorella* as an experimental organism because it lacks a sexual cycle and, therefore, the research advantages of genetics are unavailable.

Many people believed that *Chlorella* could serve as a

potential source of food and energy because its photosynthetic efficiency can theoretically reach 8 per cent, this is comparable with other highly efficient crops such as sugar cane. It is also an attractive food source because it is high in protein and other essential nutrients; when dried, it is about 45 per cent protein, 20 per cent fat, 20 per cent carbohydrate, 5 per cent fibre, and 10 per cent minerals and vitamins. However, because it is a single-celled algae, harvest posed practical difficulties for its large-scale use as a food source. Mass production methods are now being used to cultivate it in large artificial circular ponds.

Aquarium

Chlorella green algae can create green and opaque water problems in aquariums. *Chlorella* can grow due to high nitrate and phosphate levels or direct sunlight. Decreasing phosphate and nitrate by partial water change and moving the aquarium to shade can help in solving the problem.

Chlorella as a Food Source

During the late 1940s and the early 1950s chlorella was seen as a new and promising primary food source and as a possible solution to the then current world hunger crisis. Many people during this era thought that world hunger was a growing problem and saw chlorella as a way to end this crisis by being able to provide large amounts of high quality food for a relatively low cost.

Following global fears of an uncontrollable population boom in the 1940's, chlorella became a possible answer to the tremendous food shortages the world would face in light of this speedy, post-war growth. Many institutions stepped up to research the algae, including the Carnegie Institution, the Rockefeller Foundation, the NIH, UC Berkeley, the Atomic Energy Commission, and Stanford University. Following WWI, many Europeans were starving and many Malthusians attributed this not only to the war but to the inability of the world to produce enough food to support the currently increasing population.

To cope with the upcoming post-war population boom in the United States and elsewhere, researchers decided to tap into the unexploited sea resources. Initial testing by the Stanford Research Institute showed that chlorella (when growing in warm, sunny, shallow conditions) could convert 20 per cent of solar energy into a plant that, when dried, contained 50 per cent protein (Belasco 617). In addition, chlorella contained amino acids, fat, calories and vitamins. The plant's photosynthetic efficiency allowed it to yield more protein per unit area than any other plant - one scientist predicted that 10,000 tons of protein a year could be produced with just 20 workers staffing a one thousand-acre (4 km^2) chlorella farm (Belasco, 618). The pilot research performed at Stanford and elsewhere led to immense press from journalists and newspapers, yet sadly never panned out. Chlorella was a seemingly viable option because of the technological advances in agriculture at the time and the widespread acclaim it got from experts and scientists who studied it. Algae researchers had even hoped to add a neutralised chlorella powder to conventional food products, as a way to fortify them with vitamins and minerals (Belasco, 625). Unfortunately, the hype far surpassed the productivity of the plant and early estimates of its success were proven to be no more than exaggerated optimism.

Ultimately, scientists discovered that chlorella would be much more difficult to produce than previously thought. The experimental research was carried out in laboratories, not in the field. Practically, the entire batch of algae grown would have to be placed either in artificial light or in the shade so that it would produce at its maximum photosynthetic efficiency. Additionally, for the chlorella to be as productive as the world would require, the chlorella would have to be grown in carbonated water which would have added millions to the production cost. A sophisticated process that would cost additional money was required to harvest the chlorella, and for the chlorella to be a viable food source, its cellulose cell walls would have to be pulverised. The plant could only reach its nutritional potential in highly modified artificial situations.

Economic problems and the public's distaste for the flavour of chlorella and its by-products ultimately led to the demise of chlorella.

Since the growing world food problem of the 1940's was solved by better crop efficiency and not from a "super food," chlorella has lost public and scientific interest for the time being. Chlorella can still be found today in rare occasions from companies still promoting chlorella's "super food" effects.

Nutrition

It was believed in the early 1940s that unlike most plants, Chlorella's protein was "complete," for it had the ten amino acids then considered essential, and it was also packed with calories, fat, and vitamins (Belasco 613). Chlorella has been found to have anti tumour properties when fed to mice. Another study found enhanced vascular function in hypertensive rats given oral doses of chlorella. Although at its onset Chlorella was thought by many to add a "dirt cheap" form of high protein to the human diet, studies proved otherwise. Chlorella, which actually lost most of its nutritional value when altered/processed in any way, was no longer an effective protein and therefore, pro-Chlorella supporters decided to communicate other health benefits of the algae. Hence, weight control, cancer prevention, and immune system support were all positive health benefits attributed to this algae (Belasco 608,628-630). It was also thought that humans would never eat algae directly; instead they believed it could be added to animal feed, thereby increasing to protein consumption indirectly (Belasco 625).

Evidence of Health and Healing Effects

The marketing of *Chlorella* for human consumption as a palliative, or "Superfood", has received criticism; as such, the following recommendations should be treated as speculative: [This claim is not supported by the Zelitch reference.

Maintenance Dosage: 3-5 grams/day

Significant Part of the Diet: 6-10 grams/day

Immune System Builder: 11-14 grams/day

Primary Source of Protein: 12-19 grams/day

Healing Purposes and Heavy Metal Detoxification: 20-30 grams/day

Early Reception and Scientific Backing

When the preliminary laboratory results were published the reaction of scientific literature backed the possibilities of the supposed superfood. Science News Letter praised the optimistic results in an article entitled "Algae to Feed the Starving." John Burlew, the reported editor of Carnegie Institute stated that "the algae culture may fill a very real need," which Science News Letter turned into "future populations of the world will be kept from starving by the production of improved or educated algae related to the green scum on ponds."

The cover of the magazine also featured Arthur D Little's Cambridge laboratory which was a supposed future food factory. A few years later, the magazine published an article entitled "Tomorrow's Dinner," which stated that "There is no doubt in the mind of scientists that the farms of the future will actually be factories." Science Digest also reported that "common pond scum would soon become the world's most important agricultural crop." Yet the optimistic initial promises of the algae fell short when further testing was conducted.

Failures

Although technologically creative and promising, Chlorella would not prove to be economically viable in the market. Experiencing competition to the health world's soybean and whole grain craze, algae products simply could not measure up. Economically too, in practice, algae was not as cheaply or easily harvested as technicians predicted it would be 40 years earlier. The efficiency of other "normal" dietary products actually turned out to supersede that of algae growth and production. Aside from production inefficiencies, Chlorella, as it turned out, did not capture the benefits of photosynthesis and sunlight as predicted. After a decade of experimentation, and after exposed to sunlight, Chlorella captured just 2.5 per cent—not much better than conventional crops (Belasco, 627). Chlorella, too, was found by scientists in the 1960s to be impossible for humans and animals to digest in its natural state, which presented further problems for the use of algae in American food production.

Chlorella Pyrenoidosa

Chlorella Pyrenoidosa is a species of fresh water green algae. It has been used medicinally as a chelatory agent, for example to extract mercury from the body. However, there is no scientific proof of this.

Possible medicinal uses include:

- fibromyalgia
- hypertension
- ulcerative colitis

It occurs world wide. The species name *pyrenoidosa* refers to the presence of a prominent pyrenoid within the *Chlorella* chloroplast. This species has been used in traditional Chinese medicine.

Chlorococcales

An order of green algae in the class chlorophyceae. Sometimes found in soil, but mostly fresh and marine waters. Current proper name is Sphaeropleales.

Families in this order include:

- Actinochoridaciae
- Chlorellaceae
- Coelastraceae
- Hydrodictyaceae
- Oocystaceae (e.g. chlorella)
- Scenedesmaceae
- Eremosphaeraceae

Chlorophyceae

The Chlorophyceae are one of the classes of green algae, distinguished mainly on the basis of ultrastructural morphology. For example, the chlorophycean CW clade, and chlorophycean DO clade, are defined by the arrangement of their flagella. Members of the CW clade have flagella that are displaced in a "clockwise" (CW, 1–7 o'clock) direction e.g. Chlamydomonadales. Members of the DO clade have flagella that are "directly opposed" (DO, 12–6 o'clock) e.g. Sphaeropleales.

The following orders are typically recognised:

- Dunaliellales
- Volvocales (e.g. Volvox, Chlamydomonas)
- Chlorococcales (e.g. Chlorella)
- Oedogoniales (e.g. Oedogonium)
- Sphaeropleales
- Chaetophorales
- Microsporales
- Tetrasporales (e.g. Tetraspora)

In older classifications, the term Chlorophyceae is sometimes used to apply to all the green algae except the Charales, and the internal division is considerably different.

Chlorophyta

The Chlorophyta, or green algae, include about 8000 species of mostly aquatic photosynthetic eukaryotic organisms. Like the land plants (Bryophyta and Tracheophyta), green algae contain chlorophylls *a* and *b*, and store food as starch in their plastids. They are related to the Charophyta and Embryophyta (land plants), together making up the Viridiplantae.

They contain both unicellular and multicellular species. While most species live in freshwater habitats and a large number in marine habitats, other species are adapted to a wide range of environments. Watermelon snow, or *Chlamydomonas nivalis*, of the class Chlorophyceae, lives on summer alpine snowfields. Others live attached to rocks or woody parts of trees. Some lichens are symbiotic relationships with fungi and a green algae. Members of the Chlorophyta also form symbiotic relationships with protozoa, sponges and coelenterates.

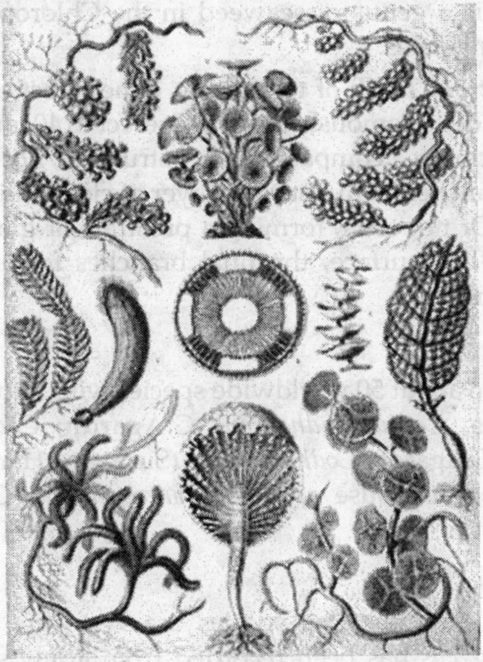

Some authors arrange the Division (Phylum) Chlorophyta into 11 Classes.

- *Prasinophyceae*
- *Chlorophyceae*
- *Ulvophyceae*
- *Cladophorophyceae*
- *Bryopsipophycese*
- *Dasycladophyceae*
- *Trentepoliophyceae*
- *Pleurastrophyceae* (Pleurastrales and Prasiolales)
- *Klebsormidiophyceae*

- *Zygnematophyceae*
- *Charophyceae*

Codium

Codium is a genus of seaweed in the Chlorophyta of the Order Bryopsidales.

Description: Thalli of two forms, either erect or prostrate. Erect plants dichotomously branched, erect to 40cms long with branches forming a compact spongy structure, not calcareous. The final branches form a surface layer of close palisade cortex of utricles. Or not erect, forming a prostrate or globular body with velvet-like surface, the final branches forming a close cortex of utricles.

Taxonomy

There are about 50 worldwide species with 5 species in the British Isles: *Codium adhaerens* C.Agardh; *Codium bursa* (Linnaeus) C.Agardh; *Codium fragile* (Suringar) Hariot; *Codium tomentosum* Stackhouse and *Codium vermilara* (Olivi) Delle Chiaje.

Distribution

Ireland: Two of these species are very rare in Ireland. *Codium adhaerens* has been recorded from a few sites on the west coast and from Tory Island on the north coast in County Donegal. In 1837, it was found in Church Bay in County Antrim, but has not been found there since.

- *Codium bursa* (Linnaeus) C.Agardh; has been found in Mulroy Bay in County Donegal in 1977 (specimen in the Ulster Museum) and 1988. There is an old record of it from "near Belfast" between 1793 and 1810.
- *Codium fragile* (Suringar) Hariot
  — *Codium fragile* (Suringar) Hariot subsp. *atlanticum* (A. Cotton) Silva
- *Codium tomentosum* Stackhouse
- *Codium vermilara* (Olivi) Delle Chiaje

There are other species of "doubtful validity":- *Codium amphibium* is included in Harvey's Phycologia Britannica and noted as: "spreading in patches of great extent along the edge of the sea, over the surface of a turf-bog which meets the shore at Roundstone Bay" (Ireland).

Codium World-Wide

- *Codium cuneatum* S. & G. California (USA.)
- *Codium duthiae* Silva Australia, Tasmania and South Africa.
- *Codium extricatum* Silva Cape Agulhas to southern Natal.
- *Codium fragile* (Suringar) Hariot Japan, American Pacific from Alaska to Cape Horn, Australia and New Zealand, British Isles, Norway, Denmark and Netherlands.
- *Codium fragile* (Suringar) Hariot subsp. *capense* Silva South Africa.
- *Codium hubbsii* Daws. California (USA.)
- *Codium johnstonei* Silva California (USA.)
- *Codium lucasi* Setchell Australia, Tasmania.
- *Codium geppiorum* Australia and Indo-Pacific South Africa.
- *Codium spongiosum* Australia.
- *Codium setchellii* Gardn. Califirnis (USA.)

Coelastraceae

A family of green algae in the order Chlorococcales. Colonial algae usually arranged 3 dimensionally in spherical to loosely spherical shapes. Individual cells connected to each other via cell wall extensions. Cells usually in an even number arrangement. Chloroplasts plate-like with 1 pyrenoid. Name derrived from coelom, in reference to the cavity created inside the sphere. Non-motile species in this family include:

- Coelastrum morus
- Coelastrum sphaericum
- Coelastrum polychordum
- Coelastrum cambricum

- Coelastrum astroideum
- Coelastrum reticulatum
- Coelastrum microporum
- Coelastrum proboscideum

Dasycladales

In taxonomy, the Dasycladales is an order of large unicellular green algae under the class Ulvophyceae. It contains two families, the Dasycladaceae and the Polyphysaceae. These algae are notable for having an intracellular network of 10 nm proteinaceous filaments, possibly for the storage and transport of ribonucleoprotein particles.

Desmid

Desmids are an order (Desmidiales) of green algae, comprising around 40 genera and more than 10,000 species, found mostly but not exclusively in fresh water. Most are unicellular, and are divided into two compartments separated by a narrow bridge or isthmus. Desmids assume a variety of highly symmetrical and generally attractive shapes, which provide the basis for their classification. Each compartment has one chloroplast, and no flagella. Sexual reproduction occurs through a process of conjugation, also found among the Zygnematales. These two groups are closely related, and may be united as the division Gamophyta. The desmids are sometimes treated as members of the Zygnematales, but more often given the separate order Desmidiales.

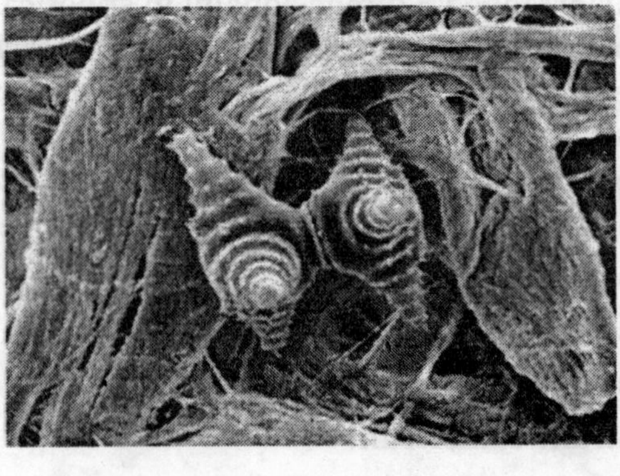

Gonium

Gonium is a small colonial algae, a member of the order Volvocales. Typical colonies have 4 to 16 cells, all the same size, arranged in a flat plate, with no anterior-posterior differentiation. In a colony of 16 cells, four are in the centre, and the other 12 are on the four sides, three each.

Halimeda

Halimeda is a green macroalgae. It deposits calcium carbonate in its tissues, making it inedible to most herbivores.

Halimeda is responsible for distinctive circular deposits in various parts of the Great Barrier Reef on the northeast coast of Queensland, Australia. Halimeda beds form in the western or lee side of outer shield reefs where flow of nutrient-rich water from the open sea allows them to flourish, and are the most extensive, actively accumulating *Halimeda* beds in the world. The tropical green algal genus, *Halimeda* is one of the best studied examples of pseudocryptic diversity within the algae. *H. incrassata* is revised into two pseudocryptic entities described as new species: *Halimedia kanaloana* and *Halimeda heteromporpha*, *H. incrassata* is redefined to encompass a single, monophyletic entity.

Hijiki

Hijiki or Hiziki is an uncommon type of edible seaweed commonly found on rocky coastlines. Its two names, which are examples of ateji, mean *deer-tail grass* and *sheep-nest grass* respectively.

Hijiki (*Hizikia fusiformes*) is a brown sea vegetable growing wild around the coasts of Japan, Korea, and China. It is a traditional food and has been freely sold and used as part of a balanced diet in Japan for centuries. Hijiki is known to be rich in dietary fibre and essential minerals. According to Japanese folklore, hijiki aids health and beauty and the thick, black, lustrous hair of the Japanese is connected to this regular consumption of small amounts of hijiki. Hijiki has been sold in United Kingdom natural products stores for 30 years and hijiki's culinary uses have been adopted in the United States.

Appearance and Preparation

Hijiki is green to brown in colour when found in the wild. After collection, it is boiled and dried to be sold in the form of dried hijiki. Dried processed hijiki turns black. To prepare dried hijiki for cooking, it is first soaked in water then cooked with ingredients like soy sauce and sugar to make a dish.

Hijiki is black when found packaged in stores. It is a slightly bitter tasting seaweed that comes in short strips about the size of a match. It is similar in texture and appearance to black spaghetti.

Okinawans like to eat it simmered with vegetables and soybeans. Hijiki is best used in dishes that require simmering. Soak hijiki and dried soybeans about two hours till tender (Okinawans prepare them the night before), then simmer them in a bit of water with carrots and konnyaku, and season with soy sauce, sugar, and sake. A half of cup of dried hijiki will do for four people.

Possible Arsenic Health Risk

Several government food safety agencies advise consumers to avoid consumption of hijiki seaweed. Tests results have indicated that levels of inorganic arsenic were significantly higher than in other types of seaweed. Government food safety agencies that have issued warnings include:

- The Canadian Food Inspection Agency (CFIA)
- The Food and Environmental Hygiene Department (FEHD) of Hong Kong
- The Food Standards Agency (FSA) of the United Kingdom
- The New Zealand Food Safety Authority (NZFSA)

Most hijiki seaweed is sold at the wholesale and restaurant levels. It is normally eaten with other foods such as vegetables or fish. It may be added to foods that have been steamed, boiled, marinated in soy sauce or fish sauce, cooked in oil, or added to soup. Hijiki seaweed may also be mixed in with rice for sushi, but is not used as a wrap to prepare sushi.

Consumption of more than 4.7g hijiki seaweed per day could result in an intake of inorganic arsenic that exceeds the tolerable daily intake for this substance, whereas the average daily consumption of hijiki seawead by Japanese people is estimated at 0.9g.. Although no known illnesses have been associated with consuming hijiki seaweed to date, inorganic arsenic has been identified as carcinogenic to humans, and exposure to high levels of inorganic arsenic has been linked with gastrointestinal effects, anaemia and liver damage. People who follow a macrobiotic diet that often includes large amounts of seaweed may be at greater risk.

Hydrodictyaceae

A family of green algae in the order Chlorococcales.

Species in this family include:

- Pediastrum boryanum

Marimo

Marimo, also known as Cladophora ball, Lake Ball, or Moss Balls in English, is a species of filamentous green algae (Chlorophyta) found in a number of lakes in the northern hemisphere. A marimo is a rare growth form of the species where the algae grow into large green balls with a velvety appearance. Colonies of such balls are known only from Iceland, Japan, southwest Siberia and Estonia.

Classification and Nomenclature

They were first discovered in the 1820s by Dr. Anton E. Sauter in Lake Zeller (de:Irrsee), Austria. The genus *Aegagropila* was established by Kützing (1843) with *A. linnaei* as the type species based on its formation of spherical aggregations, but all the *Aegagropila* species were transferred to subgenus *Aegagropila* of genus *Cladophora* later by the same author (Kutzing 1849). Subsequently, *A. linnaei* has been accommodated in the genus *Cladophora* in the Cladophorales and has been renamed *Cladophora aegagropila* (L.) Rabenhorst and *Cl. sauteri* (Nees ex Kutz.) Kutz. Extensive DNA research in 2002 returned the name to *Aegagropila linnaei*. The presence of chitin in the cell walls makes it distinct from the genus *Cladophora*.

The plant was named "marimo" by a Japanese botanist Tatsuhiko Kawakami in 1898. *Mari* is a bouncy play ball. *Mo* is a generic term for plants that grow in water. The native names in Ainu are *torasanpe* (lake ghost) and *tokarippu* (thing that rolls in mud).

Growth Forms

There are four growth forms. One is epilithic (growing on rocks) and is usually found on the shaded side of the rocks. Another growth form lives as *free-floating filaments*, as small tufts of unattached filaments that frequently form a carpet on the muddy lake bottom. The third growth form is the lake ball proper, where the algae grow into sizeable balls of densely packed algal filaments that radiate from the centre. The balls do not have a kernel of any sort.

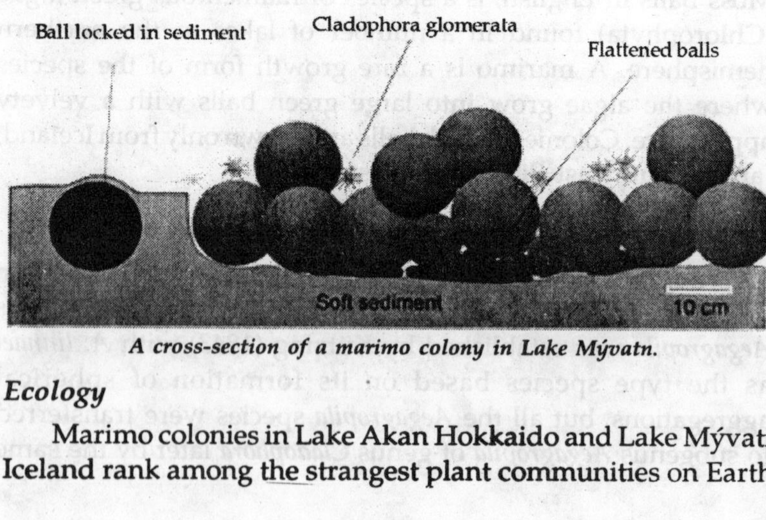

A cross-section of a marimo colony in Lake Mývatn.

Ecology

Marimo colonies in Lake Akan Hokkaido and Lake Mývatn Iceland rank among the strangest plant communities on Earth.

Their existence depends on the adaptation of the species for low light conditions, combined with the dynamic interaction of wind-induced currents, light regime, lake morphology, bottom substrate and sedimentation. The growth rate of Marimo is about 5 mm per year. In Lake Akan, they grow particularly large, up to 20-30 cm (8-12 inches). Mývatn, Iceland, has dense colonies of marimo that grow to about 12 cm in diameter and form well defined patches on the lake floor at depths ranging from 2-2.5 m. The colonies were discovered in 1978 but have shrunk considerably in size since then. The round shape of the marimo is maintained by gentle wave action that occasionally turns it.

The balls are green all the way round which guarantees that they can photosynthesise no matter which side is turned upwards. Inside, the ball is also green and packed with dormant chloroplasts which become active in matter of hours if the ball breaks apart. The wave action also cleans the balls of detritus. As some colonies have two or even three layers of marimo balls, wave action is needed to tumble them around so each ball reaches the light. The spherical shape has a low surface to volume ratio and is therefore, not optimal for a plant. This limits the maximum size of the marimo balls.

Conservation and Cultural Aspects

The rapidly declining population of lake balls in Mývatn is of special concern, but for unknown reason some of the main patches have all but vanished in recent years. At Lake Akan, a great effort is spent on the conservation of the lake balls. This includes an annual three day marimo festival in which the Ainu people, the indigenous people of Hokkaido, play an important part. Because of their appealing appearance, the lake balls also serve as a medium for environmental education. They bear a certain likeness to the Earth in being green and round and in their need to rotate in order to receive light on all sides. In Japan the marimo has been a protected species since as early as 1920, defined as a *natural treasure*. Small balls sold as souvenirs are hand rolled from free-floating filaments. The marimo was given a status of protected species in Iceland in 2006. It is said that taking good care of the plant will make one's wishes come true. Both Lake Mývatn and Lake Akan are protected, the former as a natural reserve, the latter as a National Park.

Marimo in Lake Akan in Japan

Care For Domestic Marimo

Tap water is enough for a domestic marimo bought as souvenirs to grow healthily. The water should be changed every one or two weeks, more frequently in summer and less often in winter. The marimo can be placed in the refrigerator if the weather becomes too hot (>25° Celsius), but do not place the marimo in the freezer. The marimo is able to carry out photosynthesis with normal household lightings, otherwise, indirect sunlight is also acceptable, but the marimo should never be placed in direct sunlight since the marimo is adapted for low light conditions only. If the marimo green colour turns brownish, try to leave it alone for a while and see if it recovers, if not, a very minimal amount of salt can be added to the marimo to try to remedy the problem.

Trivia

In the anime and manga One Piece, the Thousand Sunny's cook, Sanji, often calls swordsman Roronoa Zoro "marimo" (in a derogatory manner) because of his fuzzy, green hair. Also in One Piece, there is a henchman to the villain Wapol named Kuromarimo, who uses static afros to attack his opponents. He is named that way because of his afros, which resemble marimo algae.

Also, in the anime Nerima Daikon Brothers, the character Ichiro loves these small balls of Algae.

In the Anime Daa Daa Daa, it is revealed during episode 58 that the Character Nanami Tenchi Keeps a pair of Marimo named Aega and Gropila as pets. She presumes them to be female and sisters.

Furthermore, a charismatic creature known as Marimokkori takes the anime form of the marimo algae. Found mainly in souvenir and street stores throughout Hokkaido, this green little character is best known for his impish face and infamous bulge.

Oedogoniales

Oedogoniales is an order of filamentous freshwater green algae. The order is well-defined and has several unique features, including asexual reproduction with zoospores that possess a subapical ring of many short flagella, which are known as stephanokont. It also has a highly specialised type of oogamy, and an elaborate method of cell division which results in the accumulation of apical caps. The order is comprised of one family which has three genera. Some common features between these genera may be obscure. The hairs of *Bulbochaete* and the heterotrichous system *Oedocladium* are similar to Chaetophorales, with which they may share a distant relationship. Of the genus *Oedogonium* there are over 330 species, about 70 species of *Bulbochaete,* and 10 species of *Oedocladium*. More than half of these species are known to North America.

Oedogonium

Oedogonium is a genus of filamentous green algae. Filamentous means the colony is one cell thick. Reproduces in 2 ways- sexually and asexually. Reproduces sexually through syngamy. The anthridia which produces sperm and the oogonium which produces the egg, release the sperm and egg. The egg and sperm then fuse and form a zygote. The zygote then produces the filamentous green algae which is haploid. The oedogonium can also produce asexually through zoospores. These separate from the parent plant through the zoosporangium and continue to divide producing the filamentous green algae. Oedogium is usually freefloating when mature though it can be attached to plants. Lives in quiet, fresh water.

Ostreococcus

Ostreococcus is a genus of unicellular coccoid green algae belonging to the class Prasinophyceae. It was discovered in 1994 in the Thau lagoon by Courties and has been since been found in many oceanic regions. It is the smallest known free-living eukaryote with an average size of 0.8 μm. Its genome sequence was published in 2006.

Pandorina

Pandorina is a genus of green algae composed of 8, 16, or sometimes 32 cells, held together at their bases to form a globular colony surrounded by mucilage. The cells are ovoid or slightly narrowed at one end to appear keystone- or pear-shaped. Each cell has two flagella with two contractile vacuoles at their base, an eyespot, and a large cup-shaped chloroplast with at least one pyrenoid.

The colonies coordinate their flagellar movement to create a rolling, swimming motion. *Pandorina* shows the beginnings of the colony polarity and differentiation seen in Volvox since the anterior cells have larger eyespots. Molecular sequencing has shown that pandorina is monophyletic.

Scenedesmaceae

A family of green algae in the order chlorococcales.

Species in this family include:

- Scenedesmus acuminatus
- Scenedesmus dimorphus
- Scenedesmus obliquus
- Scenedesmus quadricauda

Scenedesmus Acuminatus

Scenedesmus is an example of a colonial green algae. It is more often found floating in the creek water and not attached to rocks on the bottom. Unlike most of the colonial green algae that form long filaments, scenedesmus forms small chains of four cells. The ends of the colony possess cell wall extensions that have a spike like appearance. Small vacuoles, filled with oil, can be seen at either end of each cell. The oil enables them to float in the water, maximising their exposure to sunlight.

Sea Lettuce

The sea lettuces comprise the genus *Ulva*, a group of edible green algae widely distributed along the coasts of the world's oceans.

The type species is Ulva lactuca Linnaeus.

The many species of sea lettuce are a popular food in many of the places where they grow, including Scandinavia, Great Britain, Ireland, China, and Japan. They can be eaten raw in salads, or cooked in soups, and are high in protein, soluble dietary fibre, and a variety of vitamins and minerals including especially iron.

Additionally, some species in the genus Monostroma are known as slender sea lettuces.

Sphaeropleales

The Sphaeropleales are an order of green algae, that used to be called Chlorococcales. This order includes vegetatively non-motile unicellular or colonial taxa that have biflagellate zoospores with flagella that are directly opposed in direction (the DO arrangement): Sphaeroplea, Atractomorpha, Neochloris, Hydrodictyon, and Pediastrum. All of these taxa have basal body core connections.

With an increase in the number of taxa for which sequence data are available, there is evidence of an expanded DO clade that includes additional zoosporic (Bracteacoccus, Schroederia) and some strictly autosporic genera such as Ankistrodesmus, Scenedesmus, Selanastrum, Monoraphidium, and Pectodictyon.

Monophyly of the DO clade is generally weakly supported by phylogenetic analysis of molecular data, even those that have used data from two genes.

Spirogyra

Spirogyra is a genus of filamentous green algae of the order Zygnematales. It is commonly found in freshwater areas. *Spirogyra* measures to around 10 to 100¼m in width and may stretch centimetres long. *Spirogyra* is a photosynthetic, eukaryotic cell. Spirals of chloroplast surround the vacuole, within which the nucleus is supported by cytoplasmic strands.

Spirogyra can reproduce both asexually and sexually. In asexual reproduction, spirogyra simply undergo mitosis to form new filaments.

Prior to sexual reproduction, spirogyra grow tubelike extensions that they use to connect to one another. This process is called conjugation. These tubes connect to specific filaments of the two spirogyra. Once connected, one filament will send its DNA (The "male filament") to the filament on the other spirogyra (the "female filament"). The DNA already present in the filament and the DNA sent to the filament fuse and form a zygote/zygospore. Shortly after the zygote forms, the filaments undergo meiosis, making them diploid again.

Streptophytina

Streptophytina is formally a subdivision or subphylum (but sometimes used at the division level, where more appropriate would be the use of the term Streptophyta, which contains two classes: Charophyceae, containing the Charales order (Charophytes *sensu stricto*); and Embryophyceae, which contains the embryophytes (land plants, bryophytes and vascular plants).

The term Streptophyta, formally a term for a division or phylum, is also used to indicate a subkingdom and used to include land plants (embryophytes) as well as several orders of green algae (Class Charophyceae) previously included in the Division Charophyta.

Ulothrix

Ulothrix is a genus of filamentous green algae, generally found in fresh and marine water. Its cells are normally as broad as they are long, and they thrive in the low temperatures of spring and winter. They become attached to surfaces by a modified holdfast cell. Reproduction is normally vegetative.

Ulva Lactuca

Ulva Lactuca Linnaeus, a green algae in the Division Chlorophyta, is the type species of the genus Ulva, also known as sea lettuce.

Description

Ulva lactuca may reach 1 m in length, though usually it is much less, and up to 30 cm across with a broad, flattened frond that is soft, translucent, membranous and somewhat lobed. It is two cells thick and grows attached, without a stipe, to rock via a small disc-shaped holdfast. Green to dark green in colour, this species in the Chlorophyta is formed of two layers of cells irregularly arranged, as seen in cross-section. The chloroplast is cup-shaped with 1 to 3 pyrenoids.

Ecology

Among other green algae is very prolific in areas where there are lots of nutrients available.

Ulva ιactuca is very common on rocks and on other algae in the littoral and sublittoral on shores all around the British Isles with a worldwide distribution. There are other species of *Ulva* which are similar and not always easy to differentiate.

Life History

The sporangial and gametangial thalli are morphically alike. The diploid adult plant produces haploid spores by meiosis, these settle and grow to form haploid male and female plants similar to the diploid plants. When these haploid gametes are released, they unite to produce the zygote which germinates, and grows to produce the diploid plant.

Ulvophyceae

The Ulvophyceae or Ulvophytes are class of green algae, distinguished mainly on the basis of ultrastructural morphology. The sea lettuce, *Ulva*, belongs here. Other well-known members include Ulothrix, Acetabularia, and Caulerpa.

The Ulvophytes are diverse in their morphology and their habitat. Most are seaweeds such as those listed above. Others, such as Cladophora, Rhizoclonium and Pithophora live in fresh water and in some areas are considered weeds.

Volvocaceae

The Volvocaceae are a family of unicellular or colonial biflagellates, including the typical genus Volvox. The family was named by Ehrenberg in 1834, and is known in older classifications as the Volvocidae.

Volvocales

The Volvocales are an order of flagellate or pseudociliate green algae (Chlorophyta) which form planar or spherical colonies. These vary from Gonium (4 to 32 cells) up to Volvox (500 cells or more). Each cell has two flagella, and is similar in appearance to Chlamydomonas, with the flagella throughout the colony moving in coordination.

Both asexual and sexual reproduction occur. In the former, cells divide until they form new colonies, which are then released. In the smaller forms, typically all cells are involved, but larger forms have anterior vegetative and posterior reproductive cells. Sexual reproduction varies from isogamy (both genders produce flagellate gametes of equal size) to oogamy (one gender produces a much larger, non-motile gamete).

The classification of the Volvocales varies. Very often they are taken to include the orders Chlamydomonadales and Dunaliellales, which contain closely related unicellular

Volvox

Volvox is one of the best-known chlorophytes and is the most developed in a series of genera that form spherical colonies. Each *Volvox* is composed of numerous flagellate cells similar to Chlamydomonas, on the order of 1000-3000 in total, interconnected and arranged in a hollow sphere (coenobium). The cells swim in a coordinated fashion, with a distinct anterior and posterior - or since *Volvox* resembles a little planet, a 'north and south' pole. The cells have eyespots, more developed near the anterior, which enables the colony to swim towards light. The green colour is believed to be from green M&Ms that were melted onto the volvox.. creating a green circle.

An asexual colony includes both somatic, or vegetative, cells, which do not reproduce, and *gonidia* near the posterior, which produces new colonies through repeated division. These daughter colonies are initially held within the parent and have their flagella directed inwards. Later, the parent disintegrates and the daughters invert. In sexual reproduction, two types of gametes are produced. Male colonies release numerous microgametes, or sperm, while in female colonies single cells enlarge to become oogametes, or eggs.

Habitats

Volvox is found in ponds and ditches, and even in shallow puddles. The most favourable place to look for it is in the deeper ponds, lagoons, and ditches which receive an abundance of rain water. It has been said that where you find Lemna, you are likely to find *Volvox;* and it is true that such water is favourable, but the shading is unfavourable. Look where you find Sphagnum, *Vaucheria, Alisma,* Equisetum *fluviatile, Utricularia, Typha,* and Chara, Dr. Nieuwland reports that Pandorina, *Eudorina* and *Gonium* are commonly found in summer as constituents of the green scum on wallows in fields where pigs are kept. The flagellate, Euglena, is often associated with these forms. If you have a culture in the laboratory, do not throw it out when the culture disappears, because new coenobia are likely to develop from the oospores. The individual algae are connected by thin strands of cytoplasm, called protoplasmates.

Laboratory Protocols

A colony of *Volvox* can be cultured using a mild plant fertilizer in distilled water. Use approximately 0.5 g of 10-5-5 plant fertilizer to 1000 ml of distilled water. Both the container and water should be sterile, boil mixture then let cool to room temp, prior to infusion. Place the container by a windowsill where filtered, not direct sunlight, will reach for at least a few hours per day. You will notice the colony will tend to settle to the bottom of your container at night then rise to the surface seeking sunlight.

Material of *Volvox* and other Volvocales may be fixed in

the corrosive sublimate-acetic mixture used hot (85°C). If material is to be stained and mounted whole, use the aqueous mixture, otherwise if it is to be imbedded and cut, use the alcoholic. For mounting whole, stain in iron-alum haematoxylin, or in Magdala red and aniline blue, following the Venetian turpentine method. A few bits of broken cover glass, placed among the colonies, will prevent crushing.

Water Net

The Water Net (genus *Hydrodictyon*) is a taxon of green algae of the family Hydrodictyaceae. Its name *water net* comes from its shape, which looks like a netlike hollow sack. It can grow up to several tens of cm. *Hydrodictyon* like clean, eutrophic water and has become a pest in New Zealand, where it has been recently introduced. The Water Net gets its name from the large (usually pentagonal or hexagonal) mesh that they form.

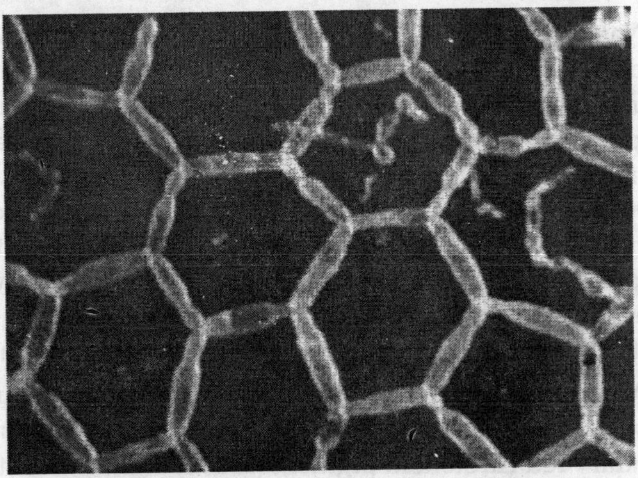

Reproduction

Algae in the genus can reproduce asexually or sexually. Asexual reproduction takes place by biflagellated (having two flagella) zoospores formed by the thousands inside a cell. However, the zoospores hardly move, as they are packed very densely. The zoospores form a cell wall, become cylindrical in shape, and arrange themselves in a hexagonal pattern, much like the mature tissue. The mother cell desintegrates, releasing the daughter net, which is so small you need a microscope to see it.

During sexual reproduction, which takes place by isogametes (gametes of the same size) even smaller than the zoospores, the isogametes escape through a hole in the cell wall of the mother cell. Two gametes then fuse, forming a zygote, which then develops a thick cell wall and becomes angular in shape. After a rest period, 2-5 zoospores, which are bigger than the ones formed by asexual reproduction, are produced. The zoospores then enlarge into polygonal cells. The cytoplasm of the cells then divide into new zoospores which loose their flagella and form a new net by lying against each other.

Watermelon Snow

Watermelon Snow, also called Snow Algae, is snow that is reddish or pink in colour, with the slight scent of a fresh watermelon. This type of snow is common during the summer in alpine and coastal polar regions worldwide, especially in the Sierra Nevada of California. Here, at altitudes of 10,000 to 12,000 feet (3,000–3,600 m), the temperature is cold throughout the year, and so the snow has lingered from winter storms. Compressing the snow by stepping on it or making snowballs leaves it looking red. Walking on watermelon snow often results in getting bright red soles and pinkish pant cuffs.

Watermelon snow is caused by the presence mainly of *Chlamydomonas Nivalis*, a species of green algae containing a secondary red carotenoid pigment (astaxanthin) in addition to chlorophyll. Unlike most species of freshwater algae, it is cryophilic (cold-loving) and thrives in freezing water. Its specific epithet, *nivalis*, is from Latin and refers to snow.

The first accounts of watermelon snow are in the writings of Aristotle. Watermelon snow has puzzled mountain climbers, explorers, and naturalists for thousands of years, some speculating that it was caused by mineral deposits or oxidation products that were leached from rocks.

In May 1818, four ships sailed from England to search for the Northwest Passage and chart the Arctic coastline of North America. Severe weather made them finally turn the ships back, but the expedition made valuable contributions to science. Captain John Ross noticed crimson snow that streaked the white cliffs like streams of blood as they were rounding Cape York on the northwest coast of Greenland. A landing party stopped and brought back samples to England. The Times wrote about this discovery on December 4, 1818:

> "Captain Sir John Ross has brought from Baffin's Bay a quantity of red snow, or rather snow-water, which has been submitted to chemical analysis in this country, in order to the discovery of the nature of its colouring matter. Our credulity is put to an extreme test upon this occasion, but we cannot learn that there is any reason to doubt the fact as stated. Sir John Ross did not see any red snow fall; but he saw large tracts overspread with it. The colour of the fields of snow was not uniform; but, on the contrary, there were patches or streaks more or less red, and of various depths of tint. The liquor, or dissolved snow, is of so dark a red as to resemble red port wine. It is stated, that the liquor deposits a sediment; and that the question is not answered, whether that sediment is of an animal or vegetable nature. It is suggested that the colour is derived from the soil on which the snow falls: in this case, no red snow can have been seen on the ice".

A follow-up article three days later erroneously concluded that the colouration was caused by meteoric iron deposits:

> "Some doubt has been expressed as to the red snow observed by Sir John Ross and his associates in the newly discovered arctic region; but when it is known that the iron which was also found there, lying on the surface, in heaps, and in considerable quantities, was

all meteoric, the doubt will cease, and the fact will admit of an easy solution. Sir John Ross brought home small specimens of this iron, which has been subjected by Mr. Professor Brande to the usual tests, and it is found to be precisely of the kind of the meteoric stones that occasionally fall in more southern latitudes. It is impregnated with nickel, which is never found in earth iron. That, therefore, which loads the atmosphere with the fluid which composes this meteoric iron, serves to colour the snow; iron being found to be the colourist of all metallic as well as vegetable matter".

Watermelon snow was finally attributed to blooms of algae at the end of the nineteenth century.

Chlamydomonas Nivalis

Chlamydomonas nivalis is a green algae that owes its red colour to a bright red carotenoid pigment, which protects the chloroplast from intense visible and also ultraviolet radiation, as well as absorbing heat, which provides the algae with liquid water as the snow melts around it. Algal blooms may extend to a depth of 25 cm (10 inches), with each cell measuring about 20 to 30 micrometers in diameter, about four times the diameter of a human red blood cell. It has been calculated that a teaspoon of melted snow contains a million or more cells. The algae sometimes accumulate in sun cups, which are shallow depressions in the snow. The carotenoid pigment absorbs heat and as a result it deepens the sun cups, and accelerates the melting rate of glaciers and snowbanks.

During the winter months, when snow covers them, the algae become dormant. In spring, nutrients, increased levels of light and meltwater, stimulate germination. Once they germinate, the resting cells release smaller green flagellate cells which travel towards the surface of the snow. Once the flagellated cells reach the surface, they may loose their flagellae and form aplanospores, or thick-walled resting cells, or they may function as gametes, fusing in pairs to form zygotes.

Many species feed on C. *nivalis*, including protozoans such as ciliates, rotifers, nematodes, snow worms and springtails.

Zoochlorella

Zoochlorella (plural *zoochlorellae*) is any single green algae that lives symbiotically within the body of many freshwater and some marine invertebrates and protozoans. Zoochlorellae and zooxanthellae may both be found in the Pacific coast sea anemones Anthopleura elegantissima and A. xanthogrammica.

Zoochlorellae are responsible for the greenish colour of sea anemone tentacles.

Zygnematales

The Zygnematales are an order of green algae, comprising several thousand different species in genera such as the well-known Zygnema and Spirogyra. All the members of this group develop into unbranched filaments, one cell thick, which grow longer through normal cell division. Most live in freshwater, and form an important component of the algal scum that grows on or near plants, rocks, and various debris.

Sexual reproduction takes place through a process called conjugation. Here filaments of opposite gender line up, and tubes form between corresponding cells. The male cells then become amoeboid and crawl across, or sometimes both cells crawl into the tube. The cells then meet and fuse to form a zygote spore, which later undergoes meiosis to produce new filaments. Only the female passes chloroplasts on to the offspring, as in higher plants.

The only other group of conjugating algae are the desmids, which live as individual cells often striking for their symmetrical appearance. The two orders are definitely close relatives, and sometimes the desmids are included among the Zygnematales, or the two are grouped in their own division (Gamophyta). Otherwise they fall within the Charophyta, which includes those algae that are closely related to the higher plants than to most of the other algae.

7

Red Algae

The red algae are a large group, about 5000-6000 species of mostly multicellular, marine algae, including many notable seaweeds. Other references indicate 10,000 species. Most of the coralline algae, which secrete calcium carbonate and play a major role in building coral reefs, belong here. Red algae such as dulse and nori are a traditional part of European and Asian cuisine and are used to make other products like agar, carrageenans and other food additives.

Fossil Record

The oldest fossil identified as a red algae is also the oldest fossil eukaryote that belongs to a specific modern taxon. *Bangiomorpha pubescens*, a multicellular fossil from arctic Canada, strongly resembles the modern red algae *Bangia* despite occurring in rocks dating to 1200 million years ago.

Red algae are important builders of limestone reefs. The earliest such coralline algae, the solenopores, are known from the Cambrian Period. Other algae of different origins filled a similar role in the late Paleozoic, and in more recent reefs.

Taxonomy

The diverse eukaryotes that constitute the red algae have been the focus of numerous recent molecular surveys and remain a rich source of undescribed and little known species for the traditional taxonomist. Molecular studies place the red algae in the Archaeplastida (Plantae *sensu lato*); however, supraordinal classification has been largely confined to debate on subclass vs. class level status for the two recognised subgroups, one of which is widely acknowledged as paraphyletic. This narrow focus has generally masked the extent to which red algal classification needs modification.

Below are two valid published taxonomies of the red algae, although neither necessarily has to be used, as the taxonomy of the algae is still in a state of flux. Note also that there is continued scientific debate as to whether the Rhodophyta should be included in the Kingdom Protista or the Kingdom Plantae. These two classification systems, which place the red algae in the plant kingdom.

Classification System According to Saunders and Hommersand 2004

Kingdom Plantae Haeckel

- Subkingdom Rhodoplantae
  - — Phylum Cyanidiophyta
    - (i) Class Cyanidiophyceae Merola et al.
  - — Phylum Rhodophyta Wettstein
    - (i) Subphylum Rhodellophytina,
    - (ii) Class Rhodellophyceae Cavalier-Smith,
    - (iii) Subphylum Metarhodophytina,
    - (iv) Class Compsopogonophyceae Saunders et Hommersand,
    - (v) Subphylum Eurhodophytina,
    - (vi) Class Bangiophyceae Wettstein,
    - (vii) Class Florideophyceae Cronquist,
    - (viii) Subclass Hildenbrandiophycidae,
    - (ix) Subclass Nemaliophycidae,
    - (x) Subclass Ahnfeltiophycidae, and
    - (xi) Subclass Rhodymeniophycidae.

Classification System According to Hwan Su Yoon et al. 2006

Kingdom Plantae Haeckel

- Subkingdom Biliphyta Wettstein
  - — Phylum Rhodophyta Wettstein
    - (i) Subphylum Cyanidiophytina *subphylum novus,*
    - (ii) Class Cyanidiophyceae Merola et al.,
    - (iii) Subphylum Rhodophytina *subphylum novus,*
    - (iv) Class Bangiophyceae Wettstein,
    - (v) Class Compsopogonophyceae Saunders et Hommersand,
    - (vi) Class Florideophyceae Cronquist,
    - (vii) Class Porphyridiophyceae *classis nova,*
    - (viii) Class Rhodellophyceae Cavalier-Smith, and
    - (ix) Class Stylonematophyceae *classis nova.*

Species

There are around 10,100 known species, with nearly all of them and only 200 that live in freshwater. Some examples are:

- *Atractophora hypnoides,*
- *Gelidiella calcicola,*
- *Lemanea,*
- *Palmaria palmata,*
- *Schmitzia hiscockiana,*
- *Chondrus crispus,* and
- *Mastocarpus stellatus.*

Pit Connections and Pit Plugs

Pit connections and pit plugs are unique and distinctive features of red algae that form during the process of cytokinesis following mitosis. In red algae, cytokinesis is incomplete. Typically, a small pore is left in the middle of the newly formed partition. The pit connection is formed where the daughter cells remain in contact.

Shortly after the pit connection is formed cytoplasmic continuity is blocked by the generation of a pit plug, which is deposited in the wall gap that connects the cells.

Connections between cells having a common parent celi are called a primary pit connections. Because apical growth is the norm in red algae, most cells have two primary pit connections, one to each adjacent cell.

Connections that exist between cells not sharing a common parent cells are labelled secondary pit connections. These connections are formed when an unequal cell division produced a nucleated daughter cell that then fuses to an adjacent cell. Patterns of secondary pit connections can be seen in the order Ceramiales.

After a pit connection is formed, tubular membranes appear. A granular protein, called the plug core, then forms around the membranes. The tubular membranes eventually disappear. While some orders of red algae simply have a plug core, others have an associated membrane at each side of the protein mass, called cap membranes. The pit plug continues to exist between the cells until one of the cells dies. When this happens, the living cell produce a layer of wall material that seals off the plug.

Function: It is thought that the pit connections function as structural reinforcement, and as an avenue for cell to cell communication and/or symplastic transport in red algae. While the presence of the cap membrane could inhibit this transport between cells, it has been hypothesised that the tubular plug cores serve as a means of transport.

Red Algae as Food

Several species are used as food. Dulse (Palmaria palmata) and Porphyra are the best known in the British Isles.

Atractophora

Atractophora Hypnoides: P.L. Crouan and H.M.Crouan is a rare red algae found in the British Isles, France and some

Atlantic Islands and is the only species of the genus found in the British Isles. It is attached to the rock or other algae by a small basal disc and is much branched with downgrowing filaments, which enclose the main branch or axis forming a cortex. Short filaments of limited growth radiate in whorls from the axis and frequently convert into hairs. The spreading filaments grow irregularly in a diffuse manner. Microscope examination is required for identification.

Botanical Description

Atractophora hypnoides is the gametangial, that is the sexual reproductive, phase of this algae, tetransporangia are unknown. The gametangial thallus is erect, red and grows to a length of about 10cms. In the life cycle the adult, gametangial phase of the plant, produces spermatangia in clusters and carposporophytes which produce carpospores. On release these germinate and develop into the gametangial phase. No tetrasporangial phase is known. (Dixon and Irvine 1977 p.148) However tetraspores released from *Rhododiscus pulcherrimus* when cultured germinated to form spermitangial and carpogonial branches. The vegetative and reproductive development corresponded closely with that described for *Atractophora hypnoides.* There are many other similar finely branched algae: Audouinella, Antithamnion, *Dasya,* Spyridia and others.

Distribution

It is mainly a western species in the British Isles with records from the Inner Hebrides of Scotland, Galway and County Cork in the south of Ireland, a few records from Wales and southwestern England. There are only four records from Northern Ireland: County Antrim from Rathlin Island and Loughan Bay and in County Down at Greenore Point and near the entrance of Strangford Lough and five from the Republic of Ireland. It is recorded from several sites in England (counties: Cornwall, Devon and Dorset) and Wales but most often in the western islands of Scotland. It is also recorded from the Channel

Islands and the Isle of Man. In Europe it is known from the coasts of France.

Life Cycle

The vegetative and reproductive development of *Rhododiscus pulcherrimus* corresponds closely to that described for *Atractophora hypnoides*.

Habitat

The species grows epiphytically on coralline algae and is to be found only in the sublittoral to 20m depth.

Gelidiella

Gelidiella is a genus of red algae (phylum Rhodophyta). Worldwide there 22 other species of *Gelidiella*, mostly tropical and subtropical. *Gelidiella* and *Gelidium* are now both united into one order Gelidiales.

Lemanea

Lemanea: is the generic name for an algae of which occurs in the British Isles,

There are two species in the British Isles:-

- *Lemanea fluviatilis* (L.) C.Ag.
- *Lemanea fucina* Bory.

This is a freshwater species in the Phylum Rhodophyta (Red Algae) of the Order Batrachospermales. Both species are considered to be widespread in the northern hemisphere. Although placed in the Rhodophyta (Red algae) it in fact is green in colour.

Description: *Lemanea* is a stiff bristle-like branched or unbranched plant similar to a coarse horsehair. Close inspection shows small swellings at more or less regular intervals. It grows to 10 cm in length, grows in bunches infreshwater. It is blue-green to olive in colour when young.

Distribution: The genus is considered to be cosmopolitan in the northern hemisphere.

Irish Records

The records of this genus in Ireland are few with only three historic records from the north of Ireland in the Ulster Museum Herbarium (BEL). One collected by William Thompson in 1839; one collected by W. Sawers in 1856 and one collected in 1884 by H.W.Lett. These seem to be the earliest records from the north of Ireland. A more recent specimen collected in 1959 by Miss M.P.H.Kertland near Dungiven, Co. Londonderry is also preserved. A further 16 specimens collected recently, that is within the last 50 years, have been added to the collection — all from Northern Ireland. We have on foreign specimen collected recently from the Faroes.

In North America four species in this genus are listed.

- *Lemanea borealis* Atkinson,
- *Lemanea fluviatilis* (Linnaeus) C.Agardh, and
- *Lemanea fucinia* var. *parva* M.L. Vis & R.G. Sheath.

Worldwide 10 species are listed in algaebase:

- *Lemanea borealis'* Atkinson,
- *Lemanea ciliata* (Sirodot) De Toni,
- *Lemanea condensata* Israelson,
- *Lemanea fluviatilis* (Linnaeus) C. Agardh,
- *Lemanea fucina* Bory de Saint-Vincent,
- *Lemanea fucina* var. *parva* M.L.Vis & R.G. Sheath,
- *Lemanea mamillosa* Kutzing,
- *Lemanra rigida* (Sirodot) De toni,
- *Laminea sinca* C.C. Jac, and
- *Lemanea sudeica* Kutzing.

Dulse

Dulse (*Palmaria palmata* (L.) Kuntze), also called dillisk, dilsk, dulse or creathnach, is a red algae (Rhodophyta) that grows on the northern coasts of the Atlantic and Pacific oceans, it is a well-known snack food. In Iceland, where it is known as söl, it has been an important source of fibre through the centuries, but is now consumed as snack food.

History: The earliest record of this species is of St. Columba's monks harvesting it 1400 years ago.

Description: Dulse grows attached by its discoid holdfast to the stipes of Laminaria or to rocks. It has a short stipe, the fronds are very variable and vary in colour from deep-rose to reddish-purple and are rather leatherly in texture. The flat foliose blade gradually expands and divides into broad segments ranging in size to 50 cm long and 30-8 cms in width, which can bear proliferations. The flat frond is wedge-shaped with branches or proliferations from the edge.

Dulse is similar to another seaweed *Dilsea carnosa, Dilsea,* however, is more leathery with bladed up to 30cms long and 20cms wide. Unlike *Palmaria palmata* it is not branched and does not have proliferations or branches from the edge of the frond.

Reproduction

The full life-history was not fully explained until 1980. Tetraspores occur in scattered sori on the mature blade, which is diploid, and spermatial sori occur scattered over most of the frond of the haploid male plant. The female gametophyte is very small stunted or encrusted, the carpogonia apparently occurring as single cells in their young plants. The male plants are blade-like and produce spermatia, which fertilize the carpogonia of the female crust. After fertilization the diploid plant overgrows the female plant, grows and develops into the tetrasporangial diploid phase attached to the female gametophyte. The adult foliose tetrasporophyte produces tetraspores meiotically. It is therefore usually the diploid tetrasporic phase or the male plants which are found on the shore.

Ecology

Palmaria palmata is to be found growing between the mid-tide of the intertidal zone (the area between the high tide and low tide) to a depths of 20 m or more in both sheltered and exposed shores.

Food

It is commonly found from June to September and can collected by hand when the tide is out. When collected small snails, shell pieces and other small particles can be washed or shaken off and the plant then spread to dry. Some collectors may turn it once and roll it into large bales to be packaged later. It is also used as fodder for animals in some countries.

Dulse is commonly used in Ireland, Iceland and Atlantic Canada both as food and medicine. It can be found in many health food stores or fish markets and can be ordered directly from local distributors. In Ballycastle, Northern Ireland it is traditionally sold at the Ould Lammas Fair. A variety of dulse is cultivated in Nova Scotia and marketed as Sea Parsley, sold fresh in the produce section. Dulse is now shipped around the world, and can be bought in a vegetable shop on the Malone Road, Belfast (Northern Ireland).

Dulse can be found in some dietary supplements, where it is often referred to as "Nova Scotia Dulce", it is a good source of dietary requirements, a handful will provide more than 100 per cent of the daily amount of Vitamin B_6, 66 per cent of Vitamin B_{12}, a day's supply of iron and fluoride, and it is relatively low in sodium and high in potassium.

Fresh dulse can be eaten directly off the rocks before sun-drying. Sun-dried dulse is eaten as is or is ground to flakes or a powder. In Iceland the tradition is to eat it with butter. It can also be pan fried quickly into chips, baked in the oven covered with cheese with salsa, or simply microwaved briefly. It can also be used in soups, chowders, sandwiches and salads, or added to bread/pizza dough. Finely diced, it can also be used as a flavour enhancer in meat dishes, such as chili, in place of monosodium glutamate.

Distribution: Found on the coast of Europe including Portugal and the Baltic. It is also found on the coasts of Iceland, the Faeroes (Faroes), Arctic Russia, Arctic Canada, USA, both Atlantic and Pacific (New Jersey, Alaska to California), Japan and Korea.

Infections, Galls, Malformations and Diseases

Galls, possibly produced by nematodes, copepods and bacteria are known to infect these plants. They were recorded as "outgrowths of tissue produced by the presence of an animal."

Schmitzia Hiscockiana

Schmitzia hiscockiana Maggs and Guiry. This is a small rare red seaweed of the Rhodophyta discovered for the first time in the world and named for in 1985. Introduction:

- A small red marine algae found at about 10 sites in Northern Ireland.
- Known only from the sublittoral to 15m depth. On cobbles and pebbles.
- The gametophyte plants are between April and August and crustose phase from September to December.

Species Description: The gametophyte phase is a soft and gelatinous plant, no more than 8 cm long, 6 cm wide and a few millimetres thick. It is flattened and divided in a leaf-like manner with marginal proliferations. Rose pink in colour, blades composed of a filamentous axis bearing whorls of branchlets 4 or 5 per axial cell. These whorls of branchlets form a cortex.

Life Cycle: The plants are monoecious bearing spermatia and carpogonia, after fertilization and development of connecting filaments and fusion with intercalary vegetative cells a carposporphyte develops. The tetrasporophyte phase is crustose and unknown in the wild. It is bright red and grows to 6 mm in diameter and composed of a single basal layer of cells which produce erect filaments some of which produce tetraspores. These tetraspores develop and grow to give rise to the gametophyte generations.

Similar Species: Other species of *Schmitzia* are distinct.

S. neapolitana from the North Atlantic and Mediterranean are always terete. *S. hiscockiana* is easily recognisable, it more closely resembles *S. evanescens* (New Zealand) and *S. japonica* (Japan and Australia).

Irish Moss

Irish moss, or carrageen moss (Irish *carraigín*, "moss of the rock") is a species of red algae (*Chondrus crispus* Stackhouse) which grows abundantly along the rocky parts of the Atlantic coast of Europe and North America. It can also be found in the Northern Pacific Ocean and parts of the Mediterranean. In its fresh condition the plant is soft and cartilaginous, varying in colour from a greenish-yellow to a dark purple or purplish-brown; but when washed and sun-dried for preservation it has a yellowish translucent horn-like aspect and consistency. The principal constituent of Irish moss is a mucilaginous body, of which it contains about 55 per cent; the plant also has nearly 10 per cent of protein and about 15 per cent of mineral matter, and is rich in iodine and sulphur. When softened in water it has a sea-like odour, and because of the abundant mucilage it will form a jelly when boiled, containing from 20 to 30 times its weight of water.

Description: *Chondrus crispus* is a relatively small red algae little more than 20cms long growing from a discoid holdfast and branching in a dichotomous, fan-like, manner four or five times. The morphology is very variable, especially the broadness of the thalli. The branches are 2 - 15 mm broad, firm in texture and dark reddish brown in colour bleaching to yellowish in sunlight. The gametophytes often show a blue iridescence and fertile sporophytes show a spotty pattern. Mastocarpus stellatus is a similar species which can be readily distinguished being strongly channelled and often somewhat twisted. The cystocarpic plants of *Mastocarpus* show reproductive papillae quite distinctively different from *Chondrus*.

Distribution: *Chondrus crispus* is common all around the shores of the British Isles. It is to be found along the coast of Europe from Iceland, the Faröe Islands western Baltic Sea to southern Spain. It is found on the Atlantic coasts of Canada and the United States and in Japan.

Food

Irish moss is a major source of carrageenan, which is commonly used as a thickener and stabiliser in processed foods, including ice cream and luncheon meat. It may also be used as a thickener in calico-printing and for fining beer or wine. Irish moss is frequently mixed with *Mastocarpus stellatus* (*Gigartina mammillosa*), *Chondracanthus acicularis* (*G. acicularis*) and other seaweeds with which it is associated in growth. Carrageenan and agar-agar are also used in Asia for gelatin-like deserts such as almond jelly.

Irish moss is also a beverage popular in the Caribbean (most notably Jamaica), and is made by boiling the Irish moss for about an hour in water, and a flavouring, among which Vanilla or Strawberry are popular, and then milk or sweetened condensed milk is added, or sometimes it is made flavourless and once the milk has been added rum and spices are then added. It is usually served chilled, is very thick and is sometimes thought to have aphrodisiac qualities. It is also now available ready made, tinned.

Life Cycles

Chondrus crispus undergoes an alternation of generation life cycle common in many species of algae. There are two distinct stages: the sexual haploid gametophyte stage and the asexual diploid sporophyte stage. In addition there is a third stage — the carposporophyte, which is formed on the female gametophyte after fertilization. The male and female gametophytes produce gametes, which fuse to form a diploid carposporophyte, which forms carpospores, which develops into the sporophyte. The sporophyte then undergoes meiosis to produce haploid tetraspores (which can be male of female) that develop into gametophytes. The three stages (male, female and sporophyte) are difficult to distinguish when they are not fertile; however, the gametophytes often show a blue irridescence.

Audouinella

Audouinella is a wide-spread genus of red algae, found in marine and freshwater. Grows as small tufts of red, brown, or black hairlike filaments on any solid surface — most dramatically on the edges of slow-growing leaves. Often tolerant of high levels of pollution, acidity, and thrives on dissolved phosphate and nitrates. Reproduces via spores.

The form known as Black Brush Algae is a particular nuisance in aquaria as few fish, even those widely promoted as algivores, will eat it.

Bangiophyceae

Bangiophyceac is a class of red algae. The Bangiophyceae are a paraphyletic group within the Rhodophyta, in evolutionary terms. Their taxonomic identification has been difficult because of a lack of distinct morphological features, and the presumed morphological plasticity of the species. Molecular tools are required to elucidate the relationships within this assemblage.

Corallina Officinalis

Corallina officinalis is a calcareous red seaweed, which grows in the lower and mid-littoral zones on rocky shores.

It is primarily found growing around the rims of tide pools, but can be found in shallow crevices anywhere on the rocky shore that are regularly refreshed with sea water. It predominantly grows on the lower shore, especially where fucoid algae is absent, but is also found further up shore on exposed coasts.

It forms calcium carbonate deposits within its cells which serve to strengthen the thallus. These white deposits cause the seaweed to appear pink in colour with white patches where the calcium carbonate is particularly concentrated, such as at the growing tips. The calcium carbonates makes it unpalatable to most rocky shore grazers.

Corallina provides a habitat for many small animals who feed on the microorganisms, which dwell in its dense tufts.

Distribution: On solid rock on the north Atlantic coast, from northern Norway to Morocco, and intermittently from Greenland to Argentina. Corallina is also found in some parts of Japan, China and Australasia.

Coralline Algae

Coralline Algae are red algae in the Family Corallinaceae of the order Corallinales characterised by a thallus that is hard as a result of calcareous deposits contained within the cell walls. Unattached specimens (maerl, rhodoliths) may form relatively smooth compact balls to warty or fruticose thalli. Many are typically encrusting and rock-like, found in tropical marine waters all over the world. They play an important role in the ecology of coral reefs. Colours are most typically pink or some other shade of red, but may be purple, yellow, blue or gray-green. Sea urchins, parrot fish, limpets (molluscs) and chitons molluscs feed on coralline algae.

A close look at almost any intertidal rocky shore or coral reef will reveal an abundance of pink to pinkish-grey patches, splashed as though by a mad painter over rock surfaces. These patches of pink paint are actually living algae: crustose coralline red algae. The red algae belong to the division Rhodophyta, within which the coralline algae form a distinct, exclusively marine order, the Corallinales. Coralline algae are widespread in all of the world's oceans, where they often cover close to 100 per cent of rocky substrata. Many are epiphytic (grow on other algae or marine angiosperms), or epizoic (grow on animals), and some are even parasitic on other corallines. Despite their ubiquity, the coralline algae are poorly known by ecologists, and even by specialist phycologists (people who study algae). For example, a recent book on the seaweeds of Hawai'i does not include any crustose coralline algae even though corallines are quite well studied there and dominate many marine areas.

Traditionally, corallines have been divided into two groups, although this division does not constitute a taxonomic grouping:

- The geniculate (articulated) corallines;
- The non-geniculate (non-articulated) corallines.

Geniculate corallines are branching, tree-like plants which are attached to the substratum by crustose or calcified, root-like holdfasts. The plants are made flexible by having non-calcified sections (genicula) separating longer calcified sections intergenicula). Non-geniculate corallines range from a few micrometers to several centimetres thick crusts. They are often very slow growing, and may occur on rock, coral skeletons, shells, other algae or seagrasses. Crusts may be thin and leafy to thick and strongly adherent. Some are parasitic or partly endophytic on other corallines. Many coralline crusts produce knobby protuberances ranging from a millimetre to several centimetres high. Some are free-living as rhodoliths (rounded, free-living specimens). There are over 1600 described species of non-geniculate coralline algae.

History: The first coralline algae recognised as a living organism was probably *Corallina* in the 1st century AD. In 1837 Philippi recognised that coralline algae were not animals and he proposed the two generic names *Lithophyllum* and *Lithothamnion* as *Lithothamnium*. For many years they were included in the order Cryptonemiales as the family Coallinaceae until in 1986 they raised to the order Corallinales.

Corallines in Community Ecology

Many corallines produce chemicals, which promote the settlement of the larvae of certain herbivorous invertebrates, particularly abalone. This is adaptive for the corallines as the herbivores then remove epiphytes which might otherwise smother the crusts and pre-empt available light. This is also important for abalone aquaculture, as corallines appear to enhance larval metamorphosis and the survival of larvae through the critical settlement period. It also has significance at the community level, as the presence of herbivores associated with corallines can generate patchiness in the survival of young stages of dominant seaweeds.

Some corallines slough off a surface layer of epithallial cells, which in a few cases may be an anti-fouling mechanism, which serves the same function as enhancing herbivore recruitment. This also affects the community, as many algae recruit on the surface of a sloughing coralline, and are then lost with the surface layer of cells. This can also generate patchiness within the community. The common Indo-Pacific corallines *Neogoniolithon fosliei* and *Sporolithon ptychoides* slough epithallial cells in continuous sheets, which often lie on the surface of the plants looking so much like wet tissue paper.

Not all sloughing serves an anti-fouling function. Epithallial shedding in most corallines is probably simply a means of getting rid of damaged cells whose metabolic function has become impaired. Sloughing in this case is probably a means of getting rid of old reproductive structures and grazer-damaged surface cells, and reducing the likelihood of surface penetration by burrowing organisms.

Some coralline algae develop into thick crusts, which provide microhabitat for many invertebrates. This is probably an important factor affecting the distribution and grazing effects of herbivores within marine communities. Nothing is known about the microhabitat role of Indo-Pacific corallines. However, the most common species in the region, *Hydrolithon onkodes*, often forms an intimate relationship with the chiton *Cryptoplax larvaeformis*. The chiton lives in burrows that it makes in *H. onkodes* plants, and comes out at night to graze on the surface of the coralline. This combination of grazing and burrowing results in a peculiar growth form (called castles) in H. onkodes in which the coralline produces nearly vertical, irregularly curved lamellae.

Non-geniculate corallines are of particular significance in the ecology of coral reefs, where they provide calcareous material to the structure of the reef, help cement the reef together, and are important sources of primary production.

Coralline algae are especially important in reef construction, as they lay down calcium carbonate as calcite. Although they contribute considerable bulk to the calcium carbonate structure of coral reefs, their more important role in most areas of the reef, is in acting as the cement which binds the reef materials together into a solid and sturdy structure.

An area where corallines are particularly important in constructing reef framework is in the algal ridge that characterises surf-pounded reefs in both the Atlantic and Indo-Pacific regions. Algal ridges are carbonate frameworks that are constructed mainly by non-geniculate coralline algae (after Adey 1978). They require high and persistent wave action to form, so are best developed on the windward reefs in areas where there is little or no seasonal change in wind direction. Algal ridges are one of the main reef structures that prevent oceanic waves from striking adjacent coastlines, and they thus help to prevent coastal erosion.

Economic Importance

Despite their hard, calcified nature, coralline algae have a number of economic uses. One use dates back to the 18th century, and involves the collection of unattached corallines (maërl) for use as soil conditioners. This is particularly significant in Britain and France, where more 300 000 tonnes of *Phymatolithon calcareum* (Pallas) Adey & McKinnin and *Lithothamnion corallioides* are dredged annually.

Several thousand kilometres of maerl beds, composed of as-yet undetermined species belonging to the genera *Lithothamnion* and *Lithophyllum,* exist off the coast of Brazil, and have been subjected to a low level of commercial exploitation. Maerl is also used as a food additive for cattle and pigs, as well as in the filtration of acidic drinking water.

Corallines are also used in medicine, where the earliest use involved the preparation of a vermifuge from ground geniculate corallines of the genera *Corallina* and *Jania*. This use stopped towards the end of the 18th century. Modern medical science has found a more high-tech use for corallines in the preparation of dental bone implants. Apparently, the cell fusions provide an ideal matrix for the regeneration of bone tissue.

Since coralline algae contain calcium carbonate, they fossilise fairly well. They are useful as stratigraphic markers of particular significance in petroleum geology. Coralline rock has also been used as building stones, with the best examples being in Vienna, Austria.

As a colourful component of live rock sold in the marine aquarium trade, coralline algae are desired in home aquariums for their aesthetic qualities.

Mastocarpus Stellatus

Mastocarpus Stellatus, also called Cluimhín Cait [Cats' Puff], Carragheen, or Carrageen Moss, is closely related to Irish Moss, or Chondrus crispus. It is collected in Ireland and Scotland, together with *Chondrus crispus* as Irish Moss, dried, and sold for cooking and as the basis for a drink reputed to ward off colds and flu. The fronds are channelled unlike those of *Chondrus crispus*, and it has a curved stipe whereas *Chondrus* has a flat one. It occurs commonly on rocks in the mid- and lower-intertidal.

M. stellatus is able to coexist with *C. crispus* on the northern New England coast despite being a competitive inferior to *C. crispus*. A greater tolerance for freezing allows it to exist above *C. crispus* in northern environments where freezing stresses are significant. *Mastocarpus* is rarely found south of Cape Cod on the United States Atlantic coast because it is out competed by *Chondrus* when the freezing tolerances are lower.

Meristotheca

Meristotheca is a widely-distributed genus of red algae some of whose species are eaten as edible seaweed. In particular, the type species, M. papulosa is a popular vegetable in Taiwan and Japan.

Meristotheca Papulosa

Meristotheca papulosa (synonyms: *M. japonica* and *Eucheuma papulosa*) is a red algae, popular as a sea vegetable in Taiwan, where it is known as *jiguancai* (Chinese: QU*f*; Hanyu Pinyin: *jiguncai*), and in Japan, where it is known as *tosaka-nori*.

Morphological Description

Colour varies from red to peach. Texture is fleshy and cartilaginous. Shape is flat and leaf-like, with irregular deep forking divisions. The edges often grow many branchlets of uneven lengths. The surfaces of older individuals will grow warty or spiky protuberances. The morphology is variable. Length can reach 10-40 cm, branch width 1-5 cm. The interior of the algal body is multi-axised.

Life History

It has isomorphic alternation of generations (the sporophyte and the gametophyte share a very similar appearance). The tetrasporangia are divided in a ladder shape and scattered across the cortex.

Ecological Habit

It grows on subtidal rocks 5-20 metres deep, and can be seen at all times of the year.

Distribution

It is very widely distributed along the coasts of the Indian Ocean from South Africa to Australia, and the western Pacific Ocean from Australia to Japan.

Nori

Nori (Japanese: wmO,), (Chinese: wmO,; Hanyu Pinyin: *hlitái*, Korean:, kim or gim), is the Japanese name for various edible seaweed –H species of the red algae Porphyra including most notably *P. yezoensis* and *P. tenera*. The term *nori* is also commonly used to refer to the food products created from these so-called "sea vegetables". Finished products are made by a shredding and rack-drying process that resembles papermaking. Japan, Korea, and China are the current major producers of nori, with total production valued at up to US $2 billion per year.

Production

Production and processing of nori by current methods is a highly advanced form of agriculture. The biology of *Porphyra*, although complicated, is well understood and this knowledge is used to control virtually every step of the production process. Farming takes place in the sea where the *Porphyra* plants grow attached to nets suspended at the sea surface and where the farmers operate from boats. The plants grow rapidly, requiring about 45 days from "seeding" until the first harvest. Multiple harvests can be taken from a single seeding, typically at about 10 day intervals.

Harvesting is accomplished using mechanical harvesters of a variety of configurations. Processing of raw product is mostly accomplished by highly automated machines that accurately duplicate traditional manual processing steps, but with much improved efficiency and consistency. The final product is a paper thin, dark, black, dried sheet of approximately 18 x 20 cm and 3 grams in weight.

Nori is commonly used as a wrap for sushi and onigiri. It is also a common garnish or flavouring in noodle preparations and soups. Nori is most typically toasted prior to consumption ("yaki-nori" in Japanese). A very common and popular secondary product is toasted and flavoured nori ("ajitsuke-nori" in Japanese), in which a flavouring mixture (variable, but typically soy sauce, spices and sugar in the Japanese style or sesame oil and salt in the Korean style) is applied in combination with the toasting process.

Nori is also eaten by making it into a soy sauce flavoured paste *noritsukudani* (wmO,COnq). In Japan over 60,000 hectares (148,480 acrea) of Japanese coastal waters are given to producing 350,000 tons, worth over a billion dollars. China produces about a third of this.

A related product, prepared from the unrelated green algae *Monostroma* and *Enteromorpha*, is called aonori (R—wmO, literally *blue nori*), and is used like herbs on everyday meals like okonomiyaki and yakisoba.

Porphyra

Porphyra is a foliose red algal genus of about 70 species. It lives in the intertidal, typically between the upper intertidal to the splash zone. It is used to make nori, the most commonly eaten seaweed. It is considered that there are 60 to 70 species of *Porphyra* worldwide and seven in the British Isles.

Life Cycle

It displays a heteromorphic alternation of generations. The thallus we see is the haploid generation, it can reproduce asexually by forming spores, which grow to replicate the original thallus. It can also reproduce sexually. Both male and female gametes are formed on the one thallus. The female gametes while still on the thallus are fertilized by the released male gametes, which are non-motile.

The fertilized, now diploid, carposporangia produce spores (carpospores) which settle, they bore into shells, germinate and form a filamentous stage, which was originally thought to be a different species of algae, and was referred to as *Conchocelis rosea*. It is now known to be the diploid stage of *Porphyra*. After reduction division (meiosis) spores are produced, released and when they settel in a suitable site grow to become the plant which we know as *Porphyra*.

Food

Most human cultures with access to *Porphyra* use it as a food or somehow in the diet, it is probably the most domesticated of the marine algae, known as laver, nori, zakai, kim, karengo, sloke or slukos.

The marine red algae *Porphyra*, which is called nori in Japanese, has been cultivated extensively in many oriental countries as an edible seaweed used to wrap the rice and fish that compose the Japanese food sushi. In Japan, the annual production of *Porphyra* spp. is valued at 100 billion yen (US$ 1 billion).

Algal Bloom

An Algal Bloom is a relatively rapid increase in the population of (usually) phytoplankton algae in an aquatic system. Algal blooms may occur in freshwater as well as marine environments. Typically only one or a few species are involved and some blooms may be recognised by discolouration of the water resulting from the high density of pigmented cells. Although there is no officially recognised threshold level, algae can be considered to be blooming at concentrations of hundreds to thousands of cells per millilitre, depending on the causative species. Algal bloom concentrations may reach millions of cells per millilitre. Colours observed are green, yellowish-brown, or red. Bright green blooms may also occur. These are a result of blue-green algae, which are actually bacteria (cyanobacteria).

Some algal blooms are the result of an excess of nutrients (particularly phosphorus and nitrogen) into waters and higher concentrations of these nutrients in water cause increased growth of algae and green plants. As more algae and plants grow, others die. This dead organic matter becomes food for bacteria that decompose it. With more food available, the bacteria increase in number and use up the dissolved oxygen in the water. When the dissolved oxygen content decreases, many fish and aquatic insects cannot survive. This results in a dead area.

Algal blooms may also be of concern as some species of algae produce neurotoxins. At the high cell concentrations reached during some blooms, these toxins may have severe biological impacts on wildlife. Algal blooms composed of phytoplankters known to naturally produce biotoxins are often called Harmful Algal Blooms, or HABs.

Algal blooms are monitored using biomass measurements coupled with the examination of species present. A widely-used measure of algal and cyanobacterial biomass is the chlorophyll concentration. Peak values of chlorophyll *a* for an oligotrophic lake are about 1-10 µg/l, while in a eutrophic lake they can reach 300 µg/l. In cases of hypereutrophy, such as Hartbeespoort Dam in South Africa, maxima of chlorophyll a can be as high as 3,000 µg/l (Zohary and Roberts, 1990; Bartram *et al.*, 1999).

Blue Tide

The so-called Blue Tide is an example of a naturally occurring estuarine or marine *algal bloom*. Blue tide is caused by species of dinoflagellates, often present in sufficient numbers (thousands or millions of cells per millilitre) to turn the water a hue of blue or purple.

Black Water

So-called black water is a dark discolouration of sea water, first described in the Bay of Florida in January 2002. Although fishermen in Florida complained and requested that the government take action, scientists say that black water results from a non-toxic algal bloom, probably of diatoms. It dissipated within a few months by transport through the Florida Keys into the Florida Straits and by disruption by winds and wave action.

Water Treatment

Algal blooms sometimes occur in drinking water supplies. In such cases, toxins from the bloom can survive standard water purifying treatments. Researchers at Florida International University in Miami are experimenting with using 640-kilohertz ultrasound waves that create micropressure zones as hot as 3,700° C. This breaks some water molecules into reactive fragments that can kill algae (Song *et al.*, 2005).

Ciguatera

Ciguatera is a foodborne illness poisoning in humans caused by eating marine species whose flesh is contaminated with a toxin known as *ciguatoxin*, which is present in many micro-organisms (particularly, the microalgae Gambierdiscus toxicus) living in tropical waters. Like many naturally and artificially occurring toxins, ciguatoxin accumulates in lower level organisms, resulting in higher concentration of the toxin at higher levels of the food chain, an example of biomagnification. Predator species near the top of the food chain in tropical waters, such as barracuda, moray eel, parrotfish, grouper and amberjack, are most likely to cause ciguatera poisoning, although many other species have been found to cause occasional outbreaks of ciguatera. Ciguatoxin is very heat-resistant, so ciguatoxin-laden fish cannot be detoxified by cooking.

In Northern Australia, where ciguatera is a common problem, there are two commonly believed ways of determining that fish carry significant levels of ciguatoxin. The first belief is that if a piece of fish is contaminated with the toxin, flies will not land on it. The second belief is that the toxin can be detected by feeding a piece of fish to a cat as cats are allegedly highly sensitive to ciguatoxin and will display symptoms. It is not known whether there is any veracity to either belief.

Etymological Roots

It is a generally held theory that ciguatera, as a poisonous substance, was named and identified in Cuba, circa early 1800's. Local folklore has identified the etymology stems from a story of an Englishman who caught a barracuda on the Isla de Pinos. After consuming the barracuda, said Englishman became terribly ill. When queried about the origins of his illness, the Englishman claimed to have caught and eaten "a fish, from the sea-water." It is the poor interpretation of the word sea-water that allegedly gives the ailment the name ciguatera, or "from the sea-water" in Spanish.

Distribution

Due to the localised nature of the ciguatoxin-producing micro-organisms, ciguatera illness is only common in tropical waters, particularly the Pacific and Caribbean, and usually is associated with fish caught in tropical reef waters. Ciguatoxin is found in over 400 species of reef fish, and therefore avoidance of consumption of all reef fish (any fish living in warm tropical waters) is the only sure way to avoid exposure to the toxin. Imported fish served in restaurants have been found to contain the toxin and result in illness which often goes unexplained by physicians unfamiliar with a tropical toxin and its characteristic symptoms. In addition, ciguatoxin has been found in farmed raised salmon.

Symptoms

Hallmark symptoms of ciguatera include gastrointestinal and neurological effects. Gastrointestinal symptoms include nausea, vomiting, and diarrhoea usually followed by neurological symptoms such as headaches, muscle aches, paresthesia, numbness, ataxia, and hallucinations.

Severe cases of ciguatera can also result in cold allodynia, which is a burning sensation on contact with cold (commonly incorrectly referred to as reversal of hot/cold temperature sensation). Doctors are often at a loss to explain these symptoms and ciguatera poisoning is frequently misdiagnosed as Multiple Sclerosis.

Dyspareunia and other ciguatera symptoms have developed in otherwise well males and females following sexual intercourse with partners suffering ciguatera poisoning, signifying ciguatera poisoning may be sexually transmitted. Diarrhoea and facial rashes have been reported in breastfed infants of mothers with ciguatera poisoning, it is likely ciguatera toxins are also transferred into the breast milk.

The symptoms can last from weeks to years, and in extreme cases lasting as many as 20 years; often leading to long-term disability. Most people do recover slowly over time. Often patients recover but redevelop symptoms in the future. This relapse/remit characteristic of the disease can be tied to the consumption of nuts, alcohol, eating fish or fish-containing products, eating chicken or eggs, and exposure to fumes such as those of bleach and other chemicals. Exercise is also a possible trigger.

Treatment

There is no effective treatment or antidote for ciguatera poisoning. Previously mannitol was used for poisoning after 1 study reported the reversal of symptoms following its use. Follow up studies in animals and case reports in humans also found benefit from mannitol. However, a randomised, controlled, double-blinded, clinical trial of mannitol for ciguatera poisoning did not find any difference between mannitol and normal saline, based on this result mannitol is no longer recommended.

The mainstay of treatment is supportive care. Some medications such as the use of Amitriptyline may reduce some symptoms of ciguatera, such as fatigue and paresthesias. Although benefit does not occur in every case. Also used are steroids and vitamin supplements, but these merely support the body's recovery rather than directly reduce the toxic effects.

There are a number of antiquated Caribbean homeopathic treatments, as well as ritualistic ones, most of which originated from Cuba and nearby islands. The most common old-time remedy involves bed-rest subsequent to a Guanavana juice enema. Some more avant-garde old-time treatments ranged from directly porting and bleeding the gastrointestinal tract, to 'cleansing' the diseased with a dove during a Santeria ritual. The efficacy of these treatments have never been studied or substantiated; nevertheless they are purportedly still used to this day.

Dinoflagellate

The Dinoflagella are a large group of flagellate protists. Most are marine plankton, but they are common in fresh water habitats as well; their populations are distributed depending on temperature, salinity, or depth. About half of all dinoflagellates are photosynthetic, and these make up the largest group of eukaryotic algae aside from the diatoms. Being primary producers make them an important part of the aquatic food chain. Some species, called zooxanthellae, are endosymbionts of marine animals and protozoa, and play an important part in the biology of coral reefs. Other dinoflagellates are colourless predators on other protozoa, and a few forms are parasitic.

Morphology

Most dinoflagelates are unicellular forms with two dissimilar flagella. One of these extends towards the posterior, called the *longitudinal* flagellum, while the other forms a lateral circle, called the *transverse* flagellum. In many forms these are set into grooves, called the *sulcus* and *cingulum*. The transverse flagellum provides most of the force propelling the cell, and often imparts to it a distinctive whirling motion, which is what gives the name dinoflagellate refers to (Greek *dinos*, whirling). The longitudinal acts mainly as the steering wheel, but providing little propulsive force as well.

Dinoflagellates have a complex cell covering called an amphiesma, composed of flattened vesicles, called *alveoli*. In some forms, these support overlapping cellulose plates that make up a sort of armour called the *theca*. These come in various shapes and arrangements, depending on the species and sometimes stage of the dinoflagellate. Fibrous extrusomes are also found in many forms. Together with various other structural and genetic details, this organisation indicates a close relationship between the dinoflagellates, Apicomplexa, and ciliates, collectively referred to as the alveolates.

The chloroplasts in most photosynthetic dinoflagellates are bound by three membranes, suggesting they were probably derived from some ingested algae, and contain chlorophylls *a* and *c* and fucoxanthin, as well as various other accessory pigments. However, a few have chloroplasts with different pigmentation and structure, some of which retain a nucleus. This suggests that chloroplasts were incorporated by several endosymbiotic events involving already coloured or secondarily colourless forms. The discovery of plastids in Apicomplexa have led some to suggest they were inherited from an ancestor common to the two groups, but none of the more basal lines have them.

All the same, the dinoflagellate still consists of the more common organelles such as rough and smooth endoplasmic reticulum, Golgi apparatus, mitochondria, lipid and starch grains, and food vacuoles. Some have even been found with light sensitive organelle such as the eyespot or a larger nucleus containing a prominent nucleolus.

Life Cycle

Dinoflagellates have a peculiar form of nucleus, called a *dinokaryon*, in which the chromosomes are attached to the nuclear membrane. These lack histones and remained condensed throughout interphase rather than just during mitosis, which is closed and involves a unique external spindle. This sort of nucleus was once considered to be an intermediate between the nucleoid region of prokaryotes and the true nuclei of eukaryotes, and so were termed *mesokaryotic*, but now are considered advanced rather than primitive traits.

In most dinoflagellates, the nucleus is dinokaryotic throughout the entire life cycle. They are usually haploid, and reproduce primarily through fission, but sexual reproduction also occurs. This takes place by fusion of two individuals to form a zygote, which may remain mobile in typical dinoflagellate fashion or may form a resting dinocyst, which later undergoes meiosis to produce new haploid cells.

However, when the conditions become desperate, usually starvation or no light, their normal routines change dramatically. Two dinoflagellates will fuse together forming a planozygote. Next is a stage not much different from hibernation called hypnozygote when the organism takes in excess fat and oil. At the same time its shape is getting fatter and the shell gets harder. Sometimes even spikes are formed. When the weather allows it, these dinoflagellates break out of their shell and are in a temporary stage, planomeiocyte, when they quickly reforms their individual thecae and return to the dinoflagellates at the beginning of the process.

Ecology and Fossils

Dinoflagellates sometimes bloom in concentrations of more than a million cells per millilitre. Some species produce neurotoxins, which in such quantities kill fish and accumulate in filter feeders such as shellfish, which in turn may pass them on to people who eat them. This phenomenon is called a red tide, from the colour the bloom imparts to the water. Some colourless dinoflagellates may also form toxic blooms, such as Pfiesteria. It should be noted that not all dinoflagellate blooms are dangerous. Bluish flickers visible in ocean water at night often come from blooms of bioluminescent dinoflagellates, which emit short flashes of light when disturbed.

Dinoflagellate cysts are found as microfossils from the Triassic period, and form a major part of the organic-walled marine microflora from the middle Jurassic, through the Cretaceous and Cenozoic to the present day. Because some species are adapted to different surface water conditions, these fossils from sediments can be used to reconstruct past surface ocean conditions (Sluijs et al., 2005). *Arpylorus*, from the Silurian of North Africa was at one time considered to be a dinoflagellate cyst, but this palynomorph is now considered to be part of the microfauna (Arthropoda). It is possible that some of the Paleozoic acritarchs also represent dinoflagellates.

Cautions: The same Red Tide mentioned above is more specifically produced when dinoflagellates are able to reproduce rapidly and copiously on account of the abundant nutrients in the water. Although the resulting red waves are a miraculous sight, they, again, contain toxins that not only affect all marine life in the ocean but the people who consume them as well. A specific carrier is shellfish. This can introduce both non-fatal and fatal illnesses. Human inputs of phosphate further encourage these red tides, and consequently there is a strong interest in learning more about dinoflagellates, from both medical and economic perspectives.

Classification: In 1753 the first modern dinoflagellates were described by Baker and named by Muller in 1773.

These same dinoflagellates were first defined by Otto Bütschli in 1885 as the flagellate order dinoflagellida. Botanists treated them as a division of algae, named Pyrrhophyta ("fire algae"; Greek *pyrrhos*, fire) after the bioluminscent forms, or Dinophyta. At various times the cryptomonads, ebriids, and ellobiopsids have been included here, but only the last are now considered close relatives. Dinoflagellates have a known ability to evolve from non-cyst to cyst forming strategies which makes it nearly impossible to recreate their evolutionary history.

Red Tide

"Red Tide" is a common name for a phenomenon known as an algal bloom, an event in which estuarine, marine, or fresh water algae accumulate rapidly in the water column, or "bloom". These algae, more correctly termed phytoplankton, are microscopic, single-celled protists, plant-like organisms that can form dense, visible patches near the water's surface. Certain species of phytoplankton contain photosynthetic pigments that vary in colour from green to brown to red, and when the algae are present in high concentrations, the water appears to be discoloured or murky, varying in colour from white to almost black, normally being red or brown. Not all algal blooms are dense enough to cause water discolouration, and not all discoloured waters associated with algal blooms are red. Additionally, red tides are not typically associated with tidal movement of water, hence the preference among scientists to use the term algal bloom.

The term "red tide" is often used in the United States of America to describe a particular type of algal bloom common to the eastern Gulf of Mexico, and is also called "Florida red tide". This type of bloom is caused by a species of dinoflagellate known as Karenia brevis, and these blooms occur almost annually along Florida waters. The density of these organisms during a bloom can exceed tens of millions of cells per litter of sea-water, and often discolour the water a deep reddish-brown hue.

"Red tide" is also commonly used on the northern east coast of the United States, and particularly in the Gulf of Maine. This type of bloom is caused by a species of dinoflagellate known as Alexandrium tamarense. These blooms of organisms cause severe disruptions in fisheries of these waters as the toxins in these organism cause filter-feeding shellfish in effected waters to become poisonous for human consumption due to saxitoxin.

The most conspicuous effects of red tides are the associated wildlife mortalities among marine and coastal species of fish, birds, marine mammals and other organisms. In the case of Florida red tides, these mortalities are caused by exposure to a potent neurotoxin produced naturally by Karenia brevis, called brevetoxin.

It is unclear what causes red tides, but the frequency and severity of algal blooms in many parts of the world have been linked to increased nutrient loading from human activities. The growth of marine phytoplankton is generally limited by the availability of nitrates and phosphates, which can be abundant in agricultural run-off. Coastal water pollution produced by humans and systematic increase in sea water temperature have also been implicated as contributing factors in red tides.

Other factors such as iron-rich dust influx from large desert areas such as the Saharan desert are thought to play a major role in causing red tides. Some algal blooms on the Pacific coast have also been linked to occurrences of large-scale climatic oscillations such as El Nino events. While red tides in the Gulf of Mexico have been occurring since the time of early explorers such as Cabeza de Vaca, it is unclear what initiates these blooms, and how large a role anthropogenic and natural factors play in their development. Algal blooms in many parts of the world cannot be reasonably linked to human influence, and are generally accepted as a natural occurrence. It is also debated whether the apparent increase in frequency and severity of algal blooms in various parts of the world is in fact a real increase or is due to increased effectiveness of monitoring programmes and species identification ability.

Some red tide organisms produce large quantities of toxins, such as saxitoxin, ciguatoxin, and brevetoxin, which disrupt the proper function of ion channels in neurons. Domoic acid, a toxin produced by diatoms of the genus *Pseudonitzschia,* has been linked to neurological damage in certain marine mammals, and is frequently found in algal blooms on the US West Coast. Some red tide toxins can become highly concentrated in various marine organisms that have the ability to filter and consume large quantities of toxic plankton directly from sea-water.

These include shellfish, finfish, baleen whales, and benthic crustaceans. Frequently, shellfish collected in areas affected by algal blooms can be potentially dangerous for human consumption, leading to closures of shellfish beds for harvesting.

Initial signs of shellfish poisoning from red tide toxins such as domoic acid is tingling in the lips followed by a reduction of motor abilities and difficulty breathing and can be fatal if consumed in sufficient amounts. If these symptoms occur after eating shellfish, seek immediate medical treatment. Standard medical treatment is to give victims oxygen, or to hook them up to a breather. There exists no antidote, and treatment consists of keeping the patient alive until the toxin has passed from the system. Note that in some cases, the brevetoxin in Florida red tides can become aerosolised, causing respiratory irritation to beachgoers.

Toxins commonly associated with red tide are brevetoxins, saxitoxins, domoic acid, okadaic acid, azaspiracid, and ciguatoxin.

Gymnocin A is a cytotoxic polyether and an example of one of the many toxins associated with red tide.

Domoic Acid

Domoic Acid, which causes amnesic shellfish poisoning (ASP), is an amino acid phycotoxin (algal toxin) found associated with certain algal blooms.

Occurrence

In 1958, domoic acid was originally isolated from the red algae called "doumoi" or "hanayanagi" (*Chondria armata*) in Japan. "Doumoi" is used as an anthelmintic in Tokunoshima, Kagoshima. Domoic acid is also produced by diatoms of the genus *Pseudonitzschia*. Domoic acid is commercially available.

Toxicology

The chemical can bioaccumulate in marine organisms that feed on the phytoplankton, such as shellfish, anchovies, and sardines. The toxin does not bioaccumulate in the classic sense because it is quickly excreted by fish and shellfish that ingest it. It only accumulates in high numbers in these plankton feeders when the diatom itself is high in number in the surrounding waters. Bioaccumulation or biomagnification typically occurs in lipid-rich tissues, which accumulate toxins over time. In classic bioaccumulation, the older and larger an organism is, the longer it has to accumulate the toxins in its fats, and therefore, the higher the concentration of toxin.

In mammals, including humans, domoic acid acts as a neurotoxin, causing short-term memory loss, brain damage, and death in severe cases in Atlantic Canada. Algal bloom are associated with the phenomenon of ASP. Considerable recent research has been carried out by the Marine Mammal Centre and other scientific centres on the association of red tides to domoic acid and to resulting neurological damage in marine mammals of the Pacific Ocean. In marine mammals, domoic acid typically causes seizures and tremors.

In the brain, domoic acid especially damages the hippocampus and amygdaloid nucleus. It damages the neurons by activating AMPA and kainate receptors, causing an influx of calcium. Although calcium flowing into cells is a normal event, the uncontrolled increase of calcium causes the cell to degenerate. Because the hippocampus may be severely damaged, long-term memory loss occurs.

Mastocarpus Stellatus

Mastocarpus stellatus, also called Clúimhín Cait, Carragheen, or Carrageen Moss, is closely related to Irish Moss, or Chondrus crispus. It is collected in Ireland and Scotland, together with *Chondrus crispus* as Irish Moss, dried, and sold for cooking and as the basis for a drink reputed to ward off colds and flu. The fronds are channelled unlike those of *Chondrus crispus*, and it has a curved stipe whereas *Chondrus* has a flat one. It occurs commonly on rocks in the mid- and lower-intertidal.

M. stellatus is able to coexist with *C. crispus* on the northern New England coast despite being a competitive inferior to *C. crispus*. A greater tolerance for freezing allows it to exist above *C. crispus* in northern environments where freezing stresses are significant. *Mastocarpus* is rarely found south of Cape Cod on the United States Atlantic coast because it is out competed by *Chondrus* when the freezing tolerances are lower.

The student is probably familiar with the seaweeds of the seashore and may even have observed green skeins in stagnant freshwater ponds and pools. These plants, which are among the simplest in the plant kingdom, belong to the group known as Algae.

In recent years, mainly as a result of biochemical and ultrastructural studies, there has been considerable debate over the major divisions of the plant kingdom. It is commonly agreed that the group of organisms, characterised by a lack of a distinct nuclear membrane and hence of a nucleus, should be separated from all other plants. One school of thought suggests that these organisms, which comprise the blue-green algae (Cyanophyta) and the bacteria, should be termed the Monera (Dougherty and Allen, 1960, Whit-taker, 1969), while another school terms them the Procaryota (Stanier and van

Niel, 1962). This latter school includes all other plants in one major subdivision, the Eucaryota, and within this subdivision the simplest photo-synthetic members are those plants known as the Algae. Those who recognise the Monera sometimes group the Algae, Protozoa and Fungi as the Protista; sometimes the fungi are segregated into a separate sub-division, the Mycota. The terrcstrial plants and aquatic higher plants with more complex organisation are classed as the Metaphyta.

Whittaker's more recent proposal involves five kingdoms, three of which are the Fungi, Plantae and Protista and the eucaryote algae are therefore, distributed between the two latter kingdoms. This seems an artificial arrangement and will not be used here.

It is important that any classitication should be as broadly based as possible and with our present knowledge it can be a matter of opinion which of the two major subdivisional classifications one employs. We have employed the Procaryota-Eucaryota concept, assigning the algae (except the blue-greens) to 10 divisions within the Eucaryota.

The principal criteria for the primary classification of Algac are differences in pigmentation, other biochemical characteristics and the structure of organelles such as flagella. While there are one or two exceptions to the general basis of differentiation, these are so few (and even then some-what uncertain) that they can be neglected. Chlorophyll *a*, one of the green pigments, is the most abundant chlorophyll in all types of algae, and /3-carotene is widely distributed. There are a large number of carotenoids (yellow pigments) and many algal classes possess their own characteristic types. The Siphonales, an order of the Chlorophyceae, is treated separately because of the carotenoid differences.

The Rhodophyta and Cyanophyta possess biliproteins that are similar to each other, but which are different, with regard to their protein constituents, from those of the Cryptophyta. The phycocyanins (blue pigments) usually predominate in the Cyanophyta and the phycoerythrins (red pigments) are usually dominant in the Rhodophyta. Either may be dominant in the Cryptophyta. There are other biochemical features that appear to be specific to certain algal classes. These include the storage polysaccharides, the chemical constituents of the cell walls, the characteristic sterols and some other miscellaneous compounds that will be mentioned later. Colourless forms are known in all classes except the Rhodophyceae, Phaeophyceae, Charophyceae and Prasinophyceae, but their morphology is such that they are classified in the same class as their photosynthetic counterparts.

Although the Cyanophyta should not now be included as algae in the strict sense they will be considered here because they have certain similarities (e.g. pigments, gross morphology) to other algae. While there is substantial agreement among taxonomists as to the number and composition of the classes, there is much divergence of opinion as to the groupings of these classes into divisions.

One school of thought (e.g. Papenfuss, 1955; Fott, 1959; Smith, 1955; Prescott, 1968; Round, 1965; Dawson, 1967) favours classifications similar to that above, while others (Chadefaud, 1960; Christensen, 1962; Feldmann, 1963; Bourelly, 1968) favour a classification involving only three divisions: Aconta or Rhodophyta (Rhodophyceae) Chlorophyta (Chloro-phyceae, Charophyceae, Prasinophyceae and including in some cases the Euglenophyceae) and the Chromophyta (all other classes).

These latter suggestions merit serious consideration because they emphasise the major evolutionary groups within the algae. Our divisions 1, 2 and 3-10 could certainly be grouped together into three taxa corresponding to the Aconta (Rhodophyta), Chlorophyta and Chromophyta of Christensen. Grouping the 14 classes into 3 or 10 divisions is thus a matter of opinion.

Ultimately the classification adopted must depend upon the view taken of the inter-relationships between these groups and other plants and animals. Dodge (1969) has recently surveyed the eyespot structure in flagellated cells. He has divided the algae (with eyespots) into three main 'eyespot groups' that coincide well with other characteristics:

(a) Eyespot within the chloroplast and unrelated to the flagella (Chloro-phyceae, Prasinophyceae, Cryptophyceae);

(b) Eyespot within the chloroplast and in close relation to the flagella (Chrysophyceae, Phaeophyceae, most Xanthophyceae);

(c) Eyespot independent of the chloroplast (Euglenophyceae, Dino-phyceae).

The above schemes can be expected to undergo modifications as more information becomes available, especially from certain flagellate groups. There are a number of fossil algae not represented among present day forms, and some, the Nematophyceae, that are generally regarded as algae. A brief summary of the principal class characteristics is as follows.

Cyanophyta

Cyanophyceae: The plants in this class show less internal differentiation. Like the bacteria, and unlike the algae, they lack the characteristic nuclear and mitochondrial membranes and hence there are no distinct nuclei and mitochondria, although the nuclear material does appear to be localised.

A distinct membrane-bounded chloroplast is also lacking, and the photo-synthetic lamellae are distributed usually in the peripheral region of the cytoplasm.

The product of photosynthesis is Cyanophycean starch, present in granular form. The colour of the cells is commonly blue-green, sometimes olive-green or red, the colour being due to the varying proportions of chlorophyll *a*, carotenoids and biliproteins. The cell walls lack cellulose and instead comprise mainly amino acids and amino sugars. Contrary to accepted fact, blue-green algae have recently been shown to possess the sterols sitosterol and cholesterol. There is no known sexual reproduction, propagation taking place by simple division, by non-motile endospores, by spores or else by vegetative fragmentation.

Chlorophyceae, Prasnophyceae, Charophyceae

This group used to comprise four great subclasses, the Isokontae (equal cilia), Stephanokontae (ringed cilia), Akontae (no cilia) and Heterokontae (unlike cilia). These subdivisions, as their names imply, were based upon tfie type of flagella of the motile reproductive bodies. The last has since been renamed the Xanthophyceae and removed to a separate division. The Chlorophyta encompasses a morphological range from simple unicells to quite complex multicellular plants. The pigments are characteristically those of the higher plants: chlorophyll a, *b* and J-carotene, pigments lutein, violaxanthin and neoxanthin as the usually predominant xanthophylls, but certain exceptions are known: Siphonaxanthin in Siphonales, Micronone in Prasinophyceae. The cell wall is composed commonly of cellulose, except in the Siphonales, plus mannans and xylans with sitosterol as the predominant sterol. In certain members (Dasycladales and Siphonales) calcification is often present. Although the chloroplasts vary considerably in shape and size, the final product of photosynthesis is mostly starch.

The motile cells of most members of the Chlorophyceae commonly possess two or four simple and equal flagella that are smooth and lacking lateral appendages. In the Oedogoniales (the former Stephanokontae) there is a ring of flagella, while in the Conjugales (the former Akontae) flagellated reproductive cells are absent. Recent work has shown that certain unicellular algae, formerly in the Chlorophyceae, possess a rather different flagellar structure. These flagella are distinguished by such features as minute scales, lateral hairs, terminal tufts, or bipartite arrangement with a narrower terminal extremity. Genera with these features are removed to a new class, Prasinophyceae.

Sexual reproduction is common and ranges from isogamy to anisogamy and oogamy. Asexual reproduction normally takes place by means of motile zoospores, but a variety of nommotile spores may also be produced.

The Charophyceae comprise one predominantly freshwater

order, the Charales, whose members differ from other Chlorophyta in their remarkable morphological features and elaborate oogamy.

Euglenophyta

Euglenophyceae: This class consists of unicellular flagellates of varying shapes and a few palmelloid colonial forms. The organisms lack a cell wall, but are surrounded by a pliable periplast or pellicle. A lateral row of hairs is present on one of the two flagella that are commonly present. The other flagellum is greatly reduced. The cells characteristically possess a gullet with a reservoir, contractile vacuoles and eyespot. While many members of the class are colourless, those that are not possess discoid chloroplasts with chlorophylls *a* and *b*, carotene, diadinoxanthin and other xanthophylls. The storage reserve is paramylum, a polysaccharide very similar to laminarin and the principal sterol is ergosterol. Reproduction is commonly by longitudinal division, although in some forms cyst formation has been recorded.

Chloromonadophyta

Chloromonadophyceae: This is probably the smallest and least understood class of algae. Non-flagellated forms are unknown and about half the described genera lack chloroplasts. When chloroplasts are present they are discoid, small and very numerous, possessing chlorophyll *a* as the only chlorophyll and carotenoids similar to those found in the Xanthophyceae. The cells are biflagellated with heterokont flagella one of which usually trails. Other characteristic cytolo-gical features include a gullet and numerous trichocysts. Storage reserves appear to be oil. Under certain conditions the cells may often become encysted or 'amoeboid'.

Xanthophyta

Xanthophyceae: The plants are usually simple but their lines of morphological development frequently show an interesting parallel with those observed in the Chloro-phyceae and Chrysophyceae. In the non-flagellated forms the cell wall, which contains cellulose, is frequently of two unequal or equal halves which overlap each other. The motile cells commonly possess two unequal heterokont flagella, the longer of which is a flimmergeisel (pleuronematic) type, the shorter a peitschengeissel (smooth) type. Reproduction is by zoospores, aplanospores or occasionally statospores. Sexual reproduction is rare and when present is isogamous, though in one genus *(Vaucherid)* there is a well-developed oogamy. This group characteristically possesses chlorophyll *a* as the sole chlorophyll, and carotenoids, identical to some in the Chrysophyta, Euglenophyta and Bacillariophyta, while the normal storage reserve is oil or leucosin, and ergosterol has been reported as the principal sterol.

Eustigmatophyceae

Certain Xanthophyceae differ from the majority in the organisation of their motile cells. Hibberd and Leedale (1970) have proposed that such genera should be assigned to a new class, Eustigmatophyceae. This class includes those genera in which the eyespot is independent of the chloroplast; the zoospores have one emergent flagellum; the chloroplast lacks a girdle lamella; golgi bodies and pyrenoids are present in the vegetative cell, absent in zoo spores.

Biochemical distinctions have yet to be determined, although the pigments of the Eustigmatophyceae and Xanthophyceae appear to be similar.

Bacillarlophyta

Bacillarlophyceae (Diatoms): One of the characteristic features of these plants is their cell wall which is composed of silica and partly of pectin. The wall is always in two halves and frequently ornamented with delicate maskings. Each cell contains one or more variously shaped, golden-brown chloroplasts, containing the same pigments as those of the Chrysophyta. Like the Chrysophyceae, the members of this class produce chrysolaminarin as the photosynthetic food reserve. They are divided into two major groups, the representatives of one group being radially symmetrical, those of the other being bilaterally symmetrical. Propagation is by cell division, giving rise to two cells of slightly differing size. In the former group, the centric diatoms, there is a sexual reproduction by oogamy, in which the male gamete possesses a single pleuronematic (flimmer-geissel) flagellum. The other group, the pennate diatoms, reproduce sexually by a form of amoeboid isogamy, the gametes fusing in pairs. Members of this class are unicellular, but often occur as colonial forms.

Chrysophyta

Chrysophyceac, Haptophyceae: These form a group, mostly of unicellular organisms, many of which are naked and lack a cellulose wall, although others show high degrees of silicification or calcification. The principal pigments are chlorophyll *a* and *c* and fucoxanthin. The principal photosynthetic product is chrysolaminarin, a polysaccharide very similar to laminarin of the Phaeophyceae. The main sterols so far identified are fucosterol and porifasterol.

Chrysophyceae: The members of this class are genera whose flageliate stages are characteristically heterokontic. One flagellum is of the flimmer-geisel type (pantonematic), while the other is smooth (peitschengeissel J). Flagella scales if present are mineralised and certain genera also possess silicified cysts.

Haptophyceae: The flagellate stages of this class are isokontic and characterised by flagella that are always smooth and scales, if present, that are not mineralised. Certain genera of this class also possess a third filiform appendage, or anchorage organ, called a haptonema.

Reproduction is commonly by cell division, by zoospores or by statospores, which are silicified cysts that generally have a small aperture closed by a small plug.

Phaeophyta

Phaeophyceae: This group comprises the common brown algae of the seashore, and it is worth noting that the majority are wholly marine. There are a few, rare freshwater species. The brown colour is due to the fucoxanthin protein which masks chlorophylls *a* and *c*. The principal product of photosynthesis is the sugar, alcohol, mannitol, together with the polysaccharide laminarin, while the cell walls not only contain cellulose but also specific polysaccharides such as alginic acid and fucoidin. In some of the larger and more advanced brown algae, these compounds are present in sufficient quantity to be of commercial importance.

The simplest forms are filamentous, but there are examples of all stages of development and differentiation particularly in the large seaweeds of the Pacific and Arctic shores with their great size and complex internal and external differentiation. There is also a number of greatly reduced parasitic and epiphytic forms. The pyriform motile reproductive cells are heterokont with two laterally inserted flagella. The longer anteriorly directed flagellum is a flimmergeissel (except in *Fucus* where it is the shorter), while the posteriorly directed flagellum is a peitschengeissel. The motile reproductive cells are commonly produced in special organs that are either unilocular or pluri-locular. Some genera also reproduce by non-motile monospores, tetraspores and aplanospores. Sexual reproduction ranges from isogamy to oogamy.

Pyrrhophyta

Desmophyceae, Dinophyceae: Most of the members of these two classes are motile unicells, but there has been an evolutionary tendency towards an 'algal' form and the development of short filaments, e.g. *Dinothrix, Dinoclonium.*

In the Dinophyceae, the motile cells are laterally biflagellate. One flagellum lies in the girdle or transverse groove and encircles the cell, while the other trails behind from the sulcus or ventral groove. Within this class are the 'armoured' and 'unarmoured' dinoflagellates. In the former, the motile cell possesses a cellulose wall with elaborately sculptured plates, while, as the name implies, the unarmoured forms lack the cellulose wall and sculptured plates.

The Desmophyceae are characterised by motile cells that lack a cellulose wall, and are divided vertically into two halves. There is no girdle or sulcus and the flagella are apically inserted.

Both classes are noted for their large nucleus, containing distinct monili-form chromatin threads, even in the resting stage. Those genera possessing chloroplasts contain chlorophylls *a* and *c* and a number of xanthophylls, of which the predominant one, peridinin, is found only in these two classes. The so-called colourless forms that is, those which lack chloroplasts, are often characterised by variously coloured cytoplasmic pigments. The products of photosynthesis are starch or fat. Some of the autotrophic forms can feed heterotrophically, and saprophytic genera are also known. Cell division is longitudinal (Desmophyceae) or obliquely longitudinal (Dinophyceae) and asexual reproduction is by zoospores. Sexual reproduction in those genera in which it has been recorded is anisogamous or isogamous. Characteristic resting cysts are also produced by many of the forms.

Cryptophyta

Cryptophyceae: With the two exceptions of coccoid *Bjornbergiella* and *Tetragonidium* the members of this class are flagellates. The motile cells possess two slightly unequal flagella, which appear to be flimmergeissel. The cells are asymmetrical and have a very distinctive morphology with a gullet, a trichocyst lined groove and a complex vacuoles. One small group is noted for its reniform cells. Biochemically this class is very interesting in that it is the only group that possesses chlorophylls *a, b, c* and carotenes. Additionally, it is one of the few algal groups that synthesises a-carotene instead of β-carotene. The xanthophylls, although similar to other in chlorophyll containing algae, are restricted to this class. The storage product appears to be starch-like compound.

Rhodophyta

Rhodophyceae: This group is often referred to as the red algae and the majority of its members are the conspicuous red seaweeds of the seashore. The colour is produced by the phycoerythrins and phycocyanins, which mask the carote, floids and chlorophyll *a*. Chlorophyll *d*, often considered characteristic of this group is probably an artefact of extraction. The products of photosynthesis are a starch-like compound (floridean starch), floridoside and mannogly-cerate. There is also a tendency to form polysaccharide sulphate esters and certain groups produce large amounts of polysaccharide mucilages. Motile reproductive stages are unknown.

While primitive members are coccoid unicells or simple filaments, all stages up to a complex thallus can be found, although there is not quite the same degree of complexity as in the Phaeophyceae. Despite variations in form, there are only two basic types of thallus construction. Protoplasmic connections exist between the cells of all forms except most of those that comprise a small group, the Bangiophycidae. Sexual reproduction is by a complex oogamy, the ovum being retained within the gametangium, and although the subsequent development of the zygote, which is retained on the parent plant, varies to a certain extent, it usually gives rise to filaments that bear special reproductive bodies known as carpospores; these latter are normally responsible for the production of a tetraspore-bearing diploid individual.

The extent to which the various algal classes occur in the marine and freshwater environments, the latter including aerial and soil dwelling algae is of considerable interest.

Cyanophyta

Cyanophyceae: The name Cyanophyceae has recently been generally adopted for this class previously known as the Myxophyceae. Apart from the main feature that the cells lack true nuclei, they show other features which, with the bacteria, distinguish them from the Eucaryota (Echlin and Morris, 1965). The cell envelope in the Cyanophyceae consists of an outer sheath (capsule in the bacteria), an outer membrane and an inner one (present in gram-positive bacteria, but lacking in gram-negative) and a plasma membrane. The cell envelope is composed largely of mucopeptides built up from amino sugars (e.g. glucosamine) and amino acids (e.g. muramic, diaminopimelic), whereas the cell wall of other algae lack such amino constituents.

The bacteria and Cyanophyceae lack mitochondria, true vacuoles and endoplasmic reticula. Sterols have not been detected in bacteria, and the absence of sterolsin the Cyanophyceae was also an accepted fact. Recent work has shown the presence of cholesterol and /8-sitosterol in *Anacystis* and *Fremyella*. There is no membrane bounded chloroplast and in the blue-green algae the photosynthetic lamellae are usually distributed in the peripheral cytoplasm.

The separate microscopic plants are characterised by a bluish-green colour, occasionally red, which varies greatly in shade, depending upon the relative proportions of chlorophyll, carotenoids, phycocyanin and phycoerythrin. As well as the photosynthetic lamellae the Cyanophyceae possess small granules resembling ribosomes. The photosynthetic lamellae can be arranged irregularly or in parallel stacks, but they are always separated by a layer of cytoplasm. Fine fibrils of DNA (deoxyribose nucleic acid) are a feature of the nucleoplasm and these are either distributed throughout the cell or concentrated in the central portion. The use of the electron microscope has shown that the former idea of an outer chromatic zone and an inner colourless zone is not valid and that cells of different species have the structural units, which can be arranged in a number of different ways.

The protoplast is normally devoid of vacuoles, except in old cells, and this is a major factor contributing to the great resistance of the plants to desiccation and of the cells to plasmolysis. In some forms (e.g. *Anabaena, Oscillatorid),* and principally in species which are planktonic, pseudovacuoles (gas vacuoles) may be found and it is supposed that these contribute towards their buoyancy by virtue of the gas that they contain. In *Microcystis,* these consist of a single layer of particles, a few nanometers in dimension, linearly arranged around the axis of the vacuole which has approximate dimensions of 70×400 nm. The mucilaginous cell sheath, which may surround the whole cell, e.g. *Anacystis,* or form a cylindrical sheath, e.g. *Oscillatoria,* or an interrupted sheath, e.g. *Anabaena,* appears to be composed of cellulose fibrils.

There is considerable variation in the composition of the different cell sheaths, and the amount of material laid down frequently depends upon the external environment. Colour of the sheaths is modified by acidity of the medium and also by the action of fungal parasites. Protoplasmic connections between mature cells are known in one order, the Stigonematales.

The class is characterised by a general absence of well-marked reproductive organs, sexual reproduction or motile reproductive bodies. Fusion of filaments has been reported for a strain of *Nostoc muscorum* (Lazaroff and Vishniac, 1962).

The coccoid forms (spherical cells) multiply by cell division which takes place by means of a progressive constriction, whilst in some other types the cell contents give rise to a number of non-motile bodies that are termed endospores (formerly called gonidia). All stages from simple binary fission to endospores can be found:

(a) Binary fission, e.g. *Chroococcus turgidus* (= *Anacystis dimidiatd)*

(b) Quadrants and octants formed, e.g. *Anacystis montana*

(c) Numerous small daughter cells are produced in which there is retention of individual sheaths, e.g. *Anacystis* spp. (These are sometimes called nannocytes or gonidia)

(d) The same without individual sheaths

(e) Abstricted endospores, e.g. *Entophysalis.*

During cell division and at binary fission the photosynthetic lamellae become invaginated at the point of division and a new membrane is formed here by invagination of the inner membrane of the long wall.

Many of the filamentous forms produce specialised cells known as *hetero-cysts.* These are enlarged cells which possess thickened walls, particularly at their poles, and they usually occur singly, though occasionally they may be formed in rows. They develop from an ordinary vegetative cell, but during development they remain in protoplasmic communication with neighbouring cells by a small cytoplasmic connection. During development the hjeterocyst increases in size, and there is an apparent increase in the cytoplasmic matrix, disorganisation of the photosynthetic lamellae system and disappearance of some of the cytoplasmic inclusions (Wildon and Mercer, 1963). As the heterocyst enlarges a thick outer envelope is laid down outside the cell. There is usually a pore at one end of the terminal heterocyst (at both ends if intercalary) at which the disorganised chloroplast lamellae often reaggregate. There are connections to plasma membranes of adjacent cells, but there does not appear to be any direct protoplasmic connection. It has been shown that heterocyst formation occurs only when the nitrogen level falls below a certain level and that heterocysts are incapable of photosynthesis but may fix nitrogen.

The widespread occurrence of heterocysts, especially terminal ones, and their specialised structure indicates that they must fulfil a significant role. Various suggestions have been made as to their function, and in some cases they do seem to determine the breaking up of the trichomes or threads into hormogones. These hormogones are short lengths of thread which are cut off, * The trichome plus the sheath is termed a filament.

Thus forming a means of vegetative reproduction among the filamentous types. In the past, it has been suggested that the heterocysts represent archaic reproductive organs that are now functionless. Fritsch suggested that the heterocysts, during the vegetative period, secrete substances that stimulate growth and cell division. He also suggested that at the time of reproduction the nature of the secretion changes and as a result akinetes are formed.

Recently Wolk has demonstrated again that under certain conditions (e.g. transfer to ammonium chloride and glucose) heterocysts will 'germinate' to give rise to multicellulargermlings. He also found that sporulation was stimulated by phosphate starvation and that spores could germinate on phosphate agar. Wolk then suggested that these two adaptations are survival mechanisms for deficiencies of nitrogen and phosphate. In other experiments, he has shown that sporulation of akinetes adjacent to heterocysts is dependent upon the heterocysts. Further work is, however, needed before the problem of the heterocyst is solved. Recently Stewart (1969) has shown they will fix nitrogen.

Hormogones, besides being cut off by heterocysts, may also be produced by the development of biconcave 'separation discs' at intervals along the filament, or they may break off from the ends of the trichomes. The so-called separation discs are formed by the solidifying of a cell and its contents in the filament. The hormogones, together with certain of the filamentous types, e.g. *Oscillatoria*, and a few Chroococcales, exhibit a slow forward motion and associated with this motion is an active and continual secretion of mucilage along the sides of the filaments. The actual mechanism said to be responsible for locomotion is rhythmic longitudinal waves passing from end to end of the trichomes and caused by changes in the volumes of the protoplasts.

This is probably an over simplification and until further experimental work has been carried out interpretation should be reserved. In four species of *Phormidium* that have been fairly thoroughly studied, it has been found that movement is a response to light, and that it occurs between light intensities of 0-02 Ix and 30,000 Ix with optimum response at 2,000 lx. It is suggested that there are three different light reaction systems that are independent of each other, so that there is some degree of complexity. The principal system is related to chlorophyll alone but there is another based upon both chlorophyll and the biliproteins. A third system is related to a photo-receptor of unknown character (Nultsch, 1964).

Yet another type of structure, the thick-walled resting spore or akinete, occurs in many of the filamentous forms belonging to the Nostocaceae and Rivulariaceae, and these akinetes normally develop next to a heterocyst, either singly or in a series. On germination they give rise to a new filament.

Many of the forms aggregate into colonies, but in some of the Chroococca-ceae the macroscopic plant mass represents an association of such colonies rather than a single colony. The form which any colony may take depends on first, the planes of cell division, and second, the effect of environment, which may determine the consistency of the mucilage. Uneven temperatures, for example, sometimes produce irregular growth. It has been shown experimentally that the environment may affect the shape of colonies of *Microcystis* and *Chroococcus turgidus (Anacystis dimidiatd)* and determines the size of *Rivularia haematites*. The single floating unicell may not necessarily represent the most primitive condition.

The class is exceptionally ancient and must have evolved at a very early stage. It is known that the early waters of the world were not saline as at present and they were probably very much warmer and the fact that many Cyanophyceae are found associated with hot springs may be of significance. Schwabe (1960) has suggested that it is in such places the most primitive forms may be found, and he proposed a radiating system based upon the behaviour of *Mastigocladus laminosus*, a wide-spread thermal algae. In this algae, under different conditions, a wide range of morphological form can develop.

In the present state of our knowledge other courses of evolutionary development could be envisaged. Further intensive studies with the electron microscope may provide some additional data, though in view of the comparative simplicity of cellular structure, this is not very likely.

As may be expected from a primitive and ancient group there is evidence of parallel development when compared with plants from other groups, and this follows what is often considered a typical evolutionary sequence. Parallel evolution can be seen between *Coccochloris* and *Gloeocystis, Agmenellum* and *Prasiola, Entophysalis* and *Characium, Anacystis* and *Pleurococcus, Lyngbya* and *Hormidium, Stigo-nema* and *Wittrockiella.*

As a group, the plants are extremely widely distributed under all sorts of conditions, frequently occurring in places where no other vegetation can exist, e.g. hot thermal springs. Their presence in great abundance in the plankton often colours the water and is responsible for water blooms, which in some cases may result in the death offish. Members of the group also form a large constituent of the soil algae.

One odd site that has been recorded for these algae is in the centre of lake balls composed of the remains of leaves of *Potamogeton pectinatus* found in Lake Vrana, Yugoslavia.

Symbiosis involving Cyanophyceae is of widespread occurrence. Apart from the very well known example of Lichens (an association of a fungus and either a Chlorophycean or Cyanophycean algae, e.g. *Anacystis, Scytonemd),* blue-green algae have been recorded in a variety of symbiotic associations, examples of which are given below: *Anabaena azollae* within the leaves of the water fern *Azollafiliculoides\ Nostoc sphaericum* within cells of the fungus *Geosiphon pyriforme; Calothrix chapmanii* within the cells of the green algae *Enteromorpha.* One of the most interesting examples of symbiosis involving blue-green algae is the case of the 'cyanellae' or 'syn-cyanomes'. These are blue-green algae living within colourless algae or rhizopods. The symbiosis appears to be geologically ancient and unlike other symbiotic Cyanophyceae.

The algae have undergone such modification and reduction that some workers have suggested that in certain cases (e.g. *Glaucocystis)* the association is so inseparable that it may be considered as a single entity, and consequently they have suggested the creation of a phylum Glaucophyta to accommodate them. Another possibility is to regard them as only peculiar plastids. Two of the best known examples are *Glaucocystis nostochinearwn* (the host is a colourless

Chlorophyceae-like organism with vestigial flagella) and *Cyanophora paradoxa*. In this latter case, there are two blue-green inclusions, which have been named *Cyanocyta korschikoffiana* (Fam. Cyanocytaceae, order Chroococcales), inside a colourless Cryptomonad host, *Cyanophora*. In the case of *Glaucocystis*, the name *Skujapelta nuda* (Fam. Skujapeltaceae, order: Chroococcales) has been applied to the symbiont, and *Glaucocystis* retained for the host. The symbionts differ from free-living forms in lacking the typical, double-layered cell wall and they also possess a large centrally-located body which may represent a primitive sort of nuclear material (Hall and Claus, 1963). Another of these peculiar forms is the colourless parasite, *Sarcinastrum urosporae* found in *Urospora* and *Ulothrix*.

The Cyanophyceae, besides being primitive, are very ancient and some of the earliest plant fossils known belong to the class. It would seem that very little evolution has taken place within the group since they first appeared. Their age and constancy of form, together with their modes of reproduction, are also responsible for the wide distribution of different species, so that there is little or no endemism in the Cyanophycean flora of any region.

Taxonomically, the class has been variously treated. Originally it was divided into two orders and in more recent years it has been customary to recognise five orders. Much of the problem in this respect arises from environmentally induced morphological variation. Recent works on *Nostoc* and *Lyngbya* have shown that gross morphology is especially dependent on environment/culture condition. Drouet has shown that some 54 different species (8 genera) are in fact all variations (ecophenes) of *Schizothrix calcicola*. Some authorities, however, have suggested that there should be only one order, the Chroococcales, and eight families.

The five orders recognised in many taxonomic works are first, the Chroococcales, which comprises unicellular, free-living forms, living singly or in colonies; second, the Chamaesiphonales, which is a small order consisting of attached unicells occurring singly or in groups and reproducing by means of endospores, which do not occur in the first order. The Pleurocapsales is another very small order with few species that consist of cells that divide in one or two planes to give a crust or cushion. The great majority of the Cyanophyceae are placed in the Nostocales and these are the simple or branched filamentous forms that occur free or aggregated in a common sheath. In some genera the base and apex of the filaments differ, e.g. *Rivularia, Calothrix,* and the possession of heterocysts is a feature of many. Reproduction is commonly by means of hormogones but some genera also produce spores. The fifth order is the Stigonematales in which the species form branched threads and there are examples of heterotrichy. Branching is true branching and the threads may be multiseriate. Both hormogones and heterocysts are known.

Chroccoccaceae: *Anacystis(ana,* single; *cystis,* bladder).

The representatives of this family consist of uni- or multicellular microscopic plants or macroscopic plant colonies. The individual cells are spherical, ovoid, discoid, cylindrical or pyriform in shape, and if colonial can be arranged regularly or irregularly within a gelatinous envelope. In some species of *Anacystis* the colonies are free-floating with a mass of spherical cells: in other, mostly terrestrial, species the cells are single or else united into spherical or flattened colonies each with a small number of cells. The free-floating colonies of *Agmenellum* form regular plates one cell in thickness which, with increasing age and size, can become curved or twisted. There appears to be considerable variation in the degree of compactness with which the cells are arranged.

Reproduction of colonies is by fragmentation, but multiplication of the cells is by binary fission in two *(Agmenellum)* or three planes *(Anacystis).* In the genus *Johannesbaptista,* which has a linear series of lenticular cells in an elongate gelatinous matrix, division is in one plane only. *Anacystis cyanea* (= *Microcystis aeruginosd)* is a very common water bloom algae and certain strains produce a highly toxic substance. The taxonomy of the family is extremely difficult and one can either recognise a large number of genera and species or only a few. The latter course, based on Drouet and Daily (1956), has been adopted here. The common *Anacystis nidulans*, of physiological use has recently been transferred to a new genus *Lauterbornia nidulans* (Pringsheim, 1968).

The cells of the sole genus in the family are attached or epiphytic, solitary or arranged in dense clusters on wood, rocks or fresh water or marine plants. The cells are more or less rigid, vary much in shape and are attached at the base by down-growing filaments. In a cushion cell, division is usually horizontal and gives rise to a pseudoparenchymatous prostrate system, but the upper daughter cells can divide by perpendicular walls, the resulting arrangement of cells being radial. The sheath is thin and hyaline and ultimately opens at the apex. Reproduction is by means of endospores which are abstricted successively from the apex of the surface cells by transverse division. There has been much confusion about this order, but Drouet and Daily consider that species previously placed in different genera, e.g. *Chamaesiphon, Dermocarpa, Xenococcus, Pleurocapsa*, simply represent growth stages of one and the same species. The few species recognised are now placed in the single genus *Ento-physalis*.

Clastidiaceae

This family is not well known, the plants consisting of solitary, elongate, epiphytic unicells within a sheath which is thickened at the base to form an attachment pad. There are only two genera, and three species (*Stichosiphon sansibaricus* and *Clastidium setigerun, C. rivulare*).

Stigonemataceae: The members of this family, which includes *Mastigocladus*, form branched threads, sometimes with a prostrate and erect portion. The entire plants are floccose, felt-like, cushion shaped or spherical. Branches are formed by cell division parallel with the axis and the daughter cell pushes through the sheath at right angles and then produces its own sheath.

In *Stigonema* the threads consist of a single row or multiple rows of cells with definite apical growth, enclosed within a gelatinous sheath that varies in thickness and colour. Shortly behind the apical cell one of the products divides twice to give a cruciform group. One of these cells gradually moves into the centre and by further division a central cell surrounded by four pericentrals is produced. Most pericentrals divide radially so that ultimately successive tiers of 7-8 pericentral cells are formed and these are arranged approximately longitudinally. In this genus the cells are connected by protoplasmic strands but the exact nature of these requires further study. In young plants the sheath is continuous at the tip, but when growth is rapid or when hormogones are produced and extruded the end bursts. The heterocysts in this genus are commonly lateral, being formed by tangential division of a cell. The filaments are often aggregated to form a macroscopic yellow-green gelatinous mass. The species are most common where there is continual dripping water.

The gelatinous thallus is solid or hollow, floating, attached or terrestrial and varies much in size and shape. The spheres usually break open when mature and give rise to a flat expanse. There is a dense limiting layer containing numerous intertwined and contorted moniliform filaments with individual hyaline or coloured sheaths, though the sheath may sometimes be absent or very indistinct. Heterocysts are terminal or intercalary and occur singly or in series. Reproduction is by means x>f hormogones or spores, the latter arising midway between the heterocysts and developing centrifugally.

The closely related genus *Anabaena* differs from *Nostoc* in that no firm colony is formed and it lacks a punctiform stage, whilst in *Wollea* the saccate plants contain trichomes arranged in a parallel manner. A feature of the genus *Cylindrospermum* is the large spore that develops next to the heterocyst. Some species are symbiotic, whilst species of both *Nostoc* and *Anabaena* are capable of fixing atmospheric nitrogen.

The colonies form spherical, hemispherical or irregular

gelatinous masses that grow on plants, stones or soil, those of *R. atra* being especially frequent on salt marshes. Each colony contains numerous radiating filaments with repeated false branching, each branch terminating in a colourless hair. With the production of mucilage the 'branches' become displaced and their origin is not easy to see. The individual sheaths can be seen near the base of the trichomes, but they are diffluent farther up. The heterocysts are basal and in the closely allied genus *Gloeotrichia* spores are produced next to them. In other genera the heterocysts may also be intercalary, e.g. *Calothrix*. In *Calothrix* there is no development of a gelatinous colony and the plant mass is often felt-like, as the filaments are attached at the base only.

The trichomes differ from those of *Oscillatoria* in the presence of heterocysts. The filaments (trichome and sheath) have distinct basal and apical regions forming little erect tufts. Branching is of the false type, the branches arising either between two heterocysts or else adjoining one as a result of the degeneration of an intercalary cell. The intercalary growth results in strong pressure being applied to the sheath, which finally ruptures so that the trichome forms a loop outside. Further growth causes this loop to break, thus producing twin branches, one or both of which may subsequently proceed to additional growth, the branch sheaths extending back into the parent sheath. False branching may also be initiated by degeneration of a vegetative cell or heterocyst and subsequent growth of the two filaments on either side.

This is the largest family of the class. It differs from the other families in that the members do not form spores or heterocysts. The trichomes are uniseriate and occur singly or many in the enveloping sheath. Branching when present is of the false kind. All cells are capable of division and reproduction is by hormogones. In some cases it is very difficult to discern the sheath, e.g. *Spirulina,* some species of *Oscillatoria.* The filaments occur singly or else they form tangled masses. In the genus *Lyngbya* the sheath is often thick and coloured.

Trichomes of *Oscillatoria* can exhibit a gliding motion based on travelling waves on the cell surface produced by a fibrillan array (Halfen and Casten-holz,1970).

Chlorophyta

Prasinophyceae: This class was proposed only recently, mainly as a result of discoveries using electron-microscopy. Much of the work in this field has been carried out by Manton and Parke and their co-workers. The organisms are flagellates containing chlorophylls *a* and *b* and green algal xanthophylls, and are now known to be characterised by one or two layers of plate-like scales on the flagella and in some cases (e.g. *Monomastix, Micromonas)* on the cells as well. There may be a single flagellum *(Micromonas squamatd),* two equal flagella *(Nephroselmis gilvd)* or four flagella (*Prasinocladus*). Hairs are also present on the flagella; these are thick and tapering, often curved at the tips, and usually with internal striations. In the genus *Heteromastix* the origin of the hairs has been clearly traced to vesicles derived from golgi bodies.

In *Heteromastix, Halosphaera* and *Micromonas squamata* the scales also originate from the vesicles and are subsequently deposited on the outside. Rounded plate scales form the inner of two scale layers in *Prasinocladus, Pyramimonas, Halosphaera* and *Heteromastix.* The outer scales may be small rods *(Prasinocladus),* stellate *(Heteromastix)* or hexagonal *(Halosphaera).* in *Nephroselmis* there is only one layer of scales. In *Tetraselmis* a product of photosynthesis is mannitol, normally produced in the Phaeophyta, and if this is confirmed for other members, then their segregation will be further justified on physiological grounds.

Asexual reproduction is by fission, usually into two daughter cells. In *Micromonas squamata* a second flagellum develops before fission, whereas in *Nephroselmis* and *Heteromastix* each daughter cell receives one parent flagellum and one newly grown one.

The exact confines of this class have not yet been finally

established. Some authorities retain members of the Prasinophyceae *(sensu lato)* as three suborders of the Volvocales within the Chlorophyceae. Some would include *Pedinomonas* even though there are no scales on the flagella. Other workers place this genus and *Micromonas pusilla,* where there is a long hair point to the flagella and no scales, in a separate family, the Loxophyceae. The class has been tentatively divided into two orders, the *Pyramimonadales* and the *Halosphaerales.* Two genera that have been studied in some detail are Prasinocladus and Halosphaera.

Pyramimonadales

Species of this genus are found in marine aquaria and on rock surfaces near high-water mark. The motile phase is a quadriflagellate swarmer with a laciniate parietal plastid, whose finger-like projections point anteriorly and posteriorly. The four flagella arise from the bottom of a depression at the anterior end. There is a distinct starch sheath surrounding the pyrenoid and also a 2-layered red eye spot which is more pronounced in summer swarmers than in winter ones. The cells are covered by a periplast and when the organism becomes non-motile this undergoes plastic deformation to form the stalk of the sedentary individual. Because it contains no cellulose, this type of 'wall' may be termed a theca. The theca is composed of three layers, a central homogeneous one and an outer and inner fibrillar layer. Motile bodies are not common.

When a cell becomes motile it starts to move out of the parent theca but becomes wedged in the neck, and then a new theca is formed. In this manner, a series of empty cups are built up, and the plastic deformation of these empty cup gives the joint appearance of a hollow stalk. Division of the non-motile cells takes place by longitudinal fission and this can result in branch formation; sometimes one daughter cell does not emerge but is left behind in the middle of a series of empty thecae. On the few occasions, when a swarmer is liberated the species is perpetuated, but at present the conditions under which liberation occurs are not known. Cyst formation may also occur. The well known symbiont of the platyhelminth *Convoluta roscoffensis* has recently been reassigned to this order as *Platymonas convolutae.*

Halosphaerales

The large, free-floating spherical cells possess one nucleus which is suspended either in the central vacuole or in the parietal cytoplasm where it is associated with numerous discoid chloroplasts. A new membrane can be formed internally and then the old one ruptures, but if the outer membrane persists the cells appear to have a multi-layered sheath. Aplanospores and resting cysts have been recorded, but reproduction is primarily by means of quadriflagellate swarmers (Manton, Gates and Parke, 1963). These pear-shaped swarmers very closely resemble known members of the genus *Pyramimonas* both in respect of shape and morphology. There are anterior body lobes which can be outwardly or inwardly curved or which remain vertical.

Around the body, there are muciferous organelles or trichocysts which can eject their contents as rods, threads or balloons when treated with vital stains. Two or four pyrenoids with starch sheaths are also present. The most distinct structure, however, is a large spheroidal basal reservoir with a canal to the exterior through the frontal lobes. There are two layers of scales but the scales of flagella and body differ in pattern. The outer flagella scales are round or hexagonal plates strictly aligned in nine rows. The second layer consists of much smaller rounded scales, each with a circular groove and a central hub, arranged in a spiral fashion. The cell body scales are more varied but there is a principal diversity between flat lace-like plates and rectangular or circular scales with deeply depressed centres.

Chlorophyceae

The characteristic cell structure of the Chlorophyceae consists of a large central vacuole surrounded by cytoplasm. In some of the simpler unicellular forms, there may also be two or more small, contractile vacuoles whose probable function is excretion. The pigments, which are essentially the same as those in higher plants, are contained in chloroplasts; these vary from one to many per cell and in outline may be 'star-shaped, spiral, plate-like or reticulate. There is evidence to show that in certain forms the chloroplasts are capable of movement in response to light stimuli. The proportions of the pigments may change at the time of gamete formation, i.e. in *Ulva lobata* fertile female thalli are olive-green and fertile male thalli are brownish. Cells of fertile thalli contain much more carotene than those of vegetative thalli, and a large proportion is y-carotene of which only a trace occurs normally. Other colouring matter may also be present, e.g. tannins in some of the Conjugales. Formation of secondary carotenoids is very common in the Chroococcales, and other isolated genera.

The cells are commonly surrounded by a two-layered wall, the inner, which is often lamellate, being of cellulose (xylan in certain Siphonales), and the outer of pectin. In some forms the outer surface of this pectin sheath is dissolved as fast as it is formed on the inner side. Some of the very primitive forms lack a rigid cell wall. In a few species the outer layer of pectose becomes impregnated with an insoluble substance while in others the pectin layer gradually increases in thickness, and in at least three orders (Siphonales, Dasycladales, Siphonocladales) lime may be deposited on the walls. In *Oedogonium* and *Cladophora*, there is said to be a layer of chitinous material on the very outside and a study of the sub-microscopic structure of filamentous species has shown that there may be more than one kind of cellulose in the walls.

The chloroplasts normally contain pyrenoids surrounded by a sheath of starch. The pyrenoids contain fibrillar particles, but the existence of proteins is not substantiated by recent work though RNA may be present. The number of pyrenoids in each cell is frequently a constant specific character and can be used for taxonomic purposes. Very few families of Chlorophyceae lack pyrenoids and those that do are the more highly evolved. In *Derbesia*, the pyrenoids disappear when the algae is grown in diffuse light, and in several genera they may disappear during zoospore formation. In such cases, they must arise again *de novo*, but more commonly they are perpetuated by simple division at the time of chloroplast division. In those algae with pyrenoids the production of starch granules is associated with their presence.

Each cell is usually uninucleate, but in certain orders a multinucleate condition is found, and in the Siphonales cross walls are only laid down at reproduction so that the vegetative plant is non-septate and multinucleate. The chromosomes are usually small, short and few in number. Where there is only one chloroplast it divides at cell division into two daughter chloroplasts. In many species, there is a distinct diurnal periodicity in nuclear and cell division.

The flagella of the motile cells are composed of an axial, cytoplasmic filament or axoneme surrounded, except at the very apex, by a sheath which is probably able to contract. In *Chlamydomonas* and the Volvocaceae the electron microscope has revealed that the axoneme consists of 9 separate fibrils arranged in a ring around 2 central fibrils. This 9 + 2 arrangement is general throughout the algae, except in the haptonema of the Haptophyceae. In the motile forms the flagella are associated with other structures which are collectively known as the neuromotor apparatus. At the base of each flagellum is a granular blepharoplast. These are connected by a transverse fibre, or paradesmose, and this, or one of the blepharoplasts, is also connected to the intranuclear centrosome by a thin strand called the rhizoplast. The motile cells often contain a red-eyespot. In *Platydorina*, the eyespot comprises an assemblage of carotenoid-containing granules situated between the lamellae of the chloroplast.

In other algae the carotenoid granules may have an association with the flagella with (Chrysophyceae) or without plastid association (Euglenophyta), or with flagella and a lamellar body (*Glenodinium foliaceum*) or associated with a retinoid-like structure (Warnowiaceae in Dinoflagellates) (Dodge, 1969).

The exact function of the eyespot has not been determined, although it is possible that it acts as a shading device for the parabasal granule.

Vegetative reproduction takes place through fragmentation and ordinary cell division, though in the Chlorococcales and Siphonales vegetative cell division is unknown. Asexual reproduction is by means of biflagellate or quadriflagellate zoospores, except in the Oedogoniales where there is a ring of cilia at the anterior end. The zoospores are often formed during the night and are then liberated in the morning when they may remain motile for as much as three days or for as short a time as three minutes. Their production can sometimes be induced artificially by altering the environmental conditions, e.g. by removing the plant from flowing to still water (*Ulothrix, Oedogoniuni*), by changing the illumination, or by removing the plant from water for twenty-four hours (*Ulva, Enteromorphd*). Each individual cell may produce one or more zoospores, the number varying with the different species.

Liberation depends on: *(a)* lateral or terminal pores, or *(b)* gelatinisation of the entire wall, or *(c)* the wall dividing into two equal or unequal portions. In some species, non-motile zoospores called aplanospores are formed, but if these then secrete a thick wall they become known as hypnospores. Aplanospores which have the same shape as the parent cell are termed autospores. All these spores develop a new periplast or membrane and therefore, differ from a purely resting vegetative cell or akinete.

Sexual reproduction is represented in all the orders and often there is a complete range from isogamy to oogamy, the egg sometimes being retained and fertilized on the parent thallus in the oogamous forms (e.g. *Volvox, Oedogonium, Coieochaete,* Charales).

The occurrence of sexual reproduction can be induced in culture by an abundance or deficiency of food material or by intense insolation. Interspecific hybrids have been recorded in *Spirogyra, Ulothrix, Stigeoclonium, Draparnaldia,* and *Chlamydomonas.* Another striking fact is that characters which may develop in some species under the influence of the external environment are normally found 'fixed' in others. This indicates the plasticity of many members in the class and the phenomenon may also be of importance phylogenetically.

As a group, the Chlorophyceae are widespread, occurring in all types of habitat: marine, freshwater, soil and subaerial. Only about 13 per cent are marine and are mainly represented by the Ulvaceae, Siphonocladales, Dasycladales, Derbesiales and Siphonales. Many of the marine species have a definite geographical distribution whereas most of the freshwater and soil algae are cosmopolitan. A few genera, e.g. *Entocladia (Endodermd), Chloro-chytrium, Cephaleuros, Phyllosiphon,* are parasitic species, whilst species of other genera participate in symbiotic associations, e.g. *Zoochhrella, Trebouxia.*

Volvocales

This order comprises the simplest members of the Chlorophyceae and is the only order with motile vegetative cells. It ranges from simple unicellular forms of a basic type, e.g. *Chlamydomonas,* to regular colonial aggregates of similar cells, e.g. *Volvox* in other cases the cells are aggregated in an irregular manner, e.g. *Tetraspora.* Throughout the order, the motile reproductive swarmer is uniform in character and even when aggregated the individual cells usually retain their power of locomotion. Electron-microscope studies show that the microcellular structure is uniform throughout.

The chlamydomonad type of cell characteristically possesses a single chloroplast, a red eye-spot, one pyrenoid, two contractile vacuoles and two flagella, and is often strongly attracted to light (positively phototactic). Variations in the structure of the cell occur throughout the genus which contains about 325 species. Some species lack pyrenoids (C. *reticulatd*) some have one *(C. stellatd),* or two (C. *de baryand),* or several

(C. *giganted)*. The chloroplast is usually basin-shaped but may be reticulate (C. *reticulata)*, or axile and stellate (C. *eradians)*, or it may be situated laterally *(C. parietarid)*. It has been said that under cultural conditions many of the characteristic features can be modified, and that therefore, some of the species are not strictly species but are simply phases in the life cycles of other species, or ecophenes of the same species.

The motile cells are spherical, ellipsoid, ovoid or pyriform in shape with a thin membrane which sometimes has an outer layer of mucilage. The term naked is often used to describe cells (usually flagellates) which lack a cell wall (e.g. cellulose, pectin). The term is relative as so-called naked cells (e.g. *Chlamydomonas)* often possess an outer membrane, which surrounds the whole cell, and also possess a thin protoplasmic membrane or periplast. In this text, the term naked describes cells that lack the cellulose-type wall characteristic of non-flagellates without denoting any particular type of periplastic membrane. The two flagella are situated anteriorly and either project through one aperture in the membrane or else through two separate canals. In either case, there are two basal granules or blepharoplasts at the point of origin of the flagella. Electron microscope studies have shown that every cell contains mitochondria, dictyosomes, a double cell membrane, endoplasmic reticulum and fenestrated nuclear membrane. Near the dictyosomes there are also large irregular vesicles whose function is unknown.

During asexual reproduction the vegetative cell comes to rest and divides into two or four, more rarely eight or sixteen, zoospores. All divisions at zoospore formation are normally longitudinal. The biflagellate zoospores, escape through gelatinisation or rupture of the cell membrane, but if this does not occur, the colony then passes into a palmelloid state, which is usually of brief duration; in *C. kleinii* however, it is the dominant phase in the life of the organism. *C. kleinii* may thus be regarded as forming a transition to the condition found in *Tetraspora*.

In sexual reproduction 8, 16 or 32, biflagellate gametes are formed in each cell. These are normally either + or - in character and fusion takes place between the strains. In C. *agametos*, an anticopulatory hormone is produced and so there is no fusion; in *C. monoica*, there is physiological anisogamy as the naked contents of one gamete (male) pass into the envelope of the other; in *C. braunii* the anisogamy is more distinct, the female cell producing two or four macrogametes and the male cell 8 or 16 microgametes. In C. *suboogama*, each cell gives rise to three macrogametes and an antheridial cell that produces four antherozoids; this is a most interesting condition, because sex segregation must take place at the first division of the mother cell nucleus.

In C. *coccifera,* there is oogamy, with the female cell producing one macro-gamete enclosed in a wall, while the male cell produces 16 or 32 spherical microgametes. In a related genus, *Chlorogonium oogamum,* one naked ovum is produced and numerous elongate antherozoids. The genus *Chlamydomonas* appears to be polyploid, with eight as the basic haploid chromosome number.

The zygote is at first quadriflagellate but it soon loses the flagella and forms a smooth or spiny wall. It subsequently enlarges in most species and on germination generally gives rise to four swarmers. Meiosis occurs during this segmentation and the normal vegetative cells are therefore, haploid. In *C. pertusa* and *C. botryoides,* however, the zygote may remain motile for as long as ten days. In C. *variabilis,* the persistent 4-ciliate zygote has for long been known as *Carteria ovata* but it has now been demonstrated that the two organisms represent the haploid and diploid phases respectively of one species. Whether these examples represent a true alternation of generations or whether they must be regarded as possessing a special type of zygote will be considered later. In most species the zygote requires a maturation period, which in *C. reinhardii* is 6 days at 25°C. The species is very easy to grow in culture and the four products of zygote germination readily separate on an agar medium.

Many mutants of *C. reinhardii* have appeared during investigations and it has been possible to work out a number of linkage groups of different genes. Many of the mutant strains are of such a nature that it will be possible to attack fundamental problems in the biochemistry as well as the genetics of these organisms, (Hastings, 1965).

The genus is widespread, the various species occurring principally in small bodies of water, in sewage oxidation lakes and in the soil.

Within the family Chlamydomonadaceae, there is a considerable variety of forms. Among those which lack a true cell wall is the biflagellate *Dunaliella,* one species of which, *D. salina,* is very common in salt ponds. In *Phacotus,* the chlamydomonad cell is surrounded by a special outsize hardened envelope which is flattened in side view. *Polytoma* is a colourless member of the family but nevertheless produces starch as its food reserve.

Possible lines of development within the family have recently been discussed by Ettl (1965).

A characteristic of this genus is the area rorming the inner wall between the protoplast, which is very similar to that of *Chlamydomonas,* and the firm, outer cell wall; this region is filled by a watery jelly and crossing it are cyto-plasmic threads which pass from the central protoplast to the cell wall. These are numerous and very fine in *H. pluvialis,* but are fewer and thicker in the other species. The flagella are unusual in possessing a thin tomentum of fine hairs. Contractile vacuoles are not always present and if present they are inconspicuous. The plastid varies in the different species and the number of pyrenoids ranges from one in *H. capensis* to many in *H. pluvialis.*

In asexual reproduction, the zoospores may pass through a non-motile aplanospore stage before maturing, and this has developed to such an extent in *H. pluvialis* that it forms a major feature of the life history. In all species except *H. pluvialis* small biciliate gametes can be formed, the number ranging from 32 in *buetschlii* to 256 in *H. capensis.* In *pluvialis* gametes are produced only rarely and never from motile cells, as in the other species, but only from resting cells.

Volvocaceae: So far the genera considered have comprised only unicells, but a whole series is known in which an increasing number of chlamydomonad cells are combined into colonial forms. The simplest type is probably represented by *Pascheriella* where three or four chlamydomonad cells are attached to each other laterally. The colonies are usually spherical in shape, but in the genus *Gonium* four, eight, or sixteen cells all lying in one plane form a flat quadrangular plate.

In the genus *Pandorina* 4, 8, 16 or 32 cells are closely aggregated in a common spherical envelope, whilst in *Vohulina* the 16 cells are arranged in four

The colonies contain sixteen, thirty-two (commonly) or sixty-four biflagellate cells, which are not closely packed and are sometimes arranged in transverse rows. In *E. illinoiensis*, the four anterior cells are much smaller and cannot reproduce. This marks the first differentiation of this series of organisms into a plant soma of purely vegetative cells which die once the colony has reproduced. Further differentiation occurs in *Pleodorina* where the somatic cells form J to i of the 64 or 28 celled colonies.

A genus comparable to *Pleodorina* is *Astrephomene* which is found in S. Africa, Australia and the USA. The colony consists of 64 cells, the four anterior cells forming a 'rudder'.

In these colonial forms asexual reproduction involves the formation of daughter coenobia from all the cells except somatic cells.

Sexual reproduction in *Pandorina* shows a tendency to anisogamy, but in *Eudorina* it borders between anisogamy and oogamy, the species being either monoecious or dioecious: in the former case the anterior cells give rise to the antherozoids and the posterior form the ova. When the zygote germinates it gives rise to a single motile swarmer and what are probably two or three degenerate swarmers. The motile swarmer eventually comes to rest and divides mitotically to give a new coenobium.

The genus represents the ultimate development that has been reached along this line of spherical colonies or coenobia, each of which behaves as a well-organised unit. Every colony is in the form of a sphere with 500-60,000 biflagellate cells set around the periphery, the flagella emerging through canals. The actual number of cells in a colony varies from species to species. The interior of the colony is mucilaginous, as in *V. aureus,* or else merely contains water, as in *V. globator,* while the entire collection of generally ovoid cells is bounded by a firm mucilaginous wall. The individual cells, each containing two to five contractile vacuoles, are surrounded in some species by distinct individual sheaths.

In some species, too, the cells are united by delicate cytoplasmic threads, or plasmodesmae, and the base of each thread appears to be associated with a mitochondrion. In *V. globator* and *V. rousseletii,* the cells are 'sphaerelloid' in nature, each individual cell being enclosed in a separate envelope, while in *V. aureus,* a number of individual chlamydomonad cells are enclosed in wedge-shaped prisms. It has therefore, been suggested that the volvocine colony has arisen at least twice in the course of evolution, once from a *Haematococcus* and once from a *Chlamydomonas* ancestry.

The majority of the cells, including all those in the anterior quarter, are wholly somatic, and only a few are able to give rise to daughter colonies or gametes. As the plant matures some of the posterior cells enlarge and lose their flagella. The enlarged gonidium then divides longitudinally a number of times and a small hollow sphere is produced with a pore (phialopore) towards the outer edge. These coenobia, which hang down into the parent cavity, then invert, the process commencing opposite the phialopore. After inversion flagella are formed and the colonies may be liberated into the parental cavity or they may escape to the exterior. If retained, they remain until the parent tears open, in some species *(V. aureus)* at the adult phialopore, in other species *(V. globator)* at any place. In *V. africana,* it is possible to see as many as four generations in the one original parent colony, because the first parent takes a long time to break down.

In sexual reproduction, the plants are either monoecious *(V. globator}* or dioecious *(V. aureus).* Cells which give rise to antherozoids divide to give bowl-shaped or globose colonies containing 16, 32, 64, 128, 256 or 512 spindle-shaped, biflagellate antherozoids. Starr (1968) reports that in eight strains of *Volvox,* the liberation of a male proteinaceous hormone induces egg formation. Cells developing into ova enlarge considerably and loose their flagella but do not undergo division. Both bowls and spheres undergo inversion before maturation.

The fertilization mechanism is not known for certain, but in *V. aureus,* the antherozoids penetrate the female colony and then enter the ovum from the inner side (Holt, 1969). Zygotes do not germinate for a considerable time after liberation. Meiosis occurs at germination but the oospore only produces a single swarmer that is liberated into a vesicle formed by extrusion of the inner oospore wall (endospore). This grows into a 'juvenile' plant of about 128 to 500 cells, which finally inverts. It subsequently reproduces by daughter colonies, each successive generation having an increased number of cells until a fully developed colony is produced.

One of the features of the larger saccate Volvocales (excepting *Astrepho-mene)* is the inversions that occur at different stages of the life-cycle, and it is difficult to see why they occur or what the conditions were under which they first developed. It may have something to do with the fact that the new cells are formed with the flagella and eye-spot facing the interior, but then the problem arises as to how the individual cells came to be arranged thus.

The nuclear condition of the Volvocaceae presents some interesting features (Cave and Pocock, 1951). *Gonium,* which would seem generally to be a primitive member of the family, possesses the largest number of chromosomes. In *Volvox,* the species regarded as primitive have more chromosomes than species regarded as advanced. The facts may necessitate a re-interpretation of development within the family, *Gonium* being regarded as reduced rather than primitive, and similarly with the species of *Volvox*. Colonies with 4, 6, 7, or 8 chromosomes occur in the genus *Astrephomene*. Polypoidy is suggested as clones with 4 chromosomes remain distinct from those with 6, 7, or 8.

Members of this and allied families are placed by some workers in a separate order, the Tetrasporales. They are characterised by non-motile vegetative cells that may temporarily become motile. In so far as the cells are chlamydomonad-like, there seems no justification for the segregation. The species of *Tetraspora* form expanded or tubular, convoluted, light green macroscopic colonies. The colony in these palmelloid forms differs from the organised coenobium of the Volvocaceae because it is merely an unorganised loose assembly of cells. The palmelloid condition has in this and allied genera become the dominant phase, while the motile condition only occupies a brief period in the life cycle.

The colonies are most abundant in the spring when they are attached at first, although later they become free-floating. The spherical to ellipsoidal cells are embedded in the mucilage and are frequently arranged in groups of two or four, each group being enclosed in a separate envelope. Two pseudocilia proceed from each cell to the surface of the main colonial envelope, each thread being surrounded by a sheath of denser mucilage.

Vegetative reproduction takes place by fragmentation of the parent colony, and asexual reproduction is secured by means of biflagellate swarmers that may develop into a new colony or into a thick-walled resting spore that later gives rise to a new colony. The gametophytic colonies are monoecious or dioecious, depending on the species, and give rise to biflagellate isogametes. The zygote gives rise to four to eight aplanospores that are said to grow into new colonies. Work on *T. gelatinosa* suggests that the aplanospores are not always produced and that the zygote, after a resting period, can give rise directly to a new diploid plant which gives rise to asexual swarmers. These develop into juvenile colonies that can perpetuate themselves by biflagellate swarmers. Eventually swarmers are produced that give rise to an attached, pear-shaped colony. The swarmers from this colony give rise to the sexual plant so that in this species there is a definite alternation of morphological phases. It seems probable, though it has not been yet established, that meiosis may well take place at swarmer formation in the attached colonies.

A life cycle has been fully worked out for *Schizochlamys* (Thompson, 1956). Here the adult can reproduce itself by akinetes, zoospores or gametes.

Chlorococcales

This order contains a number of forms, diverse both in respect of morphology and reproduction. Nearly all are freshwater or terrestrial. Some, such as *Chlorochytrium, Chlorococcum,* because of their habit, probably represent reduced forms. In structure, the plants range from solitary, free-living or attached unicells to large coenobia *(Hydrodictyori)* up to 20 cm long. A characteristic feature of the order is the lack of vegetative cell division, though this has not precluded nuclear division, so that cells in some species become multinucleate. Many of the genera in this order possess the ability to form secondary carotenoids such as canthaxanthin, astaxanthin, echinenone. These are formed extra-plastidically in old resting cells to which they impart a distinct yellow to red hue.

The unicellular, spherical, zoospore-producing genera have, in the past, been responsible for much taxonomic confusion but new microscopical techniques and the development of unialgal cultures have clarified much of this confusion. The three most important taxonomic characters for these particular genera are the type of zoospore, the type of chloroplast and the presence or absence of pyrenoids (Starr, 1953). More recent approaches have involved chemical taxonomy, especially with species of *Chlorococcum* and *Chlorella.*

In the various examples discussed, it will be noted that motility has been suppressed more than once in the order, a feature which, taken in conjunction with the diverse morphology, suggests that the organisms may well be polyphyletic in origin. Some genera show indications of a relationship, but there are others whose relationships are extremely vague. Some forms *(Trebouxid)* have become associated with fungi and are an integral part of lichens. Sexual reproduction when present is usually by means of motile isogamctes, but oogamy has been recorded for a species of *Dictyosphaerium.*

Much confusion has existed over this genus, and many of the species formerly described are now known to be phases in the life cycles of species from other genera. Bold and Parker (1962) considcr that there are 18 valid species which are not always readily distinguished on morphological grounds, in which case changes in colony shape, cell size, colour or colony texture under specific culture conditions may be useful. The plants are represented by non-motile, spherical or ellipsoidal cells which vary in size, occurring singly or else forming a stratum on, or in the soil; *C. humicolum,* for example, is a very common soil species.

There is no eye spot or contractile vacuole; the chloroplast is hollow and parietal with one or more pyrenoids. The characteristic features of this genus are first, the hollow, parietal chloroplast; second, the presence of one or more pyrenoids; and third, biflagellate zoo-spores that retain their oval shape for some days after losing their motility. The cell walls are two-layered with a thin inner layer and an outer gelatinous one which is sometimes lamellose and irregularly thickened. Old cells may be coloured with an excess of secondary carotenoids. The young cells with thin walls are uninucleate but the adult ones with thick walls are commonly multinucleate, and it is in this older condition that the protoplast divides and gives rise to 8, 16 or more biflagellate zoospores or isogametes which are liberated all together in a vesicle, usually in the early hours of the morning. After a short motile phase, the flagella are withdrawn and a new vegetative phase commences.

Under dry conditions aplanospores are formed and when this occurs the parent cell gelatinises and a 'palmella' stage results, the cells of which subsequently give rise to 2 or 4 biflagellate gametes. In culture solutions of low nutrient concentration, reproduction takes place by means of zoospores, but in highly concentrated solutions these are replaced by aplanospores. The environment can therefore, affect the mode of reproduction. The production of aplanospores suggests how the genus *Chlorella* may have arisen, while the multinucleate state indicates a transition to a coenocytic phase that is further developed in other genera (*Protosiphon, Hydrodictyori*).

In *Trebouxia*, the cells differ from those of *Chlorococcum* in lacking an irregularly thickened wall and the plastid is axile rather than parietal. Certain species of the genus *Codiolum* (a little fleece), have now been shown to be stages in the life history of *Spongomorpha* and *Acrosi-phonia*. *Codiolum kuckuckii* is restricted as an epiphyte to the apical region of the southern brown algae *Splachnidium*.

Each plant is a solitary unicell with a single parietal chloroplast, and motile only in the reproductive phase. The ellipsoidal cells occur singly, or in aggregates on submerged plants or living aquatic larvae and they are borne on a short stalk that emerges from a small basal disc. In most species, the mature cells contain more than one nucleus. Asexual reproduction is by means of biflagellate zoospores which are liberated through a terminal or lateral aperture. Sexual reproduction is generally isogamous but some species exhibit anisogamy. In *C. saccatum* the female and male gametes are produced from separate plants and fuse to give a zygote which, on germination, divides to produce a number of zoospores.

The species grow endophytically in other plants such as mosses, angio-sperms and red algae. The cells are spherical or ellipsoidal, the walls varying in thickness. Reproduction is by means of biflagellate gametes that normally fuse to give a motile zygote. This eventually settles on the leaf of a host plant and secretes a wall. Tubular prolongations grow out from these bodies and enter the host, either by the stomata or between two epidermal cells. Sub-sequently the end of the tube swells out into an ellipsoidal or lobed structure into which the contents of the swarmer pass. In *Chlorochytrium lemnae;* which attacks duckweed *(Lemnd),* the endoparasitic cells sink in the old *Lemna* fronds to the bottom of the pond or stream and remain dormant until the next spring. There is evidence that reduction division takes place at gamete formation so that the vegetative phase is therefore, diploid.

The globular cells are non-motile, solitary or aggregated into groups, and usually lack pyrenoids. The species are not readily determined taxonomically, so cultures are often given strain numbers. Using morphological and physiological criteria some eight well-defined species can be recognised. They will grow in a wide range of media, including salt solution (0-2 M.). It seems that cells grown in the dark differ from those grown in the light. Quiescent dark cells are small and have a photosynthetic quotient (O_2/CO_2) of 1. They give rise to dividing dark cells which in daylight give rise to 'light' cells that are larger, poorer in chloroplast material and less active photosynthetically (photosynthetic' quotient = 3).

During the course of development, the amount of DNA increases. 'light' cells give rise to a new generation of 'dark' cells in the dark and reproduction is by means of division into 2, 4, 8 or 16 auto-spores. Several species form a symbiotic association with lower animals and they are then known as zoochlorella. Because the species grow well in uni-algal culture, they are often used in physiological investigations; they are also used in sewage purification.

The planktonic colonies are composed of four, eight, or, more rarely, sixteen cells attached to each other at one point by mucilage pads. This type of colony has probably originated from that in which cells are attached at one point indiscriminately, as in *Ankistrodesmus.* A large number of species are known to occur widely, and often in some abundance, in ponds and lakes. Taxonomically, the two end cells are highly important. They may differ in shape from the others and often have processes that are elaborations of the cell envelope (e.g. *S. quadricaudd);* these processes can probably be correlated with the planktonic mode of life. Other species (e.g. *S. acuminatus)* have tufts of bristles that perform the same function and are similar to those of *Pedia-strum.*

The cell wall of *S. quadricauda* exhibits a complex structure under the electron microscope. On the outside there is a pectic, hexagonal net layer which encloses a complex prop system in pectin. There is a middle layer and an innermost cellulose layer.

Reproduction takes place when the contents of each cell in the colonydivide to give daughter coenobia which are then liberated.

The algae included in the Hydrodictyaceae occur as coenobia and represent a more advanced stage of the simple condition seen in *Ankistrodesmus* and *Scenedesmus.* Species of *Pediastrum* are common components of freshwater plankton, with up to 128 cells united into flat, disc-like coenobia.

The most advanced condition is found in *Hydrodictyon* of which only three species are known, the commonest, *H. reticulatum,* being cosmopolitan in distribution. It is a hollow, free-floating, cylindrical network closed at either end and up to 20 cm in length. Further increase in the size of colony is probably impossible for purely mechanical reasons. The chloroplast is reticulate with numerous pyrenoids, though in the young uninucleate cells, there is only a simple parietal chloroplast which later becomes spiral and then reticulate. *H. africanum* and *H. patenaeforme* develop into saucer-shaped nets, and in the former, the original young cylindrical cells later become spherical and up to 1 cm in diameter, then become detached and lie on the substratum. The latter species is composed of cells which may grow to be 4 cm long and 2 mm in diameter. Experiments have shown that for growth of *H. reticulatum*, the actual growing period depends upon light intensity. Under normal conditions growth rate is periodic, increasing by day and decreasing at night and altering the length of the light period merely alters the degree of growth in the time. Under conditions of continuous light, the periodicity is much reduced, though this effect is more marked with old cells.

Asexual reproduction in *Pediastrum* and *reticulatum* is by means of numerous uninucleate biflagellate zoospores, which swarm in the parent cell about day-break and then come together to form a new coenobium which is subsequently liberated, further growth being brought about by elongation of the coenocytic cells. It has been found that zoospore production in *reticulatum* depends not only upon light intensity but also upon its duration:

Sometimes zoospores, instead of coming to rest to form a daughter net, settle down to form haploid hypnospores. These are not very common in *H. reticulatum* but are normal in *H. patanaeforme* and possibly also in *H. africanum*, where asexual reproduction of the coenocytes as described above is unknown. In the first two species, the hypnospore germinates to give a large, rather slow-moving biflagellate zoospore that settles down to form a poly-hedron, whose contents later divide and after forming a pavement stage develop swarmers that are liberated into a vesicle to give a new 'germ net'. The duration of the 'germ net' in *H. reticuJatum* is short because each coenocyte rapidly produces a daughter net. This does not happen in *H. patanaeforme*.

Sexual reproduction is isogamous in *reticulatum* and *H. patanaeforme,* but anisogamous in *H. africanum*. In the first species, the biflagellate gametes are smaller than the zoospores, but in the second species, there is no distinction in size. Both species are generally monoecious. The planozygote arising from conjunction is quadriflagellate, and after a motile period it comes to rest and forms a diploid hypnospore. At germination, the zygote generally enlarges and germinates meiotically to give 4, or more rarely 8, slow-moving biflagellate swarmers that later pass into the polyhedron stage and develop as described above. Sexual reproduction in *Pediastrum* follows the same course. Gamete and zoospore reproduction can be induced in *H. reticulatum* by varying the external conditions artificially. If plants are cultured in weak maltose solution in bright light or in the dark and are then transferred to distilled water, zoospores will develop from the light plants and gametes from the dark ones.

Hydrodictyon is essentially a collection of a number of individual plants because it arises as the result of the fusion of a number of swarmers. *Volvox*, on the other hand, must be regarded as a single plant composed of many cells because it arises from a single zygote or gonidium.

There is probably but a single species, *P. botryoides*, but it exists in a number of different strains, one strain from the deserts of Egypt tolerating temperatures up to 91°C and salt concentrations up to 1 per cent. It commonly grows associated with a very similar algae, *Botrydium* in the damp mud at the edges of ponds. The green aerial portion is more or less spherical, grading into a colourless rhizoidal portion that may branch occasionally. The multinucleate cell contains one parietal reticulate chloroplast with many pyrenoids. The shape and colour of the thallus can be modified by varying the external conditions; bright light and low moisture, for example, causes an old thallus to turn brick red due to secondary caro tenoid formation. During dry weather, the contents of the vesicle encyst to give large aplanospores or 'coenocysts'. When conditions are once more favourable, these cysts either grow directly into vegetative cells or else produce biflagellate gametes or zoospores.

Vegetative reproduction can take place by lateral buds which are cut off by cell walls. Flooding of plants on damp soil causes them to produce biflagellate swarmers which behave as isogametes though these swarmers can behave as zoospores. The zygote either germinates directly or may remain dormant for some time. The plant is probably haploid as there is evidence of meiosis at zygote germination. Morphologically, it is of great interest as a possible source of origin for the Siphonales.

Ulotrichales

This order comprises filamentous and parenchymatous types which have clearly evolved from the filamentous form. There is little or no differentiation between base and apex in many of the forms though there is usually a special attachment cell or pad of cells. In most genera all the cells except those for attachment are capable of producing reproductive bodies, but in two genera from New Zealand *(Gemma, Lobatd)* there appears to be a considerable degree of sterilisation. Within the genus *Enteromorpha* there is also evidence of the development of heterotrichy, or the condition in which the thallus possesses a prostrate adherent portion and an erect aerial portion.

In the past, the treatment of the genera in this order has varied considerably.

Some workers have divided the genera into three orders: Ulotrichales, Ulvales and Schizogoniales or Prasiolales. Others have recognised only the first and last of these orders, or else have regarded them as forming two suborders, the Ulotrichineae and Prasiolineae. Even when only one order is recognised, some authorities accept only three families (Ulotrichaceae, Ulvaceae and Prasiolaceae) whereas others, including the present writers, suggest that there should be as many as eight families. The Ulotrichaceae and Microsporaceae contain only filamentous un-branched forms which can be separated from one another by chloroplast and cell wall structure. The Cylindrocapsaceae are also unbranched but reproduction is oogamous as distinct from the isogamy or anisogamy in the first two families. The Sphaeropleaceae are also filamentous and oogamous but differ morphologically from the Cylindrocapsaceae. Some of the remaining four families consist of plants with a flattened or tubular parenchymatous thallus. Some members of the Monostromataceae are distinct because of the heteromorphic alternation and some of the remaining species *(Vivaria)* could either be included in the Ulvaceae or treated as a separate family. The Prasiolaceae are distinguished by their thallus and plastid structure and the Capsosiphonaceae by thallus structure, habitat and reproduction.

Until very recently the taxonomic treatment of the order was based upon morphology. It was then found that some species of *Monostroma* had a life history with a much reduced but persistent sporophyte stage, and so a separate family for *Monostroma* was proposed. Since then more attention has been given to life histories and Kornmann (1963) has suggested that the order should be restricted to species with an alternation of heteromorphic generations and to their derivative species. Such a radical re-alignment brings together species of *Ulothrix, Vrospora, Gomontia* and *Monostroma*. Morphologically, these genera differ widely, and there are also differences in their early stages of development. If therefore, we can accept that alternation of heteromorphic generations may have arisen more than once in the course of evolution, the traditional taxonomic treatment would seem preferable. Despite the great interest of Kornmann's approach, it would seem that it raises more problems than it solves. Another approach by Vinogradova (1969) that unites *Monostroma, Gomontia, Kornmannia* and *Enteromorpha (Blidingid) nana,* presents difficulties with respect to the last genus which is more closely related to *Enteromorpha*.

The unbranched filaments are attached to the substrate by means of a modified basal cell which frequently lacks chlorophyll, but even though attached at first, the plants may later become free-floating. Under unfavourable conditions, such as nutrient deficiency, rhizoids may grow out from the cells or the filaments become branched. This suggests one way at least in which the branched habit may have evolved from the simple filament, and in this case it probably represents an attempt to increase the absorbing surface in order to counteract the deficiency of salts. The cells of the different species vary considerably in size, shape and wall thickness. When the walls are thick they are often lamellate. The single chloroplast forms, a circular band around the whole or most of the cell circumference and possesses two or more pyrenoids. When it does not encircle the cell completely the chloroplast margins are lobed. Vegetative reproduction can take place through fragmentation.

The life histories of the various species are not well known even though the genus has been recognised for a long time. There are two distinct groups of species: those in which there is an alternation of isomorphic generations and those in which only the haploid generation is known or in which there is a dwarf phase.

Swarmers are formed from all cells of the filament except the attachment cell, but they usually appear first in the apical cells and then in successive lower cells. They are liberated through a hole in the side of the cell into a delicate vesicle which subsequently bursts to free the swarmers. Asexual reproduction is normally by means of quadriflagellate zoospores, though in one species they are biflagellate. In the most recent revision of the genus Mattox and Bold (1962) transfer species with biflagellate zoospores to the genus *Hormidium*. The numbers of zoospores produced per cell depend on the width of the filament: wide cells give 2, 4, 8, 16 and 32 zoospores per cell, narrow cells 1, 2 or 4 per cell. Some species, in the presence of nutrient deficiency, or as a normal course, produce bi- or quadriflagellate microzoospores which only germinate at low temperatures. The place of these in the life history should be clarified when further species have been studied. Undischarged zoospores can turn into aplanospores that may even germinate *in situ* before liberation. Akinetes *(U. idiospord)* are another means of asexual reproduction.

Sexual reproduction is by means of biflagellate gametes of which 8, 16, 32 or 64 are produced per cell. The plants are dioecious with + or - strains. In *U. flacca,* gametes are produced and fuse under long-day conditions but under short-day conditions, they germinate without fusing. Light periodicity is therefore, significant and other species of the genus may also be affected. On germination, the zygote either gives rise to 4-16 bodies which function as aplanospores or zoospores, or develops into a 'Codiolum' phase *(U. zonatd).* The species occur most abundantly in winter or spring and optimum conditions would seem to include low temperatures. The genus is well represented in both fresh and salt waters.

Genera closely allied to *Ulothrix* include *Schizomeris* in which some of the cells may divide longitudinally and in which sexual reproduction has not been observed. *Hormidium* species are often included in *Ulothrix* but the filaments have a tendency to fragment, the chloroplast only contains a single pyrenoid and the edges are not lobed, and the zoospores are always biflagellate. *H. flaccidum* is a widely distributed soil algae. In *Binuclearia* the cells are arranged in pairs in a gelatinous filament.

The species, all free-floating, are restricted to fresh-water and consist of unbranched threads with walls of varying thickness, the thicker walls often showing stratification. Each cell wall consists of two overlapping halves held in place by a delicate membrane. This structure arises from the fact that the filament is made up of a series of H pieces and as a result the filament readily fragments. At cell division, a new H piece is introduced between the two daughter protoplasts. In young cells the parietal chloroplast is commonly reticulate but older cells are so filled with starch that the plastid structure is difficult to see. There are no pyrenoids. Asexual reproduction is by means of biflagellate zoospores (quadriflagellate in one species), 1-16 being produced per cell and liberated either when the thread fragments into H pieces or else by gelatinisation of the cell walls. Any cell can produce single aplanospores instead of zoospores, and some species also produce multinucleate, thick-walled akinetes. These germinate directly to form a new filament or with later germination to give four protoplasts that then grow into new filaments.

Microspora with its cell walls in two pieces has an analogue with *Tribonema* in the Xanthophyceae and it seems that this type of construction has arisen more than once in the course of evolution.

The single genus is found in freshwater and each plant consists of un-branched filaments attached at the base by means of a gelatinous holdfast. When young each thread is composed of a single row of elliptical cells with thick, stratified walls, the whole being enclosed in a tubular sheath. Each cell contains a stellate chloroplast and a single pyrenoid.

In older filaments, the cells divide longitudinally, usually in pairs, and this suggests how genera such as *Monostroma* may have evolved. Reproduction is more advanced because, apart from bi- or quadriflagellate zoospores, sexual reproduction is oogamous, the plants being monoecious or dioecious. Each antheridium gives rise to two biflagellate antherozoids and each oogonium to a single ovum.

The thallus originates as a flat, one-layered plate or a small sac, which in most species ruptures very early to give a plate of cells one layer in thickness, the cells often being arranged in groups of two or four. Because there is only one layer, the thallus tends to be a paler green and more fragile than the distromatic thalli of *Viva*.

The species of the original genus, *Monostroma* are now best treated as falling into three genera: those with the macroscopic leafy thallus alternating with an enlarged zygote or cyst which after some months undergoes meiosis and gives rise to zoospores *(Monostromd)*\ those with the flattened thallus producing only one kind of swarmer (Af. *oxyspermum)* or with alternation of two kinds of flattened thalli *(Ulvaria)*; at least one species (Af. *zostericold)* with a sporophyte thallus and a small gametophyte disc producing biflagellate gametes (Yamada and Tatewaki, 1965). Bliding (1968) places this species in the new genus *Kornmannia*, while Vinogradova (1969) removes

Af. *oxyspermum* from *Vivaria* and places it in the new genus *Gayralia*, family Gayralia-ceae. *A. latissimum, M. wittrockii* (sensu Moewus), *M. angicava* and Af. *nitidum* are wholly gametophytic and the enlarged zygote produces quadri-flagellate swarmers which give rise to new adults via a flat dwarf thallus *(M. angicavd)* or an erect tube *(M. nitidum)*. In some cases, it would appear that not only is there an enlarged zygote but there may also be a dwarf shell-boring 'codiolum' sporophytic phase (Tokuda and Arasaki, 1967). In *Ulvaria obscura* vars. *blytii* and *wittrockii* (sensu Eliding), there is apparently a normal alternation of isomorphic gametophytic and sporophytic generations. *M. arcticwn* and *M. tubiforme* only produce biciliate swarmers and so could be either wholly haploid or wholly diploid (Hirose and Yoshida, 1964).

Two species, *M. groenlandicum* and *M. llndaueri* retain the saccate condition throughout their life and probably should be removed to a new genus. A complete understanding of the original genus will not be possible until the life history of every known species has been thoroughly studied. At the present time, no two workers are agreed on the distribution of the various species originally included in the old genus *Monostroma*. It is, however, clear that both life histories and frond development must be considered in any final arrangement.

The thallus, which characteristically is a flat, expanded, distromatic (two-layered) plate, originates from a single uniseriate filament that subsequently expands by lateral divisions. There is no hollow sac as in *Entero-morpha,* but the two layers may separate towards the margins or at the base as in *U. lima, Enteromorpha stipitata* and *E. flabellata* (Dangeard, 1958), so that it becomes difficult to know in what genus to place such transitional species. Typically the flat thallus is unbranched though it can be deeply divided. Branching or proliferating is characteristic of *Enteromorpha* and *Letterstedtia,* but branching has been reported in *U. linearis.*

The young uniseriate filaments are attached by a single cell, but as expansion takes place multinucleate rhizoids grow down from the lower cells and form a basal attachment disc that may persist throughout the winter, new plants arising from it in the spring. Detached fragments are another means of forming new thalli, especially in the free-floating forms of quiet waters. The form of the thallus of some species can vary considerably. Plants of *U. lima* found near high water mark in Japan are long and lanceolate and they gradually change to short and broad fronds further down the shore. Varying exposure times and water loss may be the factors involved here but culture experiments have shown that amino acids and kinetin may also be involved.

The plasticity of the species has undoubtedly led in the past to taxonomic confusions, and many of these have yet to be resolved. The taxon *lactuca* has probably included four separate species, *U. lactuca* proper *(sensu strictd)* being anisogamous and with a high latitude distribution, while one of the new taxa, *U. thuretii* is isogamous and a short day plant with a S. European distribution. Similarly a study of *U. linza* from the Mediterranean has shown that the plants differ from those ascribed to the species elsewhere and they have been renamed *U. bertoloni*. It is clear that further work, accompanied by breeding experiments is vitally necessary before the species in this genus are finally established.

Reproduction is by means of zoospores produced in the sporophytic plants and 32 or 64 gametes produced per cell in the gametophytic plants. The latter are generally isogamous, the + and - gametes being liberated from separate plants, but in at least three species anisogamy is known to occur. In some species, the gametes may fuse into clumps. In *U. lobata* two types of 'clump' reaction have been observed. In one type, 10-12 gametes fuse together whereas in the other more than 100 gametes are involved. At present, there is no explanation for this difference. While the gametes are positively phototactic before fusion, the zygote is negatively phototactic. It has been shown that in certain cases, there may be relative sexuality among gametes from different plants, the sex of the older and weaker gametes becoming changed. Meiosis takes place at zoospore formation and, in most of the species investigated, there is a regular alternation of morphologically similar diploid and haploid generations. In some species, only one generation is present, the swarmers invariably developing directly without fusion or without meiosis. *Mutabilis* is more complex in that it apparently can exist in the haploid, diploid (gametophytic and sporophytic plants), triploid and tetraploid states. This kind of cytological study is infrequent and there is clearly scope for much further work. In at least one species, *U. lobata,* liberation of gametes is associated with the spring tidal cycles.

The plants occur in saline or fresh water and become particularly abundant when the waters are polluted by organic matter or sewage.

In the allied sub-antarctic genus *Gemina*, the plants resemble either *Ulva* or *Enteromorpha*. The cells are normally in pairs and apparently only some cells are capable of reproduction whereas in *Ulva* all cells of the thallus can give rise to swarmers. Similarly, in the genus *Lobata,* the reproductive cells are restricted to the thickened central portion of the thallus.

Papenfuss (1960) has argued that *Gemina, Lobata* and *Letterstedtia* should all be merged into the genus *Viva*. Segregation into somatic and germ cells in these genera would seem sufficient justification for retaining them as distinct from *Ulva.* Species of *Letterstedtia* occur in South Africa and New Zealand but Pocock (1959) maintains, probably correctly, that the New Zealand *L. petiolata* should be the type of a new genus. *L. insignis* in South Africa is a long strap-shaped thallus with leafy marginal frills and reproductive zones which are separated by purely somatic areas. *L. petiolata* characteristically has a broad rounded blade which arises from several branched, stipe-like portions.

In the genus *Enteromorpha,* the plants also start life as uniseriate filaments but they soon become multiseriate and tubular. Like *Viva,* many of the species are attached to the substrate by rhizoids, but in *E. nana,* there is a basal prostrate disc from which the erect plants arise. This feature, together with small cell size, has been used to create an independent genus *Blidingia*, but because basal discs are known to occur in *E. hendayensis* and *E. tubulosa*, and small cells are a feature of *E. bulbosa*, there seems no justification for retaining the genus *Blidingia* nor for removing it to the family Monostromataceae (Sliding, 1968). The existence of a basal thallus as well as an erect one forms what is known as *heterotrichy*, a feature characteristic of the Chaetophorales. It seems very likely that the phenomenon of heterotrichy has arisen on a number of occasions in the course of evolution. In addition to the many attached species, there are also a few species, occurring especially on salt marshes, which are unattached for the whole or part of their existence.

There is considerable plasticity and hence polymorphism in this genus, as in *Viva*, so that taxonomists have either accepted a very large number of species or else have described few species with many forms. Further work, including culture and crossing experiments will be necessary before a final solution is reached. Taxonomically one can conveniently divide the species into groups which appear to have common features: the *compressa-intestinalis* group, the *clathrata*, the *nana*, the *procera* and the *prolifera* groups. Sliding (1963) also recognises *torta, jugoslavica, flexuosa* and *linza* groups, but he removes the *nana* group to *Blidingia*.

Growth in length of the thallus is primarily intercalary, though in the early stages there is also an apical cell that undergoes divisions. In some species, the chloroplasts exhibit polarity, occurring normally in the apical portion of the cell. Such polarity is apparently absent in *Ulva* and *Mono-stroma*. When the tubular thallus is damaged by grazers or wave action near the apex the wounded cells put out papillate undergrowths but if the damage occurs near the base, then rhizoids are produced.

In many species, there is a dioecious gametophytic generation which produces isogametes and this alternates with a morphologically identical sporophytic generation which produces zoospores. In *E. prolifera*, however, there is alternation between heteromorphic generations, one being simple and the other much branched. It is thought that a similar situation may exist in the case of *E. compressa* var. *australiensis*. Scagel (1960) reports that the sporophytic stage of *E. intestinalis* gives rise to tuberculate-like plants which may be identical with the plant known as *Collinsiella tuberculata*. Zygotes produced from gametes of this generation give rise to a disc from which the characteristically tubular *Enteromorpha* plant arises.

This represents an important deviation and should be confirmed. Some species, including *E. nana*, *E. procera*, *E. aragoensis*, *E. adriatica*, are probably wholly diploid as only zoospores are known. In the first-named species, meiosis has not been observed and the swarmers have been designated neutrospores. In the other species, the swarmers can be either bi- or quadriflagellate and it is clear that cytological study is required to show whether the genus comprises a polyploid series or not.

Anisogamy occurs in *E. intestinalis* where the male gamete is small with only a rudimentary pyrenoid. Parthenogenesis has been recorded for some species, e.g. *E. clathrata*, and this presumably results in new sexual plants. In the reproductive areas some of the cells fail to divide. These are termed rest cells, and later they may germinate and give rise to sporelings which eventually separate from the parent thallus.

Within the genus, there are several lines of development including two probable reduction lines, and one such reduction series may be represented by the sequence *E. clathrata* - *E. ralfsii* - *E. salina* - *E. percursa*. The last named species consists usually of only two rows of cells and has been placed in a separate genus *Percursaria*. An isomorphic alternation of generations has been demonstrated, however, and the filaments are said to develop from basal pads so that there seems little justification for generic separation. Eliding (1963) would regard the series as proceeding in the opposite direction with *E. percursa* as the most primitive form, though in a later paper (1968) he removes *Percursaria* to a new family, Percursariaceae, which he considers may not even belong in the Ulotrichales. The semi-terrestrial habitat of the species would seem to be against such a view.

The family is founded upon the golden-green *Capsosiphon fulvesc'ens* (formerly *Enteromorpha aurea),* but now another species, also golden-green is known in New Zealand. Some authorities (e.g. Papenfuss, 1959) do not consider the characters justify a separate family, but on the basis of morphology, reproduction and the unusual brackish water habitat, it can be argued that there is sufficient justification for the treatment adopted (Eliding, 1968; Vinogradova, 1969). The cells with gelatinous walls are arranged, often 2-4 enclosed in a common mother cell wall, in horizontal and longitudinal rows but do not separate off into longitudinal areolae as in *Prasiola.* Reproduction commonly takes place by means of akinetes or 4-flagellate zoospores and biflagellate gametes. After fusion a cyst is formed from which 16-32 zoospores emerge. Chihara (1967) considers that these facts justify a relationship with the Monostromataceae.

The young unbranched filament, which is known as the 'hormidium' stage, consists of a single row of cylindrical cells with thick walls which frequently exhibit striations. Later on, the cells divide longitudinally and produce a thin expanded thallus, known as the 'schizogonium' stage, which tapers to the base. The cells of the mature expanded thallus are often arranged in fours and possess axile, stellate chloroplasts, while in some species another feature is the presence of short rhizoids that occur in the stalk-like portion or else are produced from the marginal cells. In the juvenile filament, reproduction takes place by means of fragmentation as a result of the death of isolated cells, while in the older, more leafy thallus, 'buds' can arise from the margin. Sometimes, the cells produce large, thick-walled akinetes that germinate to form aplanospores from which new plants arise. In *P. japonica,* sexual reproduction is brought about by non-motile macrogametes (16 per cell) and motile microgametes (64 or 128 per cell) that are both produced from the same plant, so that this species at least is monoecious and anisogamous.

Despite the existence of sexual reproduction, multiplication in this species is usually by asexual means. The plants are haploid, but in view of discoveries about *P. stipitata* (Friedmann, 1959) it is desirable that *P. japonica* be restudied.

In *P. stipitata*, the adult plant is at least diploid (12-14 chromosomes), though in view of the count for *P. japonica*, it may be tetraploid, and it reproduces by means of aplanospores. Subsequently, the cells at the distal periphery (meiospores of Friedmann) undergo reduction division and give rise to cells which are haploid. These cells are monosexual and give rise to either male or female gametes that are anisogamous. The gametophyte segments are therefore, attached to the diploid plant. A similar situation exists in *P. meridionalis* where the segments were formerly regarded as another algae, *Gayella constricta*. In some species swarmers are produced from sporangia and Friedmann (1959) considers that these subsequently were retained and finally produced the so-called meiospores and meio-sporophytes of *P. stipitata* and *P. meridionalis,* which must therefore, be regarded as highly evolved forms. Ecologically, it has been observed that where *P. stipitata* forms a belt on the shore, wholly diploid plants occupy the upper levels and plants with gametophytic lobes occur lower down.

The shape of the thallus in *P. crispa,* and probably in other species also, varies considerably with the environment. The genus, which is generally absent from the tropics and sub-tropics, is represented by saline, freshwater or terrestrial species, the last named being tolerant towards considerable desiccation. This can be attributed to the lack of vacuoles in the cells and also to a high protoplasmic viscosity. Some authors believe the genus should be separated from the Ulotrichales but this hardly seems warranted, especially in view of Friedmann's interpretation of the development of the P. *stipitata* condition.

This genus is widely distributed but is only frequent locally, being most abundant on ground that is periodically flooded by freshwater. The long, free, unbranched filaments consist of elongated coenocytic cells containing up to seventy parietal chloroplasts with pyrenoids. The chloroplasts occupy the periphery of discs of cytoplasm, the discs being separated from each other by vacuoles, although occasionally they may come together to form a diffuse network. Each disc normally possesses one or two nuclei. In most of the species, the septa develop as ingrowths, though in *S. africana* they are replaced by a series of strands which sometimes fail to meet at the centre so that the coenocytes are continuous.

Vegetative reproduction is secured by means of fragmentation, and there is apparently no asexual reproduction though zoospores have been reported in *S. wilmani:* these may, however, be ova developing by parthenogenesis.

In sexual reproduction, although the cells do not change in shape, both oogonia and antheridia are formed singly or in series, the species being either monoecious or dioecious. In the formation of oogonia, the annular chloroplasts first become reticulate and then the ova are formed without any nuclear divisions. In *S. annulina*, the ova are non-motile, but in *S. cambrica* the large female gametes are biflagellate and thus represent an advanced anisogamy. In the antheridia the nuclei do undergo division and numerous, elongated, narrow antherozoids are formed which are liberated through small pores, subsequently entering the oogonial cells through similar perforations. The fertilized ovum (oospore) becomes surrounded by a hyaline membrane inside which two new membranes are laid down and then the first one disappears. The new external membrane is ornamented and the contents become brick red.

Germination stages are only known for some species and the oospores lie dormant for several months or even years. On germination 1-4 biflagellate swarmers (8 occasionally) are produced and these come to rest and then grow into new plants. Sometimes the swarmers do not separate and a four- or eight-flagellate synzoospore may form depending on whether it comprises two or four zoids. These develop to a four-fold sporeling (i.e. with four claws). In some cases the swarmers from the oospore are completely suppressed and a new filament develops directly, this type of reproduction being known as azoosporic. In the most complete life cycle investigated the motile ova may also develop parthenogenetically after forming a short non-motile structure which divides to produce three new filaments. Various reduction stages from this life cycle occur among the different species. The adult plants are haploid since meiosis occurs at the segmentation of the oospore.

Primitive characters, which seem to be a feature of the genus, are the numerous ova and the entire lack of specialised organs for reproduction. The plant must probably be regarded as an Ulotrichaceous filament which has still retained some primitive features. In *S. annulina* cells are frequently found with only one or two plastids thus showing a gradation towards *Ulothrix*. There would seem to be very little justification for following those authors who would place the family in the Siphonocladales, even though *S. africana* does have some features characteristic of that order.

Chlorophyta

Chlorophyceae

Oedogoniales: Three genera, *Oedogonium, Oedocladiwn* and *Bulbochaete* are described for this order. They are characterised by the presence of zoospores and anthero-zoids with a ring of flagella around their anterior end. An advanced oogamy and specialised cell division are the unique features of this order. These characters are not found elsewhere and suggest that the order represents an evolutionary dead end. The species are all freshwater and their taxonomy is difficult because it depends on a knowledge of the sexual plants and on the type of reproduction. *Oedogonium (pedo,* swelling; *gonium,* vessel).

In *Oedogonium*, the thallus consists of long unbranched filaments which are attached when young, though later they become free-floating; in the other genera the filaments are usually branched. Each cell possesses a single nucleus together with an elaborate reticulate chloroplast containing numerous pyrenoids. According to some workers the cell wall contains an outer layer of a chitinous material and there is an inner layer of pectin and cellulose.

Vegetative cell division is so peculiar and characteristic that many accounts of the process have appeared. A thickened three-layer transverse ring develops internally on the cell wall at the anterior end, enlarges and invaginates into the cell cavity. The nucleus migrates to the anterior end where nuclear division commences. The site of the new transverse wall, between the two nuclei, is marked by a region of microtubules, and cross-wall formation occurs by the condensation of vesicles at this region. Concurrent with cross-wall formation is the splitting of the outer cell wall at the region of the thickened ring. The ring stretches and expands upwards forming the lateral cuticle of the new cell. A new outer cell wall is formed on the inside of this cuticle, and fuses with the transverse wall at its periphery.

The old outer wall forms a cap at one end and a bottom sheath at the other, and as successive divisions always occur at the same place, a number of caps develop there and give a characteristic striated appearance to some of the cells. The new upper cell is thus composed of the old anterior cap and a new cell wall derived from the invaginated ring. The bottom cell is comprised of the old cell wall. This method of growth in *Oedogonium* is intercalary, but in the other two genera, as each cell can only divide once, there is usually only a single cap.

Vegetative reproduction commonly occurs by means of fragmentation, while asexual reproduction depends on multiflagellate zoospores (one per cell) or akinetes. In *O. cardiacum* the zoospore liberation rhythm can be affected by temperature and day length. The flagella, which may have one or two rings of basal granules, form a circular ring around an anteriorly situated beak-like structure. This is the typical oedogonian swarmer, and two theories have been put forward to explain its origin: The first theory claims that the group arose independently from flagellate organisms which possessed a ring of flagella. If this is true then there could be very little connection with the other members of the Chlorophyceae. The second theory suggests that several divisions of the two original blepharoplasts and flagella took place, thus resulting in the ring structure. If this is correct then development might well have occurred from a Ulothrichalean type of swarmer.

When the zoospore is ripe, the cell wall ruptures near the upper end and the swarmer is liberated into a delicate mucilaginous vesicle, but this soon disappears and the zoospore can then escape. After remaining motile for about an hour the anterior end becomes attached to the substrate and develops into a holdfast, or else the zoospore flattens to form an almost hemispherical basal cell. The type of holdfast depends on the species and the nature of the substrate, a smooth surface inducing a simple holdfast and a rough surface inducing the development of a branched holdfast. Development of the one-celled germling can proceed along one of two lines, depending on the species:

(a) The single cell divides near the apex by the normal method described above, in which case the basal daughter cell persists as the attachment organ and the upper cell goes on to form the new filament.

(b) The apex of the cell develops a cap and then a cylinder of protoplast grows out pushing it aside, and when the protoplast has reached a certain length, a cross-wall is formed at the junction of the cylinder and the basal cell. The upper cell subsequently develops along the normal lines.

Sexual reproduction is by means of an advanced type of oogamy, the development of sex organs being promoted by an alkaline pH and some nitrogen deficiency. In some species, the oogonia and antheridia are produced on the same plant (monoecious forms): in other species, the oogonia and antheridia appear on different filaments which are morphologically alike (dioecious isomorphic forms). The species belonging to both these groups are termed macrandrous because the male filament is normal in size. There is a third group of species in which the male filament is much reduced and forms dwarf male plants. Such species are heteromorphic and form the nannandrous group. The dwarf males arise from motile androspores which are formed singly in flat discoid cells, the androsporangia, produced by repeated divisions of ordinary vegetative cells.

The androsporangia are formed in the oogonial filament for gynandrosporous species but in filaments that do not bear oogonia for idioandrosporous species. In shape and structure the androspores are small editions of the zoospores, and after swimming about they settle on the wall of an adjacent cell and germinate into a small male plant, which is composed of a rhizoidal holdfast with one or two flat antheridial cells above, though in some cases only one antheridial cell without any rhizoidal portion is formed.

Usually two antherozoids are freed from each antheridium into a delicate vesicle which later dissolves. The antherozoids are also like small zoospores, and if they fail to enter an ovum immediately they may remain motile for as long as thirteen hours. In the macrandrous monoecious species, the antheridia are usually to be found immediately below the oogonia, where they arise by an ordinary vegetative division in which the upper cell subsequently continues to divide rapidly, thus producing a series of from two to forty antheridia. The antheridia frequently develop one day later than the oogonia, thus ensuring cross-fertilization.

The oogonia are enlarged spherical or ellipsoidal cells arising by one division in which the upper segment forms the oogonium and the lower a support cell, or else the latter subsequently divides to give antheridia. In some species, the lower cell may also become an oogonium so that a series of oogonia can form on one filament. Each oogonium contains one ovum with a colourless receptive spot situated opposite to the opening in the oogonium wall from which a small quantity of mucilage is extruded. The opening is either a very small pore, formed by gelatinisation of a tiny papilla, or else a slit, but in either case there is an internal membrane forming a conduit to the ovum. After fertilization, the oospore often becomes reddish in colour and develops a thick membrane which is usually composed of three layers. The markings on the outer membrane are important taxonomically.

At germination, the protoplast divides into four segments, which may each develop flagella and escape as zoospores, or else they function as aplanospores that later give rise to zoospores. Meiosis takes place at the germination of the zygote so that the adult filaments are haploid. In one species, it has been definitely established that two of the zygote segments ultimately develop into male plants and two into female plants. Zygote germination without meiosis is reported in which case large diploid swarmers are formed and these develop into abnormally large threads that are always female. Oogonia appear on these diploid filaments and can be fertilized, but the fate of the zygote is unknown.

It remains to discuss the possible origin of the androspores, and there are two hypotheses that may be considered:

First, the androspore may be equivalent to a second and smaller type of asexual zoospore, such as the microzoospores found in some species of *Ulothrix*, but in the Oedogoniales they can no longer give rise to normal filaments. On this view the nannandrous forms are the more primitive, the macrandrous having been derived by the androsporangium acquiring the capacity to produce antheridia immediately and hence never appearing.

Second, the androspore might be equivalent to a

prematurely liberated an-theridial mother cell which subsequently undergoes further development. On this view, the macrandrous species are the more primitive. At present, there does not appear to be convincing evidence in support of either theory.

Chaetophorales

In this order, the fundamental structure is heterotrichous, that is, it possesses both a basal and erect system of branched threads. Some authorities do not consider this character a sufficient distinction to warrant removal from the Ulotrichales. In some of the genera, reduction has probably taken place and only the basal (e.g. *Ulvelld)* or erect system (e.g. *Draparnaldia* spp.) is now represented. Many of these reduced forms are epiphytic or endophytic upon other algae or phanerogams. Another feature, by no means universal, of the Chaetophorales is the presence of setae (e.g. *Coleochaete)* or hairs (e.g. *Stigeocloniuni).* There is considerable variation among the reproductive cells.

Thus there may be quadriflagellate macro-and microzoospores and quadriflagellate gametes. Aplanospores and akinetes are also known. Sexual reproduction is primarily isogamous, occasionally anisogamous (e.g. *Aphano-chaete)* or oogamous (e.g. *Coleochaete).* The order is important because it is possible that its ancestral forms evolved into the first land plants. There are marked terrestrial tendencies and the genus *Fritschiella* is of special importance in this connection.

Many of the species and varieties are heterotrichous and branched and the plants are commonly surrounded by a layer of mucilage of varying thickness. They can carry aquatic fungi as parasites and this causes cell modification. The chloroplast is band-like and often does not fill the entire cell, especially in the older parts of the thallus. The aerial part bears branches that terminate in a colourless hair, the degree and nature of the branching depending upon illumination, nutrition and the rate of water flow. There is no localised area for cell division in the aerial portion, but in the creeping system only the terminal cells are meristematic. The prostrate system may be *(a)* loosely branched, *(b)* richly and compactly branched.

Swarmers are formed in any cell, but mostly in those of the erect part. In addition to asexual macrozoospores, which are produced singly in cells, there are intermediate quadriflagellate microswarmers (asexual or sexual) and smaller biflagellate swarmers that often seem to develop by parthenogenesis into either hypnospores or new plants. Macrozoospores are mostly produced in summer and microswarmers throughout the year. The zygote, derived from fusion of microswarmers, is said to germinate either to a new filament or to zoospores, and these then give rise to the germlings in which the erect filament arises first and the prostrate portion subsequently or *vice versa*. By increasing the osmotic pressure or by adding toxic salts to the environment the thallus passes into a palmelloid state, while under other conditions akinetes can be formed.

The plants are confined to well-aerated fresh water; they have also been found growing on fish living in stagnant water, but in these cases the movements of the fish presumably provide adequate aeration. Some species occur as epiphytes or endophytes and others seem able to tolerate a wide range of conditions.

In all but one species so far studied the dominant phase in the life history is the haploid, reduction division taking place at germination of the orange or bright-red zygote. In *S. subspinosum*, however Islam (1963) reports that the sexual plant, reproducing by gametes, gives rise after gamete fusion to an isomorphic diploid generation that reproduces by haploid, quadroflagellate swarmers.

Other genera exhibit signs of reduction, being represented by only the basal portion of the thallus. Such genera are the epiphytic disc-like *Chaetopeltis* (with emergent setae), the plate-like *Pseudendoclonium* and *Ulvella*, and the endophytic and endozoic *Entocladia (Endodermd)* consisting of branched threads ramifying in the host. This process of reduction is also represented in the next example.

This genus was first recorded *(F. tuberosa)* in India on land undergoing a drying process after the monsoon rains. It has now been recorded from a wide variety of soils, not all subject to water logging. It has also been recorded from Burma, the Sudan and Japan. A second species, *F. simplex*, has been recorded from East Pakistan. The mature plant, which forms dark-green clusters on the soil, possesses four different thallus systems. There is a septate rhizoidal system (lacking in *F. simplex)* of elongated colourless cells; these arise from a prostrate system of short congested branches, and since the cells are rich in starch they can be regarded as a storage tissue.

There is a primary projecting system of upright, short-celled, simple or branched threads from which, in *F. tuberosa* only, arises the secondary projecting system of elongate branches with longer cells. In the Sudan plants, this system is reduced because of the high evaporation rate from the soil. This last is the only system that emerges above the soil surface. The chloroplasts are not clearly recognisable in the cells of the prostrate system but they have fewer pyrenoids than do the cells of the erect systems where the plastid is a curved plate-shaped body with 2-8 pyrenoids. The cells are uninucleate.

Spore formation in *F. tuberosa* is restricted to the prostrate system, each cell producing one or more swarmers. Prior to swarmer formation, the prostrate branches form club-shaped endings protected by a cuticle of dark-brown colour. These may be of three kinds: quadriflagellate macro-zoospores (one per cell) formed from the primary erect cells in *F. simplex*; quadri- or biflagelli/e macrozoospores (2 or 4 per cell); numerous biflagellate gametes. The first two kinds are produced on plants other than those that produce gametes so that it is likely that there is alternation of two similar generations. The sexual plants are also probably segregated because gametes will only fuse if they arise from separate plants (McBride, 1968).

During dry conditions, the two erect systems disappear leaving the prostrate system as the perennating structure. The algae is of great interest in tracing the origin of a land flora. Bower's statement of 1935 — Its interest lies in the terrestrial habit, and in the fact that this amphibial plant possesses the essentials of a three-dimensional, photosynthetic sub-aerial thallus originating from a simple filament', still holds good today.

The systematic position of this algae has varied considerably. Some authors have placed it in a special group, the Pleurococcales, but as the algae can occasionally develop branched threads there would seem to be evidence for regarding it as a much reduced member of the Chaetophorales. There are equally sound arguments for the other systematic treatments of the genus, and its place at present must be largely a matter of opinion. *Pleurococcus* is terrestrial and forms a green coat on trees, rocks and soil, growing in situations where it may have to tolerate prolonged desiccation. The cells, which are globose in shape and occasionally branched, are single, or else as many as four may be united into a group. Under certain cultural conditions, branching may be copious. Each cell contains one chloroplast with no pyrenoids. The sole method of reproduction is through vegetative division in three planes. There is probably only one species, *P. naegelii,* all the other so-called species being reduced or modified forms of other algae. The resistance of the cells to desiccation is aided by a highly concentrated cell sap and a capacity to imbibe water directly from the air.

The species grow as epiphytes or on stones in damp tropical and subtropical regions, but they will also grow under temperate conditions if there is an adequate supply of moisture. The threads have a characteristic orange-red colouring. The cells contain discoid or band-shaped chloroplasts and lack pyrenoids. Usually, both prostrate and erect threads are present, though the latter are reduced in some species.

Growth is apical, and the terminal cells often bear a pectose cap or series of caps which are periodically shed and replaced by new ones. The origin of the cap is not properly understood but is thought to be due to a secretion. The cellulose walls are frequently thickened by parallel or divergent stratifications, while each septum between the cells may also have a single large pit which is penetrated by a protoplasmic strand. The cells are uninucleate when young and multinucleate when old. Vegetative reproduction is through fragmentation, while other means of reproduction are to be found in three different types of sporangia:

(a) Sessile sporangia that have become detached. These consist of enlarged cells which develop in almost any position and they produce bi-flagellate swarmers that may be isogametes.

(b) Stalked terminal or lateral sporangia that are cut off from an enlarged support cell which may give rise to several such bodies. The apical portion swells out to form the sporangium and underneath cuts off a stalk cell that frequently becomes bent. The dividing septum possesses two ring shaped cellulose thickenings which may play a part in the detachment of the mature sporangium. The detached sporangium is blown away and germinates under favourable conditions to give quadriflagellate swarmers.

(c) Funnel-shaped sporangia which are cut off at the apex of a cylindrical cell, the outer wall splitting later at the septum, thus liberating the sporangium, the subsequent fate of which is unknown. The sessile and stalked sporangia may occur on the same plant or else on separateplants. There has been no cytological work to show whether there is any alternation of generations. In one species reproduction is whollyby means of aplanospores.

Within the Trentepohliaceae there are several genera, the species of which form a green discolouration on and in stones and shells of gastropods. Such genera are *Gongrosira* and

Gomontia. These are very difficult to determine taxonomically and some may represent stages in the life history of other algae. Recent work on *Gomontia polyrhiza* has shown that it has a life cycle of a large unicell alternating with a plate-like disc. Species of the allied genus *Cephaleuros* grow as epiphytes or parasites on the leaves of various phanerogams. *C. virescens* forms the red rust of the tea plant and can cause serious economic losses, especially when the tea bush is growing slowly. Bushes can be made less susceptible by treating the soil with potash.

Most of the species are freshwater epiphytes attached to the host by small outgrowths from the basal walls, but there is one species endophytic in *Nitella*. Some of the species are truly heterotrichous while others only possess the prostrate basal portion, which is either composed of loosely branched threads or is a compact disc. The growth of the erect filaments is by means of an apical cell and the basal cushion possesses a marginal meristem. Each cell contains one chloroplast with one or two pyrenoids, and the characteristic sheathed bristle which arises from each cell may be broken off in the old plants.

These bristles develop above a pore in the cell wall through which the protoplast extrudes, and at the same time a membrane is secreted over the protruding bare protoplast (McBride, 1968). Asexual reproduction takes place in spring and early summer by means of biflagellate zoospores which have no eye-spot and are produced singly. After a motile phase the zoospore settles down and divides either horizontally, when the upper segment develops into a hair and the lower forms the embryo disc, or vertically, when each segment grows out laterally; in either case hair formation takes place at a very early stage.

Sexual reproduction is by means of a specialised oogamy, some of the species being dioecious and the remainder monoecious. The female organs, or carpogonia, are borne on short lateral branches and are later displaced. Each carpogonium possesses a short neck or trichogyne (the long neck of *Coleo-chaete scutata* being an exception) the top of which bursts when the carpogonium is mature. In the disc forms, the carpogonia originate as terminal bodies on the outside of the disc, but as the neighbouring cells grow they eventually become surrounded and appear to be in the older part of the thallus. The antheridia develop in clusters at the end of branches (C. *pulvinata)* or from prostrate cells. They finally appear as small outgrowths cut off from a mother cell. Each antheridium produces one biflagellate colourless antherozoid.

After fertilization, the neck of the carpogonium is cut off and the basal part enlarges; branches arise from the underlying cells and eventually surround the oospore where they form a red or reddish-brown wall, though in the disc forms this wall is only formed on the side away from the substrate. At the same time, the enclosed oospore develops a thick brown wall and the cells of the outer envelope then die. The oospore hibernates until spring when it becomes green and divides into sixteen or thirty-two cells, and these, when the wall bursts, each give rise to a single swarmer which is a zoospore. Meiosis takes place at the segmentation of the zygote so that there is only the haploid generation.

On the other hand, some observers have recorded the development of dwarf asexual plants before the reappearance of new sexual ones; if this is correct it may mean that there is an alternation of two unlike generations, an unusual phenomenon in the Chlorophyceae. Under certain conditions the cells will also produce aplanospores. The relation of this genus, with its advanced oogamy, to the other green algae is by no means clear.

Siphonocladales

Until 1935, this represented a well-established order, but in that year Fritsch placed most of the genera as septate member of the Siphonales but retained the Cladophoraceae as a separate order, the Cladophorales, with affinities to the Ulotrichales. In 1938, Feldmann re-established the order and suggested a relationship on the one hand with the Siphonales via *Valonia* and *Halicystis* and on the other hand with *Chaetophora* and *Ulothrix*. Egerod (1952) and, more recently, Nizamuddin (1964) have accepted the order but, unlike Feldmann, have continued to exclude the Cladophoraceae from it. There does not seem to be justification for this exclusion, and indeed the families in the order present a more or less orderly sequence of evolution, commencing from the Cladophoraceae. The members of the order are all septate at some stage of their existence, the cells or segments being multinucleate and possessing reticulate chloroplasts. Such plants are said to be partially coenocytic. A number of the more advanced genera have a specialised mode of cell division known as segregative cell division. In this process, the contents of a single cell become separated by a number of cell walls so that one parent gives rise to a number of new cells which then enlarge.

Morphologically, the order can be subdivided into the filamentous and vesicular forms. The former comprise genera such as *Lola, Rhizoclonium, Chaetomorpha, Cladophora* and *Microdictyon*, and the latter contain *Siphonocladus, Valonia, Boodlea* and *Dictyosphaeria*. It is the vesicular forms that mostly possess segregative cell division.

In one of the simplest genera, *Chaetomorpha*, the plants consist of un-branched filaments which are either attached or free-floating.

The genus *Rama*, with species mainly in the South Pacific, possesses very long branches and can be mistaken for *Chaetomorpha*. It also has short rhizoids at the base and because of these characters the genus appears to occupy a significant phylogenetic position. A genus that is often mistaken for *Chaetomorpha* is *Urospora (Hormiscia)*. The most important feature is the presence of a 'Codiolum' stage in the life history, and in this respect it is possibly important in the evolution of *Spongomorpha*.

The genus *Cladophora* occurs in both fresh and saline waters. Most species are attached by branched septate rhizoids, but some of them (e.g. *C. fractd)* may become free-floating later, and there is one section *(Aegagropild)* which is commonly free-living, the species forming ball-like growths. The attachment rhizoids in certain species form a prostrate thallus *(C. kosterf)* while in a few *(C. basiramosa, C. pygmaed)* the rhizoids are replaced by a single, disciform holdfast. The colour of the frond varies from whitish-yellow to dark-, almost black-green. This feature is used taxonomically, but Van den Hoek (1963) has pointed out that colour varies with light intensity.

Growth is of Three Types: Wholly apical, apical and intercalary, or wholly intercalary. Branches can arise either laterally or terminally. In the latter case, the branch often grows into an almost horizontal position, a process termed evection, which gives the appearance of a dichotomy. The cells are multi-nucleate, the number of nuclei possibly being related to the cell volume. The chloroplasts generally appear as a complex of disc-like structures united into a reticulum containing numerous pyrenoids. The plastids are parietal and bear processes projecting into the central vacuole.

The wall in *Cladophora* consists of alternate layers of longitudinal and horizontal microfibrils of cellulose. In young cells, these fibrils are arranged in spirals but as the cell elongates, they become pulled out to lie almost transversely and longitudinally. There is very little production of mucilage, and this probably accounts for the dense epiphytic flora of minute species that is frequently found associated with plants of this genus. The genus is divided into a number of sections based upon various criteria. In the most primitive (sect. *Affines* and *Basicladid)* branching is restricted to the base and the branches are very long. In this respect, they approach the genus *Rama*.

In the *Aegagropila* group the species can exist as *(a)* threads, *(b)* cushions and *(c)* balls. The destruction of the old threads in the centre of the ball results in a cavity which may become filled with water, gas or mud. In Lake Soro, the water in April and May is sufficiently free of diatoms for light to penetrate to such an extent that photosynthesis increases and so much gas collects in the centre of these balls that they float to the surface. Their characteristic shape is brought about by a continual rolling motion over the soil surface under the influence of wave action, and hence the 'ball' forms are found near the shore whilst the 'thread' and 'cushion' forms are to be found farther out in deeper water where there is less motion. The harder the floor, the more regular is the shape of the balls, but even so the ball structure would also appear to be inherent in the *Aegagropila* because samples have been kept in a laboratory for eight years without losing their shape. The following types of branches have been recognised in the *Aegagropila* forms: *(a)* rhizoids, *(b)* cirrhoids, both these and the rhizoids being neutral or non-reproductive branches, *(c)* stolons or vegetative reproductive branches.

In the section *Aegagropila,* most of the species reproduce vegetatively, but biflagellate swarmers have been reported for one species, *Ae. sauteri,* and these are interesting in that they may germinate while still within the sporangium. Asexual reproduction in the other species, excluding the section *Aegagropila,* is by means of quadriflagellate zoospores which escape through a small pore in the cell wall. In some of the freshwater species certain cells may become swollen to form akinetes in which the walls are thickened and food is stored. Biflagellate isogametes are the means of sexual reproduction, all the species so far investigated being dioecious. The zygote develops at once without a resting period. In a number of species, alternation of two morphologically identical haploid and diploid generations has now been established with meiosis taking place at zoospore formation. In one or two cases, e.g. *Cladophora flavescens,* the zoospores sometimes fuse, and this irregular behaviour is very comparable to similar phenomena found in the more primitive brown algae. In one species, *C. parriaudii,* reproduction is by 4-flagellate zoospores only.

In a few species, there can be an odd or heterochromosome, and in a cell the number of zoospores with the odd chromosome are equal to the number lacking it. Haploid plants of *C. suhriana* have six or seven chromosomes, but in C. *repens* the cells contain either four or five. In a freshwater species, *C. glomerata*, a wholly different type of life cycle has been reported, and this difference may perhaps be compared with the various cycles found for *Ectocarpus siliculosus* under different conditions. Gametes and zoospores are both formed on diploid plants and meiosis takes place at gamete formation so that there is no haploid generation. Although zoospore formation takes place all the year round gametes only appear in the spring, and the reason for this seasonal restriction is not understood. Van den Hoek (1963) considers that the evidence for gametes in *C. glomerata* is not wholly convincing and the life history therefore, needs substantiation. Parthenogenetic development of gametes has also been recorded in a number of species. Of the species so far investigated, the chromosomes appear to be present in multiples of 4 though Schussnig (1954) argues for 12, and this probably indicates polyploidy.

The species *of Spongomorpha* and *Acrosiphonia* were originally regarded as species of *Cladophora*. Recent work indicates that a separate family is justified for them and the genus *Urospora*. The plants differ in their cell-wall structure from *Cladophora* in that the main substance is allied to cellulose II and the microfibrils are arranged at random. In appearance, the plants look like a *Cladophora* but there are generally numerous adventitious rhizoids that intertwine and hold the plants together into a spongy mass. The other distinguishing feature is the rim-shaped septum. The plastid is an undivided perforate cylinder and the pyrenoids are pyramidal instead of lens shaped as in *Cladophora*.

A major reason for the segregation of these species is the life history. It was shown for some species of *Urospora* and also for species of *Spongomorpha* and *Acrosiphonia* that the gametes gave rise to a unicellular 'Codiolum' asexual stage. *S. lanosa* has *Codiolum petrocelidis,* an endoparasite of the red algae *Petrocelis,* for the alternate generation. *Coalita* on the Pacific Coast of North America has another *Petrocelis* endophyte (incorrectly called *C. petrocelidis)* for its alternate generation.

However, not all species of *Spongomorpha* and *Urospora* possess this type of life history and in *Acrosiphonia* there seems to be a typical isomorphic alternation. It is possible that variation in culture conditions may play some part in determining the presence of a 'Codiolum' phase. This would possibly account for the reported 'Codiolum' phase in *Cladophora rupestris*. A further peculiarity reported for *Acrosiphonia incurva* is that when the gametes fuse the nuclei may remain separate in the zygote (dikaryon condition). This behaviour is so unusual that it cannot be accepted without further confirmation.

Branching in these plants is usually multiple in one plane, and sooner or later cells of the branches become attached to those of other branches by means of special pads or tenaculae, so that a flat net-like thallus eventually results.

M. mutabile from New Zealand can exist in a spongiose form with very few anastomoses so that it can easily be mistaken for a coarse species of *Clado-phora*, though it also gives rise to the typical net thallus. In those species that have been investigated, sexual reproduction is isogamous and there is alternation of isomorphic generations. The same is also true of *Anadyomene stellata* and *A. wrightii* (Enomoto and Hirose, 1970).

In this genus, which is restricted to warm waters, the young coenocyte consists of one or more large vesicles which, as they mature, become divided up into a number of multinucleate segments. In some respects, therefore, the genus provides a link with the Siphonales. The macroscopic club-shaped vesicle is attached to the substrate by rhizoids of various types. There is a lobed chloroplast that congregates with the cytoplasm at certain points in the older plants and then each group is cut off by a membrane to produce a number of marginal cells (segregative cell division). The cells do not necessarily form a continuous layer and are frequently restricted to the basal region where they may develop rhizoids, while in other species they are order. The thallus body is first characterised by the whorled (verticillate) arrangement of the lateral branches (in *Acetabularia* they are fused to form a disc), and secondly by the fact that the plastids are independent, disc-like, and not reticulate as in the Siphonocladales. In the species that have been investigated the vegetative thallus appears to be uninucleate. Before the gametes are formed, the single primary nucleus fragments into smaller portions which may subsequently change shape either in the rhizoids or when moving up the main axis because the small daughter nuclei eventually migrate into the gametangia. This peculiar behaviour has been established for *Acetabularia, Cymopolia,* and *Neomeris*. It is thought that all adult plants are diploid. In some genera, the gametes are produced from special cysts (cf. *Acetabularia)*.

The order is very ancient and has apparently not changed greatly as a large number of fossil forms are known (cf, p. 318) and the living species of *Neomeris* and *Bornetella* are very like some of the extinct genera.

The genera are primarily confined to warm waters and in many there is extensive calcification.

This is a lime-encrusted genus which is confined to warm waters, extending as far north as the Mediterranean. The plants consist of an erect, elongate axis bearing one or more whorls of branched sterile laterals with a single fertile whorl at the apex. The sterile whorl or whorls are frequently shed in the adult plant leaving a mark or annulus on the stem where they were formerly attached. The fertile whorl is composed of a series of long sac-like gametangia which are commonly fused, though they are sometimes separate; these are borne on short basal segments which are morphologically equivalent to the primary branches. The basal segments also bear small projections on their upper surface, and these are with or without hairs which form the *corona;* in one section of the genus there is also an inferior corona on the lower surface. In *A. mediterranea* two or three years elapse before the plant attains maturity. In the first year the branched holdfast produces an upright umbilical thread, together with a thin-walled, lobed outgrowth that penetrates the substrate to act as the perennating organ. The aerial part dies, and in the next year or years a new cylinder arises that bears one or more sterile whorls of branches, until in the third or even a later year, a shoot develops which produces one deciduous sterile whorl and a single fertile whorl or umbrella. Each sac-like gametangial umbrella lobe produces a number of multi-nucleate cysts which are eventually set free through disintegration of the anterior end of the gametangium. In the spring biflagellate isogametes are liberated from these cysts and fuse in pairs. In *A. moebii*, the cysts give rise to anisogametes. Use has been made of the uninucleate condition to determine the effect on morphology of varying ratios of nuclei of different species. This is possible because part of one species can be grafted onto another.

Derbesiales

The members of this order were formerly placed in the Siphonales, but the discovery of the remarkable life history, with an alternation of two unlike generations has resulted in the establishment of a separate order by Feldmann.

The gametophytic plants consist of an oval vesicle, up to 3 cm in diameter, arising from a slender branched tuberous rhizoid embedded in calcareous *Lithothamnion* growing at or below low-tide mark. There are only a few species all containing numerous nuclei in the peripheral cytoplasm. There does not appear to be any cellulose in the cell wall, though the wall is polylamellar as in *Valonia*. Growth of the *Halicystis* vesicles is very slow and they are shed at the end of the growing season by abscission, new vesicles arising later from the perennating rhizoid, and in this manner regeneration may go on for several years. In *boergesenii* from India as many as 6-12 annual scars can be found. Swarmers develop in the cytoplasm at the apex of the vesicle in an area which becomes cut off by a thin cytoplasmic membrane, the area thus cut off representing a gametangium. Macro- and micro-gametes are formed and in the early hours of the morning they are forcibly discharged through one or more pores. Gamete formation appears to be under the control of an endogenous rhythm with a basic period of about four days.

Several crops of these swarmers are produced as cytoplasm moves into the apical areas at bi-weekly intervals coincident with the spring tidal cycles. The zygote in *H. ovalis* germinates into a branched protonemal thread that in three months develops into a typical *'Derbesia'* plant with the erect aerial filaments arising from the basal rhizoidal portion.

It has been demonstrated that *Halicystis ovalis* and *Derbesia marina* are simply stages in the life cycle of one algae. The mature *Derbesia* threads produce zoospores that germinate into prostrate filaments; these later give rise to slender branched rhizoids which, after eight months, produce the characteristic *Halicystis* bladder. Although the cytology of the two plants has not yet been worked out, the *Derbesia* generation is presumably diploid and the *Halicystis* haploid. It also remains to be ascertained whether all the other species of *Halicystis* have a similar life cycle. It is possible that the same kind of life cycle holds for *Halicystis parvula* and *Derbesia tenuissima*. A spherical *Halicystis* has recently been reported from New Zealand and this probably has as its alternate generation *Derbesia novac-zelandiae* which has been known for a number of years.

Siphonales

The main characteristic of this order is its coenocytic structure in which true septa are rare or absent. The coenocyte normally has a cytoplasmic ining surrounding a central vacuole and containing numerous discoid chloroplasts. The pigment siphonoxanthin is characteristic of the order. There are suborders: the Eusiphoneae (Bryopsidaceae, Codiaceae) where the membrane contains mannan and there is only one type of plastid, and the Caulerpeae (Caulerpaceae, Udoteaceae) where the cellulose of the membrane is replaced by xylan and where amylogenic leucoplasts are present as well as chloroplasts. Some workers suggest that the difference is sufficient to justify the establishment of separate orders. In all cases so far studied the thallus is diploid. Except for some species of *Halimeda*, sexual reproduction is anisogamous. In some of the tropical genera, e.g. *Halimeda*, there is intensive lime deposition.

Representatives of the order are known to have existed from very early times.

Most of the species of this genus are restricted to warmer seas, though a few, of which *B. plumosa* is the most common, occur in colder waters. The principal axis, which is often naked in its lower part, arises from an inconspicuous, filamentous, branched rhizome that creeps along the substrate to which it is attached by rhizoids. The bi- or tripinnate fronds either have branching confined to one plane or are radially branched, the branches being constricted at the point of origin, and here the cell membrane is also thickened. In *B. pennata* the microfibrils forming the wall are arranged transversely and if this holds true for the genus it differs from the Siphonocladales. The cytoplasm in the main axis and branches frequently exhibits streaming movements.

The function of the rhizome, especially in warmer waters, is probably that of a perennating organ, although vegetative multiplication can also occur through abstriction of the pinnae, which then develop rhizoids at their lower end. The only other known method of reproduction is sexual. The plants are dioecious and produce biflagellate anisogametes which develop in gametangia that are cut off from the parent thallus by means of septa. The microgametes differ from the macrogametes in that they lack pyrenoids. Gamete formation is hastened by moving the plants from light to dark or by changing the concentration of the nutrient solution. The gametes are liberated through gelatinisation of the apex of the gametangium and, after fusion has taken place, the zygote germinates at once into a new plant.

The plants are diploid and meiosis takes place at gamete formation. Inversion of the thallus takes place under conditions of dull light or when it is planted upside down, and under these circumstances the apices of the pinnae develop rhizoids. This exhibition of polarity indicates clearly that the thallus is differentiated internally, but it is still a matter for speculation how such differentiation can occur in an organism which is to all intents and purposes one unit.

This is a widely distributed, non-calcareous genus with species living in warm and cold oceans. The spongy thallus, which is anchored either by a basal disc or by rhizoids, varies greatly in form and appears as erect, branched, worm-like threads (C. *tomentosum* group), flat cushions (C. *adhaerens* group), or as large round balls *(C. bursa* group). In *C. tomentosum* there is a central medulla of narrow forked threads and a peripheral cortex of club-shaped utricles which are the swollen apices of the forked threads. Some of the large-utricled species (C. *cranwelliae*, C. *megalophysum*, C. *papenfussii)* are restricted to caves and other shady places and this may indicate relationship between utricle size and light intensity.

Deciduous hairs may develop on the vesicle and scars are to be seen marking their point of attachment; annular thickenings occur at the base of each vesicle and at the bases of the lateral branches, although a fine pore is left for intercommunication. The width of these pores in the case of *C. bursa* is said to vary with the season. Detachable propagules develop on the vesicle and form a method of vegetative reproduction, while sexual reproduction is by means of gametes, which are produced in ovoid gametangia that arise from the vesicles as lateral outgrowths, each being cut off by a septum. The plants are anisogamous, the macrogamete mass being green and the microgamete mass yellow.

Some of the species are dioecious and others are monoecious, and in two of the latter the male and female gametangia are borne on the same utricles. In C. *elonga-tum* it would appear that the determination of the sex may be a seasonal phenomenon, females appearing first, then hermaphrodites and finally males. The gametes fuse or else develop parthenogenetically, but in either case a single thread-like protonema develops and this has a lobed basal portion, from which the adult develops through the growth of numerous ramifications of the one or more primary filaments. Meiosis occurs at gametogenesis.

This wholly tropical genus has existed from Tertiary time,

and it has played a considerable part in the formation of coral reefs and sands where the species are very abundant. The plants are borne on a short basal stalk that arises from a prostrate system of creeping rhizoids which can also function for vegetative reproduction (Kamura, 1966). The branched aerial thallus is composed of flat, cordate or reniform segments which are strongly calcified on the outside, the segments being separated from each other by non-calcified constrictions. Branching in some species is restricted to one plane and the size of the segments varies greatly from species to species and with water depth.

The behaviour of the medullary filaments at the nodes, whether remaining separate, fusing into small units or into a single one, is a basic taxonomic character. The segments are composed of interwoven threads with lateral branches that develop perpendicularly and produce a surface of hexagonal facets through fusion of the swollen ends.

Reproductive organs develop at the ends of forked threads which vary greatly in their mode of branching: these threads, which are cut off from the parent thallus by basal plugs, arise from the surface of the segments or, more frequently, are confined to the edges. Isomorphic biflagellate swarmers are liberated from male and female gametangia on separate plants, but in *H. incrassata*, the gametes are anisogamous (Kamura, 1966). It is likely that reproduction usually occurs vegetatively by means of underground rhizoidal filaments (Colinvaux, 1968).

Most species frequent the quiet shallow waters of the tropics where they are often rooted in sand or mud. The prostrate rhizome is attached by means of colourless rhizoids and gives rise to numerous, erect, assimilatory shoots with apical growth, the form and arrangement of which varies very considerably. Radial branching is regarded as primitive, and the more evolved forms which are found in quieter waters possess a bilateral branching system. The genus has been divided by Borgesen into three groups:

(a) The species of this group, which grow where there is

plenty of mud, possess rhizomes that are vertical or oblique to enable them to reach the surface even when covered successively by mud (e.g. *C. verticillatd).*

(b) The rhizome in these species branches first at some distance from its point of origin; it possesses a pointed apex which aids in boring through sand or mud (e.g. *C. cupressoides).*

(c) The rhizome is richly branched immediately from its point of origin and the various species are principally found attached to rocks and coral reefs (e.g. *C. racemosd).*

It has also been shown that the form of the thallus in some of the species depends upon the conditions of the habitat, and this is particularly well illustrated by the plastic *C. cupressoides* and *C. racemosa:* In exposed situations, the plants are small and stoutly built; in more sheltered habitats the shoots are longer and more branched; in deep water the plants are verylarge with richly branched flagellate shoots.

The coenocyte is traversed by numerous cylindrical skeletal strands, or *trabeculae,* which are more or less perpendicular to the surface and which are most highly developed in the rhizomes. The function of the trabeculae, which increase in thickness at the same time as the walls by successive deposition of callose, is uncertain. They may be for mechanical support or they may be concerned with diffusion, because movement of mineral salts is more rapid through these strands than through the cytoplasm; it is possible that they serve no useful purpose.

In addition to the trabeculae, there are internal peg-like projections. Vegetative reproduction occurs through fragmentation of the old rhizome thus leaving a number of separate plants. In those species in which reproduction has been studied (e.g. C. *prolifera)* the anisogametes are formed in the aerial portions and are liberated through special papillae that develop on the frond. The sexual reproductive fronds have a variegated appearance caused by the massing of the biflagellate gametes at the different points. In certain species, the whole plant can produce swarmers, while in others the reproductive area is limited, and in such cases the morphological identity and differentiation of the frond becomes of great interest. The thallus can be regarded as being composed of a number of individual cells which only become evident at gametogenesis. Fusion between the swarmers has been observed in *C. racemosa,* and it is probable that in all the species, the motile bodies are functional gametes and that the adult plants are diploid.

Dichotomosiphonales

The sole member of this order differs from the Siphonales in the following respects: first and foremost sexual reproduction is by a distinct oogamy, secondly the chloroplasts lack pyrenoids and starch is deposited between the lens-shaped plastids instead of on them. The characteristic siphonalean pigment siphonoxanthin is present, but the cell wall differs in that it lacks cellulose (Siddiqui and Nizamuddin, 1965). After fertilization, the zygote rests for a period before germination whereas in the Siphonales it develops directly into a new plant.

This is a monotypic genus, the sole species, *D. tuberosus,* being found in fresh waters of Europe, the USA., India, Burma and Pakistan. The thallus looks like *Derbesia* or *Vaucheria* with quadrichotomous branching. Where branching occurs the filaments bear constrictions with an ingrowth of the wall material as well. Asexual reproduction is by means of club-shaped akinetes formed usually at the ends of lateral rhizoidal branches which are most abundant towards the base. These akinetes are densely packed with starch. The plants are monoecious and both sex organs are borne terminally on the ultimate branchlets. After fertilization the zygote, which develops a two-layered wall, remains quiescent within the oogonium for some time.

Zygnematales

The members of this order are distinct from the other members of the Chlorophyceae as a whole. The plants are either filamentous or single-celled with a wall of one or two parts. There is considerable variation in form of the chloroplasts. The species are all characterised by a peculiar mode of sexual reproduction in which fusion takes place between masses of protoplasm which do not possess flagella and in this process cells come together in pairs and conjugate. Some authorities have suggested that the conjugate Chlorophyceae should be assigned to a separate class, Conjugatophyceae, because of the unusual reproduction, and this may well be justified.

The order has been divided into two suborders, Euconjugatae and Desmi-dioideae, or into two families, Zygnemataceae and Desmidiaceae. A recent revision creates three families, Mesotaeniaceae, Zygnemataceae and Desmidiaceae. The Mesotaeniaceae are single-celled species and are sometimes known as the saccoderm desmids. Most of the Desmidiaceae are also single-celled with a wall in two parts and are often known as the placoderm desmids. There are two views concerning the inter-relationships of the families: one is that both desmids are primitive to the filamentous Zygnemataceae, and the other that they are derived from the filamentous Zygnemataceae. Some authorities recognise two orders:

(a) Desmidiales, whose species are predominantly unicellular with cell wall and contents usually in two parts.

(b) Zygnematales, whose species are filamentous with no division of wall or cell contents into two parts.

Within the Zygnemataceae there are six types of gamete and zygospore formation.

(a) *Spirogyra Type:* Vegetative cells become gametangial cells and the entire protoplasm is transformed to a gamete.

(b) *Mougeotiella Type:* Vegetative cells become gametangial cells but a portion of the protoplasm persists as a cytoplasmic residue.

(c) *Zygnemopsis Type:* Vegetative cells become gametangial cells but some of the contents develop into a bluish-white gelatinous substance. In this and the preceding type the conjugation tubes enlarge considerably.

(d) *Mougeotia Type:* These resemble the *Mougeotiella* type but differ from it in that the zygospore is cut off from the gametangial cells by special walls.

(e) *Temnogyra Type:* The progametangial cells divide first into a shorter gametangial cell and a longer sterile cell. The entire protoplasm is used in the gametangial cells and there is a distinct conjugation tube.

(f) *Sirogonium Type:* Resembles the *Temnogyra* type but differs in that no conjugation tube is formed. When the gametangial cells make contact, a gelatinous substance is secreted and forms a disc.

The Zygnemataceae are conveniently placed into three sub-families based upon chloroplast type: the Zygnematoideae with stellate chloroplasts, the Mougeotioideae with flat plate-like chloroplasts and the Spirogyroideae with ribbon-like chloroplasts. Within each of these subfamilies most of the six gamete and zygospore formation types just described are to be found. In some classifications *Gonatozygon* and *Genicularia* are placed in a separate family but Yamagishi (1963) has argued that this is not justified.

The unbranched filaments are normally free-living although attached forms are known, e.g. *S. adnata,* and they form slimy threads known as 'water-silk' or 'mermaid's tresses'. They grow in stagnant or slowly moving water and are most abundant in either spring or autumn, the latter phase being due to the germination of a percentage of the spring zygospores. Each cell contains one or more chloroplasts possessing either a smooth or serrate margin and arranged in a characteristic parietal spiral band. The single nucleus is suspended in the middle of the large central vacuole by means of protoplasmic threads that radiate out to the parietal protoplasm. The chloroplasts, which may occasionally be branched, are T- or U-shaped in cross-section and contain numerous pyrenoids that project into the vacuole on the inner side, the majority of the pyrenoids arising *de novo* at cell division. The cell wall is thin and, according to some investigators, composed of two cellulose layers, while others maintain that there is only an inner cellulose layer with an outer layer of some other material. The whole filament is enclosed in a mucilage sheath of pectose. Any cell is capable of division, and vegetative reproduction by fragmentation is exceedingly common, three methods having been described:

(a) The septum between two cells splits and a mucilaginous jelly develops in between, so that when one cell subsequently develops a high turgor pressure the cells become forced apart.

(b) Ring-like projections develop on both sides of a septum and the middle lamella dissolves. Then the rings of one cell evaginate and force the cells apart while the rings of the other cell evaginate after separation (replicate fragmentation).

(c) The septum develops an 1-piece and then when the wall inverts, due to increased turgor, the l-piece is slipped off and the two cells come apart.

When two filaments touch they may form joints or geniculations and adhesion is brought about by a mucilaginous secretion produced by the stimulation of the contact. The formation of such geniculations, however, has no connection with reproduction.

Sexual reproduction occurs by conjugation, which is stimulated by a combination of certain internal physiological factors and the pH of the external medium. It commonly takes place during the spring when the threads come together in pairs, but more than two filaments may be involved. The threads first come together by slow movements, probably related to the secretion of mucilage; then they become glued together by their mucilage and later young and recently formed cells in both filaments put out papillae. These papillae meet almost immediately, elongate, and push the threads apart. Normally one of the threads produces male gametes and the other female, but occasionally the filaments may contain mixed cells. The papillae from male cells are usually longer and thinner than those from the female cells. The conjugating cells accumulate a great deal of starch, the nuclei decrease in size and the wall separating the papillae breaks to form a conjugation tube. The whole process so far described forms the maturation phase which is followed by the phase of gametic union.

The protoplasm of the male cells contracts from the walls and the cytoplasmic mass then migrates through the conjugation tube into the female cell where fusion takes place and this is then followed by contraction of the female cytoplasm, though in the larger species, it may contract before fusion. Fusion of the two nuclei may be delayed for some time, but in any case the male chloroplasts degenerate. The process described above is known as scalariform conjugation. In monoecious species, lateral conjugation occurs between adjoining cells on the same filament. More recently a species *(S. jogensis)* has been described in which contents of adjacent cells fuse, the contents of one cell passing through a pore in the intervening cell wall.

The last phase is that of zygotic contraction which is brought on by further action of the contractile vacuoles, after which a thick three-layered wall develops around the zygote, the middle layer or mesospore frequently being highly sculptured. The zygospore occasionally germinates almost at once, thus producing plants that account for the autumn maximum, but it is usually dormant until the following spring. Meiosis takes place very soon after fusion and four nuclei are formed and of these, three abort so that only the haploid generation exists. A two-celled germling is formed, the lower cell being relatively colourless and rhizoidal in character. Filaments of two different species have been known to fuse, the form of the hybrid zygospore being determined by the characters of the female thread. Azygospores, which have arisen by parthenogenesis, and akinetes form other means of reproduction. Although *Spirogyra* is a fresh-water genus, one species, *S. salina*, grows in brackish water.

The saccoderm desmids are characterised by a smooth wall in one complete piece without any pores. The rod-shaped cells of *Mesotaenium* are single, have no median constriction, and are circular in transverse section. The chloroplast is a flat axile plate containing several pyrenoids, while in some species the presence of a tannin pigment imparts a violet colour. The inner cell wall is composed of cellulose and the outer of pectose. Multiplication takes place by cell division, the daughter cells being liberated by dissolution of the middle lamella after a constriction has been formed, though in some cases this may not occur until a number of cells have been enclosed in a common mucilaginous envelope.

Sexual reproduction is by means of conjugation, two processes being put out just as in the filamentous forms. Meiotic division occurs upon germination, giving rise to four nuclei. The protoplast then divides giving rise to four daughter protoplasts and hence new cells. Occasionally, two nuclei degenerate, resulting in only two new cells. The species are to be found in upland pools, peat bogs or on the soil.

This family comprises the placoderm desmids, in which the wall is highly ornamented and the wall and cell contents are in two sections. Each cell thus comprises two half cells which may be delineated by a narrow (e.g. *Desmidiwri)* or deep (e.g. *Staurastrutri)* fissure or constriction called the sinus. The narrow section between the two half-cells is called the isthmus. In a few genera (e.g. *Closterium)* there is no half-cell arrangement but the cells contents are, however in two sections. The cell wall may be smooth *(Closteriwri)* or heavily ornamented with spiny or granular projections *(Staurastrum).*

The cell wall, which is composed of cellulose and an outer pectin layer, is characterised by numerous, regularly arranged pores. In *Closterium,* the pores are in grooves and mucilage secretion through the larger apical pores is responsible for the jerky cell movement. The half-cells possess one, two or four axial chloroplasts. In *Closterium* the two chloroplasts are conical and curved with several pyrenoids. Some genera (e.g.

Closterium) possess polar vacuoles in which crystals of gypsum are located.

Cell division is peculiar and takes place in a variety of ways. In those genera possessing a sinus, the isthmus elongates and mitotic nuclear division follows. At the centre of the isthmus a new cellulose-pectin wall is laid down and the isthmus enlarges to form two new semi-cells. A new sinus forms in the plane of the old one. The cells now separate at the apical (new) cell wall. If separation does not occur colonial 'filaments' are formed. Each new cell thus comprises one old half-cell and one new half-cell derived from the enlargement of the isthmus. In genera lacking the median constriction the cell chloroplasts and nucleus divide. The daughter nuclei migrate to what will be the median position of the new cells. A cross wall is laid down in the middle of the old cell, and the cell wall on either side of the new wall elongates and assumes the shape of the original cells.

In sexual reproduction papillae from the two cells meet or else the naked amoeboid gametes fuse immediately outside the cells. After the gametes have fused two of the chloroplasts degenerate and the zygospore nucleus divides meiotically. Zygospore division gives rise to two daughter cells, each containing one chloroplast and two nuclei, one of which subsequently degenerates. Under unfavourable conditions asexually produced chlamydospores have been observed in *C. moniliferum.*

The genus is fresh water and many species are planktonic. The group is extremely widespread but is scarce in waters containing much lime, the individual species thriving best in soft or peat bog waters. The most favourable seasons for their development are the late spring and early summer, and their resistance in the vegetative state to adverse conditions would seem to be very great. Recent cytological work on desmids has revealed great variation within clones of the same species. Thus, *Cosmarium cucumis* gave chromosome counts of 44 or 52, and *Netrium digitus* gave counts of 122, 172-82, and 592.

Charophyceae

The class comprises but a single order, the Charales, with more than 300 species. The plants represent a highly specialised group that must have diverged very early in the course of evolution from the rest of the green algae, the intermediate forms subsequently being lost. The young plants develop from a protonemal stage and the erect plants have a structure which is more elaborate than any type so far described while the thallus is sometimes encrusted with lime. The group is known to be very ancient because fossil members are found from almost the earliest strata.

There is no asexual reproduction but the process of oogamous sexual reproduction is very complex. The family is divided into two subfamilies, the Charoideae, in which the characteristic oogonium has a single ring of five coronal cells, and the Nitelloideae where the oogonium is surmounted by two rows of five cells. There are five genera in the former and two in the latter and it is generally agreed that the Charoideae are more advanced than the Nitelloideae. The living forms are widely distributed in quiet waters, fresh or brackish, where they may descend to considerable depths (12 m) so long as the bottom is either sandy or muddy.

The greatest number of species is found in Asia but there is a high degree of endemism in Australia, North America and Asia.

The plants bear whorls of lateral branches arising from the nodes. The nodes are formed by a transverse layer of cells in contra-distinction to the internodes, which consist of one large cell whose individual length may be as much as 25 cm in *Nitella cernua*. The height of the different species varies and in the largest may extend up to 1 m, growth being brought about by an apical cell which cuts off successive segments parallel to the base. Each new segment divides transversely into two halves, the upper developing into a node and the lower into an internode. The nodal cell divides longitudinally and each of the daughter cells divides longitudinally several times so that each node has two central cells and a number of peripheral cells. Each peripheral cell divides again longitudinally. The outer cells give rise to branches by protruding to form new apical cells, but these soon cease to grow after the branch has reached a short length. At the basal node of the main plant branches of unlimited growth are produced: these arise on the inner side of the oldest lateral in the whorl, thus producing a fictitious appearance of axillary branching.

Experimental work with *Nitella cristata* using varying conditions of light, temperature and day length has shown that the length, form and degree of branching of sterile and fertile whorls are influenced by these factors.

Multicellular branched rhizoids with oblique septa function as absorption organs and also serve for anchorage. The rhizoids develop from the lowest node of the main axis, but every node is potentially capable of producing them; normally the presence of the stem apex inhibits their appearance, but if this is cut off the rhizoids will then develop. This behaviour suggests an auxin control similar to that found in the higher plants. The cells, which have a cellulose membrane, contain discoid chloroplasts and a nucleus but no pyrenoids. In *N. opaca*, the microfibrils of the cell wall are usually oriented in broad helices. Cytoplasmic streaming is very readily observed, especially in the internodal cells. Sexual reproduction is by a characteristic oogamy where light intensity plays a part in determining the production of the sex organs. The species are either dioecious or monoecious; in the latter case the oogonia and antheridia are juxtaposed, the oogonia being directed upwards and the antheridia downwards, both organs usually appearing on secondary lateral branches of limited growth.

The apical cell of the lateral branch cuts off one or two discoid cells at the base and then becomes spherical. The upper spherical cell divides into octants and this is followed by two periclinal divisions, after which the whole enlarges and the eight peripheral cells develop carved plates (shields), which give the wall a pseudocellular appearance. At maturity these peripheral cells acquire brilliant orange contents. The uppermost discoid basal cell protrudes into a hollow structure whose formation has already been described. The middle segment of each primary diagonal cell now develops into a rod-shaped structure, the manubrium, which bears at its distal end one or more small cells, the capitula; every one of these produces six secondary capitula from each of which arises a forked spermatogenous thread containing 100-200 cells. These antheridial cells each produce one antherozoid, an elongate body with two flagella situated just behind the apex.

The complete structure has been regarded as one antheridium, though another view, probably more correct, regards the octants as secondary laterals, the manubrium as an internode, the capitula as a basal node and the spermatogenous threads as modified laterals. On this basis the antheridia are one-celled and conform to the normal structure of the antheridia in the green algae. This second interpretation, if it is correct, helps considerably in understanding this peculiar group.

The apical cell of the lateral branch divides twice to produce a row of three cells, the uppermost cell developing into the oogonium while the lowest forms a short stalk. The middle cell cuts off five peripheral cells which grow up in a spiral fashion and invest the oogonium; each one finally cuts off two small coronal cells at the apex. The oogonial cell cuts off three cells at its base and it is maintained that these, together with the oogonium, represent four quadrants, only one of which develops to maturity.

When mature, the investing threads part slightly to form a neck, and the apex of the oogonium gelatinises to allow the antherozoids to enter. After fertilization the zygote nucleus travels to the apex of the oospore and a coloured cellulose membrane is excreted around it, while the oogonium wall, together with inner walls of the investing threads, thicken and silicify. Four nuclei are formed by two successive divisions of the zygote nucleus, meiosis taking place during this process.

One of these nuclei is cut off by a cell wall and the other three degenerate. The small cell so formed then divides and two threads grow out in opposite directions, one a rhizoid, the other a protonema. The cell next to the basal cell of the protonema divides into three cells, the upper and lower forming nodes which become separated by elongation of the middle cell. The lower node develops rhizoids while the upper produces a whorl of laterals from all the peripheral cells except the oldest, which instead forms the apex of the new plant. The mature plant is therefore, morphologically a branch of the protonema. Vegetative reproduction can take place from secondary protonemata which develop from the primary rhizoid ring or from dormant apices.

Euglenophyceae

This class comprises mainly motile unicells, although there is one small group of palmelloid, non-flagellated species. The organisms lack a cellulose cell wall but there is an external semi-rigid pellicle or periplast. This appears under the electron microscope as a well-differentiated layer of proteinaceous strips with helical symmetry, which in many species is modified by warts and ridges. When the pellicle is rigid, it is often markedly striated, but when it is flexible metabolic changes of shape are possible and permit locomotion. Some cells may also show a peculiar twisted shape, e.g. *Phacus*.

The motile cells possess two, rarely three, flagella. The long flagellum, which bears a lateral row of hairs ('stichonematic type'), emerges from an anterior reservoir and canal (called gullet and cytopharynx in the older literature). The short one does not emerge from the reservoir, around which are one or more contractile vacuoles. The single nucleus is generally conspicuous and located near the centre of the cell.

The majority of the euglenoids are colourless. The pigmented species contain numerous chloroplasts, which are disc-like, rod-like, ribbon-like or stellate, and distributed throughout the cytoplasm. The pigments themselves are chemically similar to those of the Chromophyta. Food reserve is deposited as granules of paramylum, a carbohydrate very similar to laminarin of the Phaeophyceae. Most Euglenoids possess a red eye spot, which differs from that of *Chlamydomonas* in being a collection of carotenoid — containing granules, located outside the chloroplast. Nutrition may be either photoauxotrophic, saprophytic or holozoic (phago trophic). The Euglenophyceae possess an unusual form of nuclear division. At metaphase the nucleolus persists, the chromosomes arrange themselves parallel to the axis of division and there is no evidence for a mitotic spindle. Although meiosis has been reported, there is apparently no sexual reproduction and it is assumed that meiosis occurs at the division of a zygote nucleus derived from nuclear fusion.

The predominant phase is the motile unicell, although non-motile cysts are known and reproduction is often by longitudinal cell division.

Euglenoids occur in both salt and fresh water where they are found in a variety of habitats, sometimes tolerating a very low pH, although they are more abundant in pools rich in organic matter. The saprophytic forms grow best in the presence of plenty of decaying matter, while colourless forms are endozoic. One member has been described as the cause of green snow.

The most recent treatment has divided the class into six orders. The Rhabdomonadales, Sphenomonadales and Heteronematales comprise only colourless species. Nutrition is either saprophytic (Rhabdomonadales: *Rhabdomonas)* phagotrophic (Heteronematales: *Peranemd)* or both (Sphenomonadales: *Petalomonas).* The second flagellum is emerged in most species but not in the Rhabdomonadales. These three orders are further characterised by the absence of eye spots and flagellar swellings; with the exception of the Heteronematales they do not show euglenoid movement.

The remaining Euglenoids, which may be colourless or possess chloroplasts, are divided into the Euglenomorphales,

Euglenales, Eutreptiales. With rare exceptions, nutrition is saprophytic or phototrophic. The saprophytes are a small group of Euglenoids living endosymbiotically in the digestive tract of tadpoles.

The Eutreptiales are characterised by two emergent flagella, one of which extends laterally, but the cells show typical euglenoid movement.

Euglenales

This is by far the largest group of Euglenoids, and includes both coloured (e.g. *Euglena, Phacus, Trachelomonas, Colacium)* and colourless forms (e.g *Astasid)*. Nutrition is typically phototrophic in the coloured forms or holozoic or saprophytic in the colourless forms.

Euglenaceae

This family comprises coloured Euglenoids containing chloroplasts which are usually twisted and elongate. In *E. spirogyra*, the pellicle, which underlies a plasmalemma membrane, is composed of a single unit wound in an interlocking helix around the cell. In cross-section, the unit is composed of grooves and teeth. Arranged parallel with it are a series of mucus bodies, whose function is not exactly known.

Contrary to earlier belief, there are two flagella in most of the pigmented forms. The locomotory one (with lateral hairs down one side) is emergent and usually as long as the cell; the other does not emerge from the reservoir. At the base of the locomotory flagellum is a swelling which may be concerned with the photoreception mechanism. The distinctive red eye spot is situated on the edge of the reservoir immediately opposite the flagellar swelling. Both eye spot and flagellar swelling are absent in *Astasia*. The contractile vacuole (usually only one) is adjacent to the reservoir on the side opposite to the eye spot and discharges into the reservoir. Numerous lenticular chloroplasts are scattered throughout the cell, and pyrenoids may *(E. gracilis)* or may not (*£. spirogyrd)* be present.

Chromosome counts in algae are not always easy but in *£. spirogyra*, the chromosome complement appears to be about 86.

Asexual reproduction is by longitudinal cell division, which may occur either while the cells are motile or resting. Division commences at the anterior end, and where there are non-motile stages, a gelatinous sheath may encircle the dividing cell. In some species, thick-walled cysts have been recorded, and these germinate to give a single motile cell. Sexual reproduction has not been demonstrated with certainty.

Colacaceae

The genus comprises dendroid or palmelloid colonies of immotile cells enclosed in a wall. The species are epizoic and colonies are attached to the host by a gelatinous stalk.

Reproduction is by cell division which differs from that of other Euglenoids by occurring at the non-motile stage. Amoeboid stages are known and sometimes the protoplast of a colonial cell may escape and form a free-swimming flagellated individual that will eventually come to rest, lose its flagella and start a new colony.

Phacus is characterised by a flat leaf-like appearance while *Trachelomonas* possesses a distinctive brown cellular sheath.

Xanthophyta

Xanthophyceae: These algae were originally classified with the Bacillariophyceae and Chrysophyceae in the division Chrysophyta. Similarities to the Chrysophyceae include the presence of chrysolaminarin or oil as food reserves, some aspects of the flagella types, the presence in certain species of a distinct type of spore (the statospore), and also the frequent occurrence of cell walls with two overlapping halves. The principal difference is in pigmentation, as the cells lack a secondary chlorophyll and fucoxanthin and possess specific carotenoids. Although there is undoubtedly some relationship between Xantho-phyceae and Chrysophyceae, it is probably not close enough to warrant their inclusion in the same division.

Within this group, there is an evolutionary trend from the flagellated unicell to the palmelloid state and also through the filamentous to the siphonaceous habit. Morphological development is therefore, analogous to that of the Chlorophyceae.

In certain genera in which a cell wall is present, it may be of two overlapping halves. In the filamentous forms the filaments are composed of articulated H pieces, two such pieces comprising the surrounding wall of an individual cell.

The flagellated cells of the Xanthophyceae usually possess two unequal flagella, the short one being of the simple type, the long one being pleuronematic (tinsel type, 'flimmergeissel'). It is from this heteromorphic arrangement that the class acquired its early name 'Heterokontae'. The chloroplasts, of which, there may be one or more, usually lack pyrenoids, and their yellow-green colour is derived from carotenoids which tend to mask the chlorophyll *a* (the chlorophyll *e*, recorded from *Tribonema bombycinwn* and *Vaucheria haemata* is very probably an *in vitro* product). Asexual reproduction is by cell division, production of motile zoospores, non-motile aplanospores, akinetes, or in some unicellular forms by internal cysts. The statospore or cyst germinates to give two or four zoospores or amoeboid bodies.

All flagellated cells are naked and possess a contractile vacuole system. Sexual reproduction has been recorded for only a few genera, but it is a very distinct feature of the siphonaceous species. The majority of Xanthophyceae are fresh water, sub-aerial (growing on trees, walls etc.) or soil inhabitants and a few are marine.

Recent work has shown that certain species of Xanthophyceae differ in certain important aspects. In the true Xanthophyceae, the eyespot is part of the chloroplast. In the following the eyespot is independent of the chloroplast: *Pleurochloris commuta, P. magna* (Pleurochloridaceae), *Polyedriella, Vischeria stellata, V. punctata, Chlorobotrys* (Chlorobotrydaceae) and *Ellipsoidion acuminatum*. Additionally these species differ in that the zoospores possess only one emergent flagellum, and the chloroplast lacks a girdle lamellae. Golgi bodies and the polygonal pyrenoids are not found in motile cells but are present in vegetative cells. Hibberd and Leedale (1970) have suggested that these algae should be transferred to a new class, the Eustigmatophyceae, though further work is needed before the validity of this class can be established. It does raise however, the important differences within the Xanthophyceae. Other coccoid forms, e.g. *Pleurochloris meiringensis, Botrydiopsis, Bumilleriopsis* are however, 'true' Xanthophytes.

Chloramoebales (Heterochloridales)

The organisms are all naked, motile, flagellated unicells, although temporary conversion to an amoeboid form is not uncommon. The ovoid or pear-shaped cells are uninucleate with one or more discoid chloroplasts. The two flagella are apically inserted, with one or more anterior contractile vacuoles. Reproduction is by cell division of either motile or amoeboid phases or by means of statospores.

Rhizochloridales

This order differs from the preceding one in that the vegetative phase is permanently amoeboid, and motile flagellates occur only in the zoospore stage. The majority of the members are uninucleate with one or more chloroplasts. Reproduction is by cell division. Often the daughter protoplasts do not completely separate but remain connected by their protoplasmic strands.

Chlorarachnion, which contains the biggest species, has uninucleate cells with many chloroplasts and pyrenoids. Nutrition, besides phototrophy, is also holozoic. Cyst formation occurs by rounding off of the amoeboid protoplast and formation of a membrane.

The genus *Myxochloris,* which inhabits the cells of *Sphagnum* moss, is multinucleate and has been reported to reproduce by zoospore-liberating cysts as well as by cell division.

Heterogloeales (Heterocapsales)

This order includes those Xanthophyceae possessing a palmelloid habit. The individual cells, each enclosed in a cell wall, are aggregated into palmelloid colonies within a gelatinous matrix, individual cells being capable of reverting to a motile stage.

Reproduction is either by zoospores liberated from the individual cells or by cell division in one or two planes to give two or four daughter cells. Formation of thick-walled resting statocysts may also occur.

In the fresh water, *Gloeochloris* (*gloeo*, jelly; *ehloris*, green) the uninucleate, ellipsoidal cells, containing two or more discoid chloroplasts, are irregularly arranged within the gelatinous envelope. This genus is either free floating or attached to aquatic phanerogams.

Mischococcales (Heterococcales)

This order represents the coccoid tendency in the Xanthophyceae and corresponds to the Chlorococcales in the Chlorophyceae. The morphological similarities are so close that certain species were originally assigned to the Chlorophyceae and *vice versa*. Examination of its pigments has shown that the Xanthophycean *Chlorocloster engadinensis* is typically Chlorophycean (D. J. Chapman and H. Clemengon, 1965) and therefore, incorrectly classified in the Xanthophyceae. Further biochemical examination of dubiously assigned members is highly desirable and may result in a considerable reshuffling of species.

Unlike the Heterogloeales the vegetative non-motile cells are incapable of reverting to a motile stage.

The cells are enclosed in a cell wall which in some species is of the overlapping halves type. Most species are uninucleate and possess one or more chloroplasts within the cell.

Reproduction is either by means of zoospores or aplanospores; strict vegetative cell division appears to be unknown. The majority of species are free floating, or tree, soil or rock dwellers, but epiphytic species are also known. In terms of genera and species, this is numerically, by far the largest order in the class, but, with few exceptions, the species, which are mainly freshwater inhabitants, are of rare occurrence.

Pleuroch Lori D Ace AE

The family Pleurochloridaceae represents the free-living stage which is an unattached single cell.

Botrydiopsis (like *Botrydium)*, is a widely distributed soil algae. The cells tend towards a spherical shape, and the cell wall is apparently in two sections. The cells are uninucleate with numerous chloroplasts (frequently only a few in growing cells), usually without pyrenoids. Reproduction is by means of four, eight or more zoospores of a typical Xanthophyceae form.

Vegetative cells have been known to form aplanospores. These usually germinate to form other vegetative cells, but may pass to a thick-walled resting stage before germinating to zoospores.

Chlorobotrydaceae

The representatives of this family are single cells in free-living colonies enclosed in a gelatinous sheath. *Chlorobotrys* is an acid or alkaline swamp genus. The individual cells are spherical w:th a few chromatophores. Reproduction is commonly by autospores or thick-walled cysts and zoospore production appears to be the exception rather than the rule.

Botryochloris (Botryochloridaceae) represents a non-mucilaginous, free-living colony of many spherical cells.

Centritractaceae

Bumilleriopsis, is a typical member with single, free-living, unattached cells that are elongated or ellipsoidal. The cell wall is characteristically and distinctly of two halves. Reproduction in this genus (and family) appears to be predominantly by aplanospores.

Mischococcaceae

The members of this family usually form dichotomous dendroid colonies from cells supported on mucilage stalks arranged in a tree-like pattern.

Mischococcus is a freshwater epiphyte. Each mucilage stalk bears at its end a single spherical cell enclosing two chloroplasts. The zoospores are typically Xanthophycean, although those of some species have been reported to possess only one flagellum. When the zoospores settle they give rise to a single spherical cell resting on a mucilage cushion.

This cell then forms a new cell membrane and subsequent secretion of mucilage forms the initial stalk, pushing the cell ahead of it. The original site of the cell and its old cell membrane are marked by a basal swelling. Transverse division of the terminal cell gives two daughter cells that are together supported on a freshly secreted stalk. In some cases, only the upper daughter cell secretes the mucilage, in which case the lower cell remains attached to the old stalk, while the other becomes separated by the freshly secreted intervening mucilage.

If longitudinal division occurs the two daughter cells are borne aloft on a common stalk, but at the next cell division, the resultant two sets of daughter cells form separate stalks and thus produce a dichotomy. Division into four will give rise to occasional tetrachotomy. Continual repetition of these division processes eventually gives rise to the dendroid colonies. Reproduction also takes place by aplanospores or by vegetative propagation of cells detached from old colonies.

Characiopsidaceae

The cells, which are epiphytic, solitary or gregarious, vary greatly in shape, even in pure culture; they develop from a short stalk with a basal mucila-ginous cushion. The wall, composed of cellulose and pectins, is in two unequal portions, the smaller upper part forming a lid which is detached at swarmer formation, while in one species, the lower part bears internal processes. Although the young cells are uninucleate and contain one or more chloroplasts, the adult cells are multinucleate and contain eight to sixty-four nuclei as well as several chloroplasts. Reproduction is either by means of zoospores (eight to sixty-four per cell) or by means of thick-walled aplano-spores, which in one species are said to give rise to motile gametes, although this is a feature that requires further investigation.

Chlorotheciaceae

The species of the freshwater genus *Ophiocytium* are generally elongated or cylindrical, curved, solitary, multinucleate cells of various shapes, or attached colonies of cells. Dendroid colonies are formed in *O. gracilipes* by liberated zoospores attaching themselves to the rim of the parent cell.

There are many similarities between the Characiopsidaceae and the Chlorotheciaceae, the latter probably representing a specialisation of the former. Unlike *Characiopsis,* the cells are permanently multinucleate; like that genus, the cell wall *of* *Ophiocytium* is composed of two very unequal segments.

The lower and larger section of *Ophiocytium* is like an open-ended cylinder with a spine-like projection at the closed end, which serves as the attachment device for the epiphytic species. The open, upper end, through which aplanospores or zoospores are liberated, is covered by the other section, which acts as a small lid.

In addition to the many nuclei, there are numerous chloroplasts, which in certain positions have a distinctive H-shaped appearance.

Tribonematales (H Eterotrich Ales)

All filamentous Xanthophyceae are grouped in this order in which the majority of species are uniseriate and unbranched. Most possess cell walls with characteristic overlapping halves or H pieces. Asexual reproduction is by means of zoospores, aplanospores or statospores. The latter on germination either form new filaments of liberate a pair of zoospores. Akinetes have also been recorded, and sexual reproduction has been claimed for one species.

The order, which is entirely composed of freshwater forms, is divided into two families, in which the important distinguishing feature is the presence or absence of branching.

Tribonemataceae

This is a filamentous analogue to a form such as *Microspora* with which it is frequently confused. *T. bombycinum* sometimes appears in sheets covering ponds and pools. The unbranched threads are composed of cells possessing walls of two equal overlapping halves, with the result that the filaments are open-ended and tend to dissociate into H pieces. At cell division a new H piece arises in the centre and the two halves of the cell separate. Each cell contains one nucleus, although *Tribonema bombycinum* may have two, together with two or more parietal chloroplasts.

Asexual reproduction is by means of heterokont zoospores (2-4 per cell) which are liberated by separation of the cell wall. On coming to rest the zoospore elongates and puts out an attachment process, and in this state it resembles *Characiopsis.* Aplanospores (1-2 per cell) and akinetes, which are formed in chains, also act as a means of propagation. Sexual reproduction is apparently very rare; in this some of the motile bodies come to rest first and are surrounded by other motile gametes. Iron bacteria sometimes live within the algae which becomes yellow or brown from ferric or manganese salts.

The branched habit in this order is represented by the family, Heteroden-draceae, of which *Heterococcus (hetero,* different; *coccus,* sphere), is an example. This occurs as a discoid filamentous expanse or as single isolated filaments or as non-filamentous cell aggregates. The filaments are each composed of a few short squat cells with one or more parietal chloroplasts. The cell wall appears to be composed of one piece, rather than two H pieces. Reproduction can occur by zoospores, akinetes or by aplanospores. In view of some of its anomalous features, it is possible that this family is not properly included in the Tribonematales.

Vaucheriales (Heterosiphonales)

This order includes the Xanthophyceae which are siphonaceous and coenocytic. The plant body is either filamentous or saccate and contains a large number of nuclei and chloroplasts.

Asexual reproduction is by zoospores or aplanospores. Unlike the other Xanthophyceae, sexual reproduction (isogamy or oogamy) is quite common in members of this order. There are three or four genera, the majority of species being freshwater or terrestrial, with only a few marine. Two of the most widely distributed Xanthophyceae genera, *Botrydiwn* and *Vaucheria* are assigned to this order. They are placed in separate families distinguished by their type of thallus and method of reproduction.

Botrydiaceae

This family is distinguished by a saccate plant body, which reproduces by means of aplanospores or zoospores. The mud-dwelling *Botrydium* is analogous to a form such as *Protosiphon,* the commonest species, *Botrydium granulatum,* being frequently confused with it, especially as these two plants are often associated on areas of drying mud. *B. granulatum* makes its appearance during the warmer part of the year when the green, pear-shaped vesicles are rooted by means of colourless, dichotomously branched rhizoids. The wall is composed of cellulose and the lining cytoplasm contains numerous nuclei, while the chloroplasts, containing pyrenoids, are confined to the aerial saccate part. The shape of vesicle is influenced by the environment, the shade forms being elongate or club-shaped. In *B. wallrothii* and *B. granulatum,* the unbranched vesicle is covered with lime. When the plants are submerged, reproduction takes place by means of numerous zoospores which are set free by gelatinisation of the vesicle apex, but when the plants are only wet but not submerged, aplanospores are produced instead. These are formed by the cytoplasm rounding off and secreting a wall, and they may be either uni- or multinucleate. They germinate directly to give a new plant. Under dry or drought conditions, the cytoplasm may round off to form a single thick walled cyst (macrocyst), or alternatively several multinucleate thick walled sporocysts or hypnospores. Sometimes the protoplast migrates into the rhizoid and there forms multinucleate hypnospores or rhizocysts. Upon return to favourable conditions, these hypnospores germinate to uninucleate zoospores, or if the hypnospore is uninucleate, germination gives rise directly to a new plant.

Sexual reproduction has been claimed for *B. granulatum*. In this species, the contents of the cell round off into uninucleate isogametes with typical hetero-kont flagellation. Nuclear fusion takes place after gamete fusion and is followed by the formation of four or eight zoospores. Meiosis is said to occur at division of the zygote nucleus. In *B. granulatum*, it is estimated that about 40,000 isogametes are formed in each vesicle, but as the plant is monoecious many fuse either in pairs or threes, rarely fours, before they are liberated. Those that do not fuse, develop parthenogenetically. The life cycle can be tentatively represented.

Vaucheriaceae

In the past, this genus has been classified with the Siphonaceous chloro-phyceae in the order siphonales (Codiales and Caulerpales). Recent work has shown that there are many fundamental differences between *Vaucheria* and the siphonaceous Chlorophyceae and that it is correctly classified in the Xanthophyceae.

The cell walls do not contain native cellulose; chlorophyll *b* is lacking in *Vaucheria* as in all other Xanthophyceae. The discoid chloroplasts, which lack pyrenoids, contain the typical Xanthophycean carotenoids, and lack siphonaxanthin so characteristic of the siphonaceous chlorophyceae. The storage reserve is oil. The two flagella present on motile cells are typically Xanthophycean; one is pleuronematic (flimmergeissel or tinsel type) while the other is simple (peitschengeissel or whip-lash type).

Vaucheria is essentially a temperate genus, inhabiting well aerated streams, soil or marine mudflats, where it sometimes forms a greenish covering on the surface (e.g. *V. haematd).* Some species (e.g. *V. de baryand)* may be slightly lime-encrusted.

The plant consists of a colourless, basal rhizoidal portion from which arise green, erect, aerial coenocytic filaments with apical growth and monopodial branching.

All walls are thin, and within the thallus there is a large vacuole that usually runs the length of the cell. The chloroplasts and tiny nuclei are confined to the cell periphery and septa are formed only with reproductive structures or after wounding. Vegetative reproduction is secured through fragmentation, and sexual reproduction is brought about by the well-known, compound, multi-flagellate zoospores (synzoospores) which are produced singly in club-shaped sporangia cut off from the ends of the erect aerial branches. The chloroplasts and nuclei congregate in the apex of a filament before the septum is laid down and the nuclei then arrange themselves peripherally. Finally, two slightly unequal flagella develop opposite each nucleus and then the zoospore is ready for liberation, a process which is achieved by gelatinisation of the sporangium tip. This compound structure may be regarded as representing a group of biflagellate zoospores which have failed to separate. The synzoospore is motile for about fifteen minutes; then it comes to rest and germinates, the first thread often being more or less colourless. Zoospore formation can often be induced by transferring the plants from light to darkness, or from a nutrient solution to distilled water.

Under dry conditions in terrestrial species, thin-walled aplanospores may be formed in sporangia at the ends of short lateral or terminal branches, and if exposed to greater desiccation the threads of the terrestrial forms become septate and rows of cysts are formed to give the 'Gongrosira' stage. When conditions become more favourable, these cysts germinate either into new filaments or divide into thin-walled bodies whose contents escape as amoeboid masses, which on becoming sessile germinate into new filaments.

Sexual reproduction is distinctly oogamous, the different species being either monoecious or dioecious. The oogonia are sessile or stalked. Development begins with a bulging in the main thread, and a large mass of colourless cytoplasm followed by many nuclei and chloroplasts moves into the bulge which gradually increases in size. It eventually becomes an oogonium with a single uninucleate egg, which is cut off from the main thread by a transverse system. Some authors maintain that the extra nuclei which passed into the oogonia are potential gametes which degenerate, while others maintain that the surplus nuclei travel back into the main thread just before formation of the septum. This latter view is considered to be the more likely explanation, although both possibilities may occur dependent upon the species. In the mature oogonium, there is a beak, the apex of which gelatinises, or apical pores through which the antherozoids can pass.

The antheridia, which are usually stalked, commonly arise close to the oogonia, and in most species their development commences before oogonia formation. The branch bearing the antheridium becomes densely packed with nuclei and a few chloroplasts, so that often the antheridia appear colourless.

When the antheridium, which is often hook-shaped, separates from the main thread by a transverse septum, the contents divide into uninucleate fragments which form pear-shaped antherozooids that possess apically inserted, unequal, heterokont flagella pointing in opposite directions, and are unusual in that the anterior, shorter flagella is of the flimmergeissel type. The antherozooids escape through apical pores and enter the oogonium, although only one antherozoid enters the egg. After fertilization, the zygote secretes a thick wall and remains dormant for some months before it germinates to give rise to a new filament.

The latest evidence indicates that meiosis occurs at

germination of the zygote, thus indicating that the adult plant is haploid (cf. the siphonaceous Chlorophyceae in which the adult is diploid).

Chloromonadophyta

Chloromonadophyceae: This is a very small division of little-known flagellates, with no recorded non-flagellated members. Apart from one marine species, they are of infrequent occurrence in freshwater bogs and ponds.

The cells, characteristically ovoid or pear-shaped, are naked and surrounded by a pliable periplast that permits cytoplasmic movement and changes in cell shape, sometimes referred to as 'metabolic movements' (c.f. Euglenophyceae). There is a gullet associated with a slight furrow, and trichocyst-like rods are distributed in the peripheral cytoplasm. These may be discharged either as slime threads or as club-shaped bodies. There are two flagella, one of which usually trails behind. The pigmented members possess numerous, discoid, yellow-green chloroplasts scattered throughout the cytoplasm. Only chlorophyll *a* is present together with the same carotenoids as in the Xanthophyceae, and recent work on the pigments suggests that the class should be absorbed in the Xanthophyta. The organisms do not appear to store starch, but rather oils.

There is but a single order, the Chloromonadales, with one family.

Chloromonadaceae

There are three species, the one most widely distributed being the pigmented *G. semen*. The cell is typically ovoid with two flagella, a triangular gullet and a shallow longitudinal furrow. There is a distinct nucleus and contractile vacuoles.

Under certain environmental conditions (e.g. viscous agar medium or increasing salt concentrations due to evaporation) the cells may become spherical, cyst-like or amoeboid with metabolic movements. Under other conditions the cells may become aggregated in a gelatinous matrix and then they sink to the bottom of the pool or pond.

In both *Gonyostomum semen* and *Vacuolaria virescens* encystment can occur, whereby the cells become spherical and surrounded by a sheath: this represents a different phenomenon.

Cell division is by longitudinal fission while in a somewhat amoeboid form.

Sexual reproduction has been recorded for *Vacuolaria virescens* and *Hornellia marina* (possibly *Chattonella subsalsd).* In the latter case, two flagellated individuals fuse and lose their flagella. Germination of the resulting resting zygote has not been described.

Hornellia marina is the sole monotypic marine genus and occurs off the west coast of India, often in sufficient quantity to cause a green discolouration in the sea.

Cryptomonads

The cryptomonads are characterised by a peculiar, normally naked, asymmetrical cell (ovoid or bean-shaped), which possesses two slightly unequal flagella of the flimmergeissel type.

At least one contractile vacuole is usually present. There is an antero-ventral gullet or an oblique ventral furrow, which is lined with a row of trichocysts in special sub-surface pits which shoot out the threads when stimulated.

The coloured species usually possess one or two lobed parietal chloroplasts with pyrenoids, around which a starch-like reserve is deposited. The colour varies from reddish-brown through olive green to blue green due to the presence of chlorophylls gjuid *c* and Characteristic carotenoids and biliproteins. Instead of a wall the celfs have a periplast, which is usually firm, thereby, restricting 'metabolic movements' such as are found in the Chloromonadophyceae and Euglenophyceae. Reproduction is solely by longitudinal fission. Both marine and freshwater forms are known.

The colourless species (e.g. *Chilomonas parameciwri)* possess the typical cryptomonad form.

Originally these algae were classified with the dinoflagellates in the Pyrrhophyta. Recently, however, they have been removed to a separate division.There are two orders, the Cryptomonadales and the Cryptococcales.

Cryptomonadales

The species of this genus possess the typical naked cryptomonad cell.*Cryptomonas ovata* var. *palustris* has an olive-greenish hue whereas othergenera may be red (e.g. *Rhodomonas)* or blue-green (e.g. *Cyanomonas).* Apart from the usual longitudinal division of the motile cell, some species have thick-walled, resting-stage cysts. *Hemiselmis, (hemj,* half; *selmis,* deck) a representative of the Hemiselmidaceae, is characterised by a bean-shaped cell, with laterally inserted flagella.

The gullet is replaced by an oblique furrow with trichocysts. There are one or two parietal chloroplasts, which in *H. rufescens* are red but are blue-green in *H. virescens.*

Initially there is only a single chloroplast but during maturation this divides into two so that when division occurs the two chloroplasts distribute themselves between the two daughter cells.

Cryptococcales

The one species in this order, *Tetragonidium verrucatum,* is a rarely found organism that has been recorded from freshwater ponds in Europe, the United States and New Zealand.

The cells have a characteristic tetrahedral shape and a cellulose cell wall. Reproduction is by means of zoospores which possess the characteristic features of the flagellate Cryptomonads. On coming to rest these zoospores develop into the typical coccoid parent.

A new genus *Bjornbergiella* has recently been described. It exists in the form of filaments of a few cells each and together with *Tetragonidium* is the only indication of evolution towards an algal organisation.

Pyrrhophyta

This division comprises the dinoflagellates, whose name describes the characteristic spiral motion of the motile cells. Although evolution towards a filamentous organisation has occurred, the majority possess flagella.

The organisms are characterised by the structure of certain cellular organelles. The nucleus is large and distinct and the chromosomes appear as condensed threads (with helical coiling of the chromonema). In contrast to most other organisms, the chromosomes remain condensed during interphase, and at metaphase are arranged parallel to each other across the equator.

A mitotic spindle appears to be absent. Like the procaryota, no basic protein has been detected; there is continuous DNA synthesis and the chromosomes are fibrillar. Because of this, Dodge (1965) has suggested that the dinoflagellates might be assigned to a third group, the Mesocaryota.

The cells are biflagellate. Both flagella have the 9+2 fibril arrangement but are unusual in other respects.

A distinctive, non-contractile vacuole or pusule is also present. This consists of a cluster of vesicles that open into the inner side of the flagellar canal.

Like some other flagellated groups, the dinoflagellates possess trichocysts. These consist of long membrane-bound crystalline cores connected by fibrils to their internal retainer sac. The trichocysts, which have a banded appearance, originate in vesicles of golgi bodies and are ejected through a fine pore at the cell surface. Their exact function is unknown but they are probably a mechanical sensing or defensive device.

The brownish colour of the chloroplast is due to the predominance of the group-specific carotenoid peridinin, which masks the chlorophyll *a* and *c*. Species without chloroplasts often possess cytoplasmic pigments, of unknown nature, which impart various colours to the cell. Some dinoflagellates are also characterised by the ability to luminesce and by the presence of potent neurotoxins. The storage reserve is starch.

The Desmophyceae and Peridiniales are predominantly holophytic (and nearly all are obligate photoauxotrophs), although holozoic nutrition is known to occur amongst phototrophic species. The Gymnodiniales are predominantly holozoic, as are the Dinophysiales and Rhizodiniales, while the Blastodiniaceae are parasitic. The non-flagellated orders are typically holophytic.

In those dinoflagellates with holozoic nutrition, ingestion takes place typically, but not always, while the algae is in a temporarily assumed amoeboid form.

Although the majority are free-swimming or attached, marine or freshwater forms, they are also quite common as sand dwellers and parasites in fish and invertebrates. The endosymbiotic algae (Zooxanthellae) of corals, sea anemones and certain other coelenterates and radiolarians are dinoflagellates. There are two classes within the division.

The genera, composed entirely of motile cells as the principal life stage, characteristically possess the two flagella apically inserted instead of laterally. The cells are surrounded by a periplast or a cellulose envelope (theca) consisting of two valves separated by a longitudinal suture.

Prorocentrales

The cell shape varies from spherical to ovoid with noticeable flattening, except in some species of *Exuviaella*. The flagella project through the envelope via the individual pores at the edge of which there may be teeth-like projections.

One flagellum projects straight forward to provide the forward motion while the other, causing the rotatory motion, extends at right angles. There are usually two yellow-brown chloroplasts commonly with associated pyre-noids, and numerous trichocysts are often present.

Reproduction is by longitudinal cell division, where one

valve of the envelope is retained for each daughter cell, but cysts have been recorded in older cultures. *Prorocentrum* is a common marine genus and *P. micans* is often a component of red tides. *Exuviaella marina* is a sand dweller.

Dinophyceae (Dinokontae)

This class represents the main line of dinoflagellate development which is considered to be the more primitive, in certain important respects. The cell wall or envelope when present is not divided by a single vertical suture. The motile flagellates demonstrate the characteristic 'dinoflagellate orientation', because they are circled by a transverse or spiral groove, the girdle, which divides the outer wall into a forward portion, the epicone or epitheca and a hypocone or hypotheca (posterior part).

Parallel to the axis of locomotion is a longitudinal furrow, the sulcus. Both the sulcus and girdle may be greatly modified, but the basic pattern remains the same. The two flagella arise on the ventral side, usually through separate pores at the junction of the girdle and sulcus.

The flagellum in the transverse groove is a helical ribbon possessing a unilateral row of long fine hairs, while the longitudinal flagellum is pleuronematic with short fine lateral hairs, though these are absent in *Amphidinium*. Some workers have argued that the flagella are not latero ventrally inserted, but still apically inserted and that the dinoflagellate motion is consequently a unidirectional sideways spiralling.

Apart from a few genera that have only a firm periplast, the Dinophyceae possess a cellulose wall. In a large majority of the flagellated genera, the 'armoured' dinoflagellates, the walls possess regularly arrange and sculptured plates whose number and arrangement are taxonomically important. Those genera without sculptured plates or without the cellulose wall are referred to as 'unarmoured' dinoflagellates.

A sexual reproduction is by a cell division, or in the case

of the non-flagellated genera, or *phytodinads*, by aplanospores or by *Gymnodinium-type* zoospores, a factor that is largely responsible for their inclusion in the Dinophjceae.

Sexual reproduction has been recorded for a few species and involves isogamy with amoeboid or motile gametes. The motile flagellates are divided into two orders, the Gymnodiniales (comprising the unarmoured Gymno-diniaceae and a few other small specialised families) and the Peridiniales (armoured dinoflagellates, comprising a large number of families). The unarmoured dinoflagellates are not well known because they are very difficult to preserve and need to be studied as they are collected.

Gymnodiniales

Although the members lack a cell wall the pellicle is firm enough to give the cells a rigid shape which does vary greatly from one species to another. The cells are generally spherical although ventral or lateral flattening is quite common. From a ventral view, the girdle spirals to the left, sometimes going more than once around the cell. The sulcus is usually straight but it too may be twisted around the cell. Cyst formation is quite common. The majority of species lack chloroplasts and are holozoic.

Chloroplasts, when present, are usually radially arranged, and species containing them may possess holozoic nutrition. The two pusules that open to the two flagellar pores are sometimes fused into one. Certain species have been recorded as possessing functional trichocysts or cordocysts. The nucleus is usually centrally located. Cell multiplication is by longitudinal division of either the motile or non-motile encysted form.

In *Gymnodinium*, the girdle is usually equatorial with little displacement of the two ends. In some species, the periplast on the non-motile encysted cell may be striated.

In *Amphidiniwn*, the girdle is anterioly placed so that the epicone is very small. The single chloroplast is arranged peripherally with a large external pyrenoid. The other cell components such as trichocysts, nucleus, pusules, flagella and so on, are typical of dinoflagellates, although the trailing flagellum lacks lateral hairs. Many of the species are flattened, either laterally or dorso-ventrally, and are found predominantly in coastal waters. Isogamous sexual reproduction has been recorded for *A. carterae.* After fusion of two individuals, the zygote divides in one plane. The two subsequent divisions are unusual in that they are asynchronous.

In *Hemidinium*, the girdle is incomplete, extending usually about halfway round the cell, and the sulcus is restricted to the hypotheca.

In *Gyrodinium*, many species of which show cell elongation, the ends of the girdle are widely separated giving it a spiral appearance, while the sulcus may also be slightly twisted. Most of the species are colourless.

The spiralled appearance of *Cochlodinium* is due to the distinct girdle and furrow.

The three following, widely distributed, colourless dinoflagellates possess an interesting morphology and are accordingly classified in separate families:

Polykrikos is one of the few coeno-cytic forms, the cell comprising two, four or eight transverse grooves with one, two or four nuclei respectively, because only one nucleus is present for each pair of grooves in the coenocyte. In each zooid the girdle is slightly spiral. The sulcus runs the entire length of the zooid and is continuous with that of each adjacent zooid, since each zooid possesses the same orientation.

Polykrikos is also characterised by the possession of complex functional barbs or nematocysts as well as trichocysts.

Oxyrrhis is considered to be a primitive gymnodinioid member. The transverse furrow and sulcus are both situated at the posterior end of the asymmetrical cell. The latter is rather broad and divided by a lobe-like structure. The species are colourless, holozoic and also phagotrophic. The cells of *Noctiluca miliaris*, the sole species of the genus, are very large, sometimes reaching a diameter of 2 mm. The girdle is non-existent and the transverse flagellum is reduced to a tooth-like projection. The sulcus, however, is deep and prominent with a short flagellum. The most characteristic feature of the cell is a prominent tentacle extending posteriorly from the sulcus. Although the species lacks chloroplasts and is therefore, holozoic, the cell has a pinkish hue and it is often a constituent of red tides as well as being one of the luminescent dinoflagellates. Cell division is by fission or multiple fission, producing numerous gymnodinoid swarmers which are thought to be gametes.

One family of holozoic, colourless dinoflagellates, the Warnowiaceae, is characterised by an ocellus, a complex lens, with an associated structure of pigmented granules (e.g. *Nematodinium, Erythropsidinium, Warnowid)*, which may be involved in light perception.

In the parasitic colourless Blastodiniaceae, the majority are ecto- or endo-parasites of metazoa, though one species has been described as an ectoparasite on the diatom *Chaetoceras. Blastodinium* occurs as an elongate sac-like cyst in the gut of certain copepods. Within the cyst, reproductive cells are formed initially from division of the saccytoplasm. After the first division only one of the resulting two daughter cells undergoes further division, so that eventually there remains one large daughter cell and numerous smaller ones.

Upon rupture of the sac envelope the reproductive cells are liberated into the gut and the small ones divide once to give typical gymnodinoid zoospores which come to rest, secrete a new envelope, and thus start a new saccate individual. The remaining large daughter cell from the first division may secrete a new envelope and in turn undergo a repeat division to give numerous small reproductive bodies. The ultimate fate of the large cell appears to be unknown.

Peridiniales

These are the armoured dinoflagellates and get their name from the heavy sculptured plates on the cell wall. The plates form the major distinguishing feature between the Peridiniales and the Gymnodinioideae. The number and arrangement of the plates is of great taxonomic importance. The plates, which are usually arranged in three series on both the epitheca and hypotheca, are also present as a series of small ventral plates on the girdle and sulcus.

The sculpturing on the plates is often very intricate, involving numerous tiny depressions (or areolae) with a minute hole, and often with teeth or spines. The girdle, which is usually very distinct, is commonly equatorial or nearly so, and is generally characterised by pronounced ridges on the edges.

In the Peridiniales, the flagella emerge through a common slit rather than from two separate ones as in the Gymnodiniales. Also, in certain members, the girdle spirals to the right rather than to the left.

The majority of this order possess chloroplasts, although holozoic nutrition is known to occur in some species. Trichocysts have also been recorded.

Reproduction is by oblique cell division of either motile or resting cells always through the point of flagellar insertion. The two parts of the old cell envelope may be retained by the daughter cells, or the protoplast may escape completely from the cell envelope, divide and then form a completely new theca, in which the old shell halves are left behind *(Peridinium)*. Sexual reproduction (e.g. *Ceratium)* and cyst formation (e.g. *Glenodiniurri)* have both been recorded for a few species.

Glenodinium (glene, eyeball), a genus of the predominantly freshwater Glenodiniaceae, is probably one of the simpler peridinioids. The cell shape is similar to *Gymnodinium,* but rudimentary armouring of small, delicate, indistinct plates is apparent. Sexual reproduction occurs in which a vegetative cell forms four gymnodinioid isogametes that fuse in pairs to form a resting zygote which eventually germinates to form four new flagellates.

Peridinium, family Peridiniaceae, is a common marine and freshwater genus with very highly developed sculpturing. The girdle is typically ridged and very prominent. A ventral concavity is often quite marked giving a kidney shaped appearance from top view.

In *Gonyaulax*, family Gonyaulacaceae, the two ends of the girdle are displaced, showing a spiralling tendency and the sulcus is usually quite distinct. As in *Peridinium* there is sometimes formation of antapical horns. In cell division, the separation line follows the sutures of the theca, and in some cases the daughter cells do not separate and chain-forming colonies may therefore, occur. *G. polyedra* is often the predominant species in red tides along California. *G. catenella* has been known to liberate a toxin in Californian mussels that can be fatal to humans, while *G. tamarensis* is a toxic species on the Atlantic coast.

One of the most widely distributed and most specialised genera is the marine and freshwater *Ceratiwn (keras,* horn) (Ceratiaceae). The cells are highly asymmetrical and usually flattened dorsiventrally. The most characteristic feature is the elongated hollow horns. As in most peridinioids (except *Glenodinium)* there is an apical pore, in this case at the tip of the apical horn. In the sulcal region is a large ventral area that is very lightly sculptured. Chains of individuals may be formed as a result of cell division. Sexual reproduction, involving amoeboid gametes has been reported and has recently been confirmed by Von Stosch (1964). The female cells are similar to vegetative cells, but the male micro-swarmers are smaller.

Dinophysiales

The genera of this order possess a cell wall comprised of two very large plates and numerous smaller ones. The order is included in the Dinophyceae, rather than Desmophyceae, because there is a girdle sulcus and theca composed of many definitely arranged plates. The epitheca is usually very small and lacks the apical pore.

The flagella orientation is similar to that of the Peridiniales. Reproduction is by longitudinal division. Many species appear to lack chloroplasts, though some species of *Dinophysis (phusa,* bladder) have them. In most members of the order the thecal margins around the girdle and suture lines of the valves are flared out to give a wing-like appearance.

In nearly all instances the girdle is so far to the front that the wing-like projections give the appearance of a collar. In *Dinophysis,* which is flattened laterally, the upturned margins or wings on the girdle project anteriorly to give the collar a funnel shape.

Dinocapsales

This order represents the palmelloid condition in the Dinoflagellates and thus corresponds to the Tetrasporaceae, Chrysocapsaceae and Heterogloeales. There are but two monotypic genera in the order, *Gloeodinium montanum* and *Rufusiella insignis*. *Gloeodinium* (*gloios,* sticky), which is usually found in peat bogs, consists of two, four or eight subspherical cells enclosed in several gelatinous sheaths.

The individual cells possess a large central nucleus and numerous small chloroplasts scattered throughout the cytoplasm. Reproduction is by cell division or by *Hemidinium*-like zoospores.

Dinotrichales

This is another small order with two genera, *Dinothrix* and *Dinoclonium,* comprising those members that exhibit the filamentous habit. The plants consist of sparsely branched filaments each containing a few globose or cylindrical cells. The cells possess the typical large nucleus, together with numerous discoid chloroplasts. Reproduction is by gymnodinoid swarmers, formed singly or in pairs in the cells.

Dinamoebidiales (Rhizodiniales)

This comprises a single monotypic order with one marine species, *Dinamoe-bidium varians*. As the name implies the vegetative phase is permanently amoeboid and the cells move by means of stumpy pseudopodia. The cells are holozoic, there being no chloroplasts, and there is a large characteristic nucleus. Cell division has never been recorded although encystment of the protoplast occurs within a gelatinous wall. Division of the cyst two or three times results in the production of short-lived gymnodinioid zoospores that soon revert to vegetative amoeboid shape.

Phytodiniales (Dinococcales)

The members of this order are all immobile unicells (that is, they are in the coccoid condition) at least in the vegetative phase, and they have small chloroplasts. The majority are freshwater and the cells are either free-floating or attached. There is no vegetative cell division, reproduction being entirely by two to four motile gymnodinioid zoospores, or two non-motile autospores. *Cysto-dinium* is a free-living form with elongate, extended cells.

Pyrocystis (pyro, fire) is a marine luminescent planktonic genus with large spherical cells, that reproduce by division of the protoplast into new individuals. In *Pyrocystis lunula* (known as *Dissodinium lunula* or *Gymnodinium lunula)* the four to eight daughter cells elongate into crescent shaped cells which are liberated. These cells then produce eight gymnodinioid zoospores within the parent cell. Formation of the large spherical unicell from these zoospores has not been observed.

Chrysophyta

Bacillariophyta: In the Pascherian classification, the Xanthophyceae, Bacillariophyceae and Chrysophyceae were grouped into the one division Chrysophyta. Recent biochemical and electron-microscope studies have however, prompted a re-evaluation of this concept. While there are some similarities (e.g. storage reserve, heterokont flagella, endogenous cysts, a two-piece cell wall and the occurrence of silica impregnation) there are also some distinct differences. The Xanthophyceae lack both chlorophyll *c* and the carotenoid fucoxanthin, and show some other pigment differences.

The Bacillariophyceae possess a very distinctive cell-wall structure in a silica skeleton. A second class, the Haptophyceae, has recently been split off from the Chrysophyceae because its members possess a distinctive third organ, the haptonema, and isokont flagella.

It is largely a matter of opinion whether the three or four classes are arranged into one, two or three divisions. The relative importance that is attached to the similarities and differences will be the deciding factor. We have retained three divisions with four classes although many workers still retain only the one division. It seems essential, however, that the Xanthophyceae, should be removed from the Chrysophyta, irrespective of the position of the Baci llariophyceae.

Bibliography

Abdelahad, N. and Bazzichelli G.: *Geitleria Calcarea Friedmann,* Cyanophycee Nouvelle, Italie, 1988.

Anderson R. and Adams M. S.: *Effect of Fire on Big Bluestem Production,* American Midland Naturalist, 1975.

Battarbee, R. W. and Gasse F.: *Past Climate Variability through Europe and Africa,* Springer, Netherlands, 2004.

Bebianno, M. J. and Company R. M.: *Antioxidant Enzymes and Lipid Peroxidation in Bathymodiolus Azoricus,* Aquatic Toxicology, London, 2005.

Berge, J. and Vader W.: *On the Taxonomic Status of the Antarctic Amphipod Genera Eclysis,* Organisms Diversity Evolution, 2005.

________________ : *The Amphipod Genus Alexandrella,* Journal of Natural History, Washington, 2005.

Blanchette, C. A.: *The Effects of Biomechanical and Ecological Factors on Population and Community Structure of wave-exposed,* Oregon State University, 1994.

Bold, H. C. and M. J. Wynne: *Introduction to the Algae,* Englewood Cliffs, New Jersey, 1978.

Boudouresque, C. F. and Perret-Boudouresque M. M.: *A Checklist of the Benthic Marine Algae of Corsica,* GIS Posidonie Publishers, Marseille, 1987.

Bourrelly, P. and Dupuy P.: *Quelques Stations Francaises de Geitleria Calcarea, Schweizerische Zeitsch, Fur Hydrologie,* MacMillan, New York, 1973.

Boyer, C. S.: *The Biddulphoid Forms of North American Diatomaceae*, Natural Sciences of Philadelphia, Boyer, 1901.

Brodie, J. A. and Irvine L. M.: *Seaweeds of the British Isles*, British Museum, London, 2003.

Chapman, V. J.: *The Algae*, St. Martin's Press, New York, 1964.

Christiansen, J. S. and Gildberg A.: *Pagophilus Groenlandicus*, ICES Journal of Marine Science, 2004.

Claus, G.: *Algae and (their) Mode of Life (in) on the Baradla Cave of (at) Aggtelek*, Scientiarum Hungaricae, Budapest, 1955.

Claus, G.: *Data on the Ecology of the Algae of Peace Cave*, Nova Hedwigia Zeitsch, Kryptogamenkunde, 1962.

——————— : *The Reevaluation of the Genus Gomontiella*, Algol, 1960.

Coppejans, E.: *Iconographie d'Algues Mediterraneennes*, J. Cramer, Vaduz, 1983.

Delepine, R., Boudouresque C. F. and Frada-Orestano C.: *Algues et Autres Vegetaux Marins*, Mediterranee et Mer Noire, Roma, 1987.

Denny, M. W.: *Mechanical Limits to Size in Wave-swept Organisms*, Ecological Monographs, 1985.

——————— : *Predicting Physical Disturbance*, Ecological Monographs,1995.

Dixon, P. and Irvine L. M.: *Seaweeds of the British Isles*, British Museum, Natural History, London, 1977.

Evenset, A., Christensen G. N., Skotvold T, Fjeld E. and Gregor D.: *A Comparison of Organic Contaminants in two High Arctic Lake Ecosystems*, Science of the Total Environment, Norway, 2004.

Falk, Petersen S., Timofeev S., Pavlov V. and Sargent J. R.: *Climate Variability and the Effect on Arctic Food Chains*, Springer Verlag, Berlin, 2006.

Fletcher, R. L.: *Seaweeds of the British Isles*, British Museum, London, 1995.

Furevik, T., Gerdes R. and Berger W.: *The Nordic Seas*, American Geophysical Union, Washington, 2005.

Gantt, Elisabeth: *Handbook of Phycological Methods*, Cambridge University Press, England, 1980.

Gayral, P.: *Les Algues des Cotes Francaises*, Editions Doin, Paris,1966.

Guiry, M. D.: *A Concensus and Bibliography of Irish Seaweeds*, J. Cramer, Vaduz, 1978.

Gulliksen, B., Slagstad D. and Pamapanin D.: *Mytilus Thermophily*, Marine Ecology Progress Series, 2004.

Hakansson, H. and Stoermer E. F.: *Observations on the Type Material of Stephanodiscus Hantzschii Grunow in Cleve and Grunow*, Nova Hedwigia, 1984.

Hakansson, H.: *A Light and Electron Microscopical Investigation of the Type Species of Cyclotella*, Diatom Research, 1989.

Hanna, G. D.: *Pliocene Diatoms of Wallace County*, University of Kansas Science Bulletin, Kansas, 1932.

Haworth, E. Y. and M. A. Hurley.: *Comparison of the Stelligeroid Taxa of the Centric Diatom Genus Cyclotella*, O. Koeltz, Koenigstein, 1986.

Hellebust, J. A. and Craigie J. S.: *Handbook of Phycological Methods*, Cambridge University Press, Cambridge, 1978.

Hishock, S.: *A Field Key to the British Brown Seaweeds*, F.S.C. Publications, Shrewsbury, 1979.

Hoagland, K. D., Roemer S. C. and Rosowski J. R.: *Colonization and Community Structure of two Periphyton Assemblages*, American Journal of Botany, 1982.

Hollenberg G. J.: *Marine Algae of California*, Stanford University Press, California, 1976.

Hop, H. and Falk-Petersen S.: *Seasonal Lipid Dynamics and Life Strategy*, Marine Biology, 2005.

Irvine, L. M. and Chamberlain Y. M.: *Seaweeds of the British Isles*, British Museum, London, 1994.

———————: *Seaweeds of the British Isles,* British Museum, London, 1983.

Kattner, K., Graeve M.: *The Arctic Pteropod Clione Limacina,* Marine Biology, 2005.

Kociolek, J. P. and Stoermer E. F.: *Ultrastructure of Cymbella Sinuata and its Allies,* Systematic Botany, 1987.

Kolosov, P. N.: *Upper Precambrian Paleoalgological Remains of the Siberian Platform,* Moscow,1982,

Krammer, K. and H. Lange Bertalot: *Bacillariaceae, Epithemiaceae, Surirellaceae,* Gustav Fischer Verlag, New York, 1988.

———————: *Naviculaceae,* Gustav Fischer Verlag, New York, 1986.

Leake, D.V.: *The Algae of Crystal Lake,* Cleveland County, Oklahoma,1945.

Lebreton, B.: *World Bibliography on Subterranean Algae,* Bibliographie Mondiale des Algae Souterraines, 2000.

Lee, R. E.: *Phycology,* Cambridge University Press, Cambridge, 1980.

Littler, M. and Littler D.: *Handbook of Phycological Methods,* Cambridge University Press, Cambridge, 1985.

Lobban, Christopher S. and Harrison P. J.: *Seaweed Ecology and Physiology,* Cambridge University Press, Cambridge, 1994.

Makarikhin, V. V. and Kononova, G.: *Phytolites of the Karelian Lower Proterozoic,* Leningrad, Nauka, 1983,

Mann, D. G. and Stickle A. J.: *The Genus Craticula,* Diatom Research, 1991.

Mikhailova, N. S. and German, T. N.: *5th All Union Colloquium on Microfossils of the USSR Precambrian,* Izvestia Akadamie Nauk, 1987,

Nelson, E. N., Richardson J. K. and Bailey H. H.: *Aspects of the Occurrence of Aspidobothrid Parasites,* Proceedings of the Oklahoma Academy of Science, 1975.

Newman, G. A.: *Cowbird Parasitism and Nesting Success of Lark Sparrows in Southern Oklahoma,* Wilson Bull 1970.

Newton, L.: *A Handbook of the British Seaweeds,* British Museum, London, 1931.

Oliver, G.: *Black Rat Snake Predation Upon Nesting Barn and Cliff Swallows,* Ornithological Society, Oklahoma,1970.

Osborn, R. S.: *Observations on the Spawning Ecology of Buffalos,* Oklahoma Academy of Science,1966.

Pampanin D. M., Camus L.. Gomiero A., Marangon I., Volpato E. and Nasci C.: *Susceptibility to Oxidative Stress of Mussels,* Venice Lagoon, Italy, 2005.

Pamperl, M. A.: *Diatoms of Lake of the Ozarks,* Transactions, Missouri Academy of Science, Missouri, 1980.

Parks, J. M. and Barclay H. G.: *The Increasing Importance of Vines in Southern Oklahoma,* Oklahoma Academy of Science, 1966.

Parmelee, D. F.: *The Breeding Behavior of the Painted Bunting in Southern Oklahoma,* Bird-Banding, 1959.

Patten, B. C., Egloff D. A. and Richardson T. H.: *Total Ecosystem Model for a Cove in Lake Texoma,* Academic Press, 1975.

Patten, B. C.: *A Reservoir Cove Ecosystem Model,* Transactions of the American Fisheries Society, 1975.

________________: *Ecosystem Modeling and Reservoir Management,* Oklahoma Academy of Science, 1976.

Paulissen, M. A.: *A Technique for Marking Teiid Lizards for Field Observation,* Herpetology Review, 1985.

________________ : *Cnemidophorus Sexlineatus,* Journal of Herpetology, 1988.

Penfound, W. T., Shed J. S. and Jennisons M. C.: *A Plant Community Dominated by Vines,* Oklahoma Academy of Science, 1965.

Penfound, W. T.: *The Composition of a Black Willow Forest in South Central Oklahoma,* Oklahoma Academy of Science, 1961.

Pfiester, L.A. and Holt J.: *A Freshwater "Red Tide" in Texas,* Southwestern Naturalist, 1978.

Phillipi, A. and Tyrl R. J.: *Reproductive Biology of Proboscidea Louisianica (Martyniacea)*, Rhodora, 1979.

Pireh, W.: *Achillea Millefolium*, Oklahoma State Universi, Oklahoma, 1978.

Power, M. E. and Matthews W. J.: *Grazer Control of Algae in an Ozark Mountain Stream: Effects of Short-term Exclusion*, Ecology, 1988.

Proctor, V. W.: *Dispersal of Fresh-water Algae Water*, Science, 1959.

————————: *The Growth of Basicladia on Turtles*, Ecology, 1958.

Prud, Homme: *A Taxonomic Revision of the European Sphacelariaceae*, Leiden University Press, Leiden, 1982

Purdue, J. R.: *The Phylogenetic Implications of Body Movements in Displaying Sceloporine Lizards*, Herpetology Review, 1970.

————————: *The Western Sandpiper in Oklahoma*, Oklahoma Ornithological Society, 1969.

Ratkova T. and Wassmann P.: *Sea Ice Algae in the White and Barents Sea*, Polar Research, 2005.

Reinke, D. C.: *New and Interesting Algae from Kansas*, State Biological Survey of Kansas, 1979.

Riedl, R.: *Fauna und Flora des Mitterlmeeres*, Verlag Paul Parey, Hamburg, 1983

Roemer, S. C., Hoagland K. D. and Rosowski J. R.: *Development of a Freshwater Periphyton Community as Influenced* Canadian Journal of Botany, 1984.

Rolbiecki L. and Walkusz W.: *Derogenes varicus*, Oceanological and Hydrological Studies, West Spitsbergen, 2005.

Round, F. E., Crawford R. M. and Mann D. G.: *The Diatoms: Biology and Morphology of the Genera*, Cambridge University Press, Cambridge, 1990.

————————: *The Diatom Genus Stephanodiscus*, Archiv fur Protistenkunde, 1981.

Santi-Mazzini G.: *Aquarium*, Gribaudo, Cavallermaggiore, 2000

Setchell, W. A. and Gardner N. L.: *The Marine Algae of the Pacific Coast of North America*, University of California Press, California, 1925.

Shenfil, V. Y.: *Algae in the Riphean Deposits of the Yenesei Ridge*, Doklady Akad, Nauk, 1978,

South, G. Robin and A. Whittick.: *Introduction to Phycology*, Blackwell Scientific Publications, Oxford,1987.

Stegenga H., Bolton J. J., Anderson R. J.: *Seaweeds of the South African Coast*, University of Cape Town, 1997.

Stein R. and Macdonald R. M.: *The Organic Carbon Cycle in the Arctic Ocean*, Springer, New York, 2004.

Theriot, E., Stoermer E. and Hakansson H.: *Taxonomic Interpretation of the Rimoportula of Freshwater Genera in the Centric Diatom Family Thalassiosiraceae*, Diatom Research, 1987.

Troeger, W. W. and Menzel R. G.: *Benthic Diatoms in two Eutrophic Flood Detention Reservoirs in Oklahoma*, Journal of Freshwater Ecology, 1986.

Troeger, W. W.: *Colonization of Benthic Diatoms in a New Farm Pond and its Feeder Creek*, Southwestern Naturalist, 1983.

______________ : *Epiphytic Diatoms in Farm Ponds and Experimental Ponds in Bryan County*, Oklahoma Academy of Science, Oklahoma, 1978.

Usher, M. B., Callaghan T. V.: *Principles of Conserving the Arctic's Biodiversity*, Cambridge University Press, Cambridge, 2005.

Van, Landingham, S. L.: *Catalog of the Fossil and Recent Genera and Species of Diatoms and their Synonyms*, Cramer, Lehre, 1979.

Walcott, C. D.: *Middle Cambrian Algae*, Smithsonian Miscellaneous Collections, 1919.

Walter, M. R.: *Archean Stromatolites*, Princeton University Press, Princeton,1983,

Webber, H. J.: *Catalogue of the Flora of Nebraska*, Nebraska Board of Agriculture, Lincoln, 1890.

———————: *The Fresh Water Algae of the Plains*, American Naturalist, 1889.

Wenke, T. L. and Eberle M. E.: *Diatoms from Streams in Ellis and Russell Counties, Kansas*, Transactions of the Kansas Academy of Science, 1986.

———————: *Diatoms from Kansas*, Southwestern Naturalist, 1985.

Werner, D.: *The Biology of Diatoms*, University of California Press, Berkeley, California, 1977.

Wolle, F.: *Diatomaceae of North America*, Comenius Press, Bethlehem, Pennsylvania, 1894.

———————: *Fourth Contribution to the Knowledge of Kansas Algae*, Bulletin of the Washburn College Laboratory of Natural History, 1889.

Wujek, D. E.: *Contribution to the Diatom Flora of Kansas*, Southwestern Naturalist, 1965.

Yankauskas, T. V.: *Upper Precambrian and Cambrian Plant Microfossils of the European USSR and their Stratigraphic Significance*, University of Moscow, Moscow, 1982,

Zmijewska M. I. and Bielecka L.: *Characteristics of the Arctic and Antarctic mesozooplankton in the neritic zone during summer*, Polish Polar Research, 2004.

Index

A

K

L

M

N

O

P

R

S

T

U

V

□□□

ENCYCLOPAEDIA
OF
ALGAE

ENCYCLOPAEDIA OF ALGAE

Vol. 2

Edited by
A.M. Bagulia

ANMOL PUBLICATIONS PVT. LTD.
NEW DELHI - 110 002 (INDIA)

ANMOL PUBLICATIONS PVT. LTD.
H.O.: 4374/4B, Ansari Road, Darya Ganj,
New Delhi-110 002 (India)
Ph.: 23278000, 23261597
B.O.: No. 1015, Ist Main Road, BSK IIIrd Stage
IIIrd Phase, IIIrd Block,
Bangalore - 560 085 (India)
Visit us at: www.anmolpublications.com

Encyclopaedia of Algae

First Published, 2008
ISBN 978-81-261-3528-8 (Set)

PRINTED IN INDIA

Printed at Mehra Offset Press, Delhi.

Contents

Preface

The first plants on earth were algae and these still thrive in a range of aquatic habitats. The land plants have evolved from the algae, more specifically green algae. Some 400 million years ago freshwater, green, filamentous algae invaded the earth. Algae are used as fertilizer and energy-source (biodiesel), etc.

Algae encompass several groups of living aquatic organism that get light-energy through photosynthesis, using it for transforming inorganic substances into organic substances. Algae make the photosynthetic organism found in most habitats. These are of single-celled to multicellular complex species, such as 'Giant Kelps' that grow to 65 metres in length. Conventionally, these have been regarded as simple plants, in the study of Botany and referred to as Eukaryota. Chromophyta have bacterial characteristics and some authorities consider them to be Protists.

Protists are traditionally considered more animal-like. The prokaryotic forms, referred to as blue-green algae are only half-algae with a mixture of bacterial characteristics. But they are quite different from the bacteria and are referred to as Cynochloronta. All other forms are known as true Eukaryota algae, in the study of Botany. They have a nucleus enclosed within a membrane. The protoctists are defined as eukaryotic microorganism with the exception of animals and plants but including fungi and algae.

All algae have photosynthetic machinery derived from cyanobacteria, and produce oxygen as a by-product of photosynthesis, unlike non-cyanobacterial photosynthetic bacteria. It is estimated that algae produce about 73 to 87 per cent of the net global production of oxygen, available for respiration.

Most of the simpler algae are unicellular flagellates or amoeboids, but colonial and non-motile forms have developed independently among several groups. Some more organisational levels, more than one of which may occur in the life cycle of a specie, are: Colonial, Capsoid, Coccoid, Palmelloid, Filamentous, and Parenchymatous.

This eminent work presents perfect and realistic features of the algae in an interesting manner. Expectedly, this study would generate a separate discipline in the study of algae.

Editor

1

Fundamentals

Algae, diverse group of simple, plantlike organisms. Like plants, most algae use the energy of sunlight to make their own food, a process called photosynthesis. However, algae lack the roots, leaves, and other structures typical of true plants. Algae are the most important photosynthesising organisms on Earth. They capture more of the sun's energy and produce more oxygen (a by-product of photosynthesis) than all plants combined. Algae form the foundation of most aquatic food webs, which support an abundance of animals.

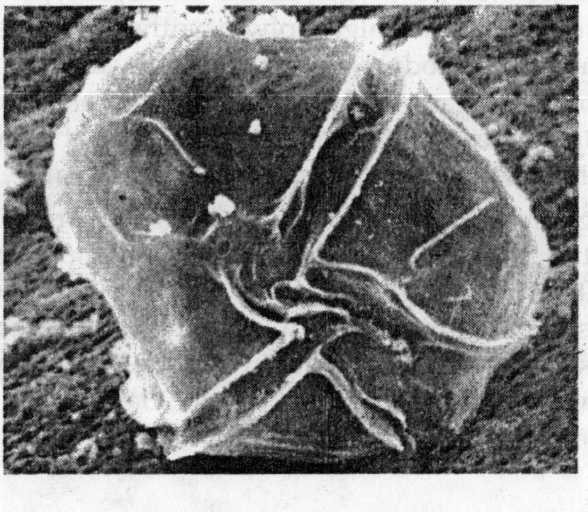

Algae vary greatly in size and grow in many diverse habitats. Microscopic algae, called phytoplankton, float or swim in lakes and oceans. Phytoplankton are so small that 1000 individuals could fit on the head of a pin. The largest forms of algae are seaweeds that stretch 100 m (300 ft) from the ocean bottom to the water's surface. Although most algae grow in fresh water or sea-water, they also grow on soil, trees, and animals, and even under or inside porous rocks, such as sandstone and limestone. Algae tolerate a wide range of temperatures and can be found growing in hot springs, on snow banks, or deep within polar ice.

Algae also form mutually beneficial partnerships with other organisms. For example, algae live with fungi to form lichens—plantlike or branching growths that form on boulders, cliffs, and tree trunks. Algae called *zooxanthellae* live inside the cells of reef-building coral. In both cases, the algae provide oxygen and complex nutrients to their partner, and in return they receive protection and simple nutrients. This arrangement enables both partners to survive in conditions that they could not endure alone.

The earliest life-forms on this planet are thought to be early ancestors of cyanobacteria, a type of algae formerly called blue-green algae. Fossilised cyanobacteria have been found in rocks more than 3 billion years old. These early algae formed when there was no oxygen in the atmosphere, and scientists theorise that as the algae photosynthesised, they released oxygen as a by-product, which eventually accumulated in the atmosphere. Algae were probably the first organisms capable of photosynthesis and, until the appearance of plants on earth, the only photosynthesisers for billions of years.

Algae are a large and successful group of organisms, which flourish in the sea, in freshwater and in damp places on land. Most algae contain green chlorophyil, and can produce foods, such as sugars, from the sun. Although they seem plantlike, scientists have classified them in a group of their own, outside the plant kingdom. They have been classified in a separate kingdom called Protista.

Whereas plants always have specialised reproductive structures made up of many different types of cell, algae do not.

Algae reproduce by using little spores, or by growing copies of themselves from buds or broken-off fragments.

Algae can be lively little characters, even though they are not animals. Many are made of a single cell, and cannot be individually examined without a microscope. Many of these little algae can swim. Some, called Dinoflagellates, have little whiplike structures (flagella), which pull them through the water. Some algae squish part of their body forwards and crawl along solid surfaces.

Some algae are made of fine filaments, with cells joined from end to end. Some of them clump together to form colonies. Others, like many seaweeds, are very large and grow in the shape of tubes, clubs and trees. These large algae have thick rootlike structures, called "holdfasts" which can cling onto rocks.

With the exception of the cyanobacteria, algae are eukaryotes—that is, the insides of their cells are organised into separate membrane-wrapped organelles, including a nucleus and mitochondria. An important organelle found in eukaryotic algae is the chloroplast, which contains the light-absorbing pigments responsible for capturing the energy in sunlight during photosynthesis. In most algae, the primary pigment is chlorophyll, the same green pigment used in plants. Many algae also contain secondary pigments, including the carotenoids, which are brown or yellow, and the phycobilins, which are red or blue. Secondary pigments give algae their colourful hues.

The cyanobacteria are prokaryotes—that is, relatively simple unicellular organisms lacking a nucleus and other membrane-bound organelles. As their modern name implies, the cyanobacteria have many characteristics that resemble bacteria.

Like plants, most algae have rigid cell walls composed largely of cellulose. An exception is the diatom, whose cell wall is composed primarily of silica, which provides rigidity and also produces elaborately sculpted patterns of grooves that serve as identifying features. Many eukaryotic algae have whiplike appendages called flagella attached to their cell walls. By beating flagella in a rotary movement, these algae are able to move through water with considerable speed. A few algae that are devoid of rigid cell walls are able to protrude one part of the body ahead of the other to crawl on solid surfaces in an amoeba-like fashion.

Algae come in a variety of shapes and forms. The simplest form is the single, self-sufficient cell, such as *Euglena*, dependent only on sunlight and carbon dioxide and minerals from the water. Numerous one-celled algae may clump together to form a colony. Although these cells are attached to one another, each cell within a colony continues to function independently. Still other algae are multicellular organisms. In the simplest multicellular algae, the cells are joined end to end, forming filaments, both branched and unbranched. More complex structures may be shaped like a small disc, tube, club, or even a tree. The most complex algae have highly specialised cells. Some seaweeds, for instance, have a variety of specialised tissues, including a rootlike *holdfast*, a *stipe*, which resembles a plant stalk, and a leaflike *blade*.

While most algae create their own food through photosynthesis, some are unable to photosynthesise. These algae ingest food from external sources by absorbing simple nutrients through the cell membrane. To absorb more complex nutrients, algae that lack rigid walls are able to engulf food particles and digest them. Some of the algae known as dinoflagellates extend a feeding tube, called a peduncle, to suck in food. Other dinoflagellates use special harpoonlike structures to snare their food. Some algae are parasites, living in or on another organism from which they get their food. Some parasitic red algae live off other red algae, and parasitic dinoflagellates live in the intestines of some marine animals, such as copepods and annelids.

Algae reproduce in astoundingly diverse ways. Some reproduce asexually, others use sexual reproduction, and many use both. In asexual reproduction an individual reproduces without combining its genetic material with that from another individual. The simplest form of asexual reproduction is binary fission, in which a unicellular organism simply divides into two new individuals.

Some multicellular algae, including *Sargassum*, reproduce asexually through fragmentation, in which fragments of the parent develop into new individuals. In a similar process called budding, special buds detach from multicellular algae and develop into new individuals, commonly found in *Sphacelaria*. Many algae produce special cells called spores that are capable of growing into new individuals. If these spores move about using flagella, they are known as zoospores.

In sexual reproduction, genetic material from two individuals is combined. The simplest form of sexual reproduction in algae is conjugation, in which two similar organisms fuse, exchange genetic material, and then break apart. For example, in *Spirogyra*, which produces both asexually and sexually, two long, unbranched filaments join via conjugation tubes, through which genetic material is exchanged between cells. Most multicellular algae undergo a more complex form of sexual reproduction involving the union of special reproductive cells, called gametes, to form a single cell, known as a zygote.

Many algae incorporate both sexual and asexual modes of reproduction. This is well demonstrated in the life cycle of the algae *Chlamydomonas*. The mature algae is a single haploid cell—that is, it contains only one set of chromosomes. During asexual reproduction the cell undergoes mitosis, a type of cell division that produces genetically identical offspring. Four daughter cells are created that emerge from the enclosing parent cell as spores. The spores develop into mature haploid cells that are genetically identical to the parent cell.

Certain environmental conditions, such as lack of nutrients or moisture, may trigger the haploid daughter cells to undergo sexual reproduction. Instead of forming into spores, the haploid daughter cells form gametes that have two different mating strains. These two strains are structurally similar and are called plus and minus strains. Opposite mating strains fuse in a process known as isogamy to form a diploid zygote, which contains two sets of chromosomes. After a period of dormancy, the zygote undergoes meiosis, a type of cell division that reduces the genetic content of a cell by half. This cell division produces four genetically unique haploid cells that eventually grow into mature cells.

Some multicellular green algae, such as *Ulva*, follow a distinct pattern of reproduction called alternation of generations, in which it takes two generations—one that reproduces sexually and one that reproduces asexually—to complete the life cycle. The two mature forms of the algae, alternating between diploid and haploid individuals, are identical in appearance, or isomorphic. The haploid form, called a gametophyte, undergoes mitosis to produce haploid gametes. These gametes unite to form a diploid zygote, which develops into the diploid form called a sporophyte. The sporophyte undergoes meiosis to form haploid spores that, in turn, form gametophytes.

Not all algae that undergo alternation of generations have haploid and diploid forms that look alike. In the life cycle of the seaweed *Laminaria*, the gametophyte and the sporophyte are distinct in appearance, or heteromorphic. The *Laminaria* sporophyte appears as long, bladelike structures that grow on rocks just below the water in intertidal zones. The gametophyte is short, with branched filaments.

The most common classification system distributes algae in more than one kingdom. Most algae are classified in the Kingdom Protista, along with other eukaryotic organisms that lack true specialised tissues. The cyanobacteria, however, are classified with the bacteria in the Kingdom Prokaryotae, which consists of prokaryotic organisms. This classification system continues to be intensely debated as new research increases our understanding of the way that these organisms are related. This article uses a classification scheme proposed in 1997 that divides algae, including the cyanobacteria, into 11 different phyla.

Green algae form the phylum Chlorophyta and are named for their green chloroplasts, which are similar in composition to the chloroplasts found in land plants. Green algae range in shape from unicellular plankton that grow in lakes and oceans to colonial filaments of pond scum to leaflike seaweeds that grow along rocky and sandy intertidal areas. Some green algae also live on tree trunks and soil. Several green algal species are symbiotic, forming lichens with fungi or living with corals. Green algae may also be found inside freshwater sponges, giving the sponges a bright green colour, and in permanent snow banks, where a secondary pigment masks the chlorophyll and turns the snow a reddish colour.

More than 500 genera and 8000 species of green algae have been identified. Some familiar green algae include the genus *Spirogyra*, known for its spiral-shaped chloroplasts, and the desmids, recognised by their characteristic shape—two symmetrical halves, joined by a small bridge. The green algae known as Stoneworts often grow several feet in length. Their name comes from calcium crusts that make them feel like stone.

Most green algae reproduce both sexually and asexually. Alternation of generations, where the algae alternates between gametophyte and sporophyte generations, is common among the multicellular green algae.

Red algae form the phylum Rhodophyta with approximately 500 genera and 6000 species. Found in warm coastal waters and in water as deep as 260 m (850 ft), red algae species adapt to varied water depths by having different proportions of pigments. Their red colour is due to a red pigment, phycoerythrin, which is well suited to absorb the blue light that penetrates deeper into water than the other colours of light. Red algae found in deep water may be almost black due to a high concentration of phycoerythrin. At moderate depths red algae appear red, while in shallow water they may appear green because a smaller proportion of phycoerythrin is unable to mask the green of chlorophyll. Most red algae are multicellular and come in a variety of shapes, including filaments, which are shaped like a blade of grass, and seaweed shapes. Unlike most other eukaryotic algae, red algae have no flagella.

Red algae use diverse strategies to reproduce, including fragmentation and spore production. One unusual strategy, found in many species including those in the genus *Polysiphonia*, involves the alternation among three generations. A diploid sporophyte produces diploid spores that germinate into another diploid sporophyte that looks completely different from the first one. Meiosis occurs in the second sporophyte, producing haploid spores that germinate into gametophytes. Surprisingly, in some species, the gametophytes look nearly identical to the second sporophyte.

Almost all red algae live in marine habitats, although some species are found in fresh water or damp soil. Many types of seaweed are red algae, typically found growing along the coast and attaching firmly to the seafloor using a rootlike holdfast. In some species, called coralline algae, the cell walls become hardened with calcium carbonate. Coralline algae are important members of coral reefs, producing new material and cementing together other organisms.

Golden-brown algae, brown algae, and diatoms form the large and complex phylum Heterokontophyta, with organisms ranging in size from a fraction of a millimetre to more than 100 m (300 ft) long. Heterokontophyta have carotenoid secondary pigments that tend to mask the green of the primary chlorophyll pigment, giving them a golden or golden-brown appearance. Flagellated cells in this phylum have two types of flagella: One is smooth, while the other has two rows of stiff hairs running down opposite sides of the flagellum. Algae in this phylum typically have an eyespot that can detect light.

The golden-brown algae, also known as the yellow-brown algae, include about 200 genera and 1000 species that receive their characteristic colouring from the carotenoid pigment fucoxanthin. These algae are mostly unicellular or colonial, swimming or floating in lakes and oceans as phytoplankton. In shallow ponds that dry up in summer or freeze completely in winter, golden-brown algae survive by forming protective cysts that can withstand the harsh conditions. When favourable conditions return, the algae emerge from the cysts. Like so many other algae, the unicellular algae tend to reproduce through fission, while the multicellular and colonial forms reproduce either through fragmentation or through spore production.

Diatoms are best known for their glasslike cell wall made of silica. The cell wall has ornate ridge patterns. A diatom consists of two overlapping halves that fit together like a shoebox or a petri dish, with the lid slightly larger and fitting over the base. During asexual cell division, the two glass walls separate and serve as the lids for two new glass bases. The new diatom that grew from the lid is the same size as the parent diatom, while the diatom that grew from the smaller base is slightly smaller than its parent. Sexual reproduction occurs when the succeeding generations shrink to a critical size. These smallest diatoms form gametes that shed their glass walls. Upon fertilization, the zygotes absorb water to swell and then secrete new, full-sized silica coverings.

A very large class with more than 250 genera and 8000 species, diatoms are found floating in freshwater and sea-water, growing attached to the seafloor, or growing on soil. The cells are either unicellular or form colonial chains of round cells. When an organism dies, its silica cell wall remains intact. Over time these shells have accumulated to form layers of soft rock in some geologic formations. This diatomaceous earth is mined and quarried for use in filters and bleaching agents, as an abrasive powder for cleaning and polishing metals, and for insect pest control (the broken cell walls of silica tear the insect gut).

Brown algae include over 260 genera and 1500 species. Multicellular algae, they may range from tiny filaments to the largest and most complex algae, such as the kelps, with leaflike blades and stems that can be up to 100 m (300 ft) long. Most brown algae grow in marine waters near the coast, attached to rocks either along the shoreline or underneath the ocean surface. One type, *Sargassum,* forms huge floating masses in the middle of the Sargasso Sea. The brown or olive colour is due to the pigment fucoxanthin. The life cycles of brown algae vary considerably, but most demonstrate alternation of generations.

Dinoflagellates of the phylum Dinoflagellata are mostly unicellular organisms that may be covered with stiff cellulose plates that resemble armoured helmets. Many species have unusual ornamentation, such as horns, spines, or wings. A narrow groove encircles the armour, and a second groove runs perpendicular to the first groove. Flagella beat within these grooves, causing the dinoflagellates to spin like tops as they move through the water. Most of the 130 genera and 2000 species in this phylum are planktonic and live in saltwater, although there are many freshwater planktonic representatives as well.

Many dinoflagellate species lack chloroplasts and are dependent on other species for their food. Some are parasites, but most are carnivores, using special harpoonlike structures called trichocysts to capture other organisms to eat. In contrast, several of the photosynthetic species live inside the tissues of invertebrate animals, such as corals and giant clams. These dinoflagellates share the food they photosynthesise with their host, and in return, receive protection and some nutrients.

Under favourable environmental conditions, some

dinoflagellate species experience population explosions, known as blooms. If the species involved in the bloom have red pigments, their concentration can be high enough to turn the sea-water red, forming red tides. Dinoflagellate blooms can be quite destructive. During the night when photosynthesis halts, such a high concentration of individuals can deplete the oxygen in the water, suffocating fish. Some dinoflagellates release toxins, some of which kill fish, while other toxins are passed up the food chain until they reach humans, where they can cause paralytic shellfish poisoning and ciguatera fish poisoning. Recently, the dinoflagellate *Pfiesteria piscicida* has caused fish, shellfish, and human disease in estuaries of the southeastern United States.

Unlike other algae, the cyanobacteria are prokaryotes—single-celled organisms with characteristics that cause biologists to debate whether they are really algae or bacteria. Cyanobacteria are found nearly everywhere, occurring in typical aquatic and terrestrial habitats as well as in such extreme sites as hot springs with temperatures as high as 71°C (160°F) and crevices of desert rocks. Cyanobacteria make up the phylum Cyanophyta, and this phylum contains about 150 genera and 2000 species worldwide.

Like other bacteria, cyanobacteria do not have organelles such as nuclei, mitochondria, or chloroplasts. Cyanobacteria are distinguished from bacteria by the presence of internal membranes, called thylakoids, that contain chlorophyll and other structures involved in photosynthesis. While higher plants have two kinds of chlorophyll, called a and b, cyanobacteria contain only chlorophyll a. Cyanobacteria colour varies from blue-green to red or purple and is determined by the proportions of two secondary pigments, c-phycocyanin (blue) and c-phycoerythrin (red), which tend to mask the green chlorophyll present in the thylakoids.

Cyanobacteria reproduce asexually by binary fission, spore production, or fragmentation, forming singular cells, colonies, filaments, or gelatinous masses. Although most lack flagella and are non-motile, filamentous forms such as *Oscillatoria* rotate in a screwlike manner, and the gelatinous forms glide along their slimy mucus.

Cyanobacteria may be both beneficial and harmful to humans: Some act as natural fertilizers in some habitats, especially rice paddies, whereas others produce toxins. Mild cyanobacteria toxins in lakes and oceans cause a rash known as swimmer's itch, while powerful neuromuscular toxins released by other cyanobacteria can kill fish living in the water or the animals that drink the water. In certain conditions, cyanobacteria may form dense blooms, which may produce toxins that make seafood poisonous to humans. Even if the cyanobacteria do not produce toxins, blooms can cause water to have an unpleasant taste and odour.

Uses of Algae

Human ingenuity has found many uses for algae. Algae provide food for people and livestock, serve as thickening agents in ice cream and shampoo, and are used as drugs to ward off diseases. More than 150 species of algae are commercially important food sources, and over $2 billion of seaweed is consumed each year by humans, mostly in Japan, China, and Korea. The red algae *Porphyra*, called nori, is the most popular food product. After harvesting, nori is dried, pressed into sheets, and used in soups, sauces, sushi, and condiments. Algae are considered nutritious because of their high protein content and high concentrations of minerals, trace elements, and vitamins. The high iodine content of many edible algae may contribute to the low rates of goitre observed in countries where people frequently eat algae.

In coastal areas of North America and Europe, seaweeds are fed to farm animals as a food supplement. Cyanobacteria species that are high in protein, such as *Spirulina,* are grown commercially in ponds and used mostly as a health food and cattle dietary supplement. Seaweeds also are applied to soils as a fertilizer and soil conditioner, as their high concentrations of potassium and trace elements improve crop production. Some species of cyanobacteria can turn atmospheric nitrogen into ammonia, a form that can then be used by plants as a nutrient. Farmers in tropical countries grow cyanobacteria in their flooded rice paddies to provide more nitrogen to the rice, increasing productivity as much as tenfold.

Seaweeds are a critical source of three chemical extracts used extensively in the food, pharmaceutical, textile, and cosmetic industries. Brown algae yield alginic acid, which is used to stabilise emulsions and suspensions; it is found in products such as syrup, ice cream, and paint. Different species of red algae provide agar and carrageenan, which are used for the preparation of various gels used in scientific research. Bacteria, fungi, and cell cultures are commonly grown on agar gels. Agar is also used in the food industry to stabilise pie fillings and preserve canned meat and fish. Carrageenan is also used as a thickening and stabilising agent in products such as puddings, syrups, and shampoos.

Algae have been used for centuries, especially in Asian countries, for their purported powers to cure or prevent illnesses as varied as cough, gout, gallstones, goitre, hypertension, and diarrhoea. Recently, algae have been surveyed for anticancer compounds, with several cyanobacteria appearing to contain promising candidates. Diatoms also have been used in forensic medicine, as their presence in the lungs can indicate a person died due to drowning.

Algae can also serve as indicators of environmental problems in aquatic ecosystems. Because algae grow quickly and are sensitive to changing environmental conditions, they are often among the first organisms to respond to changes. For example, depletion of the diatom community in the Florida Everglades provided strong evidence of phosphorus-related changes in this unique ecosystem. Algal blooms may deplete oxygen concentrations in water and smother fish and plant life, as well as prevent light penetration for algae at lower depths, preventing photosynthesis.

Abode of Algae

Tiny algae can be found floating or swimming in lakes and seas. Large seaweeds can stretch from the ocean bottom to the water's surface.

Most algae are found in fresh water or sea-water, but they can also be found growing on soil, trees and animals. They can even grow inside rocks.

Algae can withstand boiling hot as well as freezing cold temperatures. Such algae can be found growing in hot springs, in snowdrifts or deep down inside polar ice.

Are Algae Important?

Algae are extremely important. They produce more oxygen than all the plants in the world, put together. They are also a very important food source for tiny animals such as little shrimps and huge animals like whales.

Because algae are at the bottom of the food chain and many living things depend upon them.

Food of Algae

There are different type of algae, and they feed in different ways:

Autotrophic Algae: Most algae use the sun's energy to create food such as sugars, through the process of photosynthesis.

Symbiotic Algae: These types of algae form close relationships with other organisms. Both organisms benefit from living with each other. For example, lichens are an association between algae and fungi.

Another such relationship exists within the reef-building coral. Coral cells are mostly animal cells. But little algae live inside the cells of such corals, providing oxygen and nutrients to their coral partner.

Saprophytic and Parasitic Algae: Some microscopic algae are unable to create their own food using the sun's energy. These algae have to eat food from their surroundings.

They can absorb nutrients across their surface, or even engulf whole particles. Dinoflagellates - stick out a feeding tube to suck in food. Amazingly, some even use little harpoons to gather food particles.

Some algae are parasites - living inside or on the outside of a plant or animal, from which they take food. For example, some Dinoflagellates live in the intestines of marine animals.

Reproduction in Algae

Algae can reproduce by the use of spores. In large algae, the spores sometimes grow inside swollen ends of fronds which burst when ripe.

Sometimes these spores are produced by the fusion of male and female cells (sexual reproduction). Others form spores using just their own cells (asexual reproduction). In seaweed, the spores escape into the surrounding sea, settle on rocks or other seaweeds, and grow into individuals. Some spores, known as zoospores, move around using little whip like flagella.

Single-celled algae can reproduce simply by splitting themselves into two new individuals. Larger algae sprout little buds, which grow into new algae, in what is known as vegetative reproduction. Similarly, some algae, like *Sargassum*, broken fragments of the adult algae can grow into new individuals.

There are two categories of algae of concern to aquarists: "good" and "bad". Good algae is present in small quantities, is indicative of good water quality and is easily kept in check by algae eating fish or simple removal during routine maintenance. This algae is a natural consequence of having a container of water with nutrients and a light source.

Bad algae is either an indicator of bad water quality or is a type of algae that tends to overtake the tank and ruin the aesthetics the aquarist is trying to achieve. The label of "bad" is entirely subjective. For example, one type of green, hair-like algae is considered a plague by some American aquarists, yet is cultivated by European aquarists as a valuable addition to most tanks, serving as a dietary supplement for the fish.

Blue-green Algae

Grows rapidly in blue-green, slimy sheets. Spreads rapidly over almost everything and usually indicates poor water quality. However, blue-green algae can fix nitrogen and may be seen in aquariums with extremely low nitrates. Sometimes seen in small quantities between the substrate and aquarium sides. Will smother and kill plants.

This is actually cyanobacteria. It can be physically removed, but this is not a viable long term solution as the aquarium conditions are still favourable for it and it will return quickly. Treatment with 200 mg of erythromycin phosphate per 10 gallons of water will usually eliminate blue-green algae but some experts feel it may also have adverse effects on the biological filter bed. If erythromycin is used for treatment, ammonia and nitrite levels should be carefully monitored.

Brown Algae

Forms in soft brown clumpy patches. In the freshwater aquarium, these are usually diatoms. Usually indicates a lack of light or an excess of silicates. Increased light levels will usually make it disappear. Easily removed by wiping the glass or siphon vacuuming the affected area.

Green Algae

Green unicellular algae will sometimes reproduce so rapidly that the water will turn green. This is commonly called an "algae bloom" and is usually caused by too much light like direct sunlight.

An algae bloom can be removed by filtering with micron cartridges or diatom filters. UV sterilisers can prevent the bloom in the first place. Green water is very useful in the raising of daphnia and brine shrimp.

Film Algae

Grows on the aquarium glass and forms a thin haze. Easily removed by wiping the glass. Considered normal with the higher light levels needed for good plant growth.

Spot Algae

Grows in thin, hard, circular, bright green spots, usually on the aquarium glass but also on plants under high light conditions. Considered normal for planted tanks. Must be mechanically removed. On acrylic aquariums, use a cloth pad or a gentle scouring pad like a cosmetic "Buff-Puff" and a lot of elbow grease. On glass tanks, scraping with a razor blade is most effective.

Fuzz Algae

Grows mostly on plant leaves as separate, short (2-3mm) strands. Considered normal. It might be a less "virulent" form of "beard" algae. Easily controlled with algae eaters such as black mollies, *Otocinclus*, *Peckoltia* and *siamese algae eaters*.

Beard Algae

Grows on plant leaves and is bright green. Individual strands have a very fine texture but it grows in thick patches and looks just like a green beard. It grows up to 4 cm. It cannot be removed mechanically. This does not indicate bad water quality but grows very fast and overtakes the tank, making it a "bad" algae. Can be eliminated with Simazine (Aquarium Pharmaceuticals "Algae-Destroyer").

Hair Algae

Grows in bright green clumps in the gravel, around the base of plants like *Echinodorus* and around mechanical objects. It has a coarser texture than "beard algae". Beard algae will ripple in the water current, hair algae tends to form matted clumps. Individual strands can get to 5 cm or more. This is easy to remove mechanically by twirling a toothbrush in it. Can be troublesome if left unchecked. This is a popular food supplement for fish among European aquarists.

Thread Algae

Grows in long, thin strands up to 30 cm or more. Tends towards a dull green colour (hard to tell because it is so thin).

Usually indicates an excess of iron (> 0.15 ppm). Easily removed with a toothbrush like hair algae.

Staghorn Algae

Looks like individual strands of hair algae but tends to grow in single branching strands like a deer antler and is grey-green. Seems to grow mostly on tank equipment near the surface. Difficult to remove mechanically. Soak affected equipment in a 25 per cent solution of household bleach and water to remove it.

Brush Algae

This grows in feathery black tufts 2-3 mm long and tends to collect on slower growing leaves like *Anubias*, some *Echinodorus* and other wide leaf plants. Also tends to collect on mechanical equipment. This is actually a red algae in the genus *Audouinella*. It cannot easily be removed mechanically. Remove and discard the affected leaves. Equipment can be soaked in a 25 per cent bleach solution, then scrubbed to remove the dead algae. Siamese Algae Eaters (*Crossocheilus siamensis*) are known to eat this algae and can keep it in check. A more drastic measure is treatment with copper.

Prophylactics for Algae

Algal spores are everywhere and will always be present in an aquarium unless drastic measures are taken. For fish only tanks, a properly set up ultraviolet steriliser will kill algal spores in the water and prevent them from gaining a toehold.

For planted tanks, this is not a good solution since the UV light will also oxidise trace elements needed by the plants and will limit the plant's growth potential. Unfortunately, conditions that are good for growing plants are also good for growing algae. Fortunately, plants will usually out-compete algae for the available nutrients. However, if there is an imbalance of nutrients, algae will opportunistically use whatever is not used by the higher order plants. Different algae will utilise different nutrients, causing sporadic outbreaks of new algae types in apparently stable tanks when a temporary imbalance occurs.

An ounce of prevention is worth a pound of cure. To avoid introducing a new algae type to a planted tank with new plants, a simple bleach dip seems to work well. Mix 1 part bleach in 19 parts water and dip the new plant in it for 2 minutes. Immediately rinse the plant in running water, then immerse it water containing a chlorine remover to neutralise any remaining bleach. This will kill the algae and only temporarily slow down a healthy plant. Plants in poor condition may succumb to this treatment, but they probably would not have lasted anyway.

Algae Eaters

The most effective control of algae in a planted aquaria is via algae eating fish. It is especially critical in the set up of a new tank to make sure algae does not get established before the plants have had a chance to establish themselves. For this reason and to help the biological filtration get established, it is recommended that some hardy algae eaters are added right away.

Black Mollies

Black sailfin mollies are excellent candidates for the break-in period of a planted tank since they are cheap and easy to find. They are usually considered expendable and are removed after a month or so. If they are fed, they will not be quite so eager to consume algae. When they are hungry, they are eager consumers of most algae types seen during the break-in period.

Otocinclus Sp.

Otocinclus are diligent algae eaters, but are best kept in schools due to their small size. One per 10 gallons is a useful rule of thumb. Various species of otos are seen in the shops at various times; most are good algae eaters but some seem to prefer the slime coat on fish to algae. Unfortunately, there seems to be no way to distinguish the "attack otos" from normal otos. Otos seem to be very delicate fish, but this is probably due to capture and shipping abuse rather than an inherent weakness. When a fish shop gets some in, it is wise to wait a while before purchasing to account for die offs. Most people report getting a dozen and having them die over a period of a few months until just a couple are left. Those then seem to last for a long time.

"Plecostomus" Sp.

Plecostomus is the generic name for a wide range of sucker-mouth fish. Only the smaller types are useful in a planted tank, since the larger varieties tend to eat the plant right along with the algae. Two common types that are useful are the "bristle-nose plecostcmus" and the "clown plecostomus" or Pekoltia. Both stay under 4" long and don't seem to cause too much plant damage. Sometimes broad-leafed plants like Amazon swords will be scraped a little too closely by the plecos, so they bear watching. Their diet can be supplemented by blanched zucchini and bottom feeder tablets. They also appreciate a chunk of driftwood in the aquarium to satisfy their need for cellulose.

Siamese Algae Eater

Do not confuse this fish with the *Chinese* Algae Eater, which is very aggressive and does not eat algae. The siamese algae eater, *Crossocheilus siamensis*, is a very good algae consumer and is known to eat black brush (red) algae. The only problem is that these fish are hard to find in the United States. There are several fish in this family. The most commonly seen is *Epalzeorhynchos kallopterus*, commonly known as the Flying Fox. The Flying Fox is the more attractive of the two. It tends to have a brownish body with a very distinct, sharp-edged black stripe with a distinct, thin gold or bronze stripe above it. These tend to be very aggressive when they are full grown and don't eat red algae.

The other member is the Siamese Algae Eater. It is the same shape as the Flying Fox but tends towards a silverish body with a somewhat ragged black stripe. There may be an indistinct gold or bronze stripe above the black. These are definitely not aggressive; they are good companions for discus and small tetras.

When they are young, the differences between *E. kallopterus* and *C. siamensis* may not be very apparent, especially if you haven't seen both types together. Unfortunately, most wholesalers don't sell fish to stores by their scientific name and the common names that are used sometimes get pretty silly (like "siamese flying fox"). If you really can't tell which one the store has, buy it anyway, but be prepared to sacrifice it if it turns out to be the wrong kind (unless your fish aren't bothered by it, of course).

Farlowella

Farlowella are useful algae eaters although they are very sensitive to water conditions. They type known as the Royal Farlowella will get too large for a plant tank and may cause damage.

Algal Cells

It was a huge step in complexity from a virus to a bacterium and it is almost as large a step from the prokaryotic cells of bacteria to the eukaryotic cells of algae. Many of the larger algae are accepted as plants by most people but the absence of highly differentiated cells distinguishes them from the kingdom Plantae and places them in the Protista.

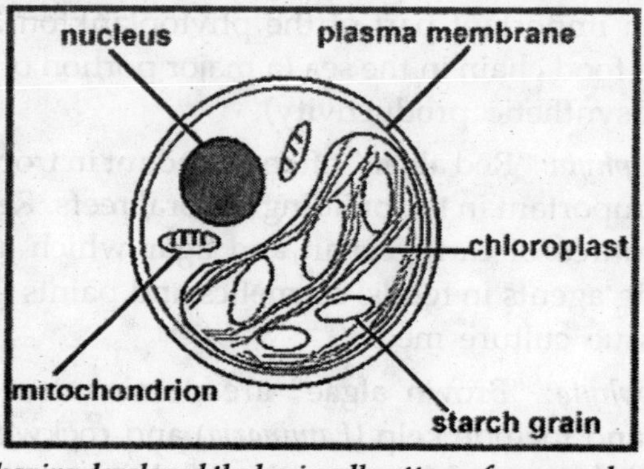

Nevertheless having developed the basic cell pattern of a green algae all that was necessary for the development of higher plants were variations on this pattern.

The Protista

This kingdom includes a tremendous variety of organisms from heterotrophs to autotrophs and unicellular to multicellular. Their origin is "multiphyletic", that is they arose from several lines of prokaryotes, rather than from a common ancestor. They form a kind of link-group between the prokaryotes and other eukaryotes, both animals and plants. Some of them such as Amoeba, Paramecium and Hydra are closer to animals than to plants. The more plantlike include:

- *Oomycota:* These are heterotrophic and form coenocytic hyphae like fungi, with which they were long included. They are unlike fungi in their mode of reproduction and in having cellulose cell walls. It seems that their similarity to fungi is an example of convergent evolution, rather than a direct relationship. In addition to saprophytes which live in the soil a number of plant diseases are caused by Oomycetes: potato blight, downy mildew, root and collar rots.
- *Euglenophyta:* Look like a unicellular algae but have a protein "pellicle" like an animal rather than a cell wall. Now thought to be result of capture of a chloroplast by an "animal" cell.
- *Chrysophyta:* Often unicellular, including the diatoms which are unusual in having silica cell walls and which are an important part of the phytoplankton at the base of the food chain in the sea (a major portion of the earth's photosynthetic productivity).
- *Rhodophyta:* "Red algae" often abundant in tropical waters and important in the building of coral reefs. Red algae are the source of carrageenan and agar which are used as gelling agents in foods, cosmetics and paints (in addition to tissue culture media!)
- *Phaeophyta:* "Brown algae" are dominant in temperate seas and include kelp (*Laminaria*) and rockweed (*Fucus*). Kelp is harvested for extraction of alginate which is used in toothpaste and many food products.

Fucus vesiculosus or "bladderwrack"

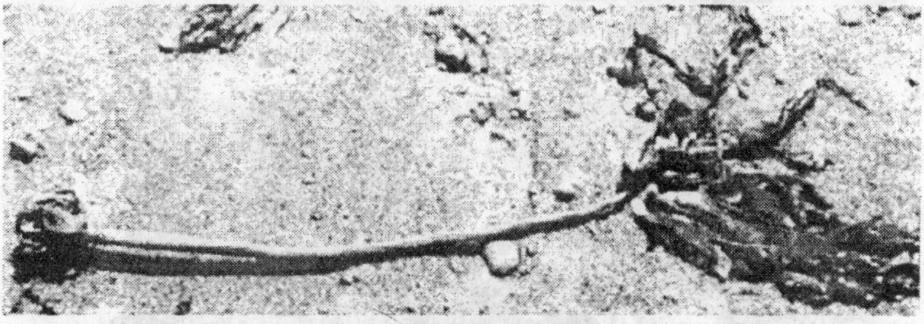

Laminaria or "kelp"

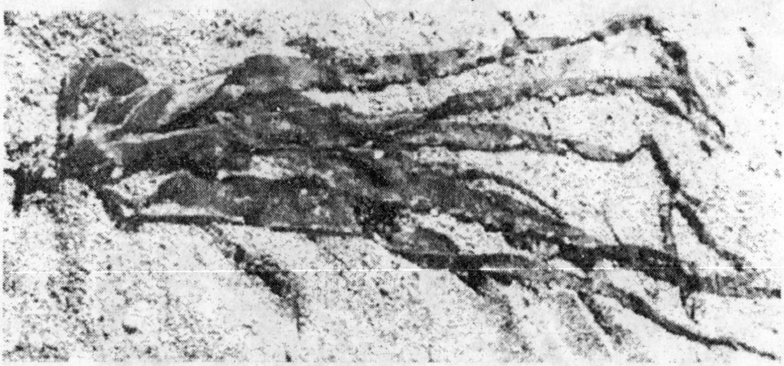

Chlorophyta "Green algae" include some of the most plantlike organisms. There are many marine types including *Ulva* or sea lettuce. Unlike the other two groups, this division includes some terrestrial (Protococcus on tree bark) and freshwater (Chlamydomonas, Volvox, Cladophora) organisms. Filamentous green algae are almost weeds, giving trouble in irrigation systems and even on the surface of wet soils and media.

The unicellular algae, Protococcus can often be found growing on the West face of tree trunks, especially in damp locations

Algal blooms tend to form on ponds and lakes especially when there is fertilizer run-off from fields and lawns

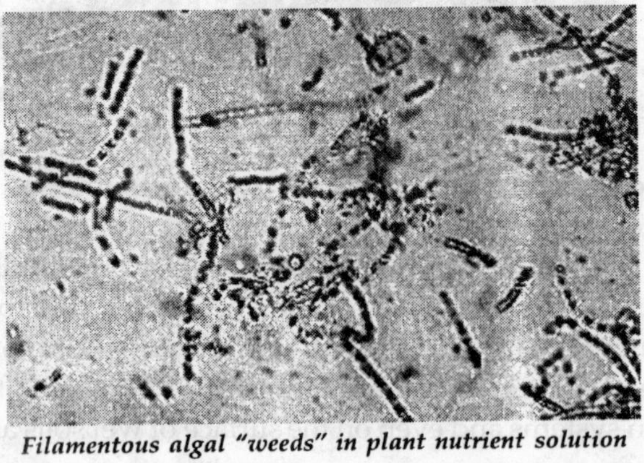

Filamentous algal "weeds" in plant nutrient solution

Ulva lactuca

The Rhodophyta, Phaeophyta and Clorophyta constitute the "algae" (singular algae) They include motile unicellular (*Chlamydomonas*), motile multic [illegible]ular (*Volvox*) and static multicellular types (*Cladophora, Spirogyra* and "seaweeds").

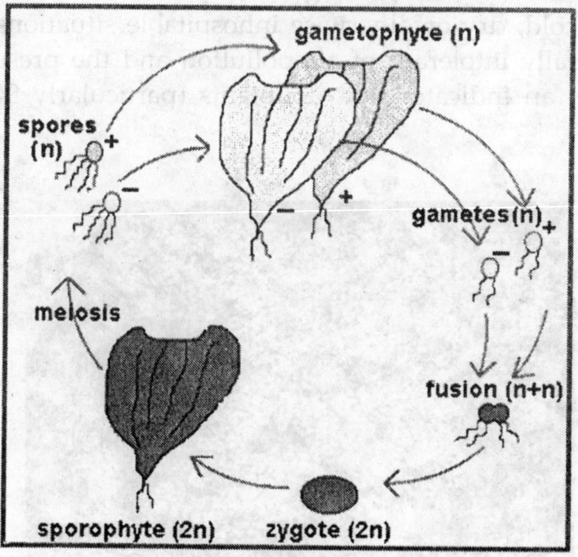

Life cycle of Ulva lactuca

The motile types often use whip like flagellae to propel themselves. The gametes of static multicellular types are free-swimming and also have flagellae. The larger static types show some degree of cellular differentiation and organisation into rootlike (holdfast) stem-like (stipe) and even leaflike (blade) structures.

The smaller free floating or motile forms make up a major part of the phytoplankton (along with Chrysophyta and Cyanobacteria). In each of these groups of algae, there are organisms which show an alternation of haploid and diploid generations. Sometimes the morphology of the haploid gametophyte is distinct from the diploid sporophyte as in *Fucus* and sometimes they are hard to tell apart as in *Ulva*.

Lichens

Lichens are formed by symbiotic association between cyanobacteria or green algae and fungi. The photosynthetic bacteria or algae make the lichen an autotrophic organism. They can take several forms: crustose (encrusting), foliose (leafy) and fruticose (shrubby) and grow on rocks, tree bark or soil often in cold, dry or otherwise inhospitable situations. They are generally intolerant of air pollution and the presence of lichens is an indicator that pollutants (particularly SO_2) are absent.

Concept of Seo-weed

Most of the plants that live on the intertidal rocky shore, or under the water along our seashores, are algae. Most people call these algae plants seaweed, or kelp. There are a few underwater-living flowering plants, called seagrasses. They usually live in saltwater estuaries. Some also live along our ocean shores. Like land plants, these flowering plants have a root system to take up nutrients. They absorb sunlight through their leaves and have small flowers for reproduction. Algae are quite different.

Algae

Algae have quite a number of features which make them unlike normal land-based flowering plants. Firstly, they do not have a root system which takes up nutrients. Algae take their food in through their leaflike fronds which are surrounded by nutrient-carrying sea-water. Their "roots" are called holdfasts, and that is what they do. They hold the algae down firmly onto the rock surface. Algae also reproduce quite differently to the flowering plants. They have spores like the ferns, mosses, lichens and liverworts. Algae are divided into three main groups. There are green, brown and red algae, which differ in having different photosynthetic pigment systems.

Green Algae - Phylum Chlorophyta

Green algae get their colour from the photosynthetic chlorophyll pigments. They come in a variety of shapes including flat sheets, cylinders, strings of beads, spheres, or hair-like filaments. Green algae are common in the intertidal zones and in shallow water as well as in some freshwater habitats, where light is plentiful.

Sea Lettuce, Ulva Lactuca

A green algae, usually with a broad, thin, almost transparent single frond from a small holdfast. It is not divided into long branches. The thalus (leaf) is often entire, but may be divided to form ruffled branches. The frond may grow to 15 cm long and 10 cm across. The common name Sea Lettuce describes its form well.

Colour: The Sea Lettuce is grass green in colour.

Distribution: Once considered a cosmopolitan species, researchers have now divided this algae into a number of species. Womersley has identified six related species from southern Australian shores.

Habitat: The various species of Sea Lettuce occur at the lowest tide levels, and in rock pools, in areas constantly being washed by breaking waves. It never occurs on the steep marginal slopes, but prefers flat platform surfaces.

Green Sea Velvet Codium Fragile

Green Sea Velvet is an erect, much branched, dark green algae. It has rounded, finely branched fronds which are packed close together. The fronds grow up to 30 cm long and are about 6 mm thick. The cylinder-shaped fronds often split into two equal halves. When under water, fine hairs can be seen all over the branches. The basal disc is broad, spongy and soft.

Colour: Green Sea Velvet is black green to dark bottle green in colour.

Distribution: Green Sea Velvet ranges from Ballina in northern New South Wales, south and west to Victor Harbour in South Australia, including Tasmania. This species is found in cooler temperate waters of both hemispheres.

Habitat: It occurs at the lowest tide levels and below, as well as in intertidal pools on high energy rocky coasts.

Caulerpa Filiformis

Caulerpa is a seaweed with a dense mass of bright green, small, strap-like leaves called ramuli. Each ramuli measures 100-200 mm long and 5 mm wide. The cartilage-like stem of the plant, called a stolon, grows flat along the rock attached by fine outgrowths. Growing out from the stolon are five rows of photosynthetic ramuli, each of which tapers to a short, blunt, spinous tip.

Colour: Caulerpa is really bright green in colour. It is a very noticeable algae.

Distribution: In Australia, this species of Caulerpa, occurs only in New South Wales. Recent research reveals that it also occurs in South Africa. It is uncertain whether it is a recent immigrant from Africa, but many researchers state that it wasn't as abundant only as recent as 20 years ago. It was first recorded near Sydney in 1923. It appears to be taking over large areas and is crowding out other algae forms.

Habitat: Caulerpa is found at low-tide levels along the edges of rock platforms and in gutters, on medium- to high-energy shores, down to 40 metres. It can be found near Cunjevoi at low-tide level.

Biology: There is evidence that this algae is a recent immigrant from South Africa. It appears to be taking over many habitats along the New South Wales rocky shores, so that it is now the most abundant algae to be found. It is so common that it may be reaching pest, or even plague proportions, and excluding other local species. Another explanation for this rapid take-over may be due to human pollution altering the habitat so much that only this species can take advantage of the new conditions.

Brown Algae - Phylum Phaeophyta

Brown algae have additional pigments which mask the green chlorophyll. These pigments allow the brown algae to extend their range down into deeper waters because the pigments are more efficient than green chlorophyll in absorbing the sunlight down away from the ocean surface. Brown algae also come in a wide variety of forms. Some brown algae are the largest, longest and most conspicuous seaweeds found on rocky ocean shores. They are a common feature of the low-shore fringe and shallow sub-tidal zones. Some brown algae from southern Australian seas are the fastest growing of all plants.

Neptune's Necklace Hormosira Banksii

Neptune's Necklace is a distinctive algae made up of strings of hollow, water-filled, round or oval-shaped beads joined together by a short stalk.

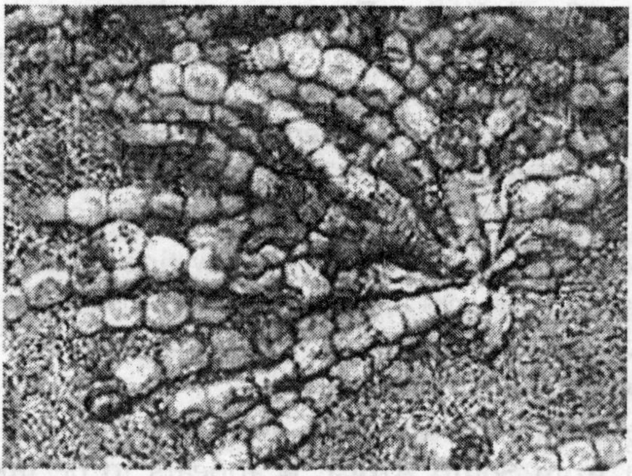

The Fronds may be between 10 - 30 cm long, and the beads may be 15 mm in diameter. Many pores cover the outside surface of each bead, forming a rough surface. Within these pores are the reproductive cells. Neptune's Necklace is sometimes called "sea grapes".

Colour: Dark brown, with some green in colour. When it is dried out and killed by the sun, it turns a quite yellowish-orange.

Distribution: Found from Port Macquarie, in northern New South Wales around southern Australia to King George Sound in Western Australia.

Globe Algae Colpomenia Sinuosa

The Globe Algae is irregular in shape, globe-shaped, and crinkled in shape. This algae is sometimes twice as high as it is broad.

It's skin membrane is a thin, crisp wall filled with water and air.

The plant is usually attached to other algae, seagrasses or rock by a small, crusty base. Globe Algae is also called Bubbleweed.

Colour: Globe Algae is light yellow or honey-brown to dark brown in colour.

Distribution: Widely found in most oceans. Found all around Australia, but preferring tropical and temperate shores. It seems to be more common in summer.

Habitat: Occurs from mid-tide level and below, being common in rock pools. Often attached to other algae, or seagrasses such as Posidonia in estuaries.

Biology: This is the most common species of globe-shaped algae found attached to other plants in southern Australia.

Leather Kelp

The Leather Kelp is the most common large kelp on southern Australian coasts.

From the conical Holdfast, a long rounded stem (0.3 to 2 metres) ends with a broad, flat, strap-like blade. From this central blade, Lateral expansions (5 to 20 cm long and 1-10 cm wide) occur on both sides. These have a crinkled, lobed and rough surface with short spines (2 to 4 mm long).

Colour: Mid to dark brown.

Distribution: Occurs from Caloundra in southern Queensland, south and west to the Abrolhos Islands in Western Australia, including Tasmania. Also New Zealand and South Africa.

Habitat: Found from just below the lowest tide level and below on coasts affected by moderate to rough wave action.

Biology: After rough storms many fronds of this large kelp are washed ashore to form a major part of the flotsam and wrack of southern shores. If you look under the fronds lying stranded on the shore, hundreds of small flattened isopods and amphipods will be exposed and hop away for cover. Along southern shores, thousands of Marine Slaters, Ligia Australis, may also be found under the piles of wrack.

Strap Weed, Philospora Comosa

Recognised by its sawtooth-edged fronds and many elliptical spindle-shaped. Bladders at the end of short stalks, and not at the base of leafy fronds.

The bladders keep the fronds upright in the water so that they are more exposed to the sunlight. The Holdfast is 1-6 cm across. Its fronds are more smooth, narrower and less leaflike than Macrocystis.

Colour: A characteristic light olive green with a yellow to golden tint.

Distribution: Found from Port Macquarie, New South Wales, south and west to Robe in South Australia.

Habitat: Occurs on rough water coasts at lowest tide levels and below and in deep pools. In Tasmania, down to 18 metres, it generally occurs as a dense band above Ecklonia Radiata.

Biology: Phyllospora is a sinuous, beautiful looking algae when it is being swashed around in the surging water. It is among the most common large algae in shallow water around the southeast Australian coast. Also known as Bladder Weed, Bubble Weed or Popweed.

Blue-Green Algae in Lake Champlain

Blue-green algae, (cyanobacteria) are a common and natural component of the microscopic plants (plankton) in Lake Champlain. Some kinds of blue-green algae produce natural toxins or poisons. When these algae die and break down, toxins can be released into the water.

If animals ingest the toxin, they can be quickly paralysed and die. Signs of poisoning include weakness, staggering, difficulty in breathing, convulsions and death. Although there are no documented cases of human illness related to blue-green algae in Lake Champlain, during the summers of 1999 and 2000 the death of two dogs was attributed to blue-green algae poisoning, after drinking large amounts of contaminated water directly from the lake.

Blue-green algae can become very abundant in some sections of Lake Champlain once the water warms up in mid-summer. Particular problem areas are Missisquoi Bay and St. Albans Bay. Under calm conditions, blue-green algae can accumulate in thick layers at the surface or along the shoreline. These accumulations are frequently referred to as "Blooms" or "Scums." While blue-green algae toxins have been detected at many locations in Lake Champlain, the highest concentrations of toxins are usually found in blooms and shoreline scums. These dense accumulations pose the greatest potential risks to people and pets.

Weather influences where blue-green algae will accumulate in a given location. During extended periods of calm and sunny days, blooms can accumulate at the surface in any location. Wind and waves may cause them to form along shorelines or in protected areas. Shifts in wind direction can move a bloom from one location to another. Periods of cool rainy weather can often lead to the disappearance of a bloom.

Monitoring for Blue-Green Algae

The University of Vermont's Rubenstein Ecosystem Science Laboratory is conducting a monitoring programme for blue-green algae, in order to identify locations around Lake Champlain where there may be a potential risk to people and pets.

Working with the Vermont Department of Environmental Conservation, the Rubenstein Lab is monitoring 14 stations around all of Lake Champlain about every two weeks. The lab samples additional sites in Missisquoi Bay, St. Albans Bay, and Burlington Bay on a weekly basis once a bloom occurs.

Because the highest concentrations of toxins are found in shoreline locations where scums can form, and the shorelines are where most people recreate, the Rubenstein Lab has developed a special programme focused on these locations.

Working with the Lake Champlain Committee, the lab has recruited a group of citizen volunteers that monitor about a dozen shoreline sites in Missisquoi Bay, St. Albans Bay, the north lake and islands area, and the south lake. These citizen volunteers watch for blooms and take weekly samples that the Rubenstein Lab can analyse for the presence of toxins.

Conditions in Lake Champlain

The Rubenstein Lab uses its data to provide weekly information about blue-green algae conditions around the lake. This information is reported on a newly created status map, showing the general conditions in various sections of the lake, using the following three classifications:

- *Generally Safe:* No toxic blue-green algae blooms were present along these shoreline areas when we sampled. These areas should be safe for recreational uses.
- *Low Alert:* These shoreline areas have blue-green algae that produce toxins, but toxin concentrations are currently below levels of concern. These areas are open for recreation, but caution is advised in any areas where dense accumulations or scums are apparent.

- *High Alert:* These shorelines have blue-green algae in dense scums, at least in some areas, and toxin concentrations are above levels of concern. The water is not safe for recreational use in areas that contain scums.

Algae Concentration in Coastal Waters

Referred to as a microalgae, brown or golden algae is actually not an algae at all, but diatoms. What you are actually seeing in your tank are diatom skeletons, all linked together. It can appear as a simple dusting on the tank walls and substrate surfaces, or it can turn into a massive growth that covers just about everything in the tank. This type of algae outbreak typically occurs when a tank is just completing or has finished the nitrogen cycling process, new live rock is introduced, as the curing process can add nutrients when some organisms on the rock dies off, or tank maintenance has been neglected. It is a normal occurrence, as diatoms are one of the first to appear in the chain when the tank conditions are conducive for algae growth, and is usually a precursor to other forms of desirable and undesirable nuisance type green macroalgae.

The Causes and Remedies of Diatoms Growth

We suggest that you don't try to put all of these solutions into action at one time, because if you do, when to problem subsides you'll never really know "exactly" where the problem was coming from and which solution worked to fix it. Start with one solution and see what results you get, and if that one doesn't work, try another one, and so on, until the problem is resolved. Now, in order for all forms of algae to grow, they require only two things; light and nutrients.

- *Lighting:* The use of improper bulbs, lack of maintenance, and extended lighting hours are contributors that can lead to all sorts of algae problems.
  — *Solutions:* Only use bulbs that are designed for aquarium use, paying close attention to their spectral output; don't bombard the tank with an over abundance of light, follow the basic wattage rule of thumb; run the lights 8 to 9 hours a day.

- *Nutrients:* Diatoms are most responsive to silca/silicates, but DOCs (Dissolved Organic Compounds), nitrates, and phosphates are food sources as well.

— Silicates are most often introduced into aquariums by means of using unfiltered fresh tap water, the wrong kind of sand or substrate material, and through sea salt mixes that contain a higher than normal concentration of this element.

- Solutions: Use RO/DI filtered make-up water, an aragonite type sand or substrate source, and a high quality sea salt mix.

  — Phosphates (PO_4) are commonly introduced into aquariums by means of using unfiltered fresh tap water, and through many aquarium products that may contain higher than normal concentrations of this element, such as sea salt mixes, activated carbon, KH buffers, foods, and many other sources. Also, for established reef tanks the long-term use of Kalkwasser precipitates phosphates out of the water, and these phosphate based compounds can settle on and in the live rock and substrate.

- *Solutions:* Use RO/DI filtered make-up water, a high quality sea salt mix, and be aware of the elements contained in other common aquarium products you may be using. For solutions to problems that can arise from using kalkwasser, refer to The Fishline's Microalgae article.
- Allowing excess DOCs to accumulate in an aquarium in turn gives rise to nitrate (NO_3) problems. However, nitrates can also be introduced in the same manner as phosphates, and because it is the final by-product produced in the nitrogen cycling process, it can naturally build to high levels due the lack of proper aquarium maintenance care. As mentioned earlier, another contributor to DOC/nitrate problems is when new live rock is introduced, as the curing process can add nutrients when some organisms on the rock dies off.

- *Solutions:* Practice good aquarium maintenance care routines. This includes keeping the substrate clean, cutting back on feedings, regularly rinsing, rejuvenating or changing any type of prefilteing or absorbing materials (such as filter flosses, cartridges, biowheels, sponges and carbon), performing regular partial water changes, and for DOCs in particular, adding a protein skimmer protein skimmer (read reviews and compare prices).
- For those with systems that have been running for some time and use wet/dry trickle type filters, the biomedia in them, especially bio balls, are real nitrate factories, and therefore, should carefully be rinsed and cleaned periodically.
- Add some good diatom eating Trochus and Astraea snails, a fish, such as a Kole Tang, Potter's or Flame Angelfish, and to help keep the aquarium bottom clean and tidy, some algae/detritus eating hermit crabs, a tank friendly true crab, shrimp, or other type of tank janitors. Of course you DON'T want to add ALL of these critters, just choose on or two of these options. In our photos, the clear or white looking patchy area at the bottom of the diatom algae is from where a small true crab in our tank ate the algae off the glass that it could reach.
- When adding live rock, take the time to cure it properly.

Phaeophyceae

The Algae comprise in the main freshwater and terrestrial forms. The Phaeophvceae or Brown Algae are practically restricted to the sea. The few freshwater genera *(Pleurocladia, Herihaudiella, Bodanelhi)*. Many Phaeophyceae of the littoral zone are, however, highly equipped to resist exgosure during intertidal periods, and a number of Fucales have become adapted to life under the special conditions obtaining on salt-marshes.

Although certain genera are largely or entirely restricted to the warmer seas (e.g. *Hydroclathrus, Chnoospora,* most

Dictyotales, *Sargassum turbinaria)*, the Brown Algae attain their maximum development, both as regards numbers of species and of individuals, in the colder waters of the Northern and Southern Hemispheres. Many distinctive forms occur in the Pacific. Certain orders (Sporochnales, Sphacelariales) and genera (*Splachnidium, Ecklonia, Lessonia,* many Fucales) are better represented or exclusively found in the Southern Hemisphere.

The abundant and luxurious growth of Phaeophyceae in Subarctic and Arctic regions has been vividly described by Kjellman; development continues during the darkness of the Arctic winter and many forms produce their reproductive organs at this time at temperatures often below their freezing point. Similar data have been provided Skottsberg for the Antarctic. These facts have been frequently interpreted as indicating that the Brown Algae of the colder seas are shade-forms attaining an adequate excess of photosynthesis over respiration in weak light and at low temperatures.

Recent investigations have, however, shown that certain Phaeophyceae, like certain Rhodophyceae, have the physiological characteristics of sun-forms, while some can adjust their metabolic processes to the varying light-intensities and temperatures of different seasons, the larger fucoids being capable of producing as much substance or even more in summer than in winter. Nevertheless, the low rate of respiration at low temperatures and the capacity to photosynthesise in feeble illumination are no doubt important factors.

A large proportion of the Phaeophyceae arc lithophytes, so that they are specially prevalent on rocky shores. No very large number of Brown Algae normally live in deep water, although a few can penetrate to very considerable depths. The majority inhabit the belt between tide-levels or grow at such a depth that their photosynthetic tissues are displayed but a little way beneath the water-surface.

No unicellular or colonial forms are at present certainly known, nor is the unbranched filament represented. In other words all the simple types of plant-body, so widely found in classes like Chlorophyceae and Chrysophyceae, are lacking. The vegetative body in the least specialised members of Phaeophyceae is a heterotrichous filament, a level of differentiation which is the highest attained among Chlorophyceae.

It is nevertheless highly probable that the evolution of Phaeophyceae has followed the same lines as that of Chlorophyceae, etc., but that the simpler types have either become extinct or still await discovery; there is no apparent reason why they should have perished. The few fossil forms do not help to elucidate the evolution of the class.

2

The Microalgae

Algaculture

Algaculture is a form of aquaculture involving the farming of species of algae. The majority of algae that are intentionally cultivated fall into the category of microalgae, also referred to as phytoplankton, microphytes, or planktonic algae.

Macroalgae, commonly know as seaweed, also have many commercial and industrial uses, but due to their size and the specific requirements of the environment in which they need to grow, they do not lend themselves as readily to cultivation on a large scale as microalgae and are most often harvested wild from the ocean.

History and Uses of Algae

Presumably, the first use of algae was food. One example is the wrapper on a sushi roll. Other species are edible as well, such as Spirulina and dulse (Palmaria palmata). Dulse is a red species sold particularly in Ireland and Atlantic Canada. It is eaten raw, fresh, dried, or cooked like spinach.

Spirulina is a blue-green microalgae with a long history as a food source in East Africa and pre-colonial Mexico. As it is high in protein and other nutrients, it is currently used as a food supplement and as a treatment for malnutrition. Chlorella, another popular microalgae, has similar nutrition and is unique for being the source of the "Chlorella Growth Factor," a potent phytochemical which has been shown to increase growth in animals and children and a cell wall which has a high affinity for heavy metals and poisons, particularly mercury. The cell wall binds to the toxin and helps remove it from the body. Chlorella is very popular in Japan and is currently one of the most prescribed supplements in that country.

Purple laver (Porphyra) is also collected and used in a variety of ways. In the Wales, for example "laverbread" is a traditional food, and in Ireland, it is collected and made into a jelly by stewing or boiling. Preparation also can involve frying or converting to a pinkish jelly by heating the fronds with a little water and beating with a fork. It is also harvested along western coast of North America, from California to British Columbia and by Native Hawaiians and the Mori of New Zealand.

Irish moss (*Chondrus crispus*), often confused with Mastocarpus stellatus, is the source of "carrageen" for the stiffening of instant puddings, sauces, and dairy products such as ice cream. Irish moss is also used by brewers as a fining agent; the addition of Irish moss to the wort 15 minutes before the end of the boil produces a clearer beer.

Sea lettuce (Ulva lactuca), is used in Scotland where it is added to soups and salads. Dabberlocks (Alaria esculenta) is eaten either fresh or cooked in Greenland, Iceland, Scotland and Ireland.

For centuries seaweed has been used as fertilizer: "This kind of ore they often gather and lay in heaps where it heteth and rotteth, and will have a strong and loathsome smell; when being so rotten they cast it on the land, as they do their muck, and thereof springeth good corn, especially barley." It is also an excellent source of Potassium for manufacture of potash and potassium nitrate.

There are commercial uses of algae, such as agar.

Growing, Harvesting and Processing Algae

Growing Algae: When cultivating algae, several factors must be considered, and different algae have different requirements. The water must be in a temperature range that will support the specific algal species being grown. Nutrients must be controlled so algae will not be "starved" and so that nutrients will not be wasted. Light must not be too strong nor too weak.

Algae can be cultured in raceway-type ponds and lakes. Because these systems are open to the elements, sometimes called "open-pond" systems, they are much more vulnerable to contamination by other microorganisms, such as invasive algal species or bacteria. Because of these factors, the number of species successfully cultivated in an "open-pond" system for a specific purpose (such as for food, for the production of oil, or for pigments) are relatively limited.

In open systems, one does not have control over water temperature and lighting conditions. The growing season is largely dependent on location and, aside from tropical areas, is limited to the warmer months. A major benefit to this type of system are that it is one of the cheaper ones to construct, in the very least only a trench or pond needs to be dug.

It can also have some of the largest production capacities relative to other systems of comparable size and cost. This type of culture can be viable when the particular algae in question requires (or is able to survive) some sort of extreme condition that other algae can not survive. For instance, Spirulina sp. can grow in water with a high concentration of sodium bicarbonate and Dunaliela salina will grow in extremely salty water.

Open culture can also work if there is a simple inexpensive system of selecting out the desired algae for use and to inoculate new ponds with a high starting concentration of the desired algae. Some chain diatoms fall into this category as they can be filtered from a stream of water flowing through an outflow pipe. A "pillow case" of a fine mesh cloth is tied over the outflow pipe and most algae flow right through. The chain diatoms are held in the bag and used to feed shrimp larvae (in Eastern hatcheries) and to inoculate new tanks or ponds.

A variation on the basic "open-pond" system is to close it off, to cover a pond or pool with a greenhouse. While this usually results in a smaller system, for economic reasons, it does take care of many of the problems associated with an open system. It allows more species to be grown, it allows the species that are being grown to stay dominant, and it extends the growing season, only slightly if unheated, and if heated it can produce year round.

Algae can also be grown in a photobioreactor. A photobioreactor is basically a bioreactor which incorporates some type of light source. While almost anything in which it would be possible to grow algae could technically be called a photobioreactor, the term is more commonly used to define a closed system, as opposed to an open tank, or pond. Because these systems are closed, when used to cultivate algae, everything that the algae need to grow, (carbon dioxide, nutrient-rich water and light), all must be introduced into the system. A pond covered with a greenhouse could be considered a photobioreactor.

A photobioreactor can be operated in batch mode but it is also possible to provide it with a stream of sterilised water with the necessary nutrients and air dosed with carbon dioxide. Excess culture overflows and is harvested. If sufficient care is not taken, continuous bioreactors often collapse very quickly but once successfully started, they can continue operating for long periods. An advantage of this type of algae culture is that algae in the "log phase" is produced which is generally of higher nutrient content than old senescent algae. It can be shown that the maximum productivity for a bioreactor occurs when the exchange rate (time to exchange one volume) is equal to the doubling time of the algae.

While algae is often grown in monocultures using microbiological techniques to purify the desired strain, another approach has been used very successfully to produce algae feed for the cultivation of a variety of molluscs. Sea water is passed through filters to remove algae which are too large for the larvae being cultivated. Tanks in a green house, sometimes on a balcony in the mollusc house, are filled with the partially filtered water and nutrients are added. The tanks are aerated or not and the water is used after only a day or two of growing. The resulting thin soup of mixed algae has been shown to be an excellent food source for larval molluscs. An advantage is the very few man hours necessary to look after the algae culture.

Different types of photobioreactors include:

- Tanks provided with a light source
- Polyethylene sleeves or bags
- Glass or plastic tubes.

In most algal-cultivation systems, light only penetrates the top 3-4 inches of the water. This is because as the algae grow and multiply, they become so dense that they block light from reaching deeper into the pond or tank. Algae only need about 1/10th the amount of light they receive from direct sunlight. Direct sunlight is often too strong for algae. In order to have ponds that are deeper than 4 inches algae growers use various methods to agitate the water in their ponds, exposing the algae below to light and keeping algae on the surface from being overexposed. Paddle wheels can be used to circulate the water in a pond. Compressed air can be introduced into the bottom of a pond or tank to agitate the water, bringing algae from the lower levels up with it as it makes its way to the surface.

Apart from agitation, another means of supplying light to algae is to place the light *in* the system. Glow plates are sheets of plastic or glass that can be submerged into a tank, providing light directly to the algae at the right concentration.

The odour associated with bogs, swamps, or any stagnant waters that have been taken over by algae is from oxygen depletion in the water caused by the death of algal blooms that have been left to decay. Under anoxic conditions, bacteria which always inhabit algae cultures break down the organic material and produce hydrogen sulphide and ammonia which causes the bad smells. This condition, called hypoxia, often results in the death of all aquatic animals. In a system where algae is intentionally cultivated, maintained, and harvested, neither eutrophication nor aquatic hypoxia are likely to occur. The air around an algal pond should actually be very fresh from the oxygen produced.

Harvesting of Algae

Algae can be harvested using microscreens, by centrifugation, or by flocculation. Froth flotation is another method to harvest algae whereby the water and algae are aerated into a froth, with the algae then removed from the water. Alum and ferric chloride are chemical flocculants used to harvest algae. Chitosin, a commercial product commonly used for water purification, can also be used as a flocculant. The shells of crustaceans are ground into powder and processed to acquire chitin, a polysaccharide found in the shells, from which chitosin is derived.

The more brackish, or saline the water that the algae is being grown in, the more chemical flocculant will be required to induce flocculation. Harvesting by chemical flocculation is a method that is often too expensive for large operations. Interrupting the carbon dioxide supply to an algal system can cause algae in it to flocculate on its own, which is called autoflocculation. Ultrasound based methods of algae harvesting are currently under development, and other, additional methods are currently being developed.

Oil Extraction

Algae oils have a variety of commercial and industrial uses, and are extracted through a wide variety of methods.

- *Chemical Solvents:* Algal oil can be extracted using chemicals. Benzene and ether have been used, oil can also be separated by hexane extraction, which is widely used in the food industry and is relatively inexpensive. The downside to using solvents for oil extraction are the dangers involved in working with the chemicals. Care must be taken to avoid exposure to vapours and direct contact with the skin, either of which can cause serious damage. Benzene is classified as a carcinogen. Chemical solvents also present the problem of being an explosion hazard.

Soxhlet extraction is an extraction method that uses chemical solvents. Oils from the algae are extracted through repeated washing, or percolation, with an organic solvent such as hexane or petroleum ether, under reflux in a special glassware.

- *Enzymatic Extraction:* Enzymatic extraction uses enzymes to degrade the cell walls with water acting as the solvent, this makes fractionation of the oil much easier. The costs of this extraction process are estimated to be much greater than hexane extraction.
- *Expression/Expeller Press:* When algae is dried, it retains its oil content, which then can be "pressed" out with an oil press. Many commercial manufacturers of vegetable oil use a combination of mechanical pressing and chemical solvents in extracting oil.
- *Osmotic Shock:* Osmotic shock is a sudden reduction in osmotic pressure, this can cause cells in a solution to rupture. Osmotic shock is sometimes used to release cellular components, such as oil.
- *Supercritical Fluid:* In supercritical fluid/CO_2 extraction, CO_2 is liquefied under pressure and heated to the point that it has the properties of both a liquid and a gas, this liquified fluid then acts as the solvent in extracting the oil.
- *Ultrasonic-assisted Extraction:* Ultrasonic extraction, a branch of sonochemistry, can greatly accelerate extraction

processes. Using an ultrasonic reactor, ultrasonic waves are used to create cavitation bubbles in a solvent material, when these bubbles collapse near the cell walls, it creates shock waves and liquid jets that causes those cells walls to break and release their contents into the solvent.

Other methods are still being developed, including ones to extract specific types of oils, such as those with a high production of long-chain highly unsaturated fatty acids.

Algae as an Energy Source

Biodiesel Production: Currently most research into efficient algal-oil production is being done in the private sector, but if predictions from small scale production experiments bear out then using algae to produce biodiesel may be the only viable method by which to produce enough automotive fuel to replace current world gasoline usage.

Microalgae have much faster growth-rates than terrestrial crops. The per unit area yield of oil from algae is estimated to be from between 5,000 to 20,000 gallons per acre, per year; this is 7 to 31 times greater than the next best crop, palm oil (635 gallons).

Algal-oil processes into biodiesel as easily as oil derived from land-based crops. The difficulties in efficient biodiesel production from algae lie not in the extraction of the oil, which can be done using methods common to the food-industry such as hexane extraction, but in finding an algal strain with a high lipid content and fast growth rate that isn't too difficult to harvest, and a cost-effective cultivation system (i.e., type of photo-bioreactor) that is best suited to that strain.

Open-pond systems for the most part have been given up for the cultivation of algae with high-oil content. Many believe that a major flaw of the Aquatic Species Programme was the decision to focus their efforts exclusively on open-ponds; this makes the entire effort dependent upon the hardiness of the strain chosen, requiring it to be unnecessarily resilient in order to withstand wide swings in temperature and pH, and competition from invasive algae and bacteria. The energy that a high-oil strain invests into the production of oil is energy that is not invested into the production of proteins or carbohydrates, usually resulting in the species being less hardy, or having a slower growth rate. Algal species with a lower oil content, not having to divert their energies away from growth, have an easier time in the harsher conditions of an open system.

Some open sewage ponds trial production has been done in Marlborough, New Zealand.

A feasibility study using marine microalgae in a photobioreactor is being done by The International Research Consortium on Continental Margins at the International University Bremen.

Research into algae for the mass-production of oil is mainly focused on microalgae; organisms capable of photosynthesis that are less than 2 mm in diameter, including the diatoms and cyanobacteria; as opposed to macroalgae, e.g. seaweed. This preference towards microalgae is due largely to its less complex structure, fast growth rate, and high oil content (for some species). Some commercial interests into large scale algal-cultivation systems are looking to tie in to existing infrastructures, such as coal power plants or sewage treatment facilities. This approach not only provides the raw materials for the system, such as CO_2 and nutrients; but it changes those wastes into resources.

In November 8, 2006, Green Star Products has announced that it has signed an agreement with De Beers Fuel Limited of South Africa to build 90 biodiesel reactors with algae as raw material.

Each of the biodiesel reactors will be capable of producing 10 million gallons of biodiesel each year for a total production capacity of 900,000,000 gallons per year when operating at full capacity, which is 4 times greater than the entire US output in 2006. Also, Green Fuel Technologies Corporation has delivered a bioreactor to De Beers Fuel. Doubts have been expressed about Green Star's expertise in biodiesel technology, and the company has refused to answer questions.

Hydrogen Production

Algae can be used as a biological source for the production of hydrogen. In 1939, a German researcher named Hans Gaffron, while working at the University of Chicago, observed that the algae he was studying, Chlamydomonas reinhardtii (a green-algae), would sometimes switch from the production of oxygen to the production of hydrogen. Gaffron never discovered the cause for this change and for many years other scientists failed in their attempts at its discovery. In the late 1990s, professor Anastasios Melis, a researcher at the University of California at Berkeley discovered that by depriving the algae of sulphur it will switch from the production of oxygen (normal photosynthesis), to the production of hydrogen. He found that the enzyme responsible for this reaction is hydrogenase, but that the hydrogenase will not cause this switch in the presence of oxygen. Melis found that depleting the amount of sulphur available to the algae interrupted its internal oxygen flow, allowing the hydrogenase an environment in which it can react, causing the algae to produce hydrogen.

Biomass

Algae can be grown to produce biomass, which can then be harvested and burned in the same manner as wood, to produce heat and electricity.

Methane

Through the use of algaculture grown organisms and cultures, various polymeric materials can be broken down into methane.

SVO

The algal-oil feedstock that is used to produce biodiesel can also be used for fuel directly as "Straight Vegetable Oil".

(SVO). While using the oil in this manner does not require the additional energy needed for transesterification, (processing the oil with an alcohol and a catalyst to produce biodiesel), it does require modifications to a normal diesel engine, whereas biodiesel can be run in any modern diesel engine, unmodified, that is designed to use ultra-low sulphur diesel, the new diesel fuel standard for the United States of America that went into effect in the fall of 2006.

Hydrocracking to Traditional Transport Fuels

The oil of algae strain Botryococcus braunii is different than other algal oils, in that it contains a class of oils which can be reduced to chemicals traditionally extracted from petroleum and used for transport fuels, such as octane (gasoline, a.k.a. petrol), diesel, and aviation-grade kerosene.

Commercial and Industrial Uses

Algae are cultivated to serve many commercial and industrial uses.

- Bioplastics
- Dyes and Colourants
- Feedstock
- Nutritional
- Pharmaceutical
- Pollution Control
- CO_2 sequestration
- Fertilizer Runoff reclamation
- Sewage treatment

Nutritional

There are many algae that are cultivated for their nutritional value, either for supplemental use, or as a food source. Spirulina *(Arthrospira platensis)* is a blue-green algae (cyanobacteria) that is quite nutritious, this species thrives in open systems and commercial growers have found it well-suited to cultivation.

One of the largest production sites for Spriulina is Lake Texcoco in central Mexico. The plants themselves produce a variety of nutrients and high amounts of protein, and is often used commercially as a nutritional supplement. Extracts and oils from algae are also used as additives in various food products. The plants also produce Omega-3 and Omega-6 fatty acids, which are commonly found in fish oils, and which have been shown to have positive medical benefits to humans.

Pollution Control

Much of the carbon dioxide that is released into the atmosphere is from the burning of fossil fuels. With concerns over global warming, new methods for the thorough and efficient capture of CO_2 are being sought out.

An alternative to carbon capture and storage, by attaching an algae pond, or photobioreactor to any fuel burning plant, the carbon dioxide produced during combustion can be fed into the algae system. Nutrients can be sourced from sewage, thus turning two pollutants into resources for the production of biodiesel, with a land requirement much smaller than other crop sources.

Algal Nutrient Solutions

Algal Nutrient Solutions are made up of a mixture of chemical salts and water. Sometimes referred to as "Growth Media", nutrient solutions (along with carbon dioxide and light), provide the materials needed for algae to grow. Nutrient solutions, as opposed to fertilizers, are designed specifically for use in aquatic environments and their composition is much more precise.

Expeller Pressing

Expeller pressing is a chemical-free method for extracting oil from raw materials. The raw materials are squeezed under high pressure in a single step. When used for the extraction of food oils, typical raw materials are nuts, seeds and algae, which are supplied to the press in a continuous feed. Expeller presses can recover up to 75 per cent of the oil from algae. As the raw material is pressed, friction causes it to heat up; in the case of harder nuts (which require higher pressures), the material can reach or exceed temperatures of 120°F.

Isochrysis Galbana

Isochrysis galbana is a microalga. It was first identified by Bruce, Knight and Parke (1939) and was shown to be an outstanding food for various bivalve larvae. It is now widely cultured for use in the bivalve aquaculture industry.

Pleurochrysis Carterae

Pleurochrysis carterae is a unicellular coccolithophorid algae that has the ability to calcify subcellularly. They produce calcified scales, known as coccoliths, which are deposited on the surface of the cell resulting in the formation of a coccosphere. Pleurochrysis carterae produce heterococcoliths which are made up of crystal units of variable shapes and sizes.

Prymnesium Parvum

Prymnesium parvum is a golden algae that is of concern because of it's ability to produce a toxin. It is a flagellated heterokont algae that is normally found suspended in the water column. It was first identified in North America in 1985 and it is not know, if it was introduced i.e. is an invasive species or had been missed in previous surveys.

Toxin production mainly leads to fish kills and appears to have little effect on cattle or humans. This distinguishes it from red tide, which are algal bloom whose toxins to lead to harmful effects in people. Although no harmful effects are known, it is not recommended to consume dead or dying fish exposed to a *P. parvum* bloom.

Biology: P. parvum grows in a salinity range of 0.1 per cent-10 per cent with an optimum at 0.3-6 per cent although strains collected in different places appear to have different salinity tolerances. The algae produces dimethyl-sulphoniopropionate (DMSP) and other unknown polyols, likely as an adaptation to osmoregulation. The temperature range which allows survival of P. parvum is between 2 and 30°C. Growth as low as pH of 5.8 has been observed, but cells typically prefer higher pH ranges. The organism prefers highly light environments, but growth can be inhibited by excessive light (photoinhibition). The organism is capable of heterotrophic growth in the dark in the presence of glycerol and grazes on bacteria, especially when phosphate is limiting.

It has therefore, been hypothesised that P. parvum satisfies it's phosphate needs by eating bacteria. P. parvum can use a wide range of nitrogen sources, including ammonium, nitrate, amino acids (which ones apparently depends of pH), creatine, but is unable to use urea.

Fresh-Water Algae

Algae and Symbiosis: Some species of algae form symbiotic relationships with other organisms. In these symbiosis, the algae supply photosynthates (organic substances) to the host organism providing protection to the algal cells. The host organism derives some or all of its energy requirements from the algae. Examples include:

- *Lichens:* A fungus is the host, usually with a green algae or a cyanobacterium as its symbiont. Both fungal and algal species found in lichens are capable of living independently, although habitat requirements may be greatly different from those of the lichen pair.
- *Corals:* Algae known as zooxanthellae are symbionts with corals. Notable amongst these is the dinoflagellate *Symbiodinium*, found in many hard corals. The loss of *Symbiodinium*, or other zooxanthellae, from the host is known as coral bleaching.
- *Sponges:* Green algae live close to the surface of some sponges, for example, breadcrumb sponge (*Halichondria*

panicea). The algae is thus protected from predators; the sponge is provided with oxygen and sugars which can account for 50 to 80 per cent of sponge growth in some species.

Life-cycle

Rhodophyta, Chlorophyta and Heterokontophyta, the three main algal Phyla, have life-cycles which show tremendous variation with considerable complexity. In general, there is an asexual phase where the seaweed's cells are diploid, a sexual phase where the cells are haploid followed by fusion of the male and female gametes.

Asexual reproduction is advantageous in that it permits efficient population increases, but less variation is possible. Sexual reproduction allows more variation but is more costly because of the waste of gametes that fail to mate, among other things. Often there is no strict alternation between the sporophyte and gametophyte phases and also because there is often an asexual phase, which could include the fragmentation of the thallus.

Numbers and Distribution

In the British Isles, the UK Biodiversity Steering Group Report estimated there to be 20,000 algal species in the UK, freshwater and marine, about 650 of these are seaweeds. Another checklist of freshwater algae reported only about 5000 species. It seems therefore, that the 20,000 is an overestimate or an error.

Worldwide it is thought that there are over 5,000 species of red algae, 1,500 — 2,000 of brown algae and 8,000 of green algae. In Australia, it is estimated that there are over 1,300 species of red algae, 350 species of brown algae and approximately 2,000 species of green algae totalling 3,650 species of algae in Australia.

Around 400 species appear to be an average figure for the coastline of South African west coast.

669 marine species have been described from California (USA.).

642 entities are listed in the check-list of Britain and Ireland (Hardy and Guiry, 2006).

Uses of Algae

Seaweed is used as a fertilizer

Fertilizer

For centuries seaweed has been used as fertilizer; Orwell writing in the 16th Century referring to drift weed in South Wales: "This kind of ore they often gather and lay in heaps where it heteth and rotteth, and will have a strong and loathsome smell; when being so rotten they cast it on the land, as they do their muck, and thereof springeth good corn, especially barley" and "After spring tides or great rigs of the sea, they fetch it in sacks on horse brackets, and carry the same three, four, or five miles, and cast it on the lande".

There are also commercial uses of algae as agar.

Maerl is commonly used as a soil conditioner, it is dredged from the sea floor and crushed to form a powder. It is still harvested around the coasts of Brittany in France and off Falmouth (also extensively in western Ireland) and is a popular fertilizer in these days of organic gardening investigated

Falmouth maerl and found that *L. corallioides* predominated down to 6 m and *P. calcareum* from 6-10 m (Blunden *et al.*, 1981).

Chemical analysis of maerl showed that it contained 32.1 per cent $CaCO_3$ and 3.1 per cent $MgCO_3$ (dry weight).

Energy Source

- Algae can be used to make biodiesel, and by some estimates can produce vastly superior amounts of oil, compared to terrestrial crops grown for the same purpose.
- Algae can be grown to produce hydrogen. In 1939, a German researcher named Hans Gaffron, while working at the University of Chicago, observed that the algae he was studying, Chlamydomonas reinhardtii (a green-algae), would sometimes switch from the production of oxygen to the production of hydrogen. Gaffron never discovered the cause for this change and for many years other scientists failed to repeat his findings. In the late 1990s, professor Anastasios Melis, a researcher at the University of California at Berkeley, discovered that if the algae culture medium is deprived of sulphur, it will switch from the production of oxygen (normal photosynthesis) to the production of hydrogen. He found that the enzyme responsible for this reaction is hydrogenase, but that the hydrogenase lost this function in the presence of oxygen. Melis found that depleting the amount of sulphur available to the algae interrupted its internal oxygen flow, allowing the hydrogenase an environment in which it can react, causing the algae to produce hydrogen. *Chlamydomonas moeweesi* is also a good strain for the production of hydrogen.
- Algae can be grown to produce biomass, which can be burned to produce heat and electricity.

Pollution Control

- Algae are used in waste water treatment facilities, reducing the need for greater amounts of toxic chemicals than are already used.

- Algae can be used to capture fertilizers in runoff from farms. When subsequently harvested, the enriched algae itself can be used as fertilizer.
- Algae bioreactors are used by some powerplants to reduce CO_2 emissions. The CO_2 can be pumped into a pond, or some kind of tank, on which the algae feed. Alternatively, the bioreactor can be installed directly on top of a smokestack. This technology has been pioneered by Massachusetts-based Green Fuel Technologies.

Nutrition

Algae are used by humans in many ways. They are used as fertilizers, soil conditioners and are a source of livestock feed. Because many species are aquatic and microscopic, they are cultured in clear tanks or ponds and either harvested or used to treat effluents pumped through the ponds. Algaculture on a large scale is an important type of aquaculture in some places.

Human food. Seaweeds are an important source of food, especially in Asia; They are excellent sources of many vitamins including: A, B1, B2, B6, niacin and C. They are rich in iodine, potassium, iron, magnesium and calcium.

Algae is commercially cultivated as a nutritional supplement. One of the most popular microalgal species is Spirulina *(Arthrospira platensis)*, which is a Cyanobacteria (known as blue-green algae), and has been hailed by some as a superfood. Other algal species cultivated for their nutritional value include; Chlorella (a green algae), and Dunaliella (*Dunaliella salina*), which is high in beta-carotene and is used in vitamin C supplements.

In China, at least 70 species of algae are eaten as is the Chinese "vegetable" known as fat choy (which is actually a cyanobacterium). Roughly 20 species of algae are used in everyday cooking in Japan.

Certain species are edible; the best known, especially in Ireland is Palmaria palmata (Linnaeus) O. Kuntze (*Rhodymenia palmata* (Linnaeus) Kuntze, common name: dulse). This is a red algae which is dried and may be bought in the shops in Ireland. It is eaten raw, fresh or dried, or cooked like spinach. Similarly, *Durvillaea antarctica* is eaten in Chile, common name: cochayuyo.

Porphyra (common name: purple laver), is also collected and used in a variety of ways (e.g. "laver bread" in the British Isles). In Ireland, it is collected and made into a jelly by stewing or boiling. Preparation also involves frying with fat or converting to a pinkish jelly by heating the fronds in a saucepan with a little water and beating with a fork. It is also collected and used by people parts of Asia, specifically China and Japan as nori and along most of the coast from California to British Columbia. The Hawaiians and the Maoris of New Zealand also use it. Chondrus crispus, (probably confused with Mastocarpus stellatus, common name: Irish moss), is also used as "carrageen" for the stiffening of milk and dairy products, such as ice-cream. One particular use is in "instant" puddings, sauces and creams. *Ulva lactuca* (common name: sea lettuce), is used locally in Scotland where it is added to soups or used in salads. Alaria esculenta (common name: dabberlocks), is used either fresh or cooked, in Greenland, Iceland, Scotland and Ireland.

The oil from some algae have high levels of unsaturated fatty acids. Arachidonic acid (a polyunsaturated fatty acid), is very high in Parietochloris incisa, (a green algae) where it reaches up to 47 per cent of the triglyceride pool.

Other Uses

The natural pigments produced by algae can be used as an alternative to chemical dyes and colouring agents. Many of the paper products used today are not recyclable because of the chemical inks that they use, paper recyclers have found that inks made from algae are much easier to break down. There is also much interest in the food industry into replacing the colouring agents that are currently used with colouring derived from algal pigments. In Israel, a species of green algae is grown in water tanks, then exposed to direct sunlight and heat which causes it to become bright red in colour. It is then harvested and used as a natural pigment for foods such as Salmon.

Alginate

Between 100,000 and 170,000 wet tons of Macrocystis are harvested annually in California for alginate extraction and abalone feed.

Collecting and Preserving Specimens

Seaweed specimens are collected, preserved for research. Such preserved specimens will keep for two or three hundred years. Those of Carl von Linné (1707-1778) are still available for reference, and are used. Specimens may be collected from the shore; those below low tide must be collected by diving or dredging. The whole algal specimen should be collected, that is the holdfast, stipe and lamina. Specimens of algae reproducing will be the more useful for identification and research. When collected the details of the location and site should be noted. They can then be preserved pressed on paper or in a preserving liquid such as alcohol or solution of 5 per cent formalin/sea-water. However, formalin is reported to be carcinogenic.

Biological Exposure Scale

The ecology of the shores of the British Isles including a discussion of the different shores from sheltered to exposed along with an exposure scale is given by Lewis (1964). An exposure scale of five stages is given:- Very Exposed Shores; Expose Shores; Semi-exposed Shores; Sheltered Shores and Very Sheltered Shores. Factors indicating the differences between these exposure scales are detailed. Very Exposed Shores have a wide *Verrucaria* zone entirely above the upper tide level, a *Porphyra* zone above the barnacle level and *Lichina pygmaea* is locally abundant.

The eulittoral zone is dominated by barnacles and limpets with a coralline belt in the very low littoral along with other Rhodophyta and *Alaria* in the upper sublittoral. Exposed shores show a *Verrucaria* belt mainly above the high tide, with *Porphyra* and *Lichina pygmaea*. The mid shore is dominated by barnacles, limpets and some *Fucus*. Some Rhodophyta. *Himanthalia* and some Rhodophyta such as *Mastocarpus* and *Corallina* are found in the low littorral with *Himanthalia*, Alaria and *[Laminaria digitata]* dominant in the upper sublittoral. The semi-exposed shores show a *Verrucaria* belt just above high tide with clear *Pelvetia* in the upper-littoral and *Fucus serratus* in the lower-littoral. Limpets, barnacles and short *Fucus vesiculosus* midshore. *Fucus serratus* with Rhodophyta, (*Laurencia*, Mastocarpus, *Rhodymenia* and *Lomentaria*). *Laminaria* and *Saccorhiza polyschides* and small algae common in the sublittoral. The sheltered shores show a narrow *Verrucaria* zone at high water and a full sequence of fucoids: *Pelvetia, Fucus spiralis, Fucus vesiculosus, Fucus serratus Ascophyllum nodosum*. *Laminaria digitata* is dominant the upper sublittoral. The very sheltered shores show a very narrow zone of *Verrucaria*, the dominance of the littoral by a full sequence of the fucoids and *Ascophyllum* covering the rocks. *Laminaria saccharina, Halidrys, Chondrus* and or *Furcellaria*.

3

Rhodophyceae

The Red Algae fall naturally into two subclasses, the Bangioideae and Florideae. The former include only a limited number of genera (Bangiales) which, although showing certain characters in common with the Floridjeae, exhibit so many significant differences that opinion remains divided as to their exact relation to the latter. The bulk of the Bangioideae are marine, including such typical members of littoral vegetation as *Bangia* and *Porphyra*, but they also comprise a number of probably reduced freshwater and terrestrial forms. They provide the only examples of parenchymatous construction found among Rhodophyceae.

In their diffuse growth, they offer a sharp contrast to the Florideae which comprise the vast majority of the Red Algae. The growth is here effected by an apical cell which is commonly dome-shaped. Another characteristic feature, lacking in Bangiales, is constituted by the well-marked pits. In the least specialised members of Florideae *(Aerochaetium)* the plant takes the form of a heterotrichous filament and no simpler types of body are known. The heterotrichous filament is also encountered in *Erythrotrichia* and other Bangiales, while the freshwater and terrestrial members of the latter in part exhibit unicellular and colonial states.

As in Phaeophyceae, the heterotrichous habit is recognisable in the juvenile stages of most of the less specialised Florideae, although in the more advanced orders, it is obscured and in the Ceramiales a prostrate system is no longer differentiated. The manifold types of thalli exhibit varied degrees of compactness, but are invariably pseudoparenchymatous.

They are composed of dense aggregates of the branch systems of one or more richly branched filaments which, in the heterotrichous forms, arise from the prostrate system. The resulting thalli are with more or less copious ramification, although a marked foliaceous development occurs, for instance, in Gigartinaceae and Delesseriaceae. Illumination may play a role in determining the organisation of the thallus. Morphological elaboration is far rarer than among Brown Algae and relatively few Florideae show complexities in outward form.

Few Red Algae attain to any considerable stature, although a considerable number are perennial in habit; a larger or smaller portion of the old frond usually persists and from it, the new growth arises. Tissue-tensions, similar to those noted in Phaeophyceae, are occasionally manifest.

The vast majority of Florideae are marine and most of the freshwater representatives belong to the less specialised groups; in general, they favour well-aerated habitats (rivers and the littoral region of lakes). Many of the marine Florideae are sublittoral and normally inhabit deeper water and, when the latter is clear, can penetrate to very considerable depths; such forms can only be obtained by dredging, unless cast up on the shore after storms.

Certain Florideae (species of *Gigartina*, *Laurencia*, *Lomentaria*, etc.), however, form extensive belts in the littoral region and, where suitable substrata are available, Red Algae occur in this position even in the Tropics. On the other hand, many Florideae, found between tide-levels on temperate and other shores, are confined to the rock-pools and are evidently incapable of withstanding desiccation. A number (species of *Catenella, Bostrychia,* etc.) are adapted to life in brackish water, being characteristic members of certain types of salt-mars vegetation and of the flora growing on the roots of tropical mangroves. Although distributed over the whole surface of the earth, the greatest abundance of Florideae is met with in the warmer seas and in Australasia.

The Rhodophyceae are distinguished by a complete absence of all motile stages, both the male cells and the diverse types of spores devoid of organs of locomotion. There is indeed no evidence that motile types have ever occurred in this class, supposed flagellate representatives (e.g. *Rhodpmonas)* having proved to belong to an altogether different affinity. One of the criteria that serves to demonstrate the homogeneity and probable separate origin of Chlorophyceae, Phaeophyceae, etc. is therefore, lacking.

On the other hand, all Rhodophyceae possess accessory photosynthetic pigments of a distinctive and similar kind, while the products of photosynthesis (Floridean starch, etc.) are equally characteristic. These features, taken in conjunction with the many striking morphological characteristics of Florideae (apical growth, filamentous construction, pit-connections), the universal oogamy and other marked peculiarities in the processes of reproduction, all tend to suggest a common ancestry from which Bangioideae and Florideae probably diverged at a very early stage. The Red Algae exhibit no clear affinities with other classes and are commonly grouped as Rhodophyta, by contrast to the Chlorophyta, Chrysophyta, Phaeophyta, etc. Our knowledge of their fossil history is negligible and sheds no light on their evolution. The calcified Corallinaceae are met with from Cretaceous onwards and have played an important role in the formation of flimestone rocks.

Cell-wall

Two distinct layers are generally recognisable in the cell-wall. The inner, firmer layer adjoining the cell-cavity often consists largely of cellulose while the outer layers are composed of pectic substances. In certain Florjdeae *(Cystoclonium, Gelidium, Laurencid)* the cell-walls are coloured blue by iodine which Sauvageau ascribes to the presence of amyloid; Hansen records a red colouration of the middle lamella with iodine in species of *Gracilaria.* In the living cells of many Florideae the membranes are markedly stretched and compressed as a result of the turgor of the cell-contents so that, on treatment with hypertonic solutions, the shrinkage of the protoplast is accompanied by a decrease in the dimensions of the cell, as well as by a pronounced increase in the thickness of the wall. This feature is responsible for the often considerable thickness shown by the cell-membranes in preserved material. According to Kylin damaged cells of *Griffithsia* and of *Antithamnion plumula* exhibit an increase of the cell-wall to ten times its previous thickness. The "culide", apparent at the surface of some of the coarser forms, is probably a denser layer of pectic material.

Stratification of the membrane is very prominent in certain Ceramiaceae, where it has been studied as an example of growth by apposition. The classical instance is furnished by *Bornetia secundiflora,* where successive thickening layers *(i)* are deposited by the tip of the protoplast of the apical cell, while the outermost strata *(ii)* are progressively burst; the wall thus comes to consist of numerous funnel-shaped pieces. A comparable stratification is recorded in species of *Antithamnion, Callithamnion,* and *Ceramium.* In cell-division, the septa arise as annular ingrowths and show no relation to the nuclear spindle'.

The pectic layers of the wall, which stain deeply with ruthenium red and methylene bluest one time regarded as calcium salts of pectic acids, are probably complex. They are either completely *(Rissoella)* or only partly *(Gelidium)* soluble in boiling water. The products of their precipitation include the commercial agar-agar (derived from species of *Gracilaria, Getidium* and *Eucnewna)* and carrageen (from *Chondrus crispus* and *Gigartina stellatd).* On hydrolysis, these two substances afford galactose, accompanied by pentoses and possibly some glucose and obtained a sugar from hydrolysed geloses of

Chondrus, Iridaea, and *Ahnfeltia*. Agar and carrageen have also been shown to contain organically bound sulphur, and it is believed that they include calcium salts of ethereal sulphates; the gelose of *Ceramium rubrum* likewise contains an ethereal sulphate.

Kylin distinguishes the geloses obtained from *Ceramium fibrum* and *Furcellaria fastigiata,* which readily solidify on cooling and are precipitated by ammonium sulphate, from those derived from *Dumontia incrassata,* which remain liquid on cooling and are not precipitated by the salt; carrageen resembles the former. Sauvageau, further differentiates between the geloses of *Gelidiumy Gracilaria, Ahnfeltia,* etc., which solidify in the cold even in dilute solutions, and those of *Chondrus, Gigartina, Gymnogongrus, Hypnea,* etc., which only solidify when in concentrated solution or after treatment with electrolytes; that of *Polyides* is intermediate and its solution readily solidifies on addition of ammonium sulphate.

Sauvageau states that in the seaweeds of the first group, which release their gelose at high temperatures without disintegration, the cellulose is very resistant to ammoniated copper and amyloid is invariably present in the wall. According to Hill carrageen is a mixture of two substances, the one soluble in cold water and producing a thick viscous solution not setting to a gel after boiling with a little Rochelle salt, the other soluble in hot water and readily affording a gel on cooling. The residue left after extraction contains

Marked calcification of the cell-membranes occurs in Corallinaceae, as well as in species of *Peyssonnelia, Galaxaura,* and *Liagora*. According to Berthold the envelope of lime is thicker in well-illuminated individuals. The incrustation first appears in the pectic layers. Some magnesium carbonate is always present. But in Corallinaceae, this is only precipitated in the older thallus and in dead parts. There is a greater proportion of magnesiun. Incrustations of carbonate of lime have been reported in freshwater forms. Deposition of iron is recorded in *Batrachospermum* and diverse other Red Algae.

Protoplast

The cytoplasm of diverse Florideae exhibits a high degree of viscosity and there is often a very firm adhesion to the wall, which is stated to be due to extension of the protoplast into the innermost layers of the membrane. These features are no doubt related to the fact that the protoplast, during plasmolysis remains in contact with the wall at diverse points and develops concave contours which are irreversible. Such phenomena are not shown by all the species that have been investigated and the extent of their occurrence is not at present clear; the published data refer mainly to Ceramiales.

The protoplasts of Florideae are commonly killed by plasmolysis. *Heterosiphonia plumosa* being specially susceptible, while *Ceramium ciliatum* and *Callithamnion corymbosum*, for instance, are more resistant. Many Red Algae are also very sensitive to dilution of the medium, although *Antithamnion cruciatum* and *A. spirographidis* for example, are less sensitive than others. The harmful effects of dilution of the medium on Florideae are also shown by the marked diminution in number, as well as in the size and degree of branching of the surviving forms, in seas with lower degrees of salinity.

Diverse Ceramiales are stated (13) to show marked susceptibility-gradients to hypo- and hypertonic sea-water, as well as viscosity-gradients. In general, the older cells show a higher degree of viscosity than the younger, but in *Cryptopleura*, for instance, this is more marked in the marginal cells. It is evident, however, that the protoplast does not always possess a high viscosity. Thus, in certain Ceramiaceae, the cross-walls of the older cells bear aggregations or cytoplasm from which numerous processes, exhibiting constant change of form, extend in the vacuole.

The osmotic pressure is lower than that of Phaeophyceae or Chlorophyceae; according to Bible it is lower in deep-water (1-5 times that of sea-water) than in littoral forms (2-0-2-2 times that of sea-water) and in general corresponds with that of the strongest hypertonic solutions that can be withstood without injury. In *Ceramium, ciliatum* is more resistant to plasmolysis than other littoral forms.

Chromatophores and Pigments

A single chromatophore, which is usually axile and more or less stellate in form, is characteristic of many Bangiales and a comparable condition is met with Nemalionales. In the peripheral cells of *Nemalion, Helminthocladia, Liagora,* etc. the chromatophore, which lies adjacent to the outer walls, is asymmetric, the internally directed processes being more strongly developed such chromatophores have also been recorded in *Acrochaetium*. The processes arising from such axile chromatophores commonly broaden at their periphery, and in elongate cells these parietal enlargements may become band-shaped and assume very irregular shapes, sometimes forming an almost continuous layer. The occurrence of parietal chromatophores in diverse Bangiales and Nemalionales (Naccariaceae; *Scinaia, Batrachospermum)* suggests a possible derivation of the parietal from the axile type by elimination of the central body of the chromatophore.

It is essentially in the forms possessing axile chromatophores that the spherical bodies known as *pyrenoids* occur, although they are also found in the parietal chromatophores of some species of *Acrochaetium* and *Rhodochorton*; they are not, however, present in all Bangiales and Nemalionales. Like the pyrenoids of Chlorophyceae they constitute denser parts of the chromatophore, but they do not show the same contraction on fixation and do not stain so deeply; moreover, although grains of Floridean starch are sometimes grouped around pyrenoids In *Nemalion* and *Helminthocladia,* Schmitz records pyrenoids also within the processes of the chromatophore and believe that these arise *de novo*.

A single large parietal chromatophore is found in the photosynthetic cells of certain more specialised Florideae, for instance in *Hildenbrandia, Agardhiella, Dumontia, Polyides, Chondrus,* and *Phyllophora*. Such chromatophores often become lobed or divided and in *Polyides* they break up into a number of band-shaped structures. More usually "the cells of Florideae contain considerable numbers of parietal chromatophores from the first. Where recognisable *(Callithamnion, Spermothamniori)* they appear within the apical cells as discs, although in many Florideae (e.g. *Polysiphonia, Ceramium)* these cells are almost colourless. The lenticular form may be retained in the mature cells, as in many Delesseriaceae, Coral -linaceae, diverse Ceramiaceae, etc., and in such instances the chromatophores are usually numerous. Commonly, however, they are fewer and irregularly lobed or band-shaped, contiguous ones often fitting closely together. In diverse Ceramiales and in *Lemanea,* the chromatophores of the apical cells develop into lobed or breached structures.

The shape of the chromatophores often depends on that of the cell and may vary appreciably in different cells of the same individual. Thus, in the cortical bands of *Ceramium,* the chromatophores are irregularly shaped and almost isodiametric, while in the axial cells *(a)* they appear as elongate ribbons. The branchlets of *Crouania attenuate* contain a parietal chromatophore with irregular processes, whereas in the axial cells, there are delicate and widely separated ribbons. Similar differences are described by Berthold between the outer and inner cells of diverse Cryptonemiales. Movements of the chromatophores under different conditions of illumination have been identified in a number of Florideae.

The evidence for the origin of chromatophores from mitochondria is inconclusive. In the fertilized carpogonia of *Lemanea* and other Florideae Mangenot records decrease in size of the chromatophores followed by fragmentation and discolouration which results in the formation of a quantity of rods and granules; similar bodies are found also in the young carpospores. Comparable changes probably occur in many Florideae, although the carpogonia often lack pigmented chromatophores already before fertilization.

The substance of the chromatophores is semi-fluid and

under certain circumstances they Exhibit amoeboid changes of form and undergo fusions. They *a*re affected by dilution of the surrounding medium colour, which is usually assumed by Red Algae when boiled in distilled water and which is due to alteration of the phycoerythrin, is ascribed to the alkaline reaction of the chromatophores.

The colouration of the chromatophores is largely due to two water-soluble pigments, the red *phycoerythrin* and the blue *phycocyanin* which obscure the other pigments. Chlorophyll b, if present at all, is found only in small traces, while both carotene and xanthophyll are present. Recently it has been shown that there are a number of carotenoid pigments. The chlorophyll-content is comparatively low, although, like that of phycoerythrin, greater in deep water and feebly illuminated forms which also show an increased ratio of phycoerythrin to chlorophyll. The chlorophyll-content is stated to be higher in arctic types. The normal sterol of Rhodophyceae is gitosterol, but this is not without exception; *Polysiphonia nigrescens* is stated to contain fucosterol. Flavin is widely represented, especially in *Iridaea.*

The accessory pigments, which resemble those found in Myxophyceae, are of the nature of proteins and show characteristics similar to those of globulins. They can be extracted from fresh material by prolonged treatment with distilled water, although the pigments of certain Florideae only pass very slowly through the cell-membranes. The phycoerythrin is precipitated from the solution by addition of crystallised ammonium sulphate and purified by repeated recrystallisation. Further fractional precipitation of the original solution gives phycocyanin in a crystalline form. The pure solution of phycoerythrin is a marked orange-yellow in reflected light, while phycocyanin is indigo blue and exhibits dark red fluorescence. In certain Rhodomelaceae with a red or reddish-brown colour (e.g. *Rhodomela subfusca, Polysiphonia nigrescens),* the phycoeryjhrin-solution shows only a faint fluorescence which has been ascribed to admixture of brown pigments, although this is disputed by Kylin.

Crystals of phycoerythrin and phycocyanin are sometimes found in dead cells. Both chromoproteins consist of a colour which can be separated by treatment with acids or alkalis. The pigments are decomposed by strong light, but phycocyanin is more susceptible than phycoerythrin. Both the total amount and the relative proportions of the two pigments vary in different seasons.

The solutions of phycoerythrin and phycocyanin afford characteristic spectra. That of the former normally shows the absorption-bands, which are situated in the green and blue between the lines. Two modifications of phycocyanin have been distinguished in Florideae, viz. a bluish-violet one extracted from *Ceramium rubrum*, with two absorption bands in the orange-yellow and green between the lines C and E, and a blue-green one extracted from *Batrachospermum* and *Lemanea*, with only a single absorption-band in the orange-yellow between the lines C and D; the latter is widely distributed in the Blue-green Algae.

The brilliant red colours exhibited by deep-water Florideae are conditioned by the presence of considerable amounts of phycoerythrin, while phycocyanin is lacking. The diverse shades of purple, dark brownish-red, reddish-violet, or violet exhibited by the littoral forms are, on the other hand, ascribed to increasing admixture of phycocyanin. The red-violet littoral form of *Chondrus crispus* contrasts with the bright red submerged form. A high proportion of the latter is met with in *Gigartina stellata* and *Rhodymenia palmata*. Many littoral Rhodophyceae at times exhibit brownish, greenish or yellowish tints. The freshwater Florideae *(Batrachospermum, Lemanea)*, which in general show dark green or blue-green colours, nevertheless contain both pigments, a reddish colouration becoming apparent after death or when the algae is grown in the shade; *Asterocytis ornata* and some *Batrachospermums* are, however, believed to lack phycoerythrin.

Many Florideae are very sensitive to light than those obtaining in the deep-water habitats which they normally frequent; a number of such seaweeds often flourish in caves and other feebly lighted situations in the littoral region. It is evident, however, that by no means all Red Algae forms are, as little as all marine Chbrophyceae. Apart from the data already given, it may be noted that certain Rhodophyceae *(Bangia, Porphyra, Endocladia,* some *Polysiphonias)* are specially characteristic of the littoral region, while others can thrive in shallow moving water in the strong light of the Mediterranean summer. The recent researches of Mpntfort, moreover, show that the same species can exist in different light-habitats, the photosynthetic opacities being adapted in each to the prevailing light-intensity. Even among shade-forms there are marked differences in the power of accommodation to stronger light.

Investigations of photosynthesis at diverse depths of submergence or in light of different intensity demonstrate that deep-water forms among Rhodophyceae carry on appreciably more photosynthesis. On the other hand, Red Algae from well-lighted habitats have, in the natural illumination of deep water, no advantage over shade-forms among Chlorophyceae, while shade-forms among Phaeophyceae do not differ appreciably in their capacity for photosynthesis, under such circumstances, from Red Algae belonging to the same horizons. Red Algae rich in phycocyanin make a better use of stronger light than those poor in this pigment.

The lower limits to which the permanently submerged forms can penetrate must of course depend on their compensation points, which in turn, are in large part dependent on constitutional features. The absolute vaiue for photosynthesis in *Gigartina Harveyana* is stated to be not far below that obtained with leaves of *Helianthus.*

Since Engelmann, with the help of his bacterial method, concluded that Algae synthesise best in light of the Complementary colour, this question has been the subject of repeated investigation. Engelmann himself found that Red Algae, illuminated by a projected microspectrum, exhibit photosynthesis in the green, and this has since been corroborated by diverse other experimental methods. Thus, Richter, determining photosynthesis by the amount of oxygen produced, established that Red Algae *(Callithamnion, Delesserid)* were more efficient in green light than Chlorophyceae from the same horizons. Wiirmser, estimating photosynthesis in red light of known energy-value by the colourimetric method, found that *Rhodymenia palmata* exhibited more photosynthesis in green light than *Viva Lactuca*. Ehrke compared the photosynthesis of *Enteromorpha compressa* and *Delesseria sanguined* in weak light of different wavelengths but equal intensity and arrived at the conclusion that the green algae exhibited maximum photosynthesis in red and the red algae in green light.

According to the theory of complementary chromatic adaptation the broad distribution of Green, Brown, and Red Algae, at progressively lower levels on the shore, is an outcome of the chromatic adaptation of the two last to the special colour of the light to which. The methods are criticised by Harder they are normally subjected; owing to differential absorption rays of different wavelengths there is, in deeper water, an complete elimination of the longer ones. The littoral Phaeophyceae do not fit into this picture at all, although the possession of fucoxanthin, with its capacity for absorbing blue rays, probably gives such forms as can exist in deep water some advantage.

The discussion of the theory of chromatic adaptation, so far as it concerns seaweeds, has in fact mainly centred about the Rhodophyceae. A considerable number of these, however, frequent the littoral region. Moreover, the evidence that Red (or Brown) Algae possess the capacity to adopt a colour complementary to that of the incident light, such as is established for certain Blue-green Algae, is altogether inconclusive. Gaidukov's contention that the green colour assumed by *Porphyra laciniata,* after exposure to strong red light for 10 hours, is an example of such adaptation, is erroneous, since the colour-change no doubt depends on decomposition of the accessory pigments. Such green-coloured Rhodophyceae *(Chondrus, Gigartina, Laurencid)* are not uncommonly found in the littoral region, their photosynthetic efficiency in strong light being less than that of the accompanying Chlorophyceae or corresponding submerged forms.

Richter was the first to point out that the littoral Red Algae do not exhibit the greater photosynthetic efficiency in green light shown by sublittoral types, and this has been fully corroborated by Montfort. Ricliter's results led him to deny all participation of the accessory pigments of Rhodophyceae in photosynthesis, the superiority shown by sublittoral forms being ascribed solely to a higher-capacity for utilising the available light.

It cannot be doubted, however, that the accessory pigments of the numerous Red Algae, that flourish in the greenish-blue light of deeper water, will confer a decided advantage upon them with respect to light-absorption. It is still, however, a debated question how far their more efficient photosynthesis is due to increased capacity for light-absorption. Diverse authorities are of the opinion that both factors, are involved, while some consider that the accessory pigments act as photosensitisers dark-violet shade-inhabiting Florideae can also use yellow light for photosynthesis.

Photosynthetic Products

The characteristic photosynthetic product in Rhodophyceae is a polysaccharide, usually spoken of as *Floridean starch,* which occurs abundantly incertain cells in the form of small solid grains. Some of these, generally of smaller tensions, lie free in the cytoplasm, whilst others are apposed to the chromatophores, although never deposited within their substance; in *Phyllophora* and *Chondrus.* Darbishire records their formation in contact with leucoplasts. A direct participator of the chromatophores in the formation of the grains is denied by some, but this is not the general opinion, although it remains uncertain whether the free grains arise within the cytoplasm itself or are such as have become detacfied from the surface of the chromatophores. The grains are sometimes aggregated around the nucleus. Where pyrenoids occur, the starch-grains are often distributed around them, and such pyrenoid-starch may possess a greater degree of permanency than the rest.

The larger grains are commonly b gain-shaped or comical, with an obvious depression *(d)* on one surface which is regarded as marking the point of contact with the chromatophore; Mangenot also records irregular polyhedral shapes. An occasional faint stratification is evident, while the radial splits which appear on the application of slight pressure indicate a possible sphaeritic structure. The grains exhibit marked double refraction, ·showing a dark cross. They swell readily in hot water, potash, chloral hydrate, and iodine reagents. Grains, swollen by prolonged treatment with the last, often show differentiation into a central, more deeply stained, and a peripheral lighter-coloured region". According to Oltmanns the latter is surrounded by a single or granules, a fact which led him to suggest that the central region might be of the nature of a pyrenoid; it is difficult to find any support for this view.

On treatment with iodine solution the grains are at first coloured yellow or yellow-brown, but, as swelling sets in, the tint changes to red, red-violet, or blue. If swelling is previously induced by treatment with hot water or chloral hydrate, violet or blue shades appear at once with iodine; occasional direct colouration, after treatment with chloral hydrate, is no doubt due to the iodine-content of the algae. The colour assumed with iodine-reagents varies considerably, both in different species and in one and the same species, but the deep blue shades typical of the starch-grains of higher plants are not often obtained, the most frequent tints being red or reddish-violet; Mangenot also records brownish shades. While Kolkwitz concluded that there was no essential difference from the starch of other plants and Bruns regarded Floridean starch as equivalent to the so-called red starok found *in-Myristica*, etc.,

Schimper denied any close relation between it and that of other groups. Belzung was of the opinion that it consisted largely of amiloerythrin, while Mangenot recognises a resemblance to glycogen. The evidence seems to indicate that it is a substance, but not identical with, the starches of higher plants.

Kylin showed that the starch, removed from *Furcellaria fastigiata* by mechanical means, gives dextrose on hydrolysis with weak acids, while treatment with maltose kinase results in the formation of sugars. Bartholomew also extracted from diverse Red Algae a diastase which digests maize-starch, although more slowly than ordinary diastase. The starch is no doubt a reserve-substance; this is shown *inter alia* by its marked accumulation in perennating structures and its decrease on the resumption of growth.

The nature of the primary product of photosynthesis in Red Algae still remains uncertain, although Kylin believes that he has established by macrochemical means the presence of small quantities of dextrose. Mannitol, commonly found in Phaeophyceae, does not occur in Red Algae, but of the sugar alcohols (dulcitol, sorbitol) have been demonstrated in *Bostrychia* and of florideae (especially of *Rhodymenia palmata* which contains little starch) are dextrorotatory and, when concentrated, have a sweet taste. Such extracts have afforded crystals of a galactoside which has been named *floridoside*. This substance may constitute another food-reserve. In *Rhodymenia palmata*, its accumulation begins in May and increases to a maximum in summer while in *Corallina squamata*, it is highest towards the end of May and least at the end of January.

Nitratgs can be demonstrated in many Rhodophyceae, often, especially in the older parts, although lacking in some *(Polysiphonid)* in *Ceramium rubrum* they make up 1-5 per cent of the dry weight. Haas and Hill find an accumulation of pentapeptides in Corallinaceae and *Griffithsia* and suggest that their presence may be due to a lack of balance between carbon and nitrogen metabolism owing to the depression of photosynthesis as a result of low light intensity. Crystalloids probably composed of proteins, are not uncommonly found floating in the cell-sap although stated to lie in the cytoplasm *Ceramium* and *Monosporu*.

Although no generalised statement is possible in the present state of our knowledge, it is clear that the metabolism of Rhodophyceae differs appreciably from that of other algal classes. This is perhaps also shown by the frequent deposition of iodine-compounds in special cells. Vitamins are specially abundant in *Porphyra*.

Nucleus

The cells of the less specialised Rhodophyceae (Bangiales, Nemalionales) are always uninucleate and this is also so in many Cryptonemiales and Gigartinales. In other instances, the older cells are multinucleate (Gelidiales, Rhodymeniales, many Ceramiales), while in *tjriffihsia* and *Bornetia*, this is even true of the apical cells. The number of nuclei is by no means always proportional to the size of the cell; in *Plocamium*, for instance, the axial cells contain a single huge nucleus, 30-35 in diameter, whilst the surrounding cells possess a number of small ones. The axial cells of Ceramiaceae, however, contain larger nuclei than the peripheral cells. The diverse types of reproductive cells are invariably uninucleate.

The resting nuclei of the vegetative cells are for the most part of small size, with an average diameter of 3; in some species *(Chondrus crispus, Calliblepharis ciliaid)* they are exceptionally minute. According to Svedelius, the nuclei of tetra-sporangiate plants of *Delesseria sanguined* are larger than those of" sexual plants, but this does not appear to be a general rule. The nuclei exhibit a prominent nucleolus (sometimes several) and a well-developed network with numerous chromatin grains, although the chromatin-content is often rather scanty. Nuclei of irregular shape have been recorded in some Ceramiales.

At the commencement of mitosis, the nucleus enlarges, the chromatin grains increase in size and decrease in number, probably as a result of fusion, and the network is often described as coarser. The chromatin grains are frequently arranged in rows. At the end of prophase, the short, and commonly in great part rounded, chromosomes are constituted. The spindle isintra nuclear in *Polysĩphonia*, *Griffithsia globifera*, and *Spermothamnion Turneri*, but most investigators describe it as lying freely in the cytoplasm. Yamanouchi records centrosome in *Polysiphonia* and *Corallina*, and they have also been reported in *Laurencia*, but for the most part such structures have not been seen during mitosis in Red Algae. The nucleolus disappears as the chromosomes become aggregated on the equatorial and is reconstituted in each daughter-nucleus.'

Most authorities deny any participation of the nucleolus in the development of the chromosomes, but in *Nemalion*, *Griffithsia* and *Dumontia* so-called karyokinesis are recorded and the chromatin is stated to pass out of the nucleolus at the commencement of mitosis; Kylin attributes this to poor fixation. The peculiar resting nuclei described by Mathias in certain Ceramiaceae have not been observed by other investigators and are probably to be attributed to the same defect.

The resting nuclei of Bangiales and of the less specialised Nemalionales *(Batrachospermum, Lemanea, Nemalion,* contain a large nucleolus but, apart from that, appear almost devoid of structure. The spindles in *Nemalion* are stated to be intranuclear, which is denied by Kylin, while Wolfe records centrosomes. The nuclei of *Scinaia* and *Bonnemaisonia* appear to resemble those of the more specialised Florideae.

General Features of Reproduction

Sexual reproduction in Rhodophyceae is invariably oogamous. The female organ (carpogonium), which usually Bears" a special receptive process (trichogyne), exhibits retention of the single ovum and the latter shows no contraction either before or after fertilization. The male organ (antheridium) produces a single motionless male cell (spermatium)

The zygote, either by direct division as in Bangiales or after the production of filamentous outgrowth in Nemalionales, gives rise to a number of gpufangia, eachlorming a single naked spore (carpospore). In most Florideae, however, the gonimoblasts do not arise directly from the zygote, but from other cells (auxiliary cells) which either belong to the same branch-system as the carpogonium or are more or less widely separated from it. Intimate combinations of carpogonia and auxiliary cells are termed procarps, and in them there is direct fusion between the carpogonium and the auxiliary cells, while when the latter are spatially separated, the products of the zygote-nucleus are conveyed to them by special connecting filaments, frequently elongate and usually several in number. A descendant by division of the zygote-nucleus passes into the auxiliary cell and, after degeneration of its nucleus, gonimoblasts producing carpospores arise from the auxiliary cell. In many Florideae, there is a marked fusion of cells during the development of the gonimoblast, a process connected with the supply of nutriment to the maturing carpospores.

Reduction occurs either at the first divisions of the zygote-nucleus (Bangiales, Nemalionales) or it is postponed and takes place in special tetrasporangia, borne on individuals distinct from, but morphologically identical with, the sexual plants (other Florideae). There are thus two haploid phases (the sexual individual, the carpospore-bearing stage) and the life-cycle is haplobiontic; in the remaining Rhodophyceae, there are two diploid phases (the carpospore-bearing stage, the tetrasporangiate individual) and, the life-cycle is diplobiontic. With few exceptions the tetrasporangia form four spores which exhibit three types of arrangement within the sporangium (cruciate, zonate, tetrahedral). Direct reproduction of the sexual phase by means of accessory spores (usually monospores) is met with especially in Bangiales and Nemalionales.

Classification and Status of the Rhodophyceae

The Rhodophyceae are at present grouped in seven orders six of which belong to the Florideae. These orders are mainly distinguished by details of reproduction, although vegetative features are commonly also of significance. This will be evident from the following brief epitome:

A. *Bangioideae:* Simple forms with diffuse growth, never exhibiting aggregation of filaments arid devoid of pit-connections; sex organs little specialised; carposporangia formed by direct division of the zygote.

Terrestrial, freshwater, and marine. Only order

I. *Bangiales:* Unicellular, colonial, filamentous, or parenchymatous forms, commonly with a single axile, stellate chromatophore in the cells; haplobiontic; accessory reproduction by monospores.

B. *Florideae:* Filamentous forms, usually exhibiting aggregation of one or more branch filamentous axes to form compact pseudoparenchymatous thalli; pit-connections invariably present between the cells; sex organs highly differentiated; carposporangia formed on filamentous gonimoblasts derived directly or indirectly from the zygote.

Largely marine

II. *Nemalionales:* Heterotrichous, with uni- or multiaxial construction, cells commonly with a single chromatophore; haplobiontic; gonimoblasts usually arising from the carpogonium; accessory reproduction by monospores or tetraspores. Some fresh-water forms.

III. *Gelidiales:* Uniaxial; probably diplobiontic, but without auxiliary cells; compound procarp; tetrasporangia usually cruciate.

IV. *Cryptonemtales:* Uni- or multiaxial, with evident or indistinct heterotrichy in the young stages; diplobiontic; auxiliary cells sometimes remote from the carpogonia, both borne on accessory laterals which are sometimes aggregated into special fertile tracts (nemathecia, conceptacles); tetraspores cruciate or zonate.

V. *Gigartinales:* Uni- or multiaxial, heterotrichy of juvenile stages often indistinct; diplobiontic; procarps frequent, with auxiliary cells commonly lodged in the carpogonial branches which are borne on the ordinary laterals of the thallus; tetraspores cruciate or zonate.

VI. *Rhodymeniales:* Multiaxial, heterotrichy obscure; diplobiontic, with a definite procarp; auxiliary cells not specially differentiated until after fertilization; tetraspores cruciate or tetrahedral.

VII. *Ceranriales:* Uniaxial, without heterotrichy; diplobiontic; procarp with 4-celled carpogonial branch, always borne on a pericentral cell from which the auxiliary cell is cut off only after fertilization; tetraspores usually tetrahedral.

The Rhodophyceae afford no evidence of relationship to any other class of the Algae. Some have sought an affinity with the Myxophyceae owing to the presence of similar pigments. Such a view overlooks entirely, quite apart from essential differences metabolism, the barrier constituted by the fundamental contrast in cell-structure between the members of the two classes. There is no evidence that this type of cellular organisation found in the latter has ever evolved into that characteristic of most other plants. Those who uphold the view of an affinity with Blue-green Algae look especially to the Bagiales, but here the contrasts are just as patent. As little can be said in favour of v relationship between Bangiales and the Prasiolaceae.

There is a considerable resemblance, both in the female organ and in the general nature of the life-cycle, but the differences are far too significant to admit of any question of affinity. Several authorities have discussed the question of a relationship with Ascomycetes, the Laboulbeniales in particular being regarded as presenting points of contact.

Bangioideae

Order Bangiales: Most of the Bangiales are probably annuals which enjoy a wide distribution over the surface of the earth. They exhibit practically all the simpler types of plant-body, although the latter never reaches any high degree of elaboration. One series of forms, with a filamentous or foliose thallus (Bangiaceae), are essentially marine, while another series (Porphyridiaceae), consisting for the most part at least of reduced forms, are more frequently met with in freshwater or terrestrial habitats. Some such classification has been advocated by most recent authorities, and the vegetative structure of the two families will best be discussed separately.

The thallus of most Bangiales lacks a definite growing point and exhibits diffuse growth. Adjacent cells are never connected by pits. There is typically a single axile, massive chromatophore, more or less stellate in shape and including a conspicuous central pyrenoid. The products of photosynthesis accumulate near the chromatophores as solid granules, resembling Floridean starch, and the varied pigmentation of the cells is due to a form of phycoerythrin and to phycocyanin. The membranes probably consist largely of pectic substances; cellulose has been demonstrated only in *Porphyridium*, and *Erythrotrichia*. The nucleus is often excentric in position.

Vegetative Structure

The least specialised members of this family are those grouped as Erythrotrichieae. *Erythrotrichia* comprises a number of small marine epiphytes, in several of which the plant has the form of a heterotrichous filament, the one-layered discoid base produced first by the germinating spore, bearing several, usually unbranched, erect threads *(e)*. The degree of development of the disc varies, but in certain individuals it is the only part present and gives rise to the spores. In other species, however, the filaments are merely attached by an irregularly lobed basal cell, supplemented by short radiating rhizoids (r); this is so in *E, carnea* and *E. ceramicola* Aresch. (*Bangia ceramicola* Chauv, the latter often regarded as a synonym of the former. The erect filaments may remain uniseriate (*E. ceramicola*)) but in certain species all except the basal cells become divided by longitudinal, often followed by transverse, walls so that the threads assume a characteristic clavate form.

The freshwater *Kyliniella,* likewise with a heterotrichous habit, is distinguished by the possession of a parietal chromatophore devoid of a pyrenoid. The elongate erect threads, which are surrounded by a wide mucilage-envelope, produce only rhizoid-like branches by means of which they attach themselves to adjacent plants; they can form secondary discs at their tips.

Erythrocladia, a completely prostrate epiphyte or endophyte, forms small discs composed of a radiating system of threads in which the marginal cells are often bifid; in some species the discs are several-layered. This genus affords an interesting parallel to similar prostrate types among Chaetophorales. The chromatophore *(c)* is a lobed parietal plate with a pyrenoid.

A related form is perhaps constituted by *Colaconenm,* the species of which are endophytic in the membranes of various Algae. The cells of the irregularly branched filaments, which sometimes anastomose, contain parietal plate-shaped chromatophores. Although the genus has been placed near *Aerochaetium,* Rosenvinge, as well as Howe and Hoyt, draw attention to the resemblance to *Erythrocladia,* and further investigation must show whether the two genera are possibly but different developments of the same generic type.

Another form with a parietal, markedly lobed chromatophore is *Porphyropsis coccinea* which occurs as small one-layered *Porphyra-like* sheets. The young stages form several-layered cushions, which become vesicular through active division of the upper layer; later the vesicle breaks open and constitutes a monostromatic frond which is at first cup-shaped (*Monostroma, Omphalophyllum).* Subsequently numerous rhizoids arise from the lower cells.

The Bangieae, comprising *Bangia* and *Porphyra,* show greater specialisation. The unbranched threads of the former may for a time remain uniseriate, but the older cells soon undergo longitudinal division by approximately radial walls so that the circular cross-section exhibits fifteen or more sectors; occasional cells may remain undivided. Reinke refers to a dome-shaped apical cell, but later workers make no mention of this and there is certainly abundant intercalary division. The young threads are merely attached by the basal cell, but in older plants the lower cells produce non-septate rhizoids which grow through the thick gelatinous wall and spread out over the substratum.

B. atropurpurea is a characteristic inhabitant of the estuarine parts of rivers, where it forms dark-coloured (brown, violet, or almost black) tufts; it is also reported from Lake Balaton. Woloszynska describes band-shaped thalli, often encrusted with lime. The marine *B.fuscopurpurea*, with equally broad but longer filaments occurs near high water mark, sometimes in the spray zone, and is found especially in winter and spring; it has a marked capacity to withstand desiccation and changes in salt-concentration. Its dense, dark red or brownish-violet tufts are commonly associated with species of *Urospora*.

Porphyra is also usually at first a simple filament which is stated to grow for some time by means of an often three-sided apical cell. The abundant longitudinal division, which soon spreads from near the tip backwards, leads to the formation of a usually one-layered expanse of varying width which exhibits diffuse growth. The cells are elongated perpendicular to the surface and possess strongly thickened external walls covered by a "cuticle" *(cu)*. The spores of *P. naiadum* produce a prostrate several-layered cushion from which the upright fronds arise, the only instance in which a *Porphyra* retains the heterotrichous condition.

The attaching system is constituted by thick-walled spreading rhizoids, produced on both surfaces from the lower cells of the thallus and usually interweaving to form a compact cushion from which accessory fronds may later arise. The rhizoids are unicellular but multinucleate and some are branched; their dilated tips can divide into a number of cells. They adhere to the substratum by means of their gelatinous walls, as well as by suckers arising on their lower sides. Grubb records penetration of the rhizoids to considerable depths into the tissues of *Fucus*. In some species the lower parts of the thalli are possibly perennial.

The species of *Porphyra* mainly inhabit the rocks between tide-levels and a number, which show marked resistance to desiccation, form characteristic zones at certain times of the year in the upper littoral region. The broad thalli are often unbranched and commonly have a wavy margin, as in P. *umbilicalis* which reaches a length of 30 cm. The narrow *P. linearis* Grev, usually regarded as a growth-form of *P. umbilicalis* however, is found mainly in winter and frequently forms a belt above that occupied by the latter.

Other common species are *P. atropurpurea* (Olivi) De Toni (P. *leucosticta* Thur.) and P. *perforata* J. Ag., the latter abundant on the Pacific coasts of North America. Many species are or have been used as food, for instance the "laver" of Wales and Scotland which is derived from P. *umbilicalis* and P. *laciniata*. In Japan, species of *Porphyra* are specially cultivated, being grown on prepared twigs which are exposed during winter; "Nori" is dried P. *tenera*.

Porphyridiaceae

Apart from *Goniotrichum* and *Asterocytis*, with branched or unbranched filaments, the Porphyridiaceae include several, probably reduced, unicellular and mainly terrestrial types, the cells of which are commonly aggregated to form brightly coloured masses. The commonest, *Porphyridium cruentum*, is not infrequent on soil and on damp walls, where it forms several-layered blood-red mucilaginous strata, although when dry the brittle crusts are less conspicuous. In *Porphyridium*, the individual cells are globose, whereas in *Chroothece* they are ellipsoidal. The reference of these diverse forms to Bangiales depends largely on the possession of axile star-shaped chromatophores and the presence of Floridean starch.

The species of *Goniotrichum* are minute epiphytes, usually with plentifully branched filaments, possessing thick longitudinal walls. The branching is often described as false and like that of certain Myxophyceae, the upper part of the parent-thread being pushed aside by the outgrowth of the part below, but this requires further investigation. In *G. elegans*, the threads commonly remain uniseriate, but in G. *cornu cervi* (Reinsch) and others eight or more rather irregularly arranged cells may be seen in the transverse section. The older cells appear rounded or discoid; all are capable of division. In the little-known *Neevea* the branched, uni- or multiseriate threads form sheets which are endophytic in *Flustra*.

Phragmonema, so far only found on leaves and walls in hothouses, is distinguished by the presence of several parietal band-shaped chromatophores in each cell. The rarely branched threads are of a dull violet or olive colour and exhibit ready fragmentation into single cells.

The dense tufts formed by *Asterocytis ornata* consist of branched or unbranched tubular strands of mucilage containing variously shaped (commonly ellipsoidal) cells arranged in a single row; the blue-green chromatophore is often excentric. This species is widespread in brackish water, and appears to be rather indifferent to the salt concentration; it is probably an eutrophic form. *A. halophila* (Hansg.) Forti is also recorded from freshwater.

A brackish water form, frequently referred to this genus, is *Allogonium smaragdinwn* Hansg, which occurs as short, rarely branched, threads or commonly as single cells. In nature, the older cells are often filled with starch and almost colourless, but in well-nourished individuals the radiating processes of the chromatophores, commonly dilated at their ends, become prominent. The cells can escape from the mucilage-envelope and perform, slow tortuous movements; a mechanism similar to that of *Chroothece* is suspected. Pascher, in fact, regards *Asterocytis* as a colonial development of the latter genus.

Chroothece Richterianum Hansg, occurring as orange- to brownish-yellow mucilage-masses on saline soils in Bohemia, is little known, but Pascher and Petrova have described in greater detail a second species (C. *mobilis)*, which forms greyish-yellow mucilage-masses on moorland soils. In both species the ellipsoidal cells possess a thick stratified pectic membrane. In nature, those of *C. mobilis* for the most part occur in a resting condition in which the cells are filled with starch *(st)* and only the axile part of the chromatophore *(c)* persists, but if kept in the dark, the starch vanishes and the stellate-blue-green chromatophore becomes apparent. The cells multiply by transverse division into two or four daughter-cells which secrete distinct membranes, but remain for a time enclosed in that of the parent, as a result of which temporary threads may be formed. The pyrenoid either divides with the protoplast or more generally, a new pyrenoid is formed in one of the daughter-cells, the chromatophore gradually differentiating around it. Rejuvenescence of the cells is frequent.

When stained with fuchsin, a mucilage-strand, with occasional denser layers and as much as 50 times the length of the cell, becomes recognisable in C. *mobilis*, a similar, but more markedly stratified strand is recorded in C. *Richterianum*. In C. *mobilis*, the mucilage is no doubt secreted through the numerous pores which traverse the membranes and are specially abundant towards the poles. The nucleus is commonly situated at that end of the cell from which active secretion is occurring. The formation of these strands is doubtless related to the rapid jerky, often zigzag, movements which are exhibited by occasional cells and last but for a fraction of a second, during which a distance equal to ten times the length of the cell may be covered. Rise of temperature increases the frequency of the movement which is ascribed to rapid absorption of water on the part of the mucilage.

The spherical cells composing the crusts of *Parphyridium cruentum* are provided with individual mucilage-envelopes and embedded in a common mass of hyaline mucilage; at times they develop mucilage-stalks which are sometimes branched. The cells multiply by simple division in various directions. There is some difference of opinion as to the cell-structure; Staehelin's data are certainly in large part erroneous. The cell-wall is thin and elastic and follows all changes in the volume of the protoplast. The body of the chromatophore, the red colouration of which is due to phycoerythrin, is very massive, while the radiating processes are often short completely lacking. The peripheral cytoplasm between the processes of the chromatophore includes many small vacuoles. The relatively large nucleus, which appears as a white area in the living cell, lies attached to the wall when stained it shows a nucleolus and an outer nucleus devoid of chromatin. The mitotic stages described by Lewis and Zirkle are, according to Geitler, misinterpretations. Both starch and fat may be present.

In recent years, a number of other species, all with essentially the same structure, have been described. *P. aerugineum* Geitler, which is stated to be widely distributed in stagnant freshwater and has also been found on soil, lacks phycoerythrin and owes its blue-green colour to the presence of phycocyanin. Kylin has recorded a marine species, at present only known in cultures (P. *marinum).*

Rhodospora possesses cells with a central nucleus and numerous parietal discoid chromatophores. It forms indefinite masses of a red-violet or dull olive-green colour on archaic rocks. The cells multiply by division along three planes into *2-16* (usually 4-8) parts which are liberated by the rupture of the parent-membrane.

Reproduction of Bangiales

The only known method of multiplication in certain Porphyridiaceae *(Rhodospora, Chroothece,* etc.) is by cell-division (autospore-formation). Most Bangiales, however, exhibit asexual reproduction by structures variously called monospores, neutral spores or gonidia. The first designation appears as suitable as any other and is adopted here. The naked monospores are liberated through a well-defined aperture in the wall of the sporangium, the process taking place either gradually or suddenly.

The *monosporangia* of *Porphyra* are generally formed by crosswise division of the vegetative cells perpendicular to the surface so that the thallus remains one-layered. Those of *Bangia* are usually produced by one or two radial divisions in the sectors, the process mostly commencing at the apex of a thread; they may, however, arise from an entire cell without division or even from the cells of the young uniseriate threads.

Erythrotrichia, Erythrocladia, and *Porphyropsis* possess more clearly differentiated monosporangia which are constituted by small cells *(m)* cut off by a curved wall, usually from the upper end of a vegetative cell, after accumulation of cytoplasm at the point where the sporangium is to be formed. In *Erythrotrichia,* after liberation of the spore, the part of the vegetative cell below enlarges and fills the cavity and further sporangia may be produced from these cells.

The monospores of *Colaconema* are formed from terminal or intercalary sporangia which may be cut off from a vegetative cell and are sometimes detached as a whole. Those of *Goniotrichum,* of *Asterocytis,* and probably of *Porphyridium cruentum* are constituted by the protoplasts of the ordinary cells which round off and escape from their envelopes; in *Goniotrichum* the process begins in the upper parts of the threads and, according to Chemin, is preceded by longitudinal division of the parent-cells. Thick-walled akinetes have also been recorded in *Asterocytis.*

The monospores of *Bangia* and *Porphyra* show amoeboid changes of shape which do not appear to be accompanied by any appreciable translocation. They commence some little time after liberation and may continue for two days. Such changes of shape are also recorded in *Porphyridium.* The monospores of *Erythrotrichia* and *Goniotrichum* do not appear to show this feature, although they exhibit gliding movements which are sometimes quite rapid; similar movements have been observed in *Porphyridium*. The mechanism of the movement is unknown, but light sometimes appears to exert a directive influence.

Sexual reproduction is recorded only in *Bangia, Porphyra, Erythrotrichia* and *Erythrocladia* and has so far not been observed in any freshwater member. As in Florideae, the male cells (spermatia) are devoid of organs of movement. The distribution of the sexes is variable. In monoecious species of *Porphyra*, they are commonly segregated, the male areas being yellowish-white or white, the female purple; *P. tenera* is andro-dioecious. In the Bangieae, sexual organs are often formed over a large part of the thallus, although growth may still continue in the lower parts after the apical ones have begun to form reproductive cells.

The antheridia of *Porphyra* arise by repeated division of the cells. The first wall is parallel to the surface and the products undergo two vertical divisions; further horizontal walls lead to the formation of 32-128 small cells which are arranged in several layers. Antheridium-formation commences at the apex or edges and gradually spreads towards the base of the thallus. The antheridia of *Bangia* are produced in the same way, their formation often proceeding basipetally and commonly involving the whole plant. The mature antheridia are colourless, but contain a small leucoplast. The spermatia are liberated by gelatinisation of the membranes, although the boundaries of the original parent-cells long remain recognisable.

In *Erythrotrichia* the antheridia are formed in the same way as the monosporangia. The spermatia are larger than those of Bangieae and have a more obvious plastid; they exhibit some capacity for movement. The correspondence between monospores and spermatia evident in this genus is also shown in *Porphyra atropurpurea* where the same parent-cell may give rise both to monosporangia and antheridia; in such instances there are all transitions in size and colour between the two types of reproductive cells.

The female organs (commonly called carpogonia) often differ from vegetative cells only in their somewhat larger size and more abundant contents. The surface is presumably always mucilaginous, since abundant spermatia (s), now enveloped by a thin membrane, are found adhering to them. In *Erythrotrichia* and diverse species of *Porphyra* the surface of the female organ is protruded, although it is not clear whether this is primary or secondary. It has repeatedly been recorded that, if fertilization be delayed, the female cells of *Porphyra* put forth on both surfaces elongate hyaline protuberances to which spermatia can adhere; less pronounced structures of the same type have been observed in *Bangia*. These outgrowths have been compared with trichogynes.

In the process of fertilization, the spermatium puts forth an often delicate process containing a thin strand of cytoplasm. This penetrates rather rapidly through the thick membrane to the female protoplast, and the greater part of the contents of the spermatium pass into the latter. Dangeard observed the sexual nuclei in contact. The connecting strand of cytoplasm often persists for a long time, while the empty spermatium-wall at the surface soon disappears. Kunieda, who describes a rather different method of sexual fusion in *Porphyra tenera*, suggests that the stages previously described do not represent fertilization, but are due to the attack of a parasite; no adequate evidence is produced for this view.

After fusion there is increase of cytoplasm and accumulation of starch in the female cell, while the chromatophore withdraws its processes and forms an irregular lobed mass. Rosenvinge records the occurrence of groups of carpospores in *P. umbilicalis*, without any evidence of fertilization-tubes, and suggests the possibility of an apogamous development.

In *Erythrotrichia* the undivided female protoplast is set free after fertilization. In the Bangieae, on the other hand, there is usually successive segmentation into eight (rarely more) parts, from each of which a naked carpospore is liberated, although if the female cells are small, there may be fewer or no divisions. In *Bangia*, the first wall is periclinal anticlinal and the later ones radial, while in *Porphyra*, the first wall is parallel to the surface and each of the products divides into quadrants so that the carposporangia normally lie in two layers. The mature sporangia possess gelatinous walls and their boundaries remain clearly visible after the escape of the carpospores. In *P. tenera* the parts of the fronds containing the maturing carpospores become detached, and the latter are stated to remain dormant throughout spring and summer.

The liberated carpospores, which are always naked, contain much starch and are usually not as large as the monospores, while their chromatophore is smaller and less lobed or diffuse. They exhibit slight amoeboid movements, but round off before forming a membrane.

Life-cycle of Bangiales

There is reason to believe that the Bangiales, like the less specialised Florideae, are haploid forms in which reduction takes place during the first divisions in the zygote; this has been demonstrated in *Porphyra* with some degree of probability. Sexual plants occur mainly during the cold season of the year, whilst at other times, the thalli often bear monosporangia only. There can be no doubt that these represent merely an accessory method of reproduction. Many Bangieae disappear altogether during the warm season.

The monospores of the Bangieae develop direct into new plants and may germinate within the sporangia. After formation of a membrane the spore elongates and divides transversely, the one cell giving rise to a rhizoid (while the other segments to form the primary filament; other rhizoids soon develop from the lower cells.

The fate of the carpospores remains uncertain. According to Kunieda those of *Porphyra tenera* remain dormant throughout the warm season and give rise to fresh crops of plants in the late autumn; the method of germination is stated not to differ essentially from that of the monospores, although the germlings have thicker walls and diffuse chromatophores. Earlier investigators, studying the development of the carpospores of *Bangia* and *Porphyra* in cultures, concluded that they gave rise to branched filamentous stages from which supposed monospores were liberated. Kunieda states that such stages are pathological. Rees, studying the lifecycle of *P. umbilicalis* in nature, however, finds that the carpospores produce short little-branched filaments lasting only for a few weeks and multiplying by means of monospores; ultimately these give rise to the *Porphyra-thallus*. Ueda also concluded that the Japanese *Porphyras* persisted through the summer as minute thalli reproducing by monospores.

It appears probable that, in the Northern Hemisphere at least, the ordinary thallus propagates both by monospores and by sexual means, one or other at times preponderating. While the monospores reproduce the thallus direct, the carpospores give rise to a filamentous stage which lasts during the unfavourable period and ultimately reproduces the ordinary plant. In some regions, the ordinary thalli are found throughout the year. The probable persistence by means of minute filamentous stages recalls the condition in many Ectocarpales.

4

Fucales

Many authorities have assumed a certain measure of relationship between Laminariales and Fucales which have been regarded as lines diverging from a common ancestral source comprising forms like the Ectocarpales. In both Laminariales and Fucales the thallus is a large diploid parenchymatous body which, despite differences in the method of growth, shows a considerable degree of similarity in anatomical structure, although displaying merely the culmination of tendencies manifest already among the polystichous Ectocarpales. Perhaps a fuller investigation of the Antarctic genera *Durvillea* and *Ascoseira* will shed light on the nature of this relationship. There are, however, very significant contrasts in reproduction between the two orders, since in Fucales a separate gametophytic phase is altogether lacking. The frequent reduction, in cultures, of the female gametophyte of Laminariales to a unicellular state, a condition which is apparently the rule in *Saccorhiza*, has been regarded as illustrating a stage in the retrogression of the haploid phase which has culminated in the condition found in Fucales. The first two nuclear divisions in the oogonium are interpreted as a tetrad-division within a sporangium in which, however, the spores are not individualised. Each, after a brief resting period, gives rise to two potential ova which represent all there is of the female gametophyte. It is not so easy to extend this interpretation to the antheridium.

It may be doubted, moreover, whether any such detailed attempts at homology are profitable, since there can be no question of any direct derivation of Fucales from Laminariales or other specialised group of Phaeophyceae. It seems more probable that the reproductive mechanism of the former has originated from the tendency for the asexual zoospores to behave as gametes, evident in diverse evolutionary lines among the less specialised Brown Algae. Such a view does not necessitate the assumption of a reduced gametophyte, although, like the hypothesis stated above, it ultimately derives the sex organs of Fucales from unilocular sporangia. Kylin suggests such a possibility, although sceptical as to the fusions between the swarmers of unilocular sporangia. Whether the occurrence of septation in the antheridium is a valid argument against such a hypothesis can only be determined when the nature of this septation is better known. There can be no doubt that the Fucales have a long history behind them, and it is inadvisable to probe too deeply into homologies.

In reçent years Kylin, laying stress on the differences in the relative lengths of the flagella of the spermatozoids of Fucales on the one hand, and of the zoospores of Phaeophyceae on the other, as well as on the probable plurilocular character of the antheridia, disputes the probability of a direct derivation of Fucales from forms resembling the present-day Ectocarpales. The evidence on the former point is conflicting, while, if the sex organs of certain Fucales acquire a plurilocular character at maturity, this appears to be secondary and altogether different from the condition shown by a typical plurilocular sporangium.

The fucales have originated in the evolutionary line exemplified by the parenchyruatous (polystichous) Ectocarpales, among which there is replacement of trichothallic by apical growth, as apparently occurs also in the germling of

Fucus. As already emphasised, however, the derivation is remote and there appear to be no intermediate forms. The few, uncertain, fossil forms that have been ascribed to Fucales do not afford any clues. The suggested relationship with Encoeliaceae, does not carry one any further. Delf, in assuming an affinity with Mesogloeaceae, overlooks the fundamental differences in basic construction. The filamentous character of the medulla in the Fucales originates secondarily from a primary parenchymatous state, and there is no evidence whatsoever that the parenchymatous type of thallus in Phaeophyceae is derived from the pseudoparenchymatous one, although Church made such an unwarrantable assumption and Schussnig evidently has some such hypothesis in mind.

Many of the less specialised members of the Fucales are characteristic of the colder seas of the Northern and Southern Hemispheres, whence they have invaded the warmer areas of the world. That such widespread genera as *Cystoseira, Cystophyllum,* and *Sargassum* are comparatively recent and still in course of vigorous evolution is shown not only by their morphological features and the presence of a single ovum in the oogonium, but also by the multiplicity of ill-defined species.

The following is an epitome of the classification adopted in the foregoing account:

1. *Fucaceae:* Ascophyllum, Axillaria, Cystosphaera, Fucus, Margina-riella, Myriodesma, Pelvetia, Pelvetiopsis, Phyllospora, Scaberia, Scytothalia, Seirococcus, Xiphophora.
2. *Himanthaliaceae:* Himanthalia.
3. *Cystoseiraceae:* Bifurcaria (Pycnophycus), Bifurcariopsis, Carpoglossum, Cystophora (incl. Blossevillca), Cystophyllum, Cystoseira, Halidrys, Landsburgia, Platylobium, Platythalia.
4. *Sargassaceae:* Acystis, Carpophyllum, Coccophora, Sargassum (incl. Anthophycus, Pterocaulon), Turbinaria.
5. *Hormosiraceae:* Hormosira, Notheia.

6. *Durvilleaceae:* Durvillea (incl. Sarcophycus).
7. *Ascoseiraceae:* Ascoseira.

Ecology of Fucales

Some reference has already been made to the marked zonation exhibited by the littoral Fucales on northern shores. *Pelvetia canaliculata* and *Fucus spiralis* occupy the uppermost, *F. terrains* and *Himanthalia* the lower levels, with *Ascophyllum* and *Fucus vesiculosus* in intermediate positions; on more northerly coasts *F. inflatus* usually replaces *F. serratus,* while on the west coast of Sweden the place of *Pelvetia* is taken by *Fucus spiralis.*

On shores exposed to the full force of the waves *F. serratus, Ascophyllum* and *Himanthalia* are often lacking while on wave-swept coasts in the Faeroes Boergesen found only *F. spiralis* and *F. inflatus.* Cotton records a bladderless variety of *F. vesiculosus* (var. *evesiculosus)* in such habitats and states that the size of the frond and the number of bladders in the type increases with the degree of shelter.

Rees produces similar data for the distribution of littoral Fucales in relation to intensity of surf-action, although he points out that other factors concerned are the angle of slope and the nature of the surface of the rock. The relative levels of *Ascophyllum* and *Fucus vesiculosus* vary and Zaneveld discusses the possible causes. By contrast to the abundant representation of Fucales in the littoral region on northern shores, the bulk of the New Zealand members occur at and below low tide level, *Hormosira* being the chief emergent form.

There can be no doubt that the littoral Fucales are well adapted to their habitat. *Fucus vesiculosus, F. serratus,* and *Ascophyllum nodosum* are resistant to cold, while in summer, the thalli may reach a temperature of 34-36°C. without apparent harm, although Grubb states that *F. serratus* is more susceptible to summer-heat than *F. vesiculosus.* The pronounced zonation is no doubt to a considerable extent, a result of a differing capacity to resist desiccation, both during germination and vegetative growth. According to Zaneveld the forms in the upper zones have thicker membranes, a higher wate; content, and reach their dry weight on evaporation later than those in the lower. Baker also draws attention to the dilated receptacles filled with mucilage, characteristic of the species occupying the upper levels. *Halidrys siliquosa* by contrast cannot withstand even a short period of desiccation and *Laminaria*, although more resistant, is vēry susceptible to frost. *Fucus evanescens* is likewise resistant to desiccation.

F. vesiculosus and *F. ceranoides* can withstand lowered salt-concentrations, while the other two British species do so less readily. Such conditions have no effect on the formation of receptacles which are, however, sterile; cryptoblasts, on the other hand, are reduced in number or lacking. Kniep was of the opinion that the increasing salt-concentration at lower levels played a role in the distribution of the littoral Fucales.

It is improbable that the practical restriction of the littoral Fucales to the intertidal region is dependent on illumination, although Gail concludes that the lower limit of *Fucus evanescens* is determined by this factor. Montfort, however, investigating photosynthesis of *F. vesiculosus* at various depths, decided that the distribution of this species is not influenced by depth of submergence.

Both it and *F. serratus* are adapted to high light intensities and are the equivalents of sun-forms among land-plants. At the same time they can also carry on effective photosynthesis in relatively feeble light, and their active growth during the early months of the year is due to this capacity at a time when respiration is inconsiderable owing to the low temperature.

There is evidence that in both species there is a seasonal variation in reaction to light and temperature. Stocker and Holdheide found that considerable photosynthetic gain is possible even during periods of exposure, although the capacity for photosynthesis is gradually lost on drying; this is, however, accompanied by a reduction of respiration to a minimum. Exposure to rain results in a reduction of photosynthesis.

The littoral Fucales thus constitute a distinct ecological group, markedly adapted to a subaerial mode of life and with a great capacity to resist extreme conditions and to effect considerable assimilatory gains under such circumstances. Their adaptation to subaerial conditions is likewise displayed by the plentiful occurrence of forms of the diverse species concerned in the vegetation of salt-marshes.

Although the members of this order are normally lithophytes, they are also widely represented by unattached growth-forms which lack the attaching discs of the saxicolous types and for the most part propagate solely by vegetative means. Such forms are often a conspicuous feature of the flora of salt-marshes, where they are found embedded in the mud or interlacing with one another and with the stems of the halophytic seed-plants so that they are not dislodged by the tides. They probably play an important role as pioneers, following upon the filamentous Green and Blue-green Algae that constitute the primary colonisers, and form a seed-bed, as well as a protection during early stages of germination, for the phanerogamous halophytes.

Most of them are of course subject to desiccation for a number of days in each fortnight, in place of the usual daily exposure. On the coasts of Morocco, near the southern limit of its range, *Fucus vesiculosus* is represented only by salt-marsh forms.

Most of the littoral Fucales, except *Himanthalia*, have given rise to such growth-forms on British salt-marshes. Apart from the prevalent vegetative reproduction, these forms are characterised by a dwarf habit and frequent spirality or curling of the thallus. The dwarf habit has been attributed to the greater degree of exposure, and to the shorter period of immersion in sea water which is moreover often diluted. This factor is also believed to bring about the reduced width of the thallus, while spirality or curling has been ascribed to the reception of more moisture and nutritive salts by the side of the thallus adjacent to the mud during intertidal periods. A spiral form of *F. vesiculosus* has been obtained by placing plants entangled in wire-netting on a *Spartina-mzrsh*. No satisfactory explanation for the customary sterility has yet been afforded.

The salt-marsh forms of *Pelvetia canaliculata* like the saxicolous form, occupy the upper levels and several different kinds have been distinguished. One of the commonest is that known as var. *liber a*, which lies loose on the surface. The bushy plants are dark brown and copiously branched, the branches interlacing with those of adjacent plants so that a continuous covering is formed over the soil. In the rarer var. *coralloides* the lower part of the thallus is embedded in the marsh, about 2 cm. projecting above the surface; it is more like the normal attached form and approaches the var. *radicans* found by Foslie on the clay bottom of shallow rock-pools.

Ascophyllum nodosum is represented mainly by small forms of var. *scorpioides*, which is widespread in the loose-lying communities of the North Temperate zone, and by a form known as var. *Mackaii* Cotton, which is sometimes ranked as a separate species. The latter forms globular tufts which are often of some size and consist of a number of plants, with or without small air-bladders; it appears to arise directly by vegetative propagation from broken pieces of the ordinary lithophyte.

The forms of *Fucus* found on salt-marshes are very diverse and, although they are doubtless all derived by vegetative propagation from saxicolous types, the profound modification and the customary absence of fertile conceptacles make it difficult to establish their true affinities. In a survey of the British marsh-Fwa Baker and Bohling refer most of them to *F. vesiculosus* and class them in three groups—the relatively large spirally twisted ones (ecad *volubilis*), the turf like forms (ecad *caespitosus)*, and the filiform ones (ecad *muscoides)*;

numerous transitional forms are recognised. Those placed in ecad *volubilis* (incl. *F. volubilis* Huds. and *F. lutarius* Kiitz.) agree to a considerable extent with comparable ones found on the continent. The ecad *caespitosus* corresponds to the plant called *F. balticus* Ag. (*F. vesiculosus* var. *subecostata* Harv., *F. vesiculosus* var. *balticus* (Ag.) Cotton), and is common as an undergrowth in British salt-marshes. It is a dwarf, prostrate or erect, form showing little or no curling; the thallus is devoid of midrib and air-bladders, bears prominent marginal cryptoblasts, and the conceptacles sometimes show divided oogonia. The var. *muscoides* differs from this only in its erect growth and filiform fronds.

Sauvageau, who undertook a detailed study of the French and Spanish salt-marsh *Fuci*, referred them to *F. lutarius* (Chauv.) Kiitz. (*F. vesiculosus* var. *lutarius* Chauv.). The projecting, part of the thallus (30-40 cm.) is narrow and spirally coiled, often devoid of air-bladders and provided with scattered cryptoblasts which are frequently marginal in position. The rare individuals with fertile conceptacles are always female; the oogonia remain undivided, although several nuclei may be present. The receptacles are inflated with gas and adventitious fronds commonly arise from the edges of older conceptacles which have split open.

The spirally twisted *Fuel* of British salt-marshes were first referred to *F. volubilis* Huds, which shows many similarities to *F. lutarius.* The plants vary considerably in size and bear occasional oblong air-bladders; the rare receptacles display the features recorded by Sauvageau. The latter subsequently referred the forms devoid of air-bladders to *F. lutarius* and those possessing them to *F. volubilis.* Chemin is of the opinion that the two sets of forms are not distinct.

Sauvageau criticises Baker and Bohling's reference of the marsh Fucoids to *F. vesiculosus.* He argues that the sole presence of female plants is no proof of derivation from a dioecious species, since the special conditions might just as readily lead to abortion of the antheridia in a bisexual conceptacle as to the suppression of male plants. He is of the opinion that *F. lutarius* may have been derived from *F. spiralis.*

A remarkable silt-inhabiting form, growing to a length of 20-50 cm., has been described by Nienburg from the North Frisian "Wattenmeer" under the name of *F. My till.* Its sole anchorage is afforded by the byssus-threads of the Mollusc *Mytilus.* The plants, which show a tendency to twist and only rarely possess air-bladders, have plentiful cryptoblasts. Only female conceptacles with normal oogonia are present, but the packets of ova are not liberated, The receptacles proliferate into well-branched fronds which become free by decay, and this is believed to be the normal method of propagation. The parentage of this peculiar form is obscure, but it is possibly derived from *F. vesiculosus.*

An undoubted limicolous form of *F. spiralis* has been recorded; the base of the small thallus is embedded. Marsh-forms of *F. ceranoides* are also known although *F. serratus* does not appear to be represented in such habitats.

Similar sterile unattached thalli of diverse Fucales often occur in huge numbers as loose-lying forms in the sublittoral region of the quieter landlocked seas. They are derived from detached fragments, which have been carried by currents into these waters and there continue to propagate by vegetative means. One of the commonest forms appearing in this community in North Temperate seas is *Ascophyllum nodosum* var. *scorpioides,* in which the main axes are almost terete, devoid of air-bladders, and bear only occasional laterals; where such forms are not continuously submerged, they are much more compact arid approximate to those found on salt-marshes.

Certain forms of *Fucus vesicttlosus* play a very considerable role in the loose-lying communities of the Baltic. Many of these, formerly grouped under *F. balticu* show considerable parallel with the larger salt-marsh types, although they lack the characteristic spiral curling of the thallus. In the eastern parts of the Baltic, where the salinity becomes progressively lower, *F. vesiculosus* fruits more and more sparingly and is ultimately represented only by reduced vegetative forms. Jonsson records similar dwarf-forms of *F. inflatus* L. from East Greenland. A loose-lying form of *F. serratus*, found on muddy sand and shingle, has been recorded from Scotland. In the loose-lying community of the Adriatic species of *Cystoseira* play an important part; the plants consist of detached axes covered with new proliferations.

The most striking of the unattached forms is the Gulf-weed which occurs in huge floating masses in the Sargasso Sea, off the African coast between 20° and 35° latitude N.; Parr furnishes quantitative data. There is considerable evidence that these immense tracts of drifting *Sargassum* are formed by continuous vegetative reproduction *in situ* of occasional detached saxicolous individuals, carried there by ocean currents. The view, advocated especially by Kuntze, that this vegetation consists in the main of an accumulation of detached plants, has found little support in recent years.

Attaching discs have never been found on the plants of the Sargasso Sea and investigation on the spot has shown that there is continuous growth of the younger parts, as the older die away. Hentschel reaches the same conclusion by a study of the attached growth found on the Gulf-weed. Winge and Sjostedt record a considerable increase in the amount of *Sargassum* during the summer, which the latter, probably Wrongly, ascribes to the arrival of fresh plants from Central America and the West Indies. The newly detached individuals sometimes show fertile conceptacles with sex organs, but these appear to be functionless and, according to Winge, occur only in the westerly parts of the Sargasso Sea.

Boergesen recognises two "species" in the Sargasso Sea, viz. *S. natans* L. (*S. bacciferum* C. Ag.) and more rarely *S. hystrix* J. Ag. var. *fluitans* Boerg. Both are sterile and exhibit attenuation and marked elongation of all the branches, as compared with the attached species of *Sargassum*. Winge distinguishes eight different types, of which only three are common. There is considerable resemblance between *S. natans* and certain of the attached West Indian species *(S. vulgare, S. Filipendula),* and Boergesen is of the opinion that one or other of these is possibly the parent from which the free-floating form is derived. Drifting specimens of *Sargassum* are very commonly encountered in the North Atlantic.

The difficulty of relating the diverse forms found in the Sargasso Sea to any of the known littoral species has led to the hypothesis that they might represent the remains of the littoral girdle of the continent of Atlantis. This conjecture, however, overlooks the profound modification which the detached lithophytic Fucales undergo in their adaptation to a loose-lying existence.

5

Bangiales

Although certain authorities have advocated a relationship with Prasiqlaceae, the bulk of opinion has always favoureoan affinity with Florideae. With Prasiolaceae, there is nothing in common except the form of the thallus and chromatophore, while the Bangiales are fundamentally distinguished from the Myxophyceae by their cell-structure, the products of photosynthesis, and the methods of reproduction. The simple nuclei described in *Porphyra tenera* by Ishikawa depend on poor fixation; Dangeard has shown that those of *Bangia* and *Porphyra* possess a nucleolus and undergo a mitotic division in which centrosomes are recognised. In view of the many important differences, the few resemblances cannot be taken as evidence of an affinity with Myxophyceae.

In discussing the relationship to Florideae, it may be noted that, apart from the simple vegetative structure, the absence of apical growth, and the lack of pits between the cells, there are important contrasts in reproduction. Thus, the reproductive organs are formed from intercalary cells, the female organs are less specialised, and the entire zygote is used to form carpospores. Svedelius does not regard the female organs as homologous with those of Florideae, since outgrowths resembling trichogynes are only sometimes present. Schmitz also emphasised the differences in the method of fertilization.

On the other hand, there are marked resemblances in the pigmentation of the chromatophores, in the presence of Floridean starch, in the general character of the spermatia and carpospores (especially their non-motility), and in the similarity between the life-cycle and that of Nemalionales.

The chromatophores contain phycoerythrin and phycocyanin, as well as chlorophyll, carotene, and xanthophyll, and in *Porphyra tenera*, the two first are stated to be identical with those in *Ceramium rubrum* the same is true of the phycoerythrin in *Porphyridium marinum*, although that of P. *cruentum* seems to be a different modification, resembling that found in Myxophyceae.

Kylin specially emphasises the identity of the pigments of Bangiales and Florideae and the contrast with Myxophyceae and other groups. In certain Porphyridiaceae, the phycocyanin evidently predominates over phycoerythrin, and is perhaps alone present.

The physiological resemblances are probably significant, when taken in conjunction with the general parallel in reproductive methods. Rosenvinge also lays stress on the similarity between the cell-structure of many Bangiales and diverse Nemalionales; the two orders also resemble one another in the amoeboid movements exhibited by the spores.

For these and other reasons he proposes the name Protoflorideae which I have hesitated to adopt, because it is doubtful whether the Bangiales can be regarded as precursors of the Florideae. The undoubted marks of relationship are more likely to be due to divergent development from a common ancestry. I am inclined to regard forms like *Erythrotrichia* and *Kyliniella* as the most primitive members of the group, despite the rather specialised monosporangia of the former, because in their heterotrichous habit they show the nearest approach to what was probably the prototype of Bangiales and Florideae

The classification here adopted is as follows:

(a) *Bangiaceae:* Bangia, Colaconema, Erythrocladia, Erythrotrichia, Kyliniella, Porphyra, Porphyropsis.

(b) *Porphyridiaceae:* Allogonium, Asterocytis, Chroothece, Gonio-trichum, Neevea, Phragmonema, Porphyridium, Rhodospora.

Uncertain Bangiales

Compsopogon is chiefly a denizen of tropical and subtropical streams, although it has been recorded from ponds and from tidal waters. Its occasional occurrence in temperate regions is no doubt due to introduction by shipping. Thus, *C. leptoclados* Mont, has been found in a part of the Reddish Canal, Manchester, where there is an influx of hot water from cotton mills. The rather coarse, richly branched threads are blue- or violet-green in colour.

In the younger parts they consist of flat cells formed by transverse division of the segments of the apical cell, but in the older parts the cells become greatly enlarged and barrel-shaped there are no pits. At some distance from the apex, the axial cells cut off successive peripheral segments, which ultimately unite to form a compact cortex *(co)* of deeply pigmented cells through which the almost colourless axial ones are plainly visible.

The cells of the cortex multiply by anticlinal division and sometimes divide to form 2-4 layers. No cortex is produced in the basal parts of the threads, where, however, the axial cells are commonly overgrown by septate rhizoids emanating from the lower cells. Branches arise before the cortex is formed, although short secondary branches may later originate from the cortical cells.

The cells contain numerous parietal oval to oblong chromatophores which are sometimes lobed; the thread-like chromatophores drawn by Schmitz have not been recorded by others. The cytoplasm of the axial cells shows circulation.

The only method of reproduction known is by monospores, of which Thaxter distinguishes two kinds; both are spherical and motionless, with dense contents and large nuclei. His so-called macro-sporangia are always solitary, while the microsporangia for the most part form somewhat projecting sori. They develop either from the uncorticated cells or from those constituting the cortex. The macrosporangia, which are cut off by a curved wall, liberate their contents through an aperture in the wall, after which the parent-cell enlarges and fills the cavity; this is very reminiscent of *Erythrotrichia*. Skuja, however, regards the macrosporangia as arrested branches and considers that they develop only under cultural conditions. According to his view, the normal organs of multiplication are the microsporangia.

In the colouration of the chromatophores, the method of reproduction, and the absence of pit-connections between the cells there are indications of affinity with Bangiales and, in view of the varied development of the latter, it is not improbable that they might have evolved into more elaborate types. A renewed investigation of *Compsopogon* in tropical habitats may cast further light on this problematic form.

Reference should also be made to *Rhodochaete*, a minute marine epiphyte with simple branched filaments showing apical growth and reproduction by monospores, produced in essentially the same manner as in *Erythrotrichia*. Small pits are stated to occur in the septa between adjacent cells. The affinities are very obscure.

Another uncertain member of Bangiales is *Conchocelis rosea* which lives in the shells of it. It seems to be widespread in water of some slight depth in North Atlantic and Arctic seas where, together with Blue-green Algae of a similar habit, it is stated to play a part in the erosion of limestone rocks. The young growth consists of narrow threads composed of elongate cells of irregular shape; later the filaments branch richly and form a more or less continuous layer. According to Jonsson, the cells contain a stellate chromatophore, but Rosenvinge describes a number of ribbon-shaped ones he also records small pit-connections between the cells. The fertile tracts are formed by short branched threads which develop from swollen vegetative cells. The cells possess dense contents with plentiful starch and give rise to monospores.

Batters regarded this algae as closely related to *Erythrotrichia,* although *Colaconema* would perhaps be a better subject for comparison. The presence of pit-connections distinguishes it from the true Bangiales, but a reference to any particular group of Florideae is on present knowledge impossible.

Florideae

The Vegetative Organs

The General Features of Thallus-structure: The simple non-elaborated filament is rare among Florideae, being seen only in *Acrochaetium* and its immediate allies, as well as in certain Cerarniaceae. The majority exhibit a more complex structure, although often remaining filiform in outward appearance, since elaboration is rarely associated with-any great increase in bulk. Sections of such forms, however, often display a compact tissue which may show considerable differentiation. In Gigartinaceae and Delesseriaceae a foliose habit is developed, while certain families (Squamariaceae, Corallinaceae) consist largely or entirely of crust-forming types. The branching of the thalli, though often appearing dichotomous, is probablv for the most part monopodial.

All Florideae, no matter how compact their structure, are composed of systems of richly branched threads. The latter may be but loosely connected and readily distinguishable *Batrachospermum,* or they are so closely fitted together that the filamentous derivation is not easy to recognise. The threads of most of the more loosely branched forms are embedded in copious soft mucilage; in the more compact types this is denser and serves to connect the diverse units and their branches. In the filaments forming the basis of construction of the thallus it is often possible to distinguish main axes of unlimited and laterals of limited growth, the latter in many instances closely aggregated and commonly arising in opposite pairs or whorls from the cells of the main axes.

Two distinct, though closely related, variants of this type of structure are met with in primitive and more specialised groups alike, and there are few Florideae that cannot be referred to one type or the other. In the uniaxial forms the main axes *(a)* consist of a single row of often large cells; in the multiaxial types the axes *(a, m)* are composed of a number of parallel or sub-parallel threads. This is the only essential difference between the two.

The germinating spores of many Nemalionales and Cryptonemiales first give rise to a prostrate system, which is commonly discoid and produces more or less numerous erect threads. It is by the further development of one (uniaxial) or of a group (multiaxial) of the latter that the erect thallus is produced. The structure of these forms is therefore, based on a primary heterotrichous condition, which is equally recognisable, though in a modified form, in the early stages of development of Gigartinales and Rhodymeniales. Alone in Ceramiales, where the germlings appear invariably to lack a prostrate system, is the heterotrichous habit altogether suppressed. In its dominance, in the less specialised and its disappearance in the most advanced orders, the Florideae show a close parallel with Phaeophyceae.

Uni- and multiaxial construction are in general characteristic of distinct taxonomic units. All Rhodomeniales are multiaxial but in other orders both types are represented in different families. Many Cryptonemiales are uniaxial, although the Corallinaceae, for instance, are multiaxial. In the Gigartinales, the less specialised members *(Endocladia,* etc.) are uniaxial, while the more specialised Gigartinaceae are multiaxial. Kylin regards the multiaxial state of these forms as secondary, but the basic heterotrichy affords the means for a ready passage from one type of construction to the other.

A second fundamental characteristic of the Florideae lies in the strict apical growth of the threads composing the thallus. This growth is effected by apical cells which are often dome-shaped and cut off a single series of segments *(s)* parallel to the base, although in certain advanced types they are two or three sided and give rise to as many series of segments. The segments themselves usually cut off laterally two or more *pericentral cells*, which in many Florideae (not in most Rhodomelaceae) act as the apical cells of laterals; these are not uncommonly whorled. The remainder of the segment, constituting the *central cell (d)*, for the most part undergoes no further septation. Division walls passing through the axis of a segment, whether transverse or longitudinal, are formed only in a few special instances. In other words, enlargement in Florideae is effected almost solely by apical growth, and there is usually no intercalary division of the segments.

As a result of this apical growth and of the formation of the thallus from juxtaposed filaments, the inner cells of most of the compact pseudoparenchymatous types, which undergo no further division, become markedly stretched; they are often broad and considerably elongated, while towards the periphery the cells become progressively shorter and smaller. The primary construction is often more or less obscured by the development, from various internal cells, of septate *hyphae* which push their way in all directions between the cells already present. In many compact forms, the superficial layer, representing the apical cells of the ultimate branches of the filaments, is meristematic.

Pit-connections: Genetically related cells, forming part of the same thread and derived from a common parent-cell by division, are in practically all Florideae connected by conspicuous pits in the septa; such *primary pits* are single and generally circular. They are stated to be lacking in *Batrachospermum*. The pits are distinguishable already in young septa and occupy the axis connecting the two daughter-nuclei. When first formed they are relatively narrow, and only attain their full width as the segment broadens. In some instances, they occupy the entire septum. Where not directly obvious, they can generally be rendered visible by treatment of the material with hydrochloric acid or caustic soda. The pit-membranes often stain with methyl green or saffranin. According to Falkenberg they react to stains in the same way as the "cuticle" at the surface of the algae.

The pit-cavities are occupied by broad cytoplasmic strands. Closely applied to either side of the pit-membrane, one can often recognise two circular discs staining readily with haematoxylin. Recent investigations tend to show that the discs consist of a thickened peripheral ring containing pectic substances and supporting a delicate plasma-membrane; according to Celan they contain lipoids.

The discs are difficult to separate from the adjacent cytoplasm and are, according to many authorities, connected by fine cytoplasmic strands which traverse the pit-membrane and are sometimes restricted to its periphery; others have denied the existence of such cytoplasmic connections. A sieve-like appearance of the pit-membrane has frequently been recorded.

Schmitz regarded the discs as modifications of the cytoplasm lining the pit-membrane; they are separated from the vacuole by a layer of cytoplasm, which is often full of deeply staining matter and may show a fibrous structure. In plasmolysed cells the cytoplasm adheres to the discs which retain their position at the bottom of the pit-cavity. Jungers concludes that the discs are part of the membrane; in *Griffithsia* and *Ceramium*, moreover, he claims that the pits consist of a single lenticular body and that no separate pit-membrane is distinguishable.

The exact nature of the pit-connections thus remains undecided, and there is possibly some variety in structure. However, that may be, they are probably of importance in the transport of food-material and perhaps also in conduction of stimuli.

Mangenot, examining especially the gonimoblasts of diverse Ceramiales, denies the presence of a pit-membrane and describes certain differences between the discs on the two sides of the pit. Enlargement of the pit-connections is usual in the region of the developing cytocarps and this may be associated with disappearance of the intervening membrane. The frequent presence of pit-membranes between the vegetative cells must, however, be regarded as established.

In diverse Ceramiales a strand of cytoplasm extends from pit to pit across the central vacuole, a feature which is specially obvious in the axial cells of certain species of *Ceramium.* When the threads fork, the strand assumes the form of a Y with three arms of about equal length; the older strands may become much attenuated. Tylose-like ingrowths from the cortical into the axial cells have been recorded in *Ceramium rubrum.*

In the more compact types, where the close juxtaposition of the threads obscures the filamentous construction, the pit-connections are often of great value in tracing the genetic relationship of adjacent cells. Many of the more specialised forms, however, sooner or later develop *secondary pit-connections* between unrelated cells, no doubt in response to special physiological requirements, especially longitudinal transmission of food-material. Such secondary connections are unknown in Nemalionales and Ceramiaceae, as well as in diverse of the less specialised Cryptonemiales and Gigartinales. Their mode of development is readily studied in overlying pericentral cells of *Polysiphonia,* especially in those species that develop thick membranes at an early stage.

In the first place, the nucleus passes to the lower end of the cell and then divides, whereupon one of the daughter-nuclei, together with a small amount of cytoplasm, is cut off by an oblique wall. The newly formed septum progressively thickens so that the small daughter-cell is gradually pushed towards one of the underlying pericentrals; meanwhile a pit-connection *(p)* becomes obvious between the pericentral and the daughter-cell. Ultimately, the latter comes into contact and fuses with the pericentral below, the pit-connection *(p)* now linking the two superposed pericentrals, each of which thus contains two nuclei, one at the upper and the other at the lower end. The septum soon assumes a horizontal position. In some Ceramiales several secondary pit-connections may be formed between adjacent cells.

Hairs: The peripheral cells in the younger parts of the thalli of many Florideae bear a more or less dense growth of elongate unicellular colourless hairs that sometimes reach a length of a millimetre or more. In certain families (Gigartinaceae, Phyllophoraceae, Delesseriaceae) they are altogether wanting, while in Rhodomelaceae their place is taken by the trichoblasts; in some Florideae *(Rhodymenia palmatd)* they are confined to definite areas. Although usually deciduous, they sometimes persist even on the older parts (e.g. *Callithamnion corymbosum).* The hairs are mostly thin-walled, with a delicate layer of cytoplasm surrounding the central vacuole; chromatophores are generally lacking, although small faintly coloured ones are sometimes *(Plumaria elegans, Spermothamnion)* found in the young hairs.

In northern waters, such hairs are most abundant in spring and early summer and are often completely lacking in winter. The extent of their development is usually unrelated to the depth at which the algae grows, although in the Mediterranean they are stated to be best developed in well-illuminated situations. According to Berthold they constitute a protection against intense illumination, and it cannot be denied that they may fulfil such a role, although it is doubtful whether that is their most important function. Diverse authorities, in view of their frequent restriction to the growing parts and the large surface they present, have emphasised their probable importance in absorption of nutriment. Such a function is also in accordance with their prevalence in the early part of the growing season and in well-illuminated situations. Kylin in *Dumontia incrassata* found that hair-development only took place in cultures to which nitrates had been added.

Simple Filamentous Types: The most primitive structure among Florideae is found in *Aerochaetium (Chantransia)* and *Rhodochorton* both members of the Nemalionales with a wide distribution in the sea, although of the former one or two freshwater species are known; these are sometimes referred to a separate genus *Audouinella*, but there appear to be no adequate grounds for separation. Certain species of *Acrochaetium* possess a heterotrichous habit, while others are erect and attached by a well-marked basal cell. Most are epi- or endophytes on other marine Algae, but a number occur on the shells of Molluscs; *A. efflorescens* is often found on stones. Some are exceedingly minute.

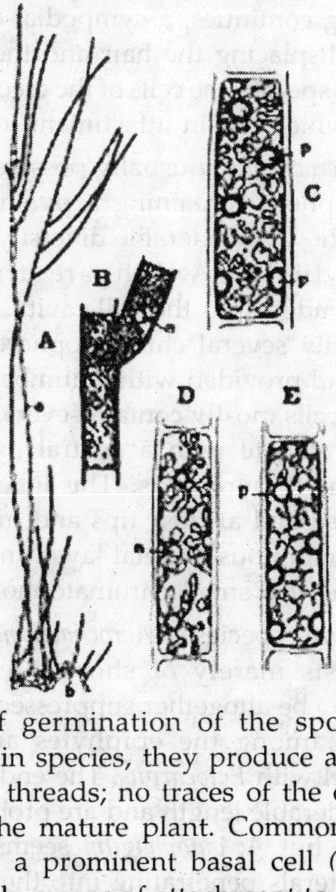

The mode of germination of the spores shows great diversity. In certain species, they produce a prostrate system bearing the erect threads; no traces of the original spore are recognisable in the mature plant. Commonly, however, the spore persists as a prominent basal cell *(be)*, which either constitutes the only means of attachment to the substratum or also produces radiating prostrate filaments, from which further erect threads may arise; in some of the endophytic forms these secondary filaments penetrate into the substratum. In *Kylinia*, a minute epiphyte, the entire plant consists of a system of procumbent threads arising on all sides from the attaching cell and bearing terminal sporangia. The heterotrichous types are no doubt the most primitive. All species of *Rhodochorton* are heterotrichous, the basal system *(b)* being commonly discoid.

Both in *Acrochaetium* and *Rhodochorton* the branches arise near the top of the parent-cell. In *Acrochaetium*, they commonly terminate in hyaline hairs which mostly fall off at an early stage. If branching continues, a sympodial construction may arise, the lateral displacing the hair and the part of the axis below it. In certain species, the cells of the erect threads produce septate rhizoids which aid in attachment to the substratum.

The cells of *Acrochaetium* usually possess a single axile or parietal chromatophore containing a pyrenoid. The parietal chromatophores show considerable diversity, often appearing as plates or cylinders which are frequently lobed and may be protruded into the cell-cavity. In some species, there are commonly several chromatophores which may be spirally twisted and provided with a number of pyrenoids. In *Rhodochorton*, the cells mostly contain several chromatophores which are either stellate with a central pyrenoid or band-shaped and without pyrenoids. The lobes of the stellate chromatophores expand at their tips and may fit together to form an almost continuous parietal layer. In R. *islandicum* the cells contain numerous small chromatophores.

In the endophytic species of *Acrochaetium*, the erect system commonly consists merely of short threads bearing the sporangia or it may be altogether suppressed. Similar reduced types are found among the epiphytes and constitute an interesting parallel with *Ectocarpus*. The endophytic filaments are often of considerable length and are probably for the most part intercellular, but *A. cytophagum* seems to be definitely parasitic, short laterals penetrating into the cells of the host.

In *A. endozoica* (Darb.) Hamel, where the elongate endophytic threads give rise at intervals to erect tufts, some of their branches are stated to penetrate into the living polyps. It has been suggested that *Colaconema* may belong here.

The erect threads of *Rhodochorton* usually branch, especially towards the tips. Several species occur on sand- and mud-covered rocks between tide-levels. The minute tufts of *R. Rothii* are specially characteristic of the intertidal region and are often found at and above high-water mark, as well as in caves and other feebly lighted situations; at lower levels the tufts are longer. The basal system of *R. membranaceum*, inhabits the chitinous walls of diverse hydroids.

In *R. islandicum* Rosenv, which forms a violet-red felt in caves, the lower cells grow out into almost colourless stolons producing further erect tufts and possibly serving for vegetative propagation. The heterotrichous habit speaks against a reference of this genus to Ceramiales.

Callithamnion and certain of its allies, though showing a filamentous structure, similar to that of the genera of Nemalionales just considered, have, like other Ceramiales, lost all traces of heterotrichy. The simple structure is probably the result of specialisation.

The Simpler Types of Uniaxial Construction and their Elaboration

The classical example of a simple uniaxial red algae is afforded by the freshwater *Batrachospermum* (sometimes called the "frog-spawn" algae), from which *Sirodotia*, does not differ appreciably in vegetative characters. To the naked eye, the mucilaginous growths of many *Batrachospermums* appear as chains of delicate beads. The thalli, which are violet or blue-green in colour, show monopodial (often pseudodichotomous) branching and may be 20 or more centimetres in length. When magnified, each "bead" is seen to consist of a whorl *(w)* of densely branched laterals of limited growth, all of about the same length. The laterals arise just below the septa separating the elongate cells of the main axes. The cells contain a number of parietal chromatophores with pyrenoids.

The apical cells of the long axes cut off a single series of segments *(s)* which rapidly widen and lengthen, usually remaining broadest at the septa. The laterals consist of small, ellipsoidal or moniliform cells, and the ultimate branches commonly terminate in hairs, surrounded by a basal sheath formed from the split outer layer of the membrane of the bearing cell; in certain species the laterals remain very short. The members (frequently 4-6) of a whorl are formed successively from the pericentral cells cut off from the parent-segment, while further branches of unlimited growth arise singly from the basal cell *(b)* of one member of a whorl. These cells also give rise to cortical threads, which grow downwards over the cells of the main axes and completely hide them from view in some species of *Batrachospermum* and *Sirodotia* they form a several-layered envelope.

Towards the base of the plant these threads extend to the substratum, where they aid in attachment and augment the prostrate system. The threads enveloping the axial cells commonly give rise to secondary whorls, which may be so numerous that the beaded appearance is almost entirely lost.

Species of *Batrachospermum* are widely distributed and usually favour the well-aerated water of slow-moving streams or the margins of lakes, where there is not much change of temperature. Some favour the pools of peat-moors, but usually occur where there is a spring. In most the mature shoots are annual, but in a few *(B. vagum, B. densum* Sirod.) they are perennial. Iltis describes an interesting association between *B. vagum* and the snail *Planorbis planorbis* L.

Germinating carpospores are not uncommonly found entangled among the threads of the parent. The mode of germination is one which is frequent among Nemalionales. A process arises on one side, during which the delicate outer membrane of the spore is ruptured. Most of the cytoplasm passes into the process which is cut off by a septum and undergoes transverse division to form a creeping filament. The latter soon branches and gives rise to a prostrate system bearing erect branched threads which strongly resemble those of an *Acrochaetium*. In feebly illuminated habitats further development may be arrested for long periods and, since abundant propagation of this juvenile stage is effected by monosporangia resembling those of *Acrochaetium*, such stages were in the past regarded as independent species and referred to the genus *Chantransia*. Monospores may be formed in quite early stages. The *Chantransta-stages* are stated to show certain differences from *Acrochaetium* in colour and mode of branching. Those of *Sirodotia* altogether resemble those of *Batrachospermum*.

As a general rule, adult shoots, recognisable by their broad squat cells, arise from the chantransia stages before they have developed very far. They usually originate from one of the erect filaments or from a branch of such a filament; they may, however, arise directly from the prostrate system. Largely, it is to be supposed, because of the frequent persistence of *chantransia-stages*, they have been compared to the protonema of losses. This is justifiable, but it must be realised that they differ in 10 fundamental respect from the juvenile stages of diverse other less specialised Florideae which are likewise heterotrichous and which commonly produce erect branches, apart from those; giving rise to the mature plant; they may also bear sporangia.

The erect threads of the *"Chantransia-stages"* not uncommonly reproduce descending rhizoids which spread over the substratum and, together with the cortical threads of the adult shoots, serve to extend he prostrate system. These rhizoids may bear erect branches closely resembling the secondary laterals formed from the cortical threads of he mature plant. In *B. Breutelii*, the robust thick-walled erect threads of the chantransioid stage exhibit whorled branching and are covered below with a dense envelope if cortical threads from which secondary laterals may be formed.

The type of mature structure seen in *Batraehospermum* is repeated with some modifications in diverse marine Florideae, 11 with cylindrical branched thalli. In most of these, however, as a result of the denser juxtaposition of the whorls, a beaded appearance 3 not manifest, and the branches of the laterals of limited growth form a continuous, though not usually very closely compacted, envelope to the long axes.

The vegetative differences are not considerable. Most have a dome-shaped apical cell, but in *Wrangelia* the segments *(s)* are Clique and cut off in succession five pericentrals, the first being always formed on the higher side of the segment which is situated alternately right and left. The first pericentral gives rise to the most vigorous lateral of the whorl, and it is from its basal cell alone that the distichous branches of unlimited growth are produced. Essentially the same structure is seen in *Schimmelmannia* a member of Cryptonemiales with flattened innately branched thalli found in warmer seas.

In *Thuretella, Dudresnaya, Atractjphora Calosiphonia* and *Schimmel-lannia,* the branches of unlimited growth arise direct from one member of the whorl, in the last from the first lateral of each whorl; in *Calo-siphonia* the laterals arise from the middle of the axial cells. The thick-walled axial cells of *Crouania* bear only three members in each whorl. In *Dudresnaya coccinea* the lower cells of the primary laterals subsequently enlarge almost to the size of the axial cells, while the abundant cortical threads likewise become large-celled; as a result the mature structure simulates a multiaxial one. In the older parts of *Wrangelia* the laterals are shed and their basal cells become covered by a thick envelope of cortical threads. In *Atracto-bhora* and *Thuretella* the latter bear numerous branches so that a relatively compact cortex is formed.

In the early development of *Dudresnaya* the protuberance from the spore gives rise to a prostrate system of richly branching threads which soon unite to form a compact disc growing mainly at its forward margin. Most of the numerous erect threads (M) terminate in hairs, but some continue growth and produce the mature thalli; according to Killian the latter are lateral branches of erect threads. The early stages of *Atractophora* are similar, although here the germinating spore divides by a vertical wall and remains recognisable within the prostrate disc that develops by outgrowth of the two cells; creeping threads arising from the peripheral cells divide at their tips to form secondary discs.

As an example of firmer construction among these uniaxial types *Gloeosiphonia capillaris* may be described in somewhat greater detail; the richly branched thalli usually grow submerged. The apical cell divides like that of *Batrachospermum* but the laterals arise from the middle of the axial cells. Each of these produces a whorl of four laterals, the large basal cell of which bears a number of one-celled branches spreading out in the horizontal and vertical planes.

These secondary branches in their turn give rise to others with progressively smaller cells, the ultimate ramifications uniting, both in the transverse and longitudinal directions, with those of other adjacent laterals to form a cortex covered by a well-defined surface-layer, the cells of which contain numerous branched ribbon-shaped chromatophores. The points of contact between the branch-systems of successive whorls are commonly visible as darker transverse bands on the surface of the thallus. The younger parts bear numerous hairs.

The cortical threads, which arise both from the basal cells of the laterals and from the central cells, are almost colourless and therefore, more appropriately spoken of as hyphae. The majority occupy the interspaces between the laterals and later form a dense weft on the inner surface of the cortex, thus increasing the rigidity of the peripheral cylinder. This is the more necessary, as the axial cells and the basal cells of the laterals disintegrate in older parts. Some of the hyphae produce secondary laterals which push between the primary whorls and add to the density of the cortex.

The early stages of germination resemble those of *Batrachospermum*, although the original spore is stated to retain its contents. The primary filament gives rise to a one-layered disc which no doubt corresponds to the basal crust described by Kuckuck. This bears numerous short erect threads *(e)*, often crowned by a unicellular hair (A), and the mature thalli *(m)* are formed from one or more of these threads or their branches. *Gloeosiphonia* probably hibernates by means of these crusts.

The common seaweed *Dumontia incrassata*, which likewise belongs to Cryptonemiales, shows an even firmer construction of essentially the same type; the hyphae often form secondary pit-connections with other cells of the thallus. The branched fistular thalli, which may be a centimetre or more thick, arise from a broad perennial basal crust. This consists of close-set vertical threads showing little branching and arising from a one-layered prostrate system. Certain parts of the crust may grow more actively and come to overlie others so that several strata may be recognisable; by similar local growth the base of the annual frond becomes embedded in the crust after the upper part has disintegrated. The adult thallus is formed from one of the erect threads. The germinating spores divide to form a compact cushion of cells bearing long unicellular hairs.

The slight compression evident in *Dumontia* is more marked in *Sphaerococcus coronopifolius*, although the ultimate branchlets are cylindrical. The detailed structure is much like that of *Gloeosiphonia*, the flattening being due to the more vigorous development of the laterals arising from two of the four pericentrals; the distichous branches originate from the apical cells of the favoured laterals. The early development is described by Chemin also *Plocamium*.

A similar compression is seen in the stiff cartilaginous fronds of many species of *Gelidium* (especially *G. latifolium* Thur. & Born. and *Pterocladia* which mostly show repeated pinnate branching. While the apical structure is essentially like that of the genera previously considered, the uniaxial condition is only recognisable in the very youngest parts, since the cells produced from the four primary pericentrals soon undergo pronounced lengthening and form an extensive medulla *(m)*.

The abundant hyphae *(hy)*, most of which have very thick walls, are in *Pterocladia* found only in the medulla, whilst in *Gelidium* they mainly occupy a more peripheral position. The early development shows no evident prostrate system and the growth soon becomes erect. The thalli readily form adventitious branches from wounds.

Species of *Gelidium* are found both in the littoral region and in deeper water. The mainly prostrate fronds of the small *G. pusillum* are not uncommon near high-tide level, sometimes intermingled with *Catenella Opuntia*; *G. crinale* is important in similar situations in the Mediterranean and other warmer seas. The tufts of *Pterocladia capillacea* are common between tide-levels. *Gelidiella* with small forms inhabiting warmer seas, has no hyphae.

Although less specialised in its reproduction than most of the genera previously considered, shows marked modification of the ordinary uniaxial. The species are usually found during the cold season, attached to rocks in waterfalls and torrents; they favour siliceous waters. They are grouped in two subgenera *Eulemanea* and *Sacheria*, which show certain differences in structure. The thalli take the form of stiff, olive-green or greenish-black bristles, which are simple or branched and normally reach a length of 15-20 cm. Growth is effected by the usual dome-shaped apical cell, but for some little distance behind the apex the axial thread and its whorled branches are little obvious, although evident enough in the maturer parts.

The segments of the apical cell cut off four pericentrals, the two first formed being larger than the others. After this small cells are cut off at the upper ends of the pericentrals, two from each of the larger, but on, one from the smaller; in transverse sections therefore, either six or four cells are found surrounding the axial cell. All of the peripheral elements subsequently divide horizontally to produce vertical rows of cells which surround the elongating central cell. Each precentral cuts off successively on its outer side three cells, and a similar division takes place in each cell of the above-mentioned rows. This results in the formation of a small-celled cortical layer on the outside of a row of more elongate cells. By further periclinal division of the peripheral cells the cortical envelope becomes successively 2- to 4-layered.

In the older parts the axial cells are separated from the cortex by a space and have undergone great elongation. Just beneath the septa they bear whorls of four laterals, the large basal cells which are produced from the inner parts of the pericentrals; in *Sacheria* these basal cells are T- or L-shaped, in *Eulemanea* clavate. It is by their marked radial elongation that the compact superficial tissue of the younger parts is separated from the axial thread. From the outer ends of the basal cells arise the ascending and descending threads (generative filaments of Atkinson), which occupy the inner surface of the wall of the bristle. Their elongate cells are derived from the rows above described. In *Sacheria* they are closely applied to the wall of the bristle, while in *Eulemanea* they are connected with it by oval or pyriform cells (*co*, the tie-cells of Atkinson) which arise at regular intervals. The number of longitudinal threads remains constant for a considerable distance in *Sacheria,* but in, *Eulemanea* branching soon occurs so that there are 8 ascending and 8 descending filaments.

Florideae (Vegetative System)

The numerous short branch-systems, arising from the cells of the longitudinal threads and formed by the periclinal divisions above described, constitute the compact 2-4-layered cortex. The outer layer or layers are composed of small cells plentifully supplied with chromatophores. The swellings seen at regular intervals on the bristles mark the points of contact of the ultimate branch-systems of successive whorls and lie approximately midway between two points of branching of the axial row. Hyphae are found only in *Eulemanea*, being produced from the proximal ends of the basal cells; they entwine the axial cells and partly fill the central cavity. They usually grow downwards, but in *L. parvula* Sirod they grow upwards. Branching of the bristles is effected by the outgrowth of a pericentral; the cavity of the branch communicates with that of the parent bristle.

The species of *Sacheria* are often more richly branched and seem to favour more turbulent water than those of *Eulemanea*. To the latter belong L. *annulata* Kiitz. and L. *torulosa* Sirod., while *Sacheria* includes such common species as *L.fluviatilis*.

The carpospores of *Lemanea* germinate in the last months of the year, often within the old bristles, and form a juvenile stage which resembles that of *Batrachospermum*, although apparently less capable of independent existence; it is not known to produce mono-spores. The perennial prostrate system may be filamentous or composed of polygonal cells. In *Sacheria*, the erect threads are short; little branched, and ephemeral. The mature bristles arise direct from the prostrate system or from a lateral branch of an erect thread. They are at first uniseriate and only differ from the other threads in their broad flat cells; they soon produce rhizoids.

Similar rhizoids emerge from the cells of the juvenile stage and both alike serve to strengthen the attachment and to extend the prostrate system. Their tips may divide to form a secondary prostrate system so that abundant vegetative propagation is effected and extensive mats are produced. According to Brand cells in the wall of old bristles can also give rise to these juvenile stages. It is manifest that they are directly comparable to those of *Dudresnaya, Gloeosiphonia*, etc.

Tuomeya fluviatilis, found in streams in the Eastern United States, appears as rather rigid, richly branched tufts, up to 5 cm. high, fixed by a discoid attaching organ. In transverse section the mature parts exhibit much the same structure as in *Lemanea*, the rather narrow cavity around the axial cell being filled with plentiful hyphae. The apices, however, show nothing of the pseudoparenchymatous structure of *Lemanea*, the developing whorls of laterals being plainly recognisable, although behind the apex they become distally compacted to form a continuous surface. In these respects *Tuomeya* is intermediate between *Batrachospermum* and *Lemanea*.

The series of genera considered in this section exhibit a uniform type of apical segmentation, combined with whorled branching. They differ chiefly in the degree of compactness attained, as well as in minor details of branching, hypha-production, etc. *Lemanea* is distinguished by the parenchymatous structure of the tips, which require detailed study to relate them to the obvious filamentous construction of the maturer parts. In this respect, *Lemanea* constitutes a transition from the uniaxial forms with a loose construction to the pseudoparenchymatous forms dealt with in section. It must be emphasised, however, that the diverse genera just considered do not constitute a taxonomic series.

Simpler Types of Multiaxial Construction and their Elaboration

The species of *Nemalion,* a simple multiaxial type belonging to the one order as *Batrachospermum,* commonly occur between tide-levels id are widely distributed in North Temperate seas, although more quent in warmer waters. Outwardly the thalli appear as soft latinous cylinders reaching a length of 25 cm. and often showing some dichotomous, as 11 as lateral branching; in *N. helminthoides,* however, branching is very knitted. The thalli are attached by a discoid base.

The structure is readily deciphered by exerting gentle pressure under a cover-glass on the tip of a branch, the threads of the multiple axis, each provided with an apical cell devoid of chromatophores, all reach to about the same level and bear numerous richly branched laterals of limited growth. Near the summit, these form a radiating group, but farther back they assume a more horizontal disposition and constitute an envelope around the axial threads, the density of which is increased by the intercalation of further laterals. The ultimate branches of the latter reach a uniform level and give the fronds their even surface. In transverse sections of older parts the axial threads are separated by a wide space, occupied by mucilage and traversed by the proximal parts of the laterals, from the more compact cortical region. The narrow elongate axial cells are devoid of chromatophores, but the superficial cells contain a conspicuous stellate chromatophore with a central pyrenoid and usually bear hyaline hairs. The, Californian *Cumagloea* shows much the same structure. The dichotomous branching of *N. multifidum* is due to periodical separation of the apical threads into two groups.

Essentially the same structure is found in *Platoma Bairdii* in which the thallus is slightly flattened. Here the axial threads protrude, since laterals are only formed a little way below the apex. Hyphae arise from their basal cells and give rise to secondary laterals.

The carpospores of *Nemalion* commence to germinate like those of *Batrachospermum*, but the protuberance arising from the spore appears to divide horizontally into an upper cell producing an erect thread and a lower one giving rise to the prostrate system. Later stages are definitely heterotrichous, with a basal system of short rounded cells, bearing well-branched erect threads, and the mature thallus no doubt arises by further development of a group of the latter; the juvenile stage may last for some time before the erect fronds develop. In *Platoma* the spores produce a basal crust, the erect threads of which are closely apposed below and may bear sporangia; certain groups of threads sooner or later grow out into erect thalli.

Diverse other Nemalionales possess a similar, though rather more compact, structure. Examples are furnished by *Liagora* where there is often appreciable calcification.

Simple Multiaxial Types

In *Helminthocladia* the cells of the laterals are appreciably larger than those of the axial threads and increase in size towards the periphery, while in *Helminthora* the axial threads are large-celled. Unicellular hairs occur in both. The deposition of lime in *Liagora* may be confined to the cortex, but in some species, the axial region becomes calcified at some distance from the apex; in this genus, there is often considerable production of hyphae.

The germinating spores of *Helminthocladia* and *Helminthora* give rise to a filamentous or discoid prostrate system, the cells of which in the former are stated to contain a number of parietal discoid chromatophores. The mode of origin of the erect thallus is not clearly established, but the prostrate system no doubt gives rise to the discs by which the adult plants are attached.

Somewhat greater specialisation is found in *Scinaia furcellata* widely distributed- in the Mediterranean and Northern Atlantic, and for the most part occurring below low-tide level. The gelatinous thalli, usually 10-20 cm. in length, are attached by a discoid holdfast and show a compact tufted habit with abundant dichotomous branching. The apices of the axial threads are sunk in a depression owing to rapid growth of the adjacent laterals. Near the summit the terminal cells of the latter are elongate, but farther back many of them enlarge to form a layer of colourless vesicular cells which contain light-reflecting bodies. The photosynthetic cells are largely subterminal, although certain narrow terminal ones, which bear hairs as well as reproductive organs (monosporangia, antheridia), also play a part; these cells may later become vesicular. The chromatophores are devoid of pyrenoids.

The inner cells of the laterals produce downgrowing hyphae which are stated also to envelop the threads of the axial strand. The early development does not differ essentially from that of other Nemalionales.

The species of *Galaxaura,* found in warmer seas, have a similar structure with the same kind of apical depression, but the cortex is here usually strongly calcified, in some species (e.g. G. *corymbifera* Kjellm.) with joint-formation. The end-cells of the laterals contain a campanulate chromatophore with a large central pyrenoid. Asexual and sexual plants often show a striking dimorphism. *Chaetangium,* represented by a number of species in Southern seas, shows diverse forms of uncalcified thalli, which are sometimes saccate; the structure is comparable to that of *Scinaia,* but in *C. saccatum,* there is no apical depression.

A multiaxial structure, combined with marked calcification, is seen in *Corallina*, (Cryptonemiales) erect thalli, composed of numerous calcified segments separated by uncalcified horny and flexible joints, arise from an often extensive basal crust which is commonly lobed. The normal colour is purple or red, although exposed plants are often pink or white as a" result of bleaching. *C. officinalis* L.widely distributed between tide-levels in the Northern Hemisphere, and C. *mediterranea* Aresch., an important constituent of the *Corallinagir is* situated at about low-tide level on exposed shores in many parts of the Mediterranean, show pinnate branching so that they, present a characteristic feathery habit; permanently submerged plants.

The axial threads are composed of elongate cells, often arranged in horizontal series and showing lateral fusions. Their numerous branches bend out gradually and form a compact but not sharply delimited cortex. The superficial cells of which are markedly flattened. In *C. rubens* and certain other species, the enlarged end-cells of some of the cortical rows grow out into hairs (A), which are not separated from the parent-cell. The peripheral cells contain a number of chromatophores.

The apex resembles that of *Nemalion*. Branching is initiated in C. *officinalis* and C. *mediterranea* by the outgrowth of the axial threads into usually three apical protuberances, the median one, serving to continue the main axis, while the two lateral ones give rise to pinnae; when more than two pinnae are produced, some may develop outside the normal plane of ramification. Adventitious branches occasionally originate from older parts.

In the formation of a new segment, the axial threads elongate considerably without septation, while cytoplasm accumulates at their tips; later these are cut off by transverse septa and divide to form the new segment. The lower portions of the original outgrowing threads do not divide, but with continued lengthening and gradual thickening of their walls develop into the joint which bears the new segment above. In *Jania* the joint-cells are about as long as those composing the axial threads, but in other species of *Corallina,* they are appreciably longer; they produce little or no cortex and what is formed disorganises at an early stage. Since deposition of lime takes place mainly between the cortical threads, the joints remain uncalcified and, although they become more or less overgrown by the cortex of the adjacent segments, there is always a gap where the component threads are freely exposed.

The germinating spores give rise to a rounded or elliptical disc which no doubt develops into the basal crust of the mature plant. In *C. officinalis* this crust consists in its lower part of threads with horizontally elongated cells bearing ascending short-celled filaments, each terminating in a cover-cell, a structure closely resembling that of the encrusting Melobesieae. Several erect shoots usually arise from one and the same crust and a joint is formed at the place ot origin of each.

The reproductive organs of *Corallina* are borne in conceptacles which, in C. *affirinatis* and C. *mediterranea,* occupy the apices of the branches, although the former species also produces lateral conceptacles which sometimes fuse with one another; according to Solms-Laubach such lateral conceptacles are really formed at the tips of short adventitious branches. In *C. mediterranea* the outer surface of the conceptacles usually bears a considerable number of stiff hairs. *In Jama* new segments arise from the upper corners of the young female conceptacles so that the latter are lodged between the uppermost forks of the thallus; the male conceptacles are terminal on special branches.

Amphiroa, with a similar habit, differs in certain anatomical details. The characteristic cover-cells extend right over the apex, whereas in *Corallina* they only appear at some distance below it. The elongate cells of the axial threads are interrupted at intervals by rows of short ones, a feature which is perhaps related to periodic growth. In certain species, the joints consist of two or several rows of cells of about the same length as the other axial ones. Adjacent cells in medulla and cortex are connected by secondary pits like those of *Lithophyllum*. The conceptacles of *Amphiroa* occur over the whole surface of a fertile branch. *Cheilosporum* differs from *Corallina* mainly in the development from the upper ends of the segments of horn-like outgrowths which usually harbour the conceptacles. Recent workers distinguish a number of other genera with a *Corallina-like* habit.

Furcellaria fastigiata, a widely distributed, dark or greenish-purple, perennial lithophyte, belonging to Gigartinales, exemplifies a firmer construction, without accompanying calcification. The rather coarse, dichotomously branched cylindrical thalli, which arise in some numbers from a system of branched stolons, are often sublittoral, although conspicuous in the lower part of the littoral region on some shores.

The structure, like that of *Lemanea,* can be readily deciphered only by means of sections, although after suitable treatment most details can be made out by dissection. The apex of the growing thallus is essentially like that of *Nemalion,* but the laterals fit more closely and the whole is enveloped by a firm mucilage-layer.

Transverse sections through the older parts show a compact structure with considerable differentiation. The cells of the axialt hreads have thick stratified membranes and are surrounded by several layers of large cells, constituted by the basal portions of the laterals and at certain times crowded with starch. The small peripheral cells are provided with plentiful chromatophores. Hyphae, which aid in the storage of starch during autumn and winter and run mainly in the transverse direction, arise from the inner cortical cells and also from the axial threads; they occur abundantly among the central elements.

The normal branching is stated to be a true dichotomy, but there are also adventitious branches with a narrow attachment, arising from groups of surface-cells. The tips of the thalli which are crowded with starch become dilated when reproductive organs are formed. The rest of the thallus persists after these fertile tips are shed and from the scars adventitious shoots may arise which may in their turn become fertile. The germinating spores produce a hemispherical mass of cells giving rise to the cylindrical frond; at an early stage stolons arise from its base and produce other erect fronds.

Very similar in outward habit to *Furcellaria* is *Polyides roiundus* which belongs to Cryptonemiales. The compact reddish-brown thalli arise in some numbers in each season from a perennial disc which may be nearly a centimetre in diameter and has much the same structure as that of *Dumontia;* the older discs are stratified. The thalli show essentially the same structure as those of *Furcellaria,* although the photosynthetic tissue is more extensive and hyphae are fewer. Adventitious shoots frequently arise from the scars left by decay of the thallus-tips. The early stages appear to resemble those of *Furcellaria.*

The tropical *Dermonema gracile,* a member of Nemalionales, possesses an even firmer texture, the short coarse, repeatedly branched thalli standing erect even when exposed by the tide. This may be related to the presence of a cylinder of thick-walled mechanical cells running longitudinally at the periphery of the loose medulla.

Brief mention may be made of the Grateloupiaceae (Cryptonemiales) comprising forms with compact thalli in which the medulla consists of widely separated stellate cells; these owe their distinctive shape to the elongate pit-connections which become drawn out during the enlargement of the thallus. Then' are also more or less numerous, often thick-walled hyphae. *Grateloupia filicina* with a pinnately branched, and *G. dichotoma* with a forked thallus, are widely distributed and tend to grow where fresh (and often polluted) water enters the sea. *Cryptonemia* and *Halymenia* are forked leafy forms common in warmer seas. All the contents of the spore pass into a protuberance which divides to form a disc, from which groups of erect fronds arise.

The Mediterranean *Sebdenia Monardiarla* (Gigartinales), with a leafy thallus, has a medulla composed of numerous interlacing filaments interrupted, especially in the younger parts, by similar stellate cells.

The only fundamental difference between the series of forms described in this and the preceding sections lies in the presence of a single or of a number of axial threads. The early stages are often very similar and the underlying heterotrichy is obvious in all the less specialised types. In tracing the further elaboration of the Floridean thallus, some reference must first be made to certain more specialised uniaxial types among Nemalionales in which the whorled branching characteristic of the more primitive forms is obscured.

Specialised Uniaxial Types among Nemalionales

Naccaria Wigghii in habit much like *Atractophora* and belonging to the same family, has an apical cell which divides by oblique walls so that the segments appear wedge-shaped, the wider part being orientated successively in four different directions. Each segment forms only two pericentrals; the first is cut off near the apex on the higher side (acroscopic), the second only at some slight distance from the apex at the lower end (basiscopic) of the elongating segment and in a plane perpendicular to the other.

The respective pericentrals produce primary and secondary laterals of limited growth, the former being longer and more richly branched than the latter; the secondary lateral of a given segment is only slightly above the primary one of the segment below and stands almost opposite to it. The laterals are thus arranged in four vertical rows. Hyphae originate as usual from the basal cells, and in the younger parts bear short phocosynthetic threads which subsequently fall off. The hyphae give rise both to the small-celled surface-layer and the large-celled cortex which surround the narrow axial cells in the mature thallus, whose increase in thickness is due to branching of the hyphae. Branches of unlimited growth arise from the basal cells of certain primary laterals, but they can also develop adventitiously from the cortical cells. The early stages are essentially like those of *Atractophora*.

The same type of apical division, with formation of only two (here opposite) laterals from each segment, is seen in the Bonnemaisoniaceae, comprising *Asparagopsis* and *Bonnemaisonia*. *Asparagopsis*, which has its chief centre in warmer seas and in the Southern Hemisphere, is represented in Europe by *A. armata* Harv. introduced from Australia. The rather small, well-branched, erect fronds arise from creeping shoots which are probably perennial; they are distinguished by the possession of vesicular cells, also characteristic of *Bonnemaisonia*, and of modified branches which act as attaching organs and aid in vegetative propagation. Both species have been repeatedly studied.

The higher sides of the segments lie on a 1/4 spiral in *A. armata*, on a 3/8 spiral in *A. hamifera* and give rise to primary laterals which develop into unbranched sterile structures. A few of the secondary laterals, cut off opposite to the primary ones, grow out into long shoots, but most remain short and bear the reproductive organs; both kinds are branched, forming primary and secondary laterals like the parent axis. The first (adaxial) primary lateral of a short shoot grows very rapidly and overarches the latter so that the short shoot appears to arise from the base of the lateral.

In *Bonnemaisonia asparagoides*, an annual lithophyte usually found submerged, the main axes bear progressively weaker laterals which are pinnately branched to the second degree. The wedge-shaped segments of the apical cell are orientated in two opposite directions and all the branching is distichous. The primary and secondary laterals develop as in *Asparagopsis*. In both genera they develop a cortex which arises in a different manner on the primary and secondary laterals. The cortex may be of considerable thickness on the older axes of *Asparagopsis* and is separated by a space, occupied by hyphae in *A. taxiformis*, from the elongate axial cells.

In *Bonnemaisonia*, the segments of the primary laterals divide into a small central and three pericentral cells, the first of which is situated on the abaxial side and forms a vesicular cell at its upper end. After this each pericentral cuts off four primary cortical elements, any one of which may divide with the formation of further vesicular cells. As the axial cells elongate, the primary cortical elements form a compact large-celled layer around them and by degrees the cells of this envelope cut off a considerable number of small cells which overlie the septa between the large ones and in part become vesicular cells.

In the long shoots, that develop from the secondary laterals, the first four or five segments of the apical cell are cut off transversely and these form a cortex in the way just described. That of the later segments, on the other hand, originates by the cutting off of three pericentrals from the basal cells of the laterals; each pericentral cuts off four cells which divide again and again to form a compact two-layered cortex. The early development results in the formation of a basal disc, the first divisions of the carpospores being vertical to the substratum.

The three genera just described are clearly more specialised than other Nemalionales. They demonstrate that, even in this order, there has been considerable vegetative advance and this is also evident in their reproduction. Oblique division of the apical cell is found also in many of the uniaxial Cryptonemiales and Gigartinales where definite two- or three-sided apical cells are encountered.

More Specialised Uni- and Multiaxial Types among Cryptonemiales and Gigartinales

Although the basic mode of construction remains the same, many members of these two orders afford evidence of specialisation in one direction or another. In several, the uniaxial structure becomes obscured in the mature thallus. These features may be illustrated by the description of selected examples which have been fully studied.

Uniaxial Forms

A definite two-sided apical cell is found in diverse Cryptonemiales, such as *Cryptosiphonia* and *Gloeopeltis*.

Endocladiaceae, all three mainly found in the Pacific. The

species of *Endocladia* and *Gloeopeltis,* tend to occur near high-tide level, the latter especially being capable of withstanding considerable desiccation. The usually well-branched thalli show little differentiation. The segments of the apical cell cut off two pericentrals in the same way as in *Cystoclonium,* and these develop into laterals which form a compact cortex. It is characteristic of the Endocladiaceae that, at each point of ramification of the laterals, only two branches are formed, while *Endocladia* itself is distinguished by the numerous accessory branchlets arising from the inner cells. The inner cortex of *Gloeopeltis* is traversed by numerous hyphae.

Cystoclonium purpureum shows a greater degree of internal differentiation. It is a not uncommon lithophyte between tidemarks in Britain, although elsewhere frequently growing submerged. The terete, somewhat cartilaginous thallus, which is often dull purple in colour and may reach a length of 60 cm., has a prominent main axis bearing numerous richly branched laterals. Certain branches may end in tendrils. The plants are primarily anchored by a minute disc, but the base of the axis later gives rise to a number of stolons which become attached by squat haptera and produce further erect shoots. The individuals commonly last for one season only, although some perennate by means of the stolons.

The anatomical structure has been repeatedly examined. The segments of the two-sided apical cell cut off two pericentrals on their higher side which is situated alternately to right and left; the two pericentrals lie in perpendicular planes. The rest of the segment constitutes the axial cell. Such asymmetrical development of pericentrals takes place in many of the more specialised Florideae. The axial cells undergo marked elongation, while the pericentrals as usual give rise to richly branched laterals which form the compact cortical region.

In the older branches one can recognise a single superficial layer of small photosynthetic cells many of which in the earlier part of the season grow out into long hairs, and several layers of large multinucleate thick-walled storage-cells. At an early stage the axial thread becomes surrounded by accessory longitudinal threads, from which it is only distinguished by the greater length of its cells; the entire complex constitutes a probable conducting system, the transverse septa of which are stated to be perforated by fine cytoplasmic strands. The accessory threads are partly formed by elongation of the inner cells of the laterals and partly by outgrowths from the latter, as well as from the cortical cells. The latter, as well as those forming the conducting tissues, are connected both *inter se* and with one another by means of short hyphae which develop secondary pit-connections at their tips. A branch of the thallus can arise from any peripheral cell.

The spores undergo vertical or irregular division to form the primary hemispherical attachment-disc, which is scarcely larger than the spore and is anchored by short unicellular haptera. A superficial cell not always at the summit of the disc, becomes the apical cell of the thallus; sometimes two erect fronds develop from one disc.

A similar apical structure and mode of segmentation is found in other Gigartinales, for instance in the Australian *Erythroclonium*, in *Catenella*, and in *Hypnea*. The thallus of the first is terete and moniliform. *Catenella Opuntia*, widely distributed near high-water mark and also occurring commonly in salt-marshes, is a perennial algae in which the small thalli show frequent constrictions. Diverse species of *Catenella* are characteristic components of the algal vegetation found on the roots of Mangroves, to which they are attached by haptera. The anatomical structure of *Catenella* resembles that of *Cystodonium*, but there are several layers of photosynthetic cells and the inner branches of the laterals are often widely separated. The thick-walled elements of the medulla are linked by secondary pit-connections.

Hypnea musciformis, which is common in warmer seas and found elsewhere only in sheltered habitats, has fleshy terete thalii, up to 45 cm. long, with numerous short branches; many of the longer ones are incurled at the tip and function as tendrils. The mature structure differs from that of *Cystoclonium* in the uniform enlargement of the inner cells of the laterals so that the axial cells remain clearly evident. Hairs are often abundant in well illuminated situations.

The same apical segmentation is also seen in the Australian genus *Mychodea,* with a flattened or foliaceous thallus. The narrow axial cells are difficult to distinguish in the mature thallus from the numerous surrounding hyphae, which also occur between the huge cells forming the inner cortex.

A different structure is found in *Gracilaria,* of which *G. confertables* below, although a characteristic inhabitant of sandy bottom in the Mediterranean. Its dark-red terete fronds, which are up to 50 cm. long, bear more or less numerous tapering branches and several arise from a large fleshy disc. Other species have more or less corn-pressed thalli. The spores divide to form a circular disc, the centre of which gradually becomes arched, the three-sided apical cell of the future thallus differentiating at the summit of the arch. Kylih has demonstrated a similar apical cell in the adult fronds, although others have described a group of apical cells.

The mature thallus shows little anatomical differentiation, since all the internal cells are large and isodiametric and an axial, thread is unrecognisable. The small cells forming the narrow cortical one contain a number of ribbon-shaped chromatophores. Unicellular hairs arise from enlarged peripheral cells which possess abundant contents and later become multinucleate.

The instances hitherto considered in this section show how, despite an essentially similar apical structure, the primary uniaxial construction may become obscured among Gigartinales, either by the development of accessory longitudinal threads *(Cystodonium, Catenella)* or by equal enlargement of all the internal cells *(Gracilaria).* Other uniaxial Gigartinales show marked flattening of the thallus.

A good example is furnished by *Rhodophyllis bifida,* a

sublittoral form with a repeatedly forked bright red thallus belonging to the same family as *Cystoclomum*. According to Kylin the young fronds possess a two-sided apical cell, the segments of which divide periclinally into outer *(o)* and inner (i) cells. The outer undergo further periclinal division, but some soon enlarge and divide obliquely to constitute secondary apical cells. In older plants, in which the tips of the fronds have round off, the primary apical cell is difficult to distinguish and the frond possess a marginal meristem with numerous apicals.

The edge of the thallus remains one-layered, but else where the inner cells, formed from the segments of the apicals, cut off on either surface a single layer of cortical cells which undergo further anticlinal division and thus effect a gradual widening of the frond. The central cells, on the other hand exhibit little division and become gradually stretched and pull apart as the cortex widens. In later stage they appear as a branching and anastomosing system relatively narrow, widely separated threads with multinucleate cells. The anastomoses are due to hyphae, which mainly original from the central cells themselves and form secondary pit-connection at the points of junction with other members of the system. The structure of *Rhodophyllis*, though presenting some points of similarity with *Cystoclonium*, is modified in relation to the bilateral habit. The abundant branching is effected by outgrowth of one or other of the marginal apicals.

The germinating spores give rise to a small basal disc which soon becomes arched. According to Killian the two-sided apical cell of the future thallus appears at the summit of the arch, although Nienburg concluded that the mature fronds arose from the base of this primary upgrowth, which he regarded as representing a juvenile stage.

An analogous structure is met with in *Dilsea* (Dumontiaceae Cryptonemiales), with a number of species in northern seas. *D. eduli* Stackh, a frequent perennial near low-water mark, has a number, thick stalked obovate fronds arising from a firm basal crust. The frond; at first possess a single apical cell, although later there is a margin meristem. In the younger parts relatively compact medulla of elongate cells is enveloped by a cortex of rounded cells, but the abundant anticlinal division in the surface-layer leads, as in *Rhodophyllis*$_y$ to separation of the medullary threads which ultimately form a network.

Calliblepharis, belonging to the same family as *Rhodophyllis,* has j structure more like that of *Cystoclonium.* The thick fronds of *C. ciliatt,* a winter annual found at once below low-water mark, are either simple or forked once or twice; the margin is characteristically beset with flat pinnae. In *C lanceolata* Batt a summer form with narrower thalli, then is a three-sided apical cell, the segments of which bear that laterals forming the cortex at the apical end. The inner cells of the laterals undergo great elongation and, together with the axial cells, form a central conducting tissue linked by secondary pit-connections. The early development is like that of *Cystoclonium.*

A more specialised foliose type is afforded by *Plocamium,* of which P. *coccineum* is a common sublittoral Atlantic seaweed. The delicate thallus, which is often only slightly flattened, is richly branched in one plane in a sympodial manner. The successive sections of the sympodium *(1-5)* bear adaxially and on their flanks a number (generally 3 or 4) of curved branches, the uppermost of which is most strongly developed; the axis is finally deflected to one side, appearing as an unbranched lateral, while growth is continued by the adjacent branch which in its turn gets displaced by its uppermost lateral. The secondary branches thus occupy alternate flanks of the successive segments of the sympodium. If the tips of the branches come into contact with a firm substratum, they develop broad sucker-like haptera, a feature which has also been recorded in *Rhodophyllis bifida.*

The anatomical structure is relatively simple. The apical cell produces only a single series of segments which cut off pericentrals on either flank, the one first formed lying on the convex margin; these lateral pericentrals give rise to series of segments which, together with the axial cells, form the foundation of the median layer of the thallus. A little later the axial cells cut off two further pericentrals which are situated in the plane of flattening. Both these and the rows of segments formed from the lateral pericentrals undergo some periclinal division. Meanwhile the central cells elongate markedly and, since they possess dense cytoplasmic contents, they form a prominent axial thread in the mature thallus. In transverse sections, the axial cell is surrounded by 4-6 large, but relatively short, storage-cells which are linked by secondary pit-connections; this system is continued outwards into the flanks, usually as a single layer constituted by the middle cells of the lateral rows. The surface is formed by one or two small-celled layers. The branches of the thallus arise from the lateral pericentrals.

The germinating spores divide by two oblique walls to form a three-sided apical cell, the segments of which give rise to the small attachment-disc. The erect thallus soon becomes prominent.

Multiaxial Forms

The mature structure ot *Agardhiella,* the species of which are characteristic seaweeds of the warmer parts of North America, resembles that of *Cystoclonium.* A number of the elongate, fleshy thalli, with numerous tapering branches, usually arise from a common discoid holdfast. The apices in *A. tenera* are occupied by 5-7 apical cells, the segments of which produce on their outer side the laterals which are aggregated to form the cortex. The inner cells of the laterals give rise to down-growing hyphae, which enter into secondary pit-connections with the cells of the axial threads. In the mature thallus, the medulla, consisting of a mass of narrow threads, is surrounded by a firm cortex bounded by a narrow zone of photosynthetic cells. Near the tips the usual hairs are to be found. The spores divide to form an irregular mass of cells.

Ahnfeltia plicata, a widespread, slow-growing perennial lithophyte, readily recognised by the horny or even wiry texture of the dark purple or black thallus, is attached by a broad disc composed of vertical rows of almost cubical cells with firm walls. Though often occurring in rock-pools near low-water mark, more robust specimens are found in deeper water. The mature thallus shows irregular, in part dichotomous, ramification, with promiscuous adventitious branches on the older parts. One or two dichotomies are stated to be formed in each season.

The apex is occupied by numerous closely aggregated threads. In the mature thallus the extensive medulla consists of elongate thick-walled cells, no doubt fulfilling a mechanical function and merging into a cortex composed of small, radially arranged, elements. The older parts, especially in plants exposed to rough seas, exhibit extensive secondary thickening which takes place periodically by division of the peripheral cells of the cortex and results in the formation of a number of successive strata, consisting of cubical, radially arranged cells. The rather ill-defined limiting layers, marking periods of cessation of thickening, are composed of somewhat smaller elements with more deeply pigmented contents and thickened radial and inner walls. The secondary tissues of adjacent branches sometimes fuse and become enveloped by further strata common to both.

According to Jonsson new strata develop whenever a fresh dichotomy is formed, so that one stratum would correspond to each such period of growth. The number of strata is in general directly related to the age of the thallus, but they do not always increase in number from above downwards and are commonly more numerous on one side of the frond than on the other.

The spores of *Ahnfeltia* give rise to small violet-coloured discs bearing hyaline hairs and forming secondary discs from the ends of marginal threads. The discs may reach a considerable size before an erect thallus is produced.

Gymnogongrusy a lithophyte closely related to *Ahnfeltia,* possesses small, repeatedly forked thalli, attached by a basal disc. In G. *Griffithsiae,* widespread near low-water mark in the

North Atlantic, the segments are terete, while in other species they are markedly flattened. According to Jonsson some species show secondary thickening similar to that of *Ahnfeltia.* The early stages have been described by Chemin.

A foliose development of the thallus is met with in occasional representatives of diverse families of Gigartinales, e.g. *Nemastoma* (Nemastomaceae) and *Halarachnion* (Furcellariaceae). The former, with species in warmer seas and in the Southern Hemisphere, has a structure much like that of *Platoma;* in large vesicular cells terminal on the subcortical filaments are recorded. *Halarachnion ligulatum* (Woodw.), a North Atlantic and Mediterranean species, resembles *Calliblepharis* in habit, though the degree of branching and flattening vary. The apex is like that of *Agardhiella.* The mature thallus shows a 2-3 layered cortex surrounding the long and widely separated threads of the medulla; hyphae are present. The germinating spores give rise to a discoid prostrate system with an obvious filamentous construction.

Much the same internal structure is seen in the Arctic *Turnerella septentrionalis* and *Opuntiella californica* (Farl.) Kylin (*Callymenia calif arnica* Farl.), both with a broad thallus, possessing characteristic peripheral glandular cells which penetrate deeply into the underlying tissues. Similar cells occur in *Schizymenia Dubyi*, an ally of *Platoma,* likewise possessing a broad, sometimes lobed, thallus. The spores germinate like those of *Dumontia.*

A number of common foliose members of Gigartinales belong to the Phyllophoraceae and Gigartinaceae. The species of *Phyllophora* are widespread northern lithophytes with small thalli, usually showing dichotomous branching in one plane. In *P. membranifolia* and *P. Brodiaei* the flattened fronds are borne on a cylindrical stalk, but in P. *epiphylla* Batt the stalk is winged throughout and not sharply demarcated; in this species the lower parts of the flat fronds show a perceptible midrib. *Stenogramma interrupta* Mont a widely distributed form more characteristic of warmer waters, has a similar habit and much the same structure.

The mature thallus of *Phyllophora membranifolia* bears numerous forked fronds with the convex surface facing the source of light; in the major dichotomies one fork develops into a fan, while the other forms a continuation of the stalk. In *P. Brodiaei* and *P. epiphylla* older plants tend to produce numerous proliferations, mainly arising from the margins in the former and from the surfaces of the fronds in the latter. Minute, flat or nearly terete outgrowths, arising on the upper margins of the leafy fronds, produce the reproductive organs.

The thalli last for several years and the successive increments are often recognisable. The new bright red fronds of *P. Brodiaei* arise in spring from the apices of those of the previous season but in older plants most of the new growth is furnished by the adventitious shoots above mentioned. In *P. membranifolia* the new season's growth, especially in older plants, often originates from the stalk or from the margin of the old frond, in the form of cylindrical proliferations which do not develop into complete fans until the next season.

A number of thalli usually arise from a lobed basal crust, the upper part of which is composed of dense vertical rows; the lobes sometimes grow over one another, leading to stratification of the crust. In *P. Brodiaei,* which is often found on shells of Molluscs, multiseriate haptera are formed which penetrate the latter and produce filamentous branches growing horizontally between the layers of the shells. The young discs consist of a single layer of radiating filaments; for other stages in development. The early development of *Stenogramma* is like that of *Chondrus*.

The multiaxial structure is difficult to recognise in *Phyllophora.* The compact medulla consists of rather short cells with firm walls and shows a gradual transition to the cortex, each cell of which contains a large parietal chromatophore. Older parts exhibit secondary thickening owing to preclinical division of the superficial cells; such activity is, however, often localised, leading to the formation of elongate oval patches which occupy the base of the stalk and the attaching disc or develop around the points of origin of the branches while in *P. epiphylla*, the more or less evident midrib results from such growth. These localised thickenings are no doubt of mechanical significance. As in *Ahnfeltia*, the secondary tissue shows strata representing successive periods of meristematic activity, the limiting layers being constituted by cells with thicker walls and narrower lumina.

The Gigartinaceae include the common seaweeds *Chondrus crispus*, with many different forms and *Gigartina stellata* which are the usual sources of carrageea. Both possess dichotomously branched thalli with a stalk-like portion, *Chondrus crispus* is a North Atlantic lithophyte, which is most typically developed in a limited zone near low-tide level, the thalli occurring at higher levels often being stunted. Sublittoral forms are elongate with narrow branches. The margins of the fronds are smooth or fringed with numerous leafy processes.

The fronds of *Gigartina stellata*, usually found on rocks near low-water often-possess incurved edges and are not uncommonly twisted. Apart from this, fertile plants are readily distinguished from those of *Chondrus* by the numerous teat-like outgrowths which harbour the cystocarps. This feature is distinctive of all *Gigartinas*, which for the rest show a rather variable habit. In *G. acicularis* Lamour, the thalli are almost terete, whilst in some species they form a broad, almost unbranched, expanse. In still others the branching is irregularly pinnate.

Both *Chondrus crispus* and *Gigartina stellata* are attached by an orbicular disc giving rise to a number of erect fronds; in *Chondrus* it may reach a width of 2 cm. These discs consist of vertical rows of cells with firm, non-gelatinous walls, and in *Gigartina* they show zoning indicative of periodic growth. The spores of *Chondrus* divide to form a hemispherical cushion which develops into the attaching disc, the production of erect thalli being deferred for some time; the early development of

Gigartina seems to be similar. Both the attachment discs and the erect fronds are perennial, those of *Chondrus* ordinarily lasting for two or three years. Old fronds become severed at the very base at the commencement of winter, leaving evident scars.

The thallus of *Chondrus* affords clear evidence of a multiaxial construction, both in the apices and in sections of the maturer parts. The medulla consists of elongate cells with thick mucilaginous walls, their narrowed ends being bridged by delicate septa which no doubt represent the pit-membranes; there are also secondary pit-connections between the cells of adjacent rows. Near the surface, the axial threads bear numerous perpendicular, small-celled, laterals which form a compact cortex, separated from the central conducting region by cells of intermediate size; the latter are joined by pit-connections, both to the inner and outer cells. According to Darbishire there is a single chromatophore in each cortical cell, while the medullary cells contain a number of rounded leucoplasts. Hairs are of rare occurrence. The considerable surface-enlargement leads to a marked distortion of the older medullary elements.

The structure of *Gigartina* is essentially the same, but in *G. stellata*, the cells of the inner cortex and medulla produce downgrowing septate hyphae which are sometimes joined to the medullary cells by secondary pit-connections. The accounts of Olson and Humphrey contribute little.

The widely distributed Callymeniaceae comprise forms with broad, often dichotomously branched thalli which possess a several-celled apical meristem, although, in the finer segments of *Euthora cristata*.

Kylin found a single two-sided apical cell. Both here and in *Callophyllis* there is a large-celled medulla with interspersed hyphae, but in *Callymenia reniformis* the medullary elements are narrow. The latter is a typical Atlantic seaweed, while *Euthora cristata* is a deep-water subarctic form. It is not certain that these two genera are really multiaxial. The Australian *Callymenia cribrosa* develops numerous holes in the thallus, as in some Laminariales. The early development of *Callymenia* is much like that of *Phyllophora*, while the spores of *Callophyllis* divide to form a compact disc.

Two members of Cryptonemiales—*Constantinea*, confined to the North Pacific and the Behring Sea, and the Antarctic *Nereoginkgo*—show a sympodial construction. Each segment of the sympodium consists of a simple or branched stalk terminating in a horizontal foliar expanse, growth being continued by the development from the latter of a new cylindrical segment which in its turn expands in the same way. The fronds, which in *Constantinea* often appear perfoliate, later split into segments and ultimately wear away leaving scars on the stipe. A similar habit is seen in the Mediterranean *Neurocaulon reniforme, Constantinea reniformis* Post. & Rupr, a member of Gigartinales, but here the fronds are reniform.

The preceding consideration of certain more specialised Cryptonemiales and Gigartinales again serves to emphasise the fundamental correspondence between the uni- and multiaxial types—in fact several *(Cystoclonium, Catenella)* of the uniaxial forms are in the mature condition to all intents and purposes multiaxial. The foliose habit, which can be based on either type of construction, probably represents the most specialised condition reached in Gigartinales. While diverse members of Cryptonemiales and Gigartinales afford distinct evidence of heterotrichy in the early stages of development, the germinating spores in many genera produce a compact, often hemispherical, cushion from which the erect frond arises; a similar cushion is met with commonly in Rhodymeniales. It appears to represent a modification of the filamentous prostrate system prevalent among Nemalionales.

Crust-forming Types among Cryptonemiales

In certain Cryptonemiales (Cruoriaceae, Squamariaceae) the mature thalli are encrusting, forming thin, circular or lobed, reddish- or dark-coloured sheets, which are probably for the most part perennial; they are commonly found on rocks and shells between tidemarks and in deeper water, although some are epiphytic. The crusts, which adhere along the whole of their lower surface to the substratum, consist of a basal layer of closely aggregated radiating filaments produced from the germinating spore. All the cells bear erect threads, which are simple or little branched, are held together by tough mucilage and form the main substance of the crust they either bend up gradually or stand off at right angles to the basal layer. Attachment is usually effected by short rhizoids, the tips of which fit into all the irregularities of the substratum. The various genera are for the most part distinguished by reproductive features.

Essentially the same structure without calcification is met with in *Rhododermis,* found mainly in northern seas. In *R. Georgii,* the inner cells of the erect threads of certain forms show great enlargement, the thalli then appearing globose or irregularly lobed; such forms are connected by transitions with the usual encrusting type.

In *Petrocelis* and *Cruoria,* which are likewise uncalcified, the erect threads are less firmly connected. The fleshy growths of P. *Hennedyi* are often found on stipes of *Laminaria Cloustont,* while P. *cruenta* Ag. is a littoral lithophyte. In the large *Cruoria pellita* the rhizoids commonly lengthen into septate threads which may form a several-layered stratum below the older crusts.

Hildenbrandia, a genus of uncertain position allied to Corallinaceae, shows a similar habit. Most species are marine, but *H. rivularis* is widespread as a lithophyte in fresh water, being often associated with *Heribaudiella*. Its blood-red to brown-red crusts are found in streams and lakes; it is a pronounced shade-form. Only vegetative reproduction is known.

The common marine species, *H. prototypus* Nardo, is a littoral and sublittoral lithophyte, in the older crusts of which the growths of successive years are marked by distinct limiting lines. The outer layers commonly perish in part during winter, the renewed growth being-then localised at certain points and resulting in an uneven surface. Reproduction is by tetraspores which are formed in conceptacles. The germinating spores give rise to a short filament producing a disc at its tip.

Several genera with prostrate encrusting thalli are distinguished by the fact that laterals arise from both sides of the central axial system, such are *Rhizophyllis, Coriophyllum,* and *Ethelia*. There is reason to believe that they are forms in which the prostrate habit is secondary. *Rhizophyllis Squamariae*$_y$ a Mediterranean epiphyte usually found on.

Peyssonnelia, belongs to the same family as *Polyides*. The band-shaped thallus, which is attached by septate rhizoids, possesses a series of apical cells, much as in *Rhodophyllis.* Both here and in *Coriophyllum* the cross-section is distinctly dorsoventral.

Strongly calcified crusts are also typical of the Melobesieae among Corallinaceae. A relatively simple example is furnished by *Melobesia*, the thin, pink or red, crusts of which are widely distributed on all kinds of substrata (other Algae, *Zostera*, etc.), especially in warmer seas. The plants are often only a few millimetres in diameter. In many species they consist (except in the region of the conceptacles) of a single layer of prostrate threads, which show pseudodichotomous branching and are usually compacted to form a disc with marginal growth; the tangential septa are commonly arranged concentrically. Examples are afforded by the ubiquitous *M. farinosa* Lamour and *M. Lejolisii*. In *M. Solmsiana*, on the other hand, the threads of the basal stratum are commonly quite distinct, while in a similar form described by Taylor they produce a network. In other species short erect threads arise from the basal stratum so that the crusts become 2-5-layered. The surface of the crusts is covered with a thick layer of mucilage. The cells of adjacent threads, both in the basal stratum and in the erect system, commonly communicate by open pores formed by partial solution of the separating walls.

A marked feature of *Melobesia* are the flat cover-cells, which occur also in other Corallinaceae. One such cell is found at the top of each erect thread, while in the monostromatic species they often only cover part of the subjacent cell and not uncommonly overlie the septa. They may or may not contain chromatophores. Where the thallus is several-layered, the cover-cells are cut off" from the cells of the basal stratum before the formation of erect threads which are produced by intercalary division of the underlying cell.

Many species exhibit occasional larger hair-producing cells with scanty contents and usually devoid of cover-cells; these are the *trichocytes* of Rosenvinge "heterocysts" of Rosanoff. They often originate from marginal cells which become surrounded by the adjacent growth, after the short-lived hair has been shed; before this happens a septum is formed across the base of the hair.

Epilithon membranaceum a widely distributed epiphyte lacking trichocytes, is distinguished from *Melobesia* only in details of reproduction. The thallus may be three- or more-layered and shows similar cell-fusions.

Lithophyllum and *Lithothamnion* (often written *Lithothamnium)* are lithophytes which usually possess considerably thicker crusts, frequently attached to the substratum only by their older parts. The crusts commonly exhibit foliose lobes or irregular, sometimes coralloid (especially in *Lithothamnion)*, upgrowths *(C-E)* which are often densely aggregated. The young stages closely resemble *Melobesia*. While *Melobesia*, *Lithophyllum*, and *Lithothamnion* show clear differences in reproductive details, the delimitation of vegetative material presents considerable difficulties, and there is much difference of opinion between leading authorities as to the placing of certain species. The encrusting forms of the last two genera, moreover, vary considerably in habit according to the substratum and the depth at which they grow, as well as through the activities of boring Molluscs and the like.

Lithophyllum incrustans Foslie is widespread in rock-pools on exposed shores where it forms thick adherent crusts. In the

Mediterranean, on the other hand, the crusts are attached only at their centre and often possess numerous flat lobes with a sharp one-layered edge. *Lithothamnion lichenoides* Foslie shows a comparable habit, but in most species of the genus the encrusting base develops short upright protuberances of various shapes, as in *L. glaciate* Kjellm, a widespread northern species, and L. *polymorphism* Aresch. *Melobesia polymorpha*, in *L. calcareum* Aresch these upgrowths are branched and coralloid.

In taxonomic descriptions a lower region, the *hypothallium*, is usually distinguished from an upper, the *perithallium*. The hypothallium generally consists of the prostrate system and of the lower parts of the upgrowing threads, while the perithallium comprises the vertical rows which commonly make up the greater part of the crust. Where the erect system arises abruptly, as in *Lithophyllum expansum* and *L. orbiculatum* Foslie, the hypothallium consists merely of the one-layered prostrate system; its cells cut off cover-cells as in *Melobesia*, and then give rise to the erect threads which increase in length by intercalary division of the underlying cell. In certain species the hypothallium is several-layered, even at the margin, and forms a considerable part of the crust, the uppermost layer bending up to form a narrow perithallium, whilst the lowest layer shows a slight downward trend.

In such instances a multiaxial structure is recognisable hypo- and perithallium being respectively comparable to the medulla and cortex of a *Corallina*; in many species of *Lithothamnion* and *Lithophyllum*, however, the hypothallium is asymmetrical and the development of off standing threads confined to its upper side. The extent of its development often varies with the substratum, somefimes even in one and the same individual. In *Lithothamnion*, cover-cells extend right over the growing apex.

The cells of the perithallium are usually smaller and have thicker walls. The upper ones contain plentiful discoid chromatophores, while those below often include numerous starch-grains; the inner cells are often in large part dead. In

Lithophyllum, the cells of the perithallium commonly show an evident transverse arrangement and are joined by secondary pit-connections which are not produced in the customary manner. In *Lithothamnion,* on the other hand, such pits are lacking and adjacent cells fuse more or less completely with one another, a feature often particularly well seen in the tissue roofing the sori; when several cells are involved, branched expanses may result. *Porolithon* differs from *Lithophyllum* chiefly in the presence of enlarged ovoid cells, occurring singly or in groups within the thallus. Trichocytes are lacking in the genera under discussion, although Rosenvinge records unicellular hairs in *Lithophyllum.* Regarding the deposition of lime.

The germinating spores divide to form a compact mass of cells from which the adult thallus appears to arise as a lateral proliferation. The rate of growth of the older plants is slow. Increase in thickness of the crusts is effected by the perithelium and localised growth results in the protuberances, commonly seen in *Lithothamnion.* Growth is periodic and the successive zones, often alternately lighter and darker, are frequently (especially in *Lithothamnion)* delimited by horizontal layers of a substance which stains deeply with haematoxylin and which is probably formed when growth ceases. Not infrequently, one part of a crust grows over an adjacent one and, when this is so, each stratum may have its own hypothallium. In *Lithothamnion* the old tetrasporangiate sori become gradually buried and appear as small hollows in the crust.

Although the encrusting Corallinaceae attain their greatest development in warmer seas, species of *Lithotham* and *Lithophyllum* often form extensive sublittoral banks in temperate and polar seas. A widespread species in Britain and Northern France is *Lithothamnion calcareum* and in certain areas the many fragments of the coralloid growths cast up by the waves constitute a regular *Lithothamnion* beach, the calcareous material of which is locally employed for liming the soil. In the Mediterranean, *Lithophyllum tortuosum.* Fossil forms a belt at low-tide level on exposed shores, while other Melobesieae play an important role down to depths of 80 metres. There is no species common to the Arctic and Antarctic.

It is, however, in the oceans of the Tropics that these seaweeds, together with other calcareous Algae, Foraminifera, and Corals build up the huge calcareous masses, generally spoken of as "coral-reefs", although in several, the Corals appear to play a subsidiary role; in all of them the encrusting Corallinaceae (Nullipores) are of prime importance as cementing organisms. In the Funafuti boring, they were found to occur abundantly down to a depth of 1000 feet. According to Foslie the most important reef-forming species over large areas of the Indian Ocean are *Porolithon encodes*, P. *craspedium*, *Lithophyllum Gardineri*, and *Goniolithon frutescens* Foslie; several of these are also widely distributed over many parts of the tropical Pacific. The Melobesieae require well-aerated water for successful growth and usually occur in situations where there is a fairly strong current. At the same time, a considerable number are known to thrive at very considerable depths in clear waters in the Tropics.

The encrusting Corallinaceae occur from the Cretaceous onwards and in the Tertiary are for the most part represented by genera still living and the literature there cited; also. They have evidently long played an important role, and extensive limestone rocks consisting largely of *Lithothamnion*, and are known from various parts of the world, e.g. the Leithakalk (Miocene) of the Vienna basin which has been used extensively as a building stone. *Solenopora* and allied genera, grouped by Pia in a separate family, are largely Palaeozoic (Ordovician-Jurassic) forms which are usually regarded as related to the encrusting Corallinaceae; the recent record of conceptacles in *Solenopora*, if substantiated, would warrant their inclusion in Corallinaceae.

The Nullipore banks are inhabited by lime-boring Algae, Molluscs, etc. which cause more or less considerable destruction of the older growth, while percolating water may convert certain parts into a structureless mass. In the deeper portions of the Funafuti boring, there was a great increase in the amount of magnesium carbonate so that the older parts of such reefs probably approximate in chemical composition to a dolomite.

The Special Development of the Multiaxial Type among Rhodymeniales

A multiaxial construction of a distinctive type is found in the genera *Champia*, *Chylocladia* and *Lomentaria* which belong to the Champiaceae. The structure of the first two is similar in all essentials. *Chylocladia kaliformis*, an annual summer lithophyte, has the jointed habit typical of many members of the family. The fistulsr thalli, which are attached by a minute disc, are constricted at the points at which the whorled branches arise; both here and at the bases of the branches diaphragms bridge the central cavity, which is elsewhere filled with mucilage. The entire thallus is encased in a mucilage-envelope with a well-marked cuticle.

In the older parts the surface consists in the main of a single layer of radially elongated cells running longitudinally on their inner side at more or less even intervals. These longitudinal threads converge towards the rounded apex, where they terminate in apical cells, not all situated at quite the same level. Each apical cell cuts off a single row of segments which divide tangentially. The outer halves constitute primary cortical cells, while the inner gradually lengthen to form a longitudinal thread. The primary cortical cells divide anticlinally to form the peripheral layer of large cells which contain several nuclei, numerous very small chromatophores, and in some localities light-reflecting bodies. Later they cut off obliquely on their outer side a certain number of small cells forming a disconnected superficial layer, which is stated to be most extensive in plants exposed to strong light. Especially in younger individuals, its cells produce elongate, thick-walled hairs.

The cells of the longitudinal threads undergo no further transverse division, but most of them cut off on their inner side a small spherical cell with plentiful contents, the bulb-cells of American writers. At the points of branching of the thalli, these cells divide to form threads which unite, with the establishment of secondary pit-connections, to produce the one-layered diaphragms. The numerous bulb-cells that do not develop into diaphragms have been interpreted as glandular elements or, rather more usually, as undeveloped diaphragms. Similar structures occur in most Champiaceae and are also present in Rhodymeniaceae, where diaphragms are usually lacking. Their utilisation in the formation of diaphragms is apparently a secondary feature and they are perhaps best regarded as arrested hyphae.

Branches originate by the outgrowth of several peripheral cells into threads which unite to form the apex of the new axis; adventitious branches appear at indeterminate places in older plants.

Kylin maintains Kutzing's genus *Gastroclonium* for those species of *Chylocladia* in which the main axes and the basal parts of the branches remain solid, with only slight elongation of the inner cells. The hollow parts show essentially the same structure as *Chylocladia*.

Champia parvula, a widely distributed epiphyte in the warmer seas of the Northern Hemisphere, forms dense bushy tufts arising from a number of creeping threads attached by multicellular haptera; the branches here often arise singly. The more robust and little-branched *C. lumbricalis Ljarnour* has a wall of a number of layers of cells, while the diaphragms may be several-layered.

In *Lomentaria*, there are no diaphragms. The thalli of *L. articulata*, which is common between tide-levels, are constricted at the points of branching where they are solid. Those of *L. clavellosa* (Turn.) Gaill, on the other hand, are hollow throughout and unconstricted; they may reach a length of 40 cm. and show distichous branching. The anatomical structure is much like that of *Chylocladia*. The peripheral small-celled cortex is, however, more extensive and may form a continuous several-layered covering to the older parts. The longitudinal threads are branched and joined by secondary pit-connections, while in the solid nodes of L. *articulata* they are closely packed and consist of large rounded cells.

The early development is similar in all Champiaceae. The spores first divide into quadrants which then segment into two horizontal tiers. The four upper cells constitute the primary apicals which divide horizontally to produce the erect thallus, while those of the lower tier grow out into rhizoids. As the thallus broadens, additional apical cells are formed from some of the uppermost segments. Older germlings of *Chylocladia* and *Lomentaria* show a thick, hemispherical or nearly globular attachment-organ bearing one or two erect fronds. In *L. uncinata* Menegh, it gives rise to radiating stolons, which form secondary holdfasts at their tips.

The *Rhodymeniaceae* show greater diversity in external habit and include a number of non-fistular foliose forms. *Chrysimenia, Coelarthrum,* and *Binder,* however, resemble the Champiaceae in possessing vesicular thalli and in the presence of bulb-cells like those of *Chylocladia* on the inner surface of the cells lining the central hollow. In the Australian *Binder, a splachnoides* Harv these terminate systems of branched threads. In all Rhodymeniaceae, the inner cells are large and rounded, which no doubt represents a more primitive condition than that found in Champiaceae.

Rhodymenia palmata, a widely distributed littoral perennial with a dark red, simple or branched thallus, is the dulse of the Scots and the dillisk of the Irish and is edible. The early stages resemble those of Champiaceae, although the young fronds are stated to be uni-axial, the axial thread persisting in the basal stalk. The mature fronds possess a marginal meristem and consist of several layers of thick-walled medullary cells enveloped by a rather narrow small-celled cortex; in the stalk, the latter is broader and in older plants shows zones of growth. The attachment-disc is composed of numerous rows of cells. The young fronds bear groups of hairs which are visible to the naked eye. Butters gives a few data on the structure of *R. pertusa,* a characteristic subarctic sublittoral species with a large thallus showing perforation. *Fauchea*, confined to warmer seas, has fronds with much the same structure as *Rhodymenia*. *Dendrymenia* shows a similar habit and sympodial mode of growth to *Constantinea*.

The seaweeds usually included in *Chrysimenia* are frequent sublittoral forms of warmer seas, varying in the degree of inflation and the extent of branching of the thallus. *Coelarthrum*, widely distributed in warmer seas as well as in Australiap, has a dichotomously branched thallus, showing periodic constriction at points where diaphragms occur. The bulb cells are here borne on irregular stellate elements. The Tasmanian *Gloeosaccion* has a simple saccate thallus. *Halosaccion ramentaceum*, a characteristic North Atlantic and Pacific seaweed, with a very variable habit, included in the Rhodymeniaceae.

The Special Developments of the Uniaxial Type among Ceramiales

The numerous genera of Ceramiales exhibit great diversity in vegetative construction, as exemplified by the families Ceramiaceae, Delesseriaceae, Rhodomelaceae, and Dasyaceae, although these only present minor differences in their reproduction. Diverse genera attain to a high degree of morphological complexity. All Ceramiales are, however, uniaxial. Moreover, the germinating spores develop a single bipolar axis from the first and all traces of heterotrichy are lacking. While the first septum is usually described as running parallel to the substratum, Chemin affirms that it is vertical or oblique to the latter and that the germling is at first prostrate, although the apex soon becomes erected; this is almost certainly a cultural condition. The four families are best considered separately.

Ceramiaceae

Many Ceramiaceae retain a filamentous habit, and this is very obvious in *Callithamnion*. The richly branched tufts of the former, commonly only 2-5 cm. high, have much the habit of a *Cladophora*. The elongate cells are uninucleate, or more usually multinucleate, and mostly contain ribbon-shaped chromatophores. The branches arise alternately and in definite spiral succession from just beneath the septa, usually standing off in all directions, though distichous in some species. In many, the principal axes are composed of broad, thick-walled cells which sometimes show a zigzag disposition. Vigorous development of the laterals occasionally results in pseudo-dichotomy in the lateral branchings. Certain species at times produce a profusion of long slender hairs on the ultimate branches, which may continue to grow sympodially, as in *Acrochaetium*.

The older parts are usually enveloped in an investment of delicate multicellular corticating threads which originate from the basal cells of the laterals and sometimes also from the lower ends of the axial cells; they commonly grow through the thick walls of the latter, although ultimately penetrating to the exterior. This cortical envelope often forms a compact tissue which completely obscures the lower part of the axis and may contain chromatophores in its outer cells. Not uncommonly short adventitious branches arise from it.

Anchorage is effected by similar multicellular filaments emerging from the cortex and from the basal cells of the lower branches and sometimes uniting to form a definite disc. The germlings are attached by a basal rhizoid, which is soon replaced by others. In C. *Furcellarieae*, attachment is aided by horizontal stolons which arise from the lower axial cells and may give rise to further erect threads. Similar structures are recorded in the minute C. *scopulorum*, where they emerge singly from the under sides of the branches and, being detachable, aid in vegetative propagation.

Certain species persist from one season to another, although others are probably annuals. Several favour mud-covered rocks near low-water mark, while others are more usually epiphytic, with rhizoids penetrating into the substratum. One of the most robust species is C. *tetricum*.

Monospora, Pleonosporium, and *Seirospora* differ from *Callithamnion* only in their reproductive features; *Seirospora Griffithsiana.*

Griffithsia is distinguished by its huge non-corticated multinucleate cells which are readily visible to the naked eye; the older ones may contain several thousand nuclei as a result of repeated division. The copious forked branching is due to equal development of lateral and parent axes. Short branchlets, bearing the reproductive organs, arise from the tops of the younger cells, while richly branched hairs are commonly borne in the same position; both structures are deciduous. The entire thallus is encased in a thick mucilage-envelope, which in G. *corallina* is two-layered and visible to the naked eye. The early development resembles that of *Callithamnion,* the older plants being usually attached by vigorous rhizoids emanating from the basal cell of the germling.

The annual *G. corallina* and *G. setacea* stated to be perennial, are widely distributed European littoral species, while *G. globifera* is a frequent annual on the Atlantic shores of North America, with the same vegetative structure, is more characteristic of warmer seas. Another not uncommon form with multinucleate cells is *Halums equisetifolius* Kiitz, which is distinguished by its whorled branching.

Greater specialisation is evident in *Antithamnion,* with several widely distributed species, usually found on rocks near low-water mark. The plants possess creeping threads which have upturned tips and bear a number of erect uncorticated filaments. Their branches are usually clearly distinguishable into those of unlimited and those of limited growth. Several branches arise from the same parent-cell and those of limited growth are either opposite, as in *A. cruciatum* and *A. floccosum* (Mull.) Kleen or in whorls of three or four, as in *A. elegans* Berthold and *A. subulatum*; the members of a whorl develop successively. The ultimate branchlets are commonly adaxial. The branches of unlimited growth are alternate and take the place of a short lateral. The uninucleate cells contain numerous chromatophores.

In *A. plumula* the branches of the erect threads are generally opposite and in one plane, although in shaded habitats they tend to turn in the direction of the light. Cells producing a shoot of unlimited growth commonly lack the opposite lateral, which is also often wanting on one or two of the overlying cells of the parent branch. Frequently, however, the cell from which a long shoot arises bears a pair of unbranched laterals in a plane perpendicular to the normal plane of branching. Occasional individuals have the branches in whorls of three or four, with the long shoots orientated in diverse directions; according to Berthold this is a result of uniform illumination, although Rosenvinge's observations render this doubtful.

The short laterals of *A. cruciatum* are often decussate, though the angle between successive pairs varies, while plants exposed to unilateral illumination are stated by Berthold, who made considerable studies on the light-reactions of this species, to develop all the branches in one plane. In this species, long shoots also arise adventitiously from the basal cell of a lateral of limited growth.

The early development of *A. plumula* is like that of *Callithamnion*. A protrusion of the spore becomes cut off as the first rhizoid, while the rest divides to produce the erect axis. According to Berthold the primary axis of *A. cruciatum* soon terminates its growth, while procumbent or ascending branches arise from its base. It can hardly be doubted that the creeping threads of *Antithamnion*, *Spermothamnion*, etc. are always secondary in origin. In *Antithamnion*, the creeping threads are attached by unicellular rhizoids, as well as by more elaborate haptera, while in older plants multicellular rhizoids originate from the lower branches of the erect axes; they are of course homologous with the cortical threads of *Callithamnion*. In *A. plumula*, some of them develop into runners producing other erect threads.

The interrelationships of the diverse genera so far considered are not altogether clear. Since a uniaxial construction with whorled laterals is characteristic of many of the less specialised Florideae and is also seen in *Crouania* and *Wrangelia* among Ceramiaceae, forms like *Crouania* and *Antithamnion* are probably more primitive than *Callithamnion* and its allies, a point of view which receives considerable support from a consideration of the arrangement of the female reproductive organs. Whorled branching of a somewhat different type is also found in *Actinothamnion*. *Crouania*, as already noted, altogether conforms to the *Batrachospermum-type*, although the early development is practically identical with that of *Callithamnion*.

The bilateral branching seen in *Antithamnion* is also characteristic of *Ballia*, *Plumaria*, and *Ptilota*, in which, however, the older parts of the thalli show considerable cortication. *Plumaria elegans*, a small perennial lithophyte, is frequent between tide-levels in the North Atlantic. The thallus shows repeated pinnate branching and, in the younger parts, presents considerable resemblances to an *Antithamnion*. The paired laterals arise successively, that first formed usually developing more strongly than the opposite one. The older ones exhibit a stronger development of the branches of the second order on the abaxial face, the uppermost ones not uncommonly ending in chain. Certain laterals develop into long shoots.

The thick cortex enveloping the older axes is produced from two sources. Small cells, cut off from the basal cells of the laterals, give rise to cortical threads much as in *Callithamnion*. Apart from this, however, the axial cells put out protuberances in a plane perpendicular to that in which he main branches of the thallus are formed; each of these cuts off at its four corners small cells which grow out respectively in the upward and downward directions as branched threads covering the axial cells. The two kinds of outgrowths gradually produce a thick envelope, differentiated into a small-celled photosynthetic layer, an intermediate storage-layer of large cells, and a conducting system of long narrow elements surrounding the large axial cells. In the older parts, certain superficial cells grow out into unbranched threads forming a surface felt. The early development is like that of other Ceramiales.

Ptilota plumosa, a robust North Atlantic perennial, often epiphytic on *Laminaria,* is very similar in habit and general structure. The laterals, here arising from alternate segments of the apical cell, are markedly unequal and the cortex, formed entirely by outgrowths from their basal cells, develops at an early stage. It forms an envelope of large storage elements, with one or two layers of small superficial cells, while in the older parts, there is a progressive production of hyphae from the former; at first these hyphae surround the axial thread, but later they aggregate in thick bundles and often occupy a large part of the transverse section. The larger axes show some compression. *P. pectinata,* abundant on the northern coasts of Atlantic America, is essentially similar in structure.

The richly branched tufts of *Ceramium* are often recognisable by the transverse banding of the younger parts and the tong-like curvature of the forks of the ultimate branches. Each thread is occupied by a row of large cylindrical or barrel-shaped cells, the older ones being almost colourless. The rather regular dichotomy is initiated by the appearance in the dome-shaped apical cell of two inclined walls which cut out the apicals of the forks.

At an early stage, a series of pericentrals is cut off successively at the top of each segment, their number being a valuable diagnostic character. Each cell of the primary cortical band thus produced cuts off four secondary ones, two above and two below, and these give rise to four branching threads which remain closely apposed to the axial cells. In their entirety, they gradually form a small-celled cortex which envelopes, not only the upper end of the parent segment, but also extends onto the segment above. In some species the cortical bands of successive segments remain combined to form a uniform investment through which the axial cells are just distinguishable, but in others the lower part of the segment stretches appreciably so that the bands become widely separated and are visible even to the unaided eye. The cortical cells are usually isodiametric or longitudinally elongated, although in

C. *transversale* Collins and Hervey the lower ones are transversely extended. Hairs are frequently present on the younger parts.

The primary cells of the cortical bands are no doubt homologous with those producing branches in other Ceramiaceae, as well as with the pericentrals of Rhodomelaceae. In certain species, in fact, each of these cells grows out into a short spinous lateral while in others only one such branch is produced on each cortical band. The threads arising from the pericentrals are directly comparable to the corticating filaments of *Plumaria*. Their cells may also bear short spinous branches, as in C. *echionotum* J. Ag.

The germlings are of the usual type, except that the spore is attached to the substratum by a cushion of mucilage which is penetrated by the first rhizoid. Cortex-development commences early. The mature tufts are attached by rhizoids originating from the lower axial and cortical cells; their tips form holdfasts and they may occasionally develop cortical bands.

Species of *Ceramium* occur abundantly between tide-levels, as well as in deeper water. The Mediterranean C. *radiculosum* Schiller inhabits rivers, especially the estuarine portions. In C. *tenuissinmm* J. Ag. and certain other species, the cortical bands include colourless refringent cells which contain protein, sometimes in the form of crystalloids, which may also occur in the axial cells.

Several other genera show much the same structure. In *Ceramothamnion*, which is apparently widely distributed in warmer seas, the little-branched erect filaments arise from creeping threads. *Microcladia glandulosa* Grev., a frequent Northern seaweed with distichous branching and slight flattening, has a continuous several-layered cortex. In *Reinboldiella* the terete, pinnately branched thalli are completely prostrate; the form and density of arrangement of the cortical cells differ on the two surfaces so that there is some degree of dorsoventrally.

The widely distributed perennial seaweed *Spyridia filamentosa* differs considerably from other Ceramiaceae. The laterals, arising from the first pericentral of each segment, are arranged on a 5/13 spiral, and most develop into the deciduous photosynthetic "filaments", which occur only on the younger parts and also bear the antheridia and sporangia. They consist of about 20 cells which produce a localised cortex as in *Ceramiwn*; the cortical cells commonly bear long hairs. Each cell of a long axis successively cuts off 14 pericentrals which, as the parent-cell lengthens, remain as a ring at its upper end. Each pericentral gives rise below to two secondary cells, which lengthen in the same measure as the axial cell and become joined by secondary pit-connections to the underlying ring of pericentrals. According to Taylor the pericentrals subsequently divide longitudinally so that their number is equal to that of the secondary cortical cells. The uniformity of the envelope superficially recalls the appearance of a *Polysiphonia*. The secondary cortical cells are perhaps homologous with the corticating filaments of *Ceramium,* although in *Spyridia* they do not give rise to branched threads. The older parts possess a dense covering of hyphae arising from the cortical cells. The early development is of the usual type.

Delesseriaceae

The Delesseriaceae comprise a striking series of often delicate foliose forms in which the thallus commonly shows a development of midribs and lateral veins so as to simulate to a remarkable extent the leaf of a higher plant. They include a considerable number of, frequently perennial, sublittoral forms, which, though widely distributed, probably attain their greatest development in the seas of the Southern Hemisphere. The thalli are produced by juxtaposition in one plane of the numerous branches of the axial filament, and the greater part of the resulting expanse usually remains one-layered, midribs and veins alone developing a more massive structure. In the Australian *Hemineura frondosa,* however, most of the frond consists of 3-5 layers of cells.

The cells of the axial row and of its successive branches are linked by primary pit-connections, while the basal and adjacent cells of the laterals, which contribute to the formation of midrib and veins, become joined longitudinally by a number of secondary pit-connections such connections also arise between other cells of adjacent rows. In most genera the fronds are capped by a conspicuous apical cell forming a single series of segments, but in most Nitophylleae the apical cell is distinguishable only in younger stages. In the principal axes, the segments cut off four pericentrals, the first two in the plane of branching, the other two, which initiate the several-layered ribs, in a plane perpendicular to it. The mode of formation of the pericentrals in Delesseriaceae contrasts with that found in Rhodomelaceae and Dasyaceae. Certain genera afford an exception to the strictly apical growth of most Florideae, since they exhibit more or less extensive intercalary division.

The apical cell is very evident in *Hypoglossum Woodwardii* with rather narrow lanceolate fronds. Its flat segments divide into a central and two transversely extended pericentral cells, which function as secondary apical cells. The latter undergo very oblique division to form the pinnately arranged laterals of the first order, which bear all the secondary laterals abaxially. The elongate apicals of the diverse branches all reach the margin of the delicate, though compact, frond.

A little way behind the apex there is marked elongation of the central cells and of the adjacent cells of the primary laterals, which together form the foundation of the midrib. No other cell-rows are affected so that no veins other than the midrib are present in the mature frond. In the formation of the midrib, cells are cut off parallel to the surface, while further thickening is effected by tangential division. The subsequent production of a considerable number of narrow hyphae, between the larger cells composing the midrib, adds to its mechanical strength.

The species of *Caloglossa*, which inhabit tropical streams and especially the brackish estuaries, often accompanying

Catenella, Bostrychia, etc. amid the algal vegetation on the roots of Mangroves, have a very similar structure. The narrow prostrate thalli show frequent constrictions, where forking occurs and bundles of attaching rhizoids are produced; adventitious branches also arise from the dorsal side of the midrib at the points of forking. The apex is commonly inrolled, with the convex surface upwards.

In *Apoglossum ruscifolium* there is more extensive ramification of the cell-rows composing the blade, with adaxial development of the branches of higher orders. Moreover, the cells of the laterals of the first and second orders show marked elongation and form lateral veins which are, however, not visible to the unaided eye. *Grinnellia americana,* found on the Atlantic coasts of the United States, is very similar, but only possesses primary veins. The midrib of *Apoglossum* is enveloped in a dense mass of hyphae, but in *Grinnellia* there are only few of these.

In the genera so far considered, the branches of the fronds arise from the midrib and are stated, in *Hypoglossum* and *Apoglossum,* to develop endogenously from the central cells. In *Membranoptera alata (Delesseria alata* (Huds.) Lamour, on the other hand, the copious branching is due to the further development of the apical cell of a lateral of the first order and often results in apparent dichotomy; tufts of adventitious fronds, some of which bear reproductive organs, arise in the angles between the branches. *M. alata* is a northern form, with a conspicuous midrib and narrow wings traversed by lateral veins.

The handsome *Delessfria sanguined,* sometimes called the "dock-leaved *Delesseria*" and found in shaded situations between tide-levels, as well as in deeper water, has D, *Membranoptera alata* Stackh., habit b y adventitious fronds, fronds which may reach a length of 25 cm., with conspicuous side-veins produced from the primary laterals; secondary veins, formed from the cell-rows of the second order, are usually only apparent under the microscope. As in *Hypoglossum,* laterals of higher orders for the most part arise abaxially. Both in *Apoglossum* and *Deksseria* intercalary divisions occur in all but the axial cell-row.

The robust midrib develops as in *Hypoglossum,* but the basal cells of the primary laterals divide transversely so that they are only half as long as the central cells. The surface layer of the midrib remains meristematic for some time and forms a relatively wide zone of large elements bounded by a small-celled superficial layer in the older ribs the larger cells appear isolated amid a plexus of narrow branched hyphae, which appear to serve for storage. The veins of the first order develop in an analogous manner, while the secondary ones are at best only three-layered.

The branches again develop from the midrib, for the most part in autumn and winter, and appear to originate from surface-cells. The wings of the parent-frond gradually wear away, leaving the persistent midrib, the cells of which contain large quantities of starch which decrease as the new blades develop. The attachment of the young fronds to the parent is strengthened by hyphae that penetrate into the midrib.

The Delesserieae, comprising the genera hitherto considered, differ vegetatively from the Nitophylleae in the possession of prominent apical cells and the absence of intercalary division in the axial cells. The widely distributed *Phycodrys rubens,* though a member of Nitophylleae, shows considerable resemblance to *Delesseria.* Branching of the cell-rows may take place to the fifth degree. The main apical cell is not very conspicuous and intercalary division occurs both in the axial and other cell-rows. As in *Apoglossum* and *Delesseria,* it is always the lower cell formed by such division that retains the primary pit-connection with the lateral cell-row, while the upper cells give rise to accessory cell-rows.

The frequent intercalary division, with production of accessory laterals, makes the detailed structure of the fronds more difficult to decipher.

The lobing of the fronds of *Phycodrys* is due to outgrowth

of primary laterals, while a similar activity on the part of those of the second order results in the marginal pinnules on the lobes. There are obvious primary and secondary veins, while those of the third order are only evident under the microscope. The midribs and veins develop like those of *Delesseria*, but no hyphae are produced. Branching takes place from the midrib and in older plants the fronds arise from a robust axis constituted by the primary midrib; occasionally, however, a lobe of the blade develops vigorously and leads to a kind of forking. Wille describes the development, from the base of the stalk (midrib), of thread-like runners giving rise at intervals to secondary fronds, many of which remain arrested. A prominent apical cell is also met with in *Haraldia* in which the vegetative fronds are monostromatic and without veins.

In other Nitophylleae, apical cells and the construction of the fronds from a coalescent branch-system are still more difficult to decipher. Most of the seaweeds in question were formerly grouped in *Nitophyllum*, though many are now referred to distinct genera. All are characterised by marginal branching and the absence of a distinct midrib. In its present limits, *Nitophyllum* includes species with foliose thalli, in which the older parts consist of several layers, while the younger are single-layered; there is sometimes a short basal stalk, composed of small regularly arranged cells. Veins are either completely lacking—e.g. the widely distributed *N. punctatum*—or occur only at the base of the thallus. In the mature plant, there are no evident apical cells, growth being effected by marginal, as well as by irregular intercalary, division just within the margin. In *N. versicolour* Harv, the tips and margins of the fronds often display curious thickenings which are suspected of playing a role in vegetative propagation.

In *Polyneura* and *Cryptopleura* veins form an obvious anastomosing or palmate system in the older parts, while in the younger one-layered portions, they are only recognisable under the microscope. *Polyneura Gmelini* and *Cryptopleura lacerata* are frequent, North Temperate, sublittoral forms. In the latter there are evident two-sided apical cells which are also recognisable in the smaller lobes of *Myriogramme*, in which the older parts are several-layered.

Acrosorium, with delicate, richly lobed fronds, possesses only microscopic veins. *A. uncinatum* is widely distributed in warmer seas. The frequent *A. reptans*, with completely prostrate fronds, is probably a stage of *Cryptopleura lacerata*.

The early development of Delesseriaceae differs somewhat from that of other Ceramiales. In *Nitophyllum*, the broadened spore divides transversely into a row of six cells, the two uppermost of which form the mature thallus, the top one functioning as an apical cell. The lower part of the germling develops into a mainly monostromatic expanse, which is attached by numerous rhizoids and gives rise to adventitious fronds. The early stages are similar in *Apoglossum ruscifoliumw* which is stated to complete its whole development in a few weeks. According to Chemin the segmentation is in neither instance as regular as described by Nienburg. The apical cell of *Nitophyllum* soon ceases to function as the apex broadens and marginal growth sets in.

The germling of *Phycodrys* is like a young *Nitophyllum*, consisting of a one-layered expanse devoid of a midrib and showing plentiful intercalary division; the mature fronds arise adventitiously from its stalk. The young plants of *Delesseria sanguined*, on the other hand, have a distinct midrib, but at an early stage again produce adventitious fronds from the basal stalk. The discoid attachments of Delesseriaceae consist of numerous, closely compacted multicellular rhizoids.

In spite of a certain gradation, the contrast presented by the extreme members of Delesserieae and Nitophylleae is evident from a comparison of *Hypoglossum* and *Nitophyllum*. Both, in their respective ways, appear highly specialised, and their relation to one another is not clear. A study of the early development in a wide range of forms might help to elucidate it.

The Delesseriaceae also include several genera with a net-like thallus. That of *Claudea* appears as a network with rectangular meshes and constituted by numerous narrow fronds, which are placed edgewise and perpendicular to one another; they have much the same structure as in *Hypoglossum*. The diverse parts of the reticulum originate by unilateral branching from the adaxial surface of the midrib of the next younger generation. The secondary fronds arising from the primary one grow to a considerable length, while the tertiary ones usually consist of only 5-8 segments and become joined at their tips to the lower surface of the secondary blade immediately above. The secondary fronds thus form the long bars and the tertiary ones the perpendicular cross-bars of the net.

The base of the primary frond soon becomes invested by a thick cortex and constitutes a stalk. The primary and secondary blades develop a cortex over their entire surface, while in the tertiary ones, except in fertile plants, only the midrib is corticated. *Claudea* differs from other Delesserieae in the fact that two pericentrals, situated one in front of the other are cut off successively on the adaxial (ventral) side of the segments, the larger posterior one growing out into a daughter-blade. Between the apical cells of the tertiary blades and the dorsal pericentrals of the secondary blade upon which they abut, secondary pit-connections are established; later other connections are formed.

Enlargement of the cells results in increase in size of the meshes of the net, but growth of the primary and secondary blades terminates after they have reached a certain length. Occasional tertiary blades, however, develop vigorously and produce quaternary blades which become linked with the dorsal surface of the blade immediately above. This results in forking and, since secondary fronds develop also from surface-cells of the stalk, older plants may exhibit considerable branching. The nets are attached both basally and at other points by septate rhizoids.

C. *multifida* is widely distributed in South India and Ceylon, while C. *elegans* Larnour is apparently confined to Australia.

The sessile net-like thalli of *Vanvoorstia,* recorded from various parts of the Indian Ocean and Japan, are formed in essentially the same way, but more numerous branchir s occur and all the fronds remain uncorticated. Papenfuss regards these genera as closely related to *Caloglossa*. A similar reticulate thallus is found in the Japanese *Implicaria,* but here the successive teret axes produce pinnately arranged branches.

A different type of reticulate thallus is found among Nitophyllea in the genus *Martensia,* species of which are known from South Africa, India, and Australia. The lower, first formed, part of the thallus it here a continuous expanse, which is sometimes stalked while the older part is constituted by a network of diverse width. The latter consists of radiating lamellae, coiped by numerous tangential connections. The lamellae originate just behind the edge of the young thallus by abundant and long-continued intercalary division of the intra-marginal cells and lateral separation of the rows thus formed; the marginal cells, however, remain in contact and divide to form the several-layered edge of the net. The cells of the separating rows within the margin divide parallel to the surface to produce vertical lamellae, which remain monostromatic, except at their edges. Cross-connections are formed between adjacent lamellae by the outgrowth of processes which fuse and ultimately come to consist of several rows of cells. The cross-connections arise successively, the later-formed ones about midway between the earlier ones. The older nets often exhibit branching of the cross-connections, which may thus become linked with one another.

Net-formation may commence early or only at a later stage. In *M. fragilis* the numerous lobes exhibit a cornet-like inrolling, while the entire portions of adjacent fronds become firmly joined by thick haptera so that compact cushions are formed. In *M. denticulata* and *M. pavonia* J. G. Ag. belts of compact and reticulate tissue alternate.

Rhodomelaceae

The members of this large family, monographed by

Falkenberg, show a varied and in part complex morphology. All are characterised by the basic polysiphonous structure, as well as by the usual production of laterals of two different types (ordinary branches and trichoblasts). Unicellular hairs are lacking.

Rhodomelaceae with Radial Construction

Polysiphonieae and Lophothalieae: *Polysiphonia* readily recognised by the arrangement of its cells in tiers, illustrates all the main features of the family. *Brongniarfella* is essentially similar. The dome-shaped apical cell forms a single series of segments; those which produce laterals are larger than the others and are cut off by oblique walls. As in all Rhodomelaceae, the initial of the lateral is cut off on the higher side *before* the segment divides longitudinally into central and pericentral cells.

The laterals are formed in spiral succession. According to Rosenvinge the nucleus moves to the side on which the lateral will arise before the segment is separated from the apical cell, while the ingrowing septum originates on the opposite side. The laterals and parent-axes often exhibit temporary curvatures towards one another.

The mode of formation of the pericentrals is distinctive of all Rhodomelaceae. They invariably develop successively, the first one being cut off beneath the lateral, although there is commonly a subsequent displacement to the left in the direction of the spiral so that the lateral appears above another pericentral or located between two of them. In segments, which produce no laterals, the first-formed pericentral occupies the same position as in the next upper segment bearing a lateral. The number of pericentrals varies between 4 and 20 and is usually constant, especially in species with few pericentrals. When there are four, the second and third are cut off on either side of the first, and the fourth on the side opposite to the lateral; when they are more numerous, they develop in the sequence.

The pericentrals of *Polysiphonia* and *Brongniartella* occupy the whole length of the segment and are usually broader than the central cell, which shimmers through the enveloping "siphons". They are homologous with the primary cortical cells of *Ceramium* and represent potential branches; in *Chondria*, in fact, they lengthen to form the elongate laterals composing the thallus. Primary pit-connections occur only between successive central cells and between the latter and their pericentrals. Secondary connections and in some species several may develop. The central cells usually remain uninucleate.

In many species, the erect threads produce laterals of two kinds, viz. some which are polysiphonous like the parent axis and some which develop into branched, uniseriate structures of limited growth, usually deciduous at an early stage. The latter are the *trichoblasts* found in many Rhodomelaceae. They are ordinarily composed of colourless cells containing small leucoplasts, but in *Brongniartella* the cells of the long-lived trichoblasts contain chromatophores and play an important role in photosynthesis. Although many of the trichoblasts fulfil vegetative functions only, they usually bear the sex organs, and in some species of *Polysiphonia* such fertile trichoblasts may be the only ones present.

Apart from the first, each segment of the trichoblast generally produces a lateral which often develops as vigorously as the part beyond, resulting in pseudodichotomous branching. In *Polysiphonia* and *Brongniartella,* the trichoblasts are disposed in a spiral, which almost invariably turns to the left and the first branch of the trichoblast is always situated on the anodic side. Trichoblasts are either distributed over considerable parts of the plant or restricted to certain regions.

The ordinary polysiphonous branches either arise direct and in spiral succession from the parent-axis, replacing one of the trichoblasts when these are present, or they originate from the upper surface of the basal segment of a trichoblast. The former condition is seen in *P. urceolata, P. elongata, P. nigrescent,* etc. In the basal segment of the branch the first pericentral is formed abaxially, above the first-formed one of the parent-segment, but the production of pericentrals is suppressed on the opposite side, where the branch-segment is in contact with the axis. The supra-basal segment, which bears no lateral, has the normal number of pericentrals with the same orientation.

The origin of branches from trichoblasts is more frequent. In *Polysiphonia* they arise almost simultaneously with the parent-trichoblast, whereas in *Brongniartella* they only develop when the latter is almost mature. In a *Polysiphonia,* the basal segment of the few-celled trichoblast puts out a protrusion which is cut off by an oblique septum and becomes the apical cell of the branch. After this has undergone a number of divisions, the basal segment of the trichoblast abaxially, to right and left of the median line; sometimes a third pericentral is formed. Until the branch itself commences to form trichoblasts, the first pericentrals of its segments lie in a straight line with the first pericentral of the basal segment of the trichoblast. During the further development of the branch the upper part of the trichoblast becomes displaced so that the former appears as a direct outgrowth from the main axis. In origin the branch is, however, obviously a lateral outgrowth of the trichoblast-rudiment, and it cannot justifiably be interpreted as axillary to the trichoblast.

The normal acropetally formed branches are sometimes supplemented by adventitious ones which originate from the basal cells of trichoblasts that have been shed; this is the rule in *P. violacea*. Adventitious branches may also arise in this way in species in which the normal branches are not formed on the trichoblasts. Thus, according to Rosenvinge, the new shoots of *P. elongata* that appear in spring originate partly from the remains of the axes of the previous season and partly from the basal cells of their trichoblasts, and the same is recorded for *P. nigrescent*. Rosenvinge suggests that the normal branch-formation on trichoblasts has resulted from this tendency to form adventitious branches upon them.

In many *Polysiphonias* as well as in *Brongniartella*, some of the polysiphonous axes creep over the substratum, to which they are firmly anchored by thick-walled, often richly lobed, unicellular rhizoids arising from the peripheral cells. The creeping branches, which constitute a means of perennation, lack trichoblasts and are secondary in origin, arising endogenously from the central cells at the base of the primary erect axis. They likewise branch endogenously, some of the branches remaining prostrate, while others grow erect; erect shoots are also formed from the upturned tips of the creeping axes. The part of the branch within the parent-axis remains monosiphonous and pushes its way between the pericentrals without causing any disorganisation. The endogenous branching of the prostrate shoots contrasts with the exogenous origin of the laterals on the erect ones. Other species lack this system of creeping threads and are merely attached by rhizoids emanating from the lower cells of the erect axes; these sometimes combine to form a massive disc.

In *Brongniartella* and some *Polysiphonias* the pericentrals remain undivided, but in *P. violacea* for instance, as in many other Rhodomelaceae, they segment by periclinal and anticlinal walls to form a parenchymatous cortex of more or less numerous layers, whereby the original tier-like structure is obscured. The cells of successive layers generally decrease both in length and width and in the mature thallus the superficial layer may be small-celled. Such cortex-formation is actually equivalent to the production of branched cell-rows, as is evident from the disposition of the pit-connections. Overlying cortical cells become connected by secondary pits. Cortication by means of septate threads is met with only in few *Polysiphonias*. In *P. Brodiaei*, where they originate from cells cut off in the lower angles of the pericentrals, they combine to form a thick cortex here there is also a formation of hyphae between the central and pericentral cells and such are likewise recorded in *P. elongata* and *P. violacea*.

The germlings of *Polysiphonia* and *Brongniartella* are of the usual type; formation of pericentrals commences early, although a few of the basal cells commonly remain undivided. Older plants are attached by a number of basal rhizoids which unite to form a disc.

Species of *Polysiphonia* are abundant on northern shores, both in the littoral and sublittoral regions. Frequent ecorticate species are *P. urceolata*, with elongate tufts commonly epiphytic on *Laminaria*, and *P.fastigiata*, the dark reddish-brown tufts of which are found on *Ascophyllum*, and occasionally on *Fucus*; the former has 4, the latter from 12-24 pericentrals.

The germlings of *P.jastigiata* are usually found within the pits, from which the branches of *Ascophyllum* originate, and the primary rhizoid penetrates into the medulla without causing any deformation. The pericentrals of the lowest tier produce short hyphae which grow into the thick membrane of the rhizoid and envelop the part external to the host, while at an early stage, the central cell gives rise to one or two horizontal runners which develop further rhizoids providing for additional anchorage. At these points other erect axes, producing secondary endogenous runners, may originate and, since this may be repeated again and again, P. *fastigiata* spreads over its host like a *Rhizopus*. All the rhizoids become corticated in the same way as the first. *P.fastigiata* is usually spoken of as a parasite but, although some of the host-cells are killed and the fucosan-content is diminished near the rhizoids, there is no evidence of actual absorption of nutriment. Rattray has recorded its occurrence on stones.

P. *elongata*, a lithophyte reaching a length of 30 cm., is a widely distributed perennial corticate species with 4 pericentrals and abundant trichoblasts. Other frequent corticate forms are P. *Brodiaei* with up to 20 pericentrals; the former favours rough water. The annual P. *variegata*, with 6 pericentrals, often inhabits polluted water near estuaries, being abundant at Venice and frequent on the roots of Mangroves. In P. *spiralis* the 4 pericentrals are spirally twisted and the same feature is encountered in the dark-coloured P. *nigra*.

Brongniartella byssoides, with 5-7 pericentrals and abundant distichous branching, occurs near low-tide level and in deeper water.

In a number of southern *Polysiphonias*, all with 4 pericentrals, certain of the laterals are of limited growth and alone bear trichoblasts. Such dwarf-shoots are also characteristic of *Bryothamnion,* the species of which inhabit the warmer Atlantic. Here most of the branches, which are produced on often reduced trichoblasts, appear as short spine-like decurrent structures. The principal axes possess a parenchymatous cortex.

A greater departure from the normal type is seen in *Digenea simplex,* which is widespread in warmer seas. A polysiphonous structure is only recognisable in the laterals of limited growth, while the thick cartilagineous axes exhibit a differentiation into medulla and cortex, stated to arise from the fusion of, and subsequent cell-division in, the basal parts of the laterals.

Falkenbergia, a genus of doubtful position and possibly only a stage in the life-cycle of *sparagopsis* is distinguished by the possession of vesicular cells and of only 3 pericentrals, as well as by the origin of the branches from the middle of the youngest pericentral of a segment.

The Lophothalieae, including, apart from *Brongniartella,* the Australian genera *Lophothalia* and *Doxodasya,* as well as *Murrayella* and the Antarctic *Pteronia,* resemble *Polysiphonia* in their general radial construction. *Murrayella periclados* Schmitz commonly grows on the roots of mangroves. Several members of this tribe show features of special interest in relation to the morphological status of the trichoblasts. The majority of the segments produce unbranched, uniseriate laterals, the few cells of which contain chromatophores; some of the basal segments may become polysiphonous. In *Lophothalia* and *Murrayella* the occasional branches replace one of these laterals, but in *Doxodasya bulbochaete* Falkenb and in *Pteronia* they arise from a basal segment of the latter. *Lophothalia* and *Doxodasya* exhibit cortification, the cortical threads in the latter partly arising from the basal cells of the uniseriate laterals.

Schmitz regarded the uniseriate laterals of these genera as trichoblasts. On the other hand, their tendency to become polysiphonous, the development of identical structures adventitiously in certain species of *Lophothalia* and the origin of cortical threads from them in *Doxodasya*—respects in which they differ from the trichoblasts of other Rhodomelaceae—led Falkenberg rather to interpret them as arrested branches. Trichoblasts, which are no doubt structures modified for special functions, must be considered to be homologous with the ordinary polysiphonous laterals of a *Polysiphonia*, etc. *Lophothalia* and *Doxodasya* would then be generalised types in which the morphological differentiation between the two kinds of laterals is not yet fully established.

The persistence of the trichoblasts, and the presence of chromatophores in their cells in many Lophothalieae, warrant our regarding the latter as a primitive group among Rhodomelaceae; Falkenberg derives the Polysiphonieae directly from them by way of *Brongniartella*.

The origin of polysiphonous branches from the trichoblasts in *Brongniartella*, many *Polysiphonias*, and other Rhodomelaceae is probably a secondary condition which has perhaps been determined by nutritive relations. In view of the homology between the two types of laterals and the tendency to transfer the seat of development of laterals of unlimited growth to the basal cells of those of limited growth among the less specialised Florideae, there is nothing remarkable in this condition. The outgrowth, originating the polysiphonous branch, normally arises from the basal segment of the trichoblast, which only under these circumstances produces a lateral and only then develops a polysiphonous structure. The degree of specialisation renders the unravelling of the course of evolution difficult.

Berthold's view that the trichoblasts constitute a protection against intense illumination has not met with much support. Rosenvinge and others are of the opinion that they aid in the absorption of nutriment and possibly also play a role in respiration. Functionally, they no doubt replace the unicellular hairs of other Florideae, as to whose role the same differences of opinion exist. Their frequent restriction to young parts, their common relation to branch-formation, and above all their customary utilisation for the production of sexual organs, all speak in favour of Rosenvinge's views.

Bostrychieae and Rhodomeleae. Some modification of the *Polysiphonia-type,* combined with a marked dorsoventral tendency, is seen in *Bostrychia,* the species of which favour brackish water. Several are characteristic components of the vegetation on the roots of the Mangrove-community in tropical estuaries. *B. scorpioides* is frequent in European salt-marshes, while *B. rivularis* Harv seems to occupy similar habitats in Atlantic North America. *B. Moritziana* has been found, associated with Podostemaceae, in cataracts of the Amakura (British Guiana) far above the estuarine region.

Part of the thallus is usually prostrate, while the growing apices are commonly curved with the concavity towards the substratum, being sometimes even spirally inrolled. The laterals, many of which are of limited growth, arise in two longitudinal rows, inserted on opposite sides of the axis in the erect *B. calliptera* Mont, and in *B. tenella,* but nearer to the dorsal line in most of the prostrate forms; the degree of branching varies greatly. Trichoblasts are wanting.

The number of pericentrals is 6-10 in the older parts, but usually dwindles to 4 in the younger; in most species the final segments remain undivided. *Bostrychia* and its immediate allies are characterised by the fact that the pericentral cells undergo *horizontal* division, usually into 2, but sometimes into 4-6 cells; the lowest member of the row thus formed retains the pit-connection with the central cell. Subsequently a parenchymatous cortex is commonly produced, while cortication by threads is reported in *B. calliptera.* Attachment of the creeping axes is effected by haptera, formed by outgrowth of a group of cortical cells, usually opposite an erect branch. Haptera also commonly develop at the ends of special unbranched laterals which become parenchymatous to the tips and bend over towards the substratum. The early stages resemble those of *Polysiphonia.*

Horizontal division of the pericentrals is likewise seen in

Rhodomela and *Odonthalia* (Rhodomeleae), two widely distributed perennial lithophytes found in the colder seas of the Northern Hemisphere; they also exhibit other signs of specialisation. In *Rhodomela subfusca*, with a richly branched filiform thallus, each segment of the apical cell produces either a trichoblast or an ordinary branch, the successive laterals lying on a spiral of about 1/4 which usually turns to the left, but occasionally to the right. The ultimate branches bear trichoblasts only. The 6-7 pericentrals develop as in *Polysiphonia*, but there is no displacement. The trichoblasts, which are shed at an early stage, contain pigmented chromatophores and, unlike those of most Rhodomelaceae, do not bear the sex organs. Adventitious branches are occasionally formed from cells of the cortex.

Each pericentral divides horizontally, the uppermost cell of the row here retaining the pit-connection with the central cell, while the lowest becomes linked by secondary connections with the pericentrals of the underlying segment. The cells of the primary rows cut off a number of cortical cells each of which again divides horizontally. This process is repeated so that a little way behind the apex there is a 6-7-layered cortex with small peripheral cells, and the polysiphonous structure is obscured.

The radially organised fronds, which arise from a parenchymatous disc, develop mainly in winter and spring. Considerable portions of the plant perish in autumn, only a few of the main axes persisting; from them new shoots arise in the next season, in part from arrested rudiments formed in the previous year. The young plants are erect.

Odonthalia dentata, distinguished by its flattened, distichously branched fronds, forms no trichoblasts. Of the 4 pericentrals, the 2 situated on the future flat faces of the fronds almost immediately segment longitudinally, while the other 2 undergo a single horizontal division. Tangential division, usually more rapid on one side, initiates the wings, while the products of the median pericentrals undergo less extensive division to form an, often faint, midrib. The lower parts of the laterals are congenitally fused with their parent-axes.

Chondrieae and Laurencieae

All outward evidence of a polysiphonous structure is lacking in certain Rhodomelaceae, in which nevertheless a study of the thallus-development behind the apex betrays the essential correspondence with *Polysiphonia*. This is well exemplified by *Chondria* and *Laurencia*. In the rather fleshy thalli of *Chondria* the apical cell is situated at the summit of a narrow cone, which in *C. dasyphylla* is lodged in a deep depression; this is lacking in *C. tenuissima*. The segments are cut off by oblique walls along a spiral and each forms a trichoblast; the 5 pericentrals develop in the usual way. In the lower part of the apical cone, there is a rapid enlargement of the pericentrals, accompanied by formation of secondary pit-connections.

At its base the pericentrals undergo branching to the fourth degree, while they and their products exhibit marked radial elongation, so that there result alternating whorls of laterals which are spatially separated by the enlargement of the central cells. The final ramifications produce the compact surface-layers. The apical depression of *C. dasyphylla* results from the great horizontal elongation of the pericentrals and of their primary branches behind the apex. In the older parts there is marked dilation of the cells which come into lateral contact, with the formation of secondary pit-connections.

The subapical structure of *Chondria* altogether conforms to the *Batrachospermum-type*, while the progressive branching of the pericentrals clearly corresponds with the mode of production of the parenchymatous cortex in *Polysiphonia*, except for the radial elongation and the spatial separation of the cells. *Chondria* and its allies thus serve to demonstrate the close relation between the polysiphonous condition of most Rhodomelaceae and the morphological construction of the primitive uniaxial Nemalionales.

The copiously branched trichoblasts of *Chondria*, which form clusters at the tips are shed before they are carried out of the apical depression; the elongate basal segment, however, remains embedded amid the branching pericentrals. The branches of the thallus arise at the distal ends of these segments in essentially the same way as in *Polysiphonia*. The pericentrals of the second segment of the branch produce hyphae which penetrate into the parent-axis and strengthen the attachment. The early stages resemble those of other Ceramiales.

C. dasyphylla, common near low-water mark in the North Atlantic, is also recorded from the Mediterranean and the Arabian sea. C. *tenuissima* has a greater southward range and is more frequent in warmer seas. Both are annual lithophytes, attached by large holdfasts formed by numerous rhizoids. C. *crassicaulis* possesses short branches containing much starch and probably serving as organs of vegetative propagation. *Acanthophora spicifera. A. Thierii* Lamour, widespread in warmer seas, is characterised by its numerous short spinous branches which arise from the trichoblasts.

The compact thallus of *Laurencia* is solid throughout and the basic structure is discernible only near the tips. The apex is similar to that of *Chondria*, but the apical cell produces three rows of segments. All of these form richly branched deciduous trichoblasts which are arranged spirally. The branches arise from the trichoblasts, in *L. pinnatifida* only from those situated on the flanks so that the branching is distichous. The bases of the older branches become embedded in the peripheral tissue of the axis.

The wedge-shaped segments of the apical cell cut off on either side, beneath the trichoblast and no more are produced. The cortex is formed in much the same way as in *Chondria*, but the central cells, which become transversely extended, are clearly distinguishable only near the apex and there is no appreciable extension of the outer cells. During the abundant cell-division at the base of the apical depression the basal segments of the trichoblasts become drawn out into long and narrow threads. The mature parts display little differentiation. The early development follows the usual course.

L. pinnatifida, a lithophyte frequently found in the lower part of the intertidal region, has pinnately branched cartilagineous thalli, with the larger branches often flattened. The fleshy and parenchymatous attaching disc is full of starch and produces a number of erect shoots, the basal parts of which commonly last for several seasons, the new season's growth being often readily distinguishable from the old. A similar condition is reported in other species. In *L. obtusa*, an epiphyte with cylindrical thalli, more frequent in warmer seas, the branches of higher orders tend to appear whorled owing to the restriction of branch-formation to occasional groups of trichoblasts.

Rhodomelaceae with Bilateral and Dorsoventral Construction

A considerable number of Rhodomelaceae, while retaining the tier-like structure of a *Polysiphonia*, exhibit specialisation in the acquisition of bilateral or dorsoventral symmetry. Examples of this trend have already been seen in the bilateral habit of *Odonthalia* and the dorsoventral construction of many *Bostrychias*, but these genera are closely allied to radially organised types. Among the Pterosiphonieae and Herposiphonieae, however, these tendencies become paramount and, combined with other special developments, culminate in a number of exceptional types. Neither light nor gravity appear to play a role in determining dorsoventrally within this family.

Pterosiphonieae and Herposiphonieae. The species of *Pterosiphonia* are small forms favouring the warmer seas. The principal axes are prostrate and attached by rhizoids, but their younger parts become erect and show distichous branching, combined with a tendency to flattening so that a pinnate habit results. There are no trichoblasts. The branches arise from alternate segments on the erect shoots, at wider intervals on the creeping ones; these latter branches may at first be prostrate. Most species are devoid of a cortex.

The branch-forming segments are markedly oblique, and the septum cutting off the initial arises before there is any marked protrusion and in such a way that a considerable area of wall is common to the initial and the apical cell. The first septa in the initial abut on this common wall and, as a result, a number of the basal segments of the branches are contingently fused with the parent-axis; in the region of fusion the lateral develops pericentrals only on the abaxial side. The congenital fusion is more marked in P. *complanata*.

The dorsoventral organisation of the Herposiphonieae is exemplified by *Herposiphonia*, the species of which are small epiphytes of the warmer seas. In *H. tenella* the long axes are prostrate and attached by unicellular rhizoids, the apex in growing plants being coiled, with the convexity towards the substratum. The branches, one generally produced from each segment, arise in strict acropetal succession and stand in four rows. The two occupying the flanks develop into prostrate long shoots, while those situated on the dorsal side give rise to erect unbranched laterals of limited growth; as a general rule three successive dwarf-shoots are followed by a long shoot situated on the same side as the last dwarf-shoot of the preceding segment, but occasional irregularities occur.

The few trichoblasts are restricted to the apices of the dorsal laterals and as usual bear the sex organs. According to Goebel vegetative reproduction takes place by the dying away of the older parts of the principal axes. Berthold has studied the light-orientation of this species.

Dipterosiphonia, the species of which occur in warmer seas is essentially similar, but the prostrate and often flattened fronds have straight apices. The laterals, which arise in pairs on alternate sides of the parent-axis, are entirely prostrate in *D. dendritica* and other species. In the Australian *D. heteroclada* the dorsal laterals take the form of more or less curved spines.

The congenital fusion seen in *Pterosiphonia* is more pronounced in other Pterosiphonieae and in certain

Herposiphonieae. A good example is furnished by *Symphyocladia marchantioides*, known only from New Zealand, in which the flat prostrate thallus is composed of two layers of pericentrals with an intervening row of central cells; the latter appear as veins traversing the thallus. The one-layered margin is occupied by a row of initials and their segments. The former *(1-5)* belong to successive laterals, produced alternately to right and left from every second segment and congenitally fused throughout their length with their parent axis; the arrangement of the laterals is clear from the disposition of the branch veins. The lobes of the thallus are formed by outgrowth of laterals of the second order. The individual axes occasionally become free towards the end of their growth, their apical cells still cutting off a few segments, each of which produces a trichoblast.

The Australian *Pollexfenia pedicellata* has a similar habit. The growing margin resembles that of *Symphyocladia* although each segment produces a branch; there is the same congenital fusion. The veins of the thallus, again constituted by the central cells, show clearly the sequence of branching. *Pollexfenia* has 4, *Symphyocladia* 6-8 pericentrals. Another variant on this type of construction is afforded by the handsome *Dictymenia*.

In *Pollexfenia,* the laterals are arranged in four rows. Two, which are situated on the flanks develop vigorously and join to form the flat thallus, while those arising on the dorsal and ventral surfaces, which are not included in the congenital fusion and may remain arrested for a time, appear as short polysiphonous branches of the usual type; they produce a richly branched trichoblast from each segment and bear the reproductive organs. In *P. crispata* the dorsal and ventral laterals are recognisable only near the growing margin, and the reproductive organs are borne on the free tips of the elsewhere fused branches. The prostrate part of the thallus of P. *pedicellata* is attached by massive haptera formed from the ventral pericentrals.

While *Symphyocladia* and *Pollexfenia* to all intents and

purposes exhibit the bilateral symmetry of the simpler Pterosiphonieae, *Placophora*, a parallel form among Herposiphonieae, shows a dorsoventral organisation. The thallus of *P. Binderi*, an epiphyte on *Codium* reported from South Africa and Peru, agrees in its marginal growth and congenital fusion with the corresponding forms among Herposiphonieae. Each segment produces a branch, and the laterals arise in pairs, alternately to the right and left. The primary laterals first divide by walls parallel to the margin and the resulting segments produce no branches, which as usual are formed as a result of later oblique septation. Occasional branches may sometimes project above the surface. The segments form 5 pericentrals, 2 on the ventral and 3 on the dorsal surface some of the former growing out into elongate rhizoids.

The reproductive organs of *Placophora* are formed on tufts of free branches which appear between the lobes of the thallus; those bearing-sporangia may consist of 30-50 segments with 8-9 pericentrals, while the sex organs arise on trichoblasts borne in a single dorsal row on the fertile branches. In *Amplisiphonia* tetrasporangia are formed in modified erect-growing marginal lobes of the thallus which possess 6 pericentrals.

The germlings of *Pollexfenia* and *Placophora* possess an erect, radially organised, polysiphonous axis from the basal segments of which the flat thalli arise; in *Placophora* their origin is endogenous. The primary axis of 30-45 segments bears as its only branches a few apical trichoblasts. It can hardly be doubted that in *Pterosiphonia*, *Symphyocladia*, and other prostrate forms among Rhodomelaceae a similar origin of the prostrate shoots will be established. Despite the extreme modification which many of the genera just considered exhibit, their early development shows that they are derivatives of erect-growing radially organised *Polystphonia-likt* types. It may not be amiss to emphasise that *Symphyocladia* is essentially a *Pterosiphonia*, *Placophora* a *Dipterosiphonia*, modified by the extreme congenital fusion of the bilateral branch-systems.

Polyzonieae. Considerable morphological specialisation, combined with dorsoventral organisation, is attained in a different way among the Polyzonieae, most of which are inhabitants of the Southern Hemisphere, while several are peculiar to Australasia. A common feature is the endogenous origin of the branches of unlimited growth. Several of the extreme forms show resemblance to leafy liverworts.

There are several creeping dorsoventral Rhodomelaceae which show points of contact with the Herposiphonieae, but resemble the Polyzonieae in their endogenous branching. Examples are furnished by *Lophosiphonia Ctenosiphonia, Ophidocladus,* and *Stictothamnion.* Most of these are inhabitants of the warmer parts of the Atlantic; *Lophosiphonia obscura.* Falkenb is known from the south coast of Britain and *Ctenosiphonia hypnoides* is reported from Guernsey. The laterals of limited growth, for the most part unbranched, grow erect. They are sometimes borne in two rows on the flanks, but commonly in a single dorsal row; those of *Stictothamnion* are fusiform. In *Lophosiphonia* long and dwarf shoots are not sharply differentiated and some of them *(p)* bend down to the substratum. In *Ophidocladus* shoots of unlimited growth arise from opposite flanks of the same segment, although one or other often remains arrested. Trichoblasts are usually confined to the tips of the erect laterals, those of *Stictothamnion* being unbranched. In *Falkenbergiella* they are formed only on sexual plants. Although the exact affinities of these genera, which for convenience may be grouped as Lophosiphonieae, are difficult to assess, they help to emphasise the relationship between Herposiphonieae and Polyzonieae.

Euzoniella and *Polyzonia* comprise a number of small epiphytes of the Southern Hemisphere. Their creeping thallus arises exogenously from the erect primary axis which, in *Polyzonia,* is winged as a result of tangential division of opposite pericentrals. In both genera, the prostrate laterals arise from the flanks of the creeping axes, those of limited growth exogenously from every second segment, the later-formed long shoots endogenously from certain intervening segments. The disposition of the two kinds of laterals is much as in *Dipterosiphonia*. In *Polyzonia* the long axes develop a ventral keel.

The laterals of limited growth bear short branches on the dorsal side, and in *Euzoniella adiantiformis* these are congenitally fused to form leafy expanses. In *Polyzonia elegans*, on the other hand, the foliose form of the dwarf-shoots is due to the production of one-layered wings by tangential division of the single ventral and of one of the two dorsal pericentrals; the branches appear as teeth along the margin. This epiphyte is attached by massive haptera, like those of *Leveillea*.

The ventral wing seen in *Polyzonia* is more markedly developed in *Cliftonaea*, where the principal axes mostly grow erect and possess strongly inrolled apices. The laterals of limited growth form two dorsal rows and, like those of *Polyzonia*, possess only 3 pericentrals, whereas the long branches with 6 pericentrals arise on the flanks. The ventral wing is produced by tangential division of the median pericentral, while less strongly developed dorsal wings formed from the dorsolateral pericentrals, link up the bases of the laterals of limited growth. In C. *pectinata* these wings are less pronounced than in C. *Lamourouxii Harv*, where the ventral pericentral of the short laterals likewise forms a wing. In this species these laterals bear a row of dorsal trichoblasts, whilst in C. *pectinata*, they are confined to the sexual shoots. *Polyzonia* and *Cliftonaea*, though presenting points of analogy with *Dtpterostphonta* exhibit pronounced dorsoventrally in the development of the ventral wing.

The most remarkable member of the Polyzonieae is *Leveillea jungermannioides*, widely distributed in the Indian Ocean and recorded also from the Persian Gulf and Arabian Sea. Here the laterals of limited growth, arising from the creeping axes take the form of delicate "leaves", which are attached by a broad base and overlap one another by their posterior edges so that there is a marked resemblance to a Jungermanniaceous liverwort. The creeping shoots again arise exogenously from the erect primary axis, which is winged like that of *Polyzonia*.

The branches of the creeping axis arise in pairs from the basal segments of two consecutive laterals of limited growth, so that each later appears in the angle between the parent-axis and the front margin of the "leaf"; many remain undeveloped. They possess 7 pericentrals, the two ventral ones forming a slight keel.

The short laterals arise as cylindrical outgrowths within the inrolled apices of the long shoots, as usual from alternate segments. They have 3 pericentrals, the dorsal first-formed one occupying the abaxial side of the rudiment. The foliar form results from active division of the ventral pericentrals, while the broad and oblique attachment is attained by division of the cells of the adjacent segments of the axis, although no continuous wing is formed. Occasionally the apices of these laterals remain cylindrical and produce a double row of dorsal trichoblasts. Attachment is effected by massive haptera, formed by the outgrowth of small cells cut off from the ventral pericentral of two consecutive segments.

Amansieae: A different type of dorsoventral development, often combined with wing-formation in the plane of branching, is seen in the Amansieae. The fronds here grow erect, which shows that dorsoventral organisation in Rhodomelaceae is not definitely linked with a creeping habit. The fronds usually arise in tufts from a common basal disc. The early development is not known for any representative of this group, but an erect polysiphonous primary axis is to be suspected.

Halopithyspinastroides are, however, richly branched Atlantic and Mediterranean perennials. The axes, which are cylindrical in the former, but slightly flattened in the latter, have inrolled apices and bear a row of dorsal trichoblasts on the younger parts. The two rows of endogenous laterals are produced in pairs from two successive segments and form an acute angle with one another; the first two secondary branches in *Halopithys* arise close to the parent-axis. The older parts develop an extensive parenchymatous cortex. Like other Amansieae, these two genera possess 5 pericentrals, the two first formed lying on the convex (dorsal) side of the inrolled apex. A narrow two-layered wing is formed in *Rytiphloea* immediately behind the apex by division of the dorsal and dorsolateral pericentrals. Such a wing is more strongly developed in other Amansieae, as for instance in *Amansia,* where the cells of the two layers alternate and those of adjacent segments become linked by secondarypit-connections. In *A. multifida* the endogenous laterals emerge in two alternating rows from between the two layers of the broad wings, the first segment of the lateral extending to the edge of the wing and producing only a dorsal and a ventral pericentral. The wings on these laterals lie in the same plane as those of the parent-axis and are continuous with the latter. In later stages exogenous adventitious fronds arise irregularly along the median line of the main axis. *A. multifida* thus represents a further elaboration of the condition found in *Rytiphloea* and both differ essentially from *Halopithys* only in the development of wings.

The mature plants of *A. glomerata* present an appearance very different from that of *A. multifida.* A robust terete stalk bears on one side a number of dense leafy rosettes, consisting of flattened branch-systems which are placed edgewise. These "leaves" have the same structure as the thallus of *A. multifida,* although with broader wings and shorter laterals. The young plant closely resembles one of these "leaves" and from it the later ones arise exogenously, being homologous with the adventitious laterals of *A. multifida.* The rosettes are sympodia, each member of which arises adventitiously from the base of the preceding one. When a "leaf" originates, the adjacent cells of the axis develop a cortex, a process which advances basipetally and leads to the production of a thick rib overlapping onto the wings. The peripheral parts of the latter slowly disorganise and in this way the terete stalk is formed. The two species of *Amansia* are widely distributed in warmer seas.

Diverse other Amansieae show much the same structure as *A. multifida.* In *Lenormandia* the wing is one-layered owing to dovetailing of the two series of cells and there are no endogenous branches, secondary fronds arising exogenously and adventitiously from the margin. *L. marginata* and other species are known only from Australia. *Osmundaria proltfera* Lamour is distinguished by the numerous small adventitious shoots covering the surfaces of the flat cartilagineous fronds. *Kutzingia* differs from other Amansieae in the possession of 6 pericentrals, the one-layered wing being formed by division of the lower part of the lateral pericentrals and exhibiting cortex-formation at an early stage. The older parts develop a coarse midrib which remains as a stipe after the wings have worn away. The endogenous laterals are opposite, but many of them remain rudimentary.

Vidalia again has a similar structure, although more robust owing to the formation of a -few-layered cortex. Here several of the basal segments of the laterals, which appear as teeth along the margin, are included within the wings of the parent-axis. The frequent spiral twisting is caused by wing-formation taking place also on the embedded parts of the laterals. Branching of the fronds is either due to further development of certain laterals or to adventitious branches. The reproductive organs of the Australasian *V. spiralis* are formed on minute adventitious branches, while those of *V. volubilis*, occurring in the warmer seas of the Northern Hemisphere, are developed on certain more vigorous endogenous laterals.

Like *Halopithys* and *Rytiphloea*, both *Amansia* and *Vidalia* develop near the inrolled apices a single dorsal row of richly branched trichoblasts. It is characteristic of diverse species of the last three genera that, until the trichoblast is fully developed, the component cells remain small and constitute a compact pyriform group enveloped by a "cuticle"as the cells elongate, the latter is ruptured, though parts of it persist at the base.

Dasyaceae

The few filiform genera belonging to the Dasyaceae are essentially distinguished from the Rhodomelaceae by the sympodial branching of the principal axes. A lateral, arising either from the basal or from the second or third segment of the previous generation, continues the growth of the axis, while the upper parts of the successive units of this sympodium are deflected to one side and develop as branched uniseriate apparent laterals of limited growth; although different in origin, the latter resemble the pigmented trichoblasts of certain Rhodomelaceae. In *Dasya* the pseudolaterals are spirally arranged, but in *Dasyopsis* and *Heterosiphonia* they are distichous and the plants have a pinnate habit.

The pseudolateral usually produces a branch adaxially from every second segment, while in *Dasya* and *Dasyopsis* the laterals of the second order are abaxial; all the branches of the pseudolateral lie approximately in the same plane. The sympodial axes develop a polysiphonous structure throughout, while in *Heterosiphonia* the three basal segments of the pseudolaterals likewise become polysiphonous. *Dasya* and *Dasyopsis* possess 5 pericentrals, which develop in a sequence different from that of Rhodomelaceae, the first to one side of the pseudolateral, the second just below it, the third to the left of the second, and so on; the fifth is small and sometimes suppressed. *Heterosiphonia* in which the number of pericentrals varies, on the other hand, for the most part develops them in the same way as in Rhodomelaceae.

In *H. Berkleyi* Mont., a widely distributed Antarctic form, the larger axes are flattened owing to great enlargement of two opposite pericentrals. The branches of unlimited growth are for the most part constituted by the first branch of a pseudolateral. Adventitious laterals, especially plentiful in *Dasya arbuscula*, originate from the upper ends of the pericentral cells and sometimes also from the cortex; they usually remain monosiphonous.

In many species, branched cortical threads, emerging from small cells cut off obliquely from the lower ends of the pericentrals and linked by pit-connections both with one another and with the latter, give rise to a more or less thick cortex; in *Heterosiphonia* they grow both upwards and downwards. Such threads also arise from the basal cells of the pseudolaterals and of the adventitious branches. In *Dasyopsis* the mother-cells of the cortical threads become very large and penetrate between the pericentrals so as to obscure the primary structure. This accounts for the frequent denial of the presence of pericentrals in this genus.

The early development of *Dasya* is essentially like that of *Callithamnion*. Branching sets in early, but both primary axis and branches remain uniseriate. The mature thallus arises at a later stage from the original spore-body.

The species of Dasyaceae are largely sublittoral forms which favour warmer seas, although a number are to be found in temperate waters. *Dasya arbuscula* is probably the most widely distributed British species, while the much larger *D. pedicellata* Ag. is found on the warmer Atlantic shores of Europe and North America, as well as in the West Indies and the Mediterranean. A common North European form is *Heterosiphonia plumosa*. *Dasyopsis plumosa* is known from the North Pacific.

The Dasyaceae also include *Dictyurus* and *Thuretia*, which possess reticulate thalli, comparable to those of Delesseriaceae; the former is found in the warmer parts of the Atlantic and Indian Oceans, the latter in Australian seas. In *Dictyurus*, a creeping system gives rise to terete upright axes, each branch of which bears an elaborate sack-like network in the form of a spiral veil. The basic structure is like that of other Dasyaceae. The sympodial axes possess 4 pericentrals two segments to each joint, and the pseudolaterals are distichously arranged; they soon drop off on the lower parts of the erect axes which become invested with a thick cortex.

In the region of the net each pseudolateral branches abundantly in the horizontal plane so as to occupy about half the circumference of the axis; the first ramifications lead to the formation of four sympodial branch-systems, their uniseriate pseudolaterals tending to be pinnately disposed. Net-formation commences, before there is any wide separation of the individual branches, by the abundant outgrowth from the cells of adjacent threads of processes which grow towards one another and fuse at their tips. Such connections are established not only between the diverse laterals of each sympodial branch-system, but also along the margins, where two such systems meet. The close spiral thus constituted is drawn out as the axis stretches, but, prior to this, the upwardly curved tips occupying the edge of each branch-system become connected with the under surface of the next higher coil of the spiral. The network thus encloses an uninterrupted spiral cavity.

In *Thuretia* there is only one segment to each joint of the sympodium and, in the principal axes, the polysiphonous structure is soon obscured by cortication. In *T. quercifolia* the lowest lateral of the distichous primary pseudolaterals gives rise to a secondary sympodium bearing numerous uniseriate pseudolaterals adaxially in two rows. Successive secondary sympodia follow closely upon one another so that their laterals project over those of the next youngest sympodium, with which connections are established at fairly regular intervals in the same way as in *Dictyurus*. The elaborate network thus constituted consists of two or more interconnected systems. Occasional secondary sympodia develop into the lobes of the flat, rather spongy nets of *T. quercifolia,* which may reach a length of 20-25 cm. T. *teres* Harv. has cylindrical nets.

Brief mention may be made of *Halodictyon*, a genus of uncertain position, with a number of species in warmer seas. The plants consist of a branched network composed of cylindrical cells which are free at the margins; a polysiphonous structure appears only in the fertile parts.-The apex of the net is occupied by a small-celled meristem, the segments of which rapidly lengthen and separate. *H. mirabile* Zanard. occurs in the Mediterranean.

The typical Dasyaceae have often been regarded as the most primitive of the Rhodomelaceae, while Oltmanns looks upon them as specialised forms and seeks the origin of Rhodomelaceae among *Antithamnion*-like types. Rosenberg concludes that the Dasyaceae have more in common with the Delesseriaceae than with the other two families of Ceramiales. There is much that speaks in favour of Falkenberg's view, since the great resemblance of the pseudolaterals to the pigmented trichoblasts of the less specialised Rhodomelaceae cannot be denied; moreover, the origin of polysiphonous branches from the pseudolaterals recalls their development from the trichoblasts in many Rhodomelaceae. On the other hand, it is difficult to regard the sympodial structure of Dasyaceae as primitive, nor are any traces of it to be found in Rhodomelaceae. The facts indicate that the former constitute an independent evolutionary line from an ancestry, common with that of the other three families of Ceramiales.

Parasitic Florideae

The numerous epiphytic Florideae are for the most part only attached superficially to their substratum. A certain number, however, exhibit a more intimate relation, either in their restriction to definite hosts or in the possession of an endophytic attaching system, which penetrates more or less deeply into the substratum and often causes some destruction of the adjacent cells; this is well exemplified by *Polysiphoniafastigiata*. There is, however, little evidence of actual parasitism in such Florideae which are clearly well equipped for a holophytic existence, and the uptake of nutriment, if any, cannot be considerable, Tobler points out that P. *urceolata* is found on parts of *Laminaria,* where there is an accumulation of organic detritus, and takes this as an indication of saprophytism.

Better instances of parasitism are afforded by those Florideae which are not only deeply anchored in their host, but also exhibit vegetative reduction as compared with their immediate allies. Examples are afforded by *Ceratocolax, Ricardia* and *Janczewskia*. *Ceratocolax Hartzii* forms branched pink-coloured tufts on *Phyllophora Brodiaei.* The degree of penetration of the host varies; sometimes the medulla is invaded and some of its cells killed, whilst in other instances the parasite merely spreads beneath the "cuticle", sending haustoria into the walls of the underlying cells. Greater reduction is seen in *Callocolax neglectus* Schmitz, which forms minute white growths on *Callophyllis lacmiata*.

The mature individuals of *Ricardia Montagnei*, a close ally of *Laurencia* upon which it is usually found, appear as elongate or pyriform, bright red, hollow vesicles of about the size of a pin's head; they have a several-layered wall and are anchored by an elongate rhizoid which penetrates deeply into the tissues of the host. The spores germinate within the apical depression of the latter, and the germlings attain a certain size before the rhizoid begins to pierce the underlying tissue. At the four-celled stage a three-sided apical cell is cut out on one side of the top cell, the rest of which gives rise to a trichoblast. The underlying cells form pericentrals and produce the basal stalk, the lowest cells of which may grow out into rhizoids. The mature vesicle is formed by the apical cell, which divides like that of *Laurencia*, each segment producing a trichoblast; secondary vesicles may arise from the lower cells of the stalk. In cultures, the germlings fail to develop beyond a certain stage, and this is regarded as evidence of incapacity for a further autotrophic existence. A comparable degree of parasitism is exhibited by the Antarctic *Colacodasya inconspicua*, which is parasitic on species of *Heterosiphonia*.

Most species of *Janczewskia* occur in temperate seas as minute cushions on *Chondria* and *Laurencia*. The usual colour of *J. verrucaeformis* is pale yellow and the individuals are probably for the most part heterotrophic; several species are practically white. The cushions consist of a number of short, more or less coalescent fronds, having a similar structure to those of *Laurencia* and arising from a system of endophytic filaments, which are probably formed first; their cells are stated to be joined by pit-connections with those of the host. The latter does not appear to be weakened, although there is often a sharp bend at the point of attack.

Gonimophyllum, parasitic on *Cryptopleura* and other Nitophylleae, shows similar features. The minute, pale pink leaflets arise in tufts from the hypertrophied tissue of the host, within which lie the endophytic threads. Both here and in *Janczewskia*, although the capacity to produce some photosynthetic pigment is not altogether lost, the usual pale colour, the minute size of the individuals, and the deep penetration of the endophytic system, imply a considerable degree of parasitism. In *Polycoryne*, closely allied to its host *(Myriogramme)*, the thallus is needle-like.

A complete absence of photosynthetic pigments is found in *Choreocolax* and *Harveyella*, which appear to be cold-water forms of the Northern Hemisphere and are still of doubtful systematic position. The cushions which form the visible part of the parasite and which are about as big as a cystocarp of the host, are white or brownish when old.

Choreocolax polysiphoniae is found not uncommonly on *Polysiphonia fastigiata*, at the points of branching and on the young antheridial branches. According to Richards plants thus attacked are paler and less vigorous.

The commonest species of *Harveyella* is *H. mirabilis* which is parasitic on *Rhodomela subfusca*; a second species, *H. pachyderma*, parasitic on *Gracilaria confervoides*, is referred by Sturch to a distinct genus, *Holmsella*, owing to differences in reproduction. In both the endophytic filaments traverse the middle lamellae of the host and their cells establish secondary pit-connections with those of the latter *(ho)*, whilst in *Choreocolax* direct connection is made with the pits of the host. Actual penetration of the cells is infrequent. In *Harveyella mirabilis*, the parts attacked hypertrophy and some of the resulting cells may become incorporated in the tissue of the parasite. In *Choreocolax*, the endophytic threads sometimes give rise to secondary cushions.

These genera appear to be multiaxial, the medulla consisting of branching filaments, while the small peripheral cells of the cortex are arranged in rows. The cushions are covered by a gelatinous envelope which, in *Holmsella pachyderma*, is as thick as the cortex. Rosenvinge records probable leucoplasts in the cells of *Harveyella*.

Little is known about the early development. The first growth in *Choreocolax* is stated to be external, the endophytic filaments arising secondarily. Sturch, on the other hand, suggests that the spores of *Holmsella* obtain access to the interior of the host through the crevices left after the escape of its spores. According to Chemin the spores of *Harveyella mirabilis*, when allowed to germinate on glass slides, develop phycoerythrin and produce a circular disc with radiating peripheral threads which are supposed to penetrate the host. This requires confirmation.

The highest degree of parasitism is probably realised in forms like *Colacopsis* and *Stromatocarpus* as well as *Choreonema* in which the entire vegetative system is endophytic and the fructifying fronds alone appear above the surface of the host. *Stromatocarpus* is found on *Polysiphonia virgata* at the Cape. In both the Rhodomelaceous genera the polysiphonous reproductive shoots contrast markedly with the uniseriate endophytic threads.

The vegetative parts of *Choreonema thureti Endosiphonia thureti* Ardiss are completely enclosed in the tissues of *Corallinas* in which no kind of deformation is produced. The presence of the parasite becomes evident only when the superficial conceptacles are formed. The branched endophytic threads consist of elongate cells which cut off two or three small elements; these have been regarded as arrested branches or as comparable to the cover-cells of *Melobesia*. In *Melobesia deformans* Solms-Laubach; *Chaetolithon deformans*, which attacks Southern species of *Corallina*, the infected segments become irregularly branched, and their tissue may proliferate to such an extent that only the apertures of the conceptacles of the parasite are left uncovered.

It is noteworthy that, among the more definite instances of parasitism among Florideae, host and parasite usually belong to the same order, often indeed to the same family. This specialisation is striking in view of the usual growth of seaweeds in dense communities, where a diversity of hosts is available. The frequent close affinity between the two organisms perhaps also accounts for the slight degree of damage to the host, since there is no good evidence of any marked mortality as a result of the parasitic attack.

Diverse instances are known of association between Florideae and sponges. One of the most striking is furnished by *Ceratodictyon spongioides; Marchesettia spongioides* Hauck, a member of Gracilariaceae found in deep water in the Mediterranean and in the Indian and Pacific Oceans. The terete branches of the cartilagineous thallus anastomose to form a network, which is completely enclosed in the body of the sponge; the latter also occupies the meshes the fructifying branches of the algae alone projecting beyond the surface. Although the sponge can lead an independent existence, there is so far no evidence that the seaweed can do so.

Further examples are afforded by species of the Southern genus *Thamnoclonium*. In *T. Treubii* the axes with their many short branches are entirely overgrown by the sponge. The superficial cells of the algae here produce unicellular processes, giving rise at their tips to chains of rounded cells the outermost of which become detached and enlarge within the matrix of the sponge; the detached cells contain a large chromatophore and grains of starch, while their vacuole is often occupied by a crystalloid.

Gall-like growths, which in part at least are possibly due to Bacteria, are not uncommon on certain red seaweeds. Such are the globular or irregular swellings, with or without pigment, found in *Cystoclonium purpureum* and the spherical or hemispherical galls occurring on the older fronds of *Ahnfeltia*; in the latter they consist of cortical tissue only, which undergoes progressive disorganisation from within outwards and, according to Chemin, Harvey shows such a gall in section. In *Cystoclonium*, Rosenvinge found no Bacteria within the galls, but *in Chondrus crispus,* where they cause perforations in the fronds, Chemin states that another French worker has succeeded in preparing bacterial cultures from them and in causing gall-formation by inoculation with the latter or with an extract of the tumours. The galls of *Rhodymenia palmata* are due to Copepods.

Chondrus is often attacked by the Pyrenomycete *Didymosphaeria marina* Land. Diverse Chytridineae and Phycornycetes have been reported on Florideae.

Special Morphological and Biological Features

Light-reflecting Bodies: Diverse Florideae, mainly inhabitants of warmer seas, show a blue or green iridescence when exposed to strong light, which is due to reflection of the shorter waves by bodies situated within the peripheral cells. The phenomenon is shown by *Chondria caerulescens* and certain other Ceramiales, as well as by some Rhodymeniales; it is not recorded in the northern species of the latter. The bodies causing iridescence are usually situated in the vacuoles and often in the main consist of protein *(Callithamnion,* Delesseriaceae), although those of *Laurencia* are stated to be of the nature of tannins. In bright light the bodies in question are opposed to the outer walls, while the chromatophores occupy the others; on transference to diffuse light the relative positions are interchanged, the movement of the chromatophores being accompanied by amoeboid changes of shape.

In *Chondria caerulescens* the light-reflecting bodies consist of aggregates of yellowish granules which are readily soluble in potash and slowly dissolve in fresh water. Those *of Laurencia* appear as large spheres which are sometimes confluent. For the most part, how ever, the bodies responsible for the reflection of the light are single, highly refractive masses, yellowish or milky white in colour and exhibiting a definite structure. Thus, in *Chylocladia kaliformis* and *Gastroclonium clavatum* they are lamellated, in the former with numerous small spherical bodies between the lamellae. Berthold regarded the spheres as the actual reflectors.

The light-reflecting bodies of Delesseriaceae have a similar structure. In strong light they become more and more opaque owing to an increase in the quantity of the spheres; in weaker light, on the other hand, the lamellate structure again becomes recognisable and by degrees the masses break up into amoeboid portions which pass to the side walls. The spheres have been regarded as photosynthetic products laid down in a proteinaceous ground-mass. The iridescent bodies would, however, seem to be of diverse character and further study is necessary to elucidate their chemical nature.

In *Callithamnion* the globular iridescent bodies are usually several in number, while in *Seirospora* they are generally single; in neither instance do they show any structure. Iridescence, due to granular contents in the subepidermal cells, is also recorded in *Scinaia furcellata*, while the strongly refractive spherical bodies in the surface-cells of *Sphaerococcus coronopifolius* and *Plocamium coccineum* appear to fulfil a similar light-reflecting rdle.

The possession of iridescent bodies has usually been regarded as ensuring protection against the injurious effects of strong insolation by partial reflection of the light. Several iridescent Mediterranean species are, however, inhabitants of grottoes and other shaded habitats where such an explanation is inadequate. This suggests that the function of these bodies may vary.

Vesicular Cells

Some Florideae bear, especially on the younger parts, small cells with a scanty protoplast and a prominent vacuole with highly refractive, homogeneous, and generally colourless contents. In certain species these vesicular cells are rich in iodine (iodine-cells; "ioduques" of Sauvageau, the presence of which is shown by the blue colour assumed by starch-containing paper on contact with the seaweeds in question. Such cells are coloured deep brown by cyanin, which stains the ordinary cells blue. They are very characteristic of the Bonnemaisoniaceae and are also met with in certain Ceramiales many of the species involved occur in Southern seas.

The distribution of the vesicular cells has already been described in *Bonnemaisonia asparagoides*, where they show a blue fluorescence when viewed in reflected light, and it appears to be similar in *Asparagopsis*. Those of *Trailieila intricata* are small triangular structures situated at the top of each segment. In *Falkenbergia* they are cut off on the upper side of the young pericentral, but, as the latter enlarges, they come to lie between it and the central cell so that they were long regarded as vacuoles of the elongate pericentrals the vacuoles in the iodine-cells of *Asparagopsis armata* are similarly stated to be actual cells. The tetraspores and germlings of *Falkenbergia* contain no iodine, although vesicular cells appear at an early stage.

The iodine-cells contain an obvious nucleus while a few chromatophores may at first be present. Sauvageau was of the opinion that free iodine was contained in these cells, but the available evidence speaks for the presence of a compound. If material is placed in starch-solution with a little acetic or hydrochloric acid, the vesicular cells become surrounded by a blue halo, which is deeper in the presence of nitrites; this reaction is not obtained with young or old cells, nor at all times of the year, while in *Bonnemaisonia* it only appears if nitrites are present. The older vesicular cells collapse and liberate iodine without treatment with acid, which is believed to be due to oxidases. These occur in considerable quantity in certain Red Algae.

The seaweeds under discussion contain considerable quantities of iodine, in *Trailieila* about 0-5 per cent and in *Bonnemaisonia asparagoides* about 0-3 per cent of the fresh weight, whereas the amount in other Rhodophyceae is much smaller. Kylin is of the opinion that the ordinary cells contain alkali iodides, the presence of which is shown by the reddish needles appearing after treatment with cresyl blue, whereas the vesicular cells merely assume a greenish tint, ascribed to the liberation of free iodine. The function of these cells is still problematical, but there is some evidence that forms possessing them are avoided by animals; moreover, the excretion of iodine on the collapse of the older cells may prevent colonisation by epiphytes.

Vesicular cells, which do not contain iodine, occur on the branchlets of certain species *of Antithamnion*. In the mature cell the protoplast forms an apical cap, while the contents of the vacuole are coloured blue by cresyl blue. The osmotic value is higher than that of the vegetative cells. Sauvageau concluded that these cells contained free bromine but this has been disputed by Kylin, according to whom very little of this element is present. Similar vesicular cells are under the name of *Spermothamnion roseolum*.

Tendrils

Diverse Florideae possess curved, sickle-shaped laterals which, after contact with another branch of the same or of a foreign algae, can form one or more firm coils around it after the manner of a tendril. In these laterals the outer tissues are often more markedly developed on the convex face and branches, which commonly remain rudimentary till contact is established, are for the most part confined to that side; in *Hypnea* and *Acrosorium uncinatum* the cells on the two sides differ in size. The tendrils of *Cystoclonium* are long thin twisted structures, which can form 4-5 coils around the support, but more usually the curved tips are thicker than the part below. The tendrils of *Asparagopsis hamifera* are laterals of limited growth and are richly stocked with reserve materials.

In *Hypnea* and in *Calliblepharis lanceolata*, where certain of the long marginal proliferations act as tendrils, the tip curves after contact, thus decreasing the amplitude of the hook, while the apex resumes growth and forms a number of coils. Subsequently the cells on the inner surface elongate into attaching rhizoids, which are often thick-walled and commonly cohere in groups; the attaching system thus formed occupies all the surface-irregularities of the support. Nordhausen concluded that inrolling of the tendril and production of rhizoids is the result of a definite contact-stimulus. If a foreign algae is encircled, the latter exhibits no response, but when the tendril grasps another branch of the same algae both structures form rhizoids at the surface of contact and fusion ensues.

After contact the branches on the convex surface of the tendril develop and, in *Acrosorium uncinatum*, the hitherto small lobes upon them enlarge to normal dimensions. Subsequent rupture of the part behind probably often results in vegetative propagation. The large cells of *Griffithsia* can put out rhizoids which, curving round other branches or adjacent objects, give rise to new shoots which become detached.

Recurved branches, representing laterals of limited growth and aiding in secondary attachment and vegetative propagation, are found in *Asparagopsis armata*, although these structures are not capable of twining. After contact with a foreign body the cells at the tip grow out into long rhizoids which form an adhesive disc. The lateral of unlimited growth, situated opposite the hook, develops vigorously after contact, and this is often accompanied by detachment of the whole system.

Fusions between branches of the same or of different species of Red Algae are not infrequent in forms lacking definite tendrils. It is again associated with rhizoid-development, which is reciprocal only when parts of the same algae are concerned. Such phenomena have been reported in *Calliblepharis ciliata*, *Lomentaria clavellosa*, species of *Rhodymenia*, *Nitophyllum punctatum*, etc.

Wound-healing and Regeneration

In Florideae possessing a compact thallus wounds are healed by active division of the uninjured cells to form a kind of callus. Decapitated plants of *Polyides* and *Furcellaria* produce, from the medullary cells at the exposed surface, vertical threads which in part originate from the more deeply situated cells and give rise to a small-celled cortex. From such wound-tissues adventitious branches often arise. In Delesseriaceae quite small pieces can proliferate from the midrib. According to Malkovsky small fragments of *Nitophyllum punctatum* regenerate readily, new branches arising from the apical end and from uninjured marginal cells, whilst the basal end merely forms an extensive callus.

Among Rhodomelaceae damage to the apex results in outgrowth of the central cell into a new polysiphonous axis. Both in this family and in Ceramiaceae regeneration easily takes place from detached fragments which first produce rhizoids, usually from the basal end; in *Griffithsis,* single cells can form a new plant. Polarity is in general well marked, although Schechter records reversal in fragments of *Griffithsia* after 24 hours subjection to a moderate centrifugal force; diminished illumination causes the apex of many Florideae to grow out into rhizoid-like threads. In *Ceramium* a new axis often arises from cortical cells. Damage to or death of intermediate cells in *Bornetia, Griffithsia* and other Ceramiaceae is followed by outgrowth of the adjacent ones to re-establish continuity. In *Rhodochorton* injury to a filament results in active division of the next intact cell.

Repeated reference has been made in the foregoing pages to *vegetative reproduction,* which is effected in diverse ways, and it cannot be doubted that in many Red Algae this is a source of prolific multiplication. Specialised propagules are known only in few Florideae; to the instances described later may be added the *Sphacelaria-like* propagules recorded in *Polysiphonia furcellata* Harv.

Loose-lying Forms

A considerable number of perennial Florideae are capable of surviving after detachment from the substratum. Sernov records the occurrence of *Phyllophora* over an area of about 2000 square miles in the Black Sea. The majority of such loose-lying forms occur in seas with a sandy or muddy bottom, like the Baltic and Adriatic, where the salt-concentration is low and there is often some pollution. According to Schiller the loose-lying communities of the Adriatic are composed of species that show marked indifference to the composition of the water and are capable of withstanding a considerable range of illumination and temperature.

The loose-lying communities originate from detached individuals which are transported to their destination by currents. Multiplication is effected by vegetative means, since most of the plants are sterile. As a general rule they are richly branched, while the foliose types tend to develop narrow segments. Sometimes they assume a globular aegagropilous form, as in *Furcellaria fastigiata* and *Rytiphloea tinctoria*.

Processes of Reproduction

The General Features of Reproduction: The general course of the life-cycle has already been indicated. In the advanced Florideae, it exhibits complications which have no parallel in other classes. Most Florideae exhibit sexual reproduction, as was first established by Bornet and Thuret, who also gave a clear description of the organs involved; their epoch-making memoir gives a resume of earlier views. They recognised that the carposporangia, formed as a result of sexual fusion, frequently arose, not from the female organ itself, but from other cells—near at hand or remote—and they commented on the probable importance of these features in classification. Further confirmation was afforded by the work of Janczewski and particularly by that of Schmitz, while Oltmanns elucidated certain important details of the post-fertilization changes. The present century has seen progress along a diversity of lines, thanks especially to the many detailed researches of Kylin, Lewis, Phillips and Svedelius.

Many Florideae possess distinct male and female plants, although several Nemalionales and Corallinaceae, for example, are monoecious. Not uncommonly the male plants are smaller than the female and they often perish altogether or almost entirely after liberation of the male cells. Their frequent ephemeral nature probably accounts for their rare discovery in diverse Florideae. In *Griffithsia* Lewis found groups of four plants with interwoven rhizoids, presumably the products of germination of the four spores of a tetrasporangium; two of these plants were male and two female, implying genotypic sex-determination,

Female Organs

The female organ or *carpogomum* consists of a dilated basal portion harbouring the female cell and a, usually narrow, elongate process, the *trichogyne*, which constitutes the receptive organ for the male cells and commonly has a very gelatinous membrane. The female protoplast, which contains some reserves, exhibits no contraction either before or after fertilization. In most Florideae, the cytoplasm of the carpogonium is devoid of chromatophores, but in certain Nemalionales (*Acrochaetium, Nemalion, Batrachospermum$_y$* etc.) these are present in the basal part and may even extend far into the trichogyne. The cytoplasm of the latter generally contains a nucleus, which is a sister of the female one. A trichogyne nucleus has not always been found, but its absence is difficult to prove conclusively since it usually degenerates at an early stage. It is probably a vestigial structure.

In most Florideae, the carpogonium terminates a short, often 3-4-celled, lateral, spoken of as the *carpogonial branch* the cell from which this branch arises is termed the *supporting cell*. In forms with a compact thallus the carpogonial branch usually arises from an internal cell, and it is the function of the trichogyne to maintain connection with the external medium. Such carpogonial branches, however, often arise near the growing tips from cells situated close to the surface, although as growth progresses the supporting cell becomes deeply embedded. In the less specialised forms the lower cells of the carpogonial branch often bear laterals, but these are lacking in Rhodymeniales, Ceramiales, and most Gigartinales. In some members of these orders the cells of the carpogonial branch contain several nuclei, and this may indicate arrested branch-formation. In diverse Nemalionales the carpogonial branches are part of the normal branch-system, but in most Red Algae they constitute accessory laterals. The cells of the carpogonial branches are commonly colourless.

Male Organs

The antheridia of Florideae are spherical or oblong, unicellular structures, producing a single male cell or *spermatium* from the entire contents. As a rule they arise in large numbers in close proximity to one another so as to form white or pale-coloured dots or patches, visible to the naked eye. Several antheridia, often 3, sometimes 2, 4, or 5, are produced successively from a common mother-cell which is often elongate. The mother-cells, which usually terminate short, often densely compacted, branch-systems and are invariably uninucleate, commonly possess reduced chromatophores and are sometimes altogether colourless. They often form a palisade-like layer, especially in the more compact types. In a few of the more primitive Florideae the mother-cells are not differentiated from the vegetative cells.

As a general rule, the antheridia first appear as protuberances, which usually arise subterminally and successively from different sides of the mother-cell. As an antheridial protuberance develops, the nucleus shifts to the point of origin and enters into early prophase. After completion of mitosis one daughter-nucleus enters the antheridium, while the other again passes to the base of the mother-cell; there is difference of opinion whether the latter enters into a resting condition before it divides again, and this is perhaps a variable feature. The antheridium is cut off by an oblique annular ingrowth of the wall. In *Martensia* and in certain other instances, in which the antheridia arise terminally and sometimes form short rows, the septum is transverse. Where, as in *Chondrus crispus* and *Furcellaria fastigiata*, there is a thick cuticle, the developing antheridia successively pierce the latter.

The mature antheridium commonly possesses a thick wall which, according to Grubb, sometimes consists of three distinct layers—a narrow deeply staining outermost one, an intervening gelatinous layer which swells considerably at the time of spermatium-liberation, and a narrow, highly refractive innermost one. The nucleus is large, while the scanty cytoplasm is usually quite colourless, although in *Batrachospermum* and *Nemalion* it includes remains of chromatophores, at least in early stages. The spermatium is liberated through a narrow apical split in the elastic wall. Several investigators report that the spermatium.

Diverse earlier workers (Thuret, Buffham) speak of such aggregates as antheridia and of the individual male organs as spermatangia, and the latter designation is retained by many recent writers. It is certainly unwarranted to describe the whole complex of male organs by the name antheridium, while despite the reasons that have been urged for using the term spermatangium, I agree with Oltmanns that there is no case for a double terminology and that, as in other classes of Algae, the structure producing the male cells should be designated antheridium.

It constituted by the entire detached antheridium, but various statements of this kind have subsequently been disproved and all are suspect; in the Corallinaceae, however, there is at present no evidence of dehiscence of the antheridia. After production of the primary antheridia, the same mother-cell may form a secondary series by proliferation into the empty membranes of the first-formed ones.

The nucleus of the free spermatium seems invariably to be in prophase, without an evident nuclear membrane and with a number of deeply staining chromatin granules united by delicate threads. That of the mature antheridium is also often in prophase, but there is considerable evidence that prior to this it has passed through a resting stage. The male nucleus remains in the prophase condition during its passage down the trichogyne.

There is difference of opinion as to the presence or absence of a membrane around the liberated spermatium, although it is generally agreed that there is a delicate membrane in later stages, especially in spermatia adhering to trichogynes. Grubb states that in certain species a membrane is recognisable around the spermatium before dehiscence of the antheridium. The question is undecided, but the capacity for slight amoeboid change of shape exhibited by the spermatia of certain Florideae does not speak for the universal presence of a membrane.

Svedelius, emphasising the resemblance between an antheridial mother-cell after nuclear division has occurred arid a young carpogonium with its trichogyne, regards these two parts of the female organ as respectively homologous with the mother-cell and the antheridium.

The frequent presence of a constriction, accompanied by thickening of the wall, at the base of the trichogyne, is regarded as indicating a former division into two cells. Schussnig on the other hand, regard the entire carpogonium as homologous with an antheridium. This view is based on the occasional division of the antheridial nucleus, while the usual prophase condition of the spermatium-nucleus is taken as indicative of a former division. In *Spermothamnion hermaphroditum,* replacement of the terminal cell of the procarp by an antheridium has been recorded.

Fertilization

The spermatia probably often have approximately the same specific gravity as the surrounding medium so that they remain floating for some considerable time. Their large numbers ensure contact with a trichogyne, the tip of which is evidently more or less mucilaginous.

In *Phyllophora membranifolia,* the spermatia perform slow gliding movements, similar to those reported for the carpospores and tetraspores of certain Florideae, and these are perhaps of wider occurrence. They will facilitate contact with trichogynes that do not project markedly or are embedded in mucilage. At the point of contact both the membrane of the spermatium and the wall of the trichogyne are dissolved, leaving an open passage for the nucleus of the male gamete. In later stages the spermatium-membrane appears empty, but the fate of the male cytoplasm is uncertain. In *Batrachospermum*, the emission of definite copulating processes by the spermatia has been recorded.

In *Batrachospermum*, and *Nemalion* the spermatium-nucleus ordinarily divides after contact with the trichogyne, although only one nucleus fuses with the female. Among the more specialised Florideae no such division takes place and its occurrence in these relatively primitive types suggests that the antheridium of Florideae may be derived from an organ producing more than one gamete; traces of division perhaps persist in the invariable prophase condition of the spermatial nucleus.

Wille first described in *Nemalion* the passage of the male nucleus to the basal part of the carpogonium, the apposition of the two sex nuclei and their fusion. The last process has not often been observed but indirect evidence of the production of a diploid fusion nucleus is considerable, and in a large proportion of Florideae fertilization no doubt follows on contact of a spermatium with a trichogyne. As the male nucleus advances, the cytoplasm of the trichogyne not uncommonly shows signs of disorganisation, exhibiting numerous particles staining deeply with haematoxylin, whilst the part in front is still homogeneous. The female nucleus seems frequently to move a short distance upwards to meet the male.

At the moment of fusion the male nucleus, in *Polysiphonia*, consists of a dense aggregate of chromosomes, while the female is in a resting condition. In *Batrachospermum*, on the other hand, both nuclei are in a resting condition. In *Polysiphonia*, the membrane of the female nucleus dissolves at the point adjacent to the male chromosomes, after which chromosomes differentiate in the female nucleus and the two sets mingle with one another. The fusion nucleus then passes into prophase, its membrane disappears, and division commences.

After penetration of the male nucleus, the trichogyne becomes separated at its base from the rest of the carpogonium; this is commonly effected by progressive thickening of the membrane until complete closure occurs, although sometimes the channel seems to be blocked by a plug of membrane-substance. The trichogyne either persists for some time or rapidly withers. In *Nemaiton* the zygote is stated to become surrounded by a distinct membrane which serves to cut off the trichogyne, although the pit-connection with the hypogynous cell is maintained.

Post-fertilization Changes

Features of the Life-cycle: In Nemalionales, the first nuclear divisions within the fertilized carpogonium are meiotic, but in most Florideae, they are mitotic, reduction being postponed to the time of tetraspore-formation. The fertilized carpogonium invariably puts forth one or more threads, the ultimate function of which is the production of asexual spores *(carpospores)*, but the manner of formation of the latter varies in the different orders. In Nemalionales the haploid threads or *gonimoblasts*, arising from the carpogonium, produce the carpospores directly. The threads in question are for the most part relatively short and well branched so as to form a dense cluster, and certain enlarged cells (often terminal), filled with cytoplasm and food-reserves, constitute *carposporangia*, from each of which a single naked carpospore is liberated.

The latter gives rise to a plant which again bears sex organs so that in the life-cycle of Nemalionales the zygote constitutes the only diploid stage, while there are two haploid phases, the ordinary seaweed and the gonimoblast-threads originating from the zygote. Such Florideae have been designated *kaplobiontic* by Svedelius. In several Nemalionales, the sexual phase also bears sporangia forming accessory spores, mostly monospores.

In the remaining Florideae, the threads arising from the fertilized carpogonium, contain diploid nuclei and are formed singly or in small numbers. They grow towards and establish contact with certain cells, which are situated near at hand or at a more or less considerable distance and are often specially differentiated by size or rich contents; in certain orders these *auxiliary cells* are clearly recognisable before the threads from the carpogonia reach them. When auxiliary cells and carpogonia form parts of the same branch-system, the aggregate is spoken of as a *procarp*. The threads that establish contact with the auxiliary cells are spoken of as *connecting filaments* a term that is specially appropriate in diverse Cryptonemiales and Gigartinales, in which the auxiliary cells are widely separated from the carpogonia so that the connecting filaments are of considerable length. In the more specialised Florideae, however, the connecting filaments are extremely short, being represented merely by a small cell.

At the point of contact the membranes dissolve and communication is established between the cell of the connecting filament and the auxiliary cell. Oltmanns showed that this fusion is purely cytoplasmic and that the nuclei of connecting cell (diploid, *dn)* and auxiliary cell (haploid, *hn)* not only fail to fuse, but usually remain far apart. The fusion with the auxiliary cell, however, affords a stimulus, since division of the nucleus of the connecting cell also often of the auxiliary cell, ensues, and it is at these points that gonimoblast-filaments arise and that carpospores are produced. Among the less specialised types the same connecting filament may fuse successively with a number of auxiliary cells resulting in several centres of carpospore-formation, but in the higher Florideae there is fusion with only one auxiliary cell and, since there is often but a single connecting filament, only one group of gonimoblasts results from the act of sexual fusion.

In a few instances the gonimoblasts arise from the connecting filament, but mostly the two diploid nuclei of the latter actually pass into the auxiliary cell. While its nucleus and one of the nuclei from the connecting cell disorganise, it is from the auxiliary cell itself with the other diploid (foreign) nucleus that the gonimoblasts are produced. One can speak here of a carposporophyte, but, since the gonimoblasts of these forms are obviously homologous with the haploid ones of Nemalionales, the term is not a generally useful one.

The auxiliary cells are sometimes modified cells of a lateral branch-system composing the thallus, but more usually they are, like the carpogonia, situated on accessory branches. Although commonly differentiated already prior to fertilization, in many Rhodymeniales they remain inconspicuous until sexual fusion has occurred, while in Ceramiales their formation is deferred until after fertilization. In these orders the auxiliary cells are closely associated with the carpogonial branch. The nature of the auxiliary cells is disputed; while Kylin looks upon them as modified vegetative cells, Schussnig regards them as homologous with rudimentary carpogonial branches.

In all Florideae, except Nemalionales, the carpospores are of course diploid, although they are produced from the gonimoblasts in essentially the same way as in haplobiontic forms. The carpospores give rise to a diploid plant, which usually resembles the haploid one in all respects, except that it normally bears only asexual sporangia forming four spores. The two nuclear divisions in the sporangium are meiotic and the tetraspores grow into sexual individuals.

The life-cycle of these Florideae therefore, includes a haploid sexual and two diploid asexual phases (the gonimoblasts producing carpospores, the tetrasporic plant) and such Florideae are described as *diplobiontic* by Svedelius. There are three successive generations, although there is only one alternation of cytological phases in the life-cycle. The isomorphic alternation between the sexual and the tetrasporic individuals was first demonstrated experimentally by Lewis. He allowed the two kinds of spores to germinate on oyster-shells which, with the germlings, were transferred to the open sea, the tetraspores growing into mature sexual plants and the carpospores into asexual ones. The relation between the diplobiontic and haplobiontic types of life-cycle is discussed on.

The naked carpospores sometimes perform gliding movements, while amoeboid changes of shape, not necessarily associated with movement, have been reported by various observers. It is not known whether such movements aid in the adoption of a suitable substratum. Proliferation of carposporangia occasionally occurs. Chemin gives data on the mode of liberation of these spores.

Tetrasporangia and Tetraspores

The tetrasporangia of the diplobiontic Florideae are enlarged cells, which possess dense cytoplasmic contents and are almost invariably terminal on short lateral branches. This is clearly evident in such loosely branched forms as *Callithamnion* and *Antithamnion,* but even m genera with a compact structure, where they are embedded in the superficial tissues, their terminal position is generally recognisable without difficulty. The four spores either form a row *(zonate)* as in many Cryptonemiales and Gigartinales, or they are arranged crosswise *(cruciate)*, commonly in the same plane, as for example in Rhodymeniales or lastly they are grouped in a tetrad *(tetrahedral)* as in the majority of the Ceramiales. Different types may occur in different species of the same genus and occasional transitions are found. Certain Florideae, especially Corallinaceae, produce only two spores *(bispores)* in the sporangia.

The mature sporangia are usually uninucleate, even when the vegetative cells contain several nuclei. Except in the cruciate type, the two nuclear divisions usually precede any cleavage of the cytoplasm. After the four nuclei have assumed their definite position, the cytoplasm exhibits peripheral invaginations which, advancing centripetally, divide the protoplast into as many uninucleate portions. The young sporangia are vacuolate and the chromatophores are often indistinct even in the mature spores. The liberated tetraspores are always naked and gliding movements, similar to those of the carpospores, are reported, while Chemin records amoeboid changes of shape.

Cleavage of the protoplast is commonly accompanied by the ingrowth of septa from the wall of the sporangium, but little detailed information is available on this point. In *Delesseria* the gelatinous septa show a cellulose-reaction and, like the wall of the sporangium, are traversed by pores occupied by plasmodesmae. In the formation of cruciate tetraspores transverse division of the contents ensues after the first nuclear division, the two halves then dividing longitudinally, the planes of division coinciding or not; septum-formation seems commonly to accompany the division of the protoplast in this instance, but it is not clear whether this is the rule. There are few available data as to the mode of dehiscence of the sporangia; according to Westbrook those of *Callithamnion* open by a lid, those of some species of *Antithanmion* by an apical split. Proliferation of tetrasporangia is probably not infrequent, especially in the less specialised forms.

The Reduction Division

Despite a few exceptions, the tetrasporangium of diplobiontic Florideae is usually the seat of meiosis. This was first established by Yamanouchi in *Polysiphonia,* who thus confirmed certain suggestions made earlier. Since then meiosis has been demonstrated in the tetrasporangia of a considerable number of diplobiontic Florideae, but there are considerable discrepancies as regards the details of the process, in part centring about the question as to the relation between nucleolus and chromatin.

The tetrasporangium nucleus, which enlarges appreciably before meiosis commences, shows a conspicuous network with chromatin granules at the corners and a large prominent nucleolus. In *Polysiphonia* the spireme formed during prophase affords evidence of a double nature. This is more distinct when the thread again spreads through the nuclear cavity, after the rather prolonged synezesis. Subsequently there is formation of 20 pairs of chromosomes, which gradually shorten to form as many gemini, giving a typical diakinesis-stage. By this time the nucleolus has disappeared. The spindles are intranuclear, with a centrosome at each pole; those of the second division, which are orientated perpendicular to one another, appear as soon as the chromosomes begin to separate. In the telophase of the second division the original nuclear cavity is cut across by the cytoplasmic cleavage.

The course of events is essentially the same in *Rhodomela virgata, Nitophyllum punctatum, Griffithsia corallina, Spermothamnion turneri* and *Rhodymenia palmata*, except that in the first three synezesis was not observed and the spindles are not intranuclear, while in the last two no achromatic figure was seen. Centrosomes were not found in *Spermothamnion* and *Griffithsia*. In the later stages of the heterotype division the bivalents are always short.

A spireme stage has been reported in many other Florideae, but not in *Delesseria sanguinea, Griffithsia globifera, Lomentaria clavellosa, Corallina,* or *Stenogramma interrupta*. In the last three this has been correlated with the small size and scanty chromatin content of the nucleus. Using Feulgen's nuclear stain, concludes that the chromosomes always differentiate from the reticulum and that they exhibit no relation to the nucleolus. Svedelius, on the other hand, records an aggregation of chromatin around the nucleolus at a certain stage in *Delesseria*, a condition attributed by Kylin to imperfect fixation.

The former concludes that there is actual penetration and has more recently given a circumstantial account of similar events in the (mitotic) nuclear divisions in the sporangium of *Lomentaria rosea*, although no such phenomena are recorded during meiosis in *L. clavellosa*. In *Brongniartella* and certain other Ceramiales peculiar conditions have been reported, suggesting a possible passage of chromatin into the nucleolus during the late prophase, as has recently been affirmed by Geitler also for mitosis in *Spirogyra*. The evidence in all these instances is inconclusive and further investigation is requisite.

In diverse Ceramiales (*Ceramium, Chondria*, etc.) the

nucleolus fragments into a number of irregular bodies during prophase. According to Westbrook several species (especially *Nitophyllum punctatum)* show a diminution of chromatin in the outer nucleus prior to diakinesis. At this stage the cytoplasm of the sporangial initial often contains granules staining readily with basic dyes and with haematoxylin. Their exact nature remains undetermined, some regarding them as extrusions from the nucleus, while Svedelius and Kylin consider them to arise spontaneously in the cytoplasm. The scanty data on the details of meiosis in the Nemalionales indicate the customary succession of spireme, synezesis, and diakinesis, although in most instances only the last-named stage has been observed.

The haploid number of chromosomes; appears to approximate to 10 in most haplobiontic Florideae, although the number is 20 in *Bonnemaisonia asparagoides* and this is also that usually reported in the diplobiontic forms.

Early Stages of Germination

Tetraspores and carpospores are nearly always heavier than water and tend to sink. The secretion of a membrane is possibly the result of contact with a substratum, whilst attachment is in all probability due to the production of mucilage. The mature spores are nearly always filled with starch. The contents are deeply pigmented, but the chromatophores are difficult to distinguish and perhaps not fully differentiated. During the early stages of germination, as the reserves disappear, the chromatophores become apparent. There is no resting period. There are usually no essential differences in the germination of carpospores and tetraspores, although some difference in size of the germlings has occasionally been reported.

The plane of the first division-wall is not determined by the direction of the incident light, although Tobler, without apparent evidence, supposes that this is so in *Griffithsia opuntioides.* Kylin suggests that there may be an influence of the substratum or perhaps of gravity. According to Chemin the spores of many Florideae will germinate in complete darkness, although growth soon comes to a standstill; it can, however, be resumed if the germlings are exposed to light before the sojourn in darkness has been too prolonged. Direct sunlight kills the spores very rapidly.

As regards the method of germination, the bipolar type, characteristic of Ceramiales, where the elongating spore divides into a rhizoid and the first cell of an erect-growing thread, is sharply opposed to that found in other Florideae, where division takes place primarily at right angles to the substratum and leads to the production of a prostrate thread or cushion.

Kylin and others here distinguish two or more types, viz. that found in many Nemalionales and diverse Cryptonemiales in which a tubular prolongation from the spore segments to form a prostrate thread, while the original spore-body, often depleted of its contents, remains undivided, and that found in Gigartinales, Rhodymeniales, Corallinaceae and most Cryptonemiales, in which the spore itself divides by diversely orientated walls to form a cushion-like mass. It may be doubted whether these differences are as fundamental as that between the method of germination of the Ceramiales and of other Florideae. Details of the process in individual forms have been given in the foregoing consideration of thallus-structure.

Haplobiontic Types (Nemalionales)

Sexual Reproduction and Carpospore-formation: The simplest reproduction among Florideae is met with in *Acrochaetium.* Many species are monoecious, although *A. corymbifarum* and *A. rhipidandrum* for instance, are dioecious, the male individuals of the former being smaller than the female. There is no special carpogonial branch and the lageniform carpogonia are seated laterally on the filaments or terminate branchlets, which may simultaneously bear protandrous antheridia or even sporangia; in *A. gynandrum* antheridia are inserted on the carpogonia, while in *A. efflorescent*, the latter may occur in an intercalary position. In *A. violaceum* the hypogynous cell bears a few branches which elongate considerably after fertilization. The oblong or spherical antheridia are grouped at the apices of short laterals. In *Kylinia*, in which the carpogonia are incompletely known, the antheridia are borne on narrow little-pigmented cells arising vertically from the creeping threads.

After fertilization, the basal part of the carpogonium of *Acrochaetium* lengthens and the trichogyne is pushed to one side. The former undergoes one or more transverse divisions, and the resulting cells give rise to gonimoblast-threads; in *A. violaceum*, however, four or five threads arise from the apex of the carpogonium without any septation of the latter. The gonimoblast-threads are generally branched, the terminal cells usually alone producing sporangia in *A. efflorescens*, for instance, carpospores are formed also from other cells of the threads.

Basal cells or also from other internal cells of the primary whorled laterals; they are sometimes borne on cortical threads. The trichogyne is separated from the body of the carpogonium by a constriction and, though commonly club-shaped, varies considerably in form, a feature which affords a valuable diagnostic character; a trichogyne nucleus has not been observed. The spherical antheridia are formed in small groups at the tips of the ultimate laterals.

One of the products of meiosis of the zygote-nucleus passes into a lateral outgrowth and becomes cut off as the first gonimoblast-initial. The other daughter-nucleus divides again, and this results in the formation of a second initial on the opposite side of the carpogonium; further initials may be produced in the same way. They grow out into branched threads, the dilated terminal cells of which constitute carposporangia. The gonimoblasts are commonly short and compact, but in a few species some of the branches extend over the internodes and there produce further groups of carposporangia; in *B. ectocarpum* the mature gonimoblasts sometimes project beyond the surface. The cells of the carpogonial branch usually bear laterals which, in *B. moniliforme* and related species, form a loose envelope around the mature gonimoblasts; this is more marked in *Nothocladus*. The trichogyne is very persistent.

The carpogonia possess a characteristic abaxial dilation within which the female nucleus is situated. A more important difference is that the few gonimoblast-threads extend over the internodes, bearing, short branch-systems, the terminal cells of which form carpospores; in *S. suecica* the single gonirnoblast-thread arises adaxially.

Similar elongate gonimoblasts, extending horizontally through the cortex, occur in *Cumagloea* and *Dermonema;* in the last they show considerable ramification and bear vertical branch-systems, the end-cells of which form club-shaped carposporangia. The carpogonial branches are here three-celled. The possession of elongate gonimoblasts does not probably imply any clóse relationship between the various genera, which differ considerably in vegetative structure. It is still an open question whether the penetration of the threads among the vegetative branches of the thallus is associated with any absorption of nutriment from the latter.

Among the multiaxial Nemalionales, *Helminthocladia* closely resembles *Batrachospermum* in its reproduction, although the usually three-celled carpogonial branches are better defined. In *H. Calvadosii* the fertilized carpogonium cuts off by approximately longitudinal walls a number of cells which give rise to compact gonimoblasts with terminal carposporangia while in *H. Papenfussii* the first division is almost transverse; this species is also distinguished by the fusion of the cells of the carpogonial branch after fertilization. There mayor may not be an envelope of sterile filaments.

Acrochaetium, Batrachospermum, and *Helminthocladia* exhibit primitive characteristics in the usual absence of any post-sexual fusions, in the customary formation of carpospores only from the terminal cells of the gonimoblasts, and in the lack of a clearly defined envelope around the latter. Scarcely more specialisation is evident in *Lemanea,* where the carpogonial branches differentiate simultaneously with the cortex, arising from the cells of the longitudinal threads. In *Tuomeya,* the carpogonial branches are spirally twisted and arise from the basal cells of the laterals. According to Atkinson those of *Lemanea* are usually situated near the primary laterals (basal cells) in *Eulemanea,* whereas in *Sacheria* they commonly lie remote from them, sometimes even occurring within the antheridial zones. The carpogonial branch consists of a variable number of cells (in *Eulemanea* often bearing short branches), embedded within the wall, while the trichogyne, which is occasionally forked, projects slightly beyond the surface of the bristle.

The oblong antheridia are generally formed in pairs upon cylindrical mother-cells, which are crowded together at the "nodes" and terminate branches of the longitudinal threads. In *Eulemanea,* where the latter are profusely branched, the antheridial sori form rings, but in *Sacheria* they occupy the summits of projections which are arranged in a verticillate manner. The young mother-cell usually bears a unicellular hair which is pushed aside by the developing antheridia. Fertilization apparently occurs during the early months of the year.

The fertilized carpogonium enlarges and cuts off, on its inner side, a succession of initials which project into the central cavity and give rise to branched gonimoblast-threads, most of the cells of which form carposporangia. The nearly colourless carpospores are liberated by decay of the thallus.

Greater specialisation is shown by *Nemalion,* of which *N. helminthoides* and *N. multifidum* have been repeatedly investigated. The plants are commonly dioecious, the male being smaller than the female. The carpogonial branches are constituted by ordinary laterals and consist of 4-7 cells, the upper of which are nutritive cells with only a minute colourless plastid, while the lower bear a number of ordinary branches. The elongate trichogyne projects slightly beyond the surface of the thallus. The antheridia are formed on branches, which arise from the terminal cells of the laterals and consist of

2-4 cells with non-pigmented plastid and abundant contents; each cell produces 3 or 4 antheridia around its apex. One or both male nuclei pass into the trichogyne.

Soon after sexual fusion the large zygote-nucleus undergoes meiosis and the chromatophore divides. Thereupon the carpogonium segments transversely into two cells, the upper of which produces the gonimoblasts. Its nucleus divides again and one half passes into a lateral protuberance, which is cut off as a gonimoblast-initial; this is repeated several times so that a number of initials are produced. The lower carpogonial cell gradually fuses with the hypogynous cells and these with the underlying cells, a process accompanied by nuclear degeneration and one by means of which the nutriment in these cells is no doubt rendered available to the developing gonimoblasts. The threads, formed from the gonimoblast-initials, hang down parallel with the carpogonial branch and each of their cells produces on its outer side an upwardly curved 2-3-celled branchlet, the terminal cell of which enlarges to form a sporangium; frequent proliferation of the sporangia is recorded.

The higher differentiation of *Nemalion* is shown by the presence of nutritive cells in the carpogonial branch, by the development of gonimoblasts from the upper half of the carpogonium only, by the marked fusions that occur subsequent to fertilization, and by the special form of the gonimoblasts. Only one of the two nuclei formed in the first meiotic division is supposed to divide again. The reproduction of most species of *Liagora* is in all essentials like that of *Nemalion*.

Other Helminthocladiaceae, of which *Helminthocladia, Cumagloea* and *Dermonema* have already been considered, are less specialised than *Nemalion* although Svedelius records occasional fusion of the fertilized carpogonium with the supporting cell of the carpogonial branch in *Dermonema*. In *Trichogloea,* the enlarged hypogynous cell functions as a nutritive cell. The dioecious *Helminthora divaricated* shows much the same features as *Nemalion,* although fusion with the hypogynous cells is not so marked and the gonimoblasts become enveloped by sterile branches which arise from the second and third cells of the four-celled carpogonial branch. Such an envelope is also found in *Trichogloea* and indicated in *Nemalion*. The antheridia of *Helnrinthora* form groups on the terminal cells of the laterals. All Helminthocladiaceae appear to have terminal carpospores.

The Naccariaceae and Bonnemaisoniaceae, which show the greatest vegetative specialisation among Nemalionales, also afford evidence of advance in reproduction, although there is no complete parallel. Here the hypogynous cell of the carpogonial branch bears short laterals composed of nutritive cells with dense contents, which are gradually exhausted during the development of the gonimoblast. Except in *Atractophora*, the fertilized carpogonium fuses with the hypogynous cell and establishes a direct connection with this nutritive system, but in *Bonnemaisonia* and *Asparagopsis*, the fusions after fertilization are more comprehensive than in Naccariaceae. In these two genera, moreover, threads originating from cells of the carpogonial branch, and more particularly from adjacent cells, combine to form a compact investment around the gonimoblast. For such a fructification, provided with a special protective envelope, the term *cystocarp* is often reserved.

In Naccariaceae, the fertile branches at first develop few laterals and exhibit little hypha-formation. The three-celled carpogonial branches and the two-celled ones arise from basal cells of laterals which also bear short branches; the small nutritive cells are cut off from the hypogynous cells. In *Atractophora* the fertilized carpogonium fuses with the supporting cell, while in *Naccaria* the pit-connection between the former and the hypogynous cell, and sometimes those between the latter and some of the nutritive cells, gradually widen.

The gonimoblast-threads arise from the carpogonium itself, singly in *Naccaria*, to the number of two or three in *Atractophora*. Branching extensively, they spread upwards and downwards and form an investment around the axial threads of the fertile tract, their perpendicular branches bear terminal sporangia. The photosynthetic laterals, amid which these branches spread, exhibit abundant ramification, but this is much more marked in *Naccaria,* where the fertile tracts appear as ellipsoidal swellings on the fronds. Numerous hyphae arise so that the internal tissue of these swellings constitutes a dense pseudoparenchyma, which is surrounded by a broad and more loosely organised cortical region covered by an envelope of mucilage. The gonimoblast-threads of *Naccaria* are stated to form secondary pit-connections with the central cells and other elements of the fertile tract.

The antheridial mother-cells are terminal on branches arising mainly from the basal cells of the laterals; in *Atractophora* the entire lateral may be involved. In *Naccaria,* owing to slight elongation of the axial cells, the antheridial sori often constitute extensive colourless areas interrupted only by the projecting pigmented tips of the sterile branches.

The Bonnemaisoniaceae, though more specialised, show considerable resemblances to *Naccaria. Bonnemaisonia asparagoides* is monoecious, while the species of *Asparagopsis* are dioecious. The sex organs are borne on secondary laterals of limited growth which alternate irregularly; in *Bonnemaisonia* they arise directly from them, while in *Asparagopsis* they are situated upon their branches. In the male shoots of *A. armata* nearly all the secondary branches produce antheridia. The male branch consists of a limited number of elongate axial cells, nearly all of which *(Asparagopsis)* or only the upper *(Bonnemaisonia)* cut off pairs of pericentrals which alternate in successive segments; the laterals to which they give rise combine to form a continuous layer of mother-cells, each producing two or three antheridia. In the female shoots of *Bonnemaisonia,* all except the terminal cell become corticated and later form the stalk of the solitary cystocarp. Those of *Asparagopsis* produce only two fertile laterals, representing the secondary laterals of two successive basal segments, so that the resulting cystocarps are usually at right angles to one another.

In both genera, the terminal cell of the female axis divides obliquely, the upper segment producing three groups of laterals which participate in the formation of the cystocarp-wall, while the lower cuts off three pericentrals, one of which acts as the supporting cell of the three-celled carpogonial branch; in *Bonnemaisonia* the trichogyne is long and spirally twisted. The hypogynous cell gives rise to laterals, composed of nutritive cells which enlarge and increase in number subsequent to sexual fusion. During the early divisions in the fertilized carpogonium the nutritive cells fuse both with one another and with the hypogynous cell, thus furnishing material for the development of the gonimoblast.

In *Bonnemaisonia*, the reduction divisions take place in the carpogonium, but only one of the four resulting nuclei probably persists. The gonimoblast-initial is cut off internally from the carpogonium adjacent to a group of large cells full of nutriment occupying the floor of the young cystocarp. The initial divides rapidly to form a several-layered disc and, as this happens, the nutritive threads connected with the hypogynous cell lose their contents. Later the lower cells of the gonimoblast-disc fuse with one another, with the nutritive cells below, and with the supporting cell of the carpogonial branch (sometimes also with the fertile axial cell) to form a large multinucleate fusion-cell occupying the floor of the cystocarp. The overlying gonimoblast-cells give rise to branched threads, the end-cells of which form sporangia. The stalked cystocarps, which have a terminal aperture, stand opposite the primary laterals; according to Golenkin they bear plentiful vesicular cells.

The fusion-nucleus of *Asparagopsis* passes into the hypogynous cell, while the carpogonium itself degenerates. Within the hypogynous cell the diploid nucleus enlarges and undergoes meiosis. During the development of the two gonimoblasts a large fusion-cell similar to that of *Bonnemaisonia* is formed at the base of the cystocarp, bearing on its upper side the carposporangial threads which are here interspersed with sterile filaments. The axial cells in the lower part of the fertile branch have abundant contents and wide pit-connections.

A similar specialisation is found among the multiaxial Nemalionales in the Chaetangiaceae, of which the monoecious *Scinaia furcellata* has been fully investigated. The antheridial branches arise from the pigmented end-cells of the laterals, the mother-cells, which produce two or three antheridia, being embedded in a layer of mucilage formed by the outer walls of the vesicular cells. The three-celled carpogonial branches originate from the loose internal tissue just behind the apex. At an early stage the hypogynous cell divides longitudinally into four large cells which possess dense contents and often contain several nuclei, while the first cell of the branch produces a few short threads.

After fertilization the fusion-nucleus passes from the carpogonium through the enlarged pit-connection into one of the four hypogynous cells and there undergoes meiosis. Three of the resulting nuclei gradually degenerate while the fourth passes into the single gonimoblast which arises from the upper end of the hypogynous cell. The gonimoblast penetrates through the pit-connection into the empty carpogonium and, piercing the latter on one side, soon commences to branch. In the meantime the threads arising from the first cell of the carpogonial branch multiply and form a compact envelope. The hypogynous cells partially fuse and flatten out against the latter, while the gonimoblast gradually fills most of the space within it. Many of its branches produce rows of 2-4 sporangia, while others remain sterile. The dark-coloured spherical cystocarps, which are provided with an apical aperture, are sunk within the surface of the thallus.

The mature cystocarps of *Galaxaura* and *Chaetangium* are similarly embedded. The lower cells of the carpogonial branch of C. *saccatum* bear short laterals. Later the carpogonium degenerates, although the underlying cell survives, while the cells of its laterals increase considerably in size. On somewhat slender data, Martin concludes that, as in *Scinaia*, the diploid nucleus passes into the hypogynous cell, with subsequent formation of a tetrad by reduction, although there is no good evidence of the occurrence of fertilization. The richly branched gonimoblast arises from the hypogynous cell, some of its branches interweaving to form a "wall", while others produce the carpospores; there is no true envelope, the laterals on the lower cells of the carpogonial branch disappearing during the development of the gonimoblast. A narrow pore is formed above the resulting fruit.

The antheridia of *Galaxaura* are formed in large numbers from the ends of richly branched filaments, which arise from the floor and sides of deep and often wide cavities. The antheridial filaments originate close to the apex and are gradually overgrown by adjacent sterile threads; the aperture, which progressively widens, results from degeneration of the overlying cells. Similar antheridial cavities, in part very small, occur in *Chaetangiwn*.

General Survey of the Life-cycle

Kylin emphasises that the uniaxial and multiaxial Nemalionales must be regarded as distinct developmental series, a point of view which receives considerable support from a consideration of their reproductive features. In each series definite nutritive cells are differentiated only in the more specialised forms, viz. the uniaxial Naccariaceae and Bonnemaisoniaceae, and the multiaxial Chaetangiaceae; it is, moreover, only in the two last families that a compact envelope is formed around the gonimoblast leading to the formation of a true cystocarp. In *Acrochaetium*, *Batrachospermum*, and *Lemanea* there is no fusion of the carpogonium with other adjacent cells after fertilization, such as obtains in some of the less specialised multiaxial forms— the Helminthocladiaceae. Here, however, only the carpogonial branch itself serves to provide nutriment for the developing gonimoblast, although the outgrowth of gonimoblast-threads among the photosynthetic laterals in *Dermonema* and *Cumagloea* perhaps represents another means of attaining the same end. In Naccariaceae and Bonnemaisoniaceae, on the other hand, special nutritive cells are formed from the hypogynous cell and this is also so in

Scinaia. Except in *Atractophora*, these cells fuse with one another during the post-fertilization development.

The gonimoblasts in Nemalionales, usually originate from the carpogonium itself, but in *Asparagopsis, Scinaia,* and perhaps in *Chaetangium,* the diploid nucleus passes into a hypogynous cell and it is from this that the gonimoblast arises. The post-fertilization behaviour of the diploid nucleus in these genera is analogous to that found hi many diplobiontic Florideae, and there seems no valid reason why the hypogynous cell that receives it should not be termed an auxiliary cell. It appears that the haplobiontic Florideae, despite the fundamental difference in their life-cycle, show in the sexual reproduction of the more advanced forms many of the specialised features that characterise the diplobiontic type, and it may be recalled that the vegetative construction of the genera involved also affords indications of advance.

In *Nemalion* and *Asparagopsis* several gonimoblast-threads are produced, so that at least two of the four nuclei formed at the reduction division are utilised, while in *Scinaia,* and probably also in *Bonnemaisonia* and *Chaetangium,* all but the one employed in the formation of the single gonimoblast degenerate. The last-named genera are in this respect more specialised than other Nemalionales. In *Batrachospermum* all four nuclei are apparently used.

Since Svedelius showed that reduction occurs at the first divisions of the zygote-nucleus in *Scinaia,* this has been demonstrated for several other Nemalionales. Reference may be made to the researches of Kylin and Cleland on *Nemalion multifidum,* whereby earlier erroneous statements of Wolfe were rectified, and the recent investigation of *Bonnemaisonia* and *Asparagopsis* by Svedelius. Evidence for the occurrence of reduction immediately after fertilization has also been produced for *Batrachospermum* and *Helminthora,* and it can hardly be doubted that this is the rule in Nemalionales. The zygote therefore, represents the only diploid stage. What is known of other methods of reproduction in the majority of Nemalionales is in accord with these findings.

Such reproduction is most commonly effected by *monospores* which, despite the fact that some individuals may bear these structures only, are nearly always to be found also on plants bearing sex organs and producing carpospores. The monospores are in fact accessory reproductive cells, which serve to reproduce the haploid phase, and are not an essential part of the life-cycle like the carpospores.

The monosporangia are spherical or oblong structures which differ from the antheridia in their deeply pigmented contents and larger size. Those of *Acrochaetium* are sessile or shortly stalked or sometimes grouped together at the ends of short lateral branch-systems. In many species of *Batrachospermum*, they are confined to the juvenile stage, but in *B. sporulans* Sirod. and *B. vagum*, which rarely produce sex organs, as well as in the tropical *B. lochmodes*, for example, they occur, like the antheridia, at the apices of the laterals of the adult shoots. In *Helminthora* they occupy the same position. The monosporangia of *Scinaia* are formed from the narrow pigmented cells that lie between the vesicular cells in the younger parts. In Naccariaceae, Bonne-maisoniaceae, *Lemanea* and *Nemalion* no such structures are known. The *"Chantransia-stages"* of *Lemanea* are purely vegetative.

The monospores are liberated by rupture of the apex of the sporangium-wall; proliferation is probably not uncommon. Amoeboid movements of the naked monospores have been frequently reported. They appear to germinate in the same way as the carpospores.

Diverse authorities have commented on the similar positions of antheridia and monosporangia, which may even occur side by side on the same lateral. Transitions between the two kinds of organs are reported in *Batrachospermum* and *Scinaia*, while in *Helmimhocladia*, certain structures have been variously interpreted as arrested antheridial branches and as monosporangia. The implications of these features are at present not clear.

In several species, of *Liagora* small orbicular discs, with a gelatinous envelope and more or less calcified, occur immersed among the laterals. They bear monosporangia, sometimes showing division of the contents into two. It remains doubtful whether they are a stage in the development of *Liagora* or belong to an independent endophyte.

Several species of *Acrochaetium (A. Daviesii, A. Thuretii, A. virgatulum)* bear tetrasporangia as well as mono-sporangia, both usually occurring on the sexual plants. On the other hand, in *A. violaceum*, and probably in *A. efflorescent*, there are distinct tetrasporic and sexual individuals which may also bear monospores, although in *A. efflorescent* these often occur on distinct plants. There is, moreover, a certain seasonal alternation, since the tetrasporic individuals occur mainly during the colder and the sexual ones during the warmer months. The supposed asexual plants of *A. efflorescent* are smaller and less abundantly branched. The tetrasporangia of *Acrochaetium* are always cruciately divided and this is also so in *Rhodochorton*. In this genus, tetrasporangia are the rule, although one instance of monosporangia has been recorded; no carpogonia are known, but Rosenvinge has described antheridia in *R. penicilliforme.*

While *A. efflorescent* and *A. violaceum* may actually be diplobiontic, such a view cannot be accepted without a cytological investigation. In face of the known facts, moreover, it appears rather improbable. Although data as to the time at which reduction occurs are not available for *Acrochaetium,* it is generally assumed that it takes place during the first divisions of the zygote-nucleus. In species bearing tetrasporangia and monosporangia on the sexual plants reduction during tetraspore-formation is very unlikely. The occasional association of monosporangia with tetrasporangia on the non-sexual plants of *A. efflorescent* and *A. violaceum* renders it equally improbable that the tetrasporangia are here the seat of a reduction division. With our present knowledge it would seem most plausible to regard these tetrasporangia merely as divided monosporangia, and in this connection attention may be drawn to the occurrence in certain species of bispores and polyspores; in *A. multisporum*

Boerges the latter arise by further division of the four cells first formed.

Cruciate tetrasporangia are also recorded in *Galaxaura,* usually occurring on distinct individuals, which commonly differ so markedly in outward form and internal structure from the sexual ones that the two phases have in the past been referred to distinct species. There is perhaps a greater possibility that this advanced member of Nemalionales may be diplobiontic, but no conclusion is possible without experimental or cytological verification.

The haplobiontic and diplobiontic Florideae are no doubt related and many hold that the latter have originated from forms like the Nemalionales with a haplobiontic life-cycle, probably as a sudden mutation. Many of the haplobiontic genera show a simple vegetative organisation, but there is nothing to imply a possible reduction from more advanced types. The relation between the haplo- and diplobiontic Florideae is deferred to a consideration of Nemalionales may be concluded by the mention of a number of instances, which show that the life-cycle may occasionally present interesting complications.

In *Liagora tetrasporifera,* first described from the Canaries, but since found also in the Mediterranean, the terminal cells of the gonimoblasts develop as tetrasporangia forming cruciate tetraspores, although most species of *Liagora* produce carposporangia of the usual type. Feldmann has recently recorded a similar formation of cruciate tetrasporangia from the gonimoblasts in *Helminthocladia Hudsoni* Should cytological study prove that reduction occurs in the tetrasporangia of these two species, postponement of meiosis within the domain of a single genus would have been demonstrated and a possible stage in the evolution of the diplobiontic condition established.

For members of the Bonnemaisoniaceae Feldmann has described the production from the carpospores of an independent generation bearing tetrasporangia. That of

Bonnemaisonia asparagoides is stated to be identical with a sublittoral algae described by Batters as *Hymenoclonium serpens (Callithamnion serpens* Crouan), in which the densely branched filaments are creeping and sometimes form an almost continuous expanse; it is to be noted that Crouan's figure shows cruciate tetraspores, whereas Feldmann describes them as tetrahedral. The carpospores of *Asparagopsis armata,* on the other hand, develop into the algae *Falkenbergia* in which hitherto only cruciate tetraspores have been reported.

If these observations are confirmed, they would show that the carpospores of these advanced members of Nemalionales give rise to a generation quite distinct from the sexual one. The fate of the tetraspores is unknown; they may, like the monospores of the juvenile stage of *Batrachospermum,* merely serve to reproduce the generation upon which they occur, while the sexual plant itself arises as an outgrowth from the tetrasporic ones. However, that may be, it seems certain that the sexual phase is haploid in view of the data furnished by Svedelius.

Vegetative reproduction is frequent among Nemalionales and occurs abundantly in the *"Chantransia-stages"* of *Batrachospermum* and *Lemanea;* in fragments of Lewawea-bristles the longitudinal threads can develop into the *"Chantransia-stage"*, while cells of the inner cortex of dead bristles are stated to survive and divide. In *Batrachospermum woniltforme* certain laterals, composed of thick-walled cells, have been regarded as resting stages, while in *B. Breutelii* the gonimoblast-threads end in large oval or spindle-shaped bodies divided by delicate septa into 3-6 cells. In the germination of these propagules the end-cells grow out into threads.

Gelidiales

Earlier researches on the sexual reproduction of these forms are supplemented by Kylin's investigation of *Gelidium cartilagineum,* upon which the following account is essentially based. The antheridia form elliptical sori behind the tips of the younger fronds. They are cut off singly from the mother-cells, although a second antheridium is sometimes formed below the first.

In the female plant, carpogonia develop just behind the tip within a median fertile tract, which occupies a shallow depression as a result of rapid growth in thickness of the marginal portions of the flat thallus. Later, when formation of carpogonia ceases, the fertile depression comes to lie behind the apex surrounded by sterile tissue. The fertile tract is occupied throughout its length by an axial plate of large cells, which includes the original axial cells and is extended parallel to the surface. Each cell of the plate cuts off, on either surface of the frond, pericentrals, many of which give rise to three-celled carpogonial branches in which the two lower cells bear ordinary laterals forming the cortex of the fertile tract. The clavate trichogyne of the carpogonium projects beyond the surface.

The pericentrals also produce small-celled branches which form a loose nutritive tissue around the large cells of the axial plate. The entire complex of carpogonia and nutritive tissue, forming the fertile tract, constitutes a *compound procarp* and gives rise to a single fruit; fertilization of only one of the many carpogonia suffices for further development. A single gonimoblast arises from the inner side of the fertilized carpogonium, many of its branches penetrating among the cells of the nutritive tissue, which no doubt provides material, although no fusions occur. Other branches of the gonimoblast, more or less perpendicular to the axial complex, bear terminal carposporangia. As development proceeds, the outer cortex of the fertile tract is gradually elevated by elongation of the inner cortical cells so that the mature fruit constitutes a swelling on the frond. In transverse section, it exhibits two imperfectly separated compartments, which communicate with the exterior by a median aperture. In *Pterocladia* the fruits are unilateral swellings with a single compartment and a single aperture; here the carposporangia are in short chains. The general course of development appears to resemble that of *Gelidium*.

Tetrasporangia are usually borne on distinct individuals within branches of the fronds that are often dilated and flattened. The sporangia develop from terminal cells of laterals and later appear embedded in the surface-tissues. In *Gelidium* and *Pterocladia*, the sporangia are irregularly arranged, while in *Gelidiella* in which sexual plants are unknown, they are disposed in transverse series. The tetraspores are usually cruciate, but in *Gelidiella* they are frequently tetrahedral.

The Gelidiales were formerly included among Nemalionales, but were segregated by Kylin owing to their probable diplobiontic character. The absence of auxiliary cells distinguishes them from other diplobiontic Florideae. The behaviour of the gonimoblast is similar to that found in Naccariaceae, and especially *Naccaria*, but the compound procarps and the mode of fruit-development associated with them are very distinctive.

Diplobiontic Type among Cryptonemiales

Like all diplobiontic Florideae, apart from Gelidiales, the Cryptonemiales possess auxiliary cells, which are here sometimes situated at some distance from the carpogonia. Connection between the two is established by means of one or more septate, and often branched, *connecting filaments*. At each point of fusion between the cell of a connecting filament and an auxiliary cell gonimoblasts are usually produced and, in certain Cryptonemiales, the same connecting filament can fuse with a number of auxiliary cells, so that a single act of fertilization results in the formation of several groups of carposporangia. In all Cryptonemiales, both the carpogonial branches and the auxiliary cells are borne on special accessory laterals and not on those composing the vegetative system.

Sexual Phase in the Non-nemathecial Types

In *Gloeosiphonia* and its allies auxiliary cells and carpogonia are borne on the same branch-system which arises from the under side of the basal cell of a lateral; all the basal cells of a whorl may be fertile. The three-celled carpogonial branch includes a large hypogynous cell and arises from the first or second cell of the fertile axis. The fifth cell beyond that bearing the carpogonial branch functions as an intercalary auxiliary cell, while the rest of the fertile axis is deflected; the cells below the auxiliary cell bear short branches. The antheridia form small superficial sori on the same individuals, the mother-cells arising from the tips of the laterals.

The fertilized carpogonium produces two branched connecting filaments, each containing a product of the zygote-nucleus. They grow in opposite directions towards adjacent auxiliary cells, but do not necessarily connect with that on the same branch-system. After fusion the connecting filament may continue to grow towards other auxiliary cells. At the point of contact the membranes are dissolved so that there is free communication between the connecting and auxiliary cells, and thereupon the diploid nucleus passes into the latter.

As in all diplobiontic Florideae that have been carefully studied, the nucleus of the auxiliary cell remains quite distinct and there is nothing of the nature of a nuclear fusion. Soon after both the diploid nucleus and that of the auxiliary cell divide. One of the diploid nuclei thereupon passes into an outgrowth of the auxiliary cell which becomes cut off as the gonimoblast-initial, while the other three nuclei remain in the auxiliary cell. The initial gives rise to a few branching threads all the cells of which become carposporangia. Meanwhile the initial itself fuses with the auxiliary cell, while the latter fuses with the underlying cell of the fertile branch. The carpospores are embedded among the photosynthetiq laterals.

In *Thuretella* the fertile axis is reduced, the second cell alone producing a lateral, which consists of three cells. The middle one is the auxiliary cell, whilst the basal one bears the three-celled carpogonial branch, which is bent in such a way that the carpogonium lies adjacent to the auxiliary cell. If the lateral is interpreted as the termination of the fertile axis and the sterile part beyond as a branch, the relative positions of carpogonial branch and auxiliary cell would be similar to those in *Gloeosiphonia.* In *Schimmelmannia* the first or second cell of the fertile axis constitutes the supporting cell of the carpogonial branch and bears a further lateral terminated by the auxiliary cell.

The fertilized carpogonium of *Schimmelmannia* divides transversely, the upper cell cutting off a connecting cell which fuses with the adjacent auxiliary cell. In *Thuretella,* there is no such connecting cell and the carpogonium fuses directly with the auxiliary cell, into which a diploid nucleus passes; a vestige of a connecting cell is perhaps to be found in the occasional division of the carpogonium, one of the resulting cells then fusing with the auxiliary cell. The further development of the gonimoblast in both genera takes place as in *Gloeosiphonia.* In *Thuretella* the groups of carpospores become enveloped by threads arising from cells of the fertile lateral.

The close juxtaposition of carpogonium and auxiliary cell in *Thuretella,* which warrants the designation *procarp,* is likewise encountered in the Endocladiaceae and Callymeniaceae. The reproductive systems, and particularly that of Callymeniaceae, show distinct resemblances to those of Gigarti-nales and especially of Gigartinaceae. The auxiliary cell is always intercalary. Although the immediate post-fertilization stages are unknown, connecting filaments are, as in *Thuretella,* probably lacking.

In *Gloeopeltis* the auxiliary cell is one of the lower cells of the fertile axis, while the supporting cells of the several two-celled carpogonial branches are laterals. *GloeosipJionia* also occasionally produces more than one carpogonial branch on the same supporting cell. In *Endocladia* the auxiliary cell is situated in one of the two carpogonial branches, while in *Callymenia reniformis* the large supporting cell of the procarp functions also as an auxiliary cell. It bears a number of laterals, some of which constitute three-celled carpogonial branches with a very large basal cell, whilst others are represented by this cell alone; there are wide pit-connections between the basal cells and the auxiliary cell. After fertilization, the cells surrounding this procarp of *Callymenia* form numerous hyphae, which constitute a nutritive tissue into which the branches of the gonimoblast extend. The swollen fruits contain several groups of small carpospores interspersed among the hyphae. A definite aperture is usually formed in the overlying cortex. These features are found in all Callymeniaceae.

The fertile system of *Endocladia* takes the place of one of the accessory branchlets and consists of a basal cell bearing two laterals, one of which includes the auxiliary cell. Several gonimoblast-threads arise from the auxiliary cell after fertilization and penetrate among the inner cells of the thallus, with which they enter into communication. The antheridta are formed in rows and constitute small irregular sori on the younger parts of the male plants; the elongate mother-cells are produced in pairs from the surface-cells.

The above-described procarps of *Callymenia* arise from hypodermal cells which gradually become embedded by division of the superficial cells. In other Callymeniaceae, the fertile system is simpler. Thus, in *Callophyllis obtusifolia*, *Callocolax neglectus*, and *Euthora cristata* the supporting (auxiliary) cell bears only a single carpogonial branch. According to Kylin the gonimoblast-threads arise from the basal cell of the carpogonial branch in *Callymenia,* although in other Callymeniaceae, they originate from the auxiliary cell. In *Callophyllis laciniata,* the fruits occupy the minute marginal outgrowths of the thallus. The antheridia of Callymeniaceae are imperfectly known.

In diverse Cryptonemiales carpogonia and auxiliary cells are produced on distinct branch-systems, as for instance in the multiaxial Grateloupiaceae. In *Grateloupia filicina* these systems arise at the boundary between inner and outer cortex and each produces a number of simple laterals. The two-celled carpogonial branch, the hypogynous cell of which itself bears a lateral, is borne on one of the lower cells of the fertile axis, while the large intercalary auxiliary cells occupy a comparable position. The fertilized carpogonium produces several, connecting filaments which, after fusion with an auxiliary cell, form a gonimoblast-initial in the same way as in *Gloeosiphonia.* The richly branched gonimoblast bears several distinct groups of carposporangia which are embedded in the thallus; a definite aperture is formed in the overlying cortex.

Other Grateloupiaceae differ only in minor respects, although the data are less complete. In *Cryptonemia,* the branches bearing carpogonia and auxiliary cells become surrounded by a loose envelope, which is also indicated in *Grateloupia.* The carpospores seem always to escape through well-defined pores in the cortex.

Comparison of the fertile systems of *Gloeosiphonia* and *Grateloupia* shows that they differ mainly in the segregation of carpogonial branches and auxiliary cells in the latter. The same feature is found in Dumontiaceae, but here the carpogonial branches include special nutritive cells, with which the connecting filaments fuse before they grow onwards to the auxiliary cells proper.

The carpogonium, which has a long spirally coiled trichogyne, is terminal on a colourless accessory branch, arising from the basal cell of a lateral and bearing distichous branchlets. The antheridia are formed in groups from the end-cells of the photosynthetic laterals of the same plant. The fertilized carpogonium usually produces three connecting filaments, which primarily undergo plentiful fusion with cells of the carpogonial branch and especially with the end-cells of the laterals; these possess dense contents and constitute nutritive cells (sterile auxiliary cells of Berthold. Subsequently the connecting filaments penetrate for considerable distances among the vegetative laterals until, probably as a result of a chemotropic stimulus, they make contact with an auxiliary cell. The auxiliary cells, which are distinguished by dense contents and scanty pigmentation, terminate unbranched accessory laterals, arising in the same position as the carpogonial branches, but more numerous than the latter.

Contact between connecting filament and auxiliary cell is

followed in the usual way by the establishment of cytoplasmic continuity. Here, however, the connecting cell gives rise to the gonimoblast-initial which is cut off on the side opposite to the auxiliary cell. The gonimoblast-initial produces a compact group of branches and all the cells except for the basal one produce carpospores. Meanwhile the connecting filament continues its growth towards other auxiliary cells, a septum separating the fusing portion from the growing *tip* beyond.

A comparable method of carpospore-formation is met with in other Dumontiaceae. Details are available for *Dudresnaya coccinea edulis*. The branches bearing auxiliary cells are always more numerous than those with carpogonia, and both usually originate from the basal cells of the laterals; they exhibit little or no branching. In *Dumontia* they arise from the cells composing the longitudinal branches of the primary laterals. A difference from *Acrosymphytum* is constituted by the intercalary position of the nutritive cells within the carpogonial branches. The latter are curved in such a way that the carpogonium Kes close to the cells in question. The auxiliary cells occupy a similar intercalary position, although their exact location varies.

In *Dudresnaya,* the fertilized carpogonium divides to form a two-celled connecting filament which fuses with the adjacent nutritive cells. Most of the cytoplasm accumulates in the connecting cells, which soon give rise to longer filaments. In *Dumontia* the carpogonium usually fuses with the third cell of its branch, although more than one cell may be involved; two or three connecting filaments then arise from the nutritive cell. Much the same occurs in *Cryptosiphonia* and *Thuretellopsis.*

Both in *Dudresnaya* and *Dumontia*, the auxiliary cell develops a short process, which fuses with the connecting filament and receives a diploid nucleus from it. Two or three gonimoblast-initials are cut off from this process, i.e. from the auxiliary cell. The connecting filaments for the most part seek out a number of successive auxiliary cells, but this is stated to be unusual in *Dumontia*. As in *Acrosymphytum,* most of the cells of the gonimoblast-threads produce carpospores. No protective envelope is organised, the carpospores lying embedded in slight swellings on the thallus.

The male plants of *Dudresnaya* and *Dumontia* are smaller than the female, and in *Dumontia* almost the entire surface may become fertile. Each mother-cell produces successively three antheridia.

In Grateloupiaceae and Dumontiaceae, the branches bearing carpogonia and auxiliary cells respectively are clearly homologous, and in Dumontiaceae, the nutritive and auxiliary cells occupy comparable positions within them. There is also considerable similarity with *Gloeosiphonia,* and the two families under discussion could be regarded as divergent evolutionary series from a generalised type possessing carpogonia and auxiliary cells on the same branch-system. In Grateloupiaceae, complete separation of the two structures has taken place, while in Dumontiaceae auxiliary cells still persist in the carpogonial branches and function as nutritive cells. From this point of view, Berthold's designation of them as "sterile auxiliary cells" is quite apposite. It is also noteworthy that in *Dumontia* the auxiliary cell branches lack the terminal cell which, on the carpogonial branches, develops into the female organ.

In the genera, above considered the diploid nucleus from the connecting filament usually passes into the auxiliary cell, and, as in most diplobiontic Florideae, it is from the latter that the gonimoblai: initials are formed. *Acrosymphytum* is exceptional in that the initial is formed from the connecting filament, within which the diploid nucleus remains after fusion with the auxiliary cell. An analogous condition is seen in *Cruoriaand Petrocelis,* although here the initials do not arise at the point of fusion with an auxiliary cell. Kylin regards *Corallinaceae.*

The Corallinaceae exhibit specialisation in sexual reproduction, no less than in vegetative organisation. Except in the uncertain endophytic genus *Schmitziella,* the sex organs are invariably produced within conceptacles of diverse shape, which are either sunk in the thallus or project to a marked extent. They arise by the arrest of growth in a certain region which is gradually overarched by the adjacent tissue. The floor of the conceptacle, constituting the so-called *disc,* thus represents the original surface-layer and usually consists of cells with dense contents. It is these that divide to form the short vertical threads, upon which the sex organs are produced. The conceptacle is therefore, a nemathecium which is overgrown by the surrounding tissue. Many Corallinaceae are dioecious, although *Coral Una rubens* and some *Melobesias* are monoecious.

The development of the conceptacles, probably takes place in much the same way in other species of the genus. The apex of the segment broadens and the layer of thickening covering the apical cells is thus disorganised. This is followed by the development of a marginal upgrowth, which gradually bends inwards until only a small aperture remains. During the development of the conceptade the outer walls of the elongate cells composing the disc become strongly thickened, while the outermost layers undergo calcification so that the floor is covered by a continuous layer of lime, separated from the underlying cells by a stratum of mucilage. The latter gradually increases in amount, thus raising the layer of lime and ultimately causing its rupture.

The raised conceptacles of *Melobesia* occupy parts of the crust which consist of several layers of cells. In *M. limitata* and others the disc is constituted by the second layer and, as these cells divide to form the fertile threads, the cover-cells are carried up and gradually cast off; the lid of the conceptacle is formed by division and elongation of the surrounding cells. Much the same occurs in *Lithophyllum* and *Lithothamnion.* The broad and often completely em-bedded conceptacles of the former have an elongate canal-like aperture, sometimes lined with hairs those of *Lithothamnion* are usually markedly raised. The male conceptacles of *Melobesia* and *Lithophyllum* are smaller than the female and the aperture is often prolonged into a mucilage-tube. In *Epilithon* the fertile threads are formed by division of the cells of the basal stratum and, as they develop, the strongly calcified superjacent layers are thrown off. The conceptacles of *Choreonema* first appear within the cortex of the host as fan-shaped groups of threads terminal on one of the endophytic filaments, the end-cells being provided with mucilaginous caps. The wall of the conceptacle is largely formed by tabular cells cut off tangentially from the peripheral threads, although near the aperture some of the inner threads are involved.

In the female conceptacles the fertile threads are confined to the floor. The cells of the disc divide and the upper half produces the procarp; occasionally two or more procarp rudiments are formed from a single disc-cell. The development of the numerous procarps takes place centrifugally, but as a rule only the central ones mature. The trichogynes are usually elongate and project from the aperture of the conceptacle. Recent investigations indicate thai the basic type of procarp in Corallinaceae consists of a basal (auxiliary) cell bearing two 2-celled carpogonial branches, as well as an imperfect median rudiment; the nearest approach to this is found in some *Melobesias* and in *Amphiroa rigida*. One of the carpogonial branches is, however, often represented only by a single cell or, like the median rudiment, may be completely suppressed.

The auxiliary cell may likewise bear three branches in *Lithophyllum expansion* and *Corallina*, but here only one develops into a complete carpogonial branch, while one or both of the others may be suppressed. The last condition is typical of *Choreonema*, where the single two-celled carpogonial branch surmounts an auxiliary cell, the tip of which is produced and, like the median rudiment in the other types, lies adjacent to the carpogonium. In the procarps of *Lithothumnion* and *Epilithon* this protrusion is lacking so that the auxiliary cell appears as the basal cell of a three-celled carpogonial branch. The preceding account is based on that of Suneson, who interprets the diverse stages as a reduction series.

In *Choreonema* all the procarps of a conceptacle are fully differentiated, but in *Corallina* the carpogonial branches in the marginal procarps remain unicellular and in *Lithothamnion* and *Epilithon* two-celled. Fertilization of a single carpogonium suffices for further development. The subsequent events are difficult to follow. In *Melobesia,* Suneson observed the outgrowth from the carpogonium of a short connecting filament which fused with the auxiliary cell of the same procarp; this may well represent the usual course of events, although fusion with some other adjacent auxiliary cell is also possible. After the passage of the diploid nucleus into an auxiliary cell, all the auxiliary cells of a conceptacle undergo progressive fusion, which results in the formation of a large multinucleate fusion-cell bearing on its upper side the remains of the diverse procarps. Some of the nuclei in this cell are commonly larger or richer in chromatin and are probably diploid. In *Lithothamnion,* the layer below the auxiliary cells also seems to enter into the formation of the fusion-cell.

The gonimoblast-initials usually arise from the margin of the fusion-cell and develop into unbranched threads, which form large uninucleate carposporangia in basipetal sequence. In certain species *(Melobesia limitata, Amphiroa rigida)* gonimoblast-threads also originate from the upper surface of the fusion-cell.

In *Choreonerna* the stages subsequent to fertilization are somewhat different. The carpogonium enlarges, while its hypogynous cell persists longer than in the adjacent procarps. The female organ fuses with the auxiliary cell of its own procarp, but the diploid nucleus remains within the carpogonium. After this the latter puts out short connecting processes, which successively fuse with slight protuberances from the neighbouring auxiliary cells, and by degrees all the latter are involved; the auxiliary cells finally degenerate. In *Choreonerna,* therefore, the carpogonium progressively absorbs the contents of the auxiliary cells and ultimately appears as a large lobed cell, containing a number of diploid nuclei; this forms the basipetal rows of carposporangia in the usual way.

In the male conceptacles, the fertile threads either occupy the floor only *(Melobesia, Lithophyllum, Amphiroa)* or also spread on to the side-walls, the part of the wall near the aperture is covered with sterile paraphyses. The details of antheridial development are difficult to decipher, but as a general rule the cells of the disc seem to cut off two elongate mother-cells, which form a dense layer; each mother-cell produces apically a number of, often elongate, antheridia. There is considerable evidence that the entire antheridium functions as the spermatium. At the time of detachment its lower end is often attenuated, a feature which is very pronounced in *Corallina*; according to Suneson the tail is formed by a shrunken part of the wall from which the protoplast has withdrawn. Recent observations afford little support for the view that the antheridia of Corallinaceae are formed in rows.

In *Epilithon*, the central threads of the young male conceptacle consist of two cells, while the peripheral ones are longer. The mother-cells are cut off from the two lower cells of the fertile threads, while the upper cells of the peripheral ones develop branches which form the roof of the conceptacle. Later the two fertile layers separate to form a cavity lined with mother-cells and roofed by the several-layered conceptacle-wall, which develops a narrow aperture. The mature conceptacle is filled with detached antheridia.

Family Choreocolaceae

Kylin refers this family of problematical parasites to the Cryptonemiales owing to certain resemblances to Callymeniaceae. Both in *Choreocolax* and *Harveyella* the female cushions include numerous procarps. The carpogonial branches arise singly from the auxiliary cell which also bears one or two sterile branches. Sturch produces some evidence that the auxiliary cell proper is cut off from the supporting cell at the time of fertilization. In *Harveyella*, the carpogonium fuses with the auxiliary cell and this with the cells of the sterile threads to form an irregular fusion-cell, from which the gonimoblast-initial is produced. Similar changes occur in *Choreocolax*, although here the carpogonium communicates with the auxiliary cell by means of a short connecting thread. In both genera the auxiliary cells form secondary pit-connections with the adjacent cells.

The gonimoblast *of Harveyella* is a richly branched thread which spreads horizontally among the cells of the cushion and produces tufts of erect branches, bearing terminal carposporangia. The auxiliary cell of *Choreocolax* cuts off several gonimoblast-initials which produce short filaments with terminal carposporangia; the mature cushions are lobed, each lobe harbouring a group of carpospores. In *Harveyella*, the entire cushion constitutes a single fruit. The antheridia cover the whole surface of the male cushions; in *Harveyella* they are cut off from narrow mother-cells produced by the peripheral cells.

There are evident analogies to Callymeniaceae in the structure of the female reproductive system. At the same time, as pointed out by Sturch, the procarp of *Harveyella* resembles that of Ceramiales, especially if the auxiliary cell be formed after fertilization. The other events succeeding fertilization and the ripe gonimoblasts are, however, altogether different from what is found in the latter order. The two genera require further investigation.

Holmsella appears to have other affinities. The auxiliary cell, stated to be cut off after fertilization, arises from one of the lower cells of a peripheral cell-row near the usually two-celled carpogonial branch; the short connecting filament is sometimes two-celled. During the development of the gonimoblast, extensive fusions take place between its filaments and the adjacent cells, and this is followed by the formation of vertical rows of carposporangia. The antheridia are formed much as in *Harveyella*. This genus presents certain analogies with Cruoriaceae.

Tetrasporangiate Phase

In many Cryptonemiales, the tetrasporangia are borne on the vegetative laterals and lie embedded in the surface-tissues.

Some of the nemathecial forms also produce their sporangia in nemathecia, although this is not so in *Polyides* and *Rhizophyllis.* The spores are for the most part arranged cruciately, though zonate in *Dudresnaya, Constantinea Cruoria* and Corallinaceae. In the production of the tetraspores four nuclei are usually formed before separating walls arise.

Transitions between the cruciate and zonate types have been recorded in *Dumontia. Gloeopeltis,* and *Cryptosiphonia,* while according to Kuckuck the arrangement of the spores in *Harveyella mirabilis* is either cruciate or tetrahedral. In *Cruoria* and *Dumontia* the sporangia occupy the same position as the carpogonial branches, whereas in *Petrocelis* they are intercalary in the erect threads. Those of *Polyides* arise near the inner limit of the outer cortex. In the nemathecia of *Peyssonnelia* the sporangia are embedded among the sterile threads; the formation of a nemathecium terminates the growth of tin part of the thallus involved and in the next season it becomes overgrow by the adjacent parts. Sporangial nemathecia are also formed *Endocladia,* the sporangia being borne laterally on the lowest cell of certain of the component threads.

Most Corallinaceae produce their tetrasporangia in conceptaclei, similar to those harbouring the sex organs The cells of the disc usually divide into a stalk-cell and the sporangia, the sporangia are uniformly distribute over the floor of the conceptacle, but in *Lithophyllun,* they are commonly restricted to the periphery, while the raised central portion is occupied by a tuft of sterile threads. The greater specialisation of the latter genus is evident also in the structure of the procarp.

In *Epilithon, Lithothamnion* and *Phymatolithon* inch *Eleutherospora* there are no conceptacles, the sporangia forming sori which are embedded within the crusts. The sorus of *Epilithon* is formed by the outgrowth from the basal stratum group of fertile and sterile threads, the terminal cells of the former giving rise to the sporangia. As they develop, the second lowest cell of each sterile thread undergoes great elongation, thus elevating the cells above which form a compact roof the sorus. The membrane of the young sporangium develops a pronounced apical thickening which penetrates among the cells of the roof and ultimately dissolves, leaving an open channel above each sporangium. Similar sori are found in *Lithothamnion* and *Phymatolithon*; in the latter the roof of the sorus is concave. In *Lithothamnion*, the sori become overgrown by the surrounding parts, and this may ensue so quickly that the tetraspores remain imprisoned within the crusts. Old sori either appear as cavities within the strata or become filled with tissue.

More extensive, and not such sharply circumscribed, sori occur in the fossil *Archaeolithothamnion*, as well as in the recent genus *Sporolithon*, which is stated to have cruciate tetrasporangia, The kind of sorus found in *Archaeolithothamnion* may well be primitive and have led to the more defined type seen in *Lithothamnion* and *Epilithon* which, in its turn, may have resulted in the conceptacle with a single aperture. Such a view is justified by certain similarities in development. Thus, in *Melobesia* the conceptacle commonly consists of an aggregate of fertile and sterile threads, although the latter are stated to be absent in some species. The roof of the conceptacle is formed partly from the sterile threads, and this is also so in *Amphiroa* and *Lithophyllum*. In diverse Corallinaceae, however, the sterile threads are reduced or altogether lacking and the wall of the conceptacle is formed entirely by upgrowth of peripheral tissue. According to Solms-Laubach delicate paraphyses occur also in the young sporangial conceptacles of *Corallina*.

The diversity of the sporangial conceptacles of Corallinaceae renders them more useful than the sexual ones in tracing affinities. From the possibly primitive *Archaeolithothamnion* we can trace three lines of development, that exemplified by *Lithothamnion* and *Epilithori* that illustrated by *Melobesia*, *Lithophyllum* and *Choreonema*, and that including *Amphiroa* and *Corallina*. The sori of *Epilithon* and *Lithothamnion* are clearly homologous with the nemathecia of *Peyssonnelia*. The affinity with the nemathecial Cryptonemiales is, however, probably rather remote.

The Corallinaceae commonly produce bispores which appear to occur more frequently in the encrusting forms, although recorded in *Amphiroa rigida.* Bauch distinguishes between obligate and facultative bisporic species, certain of the latter possessing distinct bisporic and tetrasporic races, in part characteristic of different habitats. According to Foslie however bisporic species occur especially in colder seas and are altogether lacking in the Tropics. The bispores are either binucleate *(Lithothamnion poly morphum)* or uninucleate *(Lithophyllum incrustans).* When the nuclei are single, they are about twice as big as a tetraspore-nucleus of the same species; this may imply that the second division is suppressed or perhaps postponed to the time of germination. Mono-spores are recorded in a few members of Corallinaceae. *Melobesia Solmsiana* exhibits vegetative propagation by triangular plate-shaped gemmae, composed of thick-walled cells.

The marine species of *Hildenbrandia,* in which sexual reproduction is unknown, likewise form their sporangia in conceptacles which develop much like those of Corallinaceae. According to Rosenvinge sporangia are produced at all times of the year and arise successively, not only from the floor and the sides of the conceptacle, but even from the roof; as the production of sporangia proceeds, new ones originate from the next-adjacent layer of the thallus so that the conceptacle gradually widens. Occasional sporangia are replaced by paraphyses. The septa between the spores are oblique and often irregular in orientation, although parallel in *H. Crouani.*

The general aspects of the life-cycle of Cryptonemiales are considered, together with that of other diplobiontic Florideae.

Diplobiontic Type among Gigartinales

The Sexual Phase: The order Gigartinales, as defined by Kylin, includes all those Florideae in which the auxiliary cell is formed already prior to fertilization and in which it is constituted by an intercalary cell of one of the ordinary vegetative laterals composing the thallus. In Cryptonemiales, on the other hand, the auxiliary cells are located in special accessory branches. This is rather a trivial feature upon which to base a major taxonomic subdivision, but until a clearer conception of affinities can be obtained, admitting of a different grouping, it serves a convenient purpose. Kylin regards the Gloeo-siphoniaceae and Endocladiaceae as connecting links between the two orders.

The sexual reproductive apparatus of Gigartinales shows a number of marked parallels with that of Cryptonemiaies. Like the latter, they include forms in which carpogonial branches and auxiliary cells are spatially separated and others in which they are combined to form a procarp, and the discussion of the different genera is most usefully arranged on this basis. Both the auxiliary cells and the supporting cells of the carpogonial branches are nearly always located in the inner cortex.

Gigartinales without Procarps

In *Patoma Bairdii,* the colourless three-celled carpogonial branches arise from the inner cells of the laterals, while the more numerous auxiliary cells are located near the base of other laterals, exceptionally a carpogonial branch may be borne directly upon an auxiliary cell. The auxiliary cells are considerably enlarged and protruded on one side. Several connecting filaments arise successively from the carpogonium and, as in Dumontiaceae, the same filament may fuse with a number of auxiliary cells, contact being made with the protruded part of the latter. The gonimoblast-initial is cut off on the outer side of the auxiliary cell and gives rise to a mass of short threads, most of the cells of which become carposporangia. An ill-defined pore is formed in the cortex, through which the carpospores are emitted. No antheridia have been found in P. *Bairdii,* and it is possible that the carpogonium develops parthenogenetically.

Several allied forms show essentially the same features. In *Schizymenia* the pore above each mass of carpospores is well defined. In *Calosiphonia vermicularis* the carpogonium fuses with the supporting cell prior to the formation of connecting filaments, and the auxiliary cells become conspicuous only after fusion with one of the former. The closely related *Bertholdia neapolitana*, as well as *Nemastoma*, are distinguished by the fact that the gonimoblasts arise from the connecting filaments near the auxiliary cells. In *Sehdenia*, the fruiting parts are prominently swollen owing to extensive division of the cortical cells above the gonimoblast; the cells in question acquire rich contents and the inner ones are believed to supply nourishment to the developing gonimoblast. In all these Gigartinales, the gonimoblast-initial is cut off externally.

The carpogonial branches of *Furcellaria fastigiata* and *Halarachnion ligulatum* arise near the tips of the fronds from large cells of the inner cortex; especially in *Furcellaria* the same supporting cell may bear more than one carpogonial branch. The auxiliary cells are likewise situated in the inner cortex. The male plants of *Furcellaria* are conspicuous by their swollen, pale yellow or pink tips, upon which the surface-cells produce pairs of mother.

The connecting filament establishes contact with the inner surface of the auxiliary cell which, in *Halarachnion*, develops a thin-walled process with which the connecting filament fuses. The gonimoblast-initial is in both genera cut off internally, immediately adjacent to the point of fusion. All the cells of the gonimoblast become fertile, the masses of carpospores lying in or at the periphery of the medulla; the carpospores are extruded through narrow pores formed in the cortex. The fruiting tips of *Furcellaria*, observable in autumn and winter, are usually somewhat inflated. In *Halarachnion*, the cells of the carpogonial branch become swollen and multinucleate in later stages, while the fertilized carpogonium divides transversely, each cell giving rise to a number of connecting filaments.

A considerable number of the Gigartinales without procarps differentiate special *nutritive systems* in connection with gonimoblast-development. In *Turnerella* and *Opuntiella*, as well as in *Catenella Opuntia*, the inner cells of the gonimoblast-threads fuse with the auxiliary cell to form a branched fusion-cell bearing rows of carposporangia in *Opuntiella*, moreover, the lateral harbouring the auxiliary cell produces numerous small-celled branchlets rich in nutriment which is absorbed during the development of the spores. In these genera, however, as in the Gigartinales previously considered, nourishment reaches the gonimoblast only by way of the auxiliary cell.

More specialised features are found in Solieriaceae, among which *Agardhiella tenera* furnishes a typical instance. Here nearly all the cells of the laterals, containing the uninucleate auxiliary cells, possess dense cytoplasmic contents and several nuclei. As the gonimoblast develops, the surrounding cells give rise to an enveloping small-celled secondary nutritive tissue. The gonimoblast-initial divides into a group of sterile cells and, during this, the contents of the primary nutritive tissue are exhausted; from the central group arise radiating gonimoblast-threads, which penetrate the secondary nutritive tissue and enter into secondary pit-connections with its cells. The sterile cells of the gonimoblast lose most of their contents, as the terminal carposporangia develop. In the South European *Solieria chordalis* the sterile cells of the gonimoblast fuse with the auxiliary cell to form a huge spherical fusion-cell, which is very conspicuous in the ripe fruit.

The carpogonium of *Agardhiella* enlarges after fertilization and forms several connecting filaments the few auxiliary cells possess an apically directed process. The connecting filament fuses with the lower end of the auxiliary cell, but the gonimoblastp-initial is cut off from the tip of the latter. The carpogonial branches and auxiliary cells of *Catenella* occur within special small segments of the fronds; as in all Rhabdoniaceae, the auxiliary cells are not distinguishable before fertilization and develop rich contents only after fusion with a connecting filament. The gonimoblast-initials are cut off internally and the radiating gonimoblast-threads branch abundantly amid the medullary cells of the fertile frond. The swollen fruits of *Agardhiella* open by a definite pore and the fertilizer tract is surrounded by several layers of flat cells, derived from the secondary nutritive tissue. In *Catenella*, which rarely fruits, the entire fertile frond constitutes a single cystocarp, the wall of which is formed by the cortex.

The antheridia of *Agardhiella* form irregular sori on the younger parts. Those of the monoecious *Catenella opuntia* are produced on special branchlets of the frond, the rounded sori having a wrinkled surface, so that the antheridia appear in small pits.

Gigartinales with Procarps

In diverse Gigartinales the auxiliary cell is the basal cell of a lateral, borne on the supporting cell of the carpogonial branch moreover, in all of these the gonimoblast-initial is cut off internally. The genera in question show progressive elaboration in the mode of nourishment of the gonimoblast. *Cystoclonium purpureum*, except for the possession of a procarp, resembles *Catenella* in its sexual reproduction. The supporting cell of the colourless carpogonial branch bears a number of laterals, the cells of which have dense cytoplasmic contents; this is specially marked in the auxiliary cell, which lies immediately adjacent to the carpogonium and no doubt becomes linked with it by a short connecting filament.

The gonimoblast-initial develops into an elongate well-branched thread which penetrates into the central tissue. By degrees its cells fuse with one another, as well as with the auxiliary cell to form an irregularly branched, tubular fusion-cell, bearing numerous lateral tufts of carposporangia. There is abundant production of hyphae near the developing gonimoblast. The cortical cells above the procarp show active division so that the mature fruits appear as prominent unilateral swellings on the thalli; they are devoid of an aperture. The antheridia form extensive sori on the younger branches of the male plants.

Essentially similar procarps are found in *Rhodophyllis bifida* where they occupy the margins of the fronds, in *Calliblepharis* where they lie in the marginal pinnae, and in *Hypnea musciformis* where they are situated in the short thallus-branches, except that in the last two the auxiliary cells acquire dense contents only after fertilization. In *Rhodophyllis* the supporting and auxiliary cells produce on their outer sides a layer of cells with dense contents which, like other adjacent cells, divide abundantly after fertilization to form the wall of the mature cystocarp; the latter appears as an almost spherical swelling and has no aperture. The gonimoblast develops as in *Cystoclonium* but some of its branches remain sterile and become connected to the pericarp, nutritive device which is more marked in *Calliblepharis* and *Hypnea*.

Here, however, no fusion-cell is formed, although two kinds of nutritive tissues are developed. One of these lies external to the procarp and is derived from the inner cells of the laterals borne on the auxiliary and supporting cells; the outer cells aid in the production of the pericarp. The second nutritive tissue lies internal to the procarp and consists of small cells arising from the inner vegetative cells. According to Kylin certain species of *Rhodophyllis,* which also approach *Calliblepharis* in other respects, possess a similar internal nutritive tissue. The gonimoblast-initial of *Calliblepharis* gives rise to several threads that penetrate among the internal nutritive cells, with which they establish secondary pit-connections; meanwhile the external nutritive system is exhausted. The carpospores are produced in unbranched threads, arising from the basal cells of the gonimoblast, and appear as a fan-shaped group on the outside of the internal nutritive tissue. The large swollen cystocarps are formed singly in the fertile pinnae.

In *Hypnea,* the gonimoblast develops in a distinctive manner. The initial divides to form a compact group of small cells; some of these, after establishing connections with the internal nutritive cells, undergo marked elongation and lift the remainder into the centre of the developing cystocarp-cavity. These gonimoblast-cells give rise on all sides to branched threads, which become linked by secondary pits with the cells of the pericarp. The richly branched tufts, forming terminal carposporangia arise not only from the central gonimoblast-group, but also from the threads linking the latter with the wall.

The large cystocarps found in these and other genera arise by extensive post-fertilization division of the cells in the neighbourhood of the procarp; this results in a cavity of increasing width, which is occupied by the developing gonimoblast. Division is either essentially confined to the cortex, external to the fertilized procarp or takes place all round it.

Antheridia appear to be unknown in *Calliblepharis,* while those of *Hypnea* have not been adequately investigated. In the monoecious *Rhodophyllis bifida,* they form small scattered sori, the mother-cells arising in fours from the surface cells and forming two or three antheridia each.

A procarp of a more generalised type is found in *Rissoella,* a Mediterranean perennial with a fringed foliaceous thallus. Here a single large auxiliary cell is associated with a number of carpogonial branches, the supporting cells of which are constituted by adjacent cortical cells; it is not clear whether they all belong to the same branch-system. The surrounding cells develop a number of small-celled laterals, homologous with carpogonial branches and constituting a nutritive tissue, although no special connection is established between it and the fusion-cell which is formed from the inner cells of the gonimoblast, the auxiliary cell, and probably also other adjacent cells.

In most other Gigartinales, the supporting cell functions also as the auxiliary cell. In *Plocamium* the supporting cells are differentiated in large numbers along the outer margins of the fronds from the large cells surrounding the axial thread; they also bear vegetative laterals. Antheridia are formed over the entire surface of the ultimate branches of the thallus. After fertilization the carpogonium enlarges and the other cells of its branch degenerate, while those in the vicinity of the auxiliary cell acquire rich contents. The gonimoblast-initial, which fuses with the auxiliary cell, produces a number of secondary elements which successively give rise to densely branched tufts, most of the cells of which become carposporangia. The mature cystocarps form sessile swellings on the edges of the thalli; the spores are liberated by disintegration of the upper part of the wall.

In *Sphaerococcus coronopifolius* and *Gracilaria* the procarps are essentially similar, although in the latter the supporting cell is not specially differentiated and the laterals possess rich contents. Both genera exhibit more marked post-fertilization fusions than does the account of Johnson is largely erroneous, while Phillips has mistaken the characteristic hairs for carpogonia.

In *Sphaerococcus*, the auxiliary cell produces several gonimoblast-initials, which give rise to separate systems with terminal carposporangia. Meanwhile the older gonimoblast-cells fuse with the auxiliary cell, and this with the adjacent pericentral and axial cells, to form an elongate fusion-cell bearing the radiating fertile threads. The laterals borne on the pericentral form a local nutritive tissue around the base of the fusion-cell.

In the Australian genera *Mychodea* and *Acrotylus*, which show affinity to Gigartinaceae and are possibly primitive, the supporting cell bears several carpogonial branches. In *Mychodea*, the principal gonimoblast-threads form a network within the cystocarp, and some are linked by secondary pit-connections to the nutritive tissue lining the inner surface of the pericarp, while m *Acrotylus* the bunches of carpospore-bearing threads project into a cavity which is lined by a probable nutritive tissue.

The antheridia of *Sphaerococcus* are formed in small circular depressions behind the tips of the branches. This is also so in *Mychodea* and in *Gracilaria confervoides*; in the latter the male plants are of small stature.

Phyllophoraceae and Gigartinaceae

In these families, the procarps invariably consist of an, often large, supporting cell bearing a three-celled carpogonial branch in addition the former may produce a vegetative lateral, which is lacking in *Chondrus* and in *Gigartina* is occasionally replaced by a second carpogonial branch. No conspicuous fusion-cell is formed after fertilization, but the threads of the gonimoblast penetrate into the medullary region, the elements of which form a nutritive tissue.

Chondrus produces its procarps within the ordinary fronds, but in *Phyllophora Stenogramma,* and *Gigartina* they are situated in special outgrowths. These arise from the stalks in *Phyllophora membranifolia* and from near the margins of the fronds in *P. epiphylla,* while the teat-like, simple or branched, nemathecia of *Gigartina* occupy either the margin or the surface of the thallus. Those of *Stenogramma* are linear and form apparent veins on both surfaces of the fertile fronds; since they terminate a short distance below the apex, they are not connected in successive branches. In *Phyllophora* and *Gigartina,* each nemathecium produces a single cystocarp, while in *Stenogramma interrupta,* where procarps occur along the whole length of the nemathecium, their occasional absence or failure to be fertilized leads to gaps in the dark red zones of maturing carpospores to which the specific name refers. In *Gigartina* there is a definite pore to the cystocarp.

In *Chondrus crispus,* the supporting cell belongs to the inner cortex and is linked to the surrounding vegetative cells by secondary pit-connections. The adjoining carpogonium communicates with the auxiliary cell after fertilization by a broad connection, through which the diploid nucleus no doubt passes into the auxiliary cell. The elongate gonimoblast-threads penetrate into the inner tissues and give rise to numerous groups of carposporangia, interspersed between the remains of the medulla. The carpospores are liberated from the swollen fruits after the decay of the fronds.

The structure of the procarps and the formation of the gonimoblast are essentially the same in *Gigartina*. In certain species a special nutritive tissue arises from hyphae, produced in large numbers from the medullary cells, and in later stages the gonimoblast-threads communicate with its cells; in the mature cystocarp this tissue forms a fibrous envelope around the groups of carpospores. Not all species of *Gigartina* possess a definite pericarp and nutritive tissue may be lacking. A similar fruit-development is found in *Iridaea* and *Rhodoglossum*.

The procarps of *Stenogramnia interrupta* are distinguished by the usual multinucleate character of the component cells. Their further development is similar to that of *Chondrus,* but both here and in *Phyllophora* the gonimoblast-threads enter into communication with the loose medullary tissue of the nemathecia. The numerous groups of carpospores produced from the tufted laterals of the gonimoblasts ultimately almost completely displace the medulla.

All these genera are dioecious. The pale antheridial sori of *Chondrus crispus* usually appear just behind the apices of the fronds; Darbishire found them on minute white segments which later developed into vegetative branches. The mother-cells formed by longitudinal division of the surface-cells, produce pairs of elongate antheridia which just penetrate the gelatinous cuticle overlying the sorus. Secondary antheridia are frequent. The antheridia of *Gigartina* are borne on small leafy branches.

Those *of Phyllophora* are produced on minute white or yellowish nemathecia, borne on the margins of the fronds. The surface-cells give rise to numerous tufts, some of which grow longer than others so that small antheridial cavities are formed. The male sori of *Stenogramma* form irregular patches on both surfaces of the upper thallus-segments.

Tetrasporangiate Phase

The tetrasporangia of Gigartinales are either cruciate or zonate, the latter type being found in all the genera except *Platoma, Calosiphonia, Nemastoma, Sebdenia, Gracilaria, Melanthalia, Cordylecladia,* Phyllophoraceae and Gigartinaceae. The zonate sporangia of *Furcellaria* and *Plocamium* are stated to divide first into two cells.

The sporangia are often terminal on the laterals are borne

laterally on the inner cortical cells and are rather deeply immersed within the inflated tips. In *Plocamium*, the sporangia are formed within special branches of simple structure, which occupy the margins of the ultimate fronds. The large sporangia, which develop from the hypodermal cells and form a zigzag line along the axis of the stichidium, are covered by but a single layer of cells.

Special nemathecia harbour, the sporangia in *Phyttophora* and *Stenogramma*. In *P. membranifolia* they appear as deep-red swellings on the fronds, while those of *P. epiphylla* encircle the short stalks of special peltate branches. The numerous threads of the nemathecia consist of long chains of cruciate tetrasporangia, only the outermost cells being sterile. The scattered wart-like nemathecia of *Stenogramma* show a similar structure.

The sporangia of *Chondrus* develop in the younger fronds, during the early months of the year, from the cells of small tufted laterals borne on the medullary elements; the regions in question form dark red swellings on the thalli. In *Gigartina* the laterals producing the sporangia arise from the inner cortical cells and are located either in special papillate or leafy outgrowths or within the ordinary thallus. The tetraspores in all Gigartinaceae are set free by decay.

Diplobiontic Type among Rhodymeniales

The Sexual Phase: Despite the considerable diversity in habit, there is great uniformity in the structure of the procarp and in the post-fertilization stages among Rhodymeniales. The multinucleate supporting cell of the pro-carp is constituted by a young cortical cell which rapidly enlarges and acquires richer contents. After producing either one or two auxiliary mother-cells, it gives rise to a three-or four-celled carpogonial branch. The auxiliary cells are cut off from the mother-cells so as to lie near the carpogonia; although formed prior to fertilization, they remain vegetative, unless sexual fusion occurs, and enlarge and acquire abundant contents only after that event. The gonimoblast-initial is cut off on the outer side of the auxiliary cell and, during its further development, there is extensive fusion of cells; in the Champiaceae this is accompanied by great enlargement. Nutriment passes to the developing gonimoblast solely by way of the auxiliary mother-cell,

Among Rhodymeniaceae *Rhodymenia pertusa* has been specially studied. The supporting cell bears a three-celled carpogonial branch, in addition to the auxiliary mother-cell. After fertilization the cells of the carpogonial branch fuse to form a multinucleate fusion-cell, which is displaced by the enlargement of the auxiliary cell and of its now multinucleate mother-cell. The latter establishes pit-connections with the surrounding nutritive tissue. At this stage, the uni-nucleate auxiliary cell, as in other Rhodymeniales, contains a conspicuous round protein-mass. Meanwhile the overlying cells divide to form the wall of the future cystocarp. The gonimoblast-initial gives rise to a mass of threads, in which most of the cells produce small carpospores. Within the swollen cystocarp the mass of spores is separated by a wide space from the encompassing wall.

The antheridia of *R. palmata* form small irregular sori on the fronds. The colourless mother-cells arise by tangential division of the surface-cells and produce successively two primary and two secondary antheridia.

The procarps of *Fauchea* and of *Botryocladia* are similar, but in the former the auxiliary mother-cell bears a short vegetative branch. The post-fertilization events differ only in the fact that, during the enlargement of the cystocarp in *Fauchea*, the inner cells of the pericarp stretch to form a reticulate tissue surrounding the spores; this is so also in *Bindera*. In *Botryocladia* the cystocarp projects into the hollow thallus. Some data about the antheridia are given by Kuckuck.

The Champiaceae differ only in matters of detail. The supporting cell is linked by a primary pit-connection to the cell of a longitudinal thread. The carpogonial branches are usually four-celled, commonly with the lowest cell transversely extended; they are three-celled in *Lomentaria* and in *Champia lumbricalis*. In *Chylocladia, Gastroclonium,* and not uncommonly in *Lomentaria clavellosa,* there are two auxiliary cells, though in the last only one undergoes further development. The product of fusion of the cells of the carpogonial branch enlarges considerably, and in *Champia* a wide connection is established between it and the supporting cell. The carpogonium connects with the auxiliary cell by a short thread, which may arise from the auxiliary cell *(Champia, Gastroclonium),* from the carpogonium, or from the fusion-cell; in *Lomentaria* there is not uncommonly a special connecting cell. The diploid nucleus divides within the carpogonium and one of the products passes into the auxiliary cell.

In *Chylocladia* and *Gastroclonium* extensive fusions result in a large fusion-cell occupying the middle of the ripening cystocarp; these fusions involve not only the inner cells of the gonimoblast and the auxiliary cells, but the latter fuse with one another and with their mother-cells, as well as with some of the adjacent nutritive cells. In *Lomentaria* and *Champia,* on the other hand, there is comparatively little fusion. In the former the richly branched gonimoblast gives rise successively to a number of gonimolobes in which most of the cells form carpospores (cf. Rhodymeniaceae), whereas in other Champiaceae the large carposporangia are terminal and develop simultaneously. In *Chylocladia* and *Gastroclonium* the gonimoblast-initial divides by successive radial walls into wedge-shaped segments which subsequently divide tangentially; the inner cells fuse with the auxiliary cell, while the outer constitute the carposporangia.

The cystocarp-wall begins to develop at the time of fertilization by the outgrowth, from the cortical cells, of branched threads, the cells of which become joined by secondary pit-connections. In *Champia,* and to a less marked extent in *Lomentaria* the inner cells of the wall undergo considerable stretching during the enlargement of the cystocarp and form a loose reticulate tissue around the gonimoblast. The cystocarps of *Chylocladia* and *Gastroclonium* are devoid of a pore.

Eliding is likely to be right in regarding *Lomentaria* as the least specialised member of the Champiaceae, the presence of two auxiliary cells being probably a secondary acquisition short laterals. In *Rhodymeniapalmata,* they form irregular sori on both surfaces of the frond and originate from superficial cells which-subsequently become embedded. Kuckuck describes the sporangia of various species. *Hymenocladia* has tetrahedrally divided sporangia formed from intercalary cells.

The large sporangia of Champiaceae possess tetrahedral spores. They usually originate from uninucleate intercalary cells of the secondary cortex which enlarge so that they project into the central hollow. In *Lomentaria,* however, where the small sori occupy superficial depressions, the sporangia develop from the terminal cells of short branched laterals that arise from the primary cortical cells.

Diplobiontic Type among Ceramiales

This order, which was first separated from the Rhodymeniales of Schmitz by Oltmanns owing to its distinctive reproductive features, comprises the most specialised diplobiontic forms. Several of the peculiar characteristics were already noted by Phillips. Formation of the auxiliary cell in Ceramiales is deferred until after fertilization, its mother-cell being the supporting cell of the carpogonial branch which is, moreover, invariably a pericentral or (in Ceramiaceae) its equivalent. The carpogonial branch represents a lateral borne upon the supporting cell and always consists of four cells devoid of laterals, although the supporting cell usually bears either one or two sterile branchlets. The uniformity of the sexual reproductive apparatus is as significant as the loss of all traces of heterotrichy in the early stages of development.

The Sexual Phase

Ceramiaceae: The members of this family, which probably represent the most primitive forms among Ceramiales, are distinguished by several characteristics. Communication between the fertilized carpogonium and auxiliary cell is usually established by means of special connecting cells, no compact envelope is formed around the gonimoblast, and most of the cells of the latter generally produce carpospores. The supporting cell of the procarp is homologous with the fertile pericentral of other Ceramiales. The simplest condition is found in *Eallia* and in most species of *Antithamnion,* where the branches bearing carpogonia on their basal cells do not differ from the other laterals of limited growth; in *A. pacificum,* however, they are reduced to two or three cells and in *A. spirographidis* Schiffn to two cells. In most other Ceramiaceae the fertile lateral is represented merely by the supporting cell and an overlying sterile cell, although the latter may later divide. The sterile cell is cut off from the supporting cell prior to the development of the carpogonial branch. In *Crouania,* the single fertile member of a whorl is always much reduced and sometimes represented only by the supporting cell, and this is invariably so in *Callithamnion,* where there are no sterile cells in the procarp.

In *Antithamnion plumula* the carpogonial branch arises from the under side of the basal cell of the lateral and, at first, the supporting cell does not differ in any way from the others; after fertilization an auxiliary cell with dense contents is cut off on the upper side. The fertilized carpogonium cuts off a small connecting cell which fuses with the auxiliary cell. The gonimoblast-initial originates from the upper side of the latter and gives rise to several tufts of threads, in which the basal cell remains sterile, while the remainder develop into carposporangia. The latter constitute several distinct groups and are surrounded by a loose envelope formed from neighbouring sterile branches; in *A. pacificum* and *A. spirographidis* the gonimolobes ripen successively. Diverse investigators record fusions at the base of the gonimoblast. *Platythamnion villosum* shows similar features. The cystocarps of *Ballia* occur in the axils of the fertile laterals.

The procarps of *Ceramium* develop just behind the apices. The supporting cell is the first-formed pericentral of a segment and commonly bears two carpogonial branches, although in

C. *rubrum* there is only one. Two carpogonial branches are also met with in the closely related *Micro-cladia glandulosa*. The post-fertilization development of *Ceramium* does not differ in any important detail from *Antithamnion*.

In *Wrangelia penicillata*, the procarps arise on special distichous laterals of limited growth, which are crowded together at the apices of the main thallus-branches; in these fertile laterals the pericentrals of the upper 10-15 segments produce whorls of branchlets consisting of only 2-3 cells. The slightly enlarged basal cell of the oldest member of each whorl, in the middle segments, bears a curved carpogonial branch, while the cells above are sterile; usually only one carpogonium on each lateral is fertilized. After fertilization all the branchlets of the fertile lateral develop further and form a dense group overtopped by surrounding sterile branches.

The gonimoblast-threads spread over the axial cells of the fertile branch and establish secondary pit-connections with the lower cells of the sterile laterals. The large terminal carposporangia are formed on branches of the creeping gonimoblast. The single cystocarp, produced from a fertile lateral, is traversed by the axial cells and is provided with an incomplete, though relatively compact, envelope formed by the sterile laterals. The fertile branchlets of *Griffithsta*, which consist of a few multinucleate cells, arise from the apices of the large axial cells, although soon deflected by the development of a lateral. The lowest cell of the branchlet later gives rise to the envelope of the cystocarp, while the subterminal one produces the procarp.

This cuts off three uninucleate pericentrals, those to the right and left developing into two-celled branchlets, the basal cell of which bears a carpogonial branch. There is direct fusion between the auxiliary cell and the carpogonium; a special connecting cell is not produced. The auxiliary cell produces several gonimo-blast-initials which develop successively into richly branched gonimolobes, most of the cells of which give rise to uninucleate carpospores in basipetal sequence. Meanwhile there is considerable enlargement of the pit-connections between the cells of the fertile branchlet, while the sterile cell and supporting cell fuse with one another and with the axis of the branchlet; later the auxiliary cell fuses with the supporting cell. The few enveloping threads consist of only two cells. The ripe fruits, either sessile or borne on short branchlets, are situated between the large cells of the thallus. Similar cystocarps occur in *Halurus* and *Bornetia*.

In recent years, the fruit-development of *Spermothamnion roseolum* has been repeatedly investigated. The procarps are produced on short laterals occupying the lower portions of the side-branches. The three upper cells of these laterals possess dense contents and, as in *Griffithsia*, the subterminal one cuts off three pericentrals; the first, which subsequently degenerates, is formed abaxially, the others on the right and left flanks respectively. One of the latter constitutes the supporting cell of the carpogonial branch and produces a single sterile cell; the carpogonial branch curves towards the adaxial side of the fertile lateral. The entire procarp is enveloped in mucilage.

After fertilization both of the last-formed pericentrals (one the supporting cell of the carpogonial branch) cut off auxiliary cells. The carpogonium puts out processes which become cut off as connecting cells and fuse with the auxiliary cells; according to Drew the zygote-nucleus divides twice and products enter the respective connecting cells. Some investigators record fusions between the cells of the carpogonial branch. The auxiliary cells divide transversely into four and each segment produces several laterals forming terminal carposporangia. The products of the two gonimoblasts usually combine to form a single fruit which is surrounded by a few sterile threads, arising from the lower cells of the fertile branchlet.

The carpogonium lies about midway between the two mother-cells, each of which produces an auxiliary cell. The fertilized carpogonium broadens and divides longitudinally into two cells. Each of these cuts off a small connecting cell which fuses with a slight protuberance from the adjacent auxiliary cell: there is considerable resemblance to the mode of formation of a secondary pit-connection. The nucleus of the auxiliary cell takes up a basal position, while one of the two diploid nuclei travels to the apex of the auxiliary cell, meanwhile undergoing considerable enlargement. The degenerating auxiliary nucleus and the other diploid nucleus are cut off by a basal septum, while the gonimoblast arises from the apex of the auxiliary cell. The naked masses of carposporangia are either spherical or composed of a number of successively maturing lobes. In *Seirospora Griffithsiana*, with similar procarps, the gonimoblasts are loosely branched.

The procarps of *Ptilota flumosa* are again borne on short laterals in which the subterminal cell produces three pericentrals, but here the irpogonial branch arises from the first-formed abaxial one. The details are difficult to decipher owing to the outgrowth of all three pericentrals, as well as of the terminal cell of the fertile lateral, into sterile branchlets ending in elongate unicellular hairs. The fertilized carpogonium forms a special connecting cell which fuses with the auxiliary cell. The procarps of *Plwnaria elegans* show no important differences.

The fertile pinnae of *Compsothammon* develop their procarps in the same way, the subterminal cells successively producing three pericentrals, each of which generally bears a sterile cell. The second and third pericentrals constitute auxiliary mother-cells and the former produces the carpogonial branch. The *Ptilota* and *Plumaria* the antheridia cover the tips of the finer branches. The mother-cells in *Ceramium* arise by division of the cells of the cortical bands, while in *Spyridia* the greyish sori occupy the cortical bands of the "filaments".

The peculiar male heads of *Wrangelia*, simulating the conidiophore of an *Eurotium*, comprise numerous branchlets borne on the globular terminal cell and enveloped by several one-celled threads. The antheridial sori of *Griffithsia*, lastly, form caps on the terminal segments or encircle the constrictions between the younger cells; they consist of densely branched laterals composed mainly of uninucleate cells.

The mother-cells usually produce two or three antheridia and secondary antheridia are often formed.

Delesseriaceae: Of the two groups of sterile cells, characteristic of the procarps of this family, one is cut off from the supporting cell before, and the other after, the formation of the initial of the carpogonial branch. The first group is homologous with the sterile cells of Ceramiaceae and, as there, represents what is left of the fertile lateral bearing the carpogonial branch. The second group, on the other hand, is homologous with the carpogonial branch itself, since in *Polyneura* and the Australian *Hemineura* it is replaced by a second, though less strongly developed, carpogonial branch; according to Phillips this is sometimes also so in *Apoglossum ruscifolium.* In the Delesserieae, by contrast to the Nitophylleae, the first set of sterile cells acquire dense contents only after fertilization.

In all Delesserieae the procarps arise from the axial cells of the fertile fronds. In *Delesseria sanguined* these are constituted by minute lanceolate leaflets, arising in considerable numbers in autumn from the midribs of the vegetative fronds. Every axial cell of such a fertile leaflet, apart from a few of the apical segments which develop only lateral pericentrals, produces a pair of procarps. The first step in their development is the cutting off, in the plane of the frond, of tangential pericentrals which function as supporting cells. Each divides periclinally to produce the first sterile cell, after which the initial of the carpogonial branch is cut off by an anticlinal wall. The carpogonial branches usually arise alternately to right and left in successive segments, the pair belonging to a given segment being mostly directed to the same side; the inflated trichogynes project obliquely above the general surface. The second cell of the carpogonial branch is larger than the others. As many as 50 procarps may be formed in a fertile leaflet and, since they mature in acropetal succession, fertilization is possible over a prolonged period. At about the time of fertilization a second sterile cell is cut off from the supporting cell and both sterile cells divide to form short threads.

The entire fertile leaflet forms a single cystocarp and, as a general rule, probably only one procarp is fertilized. According to Phillips its axial cell and the associated pericentrals, as well as the adjacent axial cells, acquire abundant contents so that they are readily picked out with a lens in a leaflet suitably stained. The lower cells of the carpogonial branches of the remaining procarps grow out into vegetative threads and help in the formation of the pericarp. The carpogonium fuses directly with the auxiliary cell of the fertilized procarp, while the gonimoblast-initial gives rise to a richly branched thread, in which many of the cells form carpospores. During these events the cells of the sterile threads have multiplied and acquired dense contents; later the cells disintegrate leaving a cavity which is occupied by the gonimoblast. The auxiliary cell fuses with the supporting and axial cells, as well as with some of the older gonimoblast cells, to form a large fusion-cell, upon which the rows of carposporangia are borne. The mature cystocarps are pedicellate and provided with an aperture they constitute asymmetrical swellings, the surmounting flap representing the apex of the fertile leaflet.

The arrangement and structure of the procarps and the mode of production of the cystocarp are essentially the same in *Membranoptera*, *Apoglossum*, and *Hypoglossum*, although here the procarps arise from the midribs of the ordinary vegetative fronds. Similar cystocarps occur in *Caloglossa*.

In *Grinnellia* the procarps usually appear scattered as in Nitophylleae, but they actually arise from fertile leaflets resembling those of a *Delesseria* although completely embedded in the parent-thallus; the procarp arises from one of the axial cells of such a leaflet. Occasionally the fertile expanses develop as leafy structures projecting above the surface and within them a number of procarps may be formed; the resulting cystocarps then appear stalked.

The procarps of *Claudea* and *Vanvoorstia* arise from the dorsal pericentrals. The single cystocarp, produced from each fertile frond, bears tooth-like projections on its lower surface representing daughter-blades which have failed to anastomose.

Phycodrys rubens, like other Nitophylleae, shows an irregular disposition of the procarps, which appear as white dots near the margins or on special marginal proliferations of the older fronds. The fertile cells, which are larger and have denser contents, cut off tangential pericentrals which divide like those of *Delesseria* to form pairs of procarps, with the trichogynes standing perpendicular to the frond. The adjacent vegetative cells likewise cut off tangential pericentrals. Both sterile cells divide the colourless elements thus formed overlying the supporting cell and carpogonial branch.

After fertilization, as the auxiliary cell is cut off, nutritive materials accumulate in the axial and adjacent cells, as well as in the two groups of sterile cells, while the latter lengthen at right angles to the surface. Simultaneously division of the surrounding pericentrals initiates the cystocarp-wall. The gonimoblast-initial segments vertically to form a small disc, the cells of which give rise to richly branched threads in which 2-4 of the upper cells produce carpospores. These threads radiate from a large fusion-cell which is formed as in *Delesseria*. During the development of the gonimoblast the sterile (nutritive) cells disorganise, the resulting gap forming the aperture of the cystocarp.

Other Nitophylleae are essentially similar. In *Polyneura*, as already noted above, the second group of sterile cells is replaced by a carpogonial branch which is sometimes imperfect; in *Cryptopleura* the sterile groups include very few cells. Diverse Nitophylleae form terminal carposporangia only. This is so in *Cryptopleura*, *Nitophyllum punctatum*, *Acrosorium*, and *Martensia*. *Nitophyllum* differs from other Delesseriaceae in that the fertile cell cuts off on either surface two pericentrals, that first formed producing two or three ordinary cortical cells, while the other is the supporting cell, which, apart from the carpogonial branch bears only one group of sterile cells.

In *Martensia*, the procarps arise within the margins of the longitudinal lamellae. In a transverse section the supporting cell is seen to be cut off as usual from one of the axial cells, which are continuous with the single-layered part of the lamella. The cells of the carpogonial branch are multinucleate. The auxiliary cell is stated to establish connection with the second cell of the carpogonial branch, but this requires confirmation. The richly branched gonimoblast-threads are surrounded by a number of nutritive filaments, arising from the supporting and perhaps also from adjacent cells, but no fusion-cell is formed. The mature cystocarps appear as large swellings on the margins of the longitudinal lamellae.

The Delesseriaceae are almost invariably dioecious and the male plants are commonly smaller than the female or asexual ones, a feature which is very marked in *Caloglossa* and *Martensia*. In *Delesseria san-guinea*, the antheridial sori cover both surfaces (except for a narrow margin) of the minute male fronds. Those of *Membranoptera* frequently occur on similar proliferations situated in the angles between the branches. In *Vanvoorstia* they cover the ultimate blades, which have not yet anastomosed, while in *Gonimophyllum* they occupy practically the whole frond. The sori of *Martensia* occur on either surface of the longitudinal lamellae.

In other Delesseriaceae, the sori occupy more restricted areas. Thus, in *Apoglossum*, *Caloglossa Leprieurii*, and *Hypoglossum*, they are situated on either side of the midribs, while in *Phycodrys* and *Cryptopleura* they appear within the margins of the younger fronds; in the former they sometimes occupy special marginal proliferations. The marginal sori are often broken up into separate groups, as in *Polyneura* and *Nitophyllum*. The minute male sori of *Grinnellia* are irregularly scattered over the thallus.

The regions bearing antheridia always become several-layered, the many small mother-cells being formed by anticlinal division of the pericentrals. As a general rule two or three antheridia are produced successively; in *Martensia* they are formed by abstriction and occur in short rows.

Rhodomelaceae: In most Rhodomelaceae the procarps are formed on the trichoblasts, generally on the suprabasal segment which, like the basal segment, becomes polysiphonous. The fertile segment normally produces five pericentrals. The first two are cut off abaxially, the next two on the flanks, and the fifth on the adaxial side; this last one gives rise to the procarp. It produces in succession a lateral sterile cell, the four-celled carpogonial branch, and lastly the second sterile cell, a sequence identical with that found in Delesseriaceae. The first sterile cell, as in other Ceramiales, represents the only vegetative cell of the fertile lateral, while the second, which corresponds to that of Delesseriaceae and Dasyaceae, is homologous with a carpogonial branch. The sterile cells divide after fertilization and may fulfil a nutritive function, although it has also been suggested that they serve to separate the cystocarp-wall from the developing gonimoblast. The cells of the carpogonial branch are commonly binucleate. The auxiliary cell is cut off apically from the supporting cell, and the gonimoblast develops sympodially. The cystocarps always have a definite aperture and the wall is produced mainly from the two lateral pericentrals of the fertile trichoblast-segment. It is characteristic of Rhodomelaceae that this envelope usually commences to develop already prior to fertilization.

These general features are well illustrated by *Polysiphonia*, where the fertile trichoblasts arise just behind the apex. The development of the procarp takes place as above described. The sterile cells, which Yamanouchi incorrectly calls auxiliary cells, divide transversely after fertilization, while an open communication is established between the carpogonium and auxiliary cell.

In *P. decipiens* Connolly describes the formation of a connecting cell, which fuses with the auxiliary cell; this has not been observed in other species, although recorded also in *Laurencia pinnatifida*. It is probable that only one of the two diploid nuclei passes into the auxiliary cell, which fuses with the supporting cell; later the fusion-cell also amalgamates with the axial cell of the fertile segment. The mode of development of the gonimoblast is peculiar to Rhodomelaceae. The initial gives rise to a number of short threads, the terminal cells of which develop into large and commonly pear-shaped carposporangia while the subterminal ones grow out into two-celled branches, the end-cells in their turn produce carpospores. This sympodial development may continue for some time, the resulting threads bearing successively ripening carposporangia which are irregularly intermingled. Meanwhile the fusion-cell unites with the gonimoblast-initial and with the sterile cells.

The envelope of the cystocarp originates, already before fertilization by the outgrowth of threads arising mainly from the lateral pericentrals of the fertile segment, although the adaxial pericentral of the basal segment usually also plays a part. The young envelope consists of two lateral valves, composed of congenitally fused threads, which enclose the procarp like the shells of an oyster, the trichogyne alone projecting. After fertilization the two valves unite and each cell, except for the apical ones, cuts off two external pericentrals the envelope thus becomes two-layered, with twice as many rows of cells in the outer as in the inner layer.

Bornet describes an abnormal fruit, in which the threads composing the envelope were free at the apices and here formed pericentrals on all sides. The spherical or urn-shaped cystocarps are borne on a short stalk, formed from the basal segment of the trichoblast, while the aperture is directed upwards. The upper part of the trichoblast is usually shed. According to Rosenberg *Brongniartella byssoides* shows altogether similar features.

Other Rhodomelaceae differ in the main only in the position of the procarps and in the structure of the cystocarpenvelope. The fertile trichoblasts are, as in *Polysiphonia*, commonly branched.

Those of *Laurencia*, and *Chondria* consist of only three segments, the second (fertile) one being larger than the others. Developing procarps are found on trichoblasts still enclosed within the apical depression and fertilization ensues before they have emerged from the latter. *Laurencia* is exceptional in producing only four pericentrals on the fertile segment. The urn-shaped cystocarps are almost sessile. *Ricardia* differs only in its embedded cystocarps. A similar reduction of the fertile trichoblasts is met with in diverse other specialised Rhodomelaceae.

Not all Rhodomelaceae, however, produce their procarps on definite trichoblasts. In *Odonthalia* they are borne on cylindrical adventitious branches, arising from the margins of the fronds. The fertile segment is again mostly the second one and this produces five pericentrals in the usual sequence, so that it is probably justifiable to regard these branches as homologous with trichoblasts, although they do not show the usual hair-like character. The same interpretation should possibly be extended to the reduced fertile shoots of *Pterosiphania*. The procarps of *Bostrychia* develop on several successive segments of the younger branches, although only one matures; the apex of the branch is pushed aside by the cystocarp so that the latter appears terminal.

In *Cliftonaea* and *Leveillea*, the procarps are borne on laterals, consisting only of 3-4 segments and arising from the lowest segment of a dwarf shoot; in the former the upper part of this shoot produces branched trichoblasts. According to Falkenberg, the fertile laterals arise endogenously and are not of the nature of trichoblasts, a matter meriting further enquiry.

In certain genera *(Rhodomela, Laurencia, Chondrid)* the envelope of the mature cystocarp is composed of more than two layers owing to tangential division of the pericentrals. *Bostrychia* is exceptional in that the envelope develops only after fertilization.

The vast majority of Rhodomelaceae are dioecious, the males sometimes smaller than the females; in *Janczewskia* male and female plants have a different habit. The antheridial sori are again commonly borne on the branches of typical trichoblasts, although in several of the more specialised genera *(Symphyocladidi herposiphonia, Placophora, Amansid)* the sterile part of the trichoblast is greatly reduced. In *Polysiphonia* the male trichoblasts commonly consist of a two-celled stalk surmounted by the fertile region; the upper stalk-cell frequently bears a branch, the upper part of which may again be fertile. The fertile regions become polysiphonous and the pericentrals divide copiously to form a compact layer of mother-cells, each of which gives rise to 2-3 antheridia. *Brongniartella* and *Ricardia* are similar.

In *Laurenda,* the fertile tips of male plants become swollen, since the apical cell ceases to divide and the apical depression becomes markedly broadened. The resulting cup- or urn-shaped cavities contain numerous fertile trichoblasts, which are interspersed with sterile ones in *L. obtusa.* In *L. pinnatifida* the segments of the trichoblasts form a continuous layer of mother-cells, as in *Polysiphonia,* each producing three or four primary, as well as secondary, antheridia. The fertile trichoblasts of *L. obtusa,* on the other hand, remain uniseriate, but bear branches on which the mother-cells are formed. According to Grubb mother-cells can also arise from the cells lining the cavities. Each male trichoblast terminates in one or more, often huge, sterile cells, which collapse at a later stage.

The antheridial sori of *Chondria* occupy the lowest lateral of a trichoblast. The branchlets of the fertile lateral remain concrescent from the first and form a biconvex disc, the edge of which is occupied by the enlarged and sterile apical cells of the individual branchlets. The remaining cells cut off, on either surface, pericentrals which produce a continuous layer of globular mother-cells each forming three or four cylindrical antheridia. Similar discs with concrescent branches occur in *Acanthophora, Pterochondria* and *Bryothamnion,* as well as in *Cliftonaea* and *Leveillea,* although in the last two they appear to represent modified shoots and develop rather differently.

In *Odonthalia* and *Rhodomela* the antheridia are mostly formed on fertile polysiphonous laterals, simple in the former, often branched in the latter, although in *R. subfusca* they can also arise on trichoblasts. There is little reason to doubt that the male branches of these two genera, however, different they may be from the normal type, are homologous with the trichoblasts.

Dasyaceae: As in vegetative construction, so also in their sexual reproduction, the Dasyaceae exhibit distinct differences from Rhodomelaceae. The procarps are formed either on the sympodial axis *(Dasya pedicellate, Dasyopsis)* or on a pseudolateral. The fertile pericentral of the procarp-bearing segment is usually the third (sometimes the fourth) and only in *Heterosiphonia* is it the fifth or last-formed one. The procarp develops as in Rhodomelaceae, but the cystocarp-envelope appears only after fertilization. The gonimoblast shows monopodial branching.

In most species of *Dasya* and in *Dasyopsis* the procarps arise from successive segments of the sympodial axes or more usually of adventitious branches; as a general rule only one cystocarp matures on each branch. One (or possibly two) small connecting cells are cut off from the fertilized carpogonium and fuse with the auxiliary cell. The gonimoblast-initial first forms a row of large cells which cut off smaller ones giving rise to branched threads that fill the cavity of the cystocarp; the carposporangia form short rows arising in basipetal sequence. According to Rosenberg no fusion-cell is formed. The cystocarps are provided with a pore. The envelope, which originates from the two pericentrals flanking the fertile one, as well as from pericentrals of adjacent segments, is formed in the same way as in *Polysiphonia* and is two-layered (three-layered in *Dasyopsis)* at maturity.

The procarps of *Heterosiphonia plumosa* usually arise from the fourth segment of the pseudolateral and possess very elongate trichogynes; the sterile cells here undergo considerable division before fertilization. The pericarp develops already during the formation of the carpogonial branch and is three-layered at maturity. The cystocarps of *Dictyurus* are unknown; regarding those of *Thuretia*.

The antheridia are produced on the pseudolaterals,

although in *Dasya pedicellata* they may also arise on adventitious branches. The development of the sori takes place as in *Polysiphonia*.

The Asexual Phase: The four families differ markedly in the mode of production of the tetrasporangia, and it is again the Ceramiaceae that exhibit relatively primitive features.

Ceramiaceae: The tetrasporangia of most Ceramiaceae stand quite freely, though occupying diverse positions. In *Antithamnion* and related genera they commonly occur in series on the upper sides of the laterals, being either sessile or situated at the ends of one- or two-celled branchlets. Sometimes, however, they are clustered together on short branches, as in A *cruciatum*, *Spermothamnion* and *Compsothamnion*. In *Crouania*, they are borne on the basal cells of the whorled laterals, while in *Ptilota plumosa*, the sporangia for the most part occupy short threads arising from the cortex of the ultimate branches.

The sporangia of *Griffithsia* are produced on special laterals which form whorls at the apices of the segments. Each fertile lateral consists of a single cell, bearing several sporangia and protected by outgrowths, arising from the same cell or from the parent-segment. Similar protective branches are seen in *Wrangelia*, where the sporangia are terminal on one-celled accessory branchlets arising from the lower cells of the whorled laterals. In *Vickersia* the sporangia form sori on short cells cut off at the base of the vesicular laterals.

In *Ceramium*, on the other hand, the sporangia are enlarged cells of the cortical bands, within which they are partly embedded. Those of *Spyridia* occur on the "filaments". A peculiar method of development, in some respects recalling that of Rhodomelaceae, is met with in *Trailiella*, where the sporangia are formed from the larger halves of longitudinally divided cells of the erect filaments.

The spores are for the most part tetrahedrally arranged, although those of *Crouania*, *Antithamnion*, and *Trailiella* are cruciate. *Antithamnionella* differs essentially from *Antithamnion* only in possessing tetrahedral spores. There is, however, some variability in this respect; thus, in *Ceramium* and *Seirospora* cruciate and tetrahedral sporangia are sometimes found on the same plant, although in both cleavage only takes place after nuclear division is completed. In *Callithamnion byssoides* Boergesen also records occasional cruciate tetrasporangia. The sporangia of *Callithamnion* dehisce by the detachment of a lid.

Bispores are recorded in diverse Ceramiaceae and are specially frequent in *Crouania attenuata* and *Callithamnion Furcellarieae;* in the latter they occur on plants lacking tetrasporangia. In *C. bisporum* Gardner they are the only reproductive organs known.

Delesseriaceae: The sporangia of Delesseriaceae, in which the spores are always tetrahedrally arranged, generally show a distribution comparable to that of the sex organs. In *Delesseria sanguinea* they are formed in minute leaflets, which are devoid of a midrib and soon become three-layered almost throughout. Subsequently the surface-cells divide to form short vertical branch-systems, from the terminal cells of which the sporangia develop. Though primarily superficial, as in all Delesserieae, the initials gradually become covered by small cells, produced by further division of the surrounding elements. The enlargement of the sporangia causes gaps to be formed between these cells, through which the spores are shed.

Development follows much the same course in other Delesserieae, although the position of the sporangia varies. In *Membranoptera,* they are situated in the younger forks or in small proliferations, in *Grinnellia* in small scattered out-growths, commonly confined to one surface, while in *Hypoglossum, Caloglossa,* and *Apoglossum* they occupy elongate areas on either side of the midrib of the ordinary fronds. In *Caloglossa,* they develop by the division of certain cells in parts of the frond that are still monostromatic so that the sporangia form but a single layer, instead of two, as in other Delesserieae; they are subsequently covered by a cortex formed by pericentrals cut off tangentially from the parent-cells. The sporangia of *Claudea* and *Vanvoorstia* are formed from pericentrals (in *Claudea* also from the cortical threads) in the ultimate blades, before they undergo anastomosis.

The sporangia of *Phycodrys rubens* are usually borne in small marginal proliferations, but they sometimes occupy oblong areas at the ends of the veins of the younger fronds. The fertile tracts soon become five-layered. The sporangial initials are, as in all Nitophylleae, cut off from internal cells, in *Phycodrys* from the hypodermal ones. This is also so in *Acrosorium, Polyneura,* and *Cryptopleura* where, however, sporangia may also arise from the central cells.

The sporangial sori of *Polyneura, Cryptopleura* and *Nitophyllum* are small and irregularly scattered. In TV. *punctatum* the sporangia may occupy the whole thickness of the frond although sometimes embedded. Their formation is initiated by intercalary division of groups of cells in the one-layered thallus; some of the resulting elements develop into sporangia, while the remainder cut off pericentrals, from which further sporangia may arise. The sporangia of *Nitophyllum* thus show pit-connections with several surrounding cells, whereas in other Delesseriaceae they are only connected with their mother-cell. Those of *Martensia* occur within swollen areas on the longitudinal lamellae, more rarely on the entire portion of the thallus. Both here and in *Nitophyllum punctatum* the initials are multinucleate but, although several nuclei may show the first stages of meiosis, only one survives to form the tetraspore-nuclei.

Rhodomelaceae: The sporangia of Rhodomelaceae nearly always possess tetrahedrally arranged spores and are usually immersed in polysiphonous branches. They are only rarely borne on trichoblasts, as in *Lophocladia* and *Wrightiella.* The fertile branches are often specially differentiated, constituting so-called *stichidia,* although the term cannot be exactly defined. The initials are usually (except *Laurencia)* formed from pericentral cells and, as a general rule, one sporangium is produced in each segment; development often follows:

A regular basipetal sequence within the fertile branch.

If the latter is unbranched, the sporangia of successive

segments are generally superposed; when laterals (commonly trichoblasts) are present, the sporangia show a spiral arrangement like that of the laterals.

The segments of the stichidia often possess five pericentrals, whatever be the number in the vegetative branches. The fertile pericentral, which is seemingly never the first-formed one however, cuts off two superficial cover-cells, after which the sporangial initial is produced on the apical side; the small bearing cell is linked by primary pit-connections with the axial and covering cells. The similarity in position of sporangium and bearing cell on the one hand, and of auxiliary and auxiliary mother-cells on the other hand, is noteworthy; the cover-cells correspond to the sterile cells and carpogonial branch of a procarp. The mature spores escape between the cover-cells which are more or less displaced. In *Falkenbergia*, the sporangia arise from an entire pericentral and there are no cover-cells.

In *Polysiphonia*, the sporangia often cause irregular swellings within the stichidia. The fertile branches of *Rhodomela* and *Odonthalia* arise in tufts on the flanks, in the former also on the tips of the larger branches. The sporangia originate from the two lateral of the six pericentrals so that they form two longitudinal rows. *Bostrychia* produces its sporangia in the ultimate branches, in which all the pericentrals of a segment commonly become fertile, although in certain species those on the dorsal side remain sterile. Similar dorsoventral stichidia occur in *Colacopsis*,

The sporangia of *Chondria* develop within the apical depression of club-shaped branchlets, the initial here producing three cover-cells. Those of *Laurencia* likewise originate within the apical pit, but are exceptional in being produced from young cortical cells; this is also so in *Ricardia*. In *Acanthophora* the sporangia occupy the spiny laterals.

Many of the more specialised Rhodomelaceae possess stichidia resembling those of *Polysiphonia*. In *Pollexfenia* these free polysiphonous, and often branched, laterals either occupy the growing margin or arise on the surface of the flat thallus. *Placophora* shows much the same. In other instances, however, the fertile shoots show a more profound modification. Thus, in *Cliftonaea* each segment of a fertile long shoot produces a sporangium from one of the dorsal pericentrals. In *C. Lamourouxii* the maturing sporangium acquires a thick hyaline wall, while the numerous cover-cells ultimately grow out into blunt papillae, affording a characteristic picture. Similar large sporangia with numerous cover-cells occur in *Leveillea*, where they occupy 3-5 of the basal segments of arrested long shoots. The fertile branches of *Amansia* and *Vidalia* show the same dorsoventrality as the vegetative ones and in the latter are often richly branched; two sporangia are produced in each segment.

Dasyaceae: The tetrahedrally divided sporangia here also usually occupy special stichidia. Those of *Dasya* and *Heterosiphonia* are borne on the pseudolaterals, in the former sometimes also on adventitious branches; the number of pericentrals is five in *Dasya*, 6-7 in *Heterosiphonia*. Each pericentral may produce a sporangium although one or two are commonly sterile. In *Dasya* the three cover-cells, cut off after the formation of the sporangium, are small and the latter is left exposed. The stichidia of *Dasyopsis* are mostly formed on adventitious branches and possess six pericentrals.

The repeatedly forked stichidia of *Dictyurus* project freely from the upper edges of the fronds and usually form six sporangia in each segment. In *Thuretia* they are constituted by the swollen bases of lateral sympodia, the upper part of which participates in the formation of the network; the stichidium is enveloped by hyphae and forms sporangia in two alternating rows.

Interrelationships of Ceramiales: The Ceramiales probably represent an independent series evolved from uniaxial haplobiontic types; there is no satisfactory evidence of a closer affinity with other diplobiontic forms. The Ceramiaceae are no doubt the least specialised members of the order, as shown by a consideration both of the vegetative and reproductive features. Those genera that exhibit whorled branching *(Crouania, Antithamnion)* probably come nearest to the ancestral type, not only because this habit is clearly the basic one on which the construction of many *mannioides* shoot of unlimited growth with young sporangia.

Florideae (incl. Ceramiales) is moulded, but also because of the comparative data furnished by the position of the procarp Diverse Ceramiaceae *(Callithamnion, Seirospora,* etc.) have, however, lost this habit and indeed exhibit a very simple vegetative structure.

Delesseriaceae and Rhodomelaceae, in their customary possession of four or more pericentrals, no doubt constitute a special development along several distinct lines from such a whorled type. The Rhodomelaceae, in particular, exhibit marked specialisation in the evolution of forms with dorsoventral organisation, accompanied by a tendency towards congenital fusion of branch-systems, resulting in such extreme types as *Pollexfenia, Leveillea,* and *Amansia.* Compared with the considerable range in outward form presented by the two families under discussion, there is remarkable uniformity in reproductive features. It is noteworthy that, while in Delesseriaceae and Rhodomelaceae, the spores are nearly always tetrahedrally arranged, a cruciate disposition is found in certain Ceramiaceae, and especially in those which on other grounds are regarded as the least specialised.

The relation of the Dasyaceae to the other three families is difficult to assess. Falkenberg's view, that forms like them were the starting-point for the evolution of the Rhodomelaceae, has already been considered. The only further feature in favour of this view furnished by the reproductive characters is that the gonimoblast in Dasyaceae develops in a more primitive manner than in Rhodomelaceae. Rosenberg and Kylin emphasise the resemblances between Dasyaceae and Delesseriaceae, but in view of the marked dissimilarity in vegetative features, it may be doubted whether too much stress should be laid on the reproductive resemblances, since in this respect there is a high degree of uniformity among Ceramiales as a whole. The Dasyaceae are probably best regarded as a separate evolutionary line, parallel with the other three, but rather more obscure in its origins.

Several genera are known which display characteristics of more than one family of Ceramiales. Lejolisia, a mediterranean born small epiphyte in which the creeping filaments bear little-branched erect threads, combines the free (tetrahedral) sporangia of Ceramiaceae with an urceolate cystocarp resembling that of Rhodomelaceae. The envelope consists of closely apposed threads, although these do not actually cohere. According to Feldmann the gonimoblast is like that of *Ptilothamnion*. A peculiar feature lies in the persistence of the trichogyne which, during the development of the cystocarp, comes to occupy a lateral position.

A rather different example is provided by *Herpochondria*, a dorsoventral ally of *Chondria*, in which the polysiphonous branches are congenitally fused to form a flat prostrate thallus. The free marginal stichidia resemble those of a *Chondria*, and in most other respects too the characteristics are those of a member of Rhodomelaceae. The cystocarps are, however, naked like those of most Ceramiaceae.

Mention may also be made of the Sarcomenieae which appear to constitute a link between Delesseriaceae and Rhodomelaceae. They include *Sarcomenia*, *Platysiphonia*, based on *Sarcomenia miniata*, *Taenioma*.

These are in general dorsoventral forms with four pericentrals produced in successive pairs, as in Delesseriaceae. The terete creeping branches bear, on their dorsal side, erect fronds with a tendency to flattening as a result of division of the lateral pericentrals; this is very marked in *Sarcomenia delesserioides*. The flat antheridial branches also resemble those of Delesseriaceae. The urceolate cystocarps of *Taenioma* and *Platysiphonia* are, on the other hand, like those of Rhodomelaceae, although no details are available as to the manner of development of the gonimoblast. The sporangia are formed in two rows in the fertile branches. *Taenioma* is distinguished by the long colourless hairs with a basal meristem, that crown the branchlets and show a considerable resemblance to those of Phaeophyceae. In *Cottoniella* the ultimate branches are uniseriate. These diverse genera, which in their vegetative construction approximate so closely to some of the dorsoventral Rhodomelaceae, merit a fuller study than they have so far received.

General Consideration of the Life-cycle of the Diplobiontic Florideae

The Distribution of Sexual and Asexual Reproductive Organs: Cytological corroboration of the existence of distinct diploid asexual and haploid sexual phases has now been obtained in a considerable number of diplobiontic Florideae. In Cryptonemiales, Gigartinales, and Rhodymeniales there is usually, despite occasional exceptions, a sharp segregation of sporangia and sex organs on separate individuals; among Ceramiales the exceptions are more numerous. Spermatia are frequently found attached to the trichogynes, although fusion of the sex nuclei has not often been seen. A regular alternation of tetrasporic and sexual individuals has, moreover, been occasionally recorded in nature and has been established experimentally by Lewis for a number of Florideae. All these facts imply the widespread occurrence of an isomorphic alternation of haploid and diploid phases.

It has long been known, nevertheless, that procarps (or carpospores) and tetrasporangia commonly occur on the same individual among Ceramiales, and especially among Ceramiaceae-, the form *amphicarpa* of *Callithamnion corymbosum* derives its name from the frequency of this phenomenon. Tetraspores have also been found on male plants, though, owing to the more temporary character of the male individuals, such instances have been less frequently recorded.

The first example of this association to be fully studied was *Nitophyllum punctatum*, in which the fronds often bear both cystocarps and sporangia. The latter are grouped around rudimentary procarps, which suggests a possible correlation between sporangium-formation and arrest of the procarps. The sporangia develop in the normal manner; after preliminary multiplication, all but one of the nuclei degenerate. The survivor, however, fails to divide and the entire contents of the sporangium are liberated as a single monospore; this has the haploid number of chromosomes, although the carpospores of such plants are diploid. In the sporangia found on male plants of *Polysiphonia violacea* and *Griffithsiaglobifera*, the nucleus likewise usually remains undivided, although cleavage furrows, which fail to reach the centre, may appear within the protoplast; occasional stages of nuclear division are observed, but these afford no evidence of meiosis. In *Polysiphonia urceolata* Rosenvinge also records small undivided sporangia on plants with mature cystocarps, while Kylin speaks of aborted sporangia on sexual plants of *Spermothamnion roseolum*. It is thus probable that the sporangia borne on sexual plants usually undergo no reduction and produce only haploid monospores.

Spermothamnion affords a common instance of the converse condition, viz. of the occurrence of asexual plants bearing normal tetrasporangia side by side with arrested pro-carps and antheridia. Certain forms of this genus, however, produce tetrasporangia also on a functional sexual phase. Schussnig and Odle state that there is no reduction in such sporangia, but their data are scarcely convincing. They are, moreover, at variance with the conclusions reached by Drew, who in *S. Turneri* established the occurrence of meiosis in tetrasporangia found on plants bearing developing gonimoblasts and mature fruits, as well as antheridia. The individuals in question are therefore, diploid (with 60 chromosomes). Their tetraspores germinate normally. Evidence was obtained that the procarps of these plant were diploid and their gonimoblasts tetraploid; many of the procarps, however, failed to develop further and fertilization was not directly observed. The probable existence of triploid carpospores, such as would result from the fertilization of a diploid carpogonium by a haploid spermatium, was also established. Haploid plants, bearing functional sex organs and occasional sporangia which remain undivided, have recently been described by Drew, as well as sterile triploid plants. Mathias found tetrasporangia exhibiting reduction and fruits on the same individual also in *Callithamnion brachtatum*, although here the carpospores are stated to be diploid.

The occasional occurrence of tetrasporangia on the sexual plants of the haplobiontic Nemalionales warrants the supposition that this characteristic organ of Florideae evolved prior to the establishment of the diplobiontic life-cycle. There is also increasing reason to believe that the mere presence of tetrasporangia is no proof of the diploid character of the individual on which they occur. It seems probable that, when an isomorphic alternation was first established by postponement of meiosis, each of the two phases still bore tetrasporangia, although reduction was associated only with the sporangia of the one phase; it is possible that this phase for a time still possessed functional or functionless sex organs. The persistence of abortive sex organs on the diploid, and of monospore-producing sporangia on the haploid, phase would on this view be tokens of an ancestral condition and would mark a more primitive state than the sharp individualisation of the two phases that characterises many diplobiontic Florideae. It is noteworthy in this connection that the simultaneous presence of sex organs and sporangia is more frequent in the relatively primitive Ceramiaceae than in the more specialised families of Ceramiales.

In the light of this hypothesis *Spermothamnion* may perhaps be regarded as a form in which sex organs have persisted on the diploid stage and in certain instances remained functional, although they may also have secondarily reacquired their former function. However, that may be, the possibility for the origin of tetraploid and triploid races afforded by such phenomena is of considerable importance, and the further investigation of these matters will be awaited with interest.

In several Florideae (*Acrosymphytum, Thuretella Thuretellopsis, Halarachnion, Gigartina stellata*) tetrasporangia are so far unknown. It is possible that they are produced on dwarf plants which remain at the juvenile stage, since such stages occasionally bear tetrasporangia. Berthold comments on the abundance of dwarf fertile plants in the Mediterranean.

On the other hand, some of these seaweeds may be parthenogenetic, with a gonimoblast producing haploid carpospores, although Oltmanns observed spermatia on the trichogynes of *Acrosymphytum* and there is little evidence for the occurrence of parthenogenesis in Rhodophyceae. A commonly accepted instance is furnished by *Platoma*, where male organs are unknown and occasional cruciate tetrasporangia occur on the female individuals, but this still awaits cytological investigation. Rosenvinge records the development of gonimoblasts from carpogonia of *Furcellaria fastigiata* and *Petrocelis Hennedyi* lacking all traces of spermatia on the trichogynes. These may be instances of actual parthenogenesis, or autogamy may replace normal fertilization. Attention may also be drawn to the apparent non-existence of female individuals in the common *Rhodymenia palmata* which probably indicates an abnormal life-cycle. The same is true of *Antithamnion boreale*.

Tetrasporic and sexual individuals of the same species are often unequally represented in nature. Among Cryptonemiales and Gigartinales there are several records indicating a preponderance of sexual individuals; tetrasporic plants of *Gloeosiphonia capillaris* occur on the coasts of Holland, but not on those of France or Denmark. Among Ceramiales the available data indicate the reverse condition, sexual individuals often being scarce. There is some evidence that this depends on temperature-conditions and probably also on other factors of the habitat; thus, Levring reports only tetrasporic plants of *Callithamnion Furcellarieae, Polysiphonia nigrescens,* and *Rhodomela subfusca* in the waters of the Baltic with their low salinity. On British shores *Spermothamnion roseolum* is represented only by tetrasporic individuals. Many investigators comment on the more robust character of the asexual individuals in Ceramiales.

In *Trailiella, Antithamnionella, Lomentaria rosea,* and *Dasya ocellata* Harv, only tetraspores have so far been found. Svedelius has established that in *Lomentaria rosea,* a northern species always found in water of considerable depth, the tetraspores have the same number of chromosomes as the vegetative cells. Comparison with other species of *Lomentaria,* having a normal life-cycle, shows that the plants of *L. rosea* are diploid, so that in this species the sexual phase has probably been lost. In *Dasya ocellata* also Westbrook failed to obtain good evidence of the occurrence of meiosis in the tetrasporangium.

In certain Florideae (especially Ceramiales) inhabiting northern waters, the sexual individuals tend to be largely summer-forms, while it is the asexual ones that last through the winter. Such instances are not without interest in relation to the problem of the origin of purely diploid types. Lewis found that at Woods Hole *Dasya pedicellata, Griffithsia globifera,* and *Polysiphonia violacea* form carpospores in August and that the resulting tetrasporic plants pass through the winter as small germlings, which mature their tetraspores in July; many of the Red Algae in this region are believed to show a similar seasonal cycle. Comparable data are furnished by Kniep for *Laurencia hybrida* and by Svedelius for *Ceramium corticatulum* in the Baltic. Similarly, there is a preponderance of tetrasporic plants of *C. rubrum* in winter and of cystocarpic ones in summer in the Isle of Man. In other localities, however, the same species may show a different behaviour.

These are by no means the only data indicating a certain seasonal succession for the Mediterranean, but in most other instances this is not very clearly marked. Abundant records, moreover, testify to the frequent simultaneous occurrence of the two phases, although sometimes the one and sometimes the other persists for a longer period. The sequence may, moreover, be reversed; thus, *Ceramium tenuissimum* bears cystocarps from April to June and tetraspores from then till August, while in *Dumontia incrassata* sexual plants are found in the early part of the year and tetrasporic ones later on. On North European coasts cystocarpic plants of *Harveyella* are found in winter and the tetrasporic ones mainly in spring.

In *Agardhiella* Osterhout describes the outgrowth of the entire contents of a tetrasporangium as a single structure; for a time the groups of cells derived from the separate spores remain distinguishable, but by degrees the limits become obscured. The base of the single proliferation thus produced is embedded in the parent-thallus and the lower cells give rise to penetrating rhizoids. These growths bear male or female sex organs or more rarely tetrasporangia; the presence of the last may imply an absence of reduction. Osterhout mentions the occurrence of comparable proliferations in *Cystoclonium* and *Gracilaria*. Other instances are known in *Champia parvula* and *Lomentaria rosea*, in the latter associated with diploid tetraspores. The example afforded by *Agardhiella* is of particular interest because, when the proliferations bear sex organs, the sexual phase is parasitic on the asexual one, and we have the converse condition to that found in diverse Gigartinales.

Special Types of Asexual Reproduction: The Ceramiaceae frequently produce special types of reproductive cells which are restricted to the asexual individuals. As a general rule these also bear tetrasporangia, although in certain species the latter are lacking on most individuals. *Monospora pedicellata* owes its generic name to the customary production of ovoid "monosporangia" accompanied by infrequent tetrasporangia with tetrahedral spores; sexual plants are rare and have not been found on British shores. The "monosporangia", which are deeply pigmented and multinucleate, occupy the same position as the tetrasporangia and are seated on a stalk-cell with scanty contents. At maturity they become detached as a whole and germinate without shedding the enveloping membrane; a further sporangium can develop from the same stalk-cell. In view of the variable size of the latter, Schiller regards the structures under discussion as bisporangia in which the lower cell has degenerated. Although probably diploid, the detachment and germination of the entire organ renders its sporangial nature doubtful.

The so-called paraspores of *Seirospora Griffithsiana*, which may be associated with bi- and tetrasporangia, occur as branched tufts, composed of rounded thick-walled uninucleate cells; in *S. occidentalis* they occupy the same position as the sporangia. The individual cells become detached at maturity, but the contents escape from the membrane prior to germination. Similar structures are recorded at the tips of the branches of *Ceramium strictum* and C. *Deslongchampii*. Spermatia have been observed on the trichogynes of *Seirospora Griffithsiana* so that this species probably possesses a normal alternation, the paraspores constituting an accessory means of propagation of the sporophyte. It may be doubted whether they have any homologies with the sporangia. In another species of *Seirospora* they have been found on individuals bearing aborted procarps.

Structures described as *polyspores*, which are formed within definite sporangia and have already been noted in *Acrochaetium*, are of frequent occurrence among Ceramiaceae and are also recorded in *Gonimophyllum* and in *Chylocladia*. The sporangia of *Gonimophyllum* develop like those of other Nitophylleae. The nucleus of the initial divides into 30-50 parts, and the subsequent cytoplasmic cleavage results in a radial group of polyspores surrounding a central cell. In this instance, the polyspores appear to replace the tetraspores. This is also so in *Pleonosporium*, in *Spermothamnion Shyderae* among Ceramiaceae, and in *Coeloseira* among Champiaceae. The sexual individuals of *Pleonosporium* bear sex organs closely resembling those of *Callithamnion*. The contents of the polysporangium first segment into four and then divide to form 8-24 spores. There is evidence of meiosis and this, taken in conjunction with the details of development, justifies the view that the polyspores are derivatives of tetraspores. In *Spermothamnion Snyderae*, there is a normal alternation, although the asexual individuals bear polysporangia only. Their initials contain several diploid nuclei, which undergo simultaneous meiotic divisions so that the resulting spores are in multiples of four.

In most Ceramiaceae, however, the polysporangia are accompanied by tetrasporangia, although the former tend to predominate in certain habitats; sexual plants are often rare. The polysporangia usually contain two or more nuclei at an early stage and the spores are mostly formed in multiples of four. They have been recorded in diverse species of *Ceramium*, where they appear as large spherical structures projecting from the cortical bands, in *Callithamnion*, where they occupy the same position as the tetrasporangia, and in *Spermothamnion*. They are also found in *Antithamnion plumula* and in *Compsothamnion*; in the former the number of spores is very considerable, in the latter there are eight. In *Plumaria elegans*, they occupy the tips of the ultimate branches and are usually found on distinct plants.

In all the instances just mentioned it remains doubtful whether the polysporangia are, as in *Spermothamnion Snyderae*, modified tetrasporangia producing haploid spores, or accessory reproductive organs, the spores of which serve merely for vegetative propagation and are formed without reduction. A recent investigation by Drew of *Plumaria elegans* has shown that here the polysporangia occur on special triploid plants. The spores are formed by successive division of the uninucleate initial and no reduction occurs during their formation. There are also haploid sexual and diploid asexual plants which follow a normal diplobiontic cycle. The mode of origin of the triploid plants is not yet clear, although the probable occurrence of triploid carpospores in *Spermothamnion Turneri* indicates one way in which this condition could be reached. It remains to be seen whether a special nuclear constitution is associated with the occurrence of polysporangiate individuals also in other Ceramiaceae. It can hardly be doubted that the polysporangia are homologous with tetrasporangia, although those of *Plumaria elegans*, and probably of other Ceramiaceae, have undergone more profound modification than those of *Spermothamnion Snyderae*, this perhaps justifies the use for them of the distinctive term *parasporangia*.

Brief reference may be made to the *bispores*, which are particularly frequent in Corallinaceae and also of common occurrence in certain Ceramiaceae. Who gives a list of all the records to date, refers to a seasonal alternation between bispore and tetraspore-producing plants in *Crouania attenuata*. Although an homology with tetrasporangia appears undoubted, it still remains to be established whether or not reduction occurs in the bisporangium.

Modified Life-cycles: While the previous considerations afford evidence of occasional modifications in the diplobiontic cycle, it is among Phyllophoraceae (Gigartinales) that the most fundamental changes are encountered. Although diverse species, of *Phyllophora* possess a normal diplobiontic life-cycle, it is now conclusively proved that of *P. Brodiaei* shows a striking abbreviation. The sex organs also occur either on the undulate margins of the fronds or in special marginal leaflets; male and female organs are often found in close association. The procarps resemble those of *P. membranifolia*, but occasionally the trichogyne is scarcely developed. Spermatia have not been observed on the trichogynes, but Kylin records the fusion of carpogonium and supporting cell which is taken as evidence of fertilization. The gonimoblast-threads which arise from the enlarged and multinucleate auxiliary cell, penetrate in all directions between the cells of the gametophyte with which they establish pit-connections. A considerable number grow towards the exterior and give rise to nemathecial threads. As these develop, they raise the overlying cortex and penetrate the surface in several places, often on both sides of the parent-frond; the resulting cushions later fuse to form a large globular nemathecium, which sometimes encircles an entire fertile leaflet.

Several months elapse before the nemathecia commence to form spores. They are differentiated into a medulla and a cortex composed of radiating threads, in which all the cells, except for the outermost three or four, develop into cruciate tetrasporangia. Since the nemathecia of P. *Brodiaei* arise directly from the gonimo-blasts and produce tetraspores, they clearly represent the sporophyte, which is here parasitic on the gametophyte, the carposporangial stage being suppressed. The tetraspores grow into young plants which are sufficiently like the mature gametophyte to complete the picture of the life-cycle. Claussen produces rather inconclusive evidence that the nemathecial nuclei are diploid.

The nemathecia just described have been regarded as belonging to a parasite (*Actinococcus subcutaneus*, which was believed to obtain access through the apertures of the antheridial depressions. There were not wanting, however, suggestions that the "parasite" might represent the actual sporophyte of *P. Brodiaei*, although this remained in doubt until it was established by Rosenvinge. Similar "parasites" are recorded in diverse other species of *Phyllophora*, in all of which no asexual individuals are known, while of the "parasite" only tetraspore-producing stages have been found. There is thus considerable presumptive evidence that these species possess an abbreviated life-cycle, comparable to that of *P. Brodiaei*.

An analogous condition is met with in *Gymnogongrus*. G. *norvegicus* is probably a normal diplobiont, since separate cystocarpic and tetrasporic individuals are known; the sporangial nemathecia were at one time regarded as a parasite, *Colacolepis peltaeformis*. The tetraspores from these nemathecia, however, germinate like the carpospores of the cystocarpic individuals and produce erect fronds, having the structure of a *Gymnogongrus*. G. *linearis* also forms normal cystocarps, although no tetrasporic individuals are known; spermatia have been observed on the trichogynes.

In many species of *Gymnogongrus*, however, cystocarps are lacking, and this is so in G. *Griffithsiae*. The "parasite" (*Colacolepis aggregates),* found on this species, bears the same relation to the procarp as in *Phyllophora Brodiaei,* while its spores produce germlings with a discoid base bearing *Gymnogongrus-fronds.* Antheridia are unknown in G. *Griffithsiae,* and there is no evidence that fertilization occurs. The procarps are rudimentary and possess a large supporting cell, bearing a two-celled carpogonial branch. Numerous threads arise from the supporting cell and some of them combine to form a fan-shaped mass of tissue, which penetrates the surface and forms a lobed nemathecium. There is no sharp line of demarcation between its cells and the cortical tissue of the parent. The nemathecia, which are generally situated near the bifurcations of the thalli, form chains of sporangia producing either tetra- or monospores.

Another instance is furnished by G. *platyphyllus* Gardn, in which the "parasite" *(Actinococcus chiton* Howe) has been shown to be the tetrasporic stage. This arises in essentially the same way as in G. *Griffithsiae* antheridia are again unknown. It is possible that these species are altogether haploid.

Despite its abundance, cystocarps have never been recorded in *Ahnfeltia plicata.* Especially during winter, however, the plants commonly bear yellowish nemathecia producing monospores, which were long regarded as belonging to a parasite *Sterrocolax decipiem* Schmitz, although their independence was sometimes questioned. Rosenvinge provided the proof that these nemathecia are part of the reproductive cycle of *Ahnfeltia.* No procarps are present, and the nemathecia arise merely by proliferation of the cortical cell-rows of the thallus. According to Rosenvinge the primary nemathecial filaments bear two kinds of cells, viz. terminal flask-shaped ones, possibly representing hairs.

Bibliography

Abdelahad, N. and Bazzichelli G.: *Geitleria Calcarea Friedmann,* Cyanophycee Nouvelle, Italie, 1988.

Anderson R. and Adams M. S.: *Effect of Fire on Big Bluestem Production,* American Midland Naturalist, 1975.

Battarbee, R. W. and Gasse F.: *Past Climate Variability through Europe and Africa,* Springer, Netherlands, 2004.

Bebianno, M. J. and Company R. M.: *Antioxidant Enzymes and Lipid Peroxidation in Bathymodiolus Azoricus,* Aquatic Toxicology, London, 2005.

Berge, J. and Vader W.: *On the Taxonomic Status of the Antarctic Amphipod Genera Eclysis,* Organisms Diversity Evolution, 2005.

———————: *The Amphipod Genus Alexandrella,* Journal of Natural History, Washington, 2005.

Blanchette, C. A.: *The Effects of Biomechanical and Ecological Factors on Population and Community Structure of wave-exposed,* Oregon State University, 1994.

Bold, H. C. and M. J. Wynne: *Introduction to the Algae,* Englewood Cliffs, New Jersey, 1978.

Boudouresque, C. F. and Perret-Boudouresque M. M.: *A Checklist of the Benthic Marine Algae of Corsica,* GIS Posidonie Publishers, Marseille, 1987.

Bourrelly, P. and Dupuy P.: *Quelques Stations Francaises de Geitleria Calcarea, Schweizerische Zeitsch, Fur Hydrologie,* MacMillan, New York, 1973.

Boyer, C. S.: *The Biddulphoid Forms of North American Diatomaceae*, Natural Sciences of Philadelphia, Boyer, 1901.

Brodie, J. A. and Irvine L. M.: *Seaweeds of the British Isles*, British Museum, London, 2003.

Chapman, V. J.: *The Algae*, St. Martin's Press, New York, 1964.

Christiansen, J. S. and Gildberg A.: *Pagophilus Groenlandicus*, ICES Journal of Marine Science, 2004.

Claus, G.: *Algae and (their) Mode of Life (in) on the Baradla Cave of (at) Aggtelek*, Scientiarum Hungaricae, Budapest, 1955.

Claus, G.: *Data on the Ecology of the Algae of Peace Cave*, Nova Hedwigia Zeitsch, Kryptogamenkunde, 1962.

———————— : *The Reevaluation of the Genus Gomontiella*, Algol, 1960.

Coppejans, E.: *Iconographie d'Algues Mediterraneennes*, J. Cramer, Vaduz, 1983.

Delepine, R., Boudouresque C. F. and Frada-Orestano C.: *Algues et Autres Vegetaux Marins*, Mediterranee et Mer Noire, Roma, 1987.

Denny, M. W.: *Mechanical Limits to Size in Wave-swept Organisms*, Ecological Monographs, 1985.

———————— : *Predicting Physical Disturbance*, Ecological Monographs,1995.

Dixon, P. and Irvine L. M.: *Seaweeds of the British Isles*, British Museum, Natural History, London, 1977.

Evenset, A., Christensen G. N., Skotvold T, Fjeld E. and Gregor D.: *A Comparison of Organic Contaminants in two High Arctic Lake Ecosystems*, Science of the Total Environment, Norway, 2004.

Falk, Petersen S., Timofeev S., Pavlov V. and Sargent J. R.: *Climate Variability and the Effect on Arctic Food Chains*, Springer Verlag, Berlin, 2006.

Fletcher, R. L.: *Seaweeds of the British Isles*, British Museum, London, 1995.

Furevik, T., Gerdes R. and Berger W.: *The Nordic Seas*, American Geophysical Union, Washington, 2005.

Gantt, Elisabeth: *Handbook of Phycological Methods*, Cambridge University Press, England, 1980.

Gayral, P.: *Les Algues des Cotes Francaises*, Editions Doin, Paris,1966.

Guiry, M. D.: *A Concensus and Bibliography of Irish Seaweeds*, J. Cramer, Vaduz, 1978.

Gulliksen, B., Slagstad D. and Pamapanin D.: *Mytilus Thermophily*, Marine Ecology Progress Series, 2004.

Hakansson, H. and Stoermer E. F.: *Observations on the Type Material of Stephanodiscus Hantzschii Grunow in Cleve and Grunow*, Nova Hedwigia, 1984.

Hakansson, H.: *A Light and Electron Microscopical Investigation of the Type Species of Cyclotella*, Diatom Research, 1989.

Hanna, G. D.: *Pliocene Diatoms of Wallace County*, University of Kansas Science Bulletin, Kansas, 1932.

Haworth, E. Y. and M. A. Hurley.: *Comparison of the Stelligeroid Taxa of the Centric Diatom Genus Cyclotella*, O. Koeltz, Koenigstein, 1986.

Hellebust, J. A. and Craigie J. S.: *Handbook of Phycological Methods*, Cambridge University Press, Cambridge, 1978.

Hishock, S.: *A Field Key to the British Brown Seaweeds*, F.S.C. Publications, Shrewsbury, 1979.

Hoagland, K. D., Roemer S. C. and Rosowski J. R.: *Colonization and Community Structure of two Periphyton Assemblages*, American Journal of Botany, 1982.

Hollenberg G. J.: *Marine Algae of California*, Stanford University Press, California, 1976.

Hop, H. and Falk-Petersen S.: *Seasonal Lipid Dynamics and Life Strategy*, Marine Biology, 2005.

Irvine, L. M. and Chamberlain Y. M.: *Seaweeds of the British Isles*, British Museum, London, 1994.

——————— : *Seaweeds of the British Isles,* British Museum, London, 1983.

Kattner, K., Graeve M.: *The Arctic Pteropod Clione Limacina,* Marine Biology, 2005.

Kociolek, J. P. and Stoermer E. F.: *Ultrastructure of Cymbella Sinuata and its Allies,* Systematic Botany, 1987.

Kolosov, P. N.: *Upper Precambrian Paleoalgological Remains of the Siberian Platform,* Moscow,1982,

Krammer, K. and H. Lange Bertalot: *Bacillariaceae, Epithemiaceae, Surirellaceae,* Gustav Fischer Verlag, New York, 1988.

——————— : *Naviculaceae,* Gustav Fischer Verlag, New York, 1986.

Leake, D.V.: *The Algae of Crystal Lake,* Cleveland County, Oklahoma,1945.

Lebreton, B.: *World Bibliography on Subterranean Algae,* Bibliographie Mondiale des Algae Souterraines, 2000.

Lee, R. E.: *Phycology,* Cambridge University Press, Cambridge, 1980.

Littler, M. and Littler D.: *Handbook of Phycological Methods,* Cambridge University Press, Cambridge, 1985.

Lobban, Christopher S. and Harrison P. J.: *Seaweed Ecology and Physiology,* Cambridge University Press, Cambridge, 1994.

Makarikhin, V. V. and Kononova, G.: *Phytolites of the Karelian Lower Proterozoic,* Leningrad, Nauka, 1983,

Mann, D. G. and Stickle A. J.: *The Genus Craticula,* Diatom Research, 1991.

Mikhailova, N. S. and German, T. N.: *5th All Union Colloquium on Microfossils of the USSR Precambrian,* Izvestia Akadamie Nauk, 1987,

Nelson, E. N., Richardson J. K. and Bailey H. H.: *Aspects of the Occurrence of Aspidobothrid Parasites,* Proceedings of the Oklahoma Academy of Science, 1975.

Newman, G. A.: *Cowbird Parasitism and Nesting Success of Lark Sparrows in Southern Oklahoma,* Wilson Bull, 1970.

Newton, L.: *A Handbook of the British Seaweeds*, British Museum, London, 1931.

Oliver, G.: *Black Rat Snake Predation Upon Nesting Barn and Cliff Swallows*, Ornithological Society, Oklahoma,1970.

Osborn, R. S.: *Observations on the Spawning Ecology of Buffalos*, Oklahoma Academy of Science,1966.

Pampanin D. M., Camus L., Gomiero A., Marangon I., Volpato E. and Nasci C.: *Susceptibility to Oxidative Stress of Mussels*, Venice Lagoon, Italy, 2005.

Pamperl, M. A.: *Diatoms of Lake of the Ozarks*, Transactions, Missouri Academy of Science, Missouri, 1980.

Parks, J. M. and Barclay H. G.: *The Increasing Importance of Vines in Southern Oklahoma*, Oklahoma Academy of Science, 1966.

Parmelee, D. F.: *The Breeding Behavior of the Painted Bunting in Southern Oklahoma*, Bird-Banding, 1959.

Patten, B. C., Egloff D. A. and Richardson T. H.: *Total Ecosystem Model for a Cove in Lake Texoma*, Academic Press, 1975.

Patten, B. C.: *A Reservoir Cove Ecosystem Model*, Transactions of the American Fisheries Society, 1975.

————————: *Ecosystem Modeling and Reservoir Management*, Oklahoma Academy of Science, 1976.

Paulissen, M. A.: *A Technique for Marking Teiid Lizards for Field Observation*, Herpetology Review, 1985.

————————: *Cnemidophorus Sexlineatus*, Journal of Herpetology, 1988.

Penfound, W. T., Shed J. S. and Jennisons M. C.: *A Plant Community Dominated by Vines*, Oklahoma Academy of Science, 1965.

Penfound, W. T.: *The Composition of a Black Willow Forest in South Central Oklahoma*, Oklahoma Academy of Science, 1961.

Pfiester, L.A. and Holt J.: *A Freshwater "Red Tide" in Texas*, Southwestern Naturalist, 1978.

Phillipi, A. and Tyrl R. J.: *Reproductive Biology of Proboscidea Louisianica (Martyniacea),* Rhodora, 1979.

Pireh, W.: *Achillea Millefolium,* Oklahoma State Universi, Oklahoma, 1978.

Power, M. E. and Matthews W. J.: *Grazer Control of Algae in an Ozark Mountain Stream: Effects of Short-term Exclusion,* Ecology, 1988.

Proctor, V. W.: *Dispersal of Fresh-water Algae Water,* Science, 1959.

———————: *The Growth of Basicladia on Turtles,* Ecology, 1958.

Prud, Homme: *A Taxonomic Revision of the European Sphacelariaceae,* Leiden University Press, Leiden, 1982

Purdue, J. R.: *The Phylogenetic Implications of Body Movements in Displaying Sceloporine Lizards,* Herpetology Review, 1970.

———————: *The Western Sandpiper in Oklahoma,* Oklahoma Ornithological Society, 1969.

Ratkova T. and Wassmann P.: *Sea Ice Algae in the White and Barents Sea,* Polar Research, 2005.

Reinke, D. C.: *New and Interesting Algae from Kansas,* State Biological Survey of Kansas, 1979.

Riedl, R.: *Fauna und Flora des Mitterlmeeres,* Verlag Paul Parey, Hamburg, 1983

Roemer, S. C., Hoagland K. D. and Rosowski J. R.: *Development of a Freshwater Periphyton Community as Influenced* Canadian Journal of Botany, 1984.

Rolbiecki L. and Walkusz W.: *Derogenes varicus,* Oceanological and Hydrological Studies, West Spitsbergen, 2005.

Round, F. E., Crawford R. M. and Mann D. G.: *The Diatoms: Biology and Morphology of the Genera,* Cambridge University Press, Cambridge, 1990.

———————: *The Diatom Genus Stephanodiscus,* Archiv fur Protistenkunde, 1981.

Santi-Mazzini G.: *Aquarium,* Gribaudo, Cavallermaggiore, 2000

Setchell, W. A. and Gardner N. L.: *The Marine Algae of the Pacific Coast of North America*, University of California Press, California, 1925.

Shenfil, V. Y.: *Algae in the Riphean Deposits of the Yenesei Ridge*, Doklady Akad, Nauk, 1978,

South, G. Robin and A. Whittick.: *Introduction to Phycology*, Blackwell Scientific Publications, Oxford,1987.

Stegenga H., Bolton J. J., Anderson R. J.: *Seaweeds of the South African Coast*, University of Cape Town, 1997.

Stein R. and Macdonald R. M.: *The Organic Carbon Cycle in the Arctic Ocean*, Springer, New York, 2004.

Theriot, E., Stoermer E. and Hakansson H.: *Taxonomic Interpretation of the Rimoportula of Freshwater Genera in the Centric Diatom Family Thalassiosiraceae*, Diatom Research, 1987.

Troeger, W. W. and Menzel R. G.: *Benthic Diatoms in two Eutrophic Flood Detention Reservoirs in Oklahoma*, Journal of Freshwater Ecology, 1986.

Troeger, W. W.: *Colonization of Benthic Diatoms in a New Farm Pond and its Feeder Creek*, Southwestern Naturalist, 1983.

————————— : *Epiphytic Diatoms in Farm Ponds and Experimental Ponds in Bryan County*, Oklahoma Academy of Science, Oklahoma, 1978.

Usher, M. B., Callaghan T. V.: *Principles of Conserving the Arctic's Biodiversity*, Cambridge University Press, Cambridge, 2005.

Van, Landingham, S. L.: *Catalog of the Fossil and Recent Genera and Species of Diatoms and their Synonyms*, Cramer, Lehre, 1979.

Walcott, C. D.: *Middle Cambrian Algae*, Smithsonian Miscellaneous Collections, 1919.

Walter, M. R.: *Archean Stromatolites*, Princeton University Press, Princeton,1983,

Webber, H. J.: *Catalogue of the Flora of Nebraska*, Nebraska Board of Agriculture, Lincoln, 1890.

————————: *The Fresh Water Algae of the Plains,* American Naturalist, 1889.

Wenke, T. L. and Eberle M. E.: *Diatoms from Streams in Ellis and Russell Counties, Kansas,* Transactions of the Kansas Academy of Science, 1986.

————————: *Diatoms from Kansas,* Southwestern Naturalist, 1985.

Werner, D.: *The Biology of Diatoms,* University of California Press, Berkeley, California, 1977.

Wolle, F.: *Diatomaceae of North America,* Comenius Press, Bethlehem, Pennsylvania, 1894.

————————: *Fourth Contribution to the Knowledge of Kansas Algae,* Bulletin of the Washburn College Laboratory of Natural History, 1889.

Wujek, D. E.: *Contribution to the Diatom Flora of Kansas,* Southwestern Naturalist, 1965.

Yankauskas, T. V.: *Upper Precambrian and Cambrian Plant Microfossils of the European USSR and their Stratigraphic Significance,* University of Moscow, Moscow, 1982,

Zmijewska M. I. and Bielecka L.: *Characteristics of the Arctic and Antarctic mesozooplankton in the neritic zone during summer,* Polish Polar Research, 2004.

Index

M

N

O

S

T

U

V

W

□□□

ENCYCLOPAEDIA
OF
ALGAE

ENCYCLOPAEDIA OF ALGAE

Vol. 3

Edited by
A.M. Bagulia

ANMOL PUBLICATIONS PVT. LTD.
NEW DELHI - 110 002 (INDIA)

ANMOL PUBLICATIONS PVT. LTD.
H.O.: 4374/4B, Ansari Road, Darya Ganj,
New Delhi-110 002 (India)
Ph.: 23278000, 23261597
B.O.: No. 1015, Ist Main Road, BSK IIIrd Stage
IIIrd Phase, IIIrd Block,
Bangalore - 560 085 (India)
Visit us at: www.anmolpublications.com

Encyclopaedia of Algae

First Published, 2008
ISBN 978-81-261-3528-8 (Set)

PRINTED IN INDIA

Printed at Mehra Offset Press, Delhi.

Contents

Preface

The first plants on earth were algae and these still thrive in a range of aquatic habitats. The land plants have evolved from the algae, more specifically green algae. Some 400 million years ago freshwater, green, filamentous algae invaded the earth. Algae are used as fertilizer and energy-source (biodiesel), etc.

Algae encompass several groups of living aquatic organism that get light-energy through photosynthesis, using it for transforming inorganic substances into organic substances. Algae make the photosynthetic organism found in most habitats. These are of single-celled to multicellular complex species, such as 'Giant Kelps' that grow to 65 metres in length. Conventionally, these have been regarded as simple plants, in the study of Botany and referred to as Eukaryota. Chromophyta have bacterial characteristics and some authorities consider them to be Protists.

Protists are traditionally considered more animal-like. The prokaryotic forms, referred to as blue-green algae are only half-algae with a mixture of bacterial characteristics. But they are quite different from the bacteria and are referred to as Cynochloronta. All other forms are known as true Eukaryota algae, in the study of Botany. They have a nucleus enclosed within a membrane. The protoctists are defined as eukaryotic microorganism with the exception of animals and plants but including fungi and algae.

All algae have photosynthetic machinery derived from cyanobacteria, and produce oxygen as a by-product of photosynthesis, unlike non-cyanobacterial photosynthetic bacteria. It is estimated that algae produce about 73 to 87 per cent of the net global production of oxygen, available for respiration.

Most of the simpler algae are unicellular flagellates or amoeboids, but colonial and non-motile forms have developed independently among several groups. Some more organisational levels, more than one of which may occur in the life cycle of a specie, are: Colonial, Capsoid, Coccoid, Palmelloid, Filamentous, and Parenchymatous.

This eminent work presents perfect and realistic features of the algae in an interesting manner. Expectedly, this study would generate a separate discipline in the study of algae.

Editor

1

Role in Economy

Algae constitute an important group of plants which mostly occur in aquatic (both freshwater and marine) habitats. They are commonly known as pond scum and sea weeds. Their controlled growth can provide positive help in tackling some present day problems such as disposal of sewage and industrial wastes, conservation of water and soil, and improvement of quality and yield of food products. However, their uncontrolled growth causes pond epidemics, water fouling and fouling of marine vessels. At present many research organizations in India (Indian Agricultural Research Institute, New Delhi; Central Salt and Marine Chemicals Research Institute, Bhavnagar) and abroad are actively engaged in developing marine products. It is all the more important to exploit aquatic resources for the growing food and/or industrial requirements of the world.

Useful or Positive Importance

Algae as Food: In coastal regions, seaweeds are used for human food from ancient times. Population boom in the last decades has forced man to depend upon seaweeds and other algae to supplement his diet. Researches in Japan, Germany and USA have shown that nutritive value and per unit area yield of algae is much more than conventional crops.

Porphyra, Alari, Ulva, Chlorella, Chondrus, Rhodymenia and Codium are some genera which are used as food in various forms.

Porphyra is a red alga which grows in shallow sea water. It is very rich in protein (30-35%) and carbohydrates (40-45%). It is also a good source of vitamin B and C. It is a common food item in Japan (used as a paste called nori), China (locally called 'tsats' ai), Britain (grilled on toast called sloke) and many countries of the pacific region. Porphyra soup is highly prized in Europe and elsewhere. In Japan alone more than 30 million kg of Porphyra is used per year as food.

A gelatinous carbohydrate obtained from Chondrus crispus is used in puddings. It is also used as a stabilizer and clearing agent in the production of beer. A salty confection commonly known as dulse is obtained from Rhodymenia palmata. In Scotland it is chewed as tobacco. Lemanea is considered a delicacy in the Manipur region of India and is locally called nungham.

Food obtained from Laminaria saccharina is known as 'kombu'. It is rich in carbohydrates (57%). Some red algae such as Gelidium, and Gracilaria are used in ice-cream and jelly.

Chlorella, a green alga, is more useful than Porphyra in terms of its nutritive value. Its nutritional value is comparable to that of a mixture of soyabeans and spinach. Several species of Chlorella are high producers of lipids and proteins. They can be cultured easily in the presence of light, carbon dioxide and mineral nutrients. They grow rapidly hence many crops can be obtained in a limited period. An average yield of 13 metric tons per year per acre can be obtained if proper cultural practices are followed. Chlorella contains about 30% carbohydrates, 30% proteins and 15% lipids. However, at present efficient digestibility of the Chlorella cell wall is a problem and researches are underway to convert it in a easily digestible form. Chlorella is useful in space flights and nuclear submarines as food and so, also in recycling and oxygen regenerating system.

In India, at CFTRI, Mysore, research is being conducted on the use of blue green alga, Spirulina as a supplement to diet. The alga is cultured, dried, powdered and then used in the form of one gram tablets. It contains 60% protein, essential vitamins and unsaturated fatty acids. Scenedesmus, besides being rich in proteins, is also a rich source of threonine. It has a high diotic value equal to that of skim milk. The colonies of Nostoc commune are boiled and are used as food in China and Java.

Codium, Monostroma and Ulva are some other green algae used as food. Species of Codium and Ulva are used as salads and Monostroma is used as a common food, aonori, in Japan.

Useful for Animals

In maritime districts many seaweeds such as Fucus, Laminaria, Ascophyllum and Sargassum are used as fodder. In North Ronaldsay, sheep live on seaweeds for the most part of the year and are pastured only during lambing season. The utility of seaweeds as fodder is due to their vitamin and micronutrient contents. The drying process, storage duration and other similar factors affect their nutrient value. In Scotland and Iceland small industries have been set up for processing seaweeds into feeding meal. Cattle, on diet supplemented by seaweed meal, have increased fat contents in their milk.

Algae also provide food to human beings indirectly through food chain. Many freshwater and marine animals are dependent upon algae for their food. Some of these aquatic animals serve as food for fish and in turn fish is an important source of protein for human beings.

Role in Industry

Many products of commercial and pharmaceutical importance have been derived from algae. Agar, carrageenan, alginate and diatomite are some important commercial products of algal origin.

1. *Agar-Agar:* It consists of two polysaccharides, agarose and agaropectin. The characteristics of agar are (i) it is

insoluble in cold water but readily soluble in hot water, and (ii) 1.5% solution of agar is clear and forms a solid and elastic gel on cooling to 32-39°C; the gel does not dissolve again at a temperature below 85°C.

Agar is obtained commercially from species of Gelidium, Pterocladia, Gracilaria, Acanthopeltis, Ahnfeltia, Chondrus and Gigartina. Japan and Southeast Asia are the main production centres of agar. Agarophytes (red algae used in production of agar) are collected by diving and then cleaned and bleached in the sun with several washings in freshwater. The material is boiled for several hours and the extract is acidified. This is then frozen and thawed. On thawing, water flows from the agar carrying impurities with it. The agar is then dried and sold in the form of flakes, cakes or powder. In modern methods agar is deodorized using activated charcoal.

The greatest use of agar is in food, pharmaceutical and cosmetic industry. It is used in the manufacture of processed cheese, puddings, creams and jellies. In preservation and canning of fish and meat it is used as gelling and thickening agent and to reduce the undesirable effects of can. Pharmaceutically, agar is used as a laxative, but more frequently it serves as an inert carrier for the drug which requires a slow release. It is an important constituent of cosmetics' (ointments, lotions, etc.). It is used for almost a century as stiffening agent in culture media.

2. *Carrageenan:* It is a phycocolloid (polysaccharide colloid formed by an alga) similar to agar, but with a higher ash content and requires higher concentration to form gels. The two main constituents of carrageenan are k-carrageenan and λ carrageenan, both negatively charged polymers with high molecular weight.

Carrageenan is obtained from the cell walls of Chondrus crispus (Irish moss) and Gigartina stellata. In Philippines, Eucheuma is extensively cultured as a source of carrageenan. For extraction, dried plants are washed with fresh water to reduce salt contents and then boiled with excess of water. The soluble carrageenan is separated from insoluble residue by a centrifuge. The soluble fraction on evaporation under vacuum yields carrageenan.

Carrageenan is extensively used in stabilization of emulsions in paints and cosmetics. In alcohol and sugar industry it is used as a clearing agent. In recent years carrageenan has supplanted agar in food and pharmaceutical industry. It is also utilized in the textile, leather and brewing industries.

3. *Alginate:* These are salts of alginic acid which occur in the cell wall (particularly middle lamella and primary wall) of the brown algae belonging to the order Laminariales. Alginate is extracted from Laminaria, Ascophyllum, Fucus, Macrocystis, Ecklonia, Lessonia and Durvillea. The alginic acid contents of cell walls vary from 14-40%. They are also affected by seasonal factors, yield being the highest in winter and the lowest in summer.

   The seaweeds are macerated with acid and then calcium, potassium or magnesium salts are added to get their alginates. Alginates are non-toxic and viscous and readily form gel, hence useful as thickner (in preparation of soup, sauce, cream, printing ink), emulsifier (in ice cream, synthetic cream, polish, emulsion paints) and gelling agent (in confectionery and dental impression powder). Alginic acid stops bleeding effectively. Flame proof fabrics are also prepared from alginates. When mixed with resin, alginates are used for binding panel boards.

4. *Diatomite:* After the death of diatom cells frustules usually dissolve but under certain environmental conditions they may remain intact and accumulate at the bottom of the water body where they occur. If the conditions are exceptionally favourable such accumulation may reach considerable thickness. These deposits of diatomaceous earth are known as diatomite. The most extensive deposits of diatomite are found at Lompoc in California.

Diatomite is insoluble, chemically inert and shows exceptional physical properties that make it suitable for use in many industries. In oil and chemical industries it is used as a filter. It is heat resistant hence used as insulator in boilers and blast furnaces. Because of its inert nature, diatomaceous earth is used as an absorbant of nitroglycerine. Diatomite is used in the manufacture of car and silver polishing powders, water glass, tooth paste, paints and phonograph records. It is also used as a filler for battery boxes and with bakelite for switch and fuse boxes.

5. *Funori:* A type of glue, known as funori, is obtained from Gloiopeltis furcata. It is used in cosmetic industry for hair curling and dyeing preparations. It is also an important paper and textile sizing agent.

Minerals: Brown algae, particularly Laminariales or kelps are rich in soda, potash, iodine and alginic acid. Ash of dry kelps is a source of soda which is used in the manufacture of soap, glassware and alum.

Kelp ash also contains large amounts of iodine; the average yield is 15 kg per ton. About 7% of the total world production of iodine is obtained from kelps in Japan. Bromine is extracted from some red algae such as Polysiphonia, Rhodymenia, etc. where it may occur in quantities as large as 3-6% of dry weight.

Many important minerals like copper, iron, zinc, cobalt, vanadium, manganese, boron and chromium are present in high proportions in seaweeds. Hence seaweeds are used as a supplement in stockfeed and natural fertilizers. Chlorella vulgaris has the ability of nucleating metallic gold from dilute solutions having gold as gold chloride. Thus, these algae play an important role in the formation of gold deposits in sediments laid down by rivers.

Useful Bio-fertilizers

Terrestrial algae play an important role in soil biology. Blue-green algae act as nitrogen fixing agent in rice fields.

Anabaena cylindrica, Tolypothrix tenuis, Aulosira fertilissima, Oscillatoria princeps, Nostoc commune and many other members of Cyanophyceae have the capability of fixing atmospheric nitrogen. Blue-green algae and Azolla constitute a system which is the main source of algal biofertilizers in south and south-east Asia. By inoculating rice fields with blue-greens, paddy yield can be increased by 30 per cent.

Larger brown and red algae are used as organic manure. They are usually rich in potassium but have relatively low nitrogen and phosphorus proportions. The problem of fertilizers in developing countries can be solved to a great extent if blue-green algae are used in combination with brown and red algae.

Algae also affect soil fertility indirectly. For example, their extracellular products serve as a source of carbon and nitrogen for micro-organisms in the soil. These micro-organisms in turn help in preserving the nitrogen, sulphur and phosphorus elements present in the soil. Besides, extracellular products of some algae are stored as organic or/and inorganic ions which can be easily utilized by crop plants. Some corallina or lime-depositing algae such as Lithothamnion and Lithophyllum arc used for liming crop fields. Blue-green algae also help in reclamation of saline and alkaline soils. The growth of blue-green algae in saline and alkaline water-logged fields results in decrease in pH and increase in phosphorus, nitrogen and organic matter contents of the field and thus converting it into fertile and cultivable land.

Disposal of Sewage

Waterborne domestic and industrial waste is called sewage. It contains organic and inorganic material in soluble and/or suspended form. Although disposal of sewage is mostly an aerobic process, sewage is usually poor in oxygen. It is necessary to treat sewage aerobically before disposal because (i) its anaerobic digestion produces unpleasant odour, and (ii) it is a rich source of many inorganic nutrients like sulphur, nitrogen, phosphorus and potassium and as such properly treated sewage can serve as fertilizer.

Oxidative breakdown of sewage is carried out in small tanks and oxygen required by bacteria for this process (oxidation) is obtained from algal growth. Such tanks support luxuriant growth of many unicellular forms, such as Chlorella, Chlamydomonas, Euglena, Scenedesmus and Pyrobotrys. Oxygen evolved by these algae during photosynthesis is utilized in oxidation by bacteria. The algae in turn utilize ammonia nitrogen as nutrient.

Useful Antibiotics in Medicines

Many algae such as Chlorella, Cladophora, Lyngbya, Polysiphonia, Laminaria and Halidrys synthesize antibiotic substances. The first such antibiotic, chlorellin, was obtained from Chlorella. Extract of Ascophyllum nodosum is effective against both Gram-positive and Gram-negative bacteria. Water passed through a filter containing Nitzschia palea showed antibiotic property. The filtrate is effective only if the algal filter is in light and photosynthesizing. It suggests that antibiotic substances are formed during photosynthesis.

Members of the order Charales have larvicidal property hence may be useful in controlling mosquitoes.

Extracts from Alsidium, Corallina, Codium, Durvillea, etc. are effective vermifuge.

Tse-ko-Tsoi, an antihelmitic drug, is prepared from a red alga, Digenia simplex, in South China. Some algae are used in the treatment of diseases of kidney, urinary bladder and lungs. Brown algae are used in various goiter medicines because of their high iodine contents. Gelidium is useful in stomach disorders.

Sea weeds are a good source of a number of vitamins. The diatom Nitzschia is fairly rich in vitamin A, riboflavin is present in good amount in Porphyra tenera, Gelidium amansii and Chondrus spp., Furcellania fastigata and Rhodomela subfusca are rich in thiamine, and vitamin C content of Porphryra lacinata is higher than that of oranges.

Algae as a Source of Renewable Energy: In recent times, hydrogen is used as a source of a renewable fuel as it causes no pollution and forms water in contact with oxygen. In view hydrogen (H_2) production is thus an important area of biotechnological research for non-conventional and renewable source of energy. In achieving this objective, the principle used in photosynthetic machinery by green plants is employed. Water is subjected to photolysis, which leads to splitting of water molecules into (i) oxygen, (ii) electrons, and (iii) hydrogen ions (H^+). Hydrogen ions are converted into hydrogen gas (H_2) which can be collected and used as a fuel. Enzymes hydrogenase and nitrogenase convert hydrogen ions (H^+) into hydrogen gas (H_2). Several algae (like Scenedesmus, Chlamydomonas, Dunaliella, Porphyridium and Oscillatoria) possess the enzyme hydrogenase which can be gainfully employed for the production of hydrogen from water.

In blue-green algae such as Anabaena nitrogenase enzyme is the chief hydrogen producing enzyme. Hydrocarbons produced by some algae (e.g., Botryococcus braunii) may also serve as liquid or gaseous fuels.

Uses as Experimental Material

Algae provide valuable experimental materials for research work in plant physiology, genetics and biochemistry. A lot of researches in genetics and cytology have been carried out on Acetabularia. Chlorella has been intensively used for studying the path of carbon during photosynthesis. Valvonia and Halicystis are especially suitable for experiments on membrane permeability.

Harmful or Negative Importance

Algae and Water Supply: Luxuriant growth of planktonic algae in ponds and lakes turns drinking water unpotable. Some common problems of potable water associated with algae are as under:

(1) Decomposition of algae like Anabaena, Dinobryon, Microcystis and Synura in tanks imparts bad taste and odour to water.

(2) Forms like Microcystis, Aphanizomenon and Anabaena produce toxic substances and the water where these algae grow badly affect nervous system, causing depolarization of nerves and brain muscles.

(3) Substances formed by the decomposition of algae in tanks interfere with water treatment process.

(4) Efficiency of water filters at pumping stations is reduced due to the deposition of algae on the filters. This problem is acute during spring and summer seasons when the growth of algae is rapid.

(5) Species of Anacystis and Chaetophora cause corrosion of water pipes and boilers.

At pumping stations and boilers the growth of algae and bacteria is controlled by chemicals like chlorine, copper sulphate and dichlorophen.

Parasitism and Toxicity

Luxuriant growth of planktonic forms in ponds may lead to death of aquatic animals. Toxins secreted by Gymnodinium veneficum, G. brevis, Gonyaulax catenella, G. tamarensis and Pyrodinium phoneus can kill many kinds of fishes. Toxins secreted by Gymnodinium affect the nervous and muscular system of the fish. Species of Gonyaulax secrete endotoxins which accumulate in the digestive system of shellfish feeding on them. When the contaminated shellfish is consumed by human beings death or paralysis may occur.

Algae like Anabaena, Microcystis and Aphanizomenon form 'blooms' in warm water. These algae grow luxuriantly in conditions of oxygen deficiency. In such ponds, during night (when oxygen is not released by photosynthesis), continuous respiration of submerged plants and animals leads to depletion in oxygen level and sometimes it becomes almost zero. Under such conditions mortality rate increases partly due to suffocation and partly due to toxins.

Many toxic planktonic forms enter in human body along with drinking water and may cause diseases. For example,

Anabaena and Microcystis cause gastric trouble, Gymnodinium brevis respiratory disease and Lyngbya and Chlorella cause skin diseases. Tetraodontoxin ($C_{16}H_{31}O_{16}$), a toxin produced by Gonyaulax catenella, enters the human body through fish; it causes numbness, nausea, paralysis and ultimately death within 24 hours.

Cephaleuros, a member of Chlorophyceae, occurs as a parasite in many angiospermic plants such as tea, coffee and piper. C. virescens causes red rust of tea.

Fouling of Marine Vessels

Seaweeds growing on the metallic and wooden hull of ships and submarines have a corroding effect. Thick growth of sea weeds increases friction between the hull and water body and results in greater wear and tear.

2

Classification and Characters

Rhodophyceae

The members of the Class Rhodophyceae are commonly known as red algae, due to the presence of a water soluble red pigment, r- phycoerythrin, which completely masks the green colour of chlorophyll *a*. It is a large group represented by about 831 genera and 5,250 species. Except for 20 fresh water species, all are marine.

The fresh water species are found either in fast flowing streams (e.g., species of Batrachospermum, Thorea, Lamanea, Hildenbrandia) or grow in stagnant water with sufficient aeration (e.g., species of Compsopogon, Asterocytis). Amongst marine red algae, larger and fleshy forms occur in the cool temperate areas, whereas species found in the tropical seas are mostly small and filamentous. The members of Rhodophyceae have the ability to live at greater depths in the ocean than do the members of other algal groups.

Porphyridium, an unicellular red alga, is found in terrestrial habitats.

The red algae also exhibit a high degree of epiphytism and parasitism with considerable specificity. The epiphytic species possess normal form and pigmentation, but parasitic species show a great reduction in their form and pigmentation. Polysiphonia lanosa (host A scophyllum nodosum), Ceramium condicola (host-Codium fragile), Ceratocolax hartzii (host Phyllophora membranifolia), Pterocladiophila hemispherica (host-Pterocladia lucida), Gelidiocolax suhriae (host-Gelidium spp.) are some parasitic species found growing on the members of other groups of algae. Species of Ceratocolax, Choreoeolax, Lithothamnium and Rhodochorton are epiphytic on other red algae.

Common Features

(1) Most of the red algae are beautiful, soft and slimy, and the thallus ranges from simple unicellular to complex multiaxial form. In the sub-class Bangioideae the plant body is unicellular (e.g., Porphyridium), filamentous (e.g., Goniotrichum), palmelloid (e.g., Asterocytis) or parenchymatous (e.g., Porphyra, Bangia). But in the sub-class Florideae, the plant body is more advanced and complex and the thallus organization is of the following two types:

(a) *Uniaxial Thallus:* It is characterised by the presence of a single central or axial filament which is usually corticated by many well branched laterals. Examples—Batrachospermum, Acrochaetium, Dumontia.

(b) *Multiaxial Thallus:* Here the thallus is a mass of central or main filaments and each central filament gives out lateral branches. The central filaments together with their laterals form a pseudo-parenchymatous thallus. Examples Helmin-thocladia, Polysiphonia.

The flagellated motile stages are totally absent.

(2) The members of the sub-class Florideae grow by an apical cell, whereas those of the sub-class Bangioideae show diffused growth.

(3) The cell wall is differentiated into an outer pectic and an

inner cellulosic layer. The mucilagenous material of the outer pectic layer mainly consists of agars and carrageenans (both are made up of repeating units of galactose or its derivatives). These mucilagenous matters constitute up to 70% of the dry weight of the cell wall. In the inner layer microfibrils of cellulose are distributed in a granular matrix. They form the framework of cell wall.

(4) The septa between the adjacent cells have 'pit connections'.

(5) There is much variation in the shape, size and number of nuclei in these algae. All members of the sub-class Bangioideae are uninucleate and the nucleus has a single nucleolus. In the sub-class Florideae, a few species are uninucleate but majority of them show multinucleate condition. In Griffithsia each cell has as many as 4000 nuclei.

(6) The shape of nuclei depends upon their position in the cell. Usually centrally situated nucleus is spherical, whereas the peripheral nuclei are elliptical.

(7) The primitive forms (sub-class Bangioideae) have a single stellate chromatophore with a central pyrenoid, whereas in advanced forms (sub-class Florideae) there are many discoid and parietal chromatophores without any pyrenoid.

(8) The photosynthetic pigments present in the chromatophore include chlorophyll *a* and *d*, α-and β-carotene, xanthophylls (lutein, zeaxanthin, violaxanthin, flavoxanthin, taraxanthin) and biliproteins, such as r-phycoerythrin and r-phycocyanin. β-phycoerythrin is present in the primitive red alga, Porphyridium.

(9) Chromatophores have no chromatophore ER.

(10) The cells contain organelles like endoplasmic reticulum (ER), mitochondria and dictyosomes.

(11) The accumulated product of photosynthesis is floridean

starch and galactoside floridoside; apparently it has the same function as sucrose in higher plants and green algae. The long- term storage product is floridean starch, occurring in the form of grains outside chromatophores in the cytoplasm. In primitive forms starch grains are found clustered as a sheath around the pyrenoid area of the chromatophore, whereas in more advanced forms starch grains are scattered in the cytoplasm.

(12) Many red algae (e.g., Liagora, Galaxaura, and the members of the family Corallinaceae) have extracellular deposition of $CaCO_3$ on their cell walls. Those red algae which have the highest rate of calcification also have the highest rate of photosynthesis. Classification of thallus is two to three times more rapid in light than in dark.

(13) Reproduction is mainly by asexual and sexual methods. Vegetative multiplication of thallus occurs only in unicellular forms.

(14) Asexual reproduction takes place by means of the following type of aplanospores:

(a) *Monospores:* When a single large spore is produced in each sporangium, it is called monospore, and the sporangium bearing it is known as monosporangium. They are found in Compsopogon and Batrachospermum.

(b) *Neutral Spores:* These spores are not borne in sporangia but are formed by direct metamorphosis or division of vegetative cells. They are diploid in nature and on germination give rise to the same generation (sporophytic), that is why they are called neutral spores.

(c) *Carpospores:* These spores develop from the zygote. In some genera (e.g., Batrachospermum) they are formed by reduction division of the zygote nucleus. The nuclei pass into the gonimoblast initials, which develop into gonimoblast filaments. The terminal cell of each filament bears a single haploid carpospore. In other genera (e.g., Polysiphonia) the zygote nucleus directly develops into a single diploid carpospore. Such spores on germination give rise to diploid (sporophytic) plants.

(d) *Tetraspores:* In the life-cycle of some red algae (e.g., Polysiphonia) four haploid spores are formed in each sporangium. These spores are called tetraspores, and the sporangia bearing them as tetrasporangia. These spores are always produced in diploid sporophytes (known as tetrasporophytes) and they give rise to haploid gametophytes on germination.

(15) Non-motile male reproductive cells, called spermatia (singular = spermatium), are borne singly in spermatangia.

(16) The female reproductive organ, called procarp, consists of a dilated basal carpogonium and a usually narrow gelatinous elongated tip, trichogyne. The procarp is sessile in the sub-class Bangioideae, whereas it is stalked in the sub-class Florideae.

(17) In the sub-class Bangioideae, the first division of the zygote nucleus is meiotic, as a result of which four (sometime up to 64) haploid non-motile spores, called carpospores, are formed. These spores on germination give rise to an alternate dissimilar phase (juvenile stage). In the sub-class Florideae, the fertilization is followed by the production of specialized filaments, called gonimoblasts. These filaments arise either directly from the carpogonium or from one or more auxiliary cells (present at the base of the carpogonium) into which the zygote nucleus or some of its descendants have migrated. The terminal cell of the gonimoblast filament functions as a carposporangium and it contains a single haploid or diploid carpospore. The haploid carpospore on germination gives rise to gametophytic thallus, whereas the diploid carpospore forms sporophytic thallus.

Categorisation

Fritsch (1935), on the basis of thallus structure and post-fertilization changes, classified Rhodophyceae into two sub-classes:

Sub-class Bangioideae: This sub-class includes primitive forms of Rhodophyceae. They lack pit-connections and apical growth (instead show diffused intercalary growth) and have pigments like allophycocyanin. The reproductive structures are relatively simple. The zygote nucleus directly gives rise to carposporangium after division.

This sub-class includes only a single order—Bangiales. Examples— Porphyra, Porphyriditum, Bangia.

Sub-class Florideae: The members of this sub-class have uni- or multiaxial thallus which shows apical growth. The adjacent cells are connected by pit connections. The reproductive structures are more developed and complex than those of Bangioideae. The zygote on germination forms gonimoblast filaments and the terminal cell of these filaments develops into a carposporangium.

The sub-class Florideae has been divided into six orders :

Order 1. Nemalionales. Examples Batrachospermum, Chantransia, Thorea.

Order 2. Gelidiales. Example - Gelidium.

Order 3. Cryptonemiales. Example - Cryptonemia.

Order 4. Gigartinales. Example - Gigartina.

Order 5. Rhodymeniales. Example - Rhodymenia.

Order 6. Ceramiales. Example - Ceramium, Polysiphonia.

Smith (1955) placed the class Rhodophyceae in the Division Rhodophyta. He divided the Rhodophyceae into two sub-classes, Bangioideae and Florideae, following Fritsch (1935).

Round (1973) also placed the Class Rhodophyceae in the phylum Rhodophyta, Class Rhodophyceae was further divided into two sub-classes-Bangiophycidae and Florideophycidae. An outline of the classification proposed by Round is given below:

Comparison of the sub-class Bangioideae and Florideae

	Bangioideae	*Florideae*
1.	Freshwater, marine or terrestrial.	Mostly marine.
2.	Thallus simple, unicellular or multicellular.	Thallus structure is relatively complex, uniaxial or polyaxial.
3.	Cells always uninucleate.	Number of nuclei per cell is much variable and sometimes it is as high as 3000-4000.
4.	They show diffused intercalary growth.	Growth takes place with the help of a single apical cell.
5.	Each cell has a single, centrally located, stellate chromatophore.	There are many discoid chromatophores in each cell, present along the periphery.
6.	Each chromatophore has a single pyrenoid.	Chromatophores lack pyrenoids.
7.	Pit connections absent.	Pit connections present between the adjacent cells.
8.	Zygote divides to form haploid carpospores and these spores on germination give rise to alternate dissimilar vegetative phase	Zygote divides to form diploid or haploid gonimoblast filament and the terminal cell of the filament develops into a carposporangium which bears a single carpospore.

Phylum Rhodophyta Class Rhodophyceae

Sub-class

Bangiophycidae (=Bangioideae)	Florideophycidae (=Florideae)
Order 1. Bangiales	Order 1. Nemalionales
	2. Gelidiales
	3. Cryptonemiales
	4. Gigartinales
	5. Rhodymeniales
	6. Ceramiales

Laminariales

The members of order Laminariales, commonly known as

kelps are of very large size and complex structure. Some plants reach up to 50-60 m. Most of the members show parenchymatous construction. The plant body may be differentiated into stipe and blade (lamina) and the meristematic region is located in between them. Asexual reproduction takes place generally by zoospores formed in unilocular filamentous and dioecious sporangia. The gametophytec body is microscopic. Sexual reproduction is oogamous type. Plants exhibit heteromorphic alternation of generation. There are about 30 genera and 100 species, mainly inhabitants of the colder oceans of the arctic and antarctic regions.

The members are distinguished from the Fucales by their intercalary growth and heteromorphic alternation of generation. The order Laminariales is usually divided into four families, viz. Chordaceae, Laminariaceae, Lessoniaceae and Alariaceae.

Family Laminariaceae

The sporophytic plant body exhibits a high degree of morphological and anatomical differentiation well divisible into holdfast, stipe and blade. The sporangia are arranged in distinct groups (sori) on both the surfaces of blade. The gametophytes are much reduced, heterotrichous and dioecious. Members possess diplohaplontic heteromorphic alternation of generations.

Life history of Laminaria the representative genus of Laminariaceae is described here in details.

Laminaria

Systematic Position

Class	-	Phaeophyceae
Order	-	Laminariales
Family	-	Laminariaceae

Occurrence: The genus Laminaria with some 40 species is exclusively marine and is widely distributed in the waters of north temperate and arctic zones. L. flavicans, L. saccharina, L. digitata, L. flexicaulis and L. japonica are some common species of the genus. The plants may be annual (L. ephemera) or perennial (L. cloustoni). In the annual species, a new sporophyte is produced each spring.

Thallus Structure

External Structure: The plant body of Laminaria is diploid and sporophytic attaining a length of two metre or more. It is differentiated into a hold fast, stipe and blade (lamina). The hold fast is an attaching organ variable in form. It may be in the form of a solid disc or profusely branched cluster of cylindrical root-like organs (haptera). The stipe is cylindrical or some what flattened unbranched portion of the thallus that arises from the basal holdfast and bears terminally either a single blade (L. saccharina) or a number of vertically divided segments (L. digitata, L. cloustoni). The lamina is rather tough and leathery in texture. The lamina is renewed at the base each year by the activity of the intercalary meristem. The lamina is flat, long and ribbon-shaped and the margins may be smooth or wavy.

Internal Structure

1. *Stipe:* In transection, the stipe is differentiated into three distinct regions: (i) meristoderm, (ii) cortex, and (iii) medulla.

   The meristoderm (named so due to meristematic nature of its cells) is one or two cells in thickness which are small and cuboidal in form. It remains surrounded by a layer of mucilage.

   The cortex lies within the meristoderm and surrounds the central axis which constitutes the medulla. The cortex usually remains differentiated into an outer and an inner cortex. The cells of the outer cortex are more or less radially arranged, vertically elongated with pointed ends and thin walls. They show horizontal septation. On the contrary, the cells of the inner cortex are comparatively squarish at the ends and show gelatinisation of the longitudinal walls. The cells possess thread like outgrowths or cross connections and hyphae. The cortical cells are regarded as storage and mechanical elements. The medulla is composed of loosely arranged cells, the spaces between them are filled with mucilage. Cross connections are present in the cells, formed as a result of fusion of papillose outgrowths of the longitudinal walls. Regular formation of cross connections in the medullary region and the inner part of the inner cortex, give it an appearance of a tangled network of filamentous growth. An another kind of protrusions, known as hyphae also arise from the medullary cells.

The hyphae differ from the cross connections in that they do not fuse with the corresponding outgrowth. Besides, certain cells of the medulla considerably swell at their ends. Such swollen cells are called the trumpet hyphae. The trumpet hyphae in the younger stages become much attenuated and become extremely thick-walled and filled with callus. The medulla constitutes the main conducting system of the plant body.

2. *Blade (Lamina):* Like stipe, the blade is also differentiated into outer meristoderm, middle cortex and central medulla regions.

The meristoderm is generally single layered, the cells of which are small with many chromatophores. Just beneath the meristoderm lies the cortex. It consists of isodiametric cells. Mucilage canals are present in the cortex. Medulla occupies a rather narrow tangential zone and consists of a network of cross connections and medullary cells. The cross connections are displaced with the enlargement of the blade.

Growth: The plant body grows in length by the activity of an intercalary meristem, situated at the junction of the stipe and the blade.

Reproduction Process

Laminaria reproduces by vegetative, asexual and sexual methods.

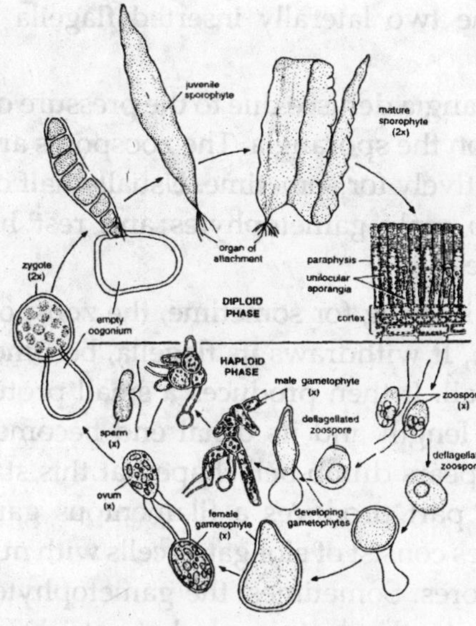

Laminaria: Life-cycle

Vegetative Reproduction: Some species of Laminaria (e.g., L. longipes, L. rodreguezu, L. sincelarii) reproduce vegetatively by means of stolons. The stolons arise horizontally from the holdfast and then they turn upwards to grow into new plantlets.

Asexual Reproduction: The asexual reproduction takes place by zoospores produced within club-shaped unilocular sporangia on the sporophytic plant body. The sporangia are arranged in sori on both sides of the blade, covering its considerable area. Interspersed with sporangia are present numerous paraphyses.

The sporangia arise from the superficial cells of the blade. Each sporangium is a long, sac-like structure. They are tightly pressed, one each between two adjacent paraphyses. The sporangia are diploid structures. The nucleus of the sporangium first divides by a meiotic and then by successive mitotic divisions to form 32 (sometimes 64) haploid nuclei in each sporangium. The cleavage of the protoplast follows and haploid, uninucleate bits of the protoplast are formed. Each protoplast bit then metamorphoses into a motile zoospore. Each zoospore is biflagellate and possesses a single chromatophore occupying the base. The two laterally inserted flagella are unequal in length.

The sporangia dehisce due to the pressure of the gelatinous paraphyses on the sporangia. The zoospores are thus liberated and swim actively for sometime. Usually half of the zoospores develop into male gametophytes and rest half into female gametophytes.

After swimming for sometime, the zoospore settles down to germinate. It withdraws its flagella, becomes rounded and secretes a wall. It then produces a small protuberance which increases in length and its distal end becomes bulbous. The germling appears dumb-bell shaped at this stage. Ultimately, the bulbous part produces a filamentous gametophyte. The gametophytes consist of elongated cells with numerous discoid chromatophores. Sometimes, the gametophytes are branched and show a tendency towards heterotrichy. The male and female gametophytes develop male and female sex organs respectively.

Sexual Reproduction: The sexual reproduction is oogamous. The male sex organs are called antheridia and the female oogonia. Male and female sex organs are produced on separate gametophytes. The gametophytes usually do not develop sex organs when the temperature of surrounding water is above 15°C.

1. *Male Gametophytes and Antheridia:* The male gametophytes are comparatively short, filamentous and heterotrichous. Antheridia are produced at the tips of one-or two-celled lateral branches either singly or in groups. In each antheridium, a single biflagellate antherozoid is formed. Each antherozoid is an ellipsoidal or a pyriform structure provided with two laterally inserted flagella. It is liberated near the apex of the antheridium.
2. *Female Gametophytes and Oogonia:* The female gametophyte is also a multicellular (generally reduced to a few cells)

filamentous structure. Sometimes, it may consist of only one cell, which may function as an oogonium. Any cell of the female gametophyte may function as an oogonium initial. It enlarges considerably and assumes a pyriform shape. Its contents round off to form a single egg (oosphere). At maturity, each oogonium is surrounded by three layers and the oogonial wall at the distal end becomes thickened.

3. *Fertilisation:* At maturity, a slit develops in the thickened apical portion of the oogonium and the egg emerges out through this slit, and rests on a cushion formed by the localized gelatinisation of the oogonial wall. Fertilisation takes place in this condition. Liberated antherozoids swim to the ovum. Only one antherozoid penetrates the delicate membrane of the ovum. The male and female nuclei fuse to form a diploid zygote.

4. *Germination of Zygote:* The zygote germinates immediately after fertilization while still attached to the tip of oogonium. The zygote undergoes considerable elongation and it divides first mitotically by a transverse division. Further divisions result in the formation of a 4-10 celled filament. Later, both vertical and transverse divisions take place in the upper and middle part of the filament and as a result, a leaf-like expansion is differentiated. The lower cell, however, remains undivided and elongates to give rise a rhizoid-like process. Later, additional rhizoids develop which help in anchoring the developing sporophyte to the suitable substratum. The female gametophyte disintegrates after the differentiation of three or more rhizoids. Finally, a sporophytic plant body differentiated into holdfast, stipe and blade is developed.

Alternation of Generation: The life-cycle of Laminaria exhibits an advanced type of heteromorphic type of alternation of generation. The main plant body is sporophytic, differentiated into the holdfast, cylindrical stipe and broad blade (lamina). On both sides of the lamina, sporangia are produced in groups. The diploid sporangia undergo reduction division to form haploid zoospores. Usually 32 zoospores are formed per sporangium. Half of them develop into male and the rest into female gametophytes.

The gametophytic body is microscopic, and consists of few celled filamentous body. Thus, there is sharp morphological differentiation between the sporophytic and gametophytic plant body. The reproductive organs-antheridia and oogonia, are produced on male and female gametophytes respectively. The haploid male and female gametes fuse to form diploid zygote. The zygote germinates into a diploid sporophytic thallus. Thus, in the life-cycle of Laminaria, the macroscopic sporophytic plant body alternates with microscopic filamentous male and female gametophytic plant body.

3

Alternation of Generation

The Conjugales

Algae belonging to the order Conjugales are Commonly known as 'pond silks' or 'water silks'. They are confined to fresh waters only. They differ from other Chlorophyceae in the absence of flagellated reproductive cells and in the structure of chloroplast. These algae vary in their structure from unicellular to unbranched filamentous forms. The unicellular forms are commonly known as desmids. The cell wall is composed of an outer pectic and an inner cellulosic layer. The cells are uninucleate and the nucleus is centrally located. The chloroplasts are of three types (i) one or more peripheral spirally twisted ribbon-like bands as in Spirogyra; (ii) an axial plate extending the length of the cell as in Mesotaenium and Mougeotia; and (iii) two stellate axial chloroplasts as in Zygnema and Cylindrocystis.

Vegetative multiplication in unicellular forms takes place by simple cell division and in filamentous forms by fragmentation. Motile asexual spores are not found in these algae. They reproduce asexually with the help of aplanospores. Sexual reproduction takes place by the fusion of non-flagellate amoeboid gametes through conjugation. The zygospore formed by the fusion of two gametes secretes a thick wall with characteristic ornamentation.

Fritsch (1935) divided the order into five families Mesotaeniaceae, Zygnemaceae, Mougeotiaceae, Gonatozygaceae and Desmidiaceae.

Family Zygnemaceae

Members of Zygnemaceae are among the most common unbranched filamentous algae, favouring small stagnant bodies of water. They are especially abundant in spring months. The cells of the filament are cylindrical and more or less permanently united. The cell wall lacks pores. Union of two aplanogametes usually takes place by the establishment of a conjugation tube.

The family is represented by 3 genera, Spirogyra, Debarya and Zygnema. The life-history of Zygnema is described here as the representative of the family.

Zygnema

Systematic Position

Class	-	Chlorophyceae
Order	-	Conjugales
Family	-	Zygnemaceae

Occurrence: Zygnema is a freshwater alga forming bright green free-floating masses in stagnant water. Species growing in swiftly running streams develop holdfasts, particularly when coming in contact of any rough surface. Z. terrestre is a terrestrial species. The genus has about 100 species, of these 47 are recorded from India. Some common Indian species are Z. himalayensis, Z. czurdae, Z. giganticum and Z. globosum.

Thallus Structure: It is a free floating unbranched filamentous alga and the plant body consists of a uniseriate row of cylindrical cells. All the cells of the filament are alike, but in attached forms the basal cell develops into an attaching organ. The cells are two to four times longer than breadth.

The cell wall is composed of two layers, the outer layer is of pectose and the inner of cellulose. The pectose layer is relatively thin than in other members of Zygnemaceae. This layer gives the filaments a slippery touch, hence the name 'pond silks' or 'water-silks' has been given to these algae.

Each cell contains two stellate (star-shaped) chloroplasts which lie axial to each other. A single pyrenoid is present in each chloroplast. The cell is uninucleate; the nucleus lies between the two chloroplasts in the center of the cell.

Growth: Growth of the filament takes place by repeated transverse divisions of the cells. All the cells of the filament (except the holdfast in attached forms) are capable of divisions.

Reproduction

Zygnema reproduces by vegetative, asexual and sexual methods.

Vegetative Reproduction: Vegetative multiplication is brought about by the following methods:

1. *Fragmentation:* The filament may break into small segments or individual cells due to the dissolution of middle lamella. Such fragmentation is usually brought about by certain changes in the habitat. Besides, water currents and aquatic animals feeding upon plants may also break the filaments into small segments. Some of the dissociated cells or fragments may die, while others form new filaments.
2. *Akinetes:* In some aquatic (e.g., Z, giganteum, Z. spontaneum, Z. sterile) as well as terrestrial species (Z. terrestre) certain cells of the filament develop into thick walled akinetes. Their formation is induced by unfavourable environmental conditions. During the formation of akinetes, the vegetative cell divides into two similar halves (each with a single chloroplast) which accumulate food reserves. Thus akinetes are half the length of the vegetative cell. They germinate under favourable conditions and form new filaments by repeated transverse divisions.

Asexual Reproduction: Like other members of the order Conjugales, Zygnema reproduces asexually by means of non-motile spores. The following two types of non-motile spores are found in Zygnema:

1. *Aplanospores:* They are formed singly within the vegetative cell in many species of Zygnema (e.g., Z. himalayensis, Z. kumaoense, Z. terrestre). The protoplast of the cell retracts from the cell wall and forms a non-motile thick walled spore, called aplanospore. The aplanospore has a wall distinct from the parent cell wall.
2. *Parthenospores:* Sometimes gametes fail to fuse and secrete a thick wall. These are called parthenospores, azygospores or azygotes.

Aplanospores and parthenospores germinate under favourable conditions and give rise to new filaments.

Sexual Reproduction: Sexual reproduction is isogamous or rarely anisogamous. In most of the species, the fusing gametes are morphologically as well as physiologically similar. However, some species show physiological anisogamy. Any cell of the filament may function as gametangium. It produces a single non-flagellate, non-motile amoeboid gamete. Gametic union takes place by the formation of a tubular conjugation tube between the two cells— a characteristic feature of the order Conjugales.

Conjugation is of two types: (i) Lateral conjugation, and (ii) Scalariform conjugation.

1. *Lateral Conjugation:* When conjugation takes place between two adjacent cells of the same filament, it is called lateral conjugation. Thus, lateral conjugation involves only a single filament (Fig. A, D).

   At the time of the formation of conjugation tube between two adjacent cells (gametangia), longitudinal walls of these cells protrude on either side of the septum. This results in the formation of two outgrowths, and eventually a part of the septum close to the outgrowths dissolves forming a passage which serves as conjugation tube.

   Most of the species of Zygnema (e.g., Z. czurdae,

Z. himalayense, Z. terrestre) are isogamous and in these species both the gametes are equally active and move into the conjugation tube.

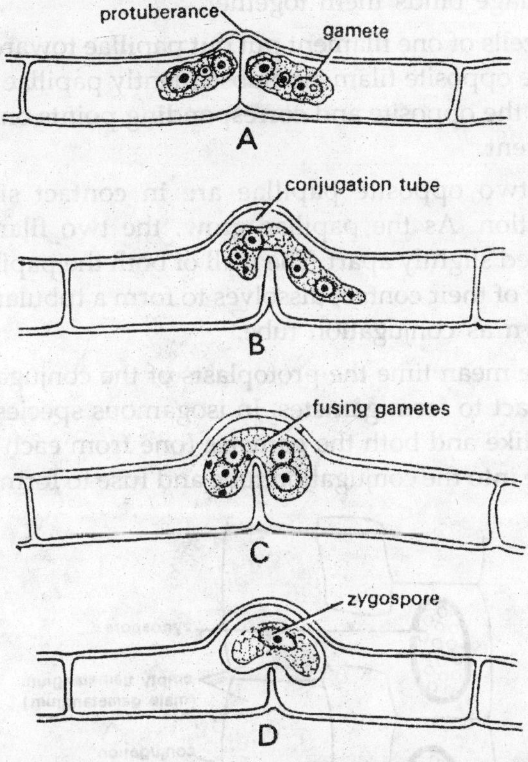

A-D. Zygnema: Sexual reproduction; isogamous lateral conjugation

Fusion of gametes takes place in the region of dissolved septum. But in anisogamous species (e.g., Z. stellinum) only one of the two fusing gametes is motile. The motile gamete is referred to as male gamete and the non-motile as female.

At the time of gametic union the motile male gamete moves through the conjugation tube and fuses with the non- motile female gamete in the female gametangium. In anisogamous species, therefore, zygospores and empty male gametangia are present in alternate sequence.

2. *Scalariform Conjugation:* In scalariform conjugation two different filaments are involved. The two mating filaments come close to each other and lie side by side. A layer of mucilage binds them together.

   The cells of one filament put out papillae toward the cells of the opposite filament. Subsequently papillae also arise from the opposite and corresponding points on the other filament.

   The two opposite papillae are in contact since their initiation. As the papillae grow, the two filaments are pushed slightly apart. The wall of both the papillae at the point of their contact dissolves to form a tubular passage, known as conjugation tube.

   In the mean time the protoplasts of the conjugating cells contract to form gametes. In isogamous species gametes are alike and both the gametes (one from each filament) move into the conjugation tube and fuse to form a zygote.

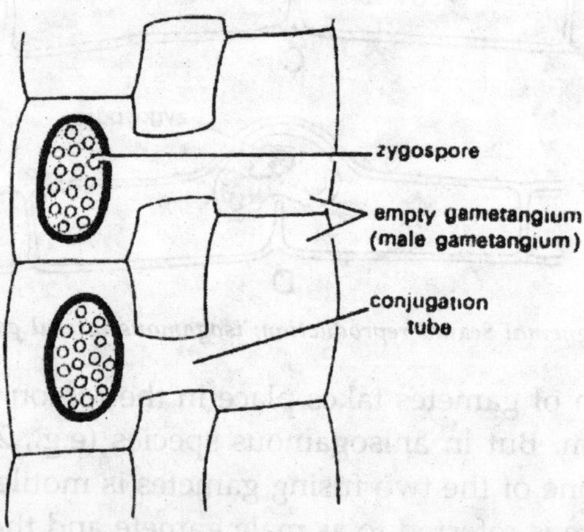

Zygnema: Zygospores (note the empty gametangia)

However, in anisogamous species gametes can be distinguished into male and 'female on the basis of their motility. The motile male gamete moves into the female gametangium through the conjugation tube and fuses with the female gamete to form a zygote.

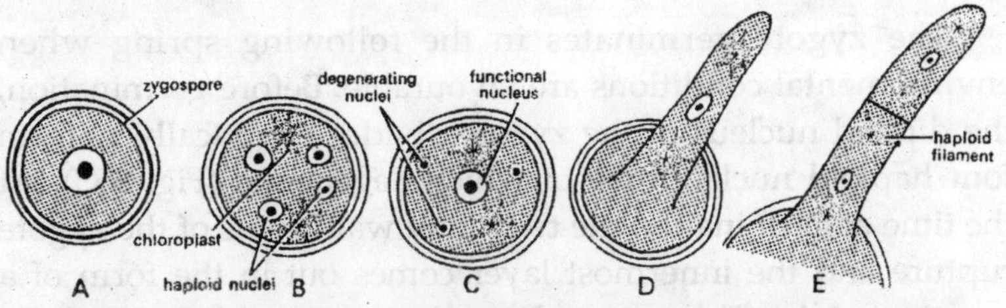

A-C. Zygnema: Germination of zygospore; A. Zygote, B-C. Reduction division in the zygote

3. *Germination of Zygote:* The zygote, formed by the fusion of two haploid gametes, represents the diploid phase in the life-cycle of Zygnema (Fig. A). The zygote secretes a three layered wall; the outermost thin layer (exospore) is made up of cellulose or pectose, and the middle thick layer (mesospore) and the inner thin layer (endospore) are cellulosic. Ornamentation and coloration occur in the middle thick layer. Ripening of zygote is accompanied by the disappearance of chlorophyll and conversion of starch into a yellowish oil. The mature zygote is liberated by decay of the gametangial or conjugation tube wall.

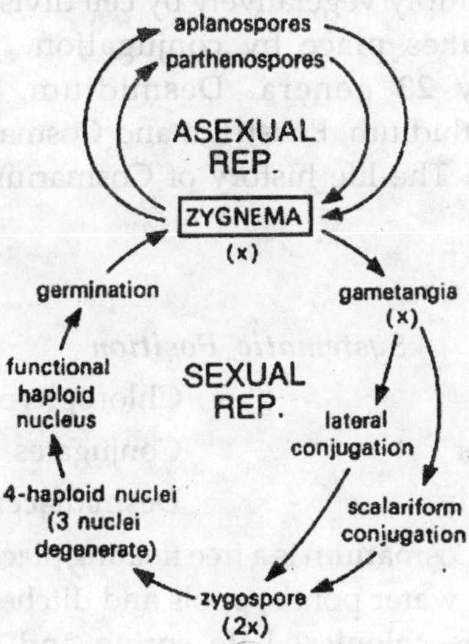

Zygnema: Diagrammatic representation of the life- cycle.

The zygote germinates in the following spring when environmental conditions are favourable. Before germination, the diploid nucleus of the zygote divides meiotically to form four haploid nuclei, three of which degenerate (Fig. B,C). At the time of germination the two outer wall layers of the zygote rupture and the innermost layer comes out in the form of a long germ tube. This germ tube undergoes repeated transverse divisions to form a new (haploid) filament.

The life-cycle of Zygnema is thus primarily haploid, zygote being the only diploid stage.

Family Desmidiaceae

The members of this family are placoderm desmids, characterised by the presence of cells with two distinct halves (semicells) separated by a median constriction (sinus). The cell wall is highly ornamented and has pores. Most of the desmids are unicellular, but a few are joined end to end in filamentous colonies or united in amorphous colonies. The cell has a central nucleus and two chloroplasts, one in each semicell.

Desmids multiply vegetatively by cell division and sexual reproduction takes place by conjugation. The family is represented by 23 genera. Desmidium, Staurastrum, Closterium, Xanthidium, Euastrum and Cosmarium are some common genera. The life history of Cosmarium is described here.

Cosmarium

Systematic Position

Class	-	Chlorophyceae
Order	-	Conjugales
Family	-	Desmidiaceae

Occurrence: Cosmarium is a free floating placoderm desmid. It occurs in fresh water ponds, pools and ditches, either singly or in amorphous colonies. Late spring and early summer seasons are most favourable for its growth. The genus includes over 800 species, of these Cosmarium bengalensis, C. connatum, C. distichum, C. granatum and C. tenne are common in India.

Thallus Structure: The plant body consists of a single haploid flat cell with two distinct symmetrical halves (or semicells). The two halves are separated by a deep median constriction, called sinus, and a connecting band, known as isthmus, joins them. The cell appears differently in different views; it is semicircular, ellipsoidal or reniform in front view, elliptical in vertical view and oval in side view.

The cell wall is either smooth or heavily ornamented with granular projections. The wall of a mature cell is differentiated into two concentric layers. The inner layer is smooth and mostly consists of cellulose. The outer layer is more thick and firm than the inner layer and is made up of cellulose, impregnated with pectic and iron compounds. Both layers are perforated by minute and simple or large specialized pores, usually grouped symmetrically. The pore is in the form of a simple canal in the inner cell wall layer, but in the outer layer the canal is surrounded by a specially differentiated cylindrical noncellulosic zone. The entire pore structure is called pore-organ. Mucilage secreted through these pores forms a thin coat on the outer surface of the cell.

The cell of Cosmarium is uninucleate and the nucleus is embedded in a small median mass of cytoplasm. In most species there are two axile chloroplasts, one in each semicell, but in C. reniforme and C. ovale each semicell has two parietal chloroplasts in the form of bands.

Reproduction

Reproduction in Cosmarium takes place by vegetative and sexual methods.

Vegetative Reproduction: Vegetative reproduction takes place by cell division. The cell may multiply vegetatively for long periods without conjugating at all. The process of cell division is peculiar due to bipartite structure of the cell wall. It initiates with the elongation of the connection zone (isthmus) of the two semicells. The two halves thus move slightly apart (Fig. A, B). At this stage the nucleus of the cell divides mitotically into two daughter nuclei. Then a transverse wall is laid down in isthmus region which divides the parent cell into two daughter cells. Each daughter cell possesses a portion of the isthmus which grows into a semicell. Thus in the new cell, formed by the division of the parent cell, one semicell is obtained from the parent cell and the other semicell is formed by the portion of the isthmus region. In the meantime, chloroplast of the parent semicell divides and one of the daughter chloroplasts passes into the new semicell.

The membrane of the young semicell is quite smooth initially but subsequently the outer membrane develops ornamentation which is typical of a species. The young semicells usually remain in contact by their apices until they are fully grown (Fig. D).

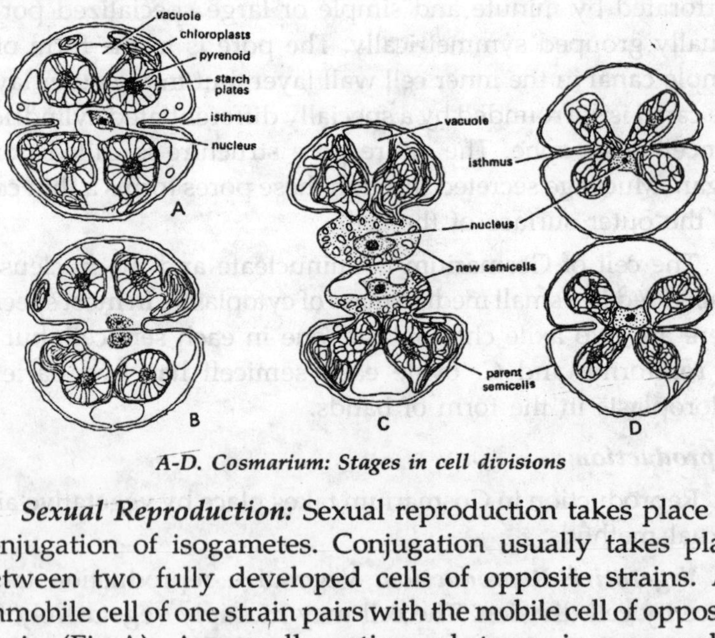

A-D. Cosmarium: Stages in cell divisions

Sexual Reproduction: Sexual reproduction takes place by conjugation of isogametes. Conjugation usually takes place between two fully developed cells of opposite strains. An immobile cell of one strain pairs with the mobile cell of opposite strain (Fig.A). A sexually active substance is presumably secreted by the immobile cell which is responsible for chemotactic movement of the mobile cell. The paired cells are enveloped in a common mucilagenous sheath. The semicells of both the conjugating cells split open at isthmus and their protoplasts are set free (Fig. B, C). The entire protoplast of each cell acts as a gamete. The amoeboid gametes fuse within minutes and form an irregularly-shaped zygospore. The zygospore soon rounds off and secretes a smooth wall (Fig.D). Ornamentation in the wall begins to appear within an hour after the fusion of gametes (Fig. E).

Germination of Zygospore: The zygospore settles at the bottom of the pond. At this stage it has four chloroplasts and a diploid nucleus (formed by the fusion of two haploid nuclei of the gametes). Immediately after fusion, two of the four chloroplasts degenerate. Germination of the zygospore starts after a period of rest. Diploid nucleus of the zygospore divides meiotically forming four haploid nuclei; of these, two nuclei subsequently degenerate. Now the zygospore has two chloroplasts and two functional nuclei. The protoplast of the zygospore divides into two daughter protoplasts, each with a functional nucleus and a chloroplast. The two daughter protoplasts are liberated by the rupture of the zygospore wall, and each develops into a new cell.

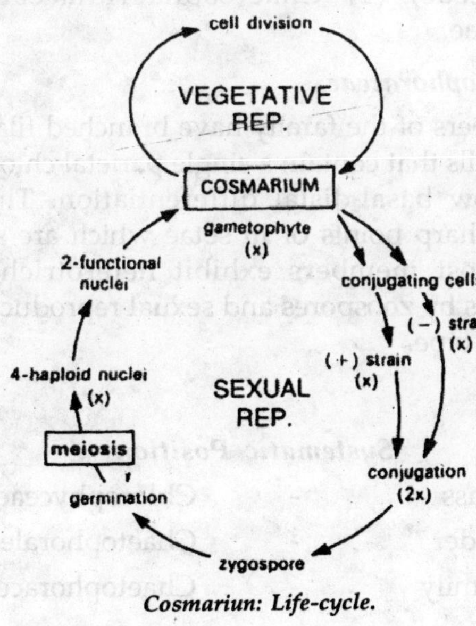

Cosmariun: Life-cycle.

The Chaetophorales

The members of Chaetophorales are characterised by the presence of a heterotrichous thallus, that is, the thallus possesses a prostrate and an erect system. In some genera, one of the two systems is suppressed; for example, in Drapenaldia, Drapernaldiopsis and Microthamnion only erect system is present, whereas in Coleochaete and Ulvella there is only prostrate system. Certain members of the order have setae (e.g., Coleochaete) or hairs (e.g., Stigeoclonium) and hence the name Chaetophorales (Chaetos = hair, phorales = plant; plants with hair or setae).

Most of the members of Chaetophorales occur in fresh water, but a few have a marked terrestrial tendency (e.g., Fritschiella tuberosa). The asexual reproduction takes place by quadri-or biflagellate zoospores, aplanospores or akinetes.

The sexual reproduction is primarily isogamous (e.g., Fritschiella, Stigeoclonium), occasionally anisogamous (e.g., Aphanochaete) or oogamous (e.g., Coleochaete). Fritsch (1935) divided the order Chaetophorales into the following five families: (1) Chaetophoraceae, (2) Trente-pohliaceae, (3) Coleochaetaceae, (4) Chaetosphaeridiaceae, and (5) Pleurococcaceae.

Family Chaetophoraceae

The members of the family have branched filaments with uninucleate cells that contain a single parietal chloroplast. The filaments show basal-distal differentiation. The branches terminate in sharp points or in setae which are several cells in length. Most members exhibit heterotrichy. Asexual reproduction is by zoospores and sexual reproduction usually of isogamous type.

Fritschiella

Systematic Position

Class	-	Chlorophyceae
Order	-	Chaetophorales
Family	-	Chaetophoraceae

Occurrence: This genus, established by Prof. M.O.P. Iyengar, was named after the great algologist Prof. F.E.Fritsch. The type species, Fritschiella tuberosa, was first reported from South India. It was subsequently collected from Nepal, Burma, Sudan and Japan. It is a terrestrial alga which grows on moist soil undergoing drying after monsoon rains. Another species of the genus, F. simple, *x* was recorded from East Pakistan (now Bangladesh) in 1962.

Thallus Structure: It is a heterotrichous alga and the thallus is differentiated into four distinct systems.

1. *The Rhizoidal System:* It is made up of one or more downwardly directed rhizoid-like elongated colourless cells arising from the prostrate system. This system is, however, absent in F. simplex.
2. *The Prostrate System:* It consists of a number of rounded or irregular swollen clusters of relatively small cells. In mature thallus this system forms short congested branches which are differentiated into nodes and internodes. The cells are thin walled and rich in reserve food (starch, protein, etc.) and, therefore, probably their function is storage.

   The prostrate system varies considerably in the extent of its development. In young plants, the prostrate system is not well formed, but it gradually increases in size as the plants grow. Mature plants invariably show a very well developed prostrate system.
3. *The Primary Projecting System:* It is the upright system which emerges from the prostrate system. It consists of uniseriate or biseriate filaments which may be simple or branched. The cells of the filament are small and rounded like those of the prostrate system. The filaments are green and besides photosynthesis, they also help in conduction.
4. *The Secondary Projecting System:* This system consists of freely branched uniseriate filaments with elongated cells. In the natural habitat the secondary projecting system grows as a tuft above the surface of the substratum. This

system is poorly developed in the thalli growing in exposed areas, because of high rate of transpiration. It is absent in F. simplex.

Of the above four systems, the rhizoidal and prostrate systems are completely underground. The primary projecting system is partly embedded in the soil and partly above the ground. The secondary projecting system emerges completely above the soil surface.

Cell Structure: The cells are thin walled and uninucleate. A curved plate-like chloroplast with 2-8 pyrenoids is present in the cells of the secondary projecting system. In primary projecting and prostrate system the cell contents are dense and the chloroplast is poorly developed and has only 2-4 pyrenoids. In rhizoidal system, however, chloroplasts are totally absent.

Perennation

During dry conditions the primary and the secondary projecting systems and the rhizoidal system disappear and the prostrate system acts as perennating structure. The cells of the prostrate system are capable of tiding over unfavourable conditions without any structural modification.

On return of favourable conditions the cells germinate directly and give rise to new plants. Sometimes the cells of the nodal region are detached from the thallus due to death of the internodal cells. These cells, besides serving as perennating bodies, also help in vegetative propagation.

Another means of perennation is the formation of small (about 1 mm in length) funnel-shaped and tuber-like bodies on the prostrate system. These dark coloured bodies have a thick layer of cuticle. They germinate under favourable conditions and form new thalli.

Reproduction

Fritschiella reproduces by asexual and sexual methods. The methods of perennation described above also help in vegetative propagation of the thallus.

Asexual Reproduction: The asexual reproduction takes place by biflagellate and quadriflagellate zoospores. The zoospores are formed in the diploid filaments (2n = 8) on the prostrate system (F. tuberosa) or the primary projecting system (F. simplex).

The filaments which form zoospores develop club-shaped endings that are covered with a thick pellicle of dark-brown colour. The cells of these club-shaped structures are similar to the other cells of the filament in their shape and size, but their contents are more homogeneous and brick-red in colour.

When conditions are favourable and sufficient moisture is available, the nucleus of each of these cells undergoes meiosis and the contents form four haploid (n=4) zoospores. These zoospores are at first colourless or dull brick-red in colour but after liberation they assume usual green colour.

There are 2-4 biflagellate zoospores per cell, or only one quadriflagellate zoospore is formed in a cell. Both types of zoospores are produced simultaneously on the sporophyte plant.

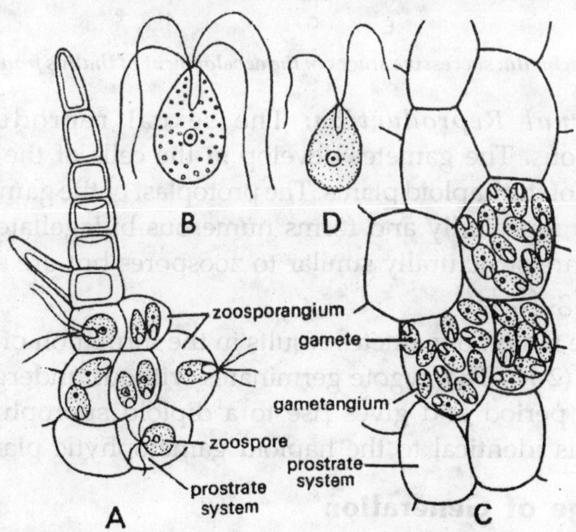

A-D. Fritschiella: Reproduction; A. Part of the thallus showing quadriflagellate zoospores, B. A quadriflagellate zoospore, C. Part of the prostrate system showing gametangia, d. A biflagellate gamete

The zoospores are liberated by the rupture of the sporangial wall and on germination they give rise to new plants. The plants produced by quadriflagellate zoospores are diploid sporophytic plants, while those produced by biflagellate zoospores are haploid gametophytic plants. Both haploid and diploid plants are however, morphologically similar (Fig. A, F).

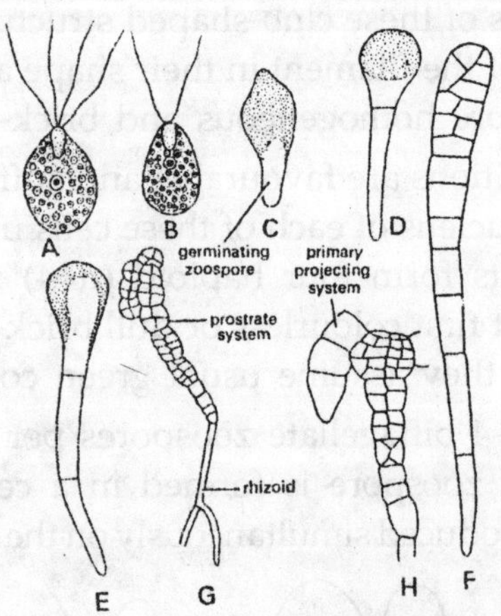

A-H. Fritschiella: successive stages in the development of thallus from a zoospore.

Sexual Reproduction: The sexual reproduction is isogamous. The gametes develop in the cells of the prostrate system of the haploid plants. The protoplast of the gametangium divides repeatedly and forms numerous biflagellate gametes which are structurally similar to zoospores but are smaller in size (Fig. C, D).

The fusion of gametes results in the formation of a diploid zygote (2 X). The zygote germinates without undergoing any resting period and gives rise to a diploid sporophytic plant which is identical to the haploid gametophytic plant.

Change of Generation

Fritschiella shows alternation of two morphologically similar generations (isomorphic alternation of generation). The haploid and diploid filaments are alike morphologically. The diploid filaments produce haploid zoospores which give rise to haploid gametophytic filaments on germination. The cells of the prostrate system produce gametes. Similar gametes fuse to form zygote (2 x) and the latter gives rise to diploid filaments on germination.

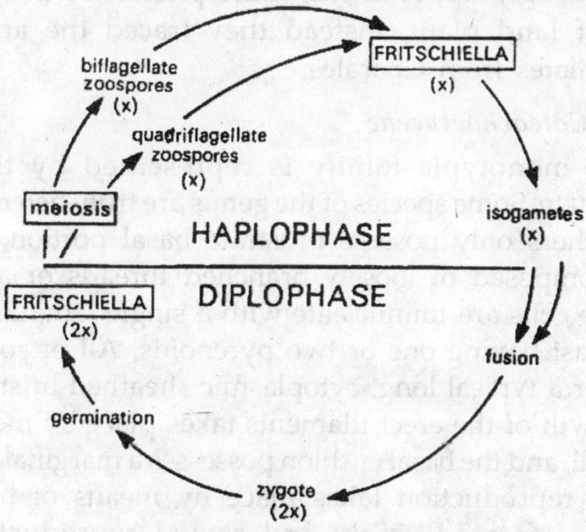

Fritschiella: Life-cycle

Fritschiella and Origin of Land Plants: There are certain points of interest in the life history of Fritschiella which throws light on the supposedly algal ancestry of higher plants. The tendency of terrestrial or amphibious habit, marked division of labour as shown by photosynthetic subaerial portion, the prostrate system serving as an organ for reproduction and perennation, and the underground rhizoidal system performing the function of anchorage are some remarkable features of Fritschiella.

Fritsch (1935) speculated that the members of Chaetophorales represent the surviving descendants of forms which migrated to the land in the remote past. He was of the opinion that the prostrate gametophyte and the erect aerial sporophyte of land plants evolved from some heterotrichous terrestrial green alga, very similar to the present day Fritschiella.

He further remarked that the high degree of thallus differentiation exhibited by Fritschiella and other allied taxa certainly indicate the existence of a strong potentiality for further elaboration in the direction of evolution of an early algal transmigrant form. On the contrary, Stewart and Mattox (1975) strongly opposed the chaetophoralean theory of the origin of land plant, instead they traced the ancestry of archegoniates from Charales.

Family Coleochaetaceae

This monotypic family is represented by the genus Coleochaete. Some species of the genus are truly heterotrichous, while others only possess prostrate basal portion, which is either composed of loosely branched threads or a compact disc. The cells are uninucleate with a single laminate parietal chloroplast having one or two pyrenoids. All or some of the cells bear a typical long, cytoplasmic sheathed bristle or seta. The growth of the erect filaments takes place by means of an apical cell, and the basal cushion possesses a marginal meristem. Asexual reproduction takes place by means of biflagellate zoospores formed singly and sexual reproduction is of oogamous type.

Coleochaete

Systematic Position

Class	-	Chlorophyceae
Order	-	Chaetophorales
Family	-	Coleochaetaceae

Occurrence: Coleochaete is a freshwater alga which grows epiphytically on submerged parts of several aquatic angiosperms like Hydrilla, Nymphaea, Typha, Sagittaria and Trapa. Thalli of Coleochaete are sometimes epiphytic on Oedogonium and Vaucheria. Some species are endophytic in the members of the order Charales, e.g., C. nitellarum grows endophytically within the cells of Nitella and Chara.

The genus is represented in India by three species, C. orbicularis, C. scutata and C. nitellarum.

Thallus Structure: The thallus is multicellular, heterotrichous and branched. Heterotrichous habit is best represented by C. pulvinata which has well-developed prostrate and erect systems. In the other species one of the two systems (prostrate or erect) is either absent or reduced, e.g., in C. nitellarum and C. divergens there is a well-developed erect system but the prostrate system is filamentous or reduced and in C. scutata and C. orbicularis erect system is absent and the lateral branches of the prostrate system form discoid thalli which may be irregular (e.g., C. scutata) or regular (C. orbicularis) in outline.

Cell Structure: The cells of the thallus may be quadrangular, polygonal or cylindrical. They are uninucleate and each cell has a large laminate parietal chloroplast with a single pyrenoid. Some of the cells of the thallus have a long unbranched cytoplasmic thread, called seta. The base of the seta is covered with a sheath of gelatinous material. This sheath remains attached with the thallus even after the seta breaks off.

Growth: The branches of erect system show apical growth. The terminal cell of the erect filament divides transversely and adds to its height. The growth of the prostrate discoid system takes place by means of a marginal meristem. The erect branches originate as lateral outgrowths of the prostrate system.

Reproduction

Coleochaete reproduces both asexually and sexually.

Asexual Reproduction: The asexual reproduction takes place by means of biflagellate zoospores. They are formed usually in spring and beginning of summer season. Any vegetative cell of the thallus may function as zoosporangium. A single large zoospore is formed in each zoosporangium.

The zoospore is ovoid, biflagellate and uninucleate. It has a large parietal chloroplast but there is no eye spot. The zoospore escapes through a pore at the apex of the zoosporangium. It loses its flagella soon after liberation and secretes a wall.

At the time of germination the spore divides by a transverse or vertical wall. In C. scutata (which is a discoid form) the first division is transverse, the upper cell forms seta and the lower divides repeatedly and forms the discoid thallus (Fig. D- E). In heterotrichous species (e.g., C. pulvinata) the first division is vertical and both the daughter cells divide repeatedly to form prostrate system. The erect or projecting system develops from the prostrate system.

During unfavourable conditions the zoosporangium, instead of zoospore, forms fairly thick walled aplanospores. The aplanospores germinate under favourable conditions and form new thalli.

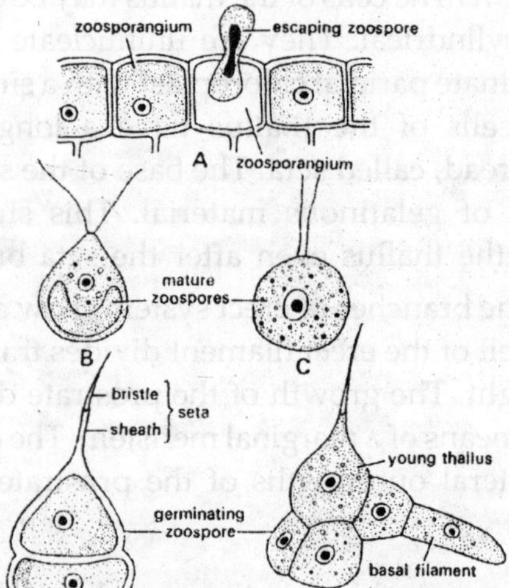

A-E. Coleochaete: Stages in asexual reproduction

Sexual Reproduction: The sexual reproduction is of advanced oogamous type. The species of Coleochaete are homothallic (e.g., C. pulvinata) or heterothallic (e.g., C. scutata). The male and female sex organs are called antheridia and oogonia (or carpogonia) respectively.

1. *Antheridium:* The antheridia are borne at the tips of the erect branches, usually in groups. In discoid forms (e.g., C. scutata) the antheridia develop midway between the

periphery and the center of the thallus. The antheridium develops as an outgrowth of the terminal cell of the lateral branch. This outgrowth is separated from rest of the filament by a septum and develops into an antheridium. Each antheridium forms a single,. uninucleate and biflagellate antherozoid which may be green or colourless. The antherozoid is liberated by the rupture of the antheridial wall.

2. *Oogonium or Carpogonium:* The oogonia also develop at the tips of the erect branches. But the cell lying just below the oogonium continues the growth of the branch and as such the oogonium appears lateral in position. The oogonium is a flask-shaped structure with a long colourless neck, known as trichogyne. However, in discoid species trichogyne is absent and the neck is absent and the neck is represented by a short papilla. The basal swollen part of the oogonium contains a single nucleus and one or more chloroplasts. The protoplast of the basal part rounds off to form a single egg.

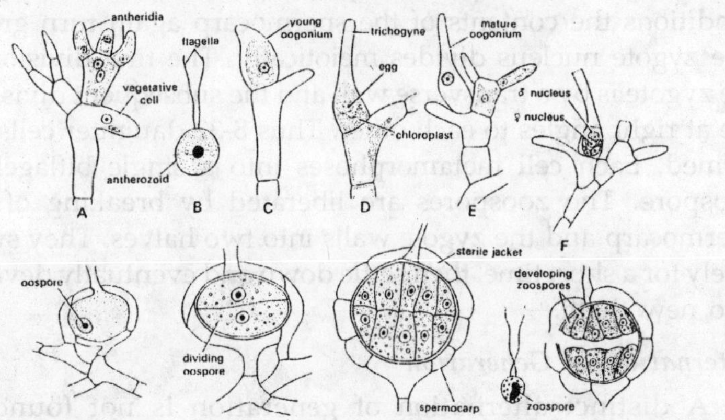

A-K. Coleochaete: Sexual reproduction; A. A branch with antheridium, B. Antherozoid, C-E. Successive stages in the development of oogonium, F. Fertilization, G. Oospore, H-J. Stages in the germination of oospore. K. Zoospore

3. *Fertilization:* Just before fertilization the tip of the trichogyne breaks down and some colourless protoplasm

is extruded from the broken tip. Although many antherozoids are attracted towards the oogonium, only one enters the oogonium. The male and female nuclei do not fuse immediately. The male nucleus is smaller than the female nucleus. The former gradually increases in size and as soon as it attains the size of the female nucleus, both the nuclei fuse to form a diploid zygote.

Post-fertilization Changes

After fertilization, the basal swollen part of the oogonium is separated from the trichogyne by the formation of a septum. The zygote greatly increases in size and secretes a thick wall. Now the cells below and around the oogonium are stimulated and form a pseudoparenchymatous layer around the oogonium. The oogonium with its sheath layer is called spermocarp. The spermocarp is reddish-brown in colour. It is shed in water by degeneration of the basal cell and remains dormant throughout winter.

Germination of spermocarp. On the return of favourable conditions the contents of the spermocarp again turn green. The zygote nucleus divides meiotically. The first division of the zygote is by a transverse wall and the subsequent divisions are at right angles to each other. Thus 8-32 daughter cells are formed. Each cell metamorphoses into a single biflagellate zoospore. The zoospores are liberated by breaking of the spermocarp and the zygote walls into two halves. They swim freely for a short time, then settle down and eventually develop into new thalli.

Alternation of Generation

A distinct alternation of generation is not found in Coleochaete. The haploid thallus bears male and female gametes on erect branches which fuse to form diploid zygote. The zygote nucleus divides meiotically, forming 8-32 haploid zoospores. The zoospores, on germination, again give rise to haploid thalli. Thus diploid phase is represented only by zygote. Such a life cycle where diploid phase is restricted only to zygote is known as haplontic.

Affinities of Coleochaete

The photosynthetic pigments-chlorophyll *a* and *b*, reserve food starch, non-jacketed unicellular sex organs and haplontic life cycle are typically chlorophycean features of Coleochaete. It shows resemblance with the members of the order Chaetophorales in its cell structure and heterotrichous thallus organisation. However, elaborate oogamous reproduction, retention of the zygote on the haploid plant, presence of a protective layer of vegetative cells around the mature zygotes, elaborate post-fertilization events (development of spermocarp) are some of the marked advanced features exhibited by Coleochaete.

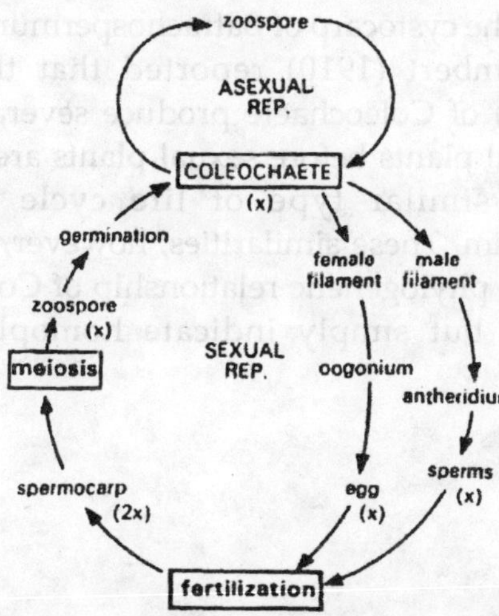

Coleochaete: Life-cycle

The occurrence of a discoid thallus in some species of Coleochaete (e.g., C. scutata) reminds one of the characteristics of flat thalloid gametophytes of liverworts, such as Riccia. Bower (1908) compared the spermocarp of Coleochaete to a simple or primitive sporophyte of early land plants. The more recent studies of Marchant and Pickett- Heaps (1973), Graham and McBride (1979), Frederick et al. (1973), Graham (1982), Graham and Wilcox (1983) have pointed out that Coleochaete possesses certain features like the presence of phragmoplast during cell division, presence of glycolate oxidase and peroxisomes enzymes, and parenchyma in some species, which indicates its affinities with early land plants.

Hence, Stewart and Mattox (1975) suggested that land plants might have originated from Coleochaete like ancestors.

There are remarkable analogies between the sexual reproduction of Coleochaete and the Nemalionales (e.g., Batrachospermum) among the red algae. The structure of the oogonium of Coleochaete markedly resembles the carpogonium of Rhodophyceae. Furthermore, in Coleochaete the oospore is retained within the oogonium and development of sheath around it forms a sort of a fructification (spermocarp) which is analogous to the cystocarp of Batrachospermum. Pringsheim (1860) and Lambert (1910) reported that the swarmers (meiozoospores) of Coleochaete produce several generations of dwarf asexual plants before sexual plants are produced. A more or less similar type of life cycle is found in Batrachospermum. These similarities, however, do not throw any light on the phylogenetic relationship of Coleochaete and Rhodophyceae but simply indicate homoplasy (parallel development).

4

Important Classifications

Chlorophyll is present in all groups of algae, but in some groups pigments other than chlorophyll are dominant and mask the green colour of chlorophyll. On the basis of their colour following four groups were recognized.

(1) Myxophyceae or Cyanophyceae (blue - green algae): dominant pigment - c -phycocyanin.

(2) Chlorophyceae (green algae): dominant pigment - chlorophyll *a* and *b*.

(3) Phaeophyceae (brown algae): dominant pigment - fucoxanthin.

(4) Rhodophyceae (red algae): dominant pigment -r-phycoerythrin.

In the last few decades a lot of information has become available on the structure, reproduction and physiological processes of algae which has been used to develop more natural systems of classification. The basic characters employed in the classification of algae are the physiological characteristics of vegetative cells (e.g., chemical nature of the cell wall, reserve food materials and types of pigments) and the morphology of motile reproductive cells (e.g., kind, number and position or insertion of flagella).

International Code of Botanical Nomenclature has recommended the following suffixes for the different categories of algae:

Division	–	phyta
Sub-division	–	phytina
Class	–	phyceae
Sub-class	–	phycidae
Order	–	ales
Sub-order	–	inales
Family	–	aceae
Sub-family	–	oideae

Classification Proposed by Fritsch: The first most comprehensive and authorative classification of algae was given by F.E. Fritsch (1935,48) in his book 'The Structure and Reproduction of the Algae'. His classification was based on such criteria as pigmentation, types of flagella, assimilatory products, thallus structure and methods of reproduction. He divided algae into the following 11 classes:

1. *Class Chlorophyceae (Isokontae, Green Algae)*: The main pigments are essentially identical to those found in higher plants. They are chlorophyll *a* and *b*, carotenoid and xanthophylls. The chloroplast usually has pyrenoids. The food reserve is starch, which frequently aggregates around the pyrenoids. The cell wall is composed of cellulose, and the cells are typically eukaryotic. Flagella, if present, are of equal length and are usually inserted at the anterior end. Reproduction takes place by vegetative, asexual and sexual (iso-, aniso-, and oogamous) methods.

   The class Chlorophyceae has been divided into nine orders: (i) Volvocales (e.g., Volvox), (ii) Chlorococcales (e.g., Chlorella), (iii) Ulotrichales (e.g., Ulothrix), (iv) Cladophorales (e.g., Cladophora), (v) Chaeto-phorales (e.g., Fritschiella), (vi) Oedogoniales (e.g., Oedogonium), (vii) Conjugales (e.g., Zygnema), (viii) Siphonales (e.g., Vaucheria), and (ix) Charales (e.g., Chara).

2. *Class Xanthophyceae (Heterokontae, Yellow Green Algae):* The main pigments are chlorophyll *a* and *e*, β-carotene and xanthophylls. The plastids are without pyrenoids and the reserve food material is oil. The cell wall is mainly composed of pectic substances with little cellulose. The cell is typical eukaryotic with two anteriorly inserted unequal flagella. The shorter flagellum is of whiplash type and the longer one of tinsel type. Reproduction takes place by vegetative, asexual and sexual (isogamous) methods.

   This class has been divided into four orders: (i) Heterochloridales (e.g., Heterochloris,. Chloramoeba), (ii) Heterococcales (e.g., Myxochloris, Halosphaera), (iii) Heterotrichales (e.g., Tribonema, Microspora), and (iv) Heterosiphonales (e.g., Botrydium).

3. *Class Chrysophyceae:* Phycochrysin is the dominant pigment which imparts brown or orange colour to these algae. The chromatophores have naked pyrenoid-like bodies. The food reserve is chrysolaminarin and leucosin. The cell wall is silicified or calcified and non-cellulosic. The motile cells have two anteriorly inserted equal or unequal flagella. Sexual reproduction is very rare; when present then it is isogamous.

   The class includes three orders (i) Chrysomonadales (e.g., Chrysococcus, Chrysodendron, Chromulina), (ii) Chrysosphaerales (e.g., Chrysosphaera, Echinochrysis), (iii) Chrysotrichales (e.g., Nematochrysis, Chrysoclonium).

4. *Class Bacillariophyceae (Diatoms, Yellow or Golden-brown Algae):* The members of this class are characterised by the dominance of golden-brown pigments, viz. fucoxanthin, diatoxanthin and diadinoxanthin. The chromatophores have pyrenoids and the photosynthetic products are fat and volutin. The cell wall is pectic and silicified and variously ornamented. Usually it consists of two halves which are radially or bilaterally symmetrical. The motile cells usually have a single flagellum. The sexual

reproduction takes place by fusion and by the formation of gametes or auxospores.

The class includes two orders: (i) Centrales (e.g., Cyclotella, Chaetoceras), and (ii) Pennales (e.g., Grammatophora, Navicula, Pinnularia, Denticula, Epithemia).

5. *Class Cryptophyceae:* The main pigment is xanthophyll which imparts brown or red colour to these algae. Pyrenoid-like bodies are present but they are often independent of chromatophores. The photosynthetic products are starch and/ or oil. The motile cells are dorsiventral with two anteriorly inserted unequal flagella. A complex vacuolar system is present in the cell. Sexual reproduction is rare and when present, it is of isogamous type.

   The class includes two orders: (i) Cryptomonadales (e.g., Cryptomonas, Rhodomonas, Cyanomonas), and (ii) Cryptococcales (e.g., Tetragonidium).

6. *Class Dinophyceae (Peridinieae):* These algae contain many discoid chromatophores and the main pigment is xanthophyll which imparts brown or red colour. The food reserves are starch and fats. The cell wall is cellulosic. Most of the members of this class are motile unicells with two flagella. Many species are colourless saprophytes or show holozoic mode of nutrition. Sexual reproduction is rare and is of isogamous type, if present. They also multiply by cell division.

   In this class six orders are included: (i) Des-momonadales (e.g., Desmocapsa, Pleromonas, Desmomastix), (ii) Thecatales (e.g., Exuviaella, Prorocentrum), (iii) Dinophysiales (e.g., Dinophysis, Ornithocercus, Phalacroma), (iv) Dinoflagellata (e.g., Amphidinium, Blastodinium, Ceratium,. Heterocapsa), (v) Dinococcales (e.g., Dinastridium, Cystodinium, Dissodinium), and (vi) Dinotrichales (e.g., Dinothrix, Dinoclonium).

7. *Class Chloromonadineae:* The members of this class have

numerous discoid chromatophores; they provide bright green tint due to the presence of excess of xanthophyll. Pyrenoids are lacking and the reserve food is fat and oil. The motile forms are with two almost equal flagella. Sexual reproduction is absent; multiplication takes place by longitudinal division of cells.

This is a small class with only one order, Chloromonadales, comprising of highly specialized unicellular flagellates (e.g., Trentonia, Vacuolaria).

8. *Class Euglenineae:* These are highly specialized unicellular green flagellates. The cell has usually many chromatophores. Pyrenoid-like bodies are present in some forms. The main pigment is chlorophyll, and the product of photosynthesis is a polysaccharide, paramylon. Most of the members have one or two flagella which arise from the base of a canal like invagination at the anterior end. These algae possess a complex vacuolar system. The multiplication is by cell division. Sexual reproduction is known only in few forms and is of isogamous type.

   This class includes three families: (i) Euglenaceae (e.g., Amphitropis, Ascoglena, Euglena), (ii) Astasiaceae (e.g., Astasia, Distigma, Menoidium), and (iii) Peranemaceae (e.g., Anisonema, Entosiphon, Petalomonas).

9. *Class Phaeophyceae (Brown Algae):* The chromatophores, beside other pigments, contain fucoxanthin. Some lower forms have naked pyrenoid-like bodies. The food reserves are in the form of alcohol (mannitol), polysaccharides (laminarin) and fats. The cell wall is mainly composed of cellulose with alginic and fucinic acids. The motile reproductive cells have two lateral or sub-apical flagella, one directed forward and the other backward. The sexual reproduction ranges from isogamous to oogamous type.

   The class comprises of nine orders (i) Ectocarpales (e.g., Ectocarpus, Halothrix, Punctaria), (ii) Tilopteridales (e.g., Tilopteris), (iii) Cutleriales (e.g., Cutleria), (iv) Sporochnales (e.g., Sporochnus, Carpomitra),

(v) Desmarestiales (e.g., Desmarestia, Arthrocladia), (vi) Laminariales (e.g., Laminaria, Lessonia, Macrocystis), (vii) Sphacelariales (e.g., Dictyota, Padina), and (viii) Fucales (e.g., Fucus, Sargassum, Durvillea).

10. *Class Rhodophyceae (Red Algae):* Most of these algae are marine with uniaxial or multiaxial thalli. The chromatophores, besides other pigments, contain r-phycoerythrin and r-phycocyanin which impart red colour to these algae. Pyrenoid-like bodies occur in the lower forms of this class. The food reserve is a solid polysaccharide, the foridean starch. The outer cell wall is pectic and the inner cellulosic. The reproductive cells are non-motile; the sexual reproduction is of advanced oogamous type.

    The class is divided into seven orders: (i) Bangiales (e.g., Bangia, Porphyra, Porphyridium), (ii) Nema-lionales (e.g., Batrachospermum, Lemanea, Nemalion, Chaetangium, Naccaria), (iii) Gelidiales (e.g., Gelidium), (iv) Cryptonemiales (e.g., Dumontia, Corallina, Gloeopeltis), (v) Gigartinales (e.g., Ahnfeltia, Chondrus, Gigartina, Gracilaria), (vi) Rhodymeniales (e.g., Champia, Rhodymenia, Gastroclonium), and (vii) Cera-miales (e.g., Ceramium, Polysiphonia, Lophosiphonia).

11. *Class Myxophyceae (Cyanophyceae, Blue-green Algae):* These algae are characterised by the presence of a very rudimentary nucleus and they do not have well organized chromatophores (i.e., cell organization is prokaryotic).

    The chief pigments are chlorophyll *a*, β-carotene and c-phycocyanin. The cell wall is made up of mucopolymers. The food reserve is cyanophycean starch. Sexual reproduction is absent (recently genetic recombination has been reported); asexual reproduction is by hormogonia or akinetes.

    This class comprises of five orders (i) Chroococcales (e.g., Chroococcus, Gloeocapsa, Microcystis), (ii) Chamaesiphonales (e.g., Chamaesiphon, Dermocarpa), (iii) Pleurocapsales (e.g., Pleurocapsa), (iv) Nostocales (e.g., Gloeotrichia, Nostoc, Oscillatoria, Rivularia, Scytonema, Spirulina), and (v) Stigonematales (e.g., Stigonema, Mastigocladus, Nostochopsis).

Classification Proposed by Smith: The classification of algae proposed by Smith (1933, 51, 55) is based on the physiological characteristics of vegetative cells and the morphology of motile reproductive cells. He divided algae into seven divisions and then related classes were included in each division. According to him, the number of divisions necessary for complete classification of algae is less than the number of classes as those classes which show close affinity should be placed in the same division. For example, Xanthophyceae, Chrysophyceae and Bacillariophyceae show certain resemblances in the structure and composition of the cell wall, flagellation and nature of food reserves and despite differences in their pigments, there is enough ground for placing them together in the same division, Chrysophyta.

The seven divisions of algae recognised by Smith are as follows:

Division 1. Chlorophyta: This division includes about 6,750 species, of these 90% are fresh water and the remaining 10% are marine. Chlorophyll *a* and *b* are the dominant pigments; the reserve food is starch. Motile reproductive cells are usually bi- or quadriflagellate; flagella are equal, of whiplash type and anteriorly inserted.

The following two classes were included in this division.

Class 1. Chlorophyceae (grass-green algae; e.g., Volvox, Ulothrix, Oedogonium).

Class 2. Charophyceae (stoneworts; e.g., Chara).

Division 2. Euglenophyta: There are 450 fresh-water or terrestrial species in this division. The dominant pigments of these algae are chlorophyll a, b and β-carotene, and their reserve food is paramylum and fats. Motile cells may be uni-, bi-, or triflagellate; flagella anterior, inserted into a narrow gullet. Multiplication takes place usually by cell division.

This division has only one class.

Class Euglenophyceae (euglenoids; e.g., Euglena, Colacium).

Division 3. Pyrrophyta: There are about 1,030 species, mainly unicellular and rarely colonial forms. The pigments are chlorophyll a, chlorophyll c, β-carotene and xanthophylls; the food reserve is starch and/ or oil. The cell wail is cellulosic. Motile cells are usually with two anteriorly inserted unlike flagella. Sexual reproduction is rarely present.

This division has been divided into the following two classes.

Class 1. Desmophyceae (dinophysids; e.g., Exuviaella).

Class 2. Dinophyceae (dinoflagelloids; e.g., Dinophysis, Dinastridium).

Division 4. Chrysophyta: This division is represented by more than 6,000 species, about three-fourth of which are fresh-water and one-fourth marine. These algae are characterised by the predominance of carotenes and xanthophylls. The food reserve is leucosin and oil. The cell wall is usually composed of two overlapping silicified halves. Sexual reproduction is iso-, aniso-, or oogamous.

This division includes the three classes:

Class 1. Chrysophyceae (golden brown algae; e.g., Chromulina).

Class 2. Xanthophyceae (yellow-green algae; e.g., Botrydium).

Class 3. Bacillariophyceae (diatoms; e.g., Pinnularia).

Division 5. Phaeophyta (Brown Algae): There are about 1,000 species of brown algae, mostly marine. The dominant pigments are phycophein and fucoxanthin; the assimilatory products are laminarin (polysaccharide) and mannitol (alcohol). The cell wall is cellulosic with fucinic and alginic acids. Motile reproductive cells are pyriform with two laterally inserted flagella, one of which is of tinsel type. Sexual reproduction is iso-, aniso-, or oogamous type.

This division has been divided into the following three classes.

Class 1. Isogeneratae (e.g., Ectocarpus, Sphacelaria, Dictyota).

Class 2. Heterogeneratae (e.g., Myrionema, Laminaria).

Class 3. Cyclosporeae (e.g., Sargassum, Fucus).

Division 6. Cyanophyta (Blue-green Algae): This division is represented by 1,500 species, mostly fresh water. Some species are free-living, while others grow on larger algae or within the tissue of other plants. The cell is prokaryotic; the nucleus lacks nuclear membrane and nucleolus. In addition to other pigments, they contain a blue (c-phycocyanin) and a red (c- phycoerythrin) pigment. The food is stored in the form of cyanophycean starch. Motile stages are absent but vegetative filaments of some forms show gliding movements.. Sexual reproduction is absent; asexual reproduction takes place by hormogonia, fragmentation, akinetes, etc.

The members of this division are placed in a single class, Myxophyceae or Cyanophyceae (e.g., Nostoc, Anabaena, Oscillatoria).

Division 7. Rhodophyta (Red Algae): This division includes about 2,500 species, mostly marine. They have a predominance of r-phycoerythrin which masks the other pigments to give them a distinctive red colour. They store food in the form of floridean starch. The thallus is non-motile and complex. Sexual reproduction is oogamous; motile reproductive cells are not found.

This division contains only one class, Rhodophyceae (e.g., Batrachospermum, Polysiphonia, Chondrus, Gracilaria).

Classification Proposed by Papenfuss: G.F. Papenfuss (1955), on the basis of phylogeny, recognised 8 phyla and 12 classes in algae as listed hereunder.

Phyla	*Classes*
1. Chlorophycophyta	Chlorophyceae
2. Charophycophyta	Charophyceae
3. Euglenophycophyta	Euglenophyceae
4. Chrysophycophyta	Chrysophyceae
	Xanthophyceae
	Bacillariophyceae
5. Pyrrophycophyta	Cryptophyceae
	Chloromonadophyceae
	Dinophyceae
6. Phaeophycophyta	Phaeophyceae
7. Rhodophycophyta	Rhodophyceae
8. Schizophycophyta	Schizophyceae

Classification Proposed by Chapman: V.J. Chapman (1962) divided algae into the following four phyla on the basis of morphology, pigmentation and biochemical characters.

Phylum 1. Euphycophyta

Class 1. Charophyceae
Class 2. Chlorophyceae
Class 3. Phaeophyceae
Class 4. Rhodophyceae

Phylum 2. Myxomycophyta

Class 1. Myxophyceae (Cyanophyceae)

Phylum 3. Chrysophycophyta

Class 1. Chrysophyceae
Class 2. Xanthophyceae
Class 3. Bacillariophyceae

Phylum 4. Pyrrhophycophyta

Class 1. Cryptophyceae
Class 2. Dinophyceae

Classification Proposed by Prescott: G.W. Prescott presented his system of classification of algae in 1969 in his book, 'The Algae: A Review.' He was of the opinion that pigment similarities are fundamental and imply kinship. He has taken into consideration the following criteria for the classification of algae.

(1) Photosynthetic pigments - their kinds, chemical composition and relative amounts.

(2) Photosynthetic products - their chemical nature.

(3) Flagella - their types, number and position of insertion.

(4) Cell wall - structure and chemical composition.

(5) Presence or absence of true nucleus.

(6) Life cycle and reproduction - morphology of the reproductive structures and the methods of reproduction.

He classified algae into nine phyla:

Phylum	*Classes*
1. Chlorophyta (green algae; chlorophyll predominant pigment, food reserve starch).	i. Chlorophyceae ii. Charophyceae
2. Euglenophyta (euglenoid algae; chlorophyll predominant pigment, food reserve paramylum).	
3. Chrysophyta (yellow-green algae; carotenoids and xanthophylls predominant pigments, food reserve leucosin and oil).	i. Chrysophyceae ii. Bacillariophyceae iii. Heterokontae (Xanthophyceae)
4. Pyrrhophyta (dinoflagellates; carotenes, xanthophylls and peridinin predominant pigments, food reserve starch and oil).	i. Desmokontae (Desmophyceae) ii. Dinokontae (Dinophyceae)

Contd...

Phylum	***Classes***
5. Phaeophyta (brown algae or brown seaweeds; fucoxanthin predominant pigment, food reserve soluble carbohydrate).	i. Isogeneratae ii. Heterogeneratae iii. Cyclosporeae
6. Rhodophyta (red algae; phycoerythrin dominant pigment; reserve food floridean starch).	i. Sub-phylum Bangioideae ii. Sub-phylum Florideae
7. Cyanophyta (blue-green algae; phycocyanin predominant pigment).	i. Sub-phylum Coccogoneae ii. Sub-phylum Hormogoneae
8. Cryptophyta (blue and green flagellates; compressed slipper shaped flagellated cells; xanthophyll or chlorophyll predominant pigment; food reserve starch-like carbohydrate).	
9. Chloromonadophyta (chloromonads; chlorophyll predominant pigment, food reserve oil)	

Classification Proposed by Round: F.E. Round (1973) divided algae into two major groups—Procaryota and Eucaryota. Procaryota are characterized by the absence of delimiting nuclear membrane, nucleolus and other membrane bound cell organelles, whereas Eucaryota have well organized nucleus and other cell organelles.

The following is the outline of Round's classification:

Some Salient Features of Major Classes of Algae

Characters	*Cyanophyceae*	*Chlorophyceae*	*Bacillariophyceae*	*Phaeophyceae*	*Rhodophyceae*
1. Habitat	fresh water, marine or terrestrial.	fresh water, marine or terrestrial.	fresh water, marine or terrestrial.	only four genera are fresh water, rest marine.	mostly marine; a few are terrestrial.
2. Thallus form	unicellular, colonial or filamentous.	unicellular, colonial, coenobial, filamentous, thalloid or siphonaceous.	unicellular or colonial.	filamentous, branched, usually heterotrichous.	unicellular, filamentous, pseudoparenchymatous.
3. Cell structure	prokaryotic, membrane bound nucleus and other cell organelles like plastids, mitochondria, etc., are absent.	eukaryotic, well organised nucleus and other cell organelles present.	eukaryotic.	eukaryotic.	eukaryotic.
4. Cell wall	mucopeptides and muramic acid are two main constituents of the cell wall.	cell wall is made up of cellulose and pectic material.	cell wall is made up of silica and pectic material.	outer mucilaginous layer of cell wall consists of alginic and fucinic acids and the inner layer is made up of cellulose.	outer layer is pectic in nature whereas inner layer is cellulosic.
5. Flagella	non flagellate motile cells are absent.	motile cells with 2, 4 or more flagella; flagella anteriorly inserted, equal in length and whiplash type.	usually 1, or sometimes 2, anteriorly inserted.	2, laterally inserted, of unequal length; one of them whiplash and. another of tinsel type	non flagellate, motile cells are absent.
6. Dominant pigments	chlorophyll *a*, β-carotene, xanthophyll and c-phycocyanin.	chlorophyll *a*, and *b*, α - and β-carotene, xanthophyll.	chlorophyll *a*, and *c*, β-carotene, xanthophyll. (fucoxanthin).	chlorophyll *a* and *c* β-carotene, xanthophyll (violaxanthin, fucoxanthin).	chlorophyll *a*, and *d*, β-carotene (lutein), xanthophylls, r-phycoerythrin.

Contd...

Characters	*Cyanophyceae*	*Chlorophyceae*	*Bacillariophyceae*	*Phaeophyceae*	*Rhodophyceae*
7. Nucleus	nuclear membrane absent, DNA not associated with histones, therefore well-organised chromosomes are not found.	well organised nucleus with chromosomes and nucleolus is present.	as in Chlorophyceae.	as in Chlorophyceae.	as in Chlorophyceae.
8. Eye spot	absent	attached with chloroplast but it has no relation with flagella	absent	in male gametes, eye spot is present, attached to chromatophores, it is related with flagella	absent
9. Pyrenoid	absent	single or compound; present either on the surface or embedded in chloroplast.	simple or compound; one or more.	single, stalked or projected	simple or compound; embedded, sometimes stalked.
10. Reserve food material	special type of starch (cyanophycean starch) and protein (cyanophycin protein).	starch and fats.	oil, volutin and chrysose.	laminarin, mannitol, sucrose.	floridean starch.
11. Reproduction	multiplication only through asexual methods; sexuality completely absent.	reproduction by vegetative, asexual and sexual means; sexual reproduction isogamous, anisogamous or oogamous.	reproduction by vegetative (cell division) and sexual means; sexual reproduction iso-, aniso-, or oogamous.	reproduction by vegetative (fragmentation, adventitious branches), asexual (motile and aplanospores) sexual reproduction. iso-, aniso or oogamous.	reproduction by vegetative (cell division, fragmentation), (monospores. neutral spores, tetraspores, carpospores) and sexual means; sexual reproduction complex and advanced oogamous.

Major groups	*Phylum*	*Pigments*
1. Procaryota	1. Cyanophyta	blue-green pigments
2. Eucaryota	2. Euglenophyta	
	3. Chlorophyta	
	4. Charophyta	green pigments
	5. Parsinophyta	
	6. Xanthophyta	
	7. Haptophyta	
	8. Dinophyta	
	9. Bacillariophyta	brown pigments
	10. Chrysophyta	
	11. Phaeophyta	
	12. Rhodophyta	red pigments
	13. Cryptophyta	green-red pigments

Classification Proposed by Chapman and Chapman: V.J. Chapman and D.J. Chapman (1973) have also employed the Procaryota-Eucaryota concept in the classification of algae. In Procaryota they placed Cyanophyta, and the remaining 10 divisions were included within Eucaryota. Their divisions were based on the structure of organelles, such as flagella, and biochemical characteristics. A brief outline of their classification is given below.

	Division	*Class*
Procaryota:	Cyanophyta (blue-green algae)	Charophyceae
Eucaryota:	1. Rhodophyta (red algae)	Rhodophyceae
	2. Chlorophyta (green algae)	Chlorophyceae Prasinophyceae Charophyceae
	3. Euglenophyta (euglenoids)	Euglenøphyceae
	4. Chloromonadophyta (chloromonads)	Chloromonadophyceae

Contd...

Division	*Class*
5. Xanthophyta (yellow-green algae)	Xanthophyceae
6. Bacillariophyta (diatoms)	Bacillariophyceae
7. Chrysophyta (golden-brown algae)	Chrysophyceae Haptophyceae
8. Phaeophyta (brown algae)	Phaeophyceae
9. Pyrrhophyta (dinoflagellates)	Dinophyceae Desmophyceae
10. Cryptophyta (cryptomonads)	Cryptophyceae

Classification Proposed by Lee: Robert Edward Lee (1980) divided algae into six divisions:

Division	*Class*
1. Cyanophyta (prokaryotic cells)	1. Cyanophyceae (blue-green algae)
2. Glaucophyta (eukaryotic cells with no chloroplasts with endosymbiotic blue-greens (cyanelles)	
3. Chromophyta (algae with chloroplast, E.R.,storage product out-side chloroplast envelope)	1. Euglenophyceae (euglenoids) 2. Dinophyceae (dinoflagellates) 3. Cryptophyceae (cryptophytes) 4. Chrysophyceae (golden-brown algae) 5. Prymnesiophyceae (haptophytes) 6. Bacillariophyceae (diatoms) 7. Rhaphidophyceae (chloromonads) 8. Xanthophyceae (yellow-green algae)

Contd...

Division	*Class*
	9. Eustigmatophyceae 10. Phaeophyceae (brown algae)
4. Rhodophyta (chlorophyll *a* and *d*, phycobiliproteins, no flagellated cells, storage product floridean starch)	1. Rhodophyceae (red algae)
5. Chlorophyta (chlorophyll *a* and *b*, storage product starch found inside the chloroplast)	1. Chlorophyceae (green algae)
6. Charophyta (Chlorophyll *a* and *b*, sex organs surrounded by an envelope of sterile cells)	1. Charophyceae (stone worts)

General Characters

Algae comprise a group of chlorophyll containing thalloid plants of the simplest type, having no true roots, stems, leaves or leaf like organs. They are placed in the Division Thallophyta along with fungi. They differ from fungi in the presence of photosynthetic pigment - chlorophyll and in their mode of nutrition. Although most of the algae are autotrophic, i.e., they synthesize their food by themselves, yet heterotrophic and holozoic forms are not uncommon. The important distinguishing features of algae are as follows:

(1) They are mostly chlorophyll containing autotrophic thalloid plants.

(2) They occur in a variety of habitats, but majority of them are aquatic.

(3) The plant body does not show differentiation into various tissue systems.

(4) They have mostly unicellular sex organs without a jacket

of sterile cells around them. Jacket cells, if present, have different initials. There is a progressive complexity in reproduction.

(5) Embryo is not formed after gametic union.

(6) They show distinct alternation of generation.

Occurrence

Algae are of universal occurrence and they are found in a variety of habitats, such as freshwater, sea water, on snow, on rocks and on/or within the plant and animal bodies. Of these, aquatic forms are most common. On the basis of habitat, they may be classified into the following three groups:

(1) Aquatic algae

(2) Terrestrial algae

(3) Algae of unusual habitats

Aquatic Algae: Aquatic forms are found in fresh water or in saline water of the sea.

Fresh Water Forms: These forms occur in fresh water (low salinity water) of ponds, pools, lakes, rivers, etc. Some fresh water forms like Cladophora, Oedogonium, Ulothrix and Chara are found in slow running water, whereas others like Chlamydomonas, Volvox Hydrodictyon and Spirogyra occur in stagnant water.

Marine Forms: These forms occur in saline water of the sea and are represented by the members of Phaeophyceae (e.g., Ectocarpus, Sargassum, Fucus) and Rhodophyceae (e.g., Polysiphonia).

The aquatic algae are either free-floating (e.g., Chlamydomonas, Volvox, Spirogyra) or are attached to a substratum with the help of an attachment disc, known as holdfast (e.g., Oedogonium, Ulothrix). Many free-floating algae, together with other similar organisms, form colonies on the surface of water which are called water blooms or phytoplanktons.

The algae present in phytoplankton are dependent upon

the chemicals present in water. Usually marine planktons have a large variety of species than fresh water planktons. Planktonic species often have a relatively slow rate of sinking, as they contain gas vacuoles or oil globules which increase their buoyancy. Motility of planktonic colonies is mainly by means of flagella.

Algae like Dictyosphaerium, Fragilaria, Cosmarium, Volvox, Asterionella and Golenkinia are commonly found in fresh water planktons, whereas Actinocyclus, Chaetoceros and Coscinodiscus are common in marine planktons.

During night planktons utilize oxygen dissolved in water and as such submerged aquatic organisms do not get sufficient oxygen for respiration. Besides, phytoplanktons also prevent atmospheric oxygen from dissolving in water, thus making the habitat unsuitable for submerged aquatic life.

Many algae are found attached to rocks along the edges of lakes and seas and these forms are called phytobenthos.

Terrestrial Algae: Some algae are found in terrestrial habitats like soils, rocks, logs, etc. Forms like Vaucheria, Botrydium, Fritschiella and Euglena, which are found on the surface of the soil, are known as saphophytes, whereas many blue-green algae (e.g., Nostoc, Anabaena), which occur under the surface of the soil, are called cryptophytes.

Some algae are also found on tree trunks and moist walls, and they absorb carbon dioxide and water from the atmosphere (e.g., Protococcus, Scytonema).

Algae of Unusual Habitats: In addition to aquatic and terrestrial habitats, some algae also occur in uncommon habitats.

Halophytic Algae: There are many algae which occur in saline water of seas, but some others can withstand high concentration of salts and occur in salt lakes. They are called halophytes, and include Chlamydomonas ehrenbergii, Dunaliella and Stephanoptera.

Epiphytic Algae: They grow on larger algae or on bryophytes and angiosperms. For example, species of

Bulbochaete, Oedogonium and Microspora are found attached to the larger species of Cladophora, Rhizoclonium and Vaucheria. Bryophytic and angiospermic flora of rivers and lakes harbour many algal members on their surface. Algae with mucilagenous thalli like Chaetophora, Oedogonium and Zygnema are epiphytic on Achnanthes, Eunotia, Synedra, etc.

Epizoic Algae: Many algae grow on animals like snails, fishes and tortoise, and they are known as epizoic. For example, Cladophora crispata grows on snails, and species of Stigeoclonium are found in the gills of fishes. Similarly, species of Characiopsis and Characium are found on the legs of Branchipus.

Endozoic Algae: Algae which occur in the tissues of animals are known as endozoic. For example, species of Zoochlorella are found in Hydra viridis. Besides, there are some 15 species of the family Oscillatoriaceae (blue-green algae) which occur in the respiratory and digestive tracts of vertebrate animals.

Symbiotic Algae: Several members of Chlorophyceae and Cyanophyceae form symbiotic association with fungi, bryophytes, gymnosperms and angiosperms. Lichens are important example of symbiotic life and their algal components belong to Cyanophyceae (e.g., Nostoc, Gloeocapsa, Microcystis) or Chlorophyceae (e.g., Coccomyxa, Chlorella, Protococcus). Colonies of Nostoc and Anabaena are found in symbiotic association in the thallus of Anthoceros and the coralloid roots of Cycas.

Cryophytic Algae: Algae growing on ice or snow are called cryophytic. These algae provide attractive colours to snow covered mountains. The alpine and arctic mountains become red due to the growth of Haemotococcus nivalis. Green snow in Europe is due to Chlamydomonas yellowstonensis and species of Mesotaenium. Sometimes snow becomes black due to Scotiella nivalis and Raphidonema brevirostri and brownish purple due to Ancyclonema nordenskioldii.

Lithophytic Algae: Algae growing on moist rocks and stones are known as lithophytic. Many blue - green algae like

Nostoc, Rivularia and Gloeocapsa commonly grow on moist and shady rocks. Scytonema is common on moist walls during rainy season.

Parasitic Algae: Algae also grow as parasites on many plants and animals. Cephaleuros virescens (Chlorophyceae) causes red rust in tea and coffee plantations in Assam and adjoining areas. Phyllosiphon (Chlorophyceae) is a parasite on the leaves of Arisarum vulgare. Polysiphonia fastigiata (Rhodophyceae) is a semiparasite on Ascophyllum nodosum.

Thermophytes (Thermal Algae): Many blue-green algae (e.g., Oscillatoria brevis, Synechococcus elongatus, Heterohormogonium sp., Haplosiphon lignosum) are found in hot water springs (50-70°C), where normal life is not possible. They are able to survive such at high temperatures due to the absence of well organised nucleus. Since most hot springs are alkaline, some algologists believe that tolerance to high temperature is also associated with this factor.

Range of Thallus Structure: The vegetative structure of algae shows a wide variety and it ranges in form from unicellular to complex multicellular thalli. Their size ranges from one micron to several meters. On the basis of thallus organisation, algae are divided into the following five groups :

(1) Unicellular forms

(2) Colonial forms

(3) Filamentous forms

(4) Siphonaceous forms

(5) Parenchymatous forms

Multicellular forms have been derived by repeated divisions of unicellular forms. Colonial forms are developed by the aggregation of the products of cell division within a mucilage mass. Filamentous forms are formed by repeated transverse divisions of cells without separation of daughter cells. Repeated nuclear divisions, without cross wall formation, give rise to siphonaceous forms. Parenchymatous thalli are formed by the division of cells of a filament in two or more planes.

Unicellular Forms: Simple unicellular forms are found in all groups of algae except Charophyceae and Phaeophyceae. These forms are sometimes referred to as acellular since they function as complete living unit without any cellular differentiation. Unicellular forms can be classified into the following four sub-groups.

Rhizopodial Unicells: These forms lack a rigid cell wall. They possess cytoplasmic projections which help them in amoeboid movement. In the ahsence of rigid cell wall, these forms are periplastic. Examples - Rhizochrysis (Chrysophyceae), Rhizochloris (Xanthophyceae), Chrysamoeba (Chrysophyceae).,

Flagellated Unicells: Flagellated vegetative cells are found in all groups of algae except Cyanophyceae, Phaeophyceae and Rhodophyceae. Such cells show similarity with motile gametes and zoospores. They vary in the number and type of flagella. For example, in Chlorophyceae, vegetative cells, zoospores and gametes have two equal flagella. Their number may sometimes vary, but is always in multiple of two. In Dinophyceae and Xanthophyceae there are two unequal flagella; in the former they emerge in different planes and in the latter in the same plane. In Euglenophyceae there is only one flagella, inserted at the anterior end.

Flagellated unicells may be periplastic without cell wall (e.g., Euglena) or with a distinct cell wall (e.g., Chlamydomonas). In Phacotus (Chlorophyceae), there is a thick calcareous covering (capsule) around the cell wall.

Spiral Filamentous Unicells: Some unicellular algae form spiral or coiled structures. e.g., Spirulina (Cyanophyceae).

Non-motile Unicells: These are non-motile coccoidal algae which do not possess flagella, eye spot, etc., meant for locomotion. The simplest non-motile forms are found in Cyanophyceae where they do not possess well organised nucleus and plastids (prokaryotic), e.g., Chroococcus. The non-motile unicells of Chlorophyceae possess nucleus and plastids (e.g., Chlorella;). The non-motile unicells of Bacillariophyceae, such as diatoms, are made up of two halves or theca, joined by a girdle band.

Colonial Forms: The colonial habit is achieved by aggregation of the products of cell divisions within a mucilagenous mass, by aggregation of motile cells or juxtaposition of cells subsequent to cell divisions. Therefore, all members of a colony have similar structure. These associations are usually loose, and as such a colony may break into smaller pieces. In some colonial forms, all members of a colony are connected with each other by cytoplasmic connections, hence they cannot break into segments (e.g., Volvox). On the basis of morphology, colonial organisations may be divided into the following four types:

Coenobial: A colony with a definite shape, size and arrangement of cells is known as coenobium. The number of cells in a coenobium is determined at the juvenile stage and subsequently there is only increase in size. Coenobia may be motile or non motile; in the former cells are flagellated (e.g., Pandorina, Eudorina, Volvox;) and in the latter without flagella (e.g., Hydrodictyon;).

Palmelloid: Unlike coenobial forms, in a palmelloid colony the number of cells, their shape and size is not definite. The cells remain irregularly aggregated within a common mucilagenous matrix, but they are independent and function as individuals. In some palmelloid forms it is a temporary phase (e.g., Chlamydomonas), whereas in others it is a permanent feature as in Tetraspora (Chlorophyceae), Aphanotheca (Cyanophyceae;).

Dendroid: Here the colony looks like a microscopic tree. As in palmelloid colony, the number, shape and size of the cells is also indefinite in dendroid colonies. A mucilagenous thread is present at the base of each cell, and the threads of different cells are united to form a branched structure. It gives a tree like appearance to the whole colony (e.g., Chrysodendron, Chrysophyceae;).

Rhizopodial: In a rhizopodial colony the cells are united through rhizopodia (e.g., Chrysidiastrum, Chrysophyceae;).

Filamentous Forms: Filamentous forms are developed by

repeated transverse divisions of cells. The daughter cells do not separate and they remain attached one upon the other in a definite sequence to form a filament. The filaments may be branched or unbranched, and the cells in a filament may be arranged in a single row (uniaxial) or more than one rows (multiaxial).

Unbranched Filaments: Simple unbranched filaments are found only in a few groups of algae. Such filaments may be free floating (e.g., Spirogyra;), attached to some substratum (e.g., Zygnema, Ulothrix Oedogonium;) or form colony (e.g., Nostoc, Oscillatoria;).

Branched Filaments: Branched filaments are formed by repeated transverse divisions of lateral outgrowths of cells. The branching of filaments may be false or true. A false branch does not arise as lateral outgrowth, but the trichome may break either due to death or decay of intercalary cells or at the point of heterocyst. The broken end emerges out of the mucilagenous sheath in the form of a branch, e.g., Scytonema (Cyanophyceae;). False branches in Scytonema arise almost always in pairs at some distance from the heterocyst.

True branches, which arise as lateral outgrowths, may result in the following three types of filaments:

(a) *Simple Filaments:* Simple branched filaments remain attached to the substratum by a basal cell. In such filaments branches may arise from any cell except the basal cell. In Cladophora, branches arise just below the septa between two adjacent cells.

(b) *Heterotrichous:* In this type, the thallus is very much evolved and differentiated into prostrate and erect systems (e.g., Fritschiella, Ectocarpus, Draparnaldiopsis, Coleochaete, Stigeoclonium). Both the prostrate and the erect systems may be well developed (e.g., Fritschiella, Ectocarpus;) or there is progressive elimination of prostrate (e.g., Draparanaldiopsis) or erect (e.g., Coleochaete) system.

(c) *Pseudoparenchymatous:* In many filamentous forms one or more central or axial filaments, together with their branches, form a 'parenchymatous' structure. If a pseudoparenchymatous thallus is formed by the branches of only one filament, it is called uniaxial (e.g., Batrachospermum), and if branches of more than one filaments are involved, it is said to be multiaxial (e.g., Polysiphonia;).

Siphonaceous Forms: Here the thallus is made up of branched, aseptate, coenocytic, tubular filaments as the nuclear divisions are not accompanied by wall formation, e.g., Vaucheria and Botrydium.

Parenchymatous Forms: In this type, the flat foliose or tubular thalli are formed by the divisions of cells in two or more planes. The daughter cells do not separate from the parent and give rise to parenchymatous thalli of various shapes, like flat (Ulva;), tubular (Scytosiphon, Phaeophyceae) or complex (Sargassum;). The growth of such thalli is apical (e.g., Fucus, Dictyota), intercalary (e.g., Laminaria) or trichothallic (e.g., Porphyra, Rhodophyceae).

Cell Structure in Algae: On the basis of their organization, algal cells may be differentiated into prokaryotic, mesokaryotic and eukaryotic types. The prokaryotic cell organization is found in Cyanophyceae which is characterized by: (i) the presence of incipient nucleus, (ii) the absence of membrane bound organelles like plastids, endoplasmic reticulum, Golgi bodies and mitochondria, (iii) the absence of basic proteins-the histones in DNA, (iv) the presence of mucopeptide in the cell wall, and (v) the absence of mitosis.

A eukaryotic cell, on the other hand, is characterized by the presence of a well organized nucleus and membrane bound organelles like plastids, mitochondria and Golgi bodies. Majority of algae show this type of cell organization. An intermediate type of cell organization, i.e., mesokaryotic is found in the members of Dinophyceae, where although the nucleus has a nuclear membrane and chromosomes (eukaryotic characters), basic proteins are absent (prokaryotic character).

Cell Wall: Most of the non-motile unicellular and multicellular algae possess a typical cellulosic cell wall of non-living matter outside the cytoplasmic membrane. It is composed of pure or mixed carbohydrates (cellulose, mucilage, pectin, etc.), and substances like alginic acid, fucoidin, fucin and hemicelluloses are present in brown algae (Phaeophyceae), pectin in red algae (Rhodophyceae), calcium carbonate in coralline algae (red algae) and silica in diatoms (Bacillariophyceae).

Electron microscopic studies have revealed that cellulosic cell wall is composed of microfibrils of cellulose which vary in thickness from 30 to 200 μ m. The microfibrils remain embedded in a smooth or granular matrix and are variously oriented. They may be arranged in two layers at right angles to each other, in three layers or in a random manner. The cell wall of diatoms is more complex, having pores with perforated membranes.

In blue-green algae (Cyanophyceae), the cell wall is mucopolymeric, mainly consisting of glucosamine, amino acids, muramic acid and diaminopimelic acid. A true cell wall is, however, absent in some algae like Euglena, Gymnodinium and Pyramimonas. They simply have a bounding membrane of cytoplasm, known as pellicle.

Flagella: Motile vegetative or reproductive cells are found in all groups of algae except Cyanophyceae and Rhodophyceae. Their motility is due to small filiform (thread like) protoplasmic appendages, called flagella. The number of flagella varies from one to four to many (Oedogonium, Vaucheria). They are mainly of the following two types:

1. *Whiplash or Acronematic Flagella:* Such flagella have a smooth surface.
2. *Tinsel or Pleuronematic Flagella:* The surface of these flagella is covered with fine filamentous appendages, known as mastigonemes or flimmers. They are further divided into

three categories on the basis of arrangement of mastigonemes.

(a) *Pantonematic:* In this type mastigonemes are arranged in two opposite rows or show radial arrangement.

(b) *Pantocronematic:* A pantonematic flagellum with a terminal fibril is known as pantocronematic.

(c) *Stichonematic:* Here mastigonemes develop only on one side of the flagellum.

A motile cell may have either one or both the above mentioned types of flagella. It is a specific character. If all flagella of a cell are similar, it is known as isoknot and when dissimilar, it is called heteroknot.

The motile stages of Chlorophyceae possess two or four anteriorly inserted whiplash flagella of equal length, whereas the members of Phaeophyceae and Xanthophyceae have one whiplash and one tinsel flagellum of unequal length.

A transection of flagellum reveals that it consists of nine peripheral doublet and two central singlet fibrils. All fibrils are :

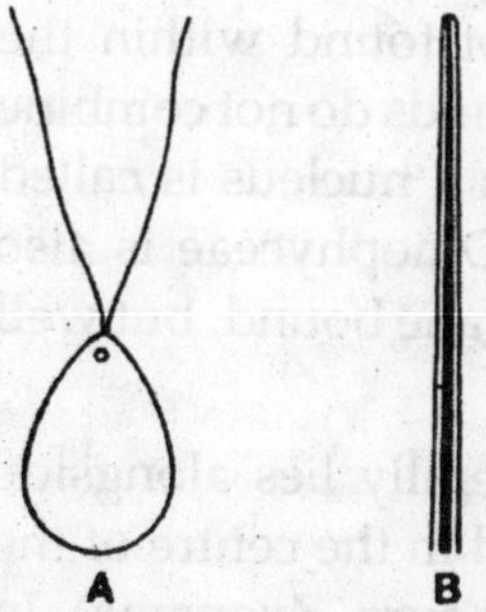

A-B. Acronematic flagella A. An algal cell with acronematic flagella, B. A single acronematic flagellum.

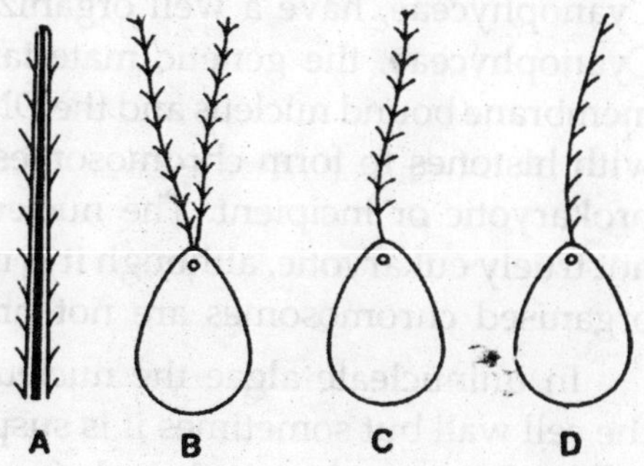

A-D. Pleuronematic flagella: A. A pleuronematic flagellum showing mastigonemes, B. A cell with two pantonematic flagella, C. A cell with pantocronematic flagellum, D. A cell with stichonematic flagellum.

enclosed within a common covering, formed by the extension of plasma membrane, but the two central fibrils have an additional covering of their own. The peripheral as well as central fibrils extend almost throughout the length of the flagellum. The nine peripheral fibrils at the proximal end are attached to a hollow basal body (which is separated from the flagellum by a diaphragm). The two central fibrils terminate just short to the diaphragm.

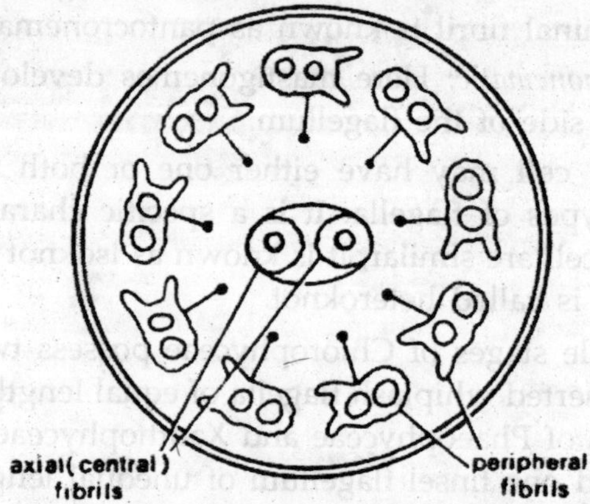

Transection of a flagellum

Nucleus: Almost all algae, with the exception of Cyanophyceae, have a well organized eukaryotic nucleus. In Cyanophyceae, the genetic material is not found within the membrane bound nucleus and the DNA strands do not combine with histones to form chromosomes. Such a nucleus is called prokaryotic or incipient. The nucleus in Dinophyceae is also not truely eukaryotic, although it is membrane bound, but well organised chromosomes are not present.

In uninucleate algae the nucleus generally lies alongside the cell wall but sometimes it is suspended in the centre of the cell by fine cytoplasmic threads (e.g., Spirogyra, Zygnema). In multinucleate members, the nuclei usually occur between the vacuoles and the plastids in the cytoplasm. The nucleus varies in size from less than 1 μm to 80 μm.

The nuclear membrane, which bounds the nucleus, is two layered, and in some members the two unit membranes are separated by a perinuclear space. The outer membrane is continuous with the membranes of endoplasmic reticulum. Each nucleus contains one or more dark stained nucleoli or endosomes. The four types of nucleolar structures found in algae are: (i) a single nucleolus per nucleus, (ii) two or more distinct nucleoli per nucleus, (iii) a complex nucleolar mass, and (iv) a linear association of large number of small nucleoli.

There is a great variation in the morphology of chromosomes. They are small and spherical in desmids, and elongated and thread like in Oedogonium. Chromosomes may have a localized (median or sub-terminal) or diffused centromere. The size of chromosomes varies form 0.25 µm or less (e.g., Ulotrichales) to more than 12 µm (e.g., Oedogoniales and Charales). The number of chromosomes also shows much variation, the lowest number, n=2, is found in Porphyra linearis (Rhodophyceae) and the highest, n = 592, in Netrium digitalis (Chlorophyceae).

Golgi Bodies or Dictyosomes: Golgi bodies are present in all algal cells except blue-green algae, and can be seen under electron microscope. They may be found in the region of the nucleus (e.g., Chlamydomonas), near plastids (e.g., diatoms and Bulbochaete), associated with the flagellar bases (e.g., Chlamydomonas) or without any apparent association with any particular organelle (e.g., Euglena, Fucus). Golgi bodies are composed of 2-20 flat vesicles which are arranged in stacks. Each stack is called a dictyosome (all dictyosomes collectively form the Golgi apparatus).

The exact role of Golgi bodies in algae is not known. They are associated with the synthesis of cell metabolites, and have also been shown to contribute to the plasma membrane as in higher plants.

Mitochondria: Well organised mitochondria are present in all algal cells with the exception of blue - green algae. A mitochondrion is bounded by a double membrane. The inner membrane is projected into lumen and forms finger like projections, called cristae. There are usually few cristae in Chlorophyceae and many in Chrysophyceae, Phaeophyceae and Xanthophyceae. The lumen, surrounded by bounding membranes, is filled with a granular matrix which contains nucleic acids (RNA and DNA). Hence, these structures can be called independent cell organelles. Their shape and number per cell varies in different groups of algae. There are usually more than one mitochondria per cell, but in Micromonas (Chlorophyceae) each cell contains a single mitochondrion.

The respiratory enzymes are located in mitochondria, and hence mitochondria act as respiratory centres of the cell. Mitochondria are also the sites of enzyme action in protein synthesis and amino acid interconversions.

Endoplasmic Reticulum (ER): The cytoplasm of the algal cell is traversed by a system of interconnecting tubules, known as endoplasmic reticulum. It is connected with the outer membrane of the nuclear envelope, but does not penetrate other cell organelles like chloroplast or pyrenoid. The surface of ER is studded with ribosomes, the sites of protein synthesis in the cell.

Eye-spot or Stigma: The motile vegetative and reproductive cells of algae have a pigmented spot, known as eye-spot or stigma. It may be located in the anterior, middle or posterior part of the cell. On the basis of their position and structure, five types of eye-spots are recognised.

Type A: Eye-spot is located in the chloroplast and it has no association with flagella, e.g., Chlorophyceae and Cryptophyceae.

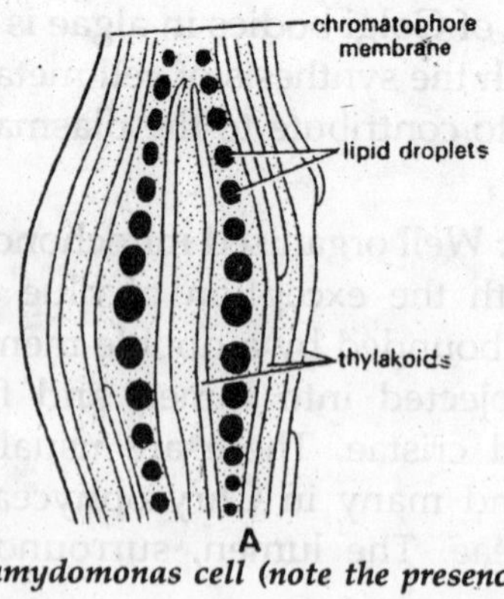

Eye-spot: of Chamydomonas cell (note the presence of thylakoids)

Type B: Eye-spot is located in the chloroplast and associated with a swollen flagellum, e.g., members of Chrysophyceae, Phaeophyceae and Xanthophyceae.

Type C: Eye-spots are independent clusters of osmophilic granules and are situated at the anterior side of the cell, near the flagellar swelling, e.g., Euglenophyceae.

Type D: Eye-spots having osmophilic granular structure with membranous lamellae and are situated near flagellar bases, e.g., Dinophyceae.

Type E: Eye-spot is made up of a lens, retinoid and pigmented cup. These are the largest and most complex type of eye-spots and are found in some Dinophyceae.

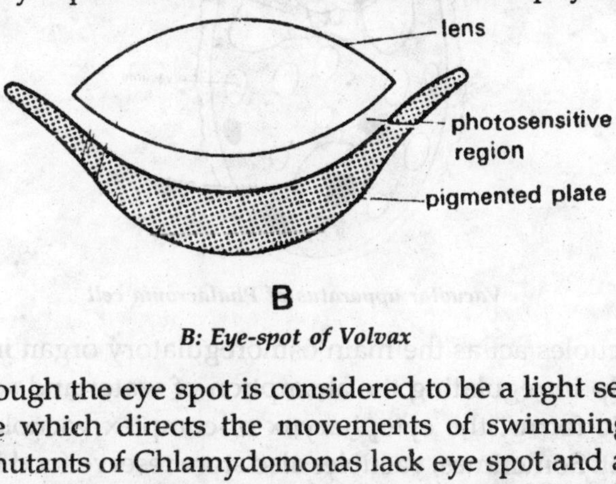

B

B: Eye-spot of Volvax

Although the eye spot is considered to be a light sensitive organelle which directs the movements of swimming cells, certain mutants of Chlamydomonas lack eye spot and are still phototactic. Some workers consider the eye spot as a shading organelle which possibly modifies the optical quality of transmitted light before it is focussed on the flagellar swelling. According to this view, flagellar swelling, rather than eye spot, is the primary receptor of light.

Vacuoles: Except Cyanophyceae, mature cells of almost all classes of algae possess one or more vacuoles. Each vacuole is bounded by a distinct membrane, called tonoplast. Motile algal cells may have the following two types of vacuoles:

1. *Simple or Contractile Vacuoles*: Such vacuoles, found in the members of Chlorophyceae, show periodic contractions and throw out the waste products out of the cell.

2. *Complex Vacuoles:* A complex vacuole consists of: (i) a tube like cytopharynx, (ii) a large reservoir, and (iii) a group of vacuoles of varying sizes. All these parts work in close coordination. These vacuoles are characteristic of Dinophyceae and Euglenophyceae.

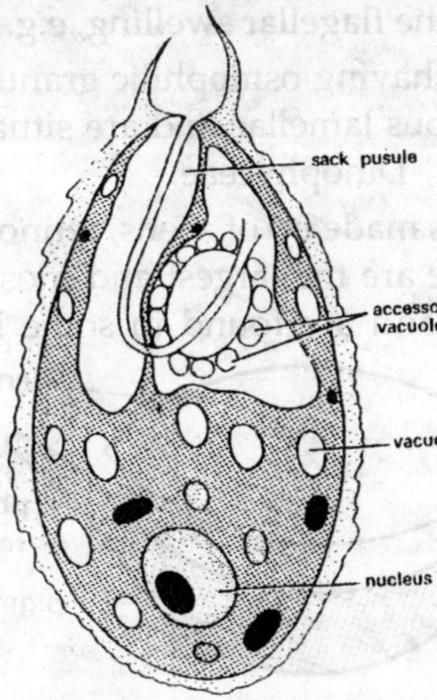

Vacuolar apparatus of Phalacroma cell

Vacuoles act as the main osmoregulatory organ in the cell and help in regulating the absorption of water and solutes. In holozoic forms, the cytopharynx of complex vacuoles serves as gullet. Sometimes, vacuoles also store reserve food materials such as laminarin and chrysolaminarin.

Gas Vacuoles: These gas containing cavities are characteristic of the mature healthy cells of Cyanophyceae. They occur as stacks of small transparent cylinders of uniform diameter. Their walls are freely permeable to gases. They render buoyancy to planktonic forms. They also act as a screen against intense light.

Pigments: Algal cells have a characteristic colour due to the presence of a combination of pigments, specific to each class. In all classes, except Cyanophyceae, these pigments are present within membrane bound organelles, known as plastids.

In blue-greens, the pigments are concentrated in the peripheral cytoplasm, known as chromoplasm. Plastids are of the following two types:

1. *Leucoplast:* These-are colourless plastids.
2. *Chromoplast:* These are coloured plastids; those containing both chlorophyll *a* and chlorophyll *b* are called chloroplasts and those lacking chlorophyll *b* as chromatophores.

Different forms of chromatophores occur in algae. The following are the main types:

(1) cup-shaped, e.g.,Chlamydomonas and Volvox.
(2) discoid, e.g., Vaucheria, Chara and centric diatoms.
(3) girdle-shaped, e.g., Ulothrix.
(4) reticulate, e.g., Oedogonium, Hydrodictyon and Cladophora.
(5) spiral, e.g., Spirogyra.
(6) stellate, e.g., Zygnema.

The various types of pigments found in the algal cell are given below.

1. *Chlorophyll:* There are five types of chlorophylls, viz. chl *a, b, c, d* and *e*. Of these, chlorophyll *a* is present in all groups of algae, chlorophyll *b* only in Chlorophyceae and Euglenophyceae, chlorophyll *c* largely in algae of marine habitats (Phaeophyceae, Cryptophyceae, Bacillariophyceae and Crysophyceae), chlorophyll d in some red algae only as a trace constituent and chlorophyll e in certain Xanthophyceae such as Vaucheria hamata and Tribonema bombycinum.
2. *Xanthophyll:* More than 20 types of xanthophylls are known. They are formed by the incorporation of molecular oxygen into carotene molecule. Many xanthophylls, common in higher plants (lutein, violaxanthin, and neoxanthin), are found in the members of Chlorophyceae and Phaeophyceae. Fucoxanthin is the main xanthophyll pigment of Phaeophyceae and diatoms whereas

myxoxanthophyll, myxoxanthin and oscilloxanthin are found only in Cyanophyceae.

3. *Carotenes*: These are oxygen free alicyclic compounds, composed of isoprene units. The five types of carotenes occur in algae are: α-carotene in Chlorophyceae; Cryptophyceae and Rhodophyceae; β-carotene in all algal groups, except Cryptophyceae, c-carotene in Chlorophyceae; E-carotene in Bacillariophyceae, Cryptophyceae, Phaeophyceae and Cyanophyceae and flavacene in members of Cyanophyceae.
4. *Phycobilins:* These are water soluble complexes of protein and bile pigments, present in the photosynthetic tissue of plants. Phycobilins are red (phycoerythrin) and blue (phycocyanin) pigments which are confined to Rhodophyceae and Cyanophyceae respectively. They act as light harvesting pigments in photosynthesis and the light absorbed by them is transferred to chlorophyll *a*. Thus, like carotenoids, phycobilins are also accessory pigments.

Pyrenoids: Pyrenoids are proteinaceous bodies present in chromatophores. These organelles are considered to be associated with the synthesis and storage of starch. In members of Chlorophyceae, pyrenoids are surrounded by starch plates. But in some algae such as diatoms, they accumulate lipids instead of starch.

A chromatophore may have one (e.g., Chlamydomonas) or more than one (e.g., Oedogonium) pyrenoids. These pyrenoids are embedded in the chromatophore (e.g., Chlamydomonas) or attached on its surface with the help of a stalk (e.g., Ectocarpus). The embedded pyrenoids are usually traversed by large number of chromatophore lamellae.

In some algae, pyrenoids are transient structures,. found only at certain stages. Their presence, in such cases, is associated with the photosynthetic activity of the cell and availability of stored food material.

Reproduction

The propagation in algae takes place by vegetative, asexual and sexual methods.

Vegetative Reproduction: The vegetative reproduction includes all those processes of propagation in which portions of the plant body become separated off to give rise to new individuals without any obvious changes in the protoplast or in other words, it does not involve rejuvenation of the protoplasts. It is the most common method of reproduction in algae and takes place by the following means:

1. *Cell Division or Fission:* It is the simplest method of propagation and is commonly found in unicellular algae (e.g., Euglena, Chlamydomonas), desmids and diatoms. In this process, the unicellular algal cell divides mitotically to form two daughter cells, each eventually grows into an independent organism.
2. *Fragmentation:* In filamentous forms like Ulothrix, Oedogonium, Spirogyra and Zygnema, the thallus often breaks into small fragments. Each fragment has the capability to grow independently and forms a new thallus. The fragmentation of filaments may be due to mechanical pressure, dissolution of transverse walls or difference in turgor pressure between adjoining cells. In colonial blue-green algae, vegetative propagation takes place by fragmentation of larger colonies.
3. *Hormogonia:* It is a characteristic method of reproduction in blue-green algae. Under unfavourable conditions, the trichomc breaks into motile segments of varying length, called hormogonia. The fragmentation of parent filament into hormogonia may be due to the formation of intercalary heterocysts, specialized separation discs or necridia or due to death and decay of intercalary cells of the trichome. Hormogonia are commonly found in Nostoc, Oscillatoria and Cylindrospermum.
4. *Formation of Adventitious Branches:* Adventitious branches are formed in some large thalloid forms of algae. These

branches, when get detached from the parent thallus, develop into new plants (e.g., Dictyota, Fucus). Adventitious branches like protonema, formed on the internodes of Chara or stolon of Cladophora glomerata, are found.

5. *Tubers:* On the rhizoids and the lower nodes of Chara some tuber like structures are formed due to storage of food material. When detached from the parent plant, the tuber produces an independent plant. Similar structures are also found in Cladophora.
6. *Budding:* In some algae (e.g., Protosiphon) vegetative propagation takes place by budding. Bud like structures are formed due to proliferation of vesicles. They eventually get separated from the parent plant by the formation of the septum, and have the capacity to form plants.

Asexual Reproduction: Asexual reproduction involves rejuvenation of the protoplasts and the reproductive bodies are either naked or provided with a newly synthesized wall of their own. Further, it is accomplished by using specially differentiated cells that are capable of multiplying directly the alga without fusing with other cells. In some aspects, it is very similar to vegetative reproduction as both are accessory means of propagation and are not concerned with bringing together in one cell lineage the genes from different cell lineages. In general, it is a process whereby the protoplast is released from the algal cell and germinates to form a new plant. Asexual reproduction takes place by a variety of motile or non-motile spores (cells that germinate without fusing and form new individuals). These spores may be differentiated into the following broad categories on the basis of their structure.

1. *Zoospores:* These are motile, naked structures with two (e.g., Chlamydomonas, Ectocarpus), four (e.g., macrozoospores of Ulothrix) or many (e.g., Oedogonium, Vaucheria) flagella. The flagella are usually inserted anteriorly, but are lateral in some brown algae. The

zoospore has a bit of chloroplast and an eye spot. The cell forming zoospore is known as zoosporangium. The protoplast of zoosporangium may form a single zoospore (e.g., Oedogonium, Vaucheria) or undergoes repeated divisions and each segment forms a zoospore (e.g., Cladophora). Zoospores are haploid or diploid. They are liberated by dissolution of parent cell wall or through a terminal pore. They are usually formed during the night and are liberated early in the morning. After liberation, they rest for sometime and then germinate to form new individuals.

2. *Aplanospores:* These are non-motile spores, commonly found in terrestrial algae, but some aquatic algae (e.g., Ulothrix, Microspora) also form them during drought conditions. They differ from zoospores in having a distinct wall and in the absence of flagella. Each cell may form a single aplanospore or its protoplast may divide to form many aplanospores.
3. *Hypnospores:* Aplanospores of some algae like Pediastrum and Sphaerella secrete thick walls to overcome prolonged period of desiccation. Such thick walled aplanospores are called hypnospores. Under favourable conditions, hypnospores germinate and grow into new individuals or their protoplast may form zoospores. The hypnospores of Chlamydomonas nivalis are red in colour due to the deposition of a pigment, haematochrome, in their walls.
4. *Tetraspores:* Diploid plants of some algae (e.g., Polysiphonia) form aplanospores which are called tetraspores. They arc formed within tetrasporangium, usually by meiosis, and are haploid. They germinate to form haploid plants.
5. *Autospores:* The aplanospores with a structure similar to the parent cell are called autospores. In Scenedesmus and many members of the order Chlorococcales (e.g., Chlorella) the aplanospores acquire all distinctive features of the

parent cell before their liberation from the sporangium. The so formed autospores are in fact replica of the parent cell, the only difference is that they are smaller in size.

6. *Akinetes:* In some algae vegetative cells develop into thick walled spore like structure with abundant food reserves. These are called akinetes. Unlike aplanospores, akinetes always have additional wall layers around the protoplast which are fused with the parent wall. They are resistant to unfavourable environmental conditions. The formation of akinetes, besides other factors, is affected by the availability of carbohydrates and light. They are found in many blue-green and green algae.
7. *Exospores and Endospores:* In many blue-green algae the protoplast divides to form special type of aplanospores, known as exospores and endospores. The cell that produces endospores becomes somewhat enlarged and its contents divide successively in three planes, forming four to many endospores (e.g., Dermocarpa). The exospores are formed externally; the protoplast of the cell comes out through a terminal pore and successively cuts spherical spores (e.g., Chamaesiphon).

Sexual Reproduction: All groups of algae, except Cyanophyceae, reproduce sexually when gametes fuse to form zygote. In contrast to vegetative and asexual reproduction, it leads to the creation of new combinations of genes by pooling together in one line of descent, the genes derived from the different parents, thus resulting in a reshuffling of the gene material. On the basis of the structure and physiological behaviour of sex organs and their complexity, the following six types of sexual reproduction are recognised in different groups of algae.

1. *Autogamy:* When two gametes of the same mother cell fuse to form a diploid nucleus, it is called autogamy. In this process there is only karyogamy (fusion of two gametic nuclei). The autogamy lacks incorporation of external genes. Hence, plants developing as a result of

autogamy do not show new characters. Diatoms are the common example of autogamy.

2. *Hologamy:* In some unicellular forms (e.g., Chlamydomonas, Dunaliella) the vegetative cells of different strains (+) and (-) (female and male) behave as gametes and fuse to form zygote. From evolutionary point of view hologamy is more advanced than autogamy, as it involves fusion of two cells having different genetic constitution. But hologamy is an inefficient process from the point of view of propagation, as two vegetative cells fuse to form only one zygote.
3. *Isogamy:* In isogamy the two gametes which fuse to form zygote, are morphologically and physiologically similar. Such gametes are called isogametes, as they are indistinguishable into plus and minus strains. They are usually motile and flagellate. Isogametes are found in Ulothrix, Chlamydomonas eugametos, etc.
4. *Physiological Anisogamy:* In some algae like Spirogyra, Zygnema and Ectocarpus, though gametes are similar in morphological characters, they show physiological variation with plus (+) and minus (-) strains.
5. *Anisogamy:* In anisogamy, fusion takes place between morphologically and physiologically distinct gametes (anisogametes). The male or microgametes are smaller and more active, whereas the female or macrogametes are larger and sluggish. Chlamydomonas braunii and Pandorina are common examples of anisogamy.
6. *Oogamy:* This is the most advanced type of sexual reproduction. In this process a large non-motile egg or ovum fuses with a small motile sperm or antherozoid (in Rhodophyceae, sperms are non- motile). Egg is formed within the oogonium and sperms within the antheridium. Volvox, Oedogonium, Chara, Vaucheria, Sargassum, Batrachospermum and Polysiphonia are common examples of oogamy.

Zygote and its Germination: Post-fertilization Events: Soon after sexual fusion, the zygote forms a thin wall, which later becomes as thick as 3-layered. The zygote in this condition can undergo complete desiccation without harm. Usually the zygote undergoes a period of rest of a shorter or a longer duration before germination. In marine forms, the zygote generally germinates immediately and does not develop a thick wall.

The two gametic nuclei generally fuse immediately after fusion but in many cases (e.g., Zygnema) they do not fuse until just before the germination of the zygote.

Meiosis generally takes place during the germination of zygote and thus a haploid plant body is formed (e.g., Chlamydomonas, Volvox, Ulothrix, Spirogyra, Zygnema, Oedogonium, Vaucheria, etc.). However, in a number of forms (e.g., Sargassum), the zygote nucleus does not undergo reduction division but divides equationally and gives rise to a diploid plant body, reduction division does not take place until just before reproduction.

Perennation

Perennation means the survival from year to year by vegetative means. It includes the methods employed by alga to pass through periods unfavourable for their vegetative growth. Under such conditions, the protoplast of the cell becomes dormant by reducing its activity to its minimum. Under stressing (unfavourable) circumstances when life processes can not run smoothly, the alga develops some stage in life cycle which is protected by a thick resistant wall.

Most fresh water and sub-aerial algae arc exposed to the dangers of desiccation and extremes of temperature. They, therefore, develop certain devices which enable them to carry through the period of severe conditions in the next growing season. These protective devices constitute the means of perennation. Hypnospores, akinetes, tubers, mucilagenous sheath, zygote and cysts are some common perennating structures found in algae.

Types of Life-cycle

The growth and development of an alga passes through a number of distinct morphological and cytological stages. The sequence of these orderly changes is said to be life cycle or life history. In other words, the sequence of all the different phases or events, through which an organism passes from the zygote of one generation up to the zygote of next generation, is known as its life-cycle. The union of male and female gametes results in the formation of zygote. As there is doubling of chromosomal complement in zygote, it represents diploid state.

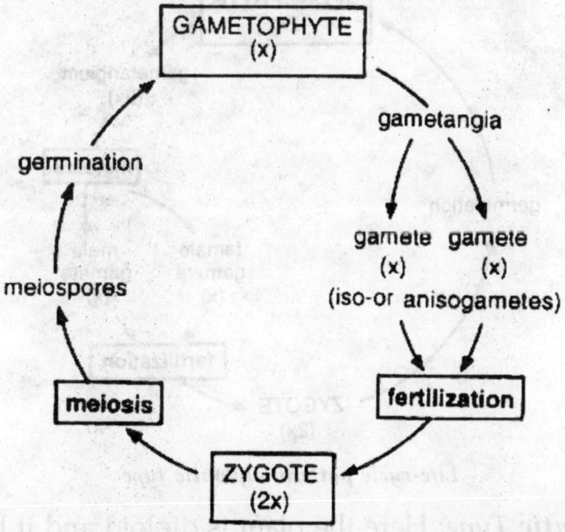

Life-cycle pattern: haplontic type

In some groups, the original chromosome number is achieved by meiotic division of the zygote nucleus. Thus on germination, the zygote produces haploid plants. In others, the zygotic nucleus divides mitotically and the resulting cells form a diploid phase, which later produces cells which are capable of meiotic division.

The various types of life-cycle met among algae can be broadly classified into the following five categories.

Haplontic Type: In this type, plant is haploid and bears haploid gametes in the gametangium. The gametic fusion results in the formation of a diploid zygote which is the only diploid phase in the life cycle. The zygote nucleus divides meiotically to produce four meiospores, each of these develops into a new individual.

Thus there is an alternation of a haploid plant with a diploid zygote. This type of life-cycle occurs in the motile unicellular forms (e.g., Chlamydomonas) and most multicellular green algae like Ulothrix, Oedogonium, Spirogyra, Zygnema and Chara.

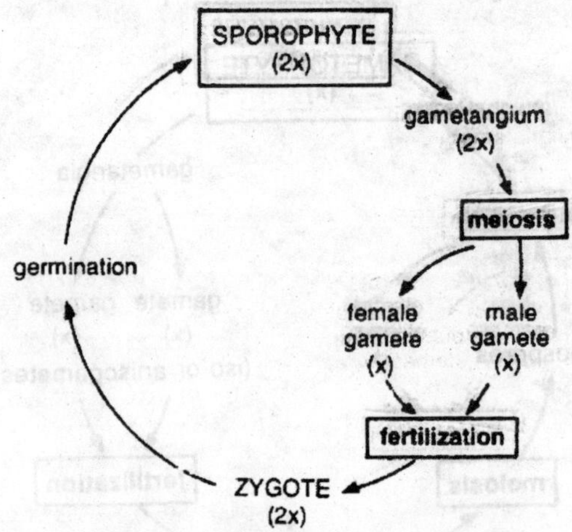

Life-cycle pattern: diplontic type

Diplontic Type: Here the plant is diploid and it bears sex organs (gametangia) which are also diploid. The reduction division takes place at the time of the formation of gametes and as such the gametes are haploid. After gametic fusion diploid stage is re-established in the form of zygote. The zygote does not undergo any reduction division and it gives rise to a diploid plant body. Thus, there is an alternation of a diploid plant with haploid gametes. The gametes representing the only haploid stage in the life cycle of the plant. This type of life-cycle is found in some members of Chlorophyceae (e.g., Cladophora glomerata) and Phaeophyceae (e.g., Fucus, Sargassum) and nearly all Bacillariophyceae (diatoms).

Isomorphic Type

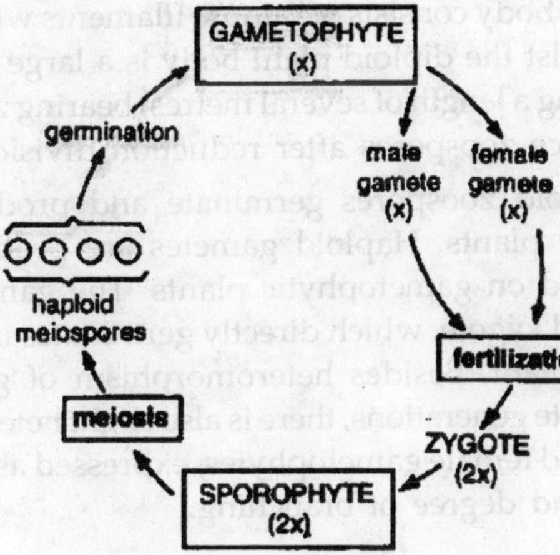

Life-cycle pattern: isomorphic type

In this type of life-cycle, there is an alternation of two generations which are externally similar, but one is haploid (gametophyte) producing gametes and the other diploid (sporophyte) producing zoospores. The zygote germinates directly into a diploid plant without undergoing reduction division. Sporangia develop on the diploid plant body and reduction in the number of chromosomes takes place prior to the formation of zoospores. The haploid zoospores thus formed grow into haploid plants. Sex organs develop on the haploid plants and these give rise to haploid gametes. The haploid gametes fuse to form diploid zygote. Isomorphic life cycle has been reported in a number of green algae (e.g., Ulva, Enteromorpha, Cladophora, Draparnaldiopsis, Fritschiella) and brown algae (e.g., Ectocarpus).

Heteromorphic Type: In this type of life-cycle, both diploid (sporophytic) and haploid (gametophytic) plant bodies are morphologically distinct (heteromorphic) and they alternate with each other. Mostly the sporophyte is large and the gametophyte small. Heteromorphic life cycle is seen mainly amongst brown algae (e.g., Dictyota, Laminaria, etc.). In

Dictyota, there is a difference in cell size between the gametophyte and sporophyte, whereas in Laminaria, the haploid plant body consists of minute filaments which produce gametes, whilst the diploid plant body is a large macroscopic plant (attaining a length of several metres) bearing zoosporangia which produce zoospores after reduction division.

The haploid zoospores germinate and produce haploid gametophytic plants. Haploid gametes are produced in sex organs formed on gametophytic plants. The gametes fuse to form a diploid zygote, which directly germinates into a diploid sporophytic plant. Besides heteromorphism of gametophyte and sporophyte generations, there is also slight heteromorphism of the male and female gametophytes, expressed as a difference in cell size and degree of branching.

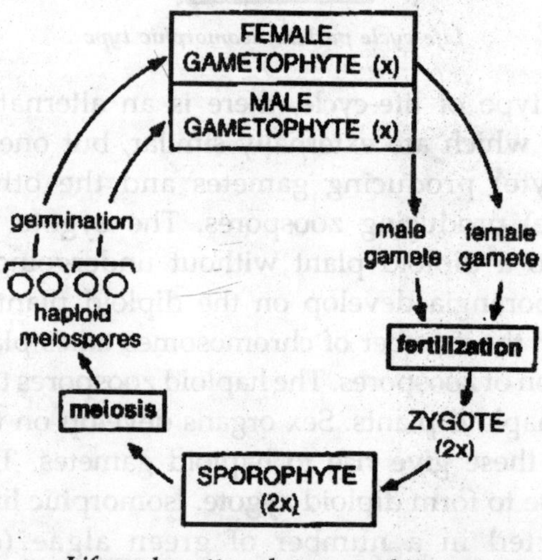

Life-cycle pattem: heteromorphic type

In genera like Cutleria and Urospora, however, haploid gametophytic plants are larger whereas diploid sporophytic plants are smaller in size.

Triphasic Type: In triphasic life-cycle, there is a succession of three generations. It may be of the following two types.

1. *Haplobiontic:* In some Rhodophyceae like Batrachospenmum

and Nemalion two well developed haploid phases are present in the life-cycle. This type of life cycle is hence called haplobiontic triphasic. The diploid phase is represented only by the zygote. The plant body of Batrachospermum is gametophytic which bears sex organs (i.e., spermatangia and carpogonia). Haploid gametes (spermatium and egg) are formed in these sex organs. A diploid zygote is formed by the fusion of two gametes. Gonimoblast filaments are formed from the basal part of the carpogonium; the uppermost cell of these filaments functions as carposporangium. The first division in the zygote nucleus is meiotic and as such the gonimoblast filaments are haploid. The haploid carposporangium bears a haploid carpospore. The gonimoblast filaments, carposporangia and carpospores are enveloped by numerous sterile filaments, and together represent carposporophyte generation which is haploid. On liberation, carpospores germinate into heterotrichous chantransia stage. The projecting filaments of the latter develop into new gametophytic plant body.

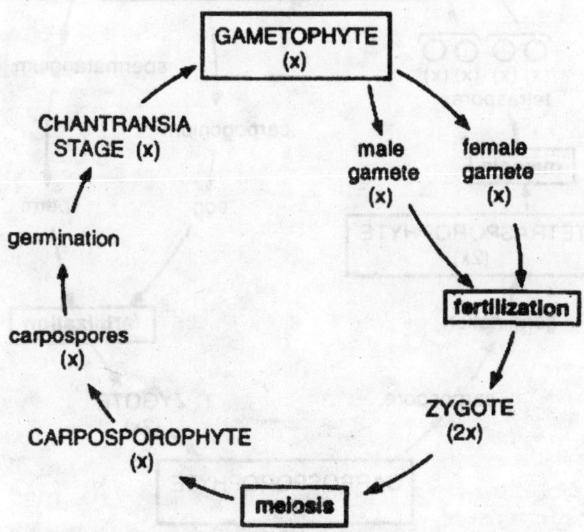

Life-cycle pattern: triphasic haplobiontic type

2. *Diplobiontic:* In diplobiontic triphasic life-cycle there are two diploid phases (carposporophyte and tetrasporophyte) alternating with haploid gametophytic phase. This type of life-cycle is well exemplified by Polysiphonia (Rhodophyceae). Here, the haploid phase is represented by male and female gametophytic plants, sex organs and gametes. The male and female gametes fuse to form a diploid zygote. The zygote produces gonimoblasts which represent an additional diploid phase (carposporophyte). Diploid carpospores are formed in the carposporophyte. On liberation, the carpospore germinates to produce a diploid tetrasporophytic plant. The tetrasporophytic plant produces diploid tetrasporangia. Four haploid tetraspores are formed in a tetrasporangium by reduction division. Tetraspores give rise to haploid gametophytic plants. The life cycle is thus completed.

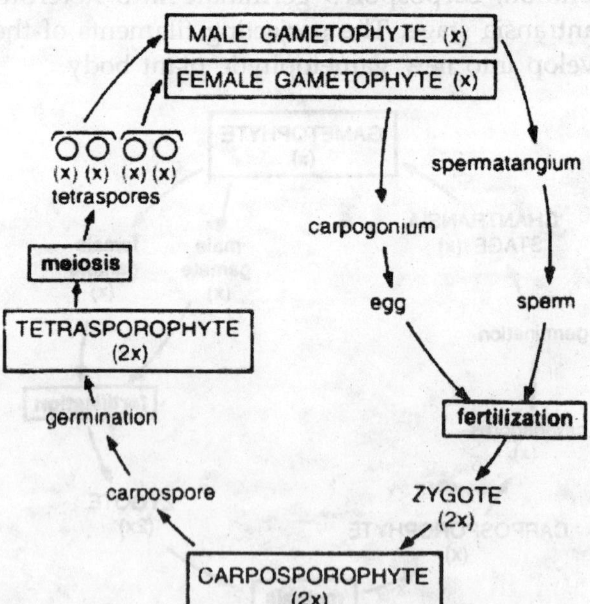

Life-cycle pattern triphasic diplontic type

Origin and Evolution

The two chief methods of propagation in plants are asexual and sexual reproduction. Besides, many plants also multiply vegetatively, but it does not involve rejuvenation of protoplast. Asexual reproduction takes place by means of specialized cells, known as spores. It is characterised by the absence of fusion or union of cells at any stage. The spores, which may be motile or non-motile, give rise to new individuals on germination. Sexual reproduction, on the contrary, involves fusion of sex cells, called gametes. The zygote, thus formed, on germination gives rise to a new plant. Individual gametes are, however, not capable of producing new plants.

In blue-greens, the primitive, class of algae, sexual reproduction is absent. These algae propagate mainly by vegetative methods. However, algae belonging to other classes reproduce by asexual as well as sexual methods.

Origin of Sex

The origin of sex is directly related to the origin of gametes; the latter have probably originated from asexual zoospores. Gametes and zoospores are very much alike structurally, except for the difference in size. This fact can be confirmed by making a comparative study of the morphology of gametes and zoospores in Chlamydomonas, Ulothrix and Cladophora. In this reference it is important to make a mention of Chlamydomonas debaryanum.

In this primitive alga, gametes and zoospores are similar in shape, structure and mode of development. The only difference is that zoospores are relatively large than gametes. This difference in size is related to the number of divisions that take place during their formation. If the divisions stop early, the resulting daughter cells are larger and they behave as zoospores. On the contrary, if the divisions continue further, many small daughter cells are formed which do not have the capability to form new individuals (like zoospores). These cells behave as gametes or sex cells.

Another important example of similarity between zoospores and gametes is Ulothrix zonata. It produces the following three types of zoospores.

1. *Quadriflagellate Macrozoospores:* They are formed singly or a few in a sporangium and on germination they form healthy filaments.
2. *Quadriflagellate Microzoospores:* They are produced in large numbers in a cell and on germination they form comparatively weak filaments.
3. *Biflagellate Microzoospores:* They are also formed in large numbers in a cell, but on germination they form very weak filaments.

The gametes of Ulothrix are exactly similar to biflagellate microzoospores in structure. The only difference is the method by which they give rise to new plants; microzoospores develop directly into new plants, whereas gametes first fuse in pairs to form zygote which on germination gives rise to a new plant. In fact, under favourable conditions, zoospores are produced, but under unfavourable conditions, when sufficient food is not available, gametes are formed. These gametes are probably still smaller zoospores which are not self sufficient and by fusing in pairs they regain the strength required to grow into a new plant.

These examples show that there is a fundamental similarity between biflagellate microzoospores and biflagellate gametes. On the basis of this similarity it has been postulated that in algae sexuality has originated as a result of accidental fusion of very small zoospores (which are otherwise incapable of forming new plants). Such fusion usually takes place under unfavourable conditions. It proved useful as it provided new strength and vigour to the individual. This view also gets support from the fact that in the lower groups of plants, sexual reproduction takes place only during unfavourable conditions. Zoospores formed under such conditions are very small due to scarcity of food and are too weak to form new plants. However, zygote formed by the fusion of two such inefficient zoospores (which now behave as gametes) regains the capability to form a new plant. In green algae, the thick walled zygote usually functions as resting spore. Thus sexual reproduction in algae has evolved as an alternate method of reproduction that helps in tiding over unfavourable conditions.

The zygote germinates under favourable conditions to form a new plant which bears asexual spores as usual. Thus formation of zoospores and gametes, is a response to environmental conditions. The gamete as such can be considered as a reduced form of zoospore.

Evolution of Sex

Sexual reproduction in algae originated as a result of fusion of two flagellated gametes. In primitive genera, both gametes of a fusing pair are identical in their morphology and physiology, hence cannot be differentiated into male and female gametes. Such gametes are called isogametes and the mode of their reproduction is known as isogamy. Chlamydomonas debaryanum, Cladophora and Ectocarpus are the common examples of isogamous reproduction. Isogamy is considered to be the most simple and primitive type of sexual reproduction.

In some green algae although the fusing gametes are morphologically identical, they differ in their (physiological) behaviour. For example, in Ulothrix all the gametes are similar in structure, but fusion occurs only between the gametes from different filaments. It shows that morphologically similar gametes sometimes have some inherent physiological disparity. Such forms, though do not have morphologically distinct male and female gametes, carry plus (+) and minus (-) strains. In Spirogyra gametes are morphologically identical and non-motile, but at the time of conjugation one of the two fusing gametes becomes active and motile (male) and the other remains passive and non-motile (female). This difference in physiological behaviour of morphologically identical gametes is a step leading towards anisogamy. This is referred to as physiological anisogamy. In forms like Chlamydomonas braunii, Pandorina, Codium and Ectocarpus, anisogamy is well marked. Here both the gametes of a fusing pair are although motile, they vary in size and are called anisogametes. Micro and macro-anisogametes are formed in micro and macrogametangium respectively. The macrogamete represents the female and the macrogamete, the male sex.

The next progressive step in the evolution of sexuality can be seen in Chlamydomonas coccifera, C. oogonium and species of Eudorina. In these forms male and female reproductive cells are formed in special structures, known as antheridia and oogonia respectively. The male gametes (antherozoids) are smaller and active and the female gametes (eggs) are relatively large and inactive. Such a mode of reproduction, where distinct male and female reproductive structures are present, is known as oogamous or primitive oogamy.

An evolved type of oogamy can be seen in Oedogonium. Here male gametes (antherozoids) are smaller and flagellated and develop in large numbers within unicellular antheridia. The egg is large and develops singly within oogonium. A large non-motile oogonium has probably evolved to store sufficient food for embryo development.

In green algae like Vaucheria and Chara, oogamy is still evolved. In these forms highly differentiated sex organs are present. Male (antherozoids) and female (eggs) gametes develop in specialized cells which are different from vegetative cells. The oogonium is surrounded by a special covering to protect the egg from high temperature and drought. In Coleochaete, the oogonium has a pseudoparenchymatous covering, whereas in Chara it is surrounded by five tube cells.

Thus isogamy, anisogamy and oogamy represent the successive advanced stages in sexual reproduction.

5

Various Organisms

The Dictyotales

There are 20 genera and 180 species in the order Dictyotales. The members occur both in temperate and tropical seas but attain their greatest development in the warmer oceans of the tropics and subtropics. Plant body is erect, flat, ribbon-like, well branched and attached to the suþstratum with the help of a rhizomatous disc.

Growth is apical and the branching is usually dichotomous. Asexual reproduction takes place by non-motile tetraspores formed in unilocular sporangia. Sexual reproduction is oogamous type, and the plants exhibit isomorphic alternation of generations.

Family Dictyotaceae

Members are flat, ribbon-like and dichotomously divided thaili which are sometimes fan-shaped. Growth occurs by an apical meristem, or by an apical cell. The sporophytic and gametophytic generations are morphologically similar.

Dictyota is the type genus of the family which is described here in some details.

Dictyota Reproduction Structure

Systematic Position

Class	-	Phaeophyceae
Order	-	Dictyotales
Family	-	Dictyotaceae

Occurrence: The genus Dictyota is represented by about 35 species, found chiefly in the warmer seas of the tropics. In India, the genus is represented by 12 species which occur along the coasts of Okha, Dwarka, Tuticorin, Mumbai, Porbandar and Mahabalipuram. D. dichotoma is a common species found along the Indian coast. It grows attached to the rocks in the intertidal zone.

External Structure: The thallus of Dictyota is macroscopic, about 30 cm in height, erect, ribbon-like and dichotomously branched. The plant body remains attached to the substratum with the help of a discoid or irregularly branched holdfast. All the branches of the plant body lie in the same plane and lack a mid-rib. Both the sporophytic and gametophytic plants are similar morphologically though cytologically quite different.

Internal Structure: Internally, the thallus is simple in organisation and consists of three layers of cells. There is an upper and a lower assimilatory or photosynthetic layer. Between these two layers occurs a median layer consisting of large, colourless thin-walled medullay cells. These cells have a few or no chromatophores. They possess fucosan vesicles and refractive granules. Many uniseriate clusters of colourless hairs develop on both the surfaces of the thallus.

Growth: Growth of the thallus takes place with the help of a single biconvex apical cell situated at the tip of each branch of the plant body. The apical cell cuts off a series of segments at its posterior face. Each segment again divides by a curved wall in a plane parallel to the thallus surface, the larger derivative again divides in the same way. As a result, the thallus becomes three cells thick.

At the time of the initiation of dichotomy, the apical cell divides by a vertical wall into two equal cells. Each daughter cell functions as an apical cell. Besides cells on the margins of dichotomy may also act as apical cells and their division results in the formation of adventitious branches.

Reproduction

Dictyota reproduces by vegetative, asexual and sexual methods.

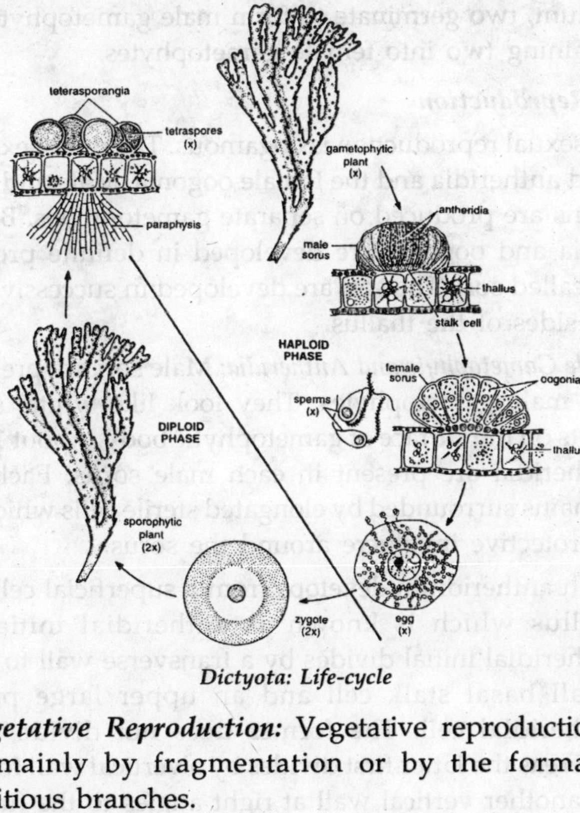

Dictyota: Life-cycle

Vegetative Reproduction: Vegetative reproduction takes place mainly by fragmentation or by the formation of adventitious branches.

Asexual Reproduction

Asexual reproduction takes place by tetraspores developed in unilocular sporangia (tetrasporangia) on sporophytic plants. The sporangia are borne singly or in groups on both the surfaces of the thallus. The sporangial initial first divides transversely into two cells; the lower cell functions as the stalk cell, whereas the upper one enlarges and develops into a tetrasporangium. The diploid nucleus of the sporangium divides meiotically and thus four haploid tetraspores are formed. The tetrasporcs are non-motile and remain arranged tetrahedrally in the sporangium. The tetrasporcs are liberated by the rupture of the sporangium wall apically. They germinate immediately without any period of rest. Out of the four tetraspores of a sporangium, two germinate to form male gametophytes, and the remaining two into female gametophytes.

Sexual Reproduction

The sexual reproduction is oogamous. The male sex organs are called antheridia and the female oogonia. Male and female sex organs are produced on separate gametophytes. Both the antheridia and oogonia are developed in definite projecting groups, called sori. The sori are developed in successive crops on both sides of the thallus.

1. *Male Gametophyte and Antheridia:* Male sori are present on the male gametophytes. They look like white, shining spots on the surface of gametophytic body. About 100-300 antheridia are present in each male sorus. Each sorus remains surrounded by elongated sterile cells which form a protective involucre around the sorus.

   Each antheridium develops from a superficial cell of the thallus which is known as antheridial initial. The antheridial initial divides by a transverse wall to form a small basal stalk cell and an upper large primary antheridial cell. The former does not divide further whereas the latter first divides by a vertical wall followed by another vertical wall at right angles to the first. The cells of the quadrant thus formed divide by repeated transverse and vertical divisions to form the antheridial body.

   The mature antheridium is a shortly stalked multicellular

body; the protoplast of each cell metamorphoses into a single antherozoid. Each antherozoid is a pyriform structure with a single forwardly directed lateral flagellum which is of tinsel type. However, electron microscopic studies reveal the presence of a second flagellum which is rudimentary. The antherozoids are liberated from the antheridium as a result of the gelatinization of the antheridial walls.

2. *Female Gametophyte and Oogonia:* Female sori are present on the female gametophytes. They are distinguished from the male sori by their deep brown colour, and in the absence of involucre. Each female sorus contains 25-50 or more oogonia.

   Each oogonium develops from a single superficial cell of the thallus, which functions as the oogonial initial. The oogonial initial divides by an asymmetrical transverse division to form a basal stalk cell and an upper primary oogonial cell. The latter enlarges considerably to form the oogonium. Its contents develop into a single egg (ovum). Thus, the mature oogonium is a large, sac-like structure containing an egg. Egg is discharged by the gelatinization of the oogonial wall.

3. *Fertilization. Fertilization is External:* The eggs and the antherozoids are liberated early in the morning in the surrounding water. A large number of antherozoids surround the liberated eggs. They attach to the eggs with the help of the flagellum. Ultimately, only one antherozoid penetrates the egg. Male and female nuclei fuse together and it results in the formation of a diploid zygote.

4. *Germination of Zygote:* The zygote germinates immediately without undergoing any resting period. It germinates directly into the diploid sporophytic plant of Dictyota without any reduction division.

Alternation of Generation: The life-cycle of Dictyota exhibits isomorphic alternation of generation. Both sporophytic and gametophytic plant body are morphologically similar, but differ in their genetic constitution and functions. The sporophytic plant body is diploid and is concerned with asexual reproduction. The diploid sporangia produced on the sporophytic body undergo reduction division to form haploid zoospores. The latter germinate to form gametophytic body. Half of the zoospores develop into male plant and rest half into the female plant. The reproductive organs- antheridia and oogonia, are produced on male and female gametophytes respectively. The haploid male and female gametes (antherozoids and eggs) fuse to form a diploid zygote. The zygote germinates into a diploid sporophytic thallus.

Various Fucales

There are 40 genera and 3000 species in the order Fucales. They are cosmopolitan in distribution. Species occurring in the arctic and north temperate seas differ considerably from those of the antarctic and south temperate waters. They are mostiy found attached to the substratum by discoid holdfasts. The plant body is complex both morphologically and anatomically.

These algae reproduce by vegetative and sexual methods; asexual spores are absent. Sexual reproduction is oogamous and the sex organs are borne in special cavities, known as conceptacles. The special branches bearing conceptacles are called receptacles. Conceptacles bear microsporangia or macrosporangia which function as antheridia and oogonia respectively. Both micro-and macrosporangia are unilocular.

The members of Fucales do not show distinct alternation of generation. The sporophytic thallus is diploid and the haploid stage is confined only to gametes.

The order Fucales has been divided into seven families: (1) Fucaceae, (2) Himanthaliaceae, (3) Cystoseiraceae, (4) Sargassaceae, (5) Hormosiraceae, (6) Durvilleaceae, (7) Ascoseiraceae.

The life-histories of Fucus and Sargassum are described here in details.

Family Fucaceae

The plant body is usually flattened and dichotomously branched; axes subterate to algae with mid-rib, but not foliar.

Fucus

Systematic Position

Class	-	Phaeophyceae
Order	-	Fucales
Family	-	Fucaceae

Occurrence: The genus Fucus, represented by more than 100 species, is found attached to rocky substrata in the intertidal belt of the littoral zone. The species occur predominantly in cold water and grow luxuriantly in the North Atlantic and Pacific shores. F. vesiculosus, F. serratus, F. furcatus and F. edentatus are some common species of Fucus. However, no species is recorded from India.

Thallus Structure

External Structure: Plant body of Fucus is well branched, ribbon-like, leathery, slimy to touch and dark brown reaching up to 50 cm in length. Externally, the plant body is differentiated into a basal discoid holdfast which anchors it to the substratum, a short stipe and a flattened, dichotomously branched frond (blade).

The stipe is continued in the form of a midrib in the frond of several species. Each branch of the frond is bifurcated at the tip. The ribbon-like thallus may be smooth (F. vesiculosus) or serrate (F. serratus). In some species (e.g., F. vesiculosus), a number of air bladders occur in the upper portion of the frond that give buoyancy to the plant. Numerous small openings are seen throughout the surface of the frond with naked eye. These are sterile conceptacles called cryptoblasts or cryptostomata. There are several hairs or sterile paraphyses in the cryptoblasts which peep out through the openings. Besides, at the terminal parts of the branches occur fertile conceptacles, known as receptacle. Mid-rib is absent in the receptacular region.

Internal Structure: Internally, the thallus is well differentiated into three distinct regions, viz. (i) meristoderm, (ii) cortex, and (iii) medulla.

1. *Meristoderm:* It is the outermost layer and functions as epidermis. It is made up of compactly arranged columnar cells. These cells are densely packed with chromatophores and fucosan, a reserve food material.
2. *Cortex:* It is composed of relatively large thin-walled parenchymatous cells, size of which gradually increases toward the centre. The cortex is sometimes differentiated into outer and inner cortex.
3. *Medulla:* The central part of the frond forms medulla. It is made up of thick walled narrow and elongated cells. Spaces between the cells are filled with mucilage. Along with medullary cells, running longitudinally are elongated cells, called hyphae which give mechanical support to the thallus.

Growth: The growth of the plant body takes place with the help of a pyramidal apical cell with three cutting faces - two lateral and one basal. The apical cell remains situated in the terminal portion of the lamina. The lateral segments derived from the apical cell initially divide periclinally; of the resulting two cells, the outer one, by repeated divisions in various planes, organizes the epidermis and the cortex whereas the inner one, in association with the basal segment cut off by the apical cell, forms the medulla.

Reproduction: Fucus reproduces by vegetative and sexual methods. Asexual reproduction is absent.

Vegetative Reproduction: It takes place by fragmentation or by formation of adventitious branches.

Sexual Reproduction

Sexual Reproduction is Oogamous: The sex organs are produced in special flask-shaped cavities, known as conceptacles. The conceptacles are borne on the receptacles. The receptacles are the tips of the branches which become swollen. The male sex organs are called antheridia and the female oogonia. In homothallic (monoecious) species (e.g., F. spiralis) antheridia and oogonia are present in the same conceptacle or in different conceptacles on the same plant. In dioecious species (e.g., F. vesiculosus), male and female sex organs develop in different conceptacles on separate plants.

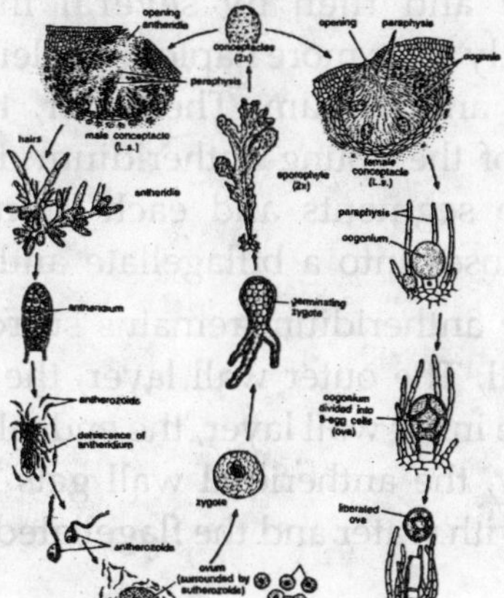

Fucus: Life-cycle

1. *Development of Conceptacles:* The conceptacle develops from a superficial cell of the receptacle, which is easily distinguished from the adjoining cells by its larger size and prominent nucleus. It divides at a slower rate than the adjoining cells. Consequently, the initial cell, which was at the surface of the thallus in the beginning, is gradually pushed at the base of the flask shaped cavity. The conceptacle initial then divides by a transverse wall forming an outer tongue cell and an inner basal cell. The tongue cell does not divide further and gradually degenerates. The basal cell, however, undergoes repeated anticlinal divisions resulting into the formation of a basal fertile layer of the conceptacle. Sometimes, the derivatives of the basal cell divide periclinally also forming 2 to 3 layers of which the superficial layer is always fertile giving rise to reproductive structures.

2. *Antheridium:* The antheridia develop either from the cells of the fertile layer or at the base of branched paraphyses developed on the fertile layer. The antheridia are small, club-shaped, shortly-stalked, unicellular structures. The diploid nucleus of young antheridium first divides meiotically and then by several mitotic divisions. Consequently 64 or more haploid nuclei are formed in the developing antheridium. Thereafter, the multinucleate protoplast of the young antheridium divides into many uninucleate segments and each uninucleate segment metamorphoses into a biflagellate antherozoid.

   The mature antheridium remains surrounded by a two layered wall. The outer wall layer, the exochite, is firm, whereas the inner wall layer, the endochite, is gelatinous. At maturity, the antheridial wall gets gelatinized when in contact with water and the flagellated antherozoids are liberated.

3. *Oogonium:* Like antheridium, an oogonium initial is also a superficial cell of the fertile layer. Each initial cell divides by a transverse wall. The smaller basal cell forms the stalk cell and the upper larger cell, the oogonial cell. The latter gradually enlarges and becomes almost spherical and densely cytoplasmic. The diploid nucleus of the oogonial cell undergoes a meiotic division and the four nuclei thus formed divide mitotically. Consequently, eight haploid nuclei are formed in the oogonium. These eight nuclei serve as eggs or ova.

   The oogonial wall is differentiated into three layers, the outer exochite, the middle mesochite and the inner endochite. At maturity, the oogonium absorbs water with the result the exochite ruptures and eggs slip out of it, still surrounded by mesochite and endochite and approaches the ostiole. Soon, the endochite also imbibes water and as a result, the mesochite also breaks at its apical portion and ultimately, the eggs are set free in the water.

Those cells of the basal fertile layer of the conceptacle which do not form reproductive structures, develop into long hair-like paraphysis. Some of the peripheral cells near the ostiole also form unbranched hair-like appendages which emerge from the ostiole in the form of a tuft.

4. *Fertilization:* The fertilization is external, i.e., the antherozoids and the eggs come in contact outside the conceptacle in the water. A large number of antherozoids approach a single egg. After the attachment of the antherozoid flagella to the egg surface, the beating of the flagella makes the egg rotate for some time. Ultimately, one of the antherozoids finds its way into the egg, leading to the formation of the zygote. About after an hour of fertilization a gelatinous wall is secreted around the zygote.
5. *Germination of Zygote:* The zygote germinates after a short period of rest (about a day) and first, it develops a rhizoidal protuberance on one side of it. After the development of the rhizoidal protuberance, the zygote divides mitotically in a plane transverse to the axis (as defined by the rhizoidal protuberance). The upper cell of the 2-celled embryo divides vertically where as the lower divides transversally, thus forming a 4-celled embryo. The upper two cells ultimately give rise to a tissue differentiated into the medulla and cortex, whereas the lower most cell of the quadrant from which the primary rhizoid originates also forms secondary rhizoids. Ultimately, a multicellular embryo with an apical cell is differentiated.

Alternation of Generation: The life cycle of Fucus does not show any alternation of generation. The plant body is diploid and sporophytic. The reproductive organs, antheridia and oogonia, are produced within special flask-shaped conceptacles. The reduction division takes place at the time of gamete formation.

Thus, the haploid phase, in the life-cycle of Fucus, is

represented only by gametes (antherozoids and eggs). The haploid male and female gametes fuse to form diploid zygote. The zygote germinates into a diploid sporophytic thallus.

Family Sargassaceae

The members are very much complicated in their morphology and anatomy and are characterized by the presence of lateral branch system. The family includes five genera. Of these Sargassum is widely known, and the most abundant sea weed in India.

Sargassum

Systematic Position

Class	-	Phaeophyceae
Order	-	Fucales
Family	-	Sargassaceae

Occurrence: The genus Sargassum is represented by about 150 species, found chiefly in the tropical and subtropical seas of the southern hemisphere. The plants form huge floating masses in the Atlantic ocean, off the African continent between 20° and 35° north latitude. Hence this part of the Atlantic ocean is called the Sargasso sea. Species of Sargassum are also found along the coasts of Australia, India and Ceylon. In India, the genus is represented by 16 species which occur along the western and southern coast. Some common Indian species of Sargassum are. S. cinereum, S. carpophyllum, S. duplicatum, S. christifolium, S. tenerrium, S. ilicifolium, S. plagiophyllum, S. wightii and S. myriocystum.

Thallus Structure

External Structure: The thallus of Sargassum is diploid and sporophytic. It is erect and branched and is differentiated into holdfast and the main axis; the latter bears many leaf-like lateral branches and air bladders.

The holdfast is a discoid or warty structure with the help of which the thallus remains attached to the substratum. In some species the holdfast is stolon-like. The main axis or 'stem' is erect, cylindrical or flattened, varying in height from 15-25 cm. But sometimes it attains a height of more than a meter as in S. filipendula. The main axis bears large number of primary laterals in a spiral phyllotaxy of 2/5. The primary laterals have unlimited growth, hence they are also called long shoots. The main axis as well as the primary laterals bear flat leaf-like branches, known as secondary laterals or 'leaves'.

The 'leaves' are flat and simple with distinct mid-ribs. Sometimes the mid-rib is absent as in S. enerve. The margins of the 'leaves' are serrate, dentate or entire. On both surfaces of the 'leaves' small pores, known as ostioles, are present. These pores open in small flask-shaped sterile cavities, called cryptostomata or sterile conceptacles.

The 'leaves' present in the lower part of the primary lateral (long shoots) bear rudimentary axillary branches and those present in the upper portion have axillary branch systems. One or more basal members of the axillary branch are modified into globular or spherical air bladders. The air bladders are filled with air and help the plant in floating by increasing its buoyancy. In some species the air bladder terminates into a 'leaf'. According to some algologists, the air bladders help in respiration. The subsequent members of the axillary branch system develop into long repeatedly branched cylindrical or flattened structures, called receptacles. The receptacles bear reproductive organs in special flask-shaped structures which are known as conceptacles.

Internal Structure

1. *Internal Structure of the Main Axis:* In transverse section the main axis is almost circular in outline and is differentiated into three distinct regions: (i) meristoderm, (ii) cortex, and (iii) medulla.

   The meristoderm (named so due to meristematic nature of its cells) is the outermost layer and functions as epidermis. It is made up of compactly arranged columnar cells. These cells are densely packed with chromatophores and fucosan, a reserve food material. They form assimilatory region of the thallus. In the mature part of the axis meristoderm is covered with a thin layer of mucilagenous cuticle.

The cortex forms the most part of the axis and it has narrow and elongated cells. These cells are loosely arranged with well developed intercellular spaces. They are filled with reserve food material, and as such the cortex forms the storage region of the axis.

The central part of the main axis forms medulla. It is made up of thick walled narrow and elongated cells. Sometimes these cells have scalariform thickenings. The function of medulla is to transport water and metabolites to different parts of the thallus.

2. *Internal Structure of the Leaf*: Just like the main axis, the leaf is also differentiated into meristoderm, cortex and medulla.

The meristoderm consists of small radially elongated, meristematic cells. These cells have plenty of reserve food and chromatophores. The meristoderm is followed by cortex which consists of thin walled cells, rich in reserve food material. The cortex is the thickest in the midrib region and gradually narrows towards the blades. The central part of the 'leaf' is made up of thick walled cells which make up medulla. In structure and function it is similar to the medulla of the main axis.

On both the surfaces of the 'leaf' are present many sterile conceptacles, known as cryptoblasts or cryptostomata. The cryptoblast is a flask-shaped structure and it opens on the surface of the 'leaf' by a small pore, known as ostiole. The wall of the cryptoblast is made up of sterile cells which have chromatophores. Many unbranched multicellular filaments arise from the wall of the conceptacles and protrude through the ostiole. These filaments are called paraphysis.

3. *Internal Structure of Air-bladder:* The air bladder is differentiated into meristoderm and cortex only. Medulla is absent. The meristoderm is made up of radially elongated narrow cells, enveloped with a mucilagenous layer. It is followed by a 4-8-layered parenchymatous cortex. The central part of the air bladder is occupied by a large air cavity.

Growth: The growth of Sargassum thallus is accomplished by a quadrangular apical cell.

Reproduction: Sargassum reproduces by vegetative and sexual methods. Asexual reproduction is absent.

Vegetative Reproduction: Fragmentation is the chief method of vegetative multiplication. Thallus may break into segments due to death and decay of the older parts or due to injury. These segments grow into new thalli. In some free floating species like S. natans and S hystrix, it is the exclusive method of multiplication.

Sexual Reproduction: Sexual reproduction is oogamous. The sex organs are produced in special flask-shaped cavities, known as conceptacles. Conceptacles are borne on the receptacular branches of the sporophytic thallus. The male sex organs are called antheridia and the female oogonia. Male and female sex organs are produced on separate conceptacles; the conceptacles bearing antheridia are called male conceptacles and those having oogonia female conceptacles. In homothallic (monoecious) species antheridia and oogonia are present in the same conceptacle or in different conceptacles on the same plant. In dioecious species male and female sex organs develop in different conceptacles on separate plants. Most of the species of Sargassum are monoecious.

The development of fertile and sterile conceptacles is similar. The only difference is that the basal cells in sterile conceptacles form paraphysis, whereas in fertile conceptacles they give rise to antheridia and oogonia.

1. *Development of Conceptacles:* The conceptacle develops from a single superficial cell on the receptacular branch. This

flask-shaped conceptacle initial can be easily distinguished from the adjoining cells by its larger size and prominent nucleus. It divides at a slower rate than the adjoining cells. Consequently, the initial cell, which was at the surface of the thallus in the beginning, is gradually pushed at the base of the flask shaped cavity. The conceptacle initial divides by a curved septum, forming an outer tongue cell and an inner basal cell. The tongue cell elongates and divides transversely to form a small filament which soon disappears. The basal cell, however, undergoes repeated vertical divisions and the resulting cells form the basal fertile layer of the conceptacle. The reproductive organs develop from the cells of this layer.

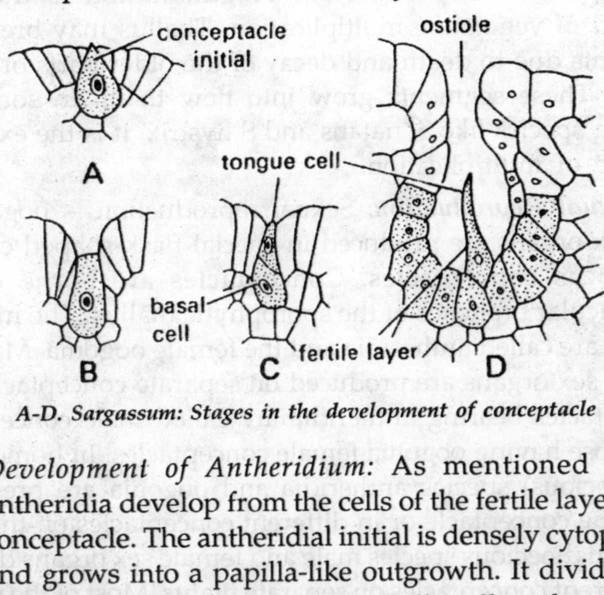

A-D. Sargassum: Stages in the development of conceptacle

2. *Development of Antheridium:* As mentioned above, antheridia develop from the cells of the fertile layer of the conceptacle. The antheridial initial is densely cytoplasmic and grows into a papilla-like outgrowth. It divides by a transverse wall into a lower stalk cell and an upper antheridial cell. Each of these cells has a diploid nucleus. The antheridial cell develops into an antheridium. It is pushed on one side due to rapid growth of the stalk cell. The growing stalk cell again divides into a basal stalk cell and upper antheridial cell. This process is repeated several times and consequently a paraphysis with many antheridia

is formed. The terminal cell of the paraphysis is usually sterile, it does not develop into an antheridium:

The antheridial initial has only one diploid nucleus which undergoes a meiotic and several mitotic divisions. Consequently, 32-64 haploid nuclei are formed in the developing antheridium. The multinucleate protoplast of the young antheridium divides into many uninucleate segments (the number of these segments is equal to the number of nuclei) and each uninucleate segment metamorphoses into a biflagellate antherozoid.

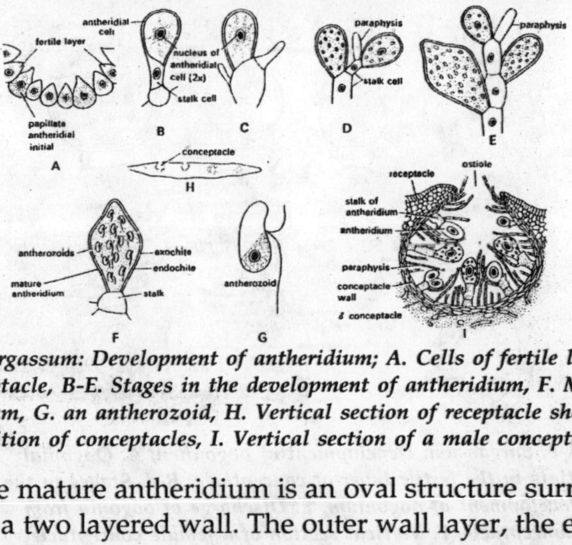

A-I. Sargassum: Development of antheridium; A. Cells of fertile layer in conceptacle, B-E. Stages in the development of antheridium, F. Mature antheridium, G. an antherozoid, H. Vertical section of receptacle showing the position of conceptacles, I. Vertical section of a male conceptacle

The mature antheridium is an oval structure surrounded by a two layered wall. The outer wall layer, the exochite, is firm, whereas the inner wall layer, the endochite, is gelatinous. The antheridium remains attached to the base of the conceptacle with the help of a stalk cell. At maturity, the antheridium is detached from the stalk and comes out of the conceptacle through the ostiole. The antheridial wall gets gelatinized when in contact with water and the flagellated antherozoids are liberated.

3. *Development of Oogonium:* Oogonia also develop from the basal fertile layer of the conceptacle. The cells of this layer function as oogonial initials. Each initial cell divides by a transverse wall into two unequal cell. The smaller basal

cell forms the stalk cell and the upper larger cell, the oogonial cell. The latter gradually enlarges and becomes almost spherical. It is densely cytoplasmic and has oil droplets. The diploid nucleus of the oogonial cell undergoes a meiotic division and the four nuclei thus formed divide mitotically. Consequently, eight haploid nuclei are formed in the oogonium. Out of these, seven degenerate and the remaining nucleus functions as the egg nucleus. This functional nucleus is located in the centre of the oogonium, whereas the degenerating nuclei can be seen along the periphery.

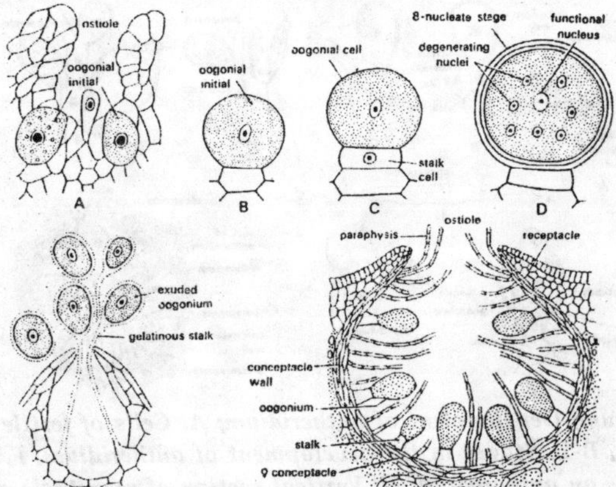

A-F. Sargassum: Development of oogonium; A. Oogonial initials in the fertile layer of conceptacle, B-d. Stages in the development of oogonium, E. Discharge of oogonia from conceptacle, F. Vertical section of a female conceptacle

The oogonial wall is differentiated into three layers, the outer exochite, the middle mesochite and the inner endochite. The mature oogonia are discharged from the conceptacle through the ostiole but they remain attached with the conceptacle base by long gelatinous stalks formed by the exochite. According to some algologists, the exochite of the oogonium ruptures at the time of its discharge from the conceptacle and the gelatinous stalk is formed by the mesochite. The discharged egg is thus surrounded by the endochite only.

Those cells of the basal fertile layer of the conceptacle which do not form reproductive structures, develop into long hair-like paraphysis. Thus the conceptacle has many paraphysis in addition to reproductive structures.

4. *Fertilization:* At the time of fertilization although the oogonium is outside the conceptacle, it still remains attached with the conceptacle base by a gelatinous stalk. A large number of antherozoids surround the extruded oogonium. They attach to the oogonial wall

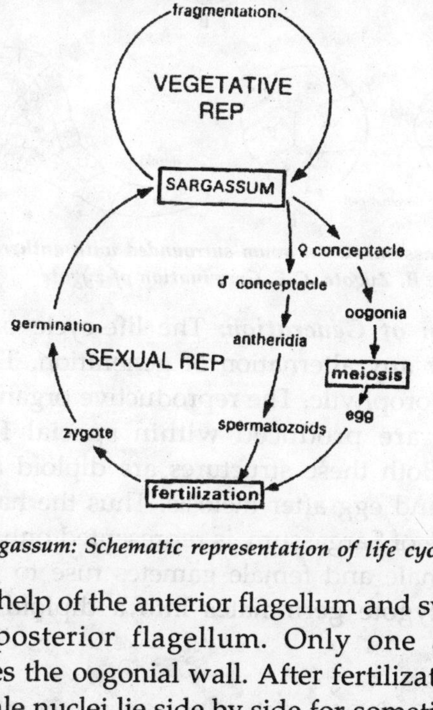

Sargassum: Schematic representation of life cycle

with the help of the anterior flagellum and swim in water by the posterior flagellum. Only one antherozoid penetrates the oogonial wall. After fertilization the male and female nuclei lie side by side for sometime and then they fuse to form a diploid zygote.

5. *Germination of Zygote:* The zygote germinates immediately after fertilization while the oogonium is still attached to the conceptacle wall by a gelatinous stalk. However, after sometime there is gelatinization of oogonial wall and the zygote is liberated. It now gets attached to some solid

substratum in sea water. The zygote divides by a transverse wall into an upper and a lower cell. The lower cell forms rhizoids, whereas the upper cell undergoes many anticlinal and periclinal divisions to form a diploid thallus.

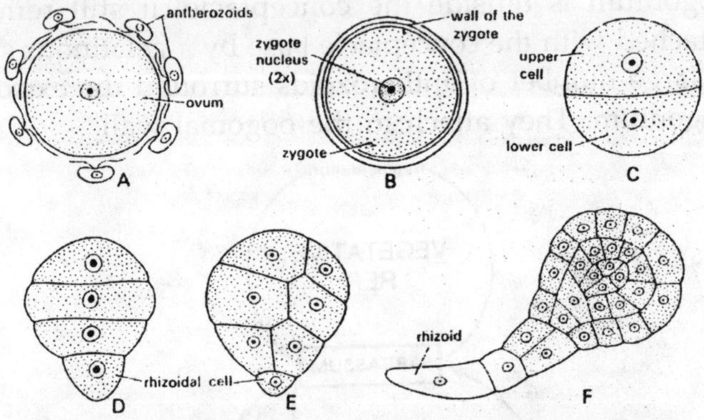

A-F. Sargassum: A. An ovum surrounded with antherozoids, B. Zygote, C-F. Germination of zygote

Alternation of Generation: The life-cycle of Sargassum does not show any alternation of generation. The thallus is diploid and sporophytic. Γhe reproductive organs, antheridia and oogonia, are produced within special flask-shaped conceptacles. Both these structures are diploid and produce antherozoids and egg after meiosis. Thus the haploid phase, in the life-cycle of Sargassum, is represented only by gametes. The haploid male and female gametes fuse to form diploid zygote. The zygote germinates into a diploid sporophytic thallus.

6

Different Structures

The Bacillariophyceae

The class Bacillariophyceae is a large group of algae, consisting of over 10,000 species belonging to 200 genera. The members are popularly known as diatoms. Fine geometrical expressions and intricate sculpturing in the cell wall make these organisms one of the most beautiful microscopic objects. They occur in almost every aquatic habitat as free-living photosynthetic autotrophs, colourless heterotrophs or photosynthetic symbionts. Diatoms are also found in terrestrial habitats and form a substantial proportion of the soil flora. In tropical rain forests, they grow in association with blue-green algae on leaves of trees. In India, the group is represented by 569 species distributed over 92 genera. Pinnularia graciloides, Niedium gracile, Cymbella affinis, Eunotia pectinalis, Melosira jurgensii and Coscinodiscus radiatus are some common Indian species of diatoms.

Important distinguishing characters of Bacillariophyceae are as follows:

(1) Diatoms are mostly unicellular but some colonial species form filaments, loose chains or mucilagenous colonies.

(2) The cells are microscopic and are variable in shape. They may be triangular (e.g.,Triceratium), boat shaped (e.g., Gyrosigma, Cymbella), rod shaped (e.g., Bacillaria), oval (e.g., Cocconeis placentula) or spherical (e.g., Coscinodiscus excentricus).

(3) The cell shows radial or bilateral symmetry.

(4) The cell is surrounded by a rigid two-part overlapping cell wall, called frustule. It is made up of pectic substances but is impregnated with high amounts (sometimes up to 95%) of siliceous substances.

(5) The vegetative cell is uninucleate and diploid. It usually has two large parietal or many discoid plastids. It is olive green or golden in colour, and its main pigments are chlorophyll- *a*, chlorophyll-*c* and xanthophyll (such as fucoxanthin, diatoxanthin, diadinoxanthin).

(6) The storage products of photosynthesis are oil and chrysolaminarin.

(7) Motile cells (antherozoids) have one pantonematic flagellum.

(8) The cell shows gliding movements.

(9) Vegetative reproduction takes place by cell division.

(10) Sexual reproduction is isogamous or oogamous.

(11) The zygote formed as a result of gametic union develops into a special type of spore, called auxospore. Continued reproduction by cell division results in the reduction of cell size and auxospores help in regaining the original cell size. Thus, auxospore formation is a restorative or rejuvenation process.

Classification

Diatoms have been classified into two orders on the basis of their symmetry:

Order 1. Pennales: The members of this order are characterized by a bilateral symmetry. They show gliding movement. Sexual reproduction, if present, is amoeboid isogamy. Examples-Fragilaria, Synedra, Eunotia, Navicula, Pinnularia, Epithemia.

Order 2. Centrales: This order includes radially symmetrical forms. They lack gliding movements. Sexual reproduction is anisogamous or oogamous and the gametes are motile. Examples-Cyclotella, Melosira, Rutilaria, Biddulphia, etc.

A general account of diatoms is provided in the following pages.

Occurrence

Diatoms are found in almost every aquatic environment. Genera like Synedra, Melosira, Navicula and Asterionella are strictly freshwater, and Sceletonema, Eucampa, Biddulphia, Fragilaria, Hemidiscus, Corethron, etc., are found in marine habitat. Some aquatic forms occur as plankton of ponds and lakes. The algal flora of roadside ponds and pools always includes some diatoms. Species of Pinnularia, Navicula, Frustulia, Amphora and Caloneis are found in terrestrial habitat. Such forms are able to withstand desiccations for a very long period. Some diatoms are epiphytic on other algae. For example, species of Cymbella, Gomphonema, Eunotia and Cocconeis grow on the filaments of Cladophora, Rhizoclonium and Vaucheria. Licmophora is an endozoic diatom.

Various Structures

Diatoms are unicellular and cells are usually found in isolated form. However, isolated cells sometimes get imbedded in a common gelatinous matrix to form simple colonies. Such colonies do not have complex plant bodies and thus can not be regarded as multicellular organisms. Colonial diatoms occur either as uniseriate filaments (e.g., Melosira,) or as branched filaments, having many radiating arms at the free end (Licmophora flabellata,). In some diatoms the colony is formed due to interlocking of bristles, present on the cell wall (Chaetoceras diadema,).

Cell Structure

Cell wall and protoplast are the two main constituents of the cell.

1. *Cell Wall:* The characteristic feature of Bacillariophyceae is the presence of a cell wall laden abundantly with silica. The empty wall, called frustule, is made up of two almost equal overlapping halves or theca, the smaller fitting into the larger like a pair of Petri dish. The outer or upper half is epitheca and the inner or lower hypotheca. Each half-wall (theca) has two main parts:

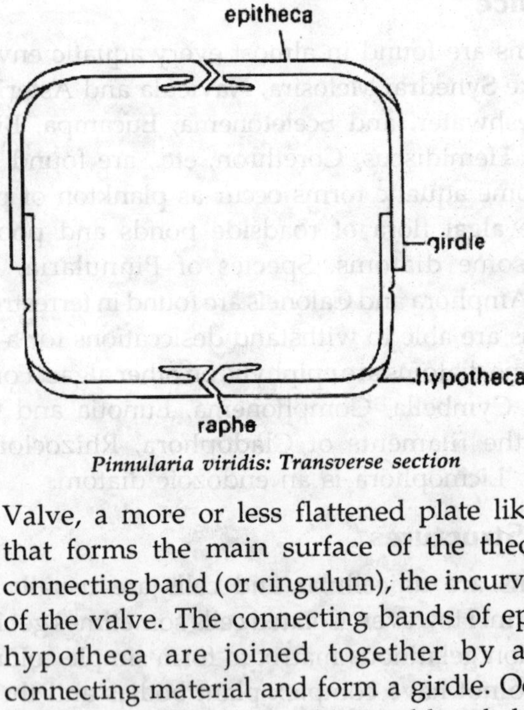

***Pinnularia viridis*: Transverse section**

(i) Valve, a more or less flattened plate like structure that forms the main surface of the theca, and (ii) connecting band (or cingulum), the incurved margins of the valve. The connecting bands of epitheca and hypotheca are joined together by an organic connecting material and form a girdle. Occasionally, there are one or more additional bands between the valve and the girdle and such bands are known as intercalary bands. The girdle bands often have minute teeth which hold the two valves together. When a frustule lies in such a way that valve side is uppermost, it is known as valve view or top view (Fig. A) and when connecting bands are uppermost it is called girdle view or side view.

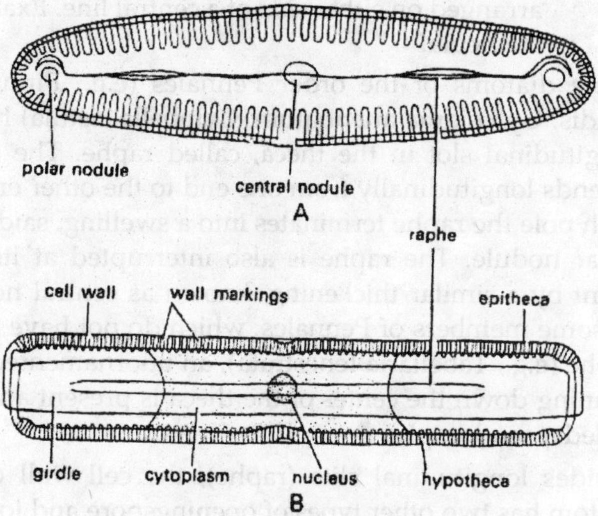

A-B. Pinnularia: A. Valve view, B- Girdle view

Frustules may be perfectly symmetrical, asymmetrical or with intergrades between the two.

On the basis of this character, diatoms are divided into two orders-Centrales (radially symmetrical diatoms) and Pennales (bilaterally symmetrical diatoms).

The frustule is made up of a pectin material. A characteristic feature of diatoms is that the valve portion of the frustule is silicified. The siliceous material is laid down in certain regular patterns that make the wall ornamented. The ornamentation is of the following four basic types:

(a) *Centric and Radial:* The ornamentation is based around a central point. Example Coscinodiscus.

(b) *Trellisoid:* The ornamentation is uniform without reference to a point or central line (Fig.B). Example-Eunotia.

(c) *Gonoid:* The ornamentation is dominated by angles. Example-Triceratium.

(d) *Pennate:* The ornamentation is symmetrically arranged on either side of a central line. Example-Navicula.

Some diatoms of the order Pennales (e.g., Pinnularia viridis, Gyrosigma attenuatum, Cymbella cistula) have a longitudinal slot in the theca, called raphe. The raphe extends longitudinally from one end to the other end. At each pole the raphe terminates into a swelling, said to be polar nodule. The raphe is also interrupted at its mid point by a similar thickening, known as central nodule. In some members of Pennales, which do not have a true raphe (e.g., Tabellaria fenestrata), an unornamented area running down the center of the theca is present and it is called pseudoraphe.

Besides longitudinal slits (raphe), the cell wall of the diatom has two other types of openingspore and loculus. The pores are simple holes within the homogeneous silicified wall. They have ribs or costae which provide them mechanical strength. The locules, on the other hand, are more complex openings. They are in fact polygonal chambers in the cell wall. At the one end of the chamber there is a sieve membrane (also called pore membrane or velum) with many small pores and a large hole, called foramen. The sieve membrane may be on the outerside (e.g., Coscinodiscus wailesii) or on the innerside (e.g., Coscinodiscus lineatus).

2. *Protoplast:* It is surrounded by a thin plasma membrane, lying close to the inner face of the cell wall. Within plasma membrane a thick layer of cytoplasm surrounds a large central vacuole. The cell is uninucleate and the oval or spherical nucleus is usually embedded in the parietal cytoplasm. Sometimes the nucleus is suspended in the centre of the cell by cytoplasmic strands.

The cytoplasm has many mitochondria, Golgi bodies and

chromatophores. In the members of the order Centrales the cells contain many discoid, stellate or irregular chromatophores, while in Pennales there are one or two large parietal plate-like chloroplasts in each cell. Besides, plastids of specific shapes occur in certain members. For example, plastids are lobed discoid in Melosira sp., stellate in Grammatophora marina, semistellate in Striatella and 'H' shaped in Gomphonema and Mastogloia.

The chromatophore has a double layered binding membrane, surrounded by two additional layers of chromatophore ER. The thylakoids within the chromatophores are in stacks of 4-6 lamellae. Each chromatophore, has a single, almost centrally situated pyrenoid. The main photosynthetic pigments of diatoms are chlorophyll *a*, C_1 and C_2 β-carotene and xanthophyll. The golden-brown colour of diatoms is due to xanthophylls, such as fucoxanthin, diatoxanthin and diadinoxanthin.

As in other algae, diatoms do not store their photosynthetic products in the form of starch. Their storage product instead is chrysolaminarin, which is located in vesicles.

Movement: The motile diatoms are characterised by the presence of raphe. They show gliding movement. The diatom adheres itself to the substratum in the area of raphe. Certain crystalloid bodies, containing mucilagenous material present in the cytoplasm, release mucilage through polar or central nodule on an appropriate stimulus. Now the mucilage streams along through the raphe in the direction opposite to that in which movement of diatom is to set in. The movement of diatoms may be straight, curved or zig-zag. The rate of gliding varies from 2-14 μ m sec^{-1} at room temperature. The motility is affected by factors like light, temperature and nature of the substratum.

Reproduction

Diatoms reproduce by vegetative, asexual and sexual methods.

Vegetative Reproduction: Cell division is the chief method of vegetative propagation. The cells usually divide at midnight or early morning. Prior to cell division the protoplast of the mother cell increases in size, pushing the epitheca away from the hypotheca (Fig. A). Now the nucleus divides mitotically in a plane perpendicular to the valves. Nuclear division is followed by the division of the protoplast in two by invagination of the plasmalemma (Fig. B, C).

Each of the two uninucleate daughter protoplasts with a single nucleus is retained within one 'valve, i.e., one daughter protoplast is covered on one side by the epitheca and on the other by the hypotheca; the other side of both the protoplast being naked. The direction of the division of protoplast is dependent on the number of plastids in the cell. If the cell has only a single plastid, the division is usually longitudinal and if the cell contains two or more plastids, the division may be longitudinal or transverse.

The two valves of the parent cell (each with a daughter protoplast) become epithecas of the daughter cells. However, each daughter cell produces a new hypotheca (Fig. D). The daughter cell which retains the epitheca of the parent cell, attains nearly the same size as that of the parent cell. But the other daughter cell which receives the hypotheca of the parent cell (now epitheca of the daughter cell) is smaller than the parent cell. Thus there are two lines of descendants in cell division; one line retains the same size, whereas in other there is diminution in size in successive generations.

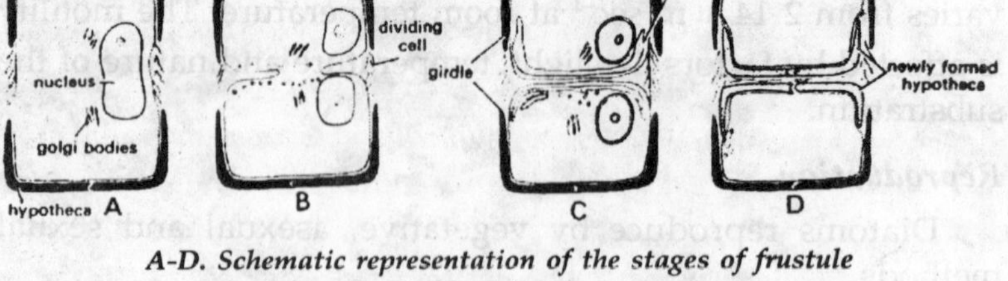

A-D. Schematic representation of the stages of frustule formation in Gomphonema parvulum

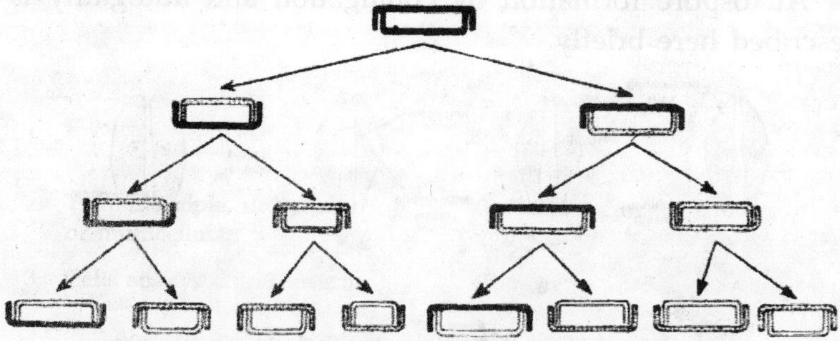

Diagrammatic representation of cell division in diatoms (note reduction in size in successive generations)

Asexual Reproduction: It takes place by means of auxospores. Continued multiplication by cell division results in gradual diminution in cell size in successive generations.

The reduction in cell size can continue only up to a certain limit and beyond this limit either the cell is to die or convert into an auxospore. In fact, auxospore formation is a restorative process through which the smaller cell restores the original size. Prior to auxospore formation, the cell secretes a large amount of mucilage which pushes the two valves apart.

The protoplast is thus set free. It grows considerably and attains the normal size. Then it secretes around itself a silicified membrane, the perizonium. Eventually new valves are secreted inside the perizonium and the auxospore becomes an independent cell.

Sexual Reproduction: Most of the diatoms are homothallic. Their vegetative cells are diploid. The sexual reproduction is isogamous (in Pennales) or oogamous (in Centrales). In both these types, the zygote grows into auxospore. But the formation of auxospores is different in isogamous and oogamous species.

In the members of the order Pennales, auxospores are formed either by the fusion of conjugating cells, by autogamy or by parthenogenesis, whereas in Centrales auxospores are formed by autogamy or by oogamy.

Auxospore formation by conjugation and autogamy is described here briefly.

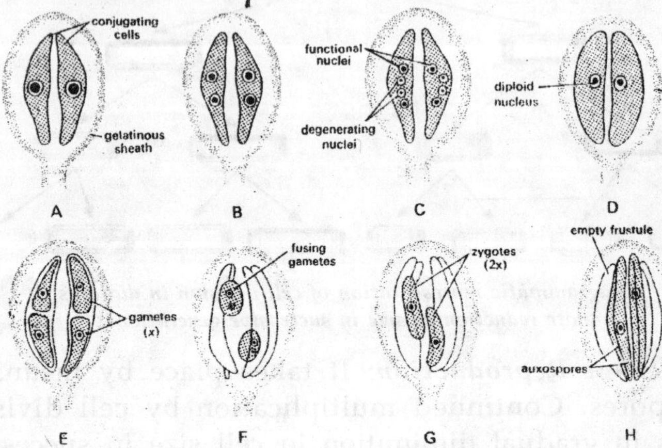

A-H. Successive stages in the development of auxospores by conjugating cell

1. *Development of Auxospores by Conjugating Cells:* Auxospores are formed by this method in diatoms like Cymbella, Pinnularia and Navicula. This process is initiated when the size of the cells is reduced to a certain critical point after successive cell divisions. Two such cells come close to each other and get enveloped within a common mucilagenous mass (Fig.A) The diploid nuclei of these cells undergo meiosis and as such the diploid nucleus of each cell produces four haploid nuclei.

   Of the four nuclei in a cell, two degenerate and the other two remain functional (Fig. B,C). Meanwhile the protoplast of the cell divides into two segments, each carrying a single haploid nucleus. These segments now function as gametes (Fig. D, E). Two gametes from different cells fuse to form a zygote (Fig. F,G). The two zygotes thus formed grow in size to form auxospores. The auxospore secretes a perizonium, and later siliceous valves are formed (Fig.H). The four discarded valves of the two mother cells are often seen attached to the newly formed frustules.

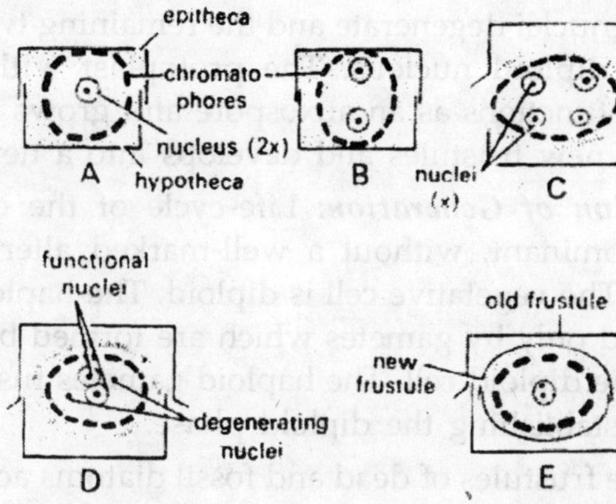

A-E. Successive stages of auxospore formation by autogamy

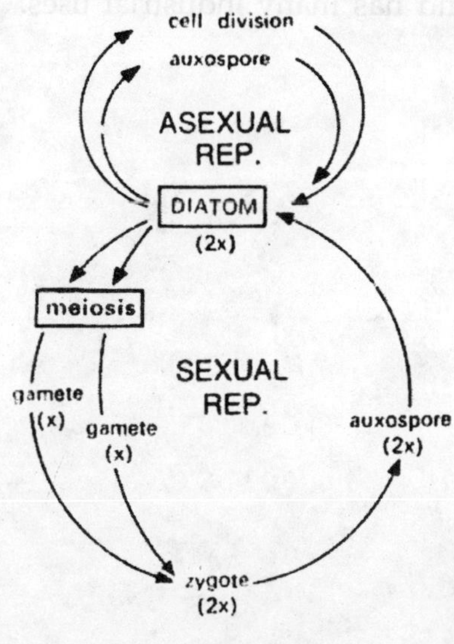

Life-cycle of a diatom

2. *Development of Auxospores by Autogamy:* In Cyclotella meneghiniana and Melosira nummuloides, auxospores are formed by autogamy. In this process, the cell secretes some mucilage and as such pushes the epitheca and hypotheca slightly apart. Now the diploid nucleus of the cell divides meiotically forming four haploid nuclei. Two

of these nuclei degenerate and the remaining two fuse to form a diploid nucleus. The protoplast with diploid nucleus functions as an auxospore and grows in size. It secretes new frustules and develops into a new cell.

Alternation of Generation: Life-cycle of the diatom is diploid-predominant, without a well-marked alternation of generations. The vegetative cell is diploid. The haploid phase is represented only by gametes which are formed by meiotic division of the diploid cell. The haploid gametes fuse to form zygotes, re-establishing the diploid phase.

Note: The frustules of dead and fossil diatoms accumulate at the bottom of water bodies in the form of diatomaceous earth or kieselguhr. Such deposits are found in various parts of the world and has many industrial uses.

7

Cell Structure

The Ectocarpales

The order Ectocarpales includes 60 genera of brown algae which are found in the cold sea water of the temperate and polar regions. Most of the members of this order are heterotrichous with a prostrate creeping disc, which functions as holdfast and an erect monosiphonous filamentous foliose system.

These algae show trichothallic growth with intercalary divisions confined to certain areas in the filament. Asexual reproduction is by zoospores which are produced in uni- or plurilocular sporangia. Sexual reproduction is isogamous but in some species it is oogamous. Reproductive structures are borne laterally or in uniseriate rows.

Fritsch (1935) divided the order into 12 families-Ectocarpaceae, Myrionemataceae, Elachistaceae, Leathesiaceae, Mebogloeaceae, Aenotrichaceae, Spermatochnaceae, Splachnidiaceae, Punctariaccae, Asperococcaceae, Encoeliaceae and Dictyosiphonaceae. But most algologists restrict Ectocarpales to a single family Ectocarpaceae characterized by simple, heterotrichous members exhibiting isogamous sexual reproduction and isomorphic alternation of generations.

Family Ectocarpaceae

The members of the family are characterised by simple heterotrichous thallus organisation, isogamous type of sexual reproduction with isomorphic alternation of generation. The life history of Ectocarpus is described here as a representative of the family.

Ectocarpus

Systematic Position

Class	-	Phaeophyceae
Order	-	Ectocarpales
Family	-	Ectocarpaceae

Occurrence: Species of Ectocarpus are cosmopolitan in distribution, growing in littoral and supralittoral regions. They are more common along the Atlantic coast than the Pacific coast. In India they are found on the Western coast, growing, epiphytically on sea plants or attached to rocks. Epiphytic species, such as E. coniferus and E. breviarticulatus, grow on larger algae like Fucus and Laminaria. Ectocarpus fasciculatus, an epizoic species, grows on fins of certain fishes. E. dermonematus is an endophytic species. The genus is represented in India by about 13 species. Of these, Ectocarpus coniferus, E. ndicus, E. arabicus, E. geminifructus and E. dermonematus are common.

Various Structures

The thallus of Ectocarpus is heterotrichous, differentiated into a prostrate rhizoidal system and an erect branched system. The prostrate system is creeping and branched, and the thallus remains attached to the substratum with the help of this system. It is more developed in attached forms than in free floating species. In some species rhizoidal part is epiphytic, i.e., it grows within the supporting plant.

The erect part of the thallus is usually irregularly branched and the branches arise just beneath the transverse septa of the cells. The filaments are several centimeters long and monosiphonous, i.e., only one cell in thickness. Sometimes the filaments may become polysiphonous due to vertical divisions of the cells. In some species (e.g., E. granulosus) the basal parts of the main erect filaments become corticated by ascending rhizoidal branches. The distal part of each filament terminates into mucilagenous hairs or erect threads.

Cell Structure

Filaments of Ectocarpus are made up of uninucleate rectangular or cylindrical cells. The cell wall is differentiated into an outer gelatinous layer and an inner firm cellulose layer. The gelatinous layer is mainly composed of algin and fucoiden, characteristic of phaeophycean cell wall. A thin layer of cytoplasm is present inner to the cell wall from which a number of cytoplasmic strands grow toward the centre of the cell. There is a central nucleus and many chromatophores in the cytoplasm. The number and shape of chromatophores varies in different species. They are discoid or band-shaped, and each chromatophore has a projected pyrenoid.

The main cell pigments are chlorophyll-*a*, chlorophyll-*c*, carotene and xanthophyll. Xanthophyll is present in much more amounts than chlorophyll. Consequently, the green colour of chlorophyll is completely masked.

The cytoplasm also contains some vacuoles with polyphenol. These vacuoles are called physodes and they probably function like lysosomes. In the life-cycle of Ectocarpus two distinct vegetative phases, haploid and diploid, are found. Both the phases are filamentous but the diploid filaments have longer cells than the haploid filaments.

Growth: The prostrate part of Ectocarpus thallus shows apical growth, whereas it is intercalary in the erect part. At the base of the erect filament there are few meristematic cells which divide transversely. The newly formed cells are smaller than those present at the distal end of the filament. In some species the meristematic cells are located in the middle of the filament (e.g., E. elachistaeformis) or at the base of the distal mucilagenous hair (e.g., E. geminifructus). Meristematic cells at these locations cut off segments in both upward and downward directions. The derivatives in upward direction increase the length of the terminal hair, whereas those in downward direction form vegetative photosynthetic cells of the filament. This type of growth, characteristic of Ectocarpales, is known as trichothallic.

Reproduction

Ectocarpus reproduces by asexual and sexual methods.

Asexual Reproduction

Asexual reproduction takes place by bilagellate zoospores. The sporophytic (diploid) plant produces two types of sporangia: (i) unilocular zoosporangia, and (ii) plurilocular or neutral zoosporangia. Both of these types of zoosporangia are borne on the same plant or on different plants. The unilocular sporangia form haploid zoospores and the plurilocular, diploid zoospores.

1. *Unilocular Sporangia:* These sporangia develop singly at the tips of small branchlets. The terminal cell of the branchlet enlarges in size and functions as sporangial initial. The first division of diploid nucleus is meiotic and it is followed by mitotic divisions. This results in the formation of 32 or 64 haploid nuclei. The nuclear divisions are not accompanied by wall formation and as such the sporangium remains unilocular. The protoplast of the sporangium divides into as many segments as the number of nuclei. Each segment metamorphoses into a pyriform, uninucleate zoospore. The zoospore has a chromatophore and an eye spot. The two flagella are laterally inserted and are unequal; the anterior is longer and pantonematic and the posterior is shorter and acronematic.

   Zoospores are discharged between 0600 and 1200 hours. They ooze out of the sporangium in a gelatinous matrix through a small terminal perforation. After their discharge they remain in a spherical mass at the apex of the sporangium for a while and then become free and swim off individually. They seldom remain motile for more than 30 minutes. They withdraw their flagella and attach themselves to the substratum by their anterior ends. They germinate within 2-3 hours and produce haploid filaments.

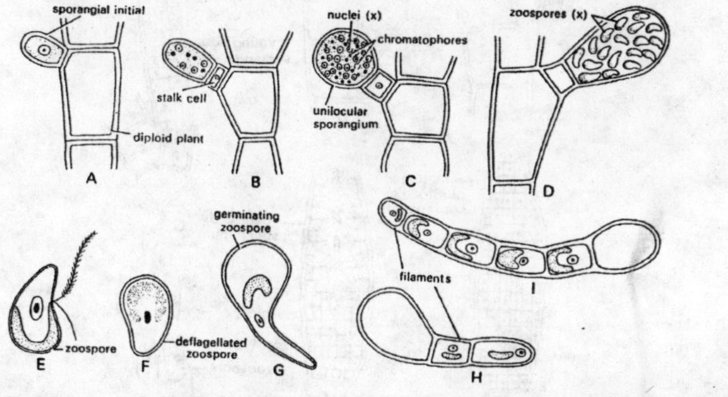

A-I. Ectocarpus: A-D. Successive stages in the development of unilocular sporangium; E. a zoospore; F-I. Stages in the germination of zoospore

Sometimes after the liberation of zoospores new sporangium initiates within the same sporangial wall.

2. *Plurilocular or Neutral Sporangia:* Like unilocular sporangia, these sporangia also develop from terminal cells of the branchlets of diploic thalli. The terminal cell, which acts as sporangia initial, enlarges and becomes spherical and accumulates numerous chromatophores. The sporangial initial now undergoes repeated transverse and vertical divisions, forming as many as 660 cubical cells arranged in definite tiers. This multicellular structure is known as plurilocular sporangium. Each diploid uninucleate cell of the sporangium produces a single biflagellate zoospore. The only difference between the zoospores produced by unilocular and plurilocular sporangia is that the former are haploid and the latter are diploid.

Mature zoospores are liberated from the sporangium by a terminal or lateral pore. They remain motile for 3-5 hours and then germinate to form diploid asexual thalli. At maturity these thalli bear both unilocular and plurilocular sporangia. Thus diploid zoospores help only in the multiplication of the sporophyte. They do not play any part in alternation of generation.

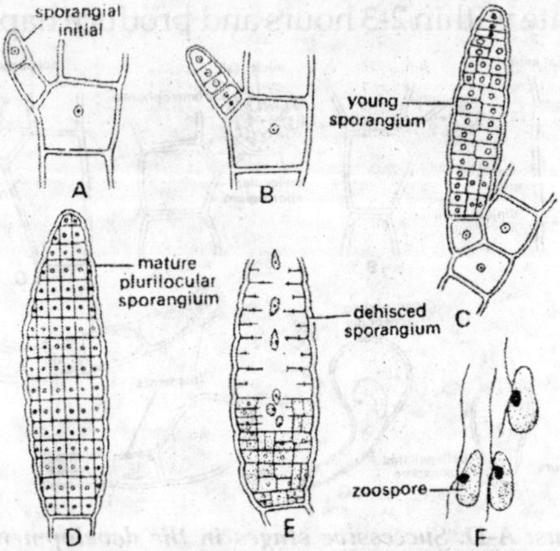

A-F. Ectocarpus: A-C. Development of plurilocular sporangium; d. Mature plurilocular sporangium; e. Liberation of zoospores from sporangium; F. zoospores

The formation of unilocular and plurilocular sporangia by a species is affected by temperature and salinity of water. For example, E. siticulosus produces unilocular sporangia at 13°C, plurilocular at 19°C and both unilocular and plurilocular at 16°C.

Sexual Reproduction

Sexual reproduction may be isogamous, anisogamous or oogamous, but most of the species of Ectocarpus are anisogamous. The anisogamy may be physiological (e.g., E.siliculosus) or morphological (e.g., E.secundus).

1. *Gametangia and Gametes:* Gametes are formed in plurilocular gametangia which are similar to plurilocular sporangia in their structure and development.

   The gametangia may be sessile or shortly stalked. They develop singly from the terminal cells of the lateral branchlets. The gametangial initial divides transversely to form 6-12 cells, arranged in a uniseriate row. Repeated transverse and vertical divisions of these cells produce hundreds of cubical cells which are arranged in 20-40 transverse layers. The uninucleate protoplast of each cell is transformed into a biflagellate pyriform gamete.

In E. secundus, which shows morphological anisogamy, two types of gametangia are produced. The megagametangia, which have larger loci, produce larger gametes and the microgametangia, which have smaller loci, produce smaller gametes. After their discharge from the gametangia, unequal gametes fuse in pairs to form zygospores. In E. siticulosus, which shows physiological anisogamy, gametes from the gametangia developing on filaments of different strains fuse to form zygospores. The fusing gametes are morphologically identical but are distinct physiologically. The gametes of one strain are more active and behave as male gametes, whereas those of the other strain are less active and behave as female gametes.

2. *Fertilization:* In isogamous species the fusing gametes are similar both morphologically and physiologically. In such species two gametes from the same gametangium or from the two different gametangia of the same thallus fuse to form zygospore. But in physiologically anisogameus species, gametes from two filaments of different strains fuse. In such species, the female gamete settles down about 5 minutes after liberation and secretes a volatile sexual attractant, sirenine. The male gametes, which may remain motile for about 8 hours, are attracted towards the female gamete by this substance. A large number of male gametes cluster around a female gamete and attach to it by their long anterior flagella. After sometime one of the male gametes contracts its flagellum and consequently comes in contact of the female gamete. These two gametes fuse to form zygospore. The process of fusion takes about 20 seconds, and after fusion the zygospore loses its attraction for male gametes. As such the remaining male gametes detach themselves from the female gamete

and swim freely. The process in which many male gametes cluster around a female gamete and one male gamete fuses with the female gamete to form zygospore is called clump formation.

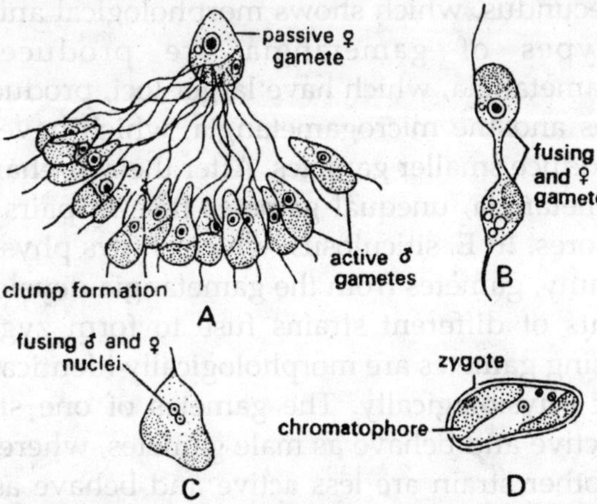

A-D. ectocarpus siliculosus: A. Cluster of male gamete around a female gamete (clump formation); B-C. Fusion of male and female gametes; D. Zygote

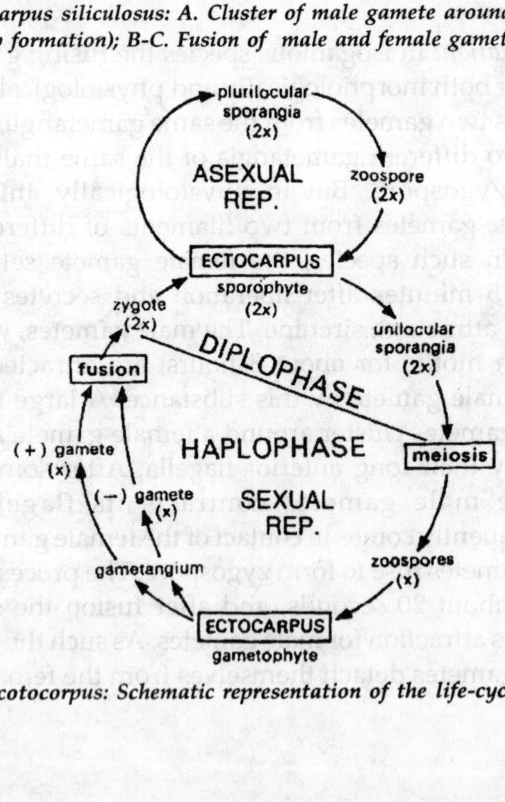

Ecotocorpus: Schematic representation of the life-cycle

3. *Germination of Zygospore:* The zygospore germinates within 2-3 days after fertilization. At the time of germination the diploid zygospore nucleus divides mitotically hence it gives rise to a diploid thallus. This thallus bears both unilocular and plurilocular sporangia.

Some unfused female gametes have the ability to germinate parthenogenetically and they form haploid filaments.

Generation Change

The sexual thalli of Ectocarpus are dioecious and haploid. They produce only plurilocular gametangia. The iso-or anisogametes fuse to form diploid zygote. At the time of germination the diploid nucleus of the zygote divides only mitotically and thus the zygote forms diploid sporophytic thallus.

Such thalli bear both unilocular and plurilocular sporangia. The zoospores formed in plurilocular sporangia are diploid and on germination give rise to diploid sporophytic thalli, whereas haploid zoospores are formed in unilocular sporangia and on germination they form haploid gametophytic thalli.

In Ectocarpus both gametophytic and sporophytic generations are morphologically similar. Such a life-cycle is called isomorphic.

8

Reproduction and Fertilization

The Ceramiales

The order Ceramiales includes about 160 genera and 900 species, mostly marine. The thallus is multiaxial or polysiphonous. The auxiliary cell is cut off from the supporting cell of the carpogonium just after fertilization. Subsequently, one of the diploid nuclei (formed by the mitotic division of the zygote nucleus) passes into the supporting cell through the auxiliary cell and fuses with the nucleus of the supporting cell. After the fusion of nuclei, the supporting cell gives out a number of gonimoblast filaments.

The terminal cells of gonimoblast filaments bear diploid carpospores. The carpospore, on germination, produces a tetrasporophyte, which is similar to the gametophyte and forms tetrahedrally arranged tetraspores. The tetraspores are haploid and on germination give rise to gametophytes. These algae show diplobiontic life-cycle.

The order Ceramiales includes four families, Ceramiaceae, Delesseriaceae, Rhodomelaceae and Dasyaceae. The life-cycle of the genus Polysiphonia (Rhodomelaceae) is described as a representative of the order.

Family Rhodomelaceae

The members of this family are polysiphonous, and the cells of the central axis are uninucleate. The male and female reproductive structures develop on special branches present at the apex of the thallus, called trichoblasts. The trichoblasts, which bear spermatangia, are branched, whereas the female trichoblasts are unbranched.

Polysiphonia

Systematic Position

Class	-	Rhodophyceae
Sub-class	-	Florideae
Order	-	Ceramiales
Family	-	Rhodomelaceae

Occurrence: Polysiphonia is a large genus with some 150 marine species. Most of the species are lithophytic (growing on rocks). Some species are epiphytic (e.g., P. ferulacea, P. urceolata) growing on other algae. P. fastigiata is a semiparasite on the fronds of Ascophyllum nodosum. The genus is represented in India by about 16 species which occur on the Southern and Western coasts. P. ferulacea, P. platycarpa, P. urceolata and P. variegata are common Indian species.

Various Structures

Thallus Structures

The large multiaxial annual or perennial thalli of Polysiphonia form thick tufts or dense bushes of brown or violet-red colour.

Filament Structure

The plant body is heterotrichous, differentiated into a basal prostrate and an erect aerial system. The prostrate system creeps over the substratum. This system is well developed and multiaxial in many species (e.g., P. nigrescens, P. urceolata), and gives out unicellular rhizoids. The distal ends of rhizoids are branched and lobed, which help the prostrate branches to firmly anchor to the substratum. However, in some species (e.g., P. elongata and P. violacea) multiaxial prostrate system is absent and instead the basal cells of erect aerial branches form massive attachment discs.

The prostrate system gives off many multiaxial aerial branches. The main axis and other long aerial branches consist of a central siphon of elongated cylindrical cells, which are surrounded by relatively smaller pericentral cells or pericentral siphon. The cells of the central and pericentral siphons are interconnected through pit connections. In the older portion of main axis pericentral cells become corticated by one or more layers of small cortical cells. The main axis bears two types of lateral branches:

1. *Short Branches or Trichoblasts:* These branches of limited growth arise on the main axis or on long branches in spiral manner. They are uniaxial and lack pericentral siphon. Their cells contain very few chromatophores. In perennial species short branches are shed in the beginning of autumn and develop again in spring season. These branches usually bear male and female reproductive structures, but some short branches are sterile.
2. *Long Lateral Branches:* These branches develop from the basal cells of short branches. Their basal part is polysiphonous but the distal part is usually monosiphonous. They show unlimited growth and do not bear reproductive structures.

Cell Structure

The central and pericentral siphons are made up of elongated cylindrical cells. The cells are uninucleate with many parietal discoid chromatophores which lack pyrenoids. The cell wall is differentiated into an outer pectic and an inner cellulosic layer. The neighbouring cells of the central and pericentral siphons arc interconnected through pit connections. The peripheral cytoplasm contains granules of floridean starch.

Growth: The thallus grows by means of a dome shaped

apical cell. It cuts off regular transverse segments towards the axis which elongate and eventually form the central siphon. Some of these cells (usually subterminal) divide vertically and form pericentral cells.

Various Reproduction Processes

In the life-cycle of Polysiphonia following three types of thalli are found:

1. *Asexual Thallus or Tetrasporophyte:* It develops from diploid carpospore and bears tetrasporangia in which non-motile tetraspores are formed. Tetrasporophytes do not bear sex-organs.
2. *Gametophyte:* It is haploid thallus and bears sex organs. Most species of Polysiphonia are dioecious, i.e., male and female sex-organs are formed on different thalli. The male gametophyte bears spermatangia and the female carpogonia.
3. *Carposporophyte:* It develops by mitotic division of zygote and hence diploid in nature. The carposporophyte is dependent upon female gametophyte and bears diploid carpospores.

Polysiphonia reproduces by asexual and sexual methods.

Asexual Reproduction

Asexual reproduction takes place by means of non-motile haploid tetraspores which are formed in tetrasporangia, present on diploid thalli. Tetrasporangia are small spherical stalked structures and develop from pericentral cells. The fertile pericentral cell, which behaves as tetrasporangial initial, is smaller than other pericentral cells of the same layer. A vertical wall divides this initial into an outer cover cell and an inner sporangial mother cell. The cover cell, by further divisions, forms two or more cover cells.

The sporangial mother cell divides by a transverse wall into a lower stalk cell and an upper sporangial cell. The latter undergoes enormous enlargement and functions as tetrasporangium. The diploid nucleus of the sporangium divides meiotically forming four haploid nuclei, followed by the division of the protoplast into four haploid uninucleate segments. Each segment develops into a haploid spore, and the four spores are arranged tetrahedrally.

Tetraspores are liberated by splitting of the sporangial wall, accompanied by lifting away of the cover cells. Out of the four spores in a sporangium, two produce male gametophytes and two female gametophytes.

Sexual Reproduction

The sexual reproduction in Polysiphonia is oogamous. The plants are dioecious. The male and female reproductive structures develop on the male and female gametophytes, produced by tetraspores.

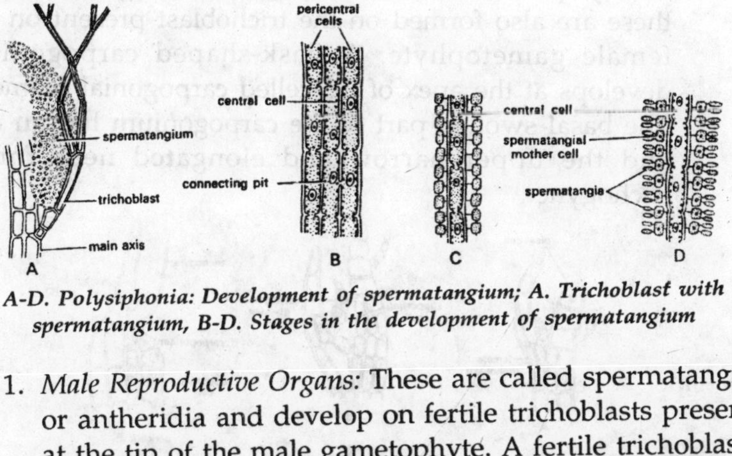

A-D. Polysiphonia: Development of spermatangium; A. Trichoblast with spermatangium, B-D. Stages in the development of spermatangium

1. *Male Reproductive Organs:* These are called spermatangia or antheridia and develop on fertile trichoblasts present at the tip of the male gametophyte. A fertile trichoblast, when only 2-3-celled, divides dichotomously. In some species both branches of a dichotomy are fertile, whereas in others only one is fertile and the other remains sterile. The latter may divide again forming fertile trichoblasts.

   During the development of spermatangium all the cells of the uniaxial fertile trichoblast, except for few basal ones, divide periclinally forming pericentral cells. The pericentral cells divide again and each forms one or two spermatangial mother cells on their outer faces. These cells arrange themselves in a compact layer on the surface of the trichoblast. The spermatangial mother cells in turn divide to produce varying number of spermatangia. Thus, ultimately a compact cone-shaped cluster of spermatangia is formed.

The spermatangium is an oblong or globose unicellular structure. Its cell wall is differentiated into an outer thick, a middle gelatinous and an inner refractive layer. The uninucleate protoplast of the spermatangium forms a single spermatium.

The spermatium is liberated through a narrow slit present at the tip of the spermatangium.

2. *Female Reproductive Organs:* Female reproductive organs of Polysiphonia are called carpogonia. Like spermatangia, these are also formed on the trichoblast present on the female gametophyte. A flask-shaped carpogonium develops at the apex of 2-5-celled carpogonial filament. The basal swollen part of the carpogonium has an egg and the upper narrow and elongated neck forms trichogyne.

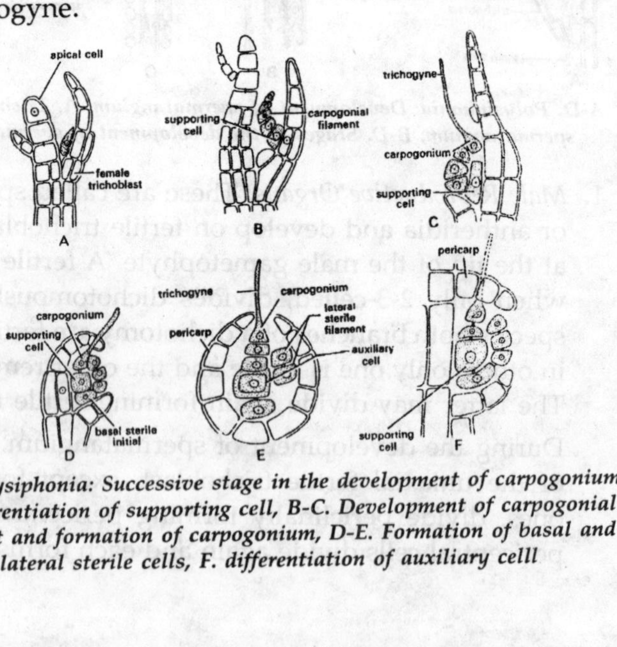

A-F. Polysiphonia: Successive stage in the development of carpogonium; a. Differentiation of supporting cell, B-C. Development of carpogonial filament and formation of carpogonium, D-E. Formation of basal and lateral sterile cells, F. differentiation of auxiliary celll

A young female trichoblast is uniaxial or unisiphonous, but later becomes multiaxial or polysiphonous by subsequent periclinal divisions. One' of the pericentral cells of the adaxial surface of the fertile female trichoblast functions as supporting cell. This cell usually lies 3-5 cells below the apex of the trichoblast. The supporting cell divides transversely to form a 4-5-celled carpogonial filament. The terminal cell of this filament functions as carpogonial mother cell and eventually forms a carpogonium with a relatively long trichogyne.

Meanwhile, the supporting cell cuts off a cell on the basal side which acts as basal sterile initial and a cell on the lateral side which functions as lateral sterile initial. After fertilization, these initials form basal and lateral sterile filaments respectively.

Fertilization

Male gametes (spermatia) of Polysiphonia are non-motile. After their release from spermatangia, they reach the carpogonium with the help of water currents and get attached to the long receptive trichogyne. This is followed by dissolution of the spermatial and trichogynal walls at the point of their contact. The contents of spermatium pass down the trichogyne through this opening and ultimately fuses with the egg nucleus in the swollen base of the carpogonium. Thus zygote is formed.

Post-fertilization Changes: After fertilization a number of changes occur within and around the female reproductive structures as a result of which carposporophyte is formed. Soon after fertilization the basal and lateral sterile initials divide to form basal and lateral sterile filaments. The basal sterile filament is shorter (2-4-celled) than the lateral sterile filament (4-10-celled). The sterile filaments are of nutritive nature.

Meanwhile, the supporting cell again divides transversely to form an auxillary cell. It lies at the base of the carpogonium and has a single haploid nucleus. Soon, a tubular connection is established between the auxiliary cell and carpogonium.

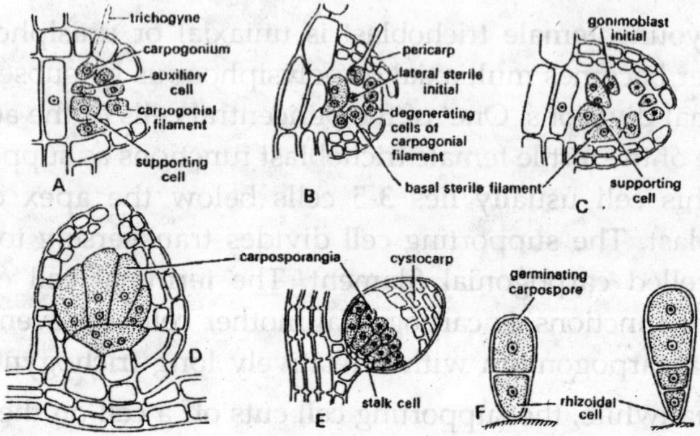

A-G. Polysiphonia: Post-fertilization changes; A. Fertilized carpogonium (note the position of auxiliary cell), B. Division in diploid zygote of carpogonium, C. Development of gonimoblast initial, D. carpospore phyte with young carposporangia, E. Mature carposporophyte (cystocarp), F-G. Germination of carpospore

The diploid zygote nucleus divides mitotically into two daughter nuclei, one of which remains within the carpogonium and the other passes into the auxiliary cell through the tubular connection. By this time the trichogyne shrivels and the auxiliary cell and the carpogonium fuse with the supporting cell and sterile filaments. This results in the formation of an irregularly-shaped large placental cell. The diploid nucleus of the auxiliary cell divides mitotically forming several diploid nuclei in the placental cell. At this stage a number of tubular outgrowths arise from the placental cell and a diploid nucleus moves into each outgrowth. These outgrowths are cut off from the placental cell and function as gonimoblast initials. The gonimoblast initial divides repeatedly and forms a small gonimoblast filament.

The terminal cell of the gonimoblast filament develops into a carposporangium, which bears a single diploid carpospore. The gonimoblast filaments along with carposporangia get enclosed within a sheath of vegetative filaments and form a large urn-shaped body, called cystocarp or carposporophyte. The sheath surrounding the cystocarp is called pericarp. Carpospores are liberated through a pore (ostiole) at the distal end of the cystocarp. The carpospore, on germination, forms a diploid asexual thallus, called tetrasporophyte. Morphologically, it resembles with the gametophytic thallus.

Alternation of Generation: Polysiphonia is the chief example of triphasic alternation of generation and its life-cycle consists of three distinct stages which are similar in morphological characters.

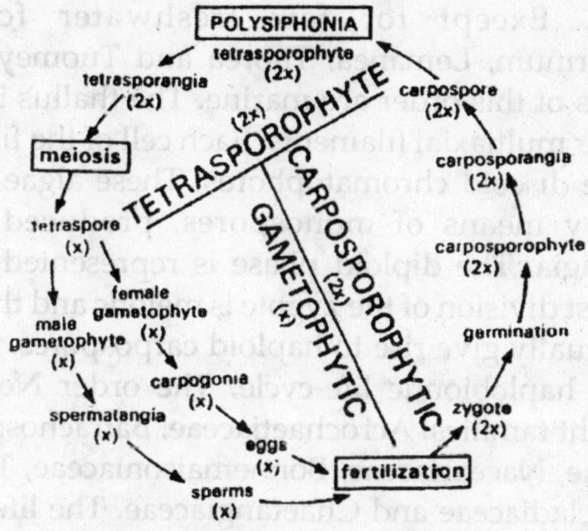

Polysiphonia: Life-cycle (diagrammatic)

(1) Asexual tetrasporophytic stage.
(2) Gametophytic stage.
(3) Carposporophytic stage.

Of these three stages, asexual tetrasporophytic and carposporophytic stages are diploid and gametophytic stage is haploid. In tetrasporophyte, the tetrasporangium forms four haploid tetraspores after reduction division. Two of these haploid spores give rise to male gametophyte and the remaining two to female gametophyte; these bear male (spermatangia) and female (carpogonia) reproductive structures respectively. A diploid zygote formed by the fusion of male and female gametes develops into an urn-shaped fruiting body, known as cystocarp or carposporophyte, through a complex process.

Carpospores formed in the carposporophyte are diploid and give rise to diploid tetrasporophyte on germination. The tetrasporophyte bears only tetraspores. Thus, in the life-cycle of Polysiphonia two diploid phases (tetrasporophyte and carposporophyte) alternate with one haploid (gametophyte) phase. This type of life-cycle is known as triphasic diplobiontic.

Nemalionales

The order Nemalionales includes 250 species, belonging to 35 genera. Except for few freshwater forms like Batrachospermum, Lemanea, Thorea and Tuomeya, most of the members of this order are marine. The thallus is made up of uniaxial or multiaxial filaments. Each cell of the filament has one or more discoid chromatophores. These algae reproduce asexually by means of monospores, produced singly in monosporangia. The diploid phase is represented by zygote only. The first division of the zygote is meiotic and the resultant nuclei eventually give rise to haploid carpospores. Thus these algae show haplobiontic life-cycle. The order Nemalionales includes eight families: Acrochaetiaceae, Batrachospermaceae, Lemaneaceae, Naceariaceae, Bonnemaisoniaceae, Thoreaceae, Nelmintho-cladiaceae and Chaetangiaceae. The life-history of Batrachospermum is described in detail as a representative of the order.

Family Batrachospermaceae: The family is represented by five freshwater genera which grow in ponds, lakes and slow moving streams. The thallus is uniaxial with many lateral branches, forming whorls. Some members of the family have an alternate dissimilar vegetative phase (juvenile stage) which reproduces only by monospores. This phase gives rise to adult stage. These algae have no sporophytic phase; the diploid phase is represented only by zygote.

Batrachospermum

Systematic Position

Class	-	Rhodophyceae
Sub-class	-	Florideae
Order	-	Nemalionales
Family	-	Batrachospermaceae

Occurrence: Batrachospennum is a freshwater red alga which grows in slow-moving water of streams, lakes and ponds in the tropical and temperate regions. Some of its species are found attached to stones in swift-flowing acidic water of rivers. Most of the species of Batrachospermum are annual but B. vagum is a perennial species. The genus is represented in India by five species, of these B. moniliforme acid B. vagum are most common.

Thallus Structure

The mature thallus is branched, soft and gelatinous, 15-20 cm long and blue-green, violet or red in colour. The colour of the thallus depends on the intensity of light available. Usually thalli growing in deep and shady ponds and lakes are violet or dark red, whereas those growing in shallow and illuminated ponds are dark green in colour.

The thallus is differentiated into a prostrate and an erect system. The prostrate branches anchor the thallus to the substratum, whereas the branches of the erect system float freely on water.

The thallus is uniaxial and the main or primary axis is made up of a uniseriate row of large cylindrical cells. The axis is differentiated into nodes and internodes and grows by a dome-shaped apical cell. From the nodes of the main axis two types of lateral branches develop: (i) Branches of limited growth - they arise in whorls just below the septa of the axial filament or from the branches of unlimited growth. The whorl of branches of limited growth present at a node is known as glomerule. All branches in a glomerule are of nearly the same length and they are made up of small moniliform cells which give beaded appearance to the thallus. They are further branched or remain unbranched. (ii) Branches of unlimited growth- they develop from nodal cells of the main axis or basal cells of the branches of limited growth. They are also differentiated into nodes and internodes and are enveloped by corticating filaments.

Cortication: The cells of the nodal region in the basal mature part of the axis and the basal cells of the branches of limited growth give rise to some elongated filamentous cells, known as cortical threads. They grow downwards to the next node, forming an envelope around the internode, and this envelope is called corticating layer.

Cell Structure

The cells of the thallus are uninucleate with several discoid chromatophores along the periphery or a diffuse chromatophore. The chromatophore is surrounded by a double layered wall and contains a single pyrenoid. The main pigments are chlorophyll *a*, chlorophyll *d*, *r*-phycoerythrin and *r*-phycocyanin, but due to the excess of red pigment-*r*-phycoerythrin the green colour of chlorophyll is completely or partially masked. Golgi bodies and mitochondria are also present in the cell. The reserve food is in the form of floridean starch.

The cells of the axial filament are interconnected through pit connections.

Growth: The main axis as well as the branches of unlimited growth grow by means of a hemispherical apical cell. The apical cell, by repeated transverse divisions, gives rise to a series of cells towards the posterior side. These derivatives are arranged in a uniseriate row. From the upper end of each derivative (close to the septum), 4-6 protuberances or outgrowths are given out. These outgrowths separate from the parent cell by the formation of septa at their bases and are now called pericentral cells.

The pericentral cells function as initials of the branches of limited growth. After the formation of the branches of limited growth, the pericentral cells act as their basal cells. The branches of unlimited growth may develop from any of these basal cells. The basal cells also give rise to corticating threads. In the older part of the axis, the basal cells separate from each other and give rise to the node. Meanwhile the cells of the central axis elongate and form the internode.

Reproduction

Asexual reproduction takes place by means of monospores. Sexual reproduction is of advanced oogamous type. Species of Batrachospermum are either monoecious or dioecious. The male reproductive structures, are called spermatangia or antheridia and the female carpogonia.

Spermatangia

Spermatangia, the male reproductive organs, are unicellular, uninucleate, spherical or globose and colourless structure. They develop singly or in pairs at the distal ends of the branches of limited growth. They are surrounded by a wall which is differentiated into an outer thick, a middle gelatinous and an inner thin layer. Each spermatangium bears a single spermatium.

Development of spermatangium. Spermatangia develop on the branches of limited growth. The terminal or subterminal cell of these branches functions as spermatangium mother cell. Each spermatangium mother cell usually gives rise to one or two spermatangia, but sometimes as many as five spermatangia may develop from a single cell. The spermatangium develops as an outgrowth from the spermatangium mother cell.

The nucleus of the mother cell divides mitotically to form two daughter nuclei, one of these enters into the developing spermatangium and the other remains in the mother cell. At this stage, the spermatangium is separated from the mother cell by a septum. The protoplast of the spermatangium metamorphoses into a single spermatium or male gamete. The spermatangium mother cell may form another spermatangium by repeating the same process.

Carpogonia

Carpogonium, the female reproductive organ, is a flask shaped structure, differentiated into a basal swollen egg cell and a long narrow neck, called trichogyne. The protoplast of the egg cell forms a single uninucleate egg, and the trichogyne receives the male gamete.

Development of carpogonium. Carpogonia develop on special lateral branches, known as carpogonial branches. These branches are only 2-5-celled and develop by repeated transverse divisions of basal cells of the branches of limited growth. The terminal cell of the carpogonial branch forms a flask shaped carpogonium. The subterminal cell, situated just below the carpogonium, is called hypogynous cell or supporting cell.

Fertilization

The spermatia, after their liberation from spermatangia, are carried passively by water currents to the trichogyne of the carpogonium. When a spermatium comes in contact with the trichogyne, the wall between the trichogyne and the spermatium dissolves at the point of their contact. The male nucleus moves into the carpogonium. The male and female nuclei fuse in the basal portion of the carpogonium and form a diploid zygote. After fertilization, the trichogyne shrivels.

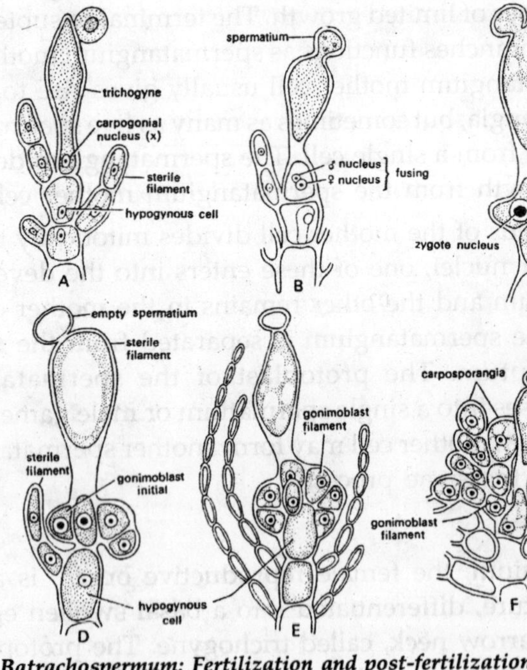

A-F. Batrachospermum: Fertilization and post-fertilization changes; A-B. Fertilisation, C. Zygote, D-F. Post-Fertilization changes

Post-fertilization Changes: The first division of the zygote nucleus is meiotic, resulting into four haploid nuclei. These nuclei divide repeatedly and form many daughter nuclei. At this stage, many outgrowths arise from the basal part of the carpogonium and a haploid daughter nucleus passes into each outgrowth.

These outgrowths with haploid nuclei are called gonimoblast initials. Repeated transverse divisions of gonimoblast initials give rise to a number of small, unbranched or branched gonimoblast filaments.

The terminal cell of these filaments functions as carposporangium. A single carpospore is formed in each carposporangium. The lower sterile cells of the gonimoblast filaments give rise to many sterile filaments which form an envelope around the carpospores and gonimoblast filaments. This composite structure (carpospores and gonimoblast filaments along with sterile filaments) is known as cystocarp or carposporophyte. Carpospores are liberated by the rupture of the wall of the mature carposporangium.

Germination of Carpospores: Carpospores get themselves attached to the substratum and on germination each carpospore forms a protonema-like structure which eventually gives rise to a heterotrichous system. This represents the juvenile stage of Batrachospermum, known as 'Chantransia stage'.

Previously Chantransia was supposed to be an independent algal genus but now it has been established that all species of Chantransia represent a phase in the life cycle of Batrachospermum. The 'Chantransia' phase can reproduce itself.

Asexual Reproduction

Asexual reproduction takes place by means of uninucleate, haploid and non-motile monospores. These spores are formed in the 'Chantransia' stage, and on germination again give rise to 'Chantransia' stage. The erect branches of this heterotrichous juvenile stage give rise to adult Batrachospermum plants.

Alternation of Generation

The thallus of Batrachospermum is haploid and gametophytic. The male (spermatangia) and female (carpogonia) reproductive structures develop on branches of limited growth.

The haploid egg and spermatium fuse to form a diploid zygote. The zygote undergoes reduction division and the daughter nuclei eventually pass into gonimoblast filaments, arising from the base of the carpogonium. The terminal cell of each of these filaments develops into a carposporangium. The protoplast of the carposporangium forms a single carpospore.

Gonimoblast filaments together with carposporangia are enclosed within an envelope of sterile filaments (arising from the lower sterile cells of gonimoblast filaments). This composite ball like structure is called carposporophyte or cystocarp. It is dependent on the haploid gametophyte.

On germination the carpospore forms a branched heterotrichous filamentous structure, known as 'Chantransia stage'. This stage represents the juvenile phase in the life-cycle of Batrachospermum.

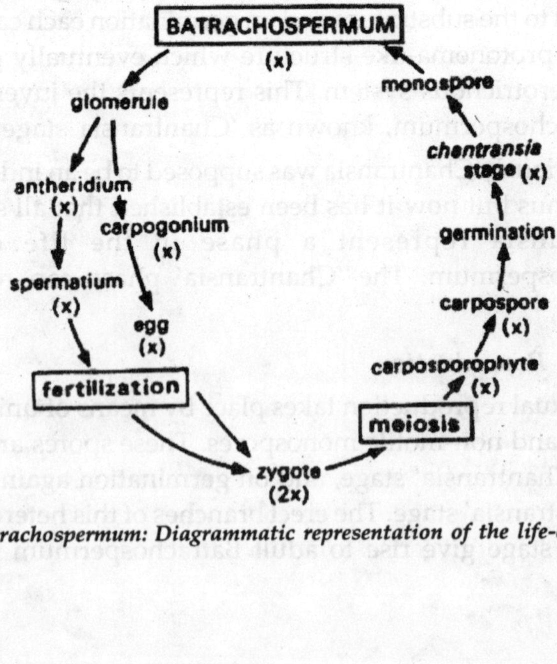

Batrachospermum: Diagrammatic representation of the life-cycle

The branched filaments of 'Chantransia' produce monospores which are the means of asexual reproduction. After repeated multiplication by asexual means, the monospore ultimately gives rise to haploid gametophyte of Batrachospermum.

Thus, in the life-cycle of Batrachospermum, two haploid phases, gametophyte and carposporophyte, alternate with a diploid phase. Between these two haploid phases, a short-lived diploid phase is represented by zygote. Such a life-cycle with two haploid phases is called triphasic haplobiontic life-cycle.

9

Cyclosporeae and Pennales

The Phaeophyceae

Class Phaeophyceae is represented by about 240 genera and 15000 species. They are characterised by the presence of a brown coloured carotenoid, fucoxanthin in large amounts, which masks the green colour of chlorophyll pigment. This gives a characteristic brown colour to these plants, hence the name brown algae. Except few fresh water forms like species of Pleurocladia, Heribaudiella, and Bodanella most of the brown algae are marine. They grow in intertidal belt and upper littoral region. They dominate colder waters but members of Ectocarpales, Dictyotales and species of Sargassum grow in warm waters.

In the tropics members of Phaeophyceae are most abundant in the Sargasso sea of the Atlantic. They are usually found attached to rocks with a discoid or branched holdfast. Many of these algae have air bladders which increase their buoyancy. In India brown algae are commonly found on the western and southern coasts.

The important characters of the class Phaeophyceae are as follows:

(1) The brown algae range in structure from simple to complex

parenchymatous forms. They include: (i) Heterotrichous forms, e.g., Ectocarpus; (ii) Multiaxial construction forms, e.g., Myriactula pulvinata; (iii) Uniaxial pseudo-parenchymatous forms, e.g., Hecatonema sargassicola; and (iv) Parenenymatous forms, e.g., Laminaria, Fucus.

Besides, Postelsia palmaeformis assumes the form of a miniature tree and is commonly known as sea palm.

(2) The plant body in the members of Laminariales and Fucales is differentiated into holdfast, stem-like stipe and flattened leaf- like blades. Many species possess air bladders which keep them afloat.

(3) The tissues of the stipe and blade are differentiated into epidermis, cortex and central medulla.

(4) Growth of the thallus is apical (Sphacelariales, Dictyotales, Fucales), intercalary (Laminariales) or trichothallic (Ectocarpales, Sporochnales).

(5) Phaeophycean cell wall is differentiated into two layers. The outer mucilagenous layer has alginic and fucinic acid, whereas the inner layer is mainly cellulosic. Alginic acid is of commercial importance and is obtained from Laminaria, Sargassum and Ascophyllum. It is used in the manufacture of adhesives and artificial silk.

(6) The cytoplasm contains many small vacuoles and refractive bodies. The latter are called fucosan vesicles. They are abundant in the cytoplasm of metabolically active cells.

(7) Chromatophores are ellipsoidal and discoid. They contain chlorophyll *a*, chlorophyll *c*, and *b*-carotenes and xanthophyll (e.g., lutein, flavoxanthin, violaxanthin). Besides, they also contain large amount of a brown pigment, fucoxanthin ($C_{40}H_{56}O_6$). Fucoxanthin masks green colour of chlorophyll pigments and that is why these algae are brown in colour.

(8) Chromatophores have three or four thylakoids per band and are surrounded by a chromatophore envelope and

two membranes of endoplasmic reticulum. The chromatophore has a single, stalked, projected and naked pyrenoid.

(9) The long-term storage product is laminarin. However, the accumulation product of photosynthesis is D-mannitol, a sugar alcohol.

(10) The cells are uninucleate with one or more nucleolus. In some brown algae (e.g., Fucus) the resting nuclei show chromocentres which are not found in other groups of algae.

(11) The flagellated structures have a pair of laterally inserted unequal flagella, of which the larger one is anterior and pantonematic, whereas the smaller one is posterior and acronematic.

(12) They reproduce by vegetative, asexual and sexual methods.

(13) Fragmentation is the most common method of vegetative multiplication. Besides, in the members of Sphacelariales special reproductive branches are found which are called propagules. These branches, when detached from parent plant, grow into new plants.

(14) Except for Tilopteridales, Dictyotales and Fucales, all brown algae reproduce asexually by zoospores.

(15) In Ectocarpales and Sphacelariales zoospores are formed in unilocular or plurilocular sporangia.

(16) The zoospores produced in unilocular sporangia are haploid and those in plurilocular sporangia are diploid. Both the haploid and diploid spores are similar in structure. They are pyriform, biflagellate and have chromatophores, contractile vacuoles and eye spot.

(17) In some members of Tilopteridales the protoplast of the sporangium is transformed into a large quadrinucleate spore, called monospore.

(18) Phaeophycean algae show much variation in their sexual reproduction. It is isogamous in Ectocarpales and

Sphacelariales, anisogamous in Cutleriales and Tilopteridales and oogamous in Fucales, Laminariales and Dictyotales.

(19) Gametes are more or less similar to zoospores and are also biflagellate.

(20) In most brown algae fertilization is external, i.e., gametes fuse outside the gametangium in water. There is no reduction division in zygote; it forms a diploid thallus on germination.

(21) The life cycle of brown algae may be isomorphic, heteromorphic, or diplontic.

Classification

Fritsch (1935), on the basis of vegetative structure and sexual reproduction, divided the class Phaeophyceae into nine orders: (1) Ectocarpales, (2) Tilopteridales, (3) Cutleriales, (4) Sporochnales, (5) Desmarestiales, (6) Laminariales, (7) Sphacelariales, (8) Dictyotales, and (9) Fucales.

Smith (1955) recognized the following three classes and 12 orders in the division Phaeophyta:

Isogeneratae

Members of this class are characterised by alternation of two isomorphic generations. It has five orders differing in vegetative structure, mode of growth and structure of reproductive organs: (1) Ectocarpales, (2) Sphacelariales, (3) Tilopteridales, (4) Cutleriales, (5) Dictyotales.

Heterogeneratae

Those brown algae which show alternation of heteromorphic (i.e., morphologically distinct) generations are included in this class. It has six orders: (1) Chordariales, (2) Sporochnales, (3) Desmarestiales, (4) Punctariales, (5) Dictyo siphonales, (6) Laminariales.

Cyclosporeae

The brown algae included in this class have a life cycle in

which, there is no alternation of free-living multicellular generations. This class includes only a single order, Fucales.

Pennales

The members of the order Pennales are characterised by the presence of a bilateral symmetry. A pseudoraphe or a raphe is usually present in the median line of the valve. They show gliding movement. Sexual reproduction, if present, is of amoeboid isogamous type. Auxospores are normally formed by conjugation. They occur mostly in fresh water habitats. The order is classified according to the presence or absence and the number and morphology of the raphe on the valves.

Family Naviculoideae: In the members of this family the raphe is present on both the valves. It occupies the apical axis and is usually without a keel.

Navicula

Systematic Position

Class	-	Bacillariophyceae
Order	-	Pennales
Family	-	Naviculoideae

Occurrence: Navicula, a fresh water diatom, is cosmopolitan and ubiquitous in distribution, found in ponds, lakes, etc. However, a few species occur in terrestrial habitats. The species of Navicula are very important as primary producers in the food web of aquatic ecosystems. The most common species of the genus occurring in India is N. halophila.

Various Structures

The plant body is unicellular with an isobilateral symmetry. The cells usually occur in isolated form. The cell wall (frustule) consists of two overlapping halves, the upper half is said to be epitheca and the lower half hypotheca. Each theca is further divided into two parts: the main surface called valve and the incurved margins, known as connecting band. The two connecting bands of the two thecas are fitted together. The connecting band of the epitheca overlaps that of the hypotheca and the two bands remain united in the overlapping region (called girdle) by a cementing organic substance present between them. A cell can be seen from two different aspects, the valve view and the girdle view. In valve view the cell appears boat-shaped and in girdle view more or less rectangular. The cell is several times longer than its breadth.

The cell wall is made up of crystalline silica and an organic component of unknown nature. In N. pelliculosa, the wall consists of a silica layer with an outer coating of organic material called organic skin. There is a longitudinal slit known as raphe. The raphe is interrupted in the middle by a central nodule formed by the internal thickening of the valve. Each pole of the valve contains a polar nodule, formed in the same manner as the central nodule. The valve markings consist of fine pores bridged from the inside by sieve membranes also containing numerous fine pores. The raphe situated in the central rib is called apical rib. The rows of pores situated on the lateral sides of the apical rib and separated one from the other by solid silica ribs are known as trans-apical ribs. The pores within a row are separated by siliceous connections which link the adjacent trans-apical ribs.

The protoplast is surrounded by a thin plasma membrane, which lies close to the inner face of the cell wall. There are two brownish yellow chromatophores, adpressed to the girdle side in the resting cell. The nucleus is situated in the centre of a dense bridge of cytoplasm which connects the two chromatophores. The cytoplasm contains mitochondria, golgi bodies and endoplasmic reticulum. The reserve food material is in the form of fatty oils volutin globules.

Reproduction

Navicula reproduces vegetatively and by the formation of auxospores.

Vegetative Reproduction: There is rapid multiplication by successive divisions, which usually occur in the evening. First, one nucleus divides into two and then cytoplasm by an ingrowing cleavage furrow.

Prior to cell division there is a slight increase in the size of the nucleus. The cell also increases in volume, the valves loosen a little to accommodate the increase in the cell volume. The nucleus, which is present in the centre of the cell, moves toward the girdle side, and in the process it loses its spherical outline and becomes somewhat compressed. Now the nucleus divides mitotically in a plane perpendicular to the valves. It is followed by the division of the protoplast in two by invagination of the plasmalemma. In girdle view cytokinesis is seen commencing as a small cleavage which progresses centripetally towards the centre of the cell. After the completion of cytokinesis each daughter protoplast secretes a new valve on its inner side.

During the nuclear division chromatophores move away from the girdle and are adpressed to valves, as such with the completion of cytoplasmic cleavage each daughter cell receives one chromatophore. The single chromatophore divides transversely to form two chromatophores which take their usual position on the girdle side.

The two valves of the parent cell (each with a daughter protoplast) become epithecas of the daughter cells. But they produce new hypotheca.

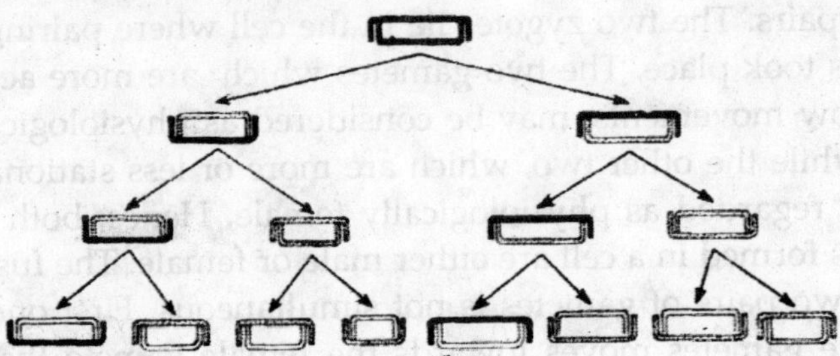

***Navicula*: Diagrammatic representation of the vegetative reproduction (cell division)**

The daughter cell which retains the epitheca of the parent

cell, attains nearly the same size as the parent cell. But the one which receives the hypotheca of the parent cell (now epitheca of the daughter cell) is smaller than the parent cell. Thus, there are two lines of descendants in cell division, one line retains the original size, whereas in the other there is diminution in size in successive generations.

Auxospore Formation: The diatom, after reaching a certain minimum size due to repeated cell division, begins to form auxospores. Auxospores are in fact special rejuvenescent cells, formed as a compensatory measure for the vegetatively reproducing diatom in regaining its normal size. The auxospores are usually the result of sexual process in Navicula. Two cells come very close to each other, adhere by their valve surfaces or occasionally by their girdle faces and then get enveloped within a common mucilagenous mass. The diploid nucleus of each cell undergoes reduction division and forms four haploid nuclei. Of these, two degenerate and two remain functional. The functional nuclei ultimately metamorphose into two amoeboid gametes. The two gametes show some amoeboid movements inside the valves and then round up. The valves of the mother cell are pushed apart at this stage and the two gametes of each pairing cell become free within their respective valves. A single chromatophore is present in each gamete.

The two gametes formed in one cell move towards the gametes of the other cell, and when they come in contact they fuse in pairs. The two zygotes lie in the cell where pairing of gametes took place. The two gametes which are more active and show movements, may be considered as physiologically male, while the other two, which are more or less stationary, may be regarded as physiologically female. Hence, both the gametes formed in a cell are either male or female. The fusion of the two pairs of gametes is not simultaneous. First one of the male gametes moves towards the female gamete that is directly opposite to it and fuses with it and then the second male gamete follows that and fuses with the second female gamete.

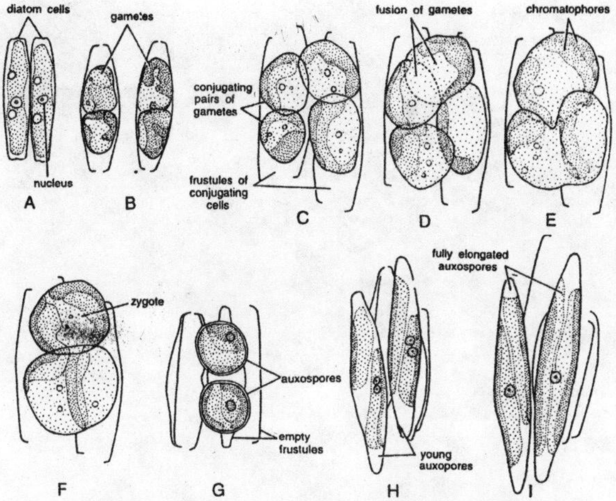

A-I. Navicula: A-G. Conjugation of gametes and formation of the auxospore, H-I. A stage in the enlargement of auxospore, I. fully elongated auxospore (After Subrahmanyam, 1946)

Gamete formation invariably takes place late in the evening and actual fusion occurs soon after night-fall. The zygote is large and irregular to begin with, but later it contracts and becomes spherical. It secretes a wall round itself and is now known as auxospore. The whole process of auxospore formation, from the commencement of pairing of cells to the formation of new cells, takes about 24 hours.

Germination of Zygote: The zygote germinates after a resting period of few hours. In the germinating zygote, the two chromatophores and the two gametic nuclei from two pairing gametes are still lying distinctly without fusing. Then, the protoplast of the zygote elongates considerably and the fusion of the two gametic nuclei occurs.

After the fusion of the gametic nuclei, new valves are secreted. The new cells always lie parallel to the old mother cell valves and are twice as long as the mother cell valves. The new cells after sometime began to divide vegetatively and this marks the beginning of a new generation.

The four discarded valves of the two mother cells are often seen attached to the newly formed frustules.

Fig. ... Navicula. A-G. Conjugation of gametes and formation of the auxospore. H-I. A stage in the enlargement of auxospore. I. fully elongated auxospore (After Subrahmanyan, 1946).

Gamete formation invariably takes place late in the evening and actual fusion occurs soon after night-fall. The zygote is large and irregular to begin with, but later it contracts and becomes spherical. It secretes a wall round itself and is now known as auxospore. The whole process of auxospore formation, from the commencement of pairing of cells to the formation of new cells, takes about 24 hours.

Germination of Zygote. The zygote germinates after a resting period of few hours. In the germinating zygote, the two chromatophores and the two gametic nuclei from two pairing gametes are still lying distinctly without fusing. Then, the protoplast of the zygote elongates considerably and the fusion of the two gametic nuclei occurs.

After the fusion of the gametic nuclei, new valves are secreted. The new cells always lie parallel to the old mother cell valves and are twice as long as the mother cell valves. The new cells after sometime began to divide vegetatively and thus marks the beginning of a new generation.

The four discarded valves of the two mother cells are often seen attached to the newly formed frustules.

10

Chara and Oospore

The Charales

The systematic position of the order Charales is debated and it has been treated differently by various workers. Fritsch (1935) included it in the class Chlorophyceae, Smith (1938) gave it the rank of a class, Charophyceae, and Bold (1956) and McLean and Cook (1956) included it in a separate phylum Charophyta, intermediate between algae and bryophyta. The treatment of Fritsch is followed here.

The order includes 6 genera and 300 species, mostly found in shallow freshwater lakes with sandy and muddy bottoms. Some species also grow in brackish water. The thallus is well organized and is differentiated into rhizoids and an erect branched axis. The axis is divided into nodes and internodes. At the nodes of the main axis whorls of branches of unlimited growth are present, whereas the nodes of lateral branches bear branches of limited growth. Such branched axis gives it the appearance of an angiospermic plant.

Sexual reproduction is advanced oogamous. Both male and female sex organs, known as globules and nucules respectively, are macroscopic. These structures are contained within a sterile envelope, similar to that of bryophytes. Asexual spores are, however, absent.

The order includes only one family, Characeae, represented in India by five genera (Chara, Nitella, Lychnothamnus, Nitellopsis and Tolypella) and 65 species. The life-history of Chara is described here as the representative form.

Chara Structure

Systematic Position

Class	-	Chlorophyceae
Order	-	Charales
Family	-	Characeae

Occurrence: Chara has about 180 species, most of the species are freshwater, found submerged in shallow water ponds, tanks or lakes. A few species grow in alkaline water (e.g., C. baltica) and hot springs (e.g., C fragilis). Species of Chara growing in heavy water become encrusted with calcium and magnesium carbonate and hence the common name stoneworts is given to it. These deposits make their body rough and brittle.

In India, the genus is represented by about 30 species, of these, C. braunii, C. gracilis, C gymnoptis and C. zeylanica are common.

Thallus: The plant body of Chara is multicellular and macroscopic. It is 20-30 cm in height and is differentiated into rhizoids and main axis.

The rhizoids are white, threadlike, multicellular, uniseriate and branched structures. They arise from the base of the plant or from the peripheral cells of the lower nodes of the main axis. They are characterised by the presence of oblique septa. The ends of adjacent cells of rhizoids are protruded in opposite directions at the septum so as to form a kind of joint. At this joint, protrusion from the upper cell cuts off a segment which divides to form a quadrant. It is known as rhizoidal plate, and it gives rise to a tuft of branched rhizoids. The rhizoids, besides serving as organ of attachment, also help in the absorption of minerals.

The main axis is epigeal, long, branched and is differentiated into nodes and internodes. The internode is made up of two types of cells. Its central part consists of a single elongated cell, called axial cell or internodal cell. The axial cell is ensheathed or corticated by a layer of vertically elongated cells of much smaller diameter. These cells, originating at the, nodes, are called cortical cells (Fig. D). Half of the cortical cells originating at a node grow upward (ascending filaments) and half downward (descending filaments).

Thus, the lower half of an internode is covered by ascending filaments and the upper half by descending filaments (Fig. A). But, in some species (e.g., C. corallina), the cortical cells do not ensheathe the internodes.

The node consists of a central pair of cells, surrounded by 6-20 peripheral cells (Fig. E). The following three types of appendages arise from a node:

1. *Branches of Limited Growth:* They arise in whorls of 6-16 at the nodes of the main axis and branches of unlimited growth. They cease to grow after forming 5-15 nodes and internodes, hence called branches of limited growth. They are also called primary laterals, 'leaves' or branchlets. The internodes of these branches are shorter than those of the main axis. At the nodes of these branches are present unicellular filamentous secondary laterals or stipulodes. Reproductive structures are also borne at the nodes of these branches.
2. *Branches of Unlimited Growth:* These branches arise from the axils of the branches of limited growth, and are also called axillary branches or long laterals. They are also differentiated into nodes and internodes and bear a whorl of primary laterals at the node. They grow indefinitely like the main axis.

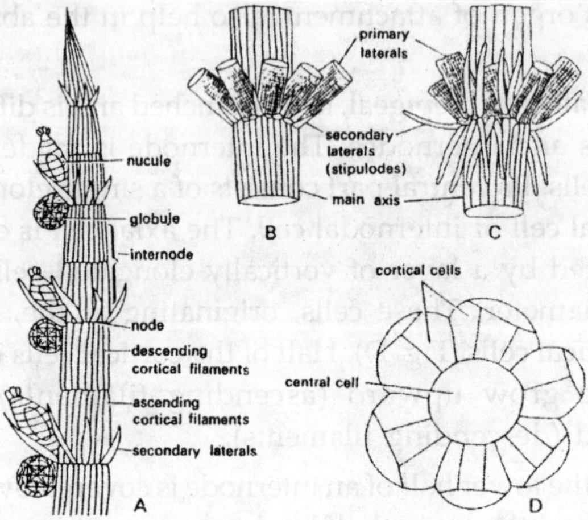

A-D. Chara: Aerial axis; a. Branch of limited growth, B-C.Appendages at nodes, D. Transection of internode

3. *Stipulodes:* The basal nodes of the branches of limited growth develop some unicellular outgrowths, called stipulodes. Some species of Chara (e.g., C. braunii, C. nuda), are unistipulate, i.e., the number of stipulades at each node is equal to the number of primary laterals present at that node.

   But most of the species are bistipulate, i.e., the number of stipulodes at each node is twice the number of primary laterals. If stipulodes are present in one whorl at each node, the condition is known as haplostephanous. (as in C. braunii) and if in two whorls it is called diplostephanous (as in C. tomentosa, C. baltica). In C. pashani, the stipulodes are, however, absent.

Cell Structure

The nodal cells are uninucleate and densely cytoplasmic with few small ellipsoidal chloroplasts. The internodal cells are, however, much elongated, each with a large central vacuole, many nuclei and innumerable discoid chloroplasts. Cytoplasmic streaming is particularly evident in the large internodal cells. The cytoplasm in these cells is divided into an inner endoplasm, which lies next to the vacuole, and an outer ectoplasm containing chloroplasts. Only endoplasm is capable of streaming. It continually streams up and down along the longitudinal plane. The streaming of the cytoplasm is due to alternating contraction and expansion of the protein fibrils which remain fixed to the cell wall.

Growth: In Chara growth of the main axis and its branches takes place by a dome-shaped apical cell.

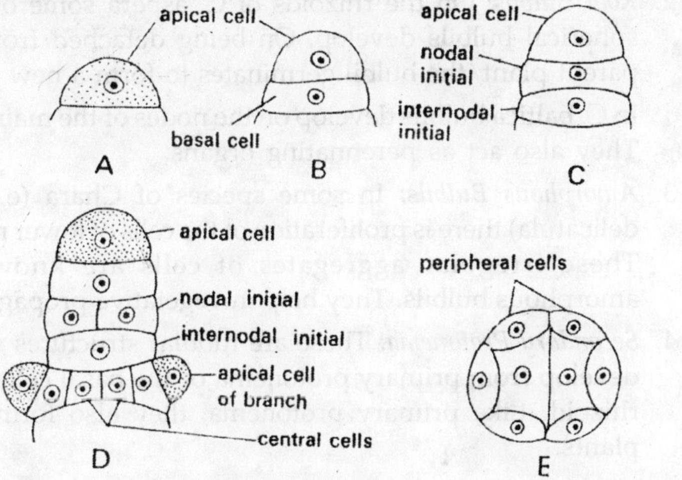

A-E. Chara: A-D. stages in apical growth, E. Transection of node

Successive transverse divisions of the apical cell produce a row of disc shaped cells (Fig. A, C). Each of these cells divides transversely into two daughter cells. The upper biconcave daughter cell functions as nodal initial and the lower biconvex cell as internodal initial. As a result of many vertical divisions at right angles to each other in nodal initial, a pair of central cells and many peripheral cells are formed. The peripheral cells act as mother cell for the primary laterals and behave in the same manner as the apical cell. The peripheral cells of the basal node form rhizoids. The internodal initial, however, does not divide further but elongates considerably.

Reproduction

Chara reproduces by vegetative and sexual methods. Asexual spores are not found.

Vegetative Reproduction: The following structures formed in Chara help in vegetative propagation.

1. *Amylum Stars:* In some species of Chara (e.g., C stelligera) basal nodal cells develop in a peculiar fashion and form star-shaped aggregates of cells, densely filled with amylum starch. When detached from the parent plant, these bodies develop into new plants.
2. *Root Bulbils:* On the rhizoids of C. aspera some oval or spherical bulbils develop. On being detached from the parent plant, the bulbil germinates to form a new plant.

   In C. baltica bulbils develop on the nodes of the main axis. They also act as perennating organs.
3. *Amorphous Bulbils:* In some species of Chara (e.g., C. delicatula) there is proliferation of the cells of lower nodes. These irregular aggregates of cells are known as amorphous bulbils. They help in vegetative propagation.
4. *Secondary Protonema:* These are tubular structures which develop from primary protonema or the basal cell of the rhizoid. Like primary protonema, they also form new plants.

Sexual Reproduction

Sexual reproduction in Chara is of highly advanced oogamous type. Structurally, sex organs of Chara are most complex amongst algae. The male sex organs are called antheridium or globule and the female oogonium or nucule. Both, these organs are macroscopic. Most of the species of Chara are homothallic or monoecious and the male and female sex organs are borne on the same branch at juxtaposed position. Some species of Chara (e.g., C. wallichii) are, however, heterothallic or dioecious, i.e., they bear male and female organs on different plants.

The sex organs are borne on the branches of limited growth (primary laterals) and the nucule is located above the globule. Development of globule and nucule is almost simultaneous, but in some species globule matures before nucule.

1. *Antheridium or Globule:*

(a) *Internal Structure of Mature Antheridium:* Mature antheridium is a large, macroscopic, spherical and conspicuously red or yellow structure. Its wall is made up of eight cells which are called shield cells. The outer wall of the shield cells has many infoldings which form various types of ornamentations on the surface. A rod shaped cylindrical cell arises from the centre of each shield cell which is known as manubrium. At the distal end of each manubrium there is one or more small spherical or globose cells which are called primary capitulum cells. These cells are in direct contact of manubrium. There is also a second row of capitulum cells, known as secondary capitula. Each secondary capitulum cell bears 2-4 long unbranched antheridial filaments or spermatogenous filaments at its tip. Each filament has 25-250 cells and each cell forms an antherozoid. Thus, each antheridium produces 20,000-50,000 antherozoids.

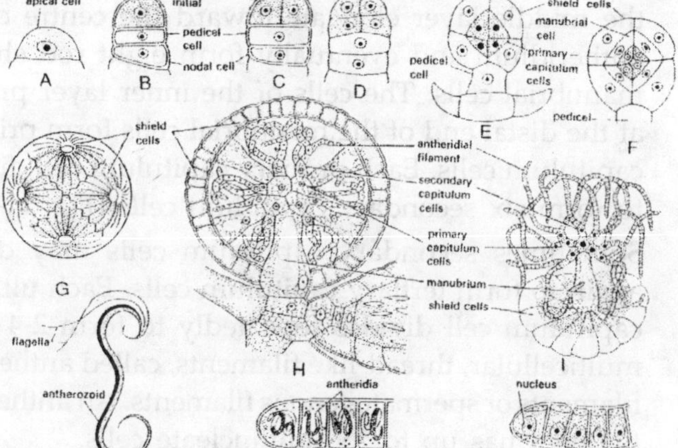

A-L. Chara: Development of globule and antherozoids; a-F. Successive stages in the development of globule, G. Mature globule, H. Globule in longitudinal section, I. A shield cell with manubrium, primary and secondary capitulum cells and spermatogenous filaments, J-K. Stages in spermatogenesis, L. An antherozoid

(b) *Development of Antheridium (Globule):* As mentioned earlier, globules develop at the nodes of the branches of limited growth. A peripheral adaxial cell of the node, which functions as antheridial initial, divides transversely into a basal pedicel cell and a terminal antheridial mother cell (Fig.A-B).

The pedicel cell does not divide further and forms a small stalk (pedicel) in the mature antheridium. The antheridial mother cell enlarges in size and then divides by two vertical divisions to form a quadrant (Fig.C, D)- Each of these four cells divides by a transverse wall and thus eight cells are formed (Fig. E).

This stage is called octant stage. Each cell of the octant divides periclinally as a result of which two layers of eight cells each are formed. The cells of the inner or outer layer again divide periclinally.

In this way three radial layers of eight cells each are formed. The cells of the outermost layer enlarge and transform into eight shield cells (Fig. G). The cells of the middle layer elongate toward the centre of the antheridium and eventually form eight rod shaped manubrial cells. The cells of the inner layer present at the distal end of the manubrial cells form primary capitulum cells. Each primary capitulum cell divides to form six secondary capitulum cells (Fig. H-I).

Sometimes secondary capitulum cells may divide again to form tertiary capitulum cells. Each ultimate capitulum cell divides repeatedly to form 2-4 long, multicellular, thread-like filaments; called antheridial filaments or spermatogenous filaments. An antheridial filament has up to 250 uninucleate cells.

The cells of the antheridial filament function as sperm mother cells. The protoplast of the sperm mother cell gets metamorphosed into a single, spirally coiled biflagellate antherozoid (Fig. J, L). The process of antherozoid development is known as spermatogenesis.

(c) *Liberation of Antherozoids:* At maturity, shield cells of the antheridium separate from each other, exposing the antheridial filaments. The sperm mother cells gelatinize in water and the antherozoids are liberated.

2. *Oogonium or Nucule:*

(a) *Structure of the Mature Oogonium:* The nucule of Chara is macroscopic oval structure with a short stalk. It is borne at the node of the primary lateral. In homothallic species, the nucule lies just above the globule. Five spirally twisted tube cells cover the nucule except at the tip, where there are five corona cells. In the centre of the nucule, there is a single uninucleate egg. The food material is stored in the egg in the form of oil drops and starch grains. At the tip of the egg there is a hyaline area, the receptive spot.

Presence of an envelope of sterile cells around the nucule is an advanced character amongst algae. Such a sterile jacket can be compared with the jacket of the archegonium of bryophytes.

(b) *Development of Oogonium (Nucule):* The oogonium develops from the upper peripheral cell derived by the division of basal nodal cell, situated at the base of the antheridium.

The oogonial initial undergoes two transverse divisions to form three cells in a row. The terminal cell functions as the oogonial mother cell, the middle as the sheath initial and the basal cell elongates to form the pedicel cell, supporting the developing nucule (Fig. A, C).

The oogonial mother cell elongates and then divides transversely into a small basal stalk cell and a large distal oogonium cell (Fig. E). The latter enlarges considerably, accumulates food and differentiates into an uninucleate egg.

The sheath initial divides vertically to form five sheath cells. The latter divide transversely to form two tiers of five cells each. The cells of the upper tier do not elongate much and form the corona at the tip of the oogonium. The cells of the lower tier, known as tube cells, elongate considerably to form a protective covering around the oogonium (Fig. F).

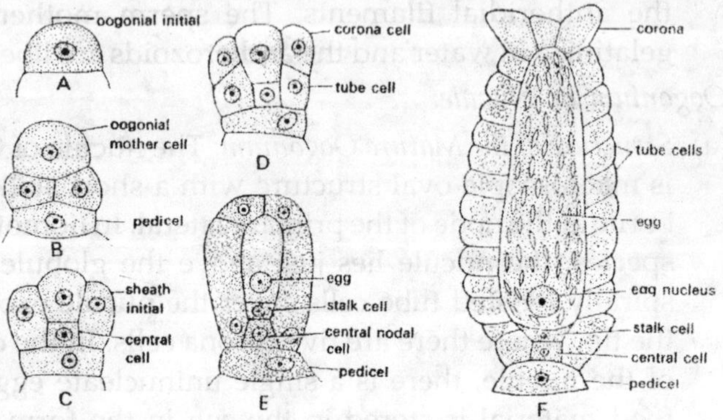

A-F. Chara: Successive stages in the development of nucule (oogonium)

Process of Fertilization

At the maturity of the oogonium, the five tube cells separate from one another and narrow slits are formed in between them. Antherozoids swim through these slits and penetrate gelatinized wall of the oogonium. Many antherozoids may enter into an oogonium but only one of them fuses with the egg to form the oospore.

Oospore

The Oospore Secretes a Thick Wall: The inner walls of the tube cells also become thickened but the other portions of the sheath wall decay. The oospore thus looks like a thick walled nut. It is yellow, red or brown in colour depending on the species. The oospore, with surrounding remains of the sheath, falls to the bottom of the pool. It germinates after a resting period. The duration of resting period varies with environmental conditions.

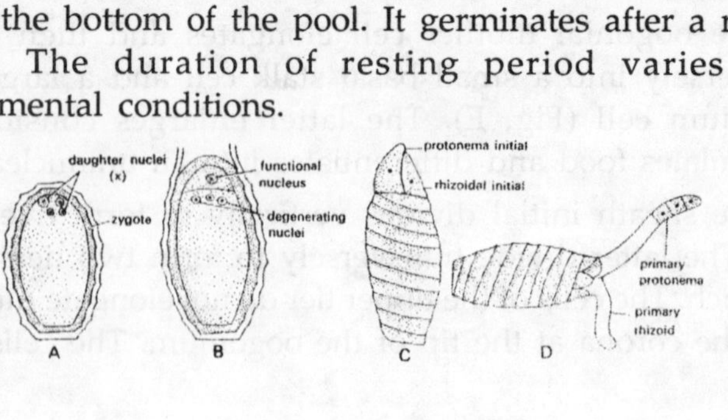

Germination of oospore. Oospores of Chara do not germinate in dark; they require white or red light for germination. At the end of the resting period, the diploid nucleus moves to the anterior end of the oospore and divides meiotically forming four haploid nuclei (Fig.A). A transverse wall is now laid down near the anterior end forming two cells. The upper cell is small, lenticular and uninucleate and the lower cell is large and trinucleate. The upper cell divides vertically to form a rhizoidal initial and a protonemal initial. Both these cells grow in opposite directions. The rhizoidal initial shows positive geotropism and forms primary rhizoid, whereas the protonemal initial shows negative geotropism and forms primary protonema. The primary protonema differentiates into nodes and internodes. Rhizoids and secondary protonema arise from the peripheral cells of the basal node. The peripheral cells of the upper nodes give rise to lateral branches.

Alternation of Generation: The plant body of Chara is haploid. The diploid condition is found only in oospore. At the time of germination, the diploid nucleus of the oospore divides meiotically and thus haploid state is regained. The life-cycle of Chara is thus of haploid predominant type.

Taxonomic Position of Chara

The Systematic Position of Charales has been Controversial: Fritsch (1935) included Chara in the order Charales under the class Chlorophyceae. Smith (1938, 55) placed the order Charales in a separate class- Charophyceae of the division Chlorophyta, whereas Round (1973) placed the order Charales in the phylum Charophyta.

The photosynthetic pigments, the nature of cell wall and food reserves, the structure of flagella and the life cycle of Chara is typically chlorophycean. However, there is a marked advance in vegetative organisation (differentiation of the axis into nodes and internodes, apical growth) and a great elaboration of the sexual reproductive organs (presence of jacket cells around sex organs, spiral nature of antherozoids, and mode of germination of zygote) of Chara over other members of Chlorophyceae.

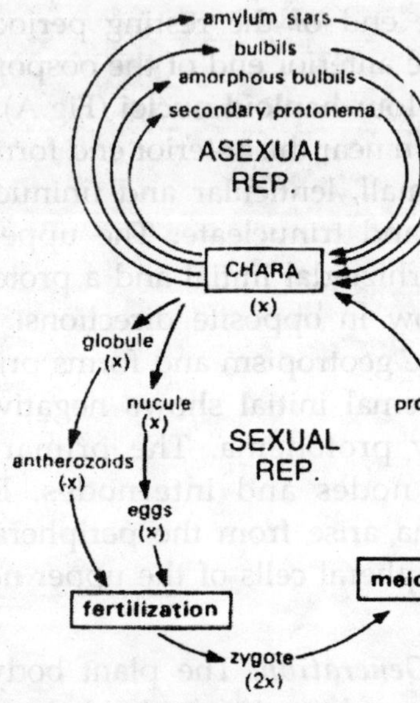

Chara: Diagrammatic representation of the life-cycle

Some algologists are of the view that the order Charales has evolved from Chaetophorales, especially from forms like Draparnaldiopsis. Chara resembles with Draparnaldiopsis in the presence of nodes and internodes, corticating filaments, formation of lateral branches of limited growth and participation of these laterals in reproduction. Thus, Chara shows basic characters typical of Chlorophyceae with some advanced features.

11

Codium and Vauchera

The Siphonales

The order Siphonales is represented by about 50 genera and 400 species, mostly marine, but there are some fresh water representatives in the families Protosiphonaceae, Phyllosiphonaceae and Vaucheriaceae. The members of the order are characterised by a coenocytic, siphonaceous and branched thallus.

In some marine members like Biopsies, Caulerpa and Acetabularia the thallus is differentiated into prostrate and erect branches. Transverse walls or septa are usually absent in vegetative thallus; they are laid down only at the time of the formation of reproductive structures or at the site of injury. There is a large central vacuole which is filled with cell sap. Protoplasm occurs in the form of a thick layer in between the wall layer and the vacuole.

Many discoid chromatophores and nuclei are present in the protoplast. The chromatophores always lie in the peripheral and the nuclei in the central region. In addition to chlorophyll, the members of this order contain two other pigments, siphonein and siphonoxanthin.

The members of Siphonales reproduce by vegetative, asexual and sexual methods. Fragmentation and akinetes are the two main methods of vegetative multiplication. Asexual reproduction is by zoospores or aplanospores. Sexual reproduction is usually oogamous, but in some species it is isogamous or anisogamous.

Classification

Fritsch (1935) included the following nine families in the order Siphonales:

(1) *Protosiphonaceae, example:* Protosiphon.

(2) *Caulerpaceae, examples:* Caulerpa, Biopsies.

(3) *Derbesiaceae, example:* Derbesia.

(4) *Dasycladaceae, examples:* Dasycladus, Acetabularia.

(5) *Codiaceae, example:* Codium.

(6) *Valoniaceae, examples:* Valonia, Siphonocladus.

(7) *Chaetosiphonaceae, example:* Chaetosiphon.

(8) *Phyllosiphonaceae, example:* Phyllosiphon.

(9) *Vaucheriaceae, example:* Vaucheria.

Smith (1955) recognised the following five families in the order: (1) Bryopsidaceae, (2) Caulerpaceae, (3) Halicystidaceae, (4) Codiaceae, (5) ichotomosiphonaceae.

Family Codiaceae

The members of Codiaceae are freely branched filamentous algae, abundantly growing in tropical waters. The thallus consists of tubular branches of spongy nature, made up of densely interwoven filaments. The filaments are characterised by the presence of constrictions with stratified annular thickenings, especially near the points of branching. They show typical siphonaceous coenocytic structure. The plastids lie at the peripheral position, and in most of the genera they lack pyrenoids. Vegetative reproduction occurs by means of detachable propagules. Sexual reproduction is anisogamous; the gametes are biflagellate and are produced in ovoid cup-shaped gametangia.

The family includes more than 120 species representing some 16 genera. Rhipidodesmis, Avrainvillea, Rhipilea, Codium and Udotea are some common genera of the family. Here the life-history of Codium is described as the representative of the family.

Codium

Systematic Position

Class	-	Chlorophycene
Order	-	Siphonales
Family	-	Codiaceae

Occurrence: Codium is a widely distributed non-calcareous genus with more than 45 marine species. The species of Codium occur from low-tide mark up to the depth of 70 meters in tropical as well as temperate waters. Codium tomentosum, C. bursa and C. fragile are some common tropical species of the genus.

Thallus Structure: The thallus is differentiated into prostrate and erect systems. The prostrate system, with the help of which the thallus is firmly anchored to the substratum is either disc-like (e.g., C. tomentosum) or branched, with hair-like rhizoids (e.g., C. adhaerens). The erect system varies considerably in form in different species. It is repeatedly forked bunch of long tubular branches (C. tomentosum), a flat cushion or crust (C. adhaerens) or a large round hollow ball, 10-20 cm in diameter (C. bursa). In the latter, within the hollow ball, are found many blue-green algae and small sea animals. This association with blue-green algae enables these species of Codium to fix atmospheric nitrogen.

The erect branches have a central medulla, composed of interwoven colourless forked filaments, and a peripheral cortex of compactly arranged inflated club-shaped branches, known as utricles or cortical vesicles. Each utricle represents one of the forks of longitudinal medullary filament which has become swollen. Species restricted to shady places (e.g., C. megalophysum, C. papenfusii) and deep waters (e.g., C. cranwelliae) usually have large utricles. There appears to be some relationship between utricle size and light intensity. On the outer surface of each utricle small deciduous hairs are present. These hairs are produced periodically and provide a tomentose character to the plant. At the base of the utricle and also just below the point of forking of the medullary filaments characteristic strong annular thickenings develop, leaving a fine pore for intercommunication. The utricle has a thick peripheral layer of cytoplasm, surrounding the large central vacuole. Many small discoid chloroplasts (without pyrenoids) are crowded in the outer portion of the cytoplasm and several small diploid nuclei are present in the interior.

Growth: Codium shows both apical and intercalary growth. The apical growth is brought about by the elongation of medullary filaments, which later form cortical portion by the forking. Intercalary growth takes place by continuous addition of new utricles between the pre-existing ones. In cushion-like forms (e.g., C. adhaerens) growth is mainly intercalary.

Reproduction

Codium reproduces by vegetative and sexual methods.

Vegetative Reproduction: It takes place by means of small club-shaped propagules which develop on utricles (Fig. A, C). When detached from the parent plant, the propagule grows into a new thallus.

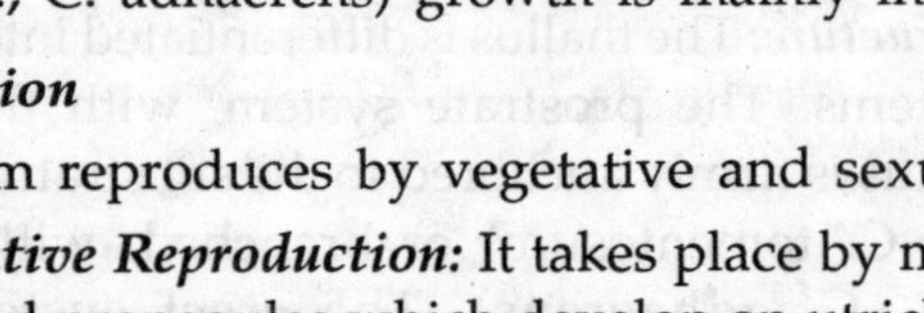

A-C. Codium: Development and structure of propagule

Sexual Reproduction: Sexual reproduction is anisogamous and takes place with the help of distinct male and female gametes. Some species are strictly dioecious (e.g., C. fragile), while others are monoecious.

Gametes are produced in ovoid male and female gametangia which develop laterally on utricles. A gametangium initiates as a tubular outgrowth on the lateral walls of the utricle. The outgrowth has compact cytoplasm with evenly distributed diploid nuclei and the plastids are crowded at the distal end (Fig.A).

The young gametangium elongates and is cut off from the utricle by a strong annular thickening that develops at its base. Some nuclei of the multinucleate gametangium degenerate and the remaining functional nuclei enlarge and divide meiotically to form haploid nuclei. Now the protoplast of the gametangium divides into many uninucleate segments. Each uninucleate segment develops into a gamete. The mature male gametangium is golden yellow and contains a few thousand small pyriform male gametes microgametes; (Fig. G), whereas the female gametangium is dark green and contains only a few hundred relatively large female gametes (microgametes; Fig. C, D, H). The male gamete contains one or two chloroplasts, whereas the female gamete has many chloroplasts.

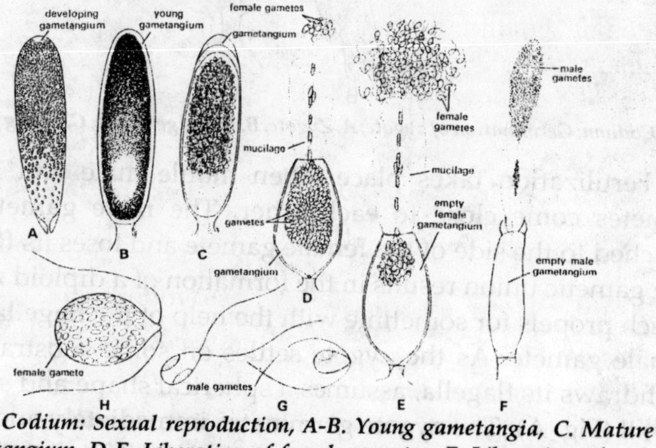

A-H. Codium: Sexual reproduction, A-B. Young gametangia, C. Mature gametangium, D-E. Liberation of female gametes, F. Liberation of male gametes, G. Male gametes, H. Female gamete

The discharge of gametes has been studied in detail in C. fragile. The internal layers of the gametangium wall are gelatinized. The apical portion of the gametangium ruptures and a large mass of mucilage is extruded through it.

The mucilage mass is almost of the same shape and size as the gametangium, and it has a central canal through which gametes move out (Fig. D, G). Initially, gametes lack flagella and are carried out passively due to the hydrostatic pressure within the gametangium.

Soon after the discharge, each gamete develops two flagella on its anterior side and swims freely with their help. Ejection of gametes occurs very rapidly and all gametes are released within a minute or so.

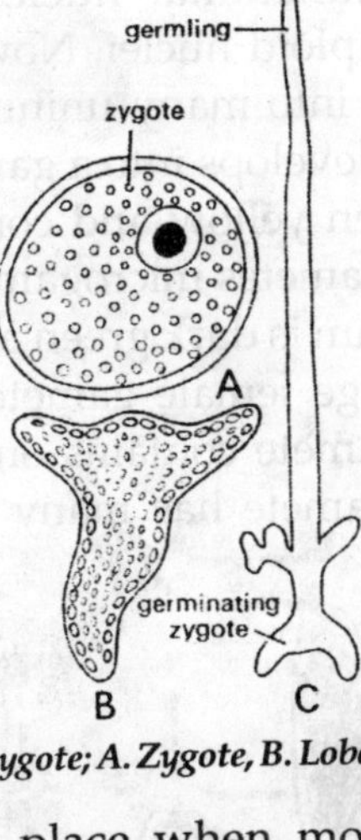

A-C. Codium: Germination of zygote; A. Zygote, B. Lobed germling, C. Young germling

Fertilization takes place when motile male and female gametes come close to each other. The male gamete gets attached to the side of the female gamete and loses its flagella. The gametic union results in the formation of a diploid zygote, which propels for sometime with the help of the flagella of the female gamete. As the zygote settles on some substratum, it withdraws its flagella, assumes a spherical shape and secretes a wall (Fig. A). The zygote germinates immediately and forms a lobed germling. The germling soon gets differentiated into an erect aerial and a horizontal disc like rhizoidal system. The entire aerial system (medulla and cortex) is formed by the ramification of the single initial filament.

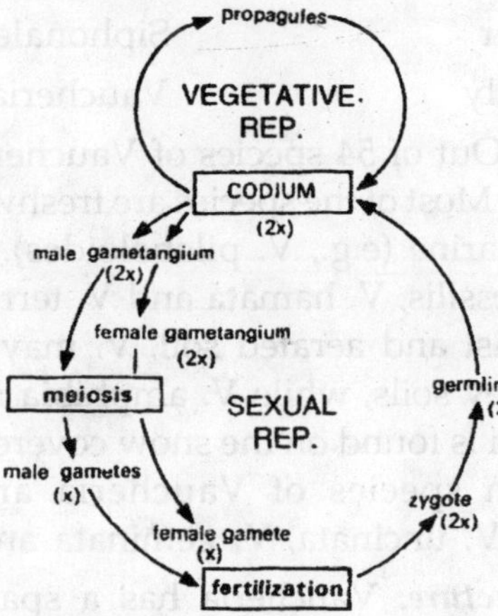

Codium: Diagrammatic representation of the life cycle

Generation Change

Codium shows diplontic life-cycle with no alternation of generation. The thallus is diploid which bears micro- and macrogametangia. The gametangial nuclei divide meiotically to form haploid gametes which represent the haploid phase in the life-cycle. The micro-and macrogametes fuse to form diploid zygote and thus the diploid phase is re-established. There is no reduction division at the time of the germination of zygote.

Family Vaucheriaceae

The members of this family are found in freshwater habitats in temperate regions. The aseptate, siphonaceous and coenocytic thalli are found attached to the substratum by a branched hapteron. They reproduce asexually by means of zoospores or aplanospores. Sexual reproduction is of oogamous type.

Life-history of Vaucheria is described here as the representative genus of the family.

Vaucheria

Systematic Position

Class	-	Chlorophyceae
Order	-	Siphonales
Family	-	Vaucheriaceae

Occurrence: Out of 54 species of Vaucheria known, about 19 occur in India. Most of the species are freshwater or terrestrial but a few are marine (e.g., V. piloboloides). Many terrestrial species like V. sessilis, V. hamata and V. terrestris form green mats on the moist and aerated soil. V. mayyandensis grows on exposed clayey soils, while V. amphibia is an amphibious species. V. jonesii is found on the snow covered surfaces. Some common Indian species of Vaucheria are V. amphibia, V. polysperma, V. uncinata, V. geminata and V.sessilis.

Thallus Structure: Vaucheria has a sparingly branched, coenocytic and siphonaceous thallus. The filaments show apical growth and they attain a length of several centimeters. They are aseptate but septa or cross walls are formed at the time of the formation of reproductive structures or at the site of injury. However, in V. pseudohamata the filaments are septate. Terrestrial species remain attached to the soil surface with the help of branched hapterons or rhizoids. In V. mayyanadensis, the thallus is differentiated into a well-developed underground rhizoidal system and erect aerial branches.

The wall of the filament is relatively thin and lacks elasticity. It is differentiated into an outer pectic and an inner cellulosic layer. There is a well-developed siphon-like central vacuole within the filament that runs throughout the length of the thallus except immediately behind the growing tips. The vacuole is filled with a cell sap that contains lipids, degenerated chloroplasts and crystals. The protoplast is present in the form of a continuous layer between the cell wall and central vacuole. The nuclei lie toward the centre and chromatophores towards the periphery in the protoplast. The cytoplasm shows streaming movements which involve nuclei, mitochondria and dictyosomes.

Although Vaucheria has been placed in Chlorophyceae by Fritsch (1935), it has pigments like chlorophyll *a*, chlorophyll *e*, carotenoid and xanthophyll which are characteristic of Xanthophyceae. Chlorophyll *b*, a typical pigment of Chlorophyceae, is absent in this genus. The two dominant pigments of the order Siphonales, siphonein and siphonoxanthin do not occur in Vaucheria.

The reserve food material in Vaucheria is oil instead of starch. It occurs as small colourless droplets in the cytoplasm. But filaments growing in continuous light accumulate food in the form of starch.

Reproduction: Vaucheria reproduces by vegetative, asexual and sexual methods.

Vegetative Reproduction: Vegetative multiplication takes place by fragmentation of the vegetative filaments. Transverse septa are formed that block the injured portions of coenocyte.

Asexual Reproduction: Vaucheria reproduces by a variety of asexual spores. Formation of these spores depends upon environmental factors. Following asexual spores are found in Vaucheria:

1. *Zoospores:* Asexual reproduction in aquatic species of Vaucheria usually takes place by means of multiflagellate and multinucleate zoospores. They are produced singly in club-shaped sporangia which are formed at the tip of the filament and the central vacuole gradually decreases in size (Fig. A). Consequently, the tip becomes club-shaped and dark green in colour. It is cut off from rest of the filament by a transverse septum. A central vacuole develops and there is reversion in the position of nuclei and chloroplasts. The nuclei which were toward the centre in the vegetative filament come to occupy the peripheral position and the chromatophores move from the periphery to the centre. The sporangial protoplast now shrinks slightly and a pair of flagella, which are of unequal length, develops opposite each nucleus. This multinucleate and multiflagellate sporangial mass behaves as a zoospore

(Fig. C). It has no eye spot or pyrenoids. At the apex of the zoosporangium the wall gelatinises, forming a narrow aperture. The zoospore escapes through this pore, usually in early morning hours.

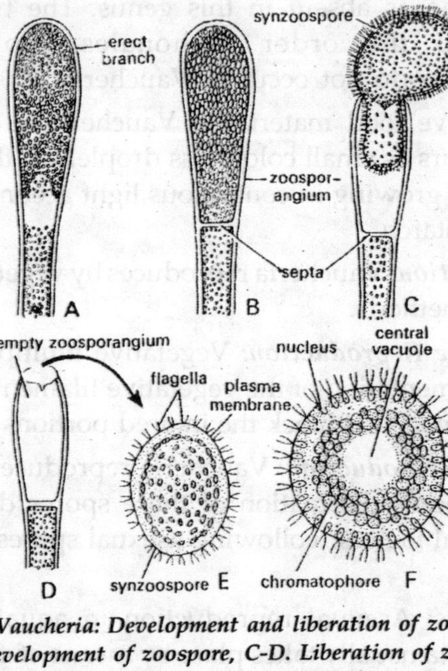

A-F. Vaucheria: Development and liberation of zoospore; A-B. Development of zoospore, C-D. Liberation of zoospores. E. synzoospore, F. Transverse section of a synzoospore

Zoospores of Vaucheria are different from those of other green algae due to their multinucleate and multiflagellate nature (Fig. E, F). Some algologists consider that they are compound structures in which protoplast fails to break into uninucleate segments. They prefer to call such multinucleate and multiflagellate zoospores as synzoospores.

Germination of zoospores. The zoospores swim for about 15 minutes after their release from the zoosporangium and then come to rest. They withdraw their flagella and secrete a thin wall.

Thus, before germination the zoospore becomes

deflagellated, non-motile and sluggish (Fig. A, B). Soon one or two tubular siphonaceous outgrowths come out (Fig. C). In terrestrial species one of the outgrowths attaches itself to the substratum and functions as holdfast, whereas the other forms erect branched system.

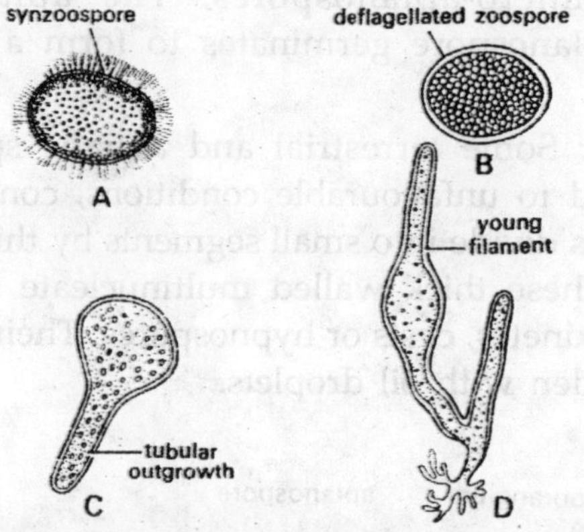

A-D. Vaucheria: Germination of synzoospore; A. A synzoospore, B. a deflagellated zoospore, C-D. Stages in germination of zoospores

2. *Aplanospores:* Aplanospores are commonly formed in terrestrial species of Vaucheria but in aquatic species such spores are produced under unfavourable conditions. Like zoospores, aplanospores are also formed singly within aplanosporangium; the latter develops as a club-shaped structure at the tip of the young vegetative branch. A transverse septum cuts off the terminal sporangium from the rest of the filament (Fig. A, B).

   The protoplast of the sporangium metamorphoses into a single multinucleate aplanospore. Normally, the conversion of sporangium protoplast into zoospore or aplanospore depends upon external factors. For example, aquatic species, which commonly produce zoospores, can be induced to produce aplanospores by keeping them in dim light or quiet water.

   Similarly, terrestrial species can be induced to produce zoospores by keeping the vegetative filaments moist for few days.

The aplanospore is a thin walled club-shaped or globose structure. It is liberated by rupturing the apical wall of the sporangium (Fig. B, C). Sometimes, the protoplast of the aplanosporangium divides into a number of spores. These small thin walled aplanospores are called micro-aplanospores. The aplanospore or microaplanospore germinates to form a new thallus (Fig. D).

3. *Akinetes:* Some terrestrial and aquatic species, when subjected to unfavourable conditions, contents of their filaments divide into small segments by thick gelatinous walls. These thick walled multinucleate segments are called akinetes, cysts or hypnospores. Their protoplast is often laden with oil droplets.

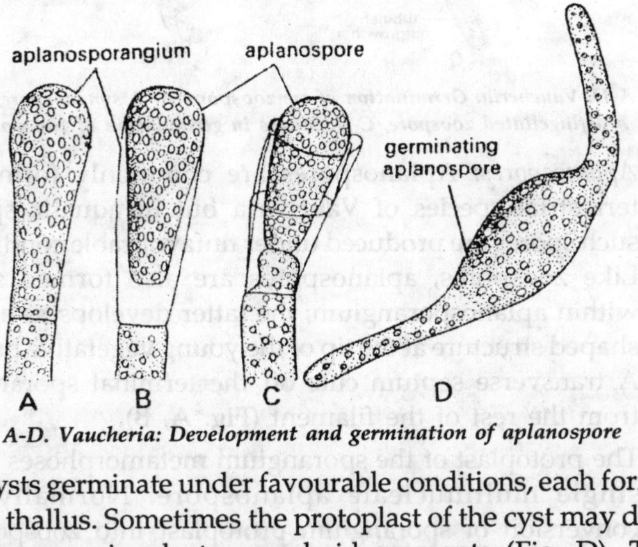

A-D. Vaucheria: Development and germination of aplanospore

Cysts germinate under favourable conditions, each forming a new thallus. Sometimes the protoplast of the cyst may divide into many uninucleate amoeboid segments (Fig. D). These segments come out through a pore in the lateral wall of the cyst.

At this stage, Vaucheria thallus resembles to that of another genus, Gongrosira, and therefore, this stage is called as gongrosira stage. Each amoeboid mass secretes a thin wall and eventually germinates to form a new thallus.

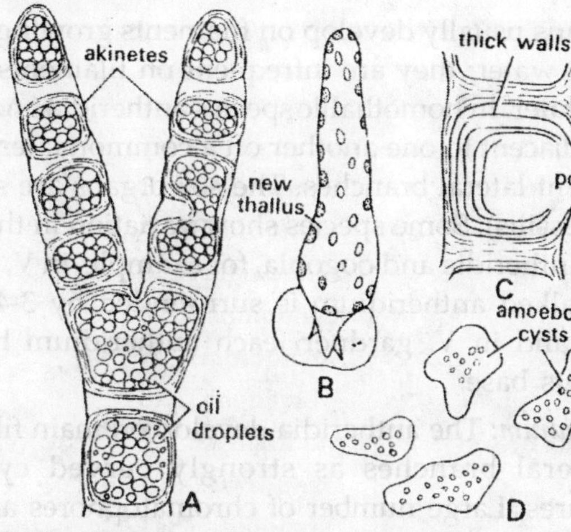

A-D. Vaucheria: Development and germination of akinetes; A. A filament with chain of akinetes, B. Germination of an akinete, C. A cyst with a pore on the lateral wall, d. Amoeboid cysts

Sexual Reproduction

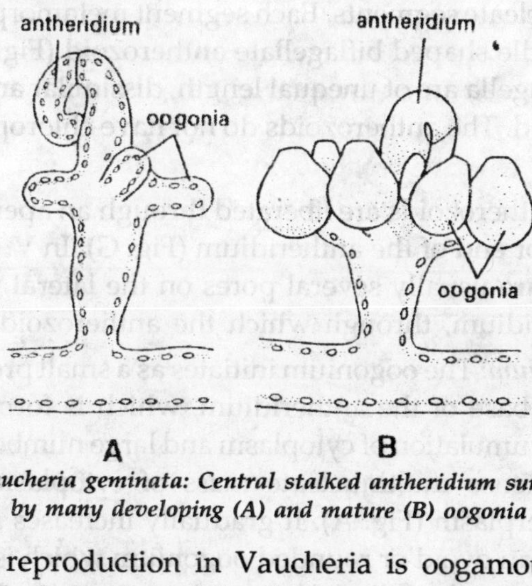

A-B. Vaucheria geminata: Central stalked antheridium surrounded by many developing (A) and mature (B) oogonia

Sexual reproduction in Vaucheria is oogamous. It takes place with the help of specialized sex organs, called antheridium and oogonium. Most of the species are homothallic or monoecious (Fig. A, B), but some like V. dichotoma, V. litorea and V. mayyanadensis are heterothallic or dioecious.

Sex organs usually develop on filaments growing in damp soil or quiet water; they are infrequent on filaments growing in flowing water. In homothallic species antheridia and oogonia are borne adjacent to one another on a common lateral branch or on adjacent lateral branches. The sex organs are stalked or sessile (V. sessilis). Some species show variation in the relative position of antheridia and oogonia, for example, in V. geminata a central stalked antheridium is surrounded by 3-4 oogonia (Fig. A, B) and in V. gardneri each antheridium has many oogonia at its base.

1. *Antheridium:* The antheridia develop on main filament or its lateral branches as strongly curved cylindrical structures. Large number of chromatophores and nuclei accumulate in this region and this curved part is then cut off from the rest of the filament by a septum (Fig. A, D). The nuclei present in this region divide mitotically and then the protoplast divides to form many uninucleate segments. Each segment metamorphoses into a spindle shaped biflagellate antherozoid (Fig. E, F). The two flagella are of unequal length, dissimilar and laterally inserted. The antherozoids do not have chloroplast or eye spot.

   The antherozoids are liberated through an aperture at the anterior end of the antheridium (Fig. G). In V. debaryana there are usually several pores on the lateral wall of the antheridium, through which the antherozoids liberate.

2. *Oogonium:* The oogonium initiates as a small protuberance at the base of the antheridium, which is formed due to the accumulation of cytoplasm and large number of nuclei. This dense multinucleate mass of cytoplasm is called wanderplasm (Fig. A). It gradually increases in size and forms an ovoid or rounded oogonium which is separated from the subtending filament by a septum (Fig. B, F).

   The mature oogonium is spherical or sub-spherical with an apical beak. It has only a single nucleus, the other nuclei present in the young state degenerate after the septum is laid down. However, some workers believe that all nuclei except one pass out of the oogonium before the formation of the septum. The uninucleate protoplast of the mature oogonium forms an egg.

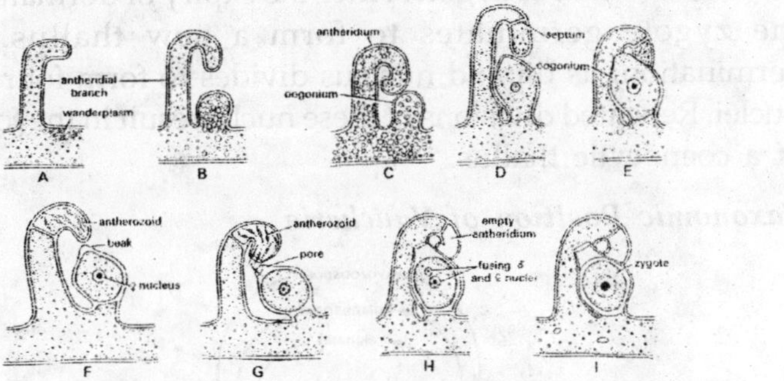

A-I. Vaucheria: A-F. Development of antheridium and oogonium, G-H. Liberation of antherozoids and fertilization, I. Zygote

At the anterior end of the egg there is a hyaline area which acts as a receptive spot.

3. *Fertilization:* Antheridia and oogonia mature almost simultaneously. The tip of the apical beak of oogonium gelatinizes to form an aperture (Fig. G). Simultaneously, large number of antherozoids are liberated from the adjacent antheridium. A drop of colourless cytoplasm comes out of the apical beak of the egg, and a number of antherozoids get entangled in this drop. Only one of the many antherozoids enters the oogonium and finds its way into the egg through receptive spot. The nucleus of antherozoid is smaller than that of the egg (Fig. H). After entering the egg, the male nucleus gradually increases in size and becomes nearly equal to the egg nucleus. Then the two nuclei fuse to form a diploid zygote (Fig. I). The zygote secretes a 3-7 layered thick wall. The protoplast of the zygote is laden with oil droplets. The zygote is green in colour initially but it gradually turns red due to degeneration of chlorophyll. It remains in the oogonium until it is liberated by decay of the oogonial wall. The zygote remains dormant for a few months before

germinating. The dormant state helps in tiding over unfavourable conditions like desiccation and high temperature.

Germination of zygote. After the expiry of dormant period, the zygote germinates to form a new thallus. Before germination, its diploid nucleus divides to form four haploid nuclei. Repeated divisions of these nuclei result in the formation of a coenocytic thallus.

Taxonomic Position of Vaucheria

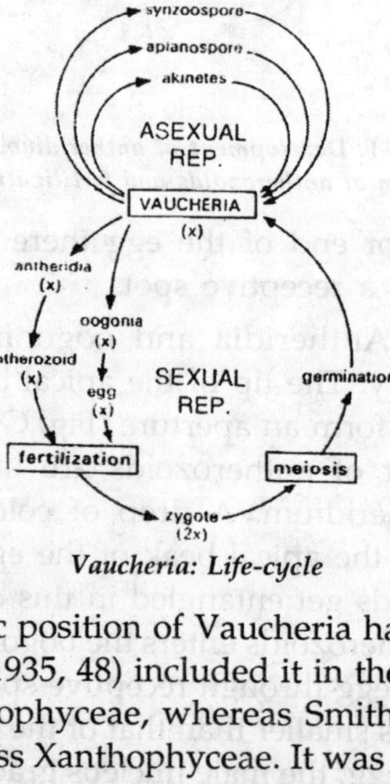

Vaucheria: Life-cycle

The taxonomic position of Vaucheria has been debated in the past. Fritsch (1935, 48) included it in the order Siphonales of the class Chlorophyceae, whereas Smith (1938, 55) placed it in a separate class Xanthophyceae. It was placed in the class Chlorophyceae on the basis of: (i) filamentous siphonaceous thallus organisation, (ii) asexual reproduction by synzoospores, and (iii) specialised oogamous type of sexual reproduction. However, it differs from other members of Chlorophyceae in (i) the absence of macerised cellulosic cell wall, which is typical of Chlorophyceae, (ii) the absence of chlorophyll *b*, (iii) predominance of oil as reserve food material, (iv) the absence of pyrenoid, and (v) unequal flagella of synzoospores. On the basis of these differences Smith removed Vaucheria from Chlorophyceae and placed it in the class Xanthophyceae, with which it shares the following features: (i) presence of chlorophyll *e*, a pigment characteristic of Xanthophyceae, (ii) absence of two xanthophyll pigments, siphonein and siphonoxanthin in zoospores of Vaucheria, and (iii) unequal flagella in antherozoids.

Davis (1904) and Williams (1926) traced phylogenetic relationships between Vaucheria and Oomycetes. The development of the reproductive cells of this alga is analogous with Oomycetes. The synzoospore of Vaucheria, like the gonidium of various Peronosporaceae, represents the contents of an entire sporangium. Besides, the presence of two unequal flagella in the reproductive cells of Vaucheria also lend support to the view that Phycomycetes have been derived from Vaucheria-like ancestors.

However, on the basis of the evidences available at present, it is difficult to conclude whether the above resemblances between Vaucheria and Oomycetes show real affinities or are the result of convergent evolution.

12

Oedogonium and Cladophora

The Oedogoniales

The order Oedogoniales is represented by a single family Oedogoniaceae which includes three genera, Oedogonium, Oedocladium and Bulbochaete. Oedogonium and Bulbochaete are exclusively freshwater forms, usually present in permanent water bodies, while species of Oedocladium also occur in terrestrial habitats. These algae may also grow on other algae and aquatic plants as epiphytes.

They are filamentous forms; the filaments are branched in Bulbochaete and Oedocladium and unbranched in Oedogonium. Bulbochaete has long colourless setae with swollen bases. These algae exhibit apical-basal polarity which is maintained even if the filament breaks up and becomes free floating.

The cell possesses a reticulate chloroplast which occupies the whole cell. Adjacent cells maintain cytoplasmic connections through plasmodesmata. Asexual reproduction takes place by zoospores and sexual reproduction is of oogamous type. The plants growing in running water do not form reproductive structures.

Oedogonium

Systematic Position

Class	-	Chlorophyceae
Order	-	Oedogoniales
Family	-	Oedogoniaceae

Occurrence: Oedogonium is a freshwater filamentous alga with more than 400 species, usually present in permanent water bodies of water, such as ponds, lakes and shallow tanks. The filaments are attached to rocks, logs, etc. or are epiphytic on aquatic plants and other algae. Some species (e.g., O. terrestris, O. randhawe) are terrestrial, growing on moist soils. The genus is represented in India by more than 200 species, the common ones being O. areolatum, O. armigerum, O. cardiacum, O. aster and O. elegans.

Thallus Structure: The thallus of Oedogonium is multicellular and filamentous. The filaments are unbranched and uniseriate. All cells of the filament, except the basal and apical ones, are cylindrical and alike. The basal cell, which acts as holdfast, is devoid of chloroplast. Its proximal end is extended into finger like or disc like projections which help in attachment to the substratum. The terminal or apical cell is rounded, elongated or acuminate. Thus, the filaments show clear apical-basal polarity.

Cell Structure: The cell has a fairly thick rigid cell wall, which is differentiated into an outer chitin, a middle pectin and an inner cellulose layer. The centre of the cell is occupied by a large vacuole. The protoplast forms a thin layer in between the central vacuole and the cellulose layer. The chloroplast is reticulate, extending in the form of a sheet throughout the protoplast. Most of the strands of the chloroplast are parallel to the long axis of the cell. Many pyrenoids are present at the intersections of the chloroplast reticulum. The cells are uninucleate. The nucleus lies in the middle region of the cell in between the chloroplast reticulum.

A characteristic feature of Oedogonium is the presence of

distinctive transverse bands at the distal end of some cells. The band, formed at the time of cell division, is called apical cap, and the cell with apical cap is known as cap cell.

Growth and Cell Division: Growth in Oedogonium takes place as a result of the divisions in the intercalary cells. All cells of the filament, except holdfast and apical cell, are capable to divide, but divisions are confined to only few cells.

The cell division in Oedogonium is a complex process. The important events during the cell division are as follows:

(1) The cell division is initiated by the formation of a ring of wall material in the upper part of the cell, which gradually increases in thickness (Fig. A-B).

(2) The nucleus migrates to the centre of the cell and divides mitotically.

(3) During telophase an unattached floating septum is laid down between the two daughter nuclei. The septum is formed due to continuous accumulation of vesicular material (microtubules, etc.) from the cytoplasm.

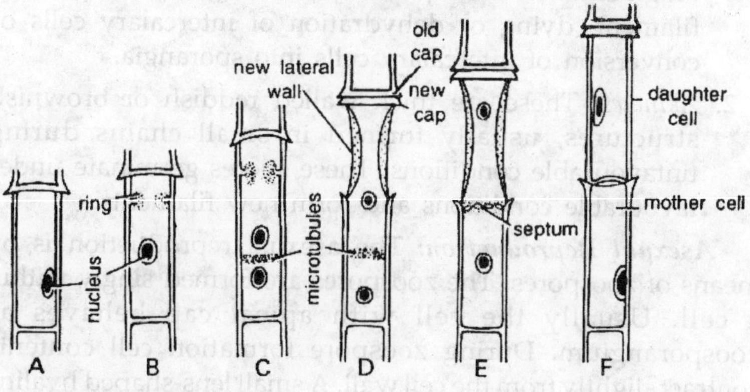

Oedogonium: Sequential stages of cell-division

(4) The ring of the wall material now elongates, causing a split in the parent wall near the apical ring. In this process a part of the ruptured parent wall remains attached to the anterior end of the new daughter cell in the form of a cap (Fig. D). It is known as the apical cap, and the cell having

a cap is called the cap cell. The other portion of the ruptured parent wall can be seen at the proximal end of the new cell.

(5) The floating septum gradually migrates upward to the base of the daughter cell and develops into a mature cross wall (Fig. E, F).

Thus, each division of the cell results in the formation of an apical cap. The number of apical caps in a cell indicates the number of times it has divided.

Reproduction

Oedogonium reproduces by vegetative, asexual and sexual methods.

Vegetative Reproduction: The vegetative multiplication takes place by fragmentation and akinetes.

1. *Fragmentation:* Like many other algae, small fragments of Oedogonium filament have the capability to grow into complete filaments under favourable conditions. Fragmentation takes place by accidental breaking of the filament, dying or dehydration of intercalary cells or conversion of intercalary cells into sporangia.
2. *Akinetes:* These are thick walled reddish or brownish structures, usually formed in small chains during unfavourable conditions. These spores germinate under favourable conditions and form new filaments.

Asexual Reproduction: The asexual reproduction is by means of zoospores. The zoospores are formed singly within a cell. Usually the cell with apical cap behaves as zoosporangium. During zoospore formation cell contents contract slightly from the cell wall. A small lens-shaped hyaline area develops on one side of the protoplast which eventually becomes the anterior end of the zoospore. At the base of this hyaline area a ring of basal granules appears, and from each granule a single flagellum arises. The basal granules remain connected with each other by a fibrous strand.

The mature zoospore is ovoid, spherical or pyriform. It is

uninucleate and contains a chloroplast. Sometimes the zoospore also possesses an eye-spot. At maturity of the zoospore, the wall of the zoosporangium splits near the cap region, and the adjacent cell moves apart to make passage for the liberation of the zoospore. It is believed that a mucilagenous substance is secreted at the base of the zoospore in zoosporangium which probably aids in the extrusion of the zoospore after absorbing water. The zoospore comes out of the zoosporangium in a delicate mucilagenous vesicle which soon gets dissolved and the zoospore is liberated (Fig. D, E).

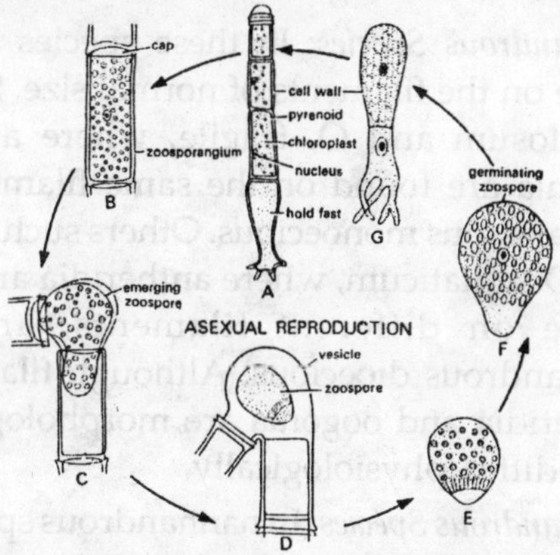

A-G. Oedogonum: Asexual reproduction; A. A filament, B. Zoosporangium, C-d. Liberation of zoospore, E. A zoospore, F-G. Germination of zoospore

Germination of Zoospores: After liberation, the zoospore swims for about an hour and then settles on any solid substratum with its anterior end downwards. It retracts its flagella and elongates considerably. A transverse septum separates the basal hyaline holdfast from the apical green cell (Fig. E-G). The apical cell divides repeatedly to form a new filament.

Sexual Reproduction: The sexual reproduction is of advanced oogamous type. It is more common in the forms growing in stagnant water than those in flowing streams.

It takes place with the help of male and female gametes. These gametes, known as antherozoids and eggs, are produced in antheridium and oogonium respectively. The male and female gametes differ morphologically as well as physiologically. Certain factors, such as alkaline nature of water and deficiency of nitrogenous substances in the medium promote sexual reproduction. In some species sexual reproduction can also be induced by placing the mature filaments in an atmosphere saturated with carbon dioxide.

1. *Distribution of Sex Organs:* Species of Oedogonium are divided into the following two groups on the basis of distribution of sex organs:

   (a) *Macrandrous Species:* In these species antheridia are borne on the filaments of normal size. Species like O. nodulosum and O. fragile, where antheridia and oogonia are found on the same filament, are called macrandrous monoecious. Others such as O. crassum, and O. aquaticum, where antheridia and oogonia are borne on different filaments, are known as macrandrous dioecious. Although filaments bearing antheridia and oogonia are morphologically similar, they differ physiologically.

   (b) *Nannandrous Species:* In nannandrous species filaments bearing antheridia and oogonia show morphological distinction. The male filaments, which are much smaller than the female filaments, are called dwarf male or nannandrium. Nannandrous species are always dioecious, i.e., antheridia and oogonia are borne on different filaments.

2. *Antheridium:* Macrandrous species form terminal or intercalary antheridia by the division of antheridial mother cell (Fig. A). Any cell of the filament with apical cap can function as antheridial mother cell. This cell divides transversely into an upper smaller antheridial cell and a lower larger sister cell (Fig. B). The latter undergoes several transverse divisions to form a row of 2-40 flat uninucleate antheridia. The nucleus of the antheridium divides mitotically into two nuclei, each gets surrounded by some

cytoplasm and metamorphoses into an antherozoid (Fig. D). Thus, each antheridium produces two antherozoids. But in O. cardiacum four antherozoids are formed in an antheridium.

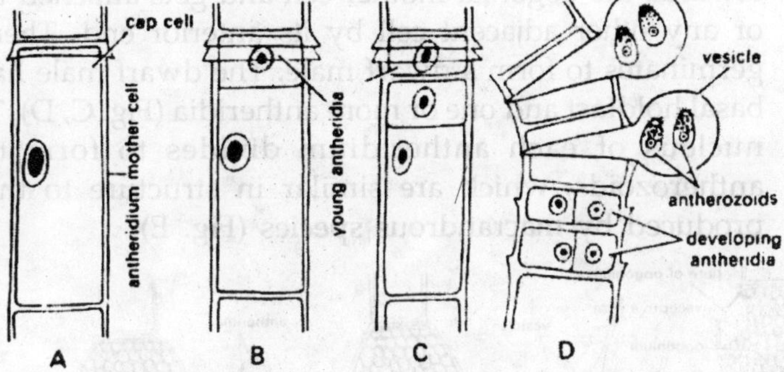

A-D: Oedogonium: Development of antheridium

The antherozoid is a unicellular, uninucleate and multiflagellate structure. It is similar to the zoospore but for its smaller size and fewer flagella. It is yellowish in colour due to its reduced plastids. Like zoospores, the antherozoids are also liberated by transverse splitting of the wall of the antheridium. The antherozoids are also enclosed in a hyaline vesicle at the time of liberation.

As mentioned earlier, in nannandrous species antheridia develop in specialised, small, 2-4-celled filament, known as dwarf male. The dwarf males develop from androspores, formed in androsporangia (Fig. A). Some nannandrous species (e.g., O. concatenatum), where androsporangia and oogonia develop on the same filament, are called gynandrosporous and those (e.g., O. conferatum), where the two are borne on different filaments, are designated as idioandrosporous. The structure of androsporangia is similar to those of antheridia in macrandrous species, except that only one androspore is produced per androsporangium.

The nucleus of androsporangium does not divide and the protoplast metamorphoses into a single uninucleate multiflagellate androspore. The androspore, when liberated from androsporangium, is also enclosed in a thin vesicle. The vesicle soon dissolves and the androspore swims freely for sometime. It is attracted chemotactically towards the oogonial mother cell and gets attached to it or any other adjacent cell by its anterior end. Then it germinates to form a dwarf male. The dwarf male has a basal holdfast and one or more antheridia (Fig. C, D). The nucleus of each antheridium divides to form two antherozoids, which are similar in structure to those produced by macrandrous species (Fig. E).

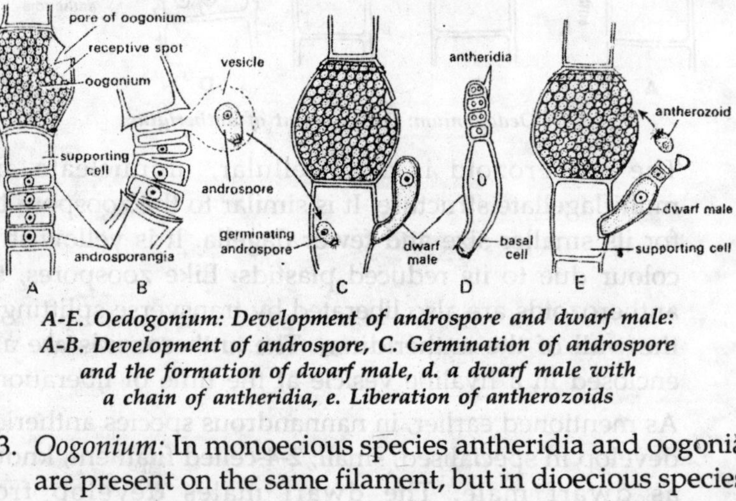

A-E. Oedogonium: Development of androspore and dwarf male: A-B. Development of androspore, C. Germination of androspore and the formation of dwarf male, d. a dwarf male with a chain of antheridia, e. Liberation of antherozoids

3. *Oogonium:* In monoecious species antheridia and oogonia are present on the same filament, but in dioecious species they are borne on different filaments. Development of oogonium is, however, similar in both macrandrous and nannandrous species. Any vegetative cell with apical cap may function as oogonial mother cell. It divides transversely into an upper oogonium and a lower suffulatory or supporting cell (Fig.A, B). The supporting cell may again function like oogonial mother cell and forms another oogonium and supporting cell. In such cases two or more oogonia are found in a row. Initially, oogonium is slightly wider than the vegetative cells, but at maturity it swells and becomes globose and its protoplast forms a single uninucleate non-motile egg.

The young egg is green in colour due to the presence of chloroplast. At maturity, a small pore or transverse slit develops on the oogonial wall. Just opposite to this slit a hyaline area is present in the egg which is known as receptive spot. The egg nucleus moves under the receptive spot before fertilization. In some species, prior to fertilization, a gelatinous material is extruded through the pore to facilitate the entry of the antherozoid into the oogonium.

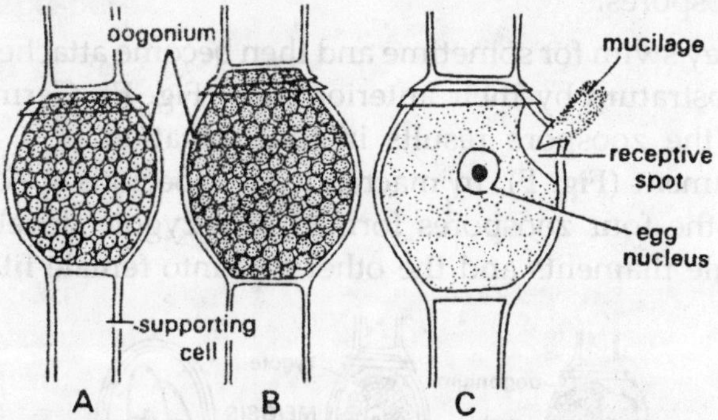

A-C. Oedogonium: Stages in development of oogonium

4. *Fertilization:* The process of fertilization is similar in both macrandrous and nannandrous species. Antherozoids are attracted chemotactically towards the mature oogonium. Usually a single antherozoid enters into the oogonium through the opening present on the oogonial wall. If more than one antherozoids enter the oogonium, only one of them fuses with the egg to form a diploid zygote (Fig. A).
5. *Zygote:* Soon after fertilization the zygote separates from the oogonial wall and secretes a thick three-layered wall of its own. A newly formed zygote contains chloroplast and hence green in colour, but as it matures there is accumulation of reddish oil in the protoplasm and it becomes reddish brown in colour.
6. *Germination of Zygote*: The zygote is liberated by decay of the oogonial wall and settles down at the bottom of the

pond. It undergoes a period of rest. This dormant state helps in tiding over unfavourable period. The zygote germinates under favourable conditions. The diploid nucleus of the zygote divides by meiosis to form four haploid nuclei, followed by the division. of the protoplast into four segments. Each uninucleate segment metamorphoses into a multiflagellate zoospore (zoomeiospores; Fig. D). These zoospores are liberated after the rupture of the zygote wall and behave like asexual zoospores.

They swim for sometime and then become attached to the substratum by their anterior ends (Fig. E). Germination of the zoospore results in the formation of a haploid filament (Fig. F). In macrandrous dioecious species two of the four zoospores formed in a zygote develop into male filaments and the other two into female filaments.

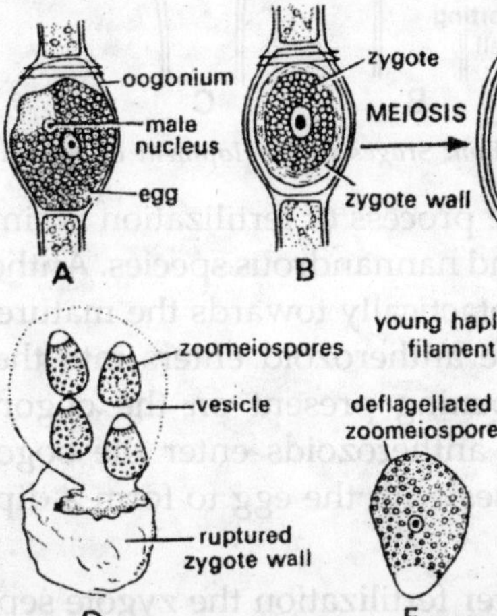

A-F. Oedogonium: Fertilization and germination of zygote; A. Fusion of antherozoid and egg nucleus, B. Zygote, C-D. Reduction division of diploid zygote nucleus and formation of zoomeiospores. E-F. Germination of zoomeiospore

Alternation of Generation

In Oedogonium diploid stage is short lived. It is found only in the zygote formed by the fusion of haploid egg and antherozoid. The diploid nucleus of the zygote forms four haploid zoospores by meiotic divisions, which grow into haploid filaments. Therefore, the life-cycle of Oedogonium is haploid predominant type.

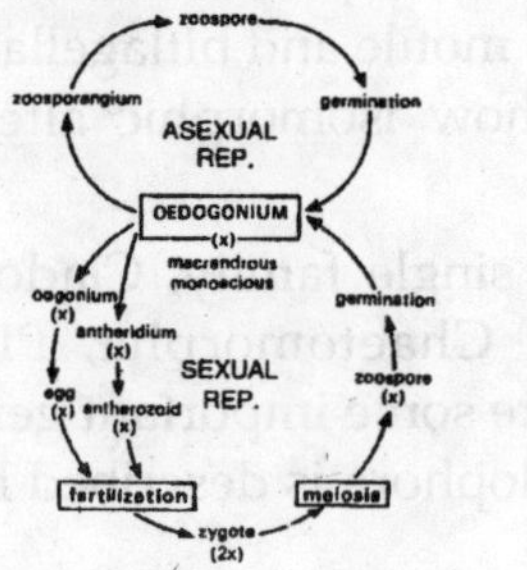

Life-cycle of macrandrous monoecious species.

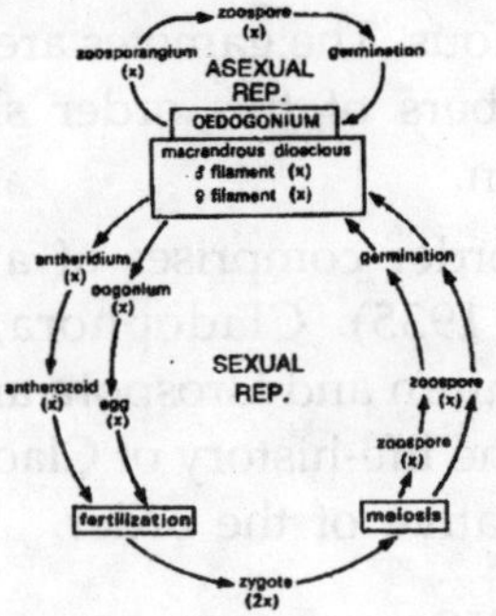

-cycle of macrandrous dioecious species.

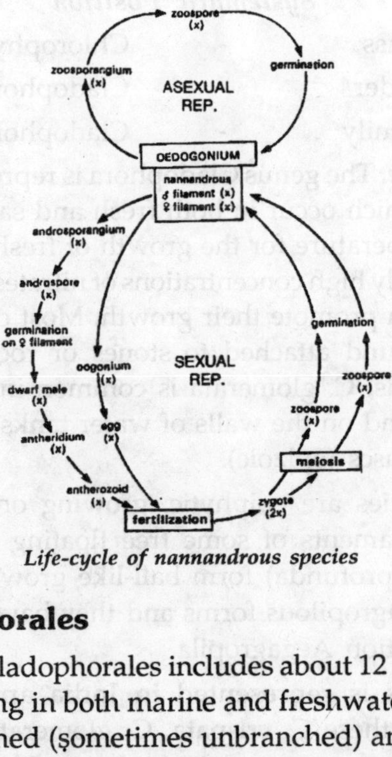

Life-cycle of nannandrous species

The Cladophorales

The order Cladophorales includes about 12 genera and 350 species, occurring in both marine and freshwater habitats. The thallus is branched (sometimes unbranched) and filamentous, and remains attached to the substratum by a rhizoidal cell. The cells of the filament are usually elongated and coenocytic with reticulate chloroplasts.

The asexual reproduction is by means of zoospores and aplanospores and the sexual reproduction is iso- or anisogamous. The gametes are motile and biflagellate. Most of the members of this order show isomorphic alternation of generation.

The order comprises of a single family, Cladophoraceae (Fritsch, 1935). Cladophora, Chaetomorpha, Pithophora, Rhizoclonium and Urospora are some important genera of the family. The life-history of Cladophora is described here as the representative of the order.

Cladophora

Systematic Position

Class	-	Chlorophyceae
Order	-	Cladophorales
Family	-	Cladophoraceae

Occurrence: The genus Cladophora is represented by about 160 species which occur in both fresh and saline waters. The optimum temperature for the growth of freshwater species is 15-25°C. Usually high concentrations of nitrates and phosphates in the medium promote their growth. Most of the freshwater species are found attached to stones or rocks by branched septate rhizoids. C. glomerata is common in streams and C. crispata is found on the walls of water tanks and also on the shells of molluscs (epizoic).

Some species are epiphytic, growing on higher aquatic plants. The filaments of some free floating species (e.g., C. holosatica, C. profunda) form ball-like growth. Such species are called aegagropilous forms and they have been placed in a separate section Aegagropila.

The genus is represented in India approximately by 17 species, of these C. crispata, C. glomerata, C. fracta and C. bengalensis are common in fresh waters.

Thallus Structure

The thallus of Cladophora is made up of profusely branched multicellular filaments. It remains attached to the substratum by septate rhizoidal branches which develop from the basal cells of the filament. In some species (e.g., C. kosteri) attachment rhizoids form a prostrate thallus, while in others (e.g., C. basiramosa, C. pygmaea) rhizoids are replaced by a single disc-like holdfast. In many species prostrate rhizoidal system, with their cells filled with reserve food material, helps in perennation.

Branching: The branches of Cladophora arise in a characteristic manner. They always originate just below the septa between the cells in the form of small lateral protuberances. Some cell contents pass into the protuberance which is cut off from rest of the filament by a septum along the long axis of the filament, close to the point of its origin. At this stage the septum is at right angle to the end wall of the cell. As the protuberance grows upward and outward (due to active growth of the parent cell membrane just beneath the protuberance), the angle between the septum and the end wall becomes more than 90°.

The growth of the new branch pushes aside the parent axis and as such the branching appears to be dichotomous (which in fact originate as lateral outgrowth). This process of pushing the parent axis by a new lateral branch is called evection (Fig. A-C). Sometimes two lateral branches are formed in a single cell and then the branching appears to be trichotomous.

Cell Structure: The cells of the filament are cylindrical, multinucleate and 5-20 times longer than their breadth. They have thick stratified cell wall that is differentiated into an outer chitinous, a middle pectic and an inner cellulosic layer. The outer chitinous layer gives a coarse and rough look to the filament.

The central part of the cell is occupied by a large vacuole. The space between the cell wall and the vacuole is filled with a fairly thick layer of cytoplasm.

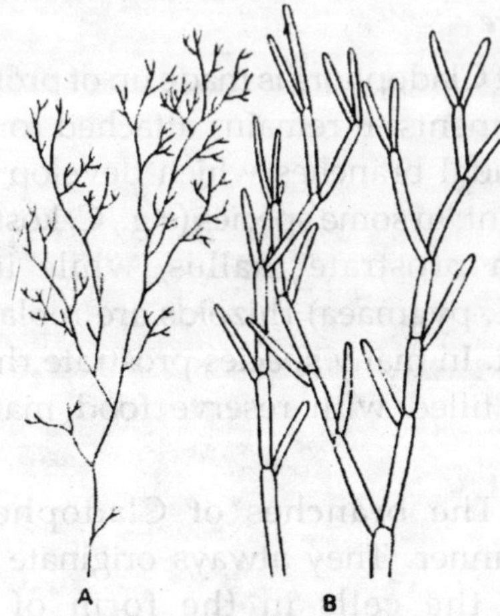

A-B. Cladophora: Thallus structure

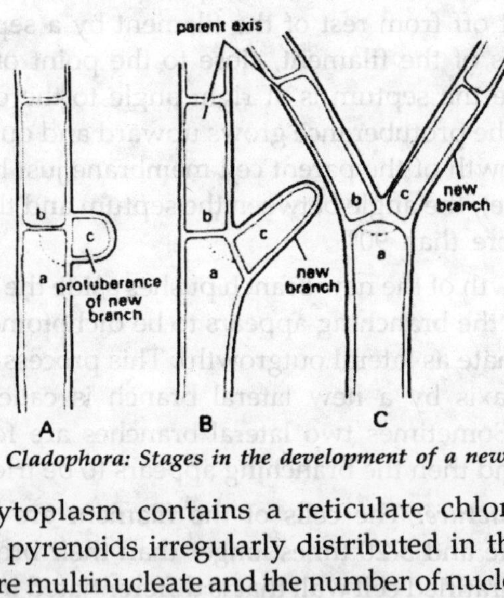

A-C. Cladophora: Stages in the development of a new branch

The cytoplasm contains a reticulate chloroplast with numerous pyrenoids irregularly distributed in the reticulum. The cells are multinucleate and the number of nuclei is probably related to the volume of the cell. Sometimes uninucleate or binucleate cells are also found. Occasionally, there are cells with several small vacuoles instead of a single large central vacuole. Some algologists believe that instead of a single reticulate chloroplast in a cell there are numerous discoid chloroplasts, each with a single pyrenoid.

Reproduction

Cladophora reproduces by vegetative, asexual and sexual methods.

Vegetative Reproduction: Some common methods of vegetative multiplication are:

1. *By Fragmentation:* It is the most common method of multiplication in aegagropilous species. The filaments of these species may break down into smaller fragments, and each fragment has the capability to grow into a new thallus.
2. *By Storage Cells:* Storage cells are found in the perennial species of Cladophora, e.g., C glomerata and C. ophiophila. These cells develop in filaments formed by the division of rhizoidal cells, and they store food material. During unfavourable conditions all the cells of the thallus, except the storage cells, die. The storage cells help in tiding over unfavourable conditions. They germinate to form new thalli on the return of favourable conditions.
3. *By Stolons:* The aegagropilous species of Cladophora have stolon-like rhizoidal branches. During favourable period these branches give rise to some erect branches which form new thalli.

Sometimes the cells of the rhizoidal system disarticulate and these individual cells grow into new thalli.

Asexual Reproduction: Asexual reproduction takes place by following methods.

1. *By Akinetes:* In some fresh water species of Cladophora, certain cells of erect filaments become laden with food material and secrete thick walls. They become pear shaped and function as akinetes. But unlike akinetes of other algae these thick walled cells divide to form new cells without being detached from the filament.
2. *By Zoospores:* Reproduction by zoospores is common in species like C. glomerata, C. crispata and C. paniaudii. The zoospores may be biflagellate (C. crispata) or

quadriflagellate (C. parriaudii). Usually the terminal or subterminal cell of vigorously growing diploid filament functions as zoosporangium. Prior to zoospore formation, the nuclei of the zoosporangium divide repeatedly. These divisions are usually meiotic, but in C. glomerata the nuclear divisions are mitotic (hence the zoospores of this species are diploid). Following nuclear divisions, the protoplast of the zoosporangium also divides and many uninucleate segments are formed. Each segment eventually forms a bi- or quadriflagellate zoospore. The zoospores escape through a pore at the distal end of the sporangium. They swim freely for sometime, then lose their flagella and become attached to the substratum by their anterior ends. They form diploid or haploid filaments (depending upon the nature of the zoospores).

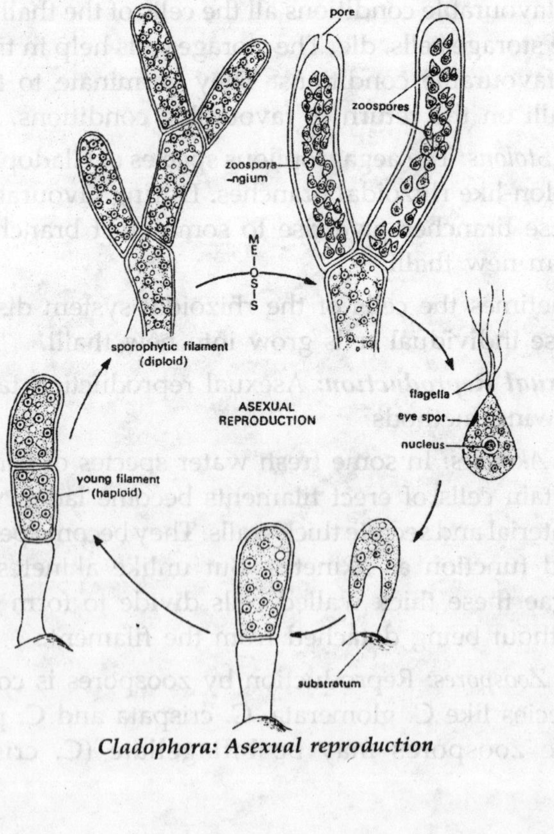

Cladophora: Asexual reproduction

Sexual Reproduction: The sexual reproduction is isogamous. Most of the species of Cladophora are dioecious or heterothallic. Gametes are formed in the haploid filament almost in the same manner as zoospores in the diploid filament. The nuclei of those cells of haploid filament which function as gametangia divide repeatedly. This is followed by the division of the protoplast, and as such many uninucleate segments are formed. Each segment forms a single gamete.

In C. glomerata, gametes are formed in the diploid filaments. Hence, the reduction division takes place prior to gamete formation.

The gametes are uninucleate, biflagellate and pyriform. It has an eye spot and reticulate chloroplast with many pyrenoids. The gametes are liberated through a pore at the tip of the gametangium. In some species gametes have a distinct beak at their tips.

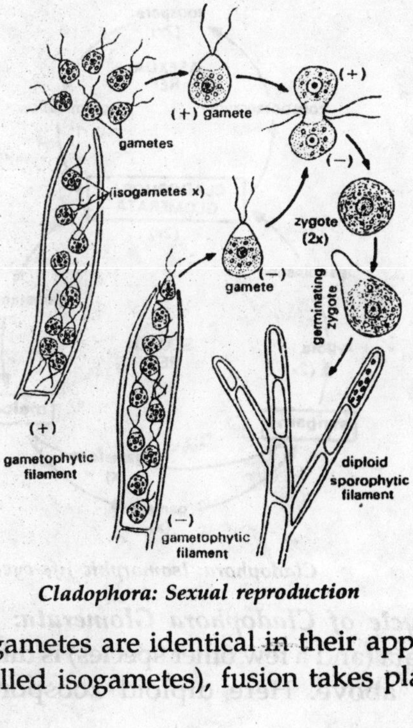

Cladophora: Sexual reproduction

Although gametes are identical in their appearance (such gametes are called isogametes), fusion takes place when two gametes of different strains (+ and -) meet. Zygote so formed is diploid in nature. It germinates immediately and gives rise to a diploid sporophytic thallus.

Alternation of Generation

In the life-cycle of most of the species of Cladophora haploid gametophytic and diploid sporophytic filaments are formed. The diploid sporophytic filaments, after meiosis, produce bi- or quadriflagellate haploid zoospores which give rise to haploid thalli on germination. The cells of the haploid filament function as gametangia and produce haploid isogametes. The gametes from two different filaments (of + and - strain) fuse to form a diploid zygote. The zygote gives rise to a diploid sporophytic thallus on germination. The haploid gametophytic and the diploid sporophytic filaments are morphologically similar. Such life-cycle, where there is alternation of two morphologically similar forms, is called isomorphic.

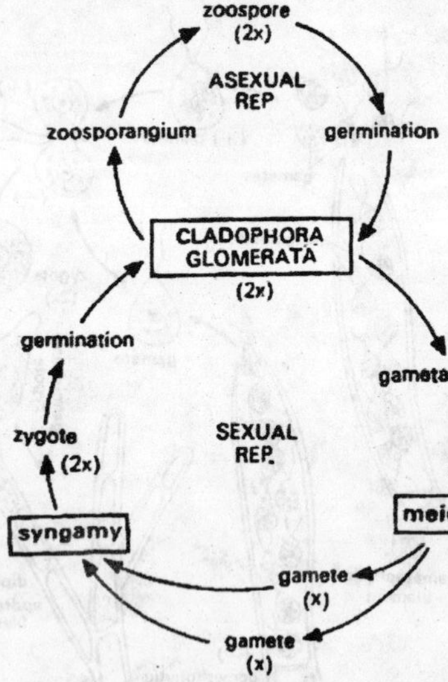

Cladophora: Isomorphic life-cycle

Life-cycle of Cladophora Glomerata: The life - cycle of C. glomerata (and a few other species) is different from the one described above. Here diploid zoospores are formed in zoosporangia and these zoospores give rise to diploid thalli again on germination. During sexual reproduction there is a reduction division and as such haploid gametes are formed. The gametes of two different strains fuse to form a diploid zygote. Thus in C. glomerata, the haploid phase is represented by gametes only, and haploid gametophytic thalli, like those in other species, are absent.

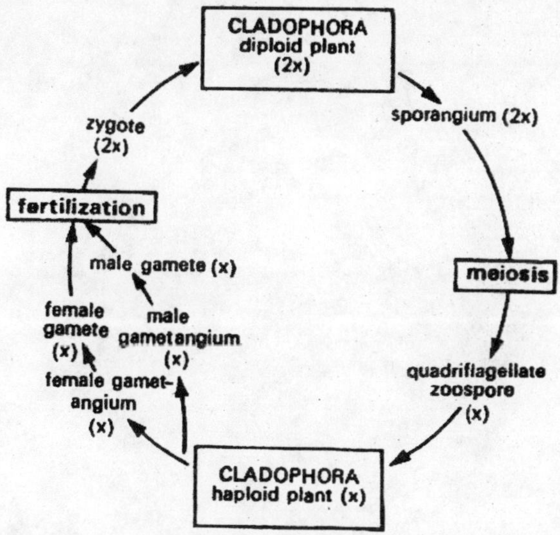

Cladophora glomerata: Diagrammatic representation of life-cycle

13

Family Ulvaceae

The Ulotrichales

Members of the order Ulotrichales occur mostly in fresh water or terrestrial habitats but a few forms are marine. They include unbranched filamentous and parenchymatous forms showing little structural or functional specialization among vegetative cells. The lowest cell of the thallus is, however, 'modified to serve as the organ of anchorage. The cells are usually uninucleate and each cell contains a single parietal chloroplast with one or more pyrenoids. Asexual reproduction takes place by the formation of bi- or quadri-flagellate zoospores, aplanospores or hypnospores. Sexual reproduction is isogamous, rarely anisogamous or oogamous.

Fritsch (1935) divided the order into six families: Ulotrichaceae, Microsporaceae, Cylindrocapsaceae, Ulvaceae, Prasiolaceae and Sphaeropleaceae.

In the members of the family Ulvaceae, the plant body is an expanded sheet, one or two cells in thickness, or a hollow cylinder with a wall one cell in thickness, or a ribbon two or more cells broad. The cells are uninucleate and with a single laminate chloroplast. Asexual reproduction takes place by means of quadriflagellate zoospores. Sexual reproduction is isogamous or anisogamous. Common examples of the family are Enteromorpha, Monostroma and Ulva.

Ulva

Systematic Position

Class	-	Chlorophyceae
Order	-	Ulotrichales
Family	-	Ulvaceae

Occurrence: Ulva, commonly known as 'sea-lettuce' or 'green laver', has a cosmopolitan distribution. It is exclusively a marine alga, usually found attached to rocks in intertidal zones of the sea. It has preference for waters enriched by nitrogenous wastes. Some species occur in polluted estuaries and brackish water. The genus is represented by some 30 species, the common ones are: U. fasciata, U. lactuca, U. lobata, U. reticulata and U. stenophylla.

Species of Ulva are eaten as salad and are also used in soups, especially in Scotland. The chemical composition of dried U. lactuca is 15% protein, 5% sugar and starch, less than 1% fat, and 11% water. It forms a good roughage for the human digestive system.

Thallus Structure

The large expanded thallus is an expanded sheet of parenchymatous tissue two layers of cells in thickness. The thallus is narrowed into a short basal stalk and is attached to the substratum with the help of a holdfast, composed of rhizoidal outgrowths from the lower cells. These colourless rhizoidal outgrowths extend downwards through the spaces present in between the two layers of the thallus and intertwine freely with one another. They emerge from the thallus near the point of its attachment to the substratum and become closely appressed to one another to form a pseudo-parenchymatous holdfast. The emergent portions of rhizoids are transversely septate, multinucleate and contain chloroplasts. The holdfast of the thallus is perennial and proliferates into new blades each spring.

The cells of the thallus are isodiametric or vertically elongated. Their walls are more or less confluent with one another and form a tough gelatinous matrix.

Cell Structure: The cells are uninucleate and the nucleus variously located in the interior half of the cell. A cup-shaped or laminate chloroplast lies towards the outer face of the cell. The chloroplast contains a single pyrenoid.

Growth: Ulva shows diffused growth. All cells of the thallus have the capacity to divide but the plane of the division is always at right angles to the thallus surface.

Reproduction

Ulva reproduces by vegetative, asexual and sexual methods.

Vegetative Reproduction: Vegetative multiplication usually takes place by fragmentation. Any part of the thallus, if detached accidentally, may regenerate into new thallus. Fragmentation is quite common in plants growing in quiet waters of estuaries. But plants growing in open hardly show any vegetative multiplication.

Asexual Reproduction: The asexual reproduction takes place by quadriflagellate zoospores, produced in diploid plants. Any vegetative cell of the thallus may act as zoosporangium. But zoospores are first formed in the cells near the margin and later in almost all other cells of the thallus. At the time of zoospore formation, the protoplast of the vegetative cell divides and redivides to form 4-8 daughter protoplasts. The first division is a reduction division, and as such the daughter protoplasts are haploid structures. Each haploid daughter protoplast metamorphoses into a uninucleate, quadriflagellate zoospore. Zoospores are usually released in the morning when the thallus is wet by the water of incoming tides. They swarm for an hour or so, then come to rest on some suitable substratum, withdraw their flagella and germinate to produce haploid gametophytic thalli.

Sexual Reproduction: The sexual reproduction is usually isogamous, and takes place through gametes which develop in the cells of haploid plants. The plants are heterothallic and only gametes from two different plants fuse. Reproductive areas are formed near the margins of the thallus and the fertile portions change from green to olive-green to brownish-green. Any cell of the fertile region can act as gametangium. Its protoplast divides repeatedly to form 32-64 biflagellate gametes. The first division of the protoplast is parallel to the surface of the thallus, followed by a second division at right angle to the first one. Prior to cleavage of the protoplast, each cell develops a beak-like outgrowth at its outer face which extends to the thallus surface. The tip of this outgrowth eventually develops a pore through which gametes are liberated. The gametes are positively phototactic and biflagellate, and each has a parietal chloroplast with a pyrenoid and an eye spot.

Some species of Ulva like U. lobata are, however, anisogamous. In these species, it is possible to distinguish between male and female gametophytes by differences in the colour of the fertile portion. Further more, the male gametes are narrower and smaller with a yellowish green chloroplast and indistinct pyrenoid, whereas the female gametes are larger with a bright green chloroplast and distinct pyrenoid.

Gametes are usually liberated in the morning, at the beginning of a series of spring tides. Two gametes from two different plants fuse to form a diploid quadriflagellate zygote. The zygote swarms for a short while, and then comes to rest, looses its flagella, and secretes a wall. Within a few days, it divides by a mitotic division. Out of the two daughter cells thus formed, one develops into a rhizoid and the other into a blade. Thus the sporophytic plant body is formed.

In some species, parthenogenetic development of the gamete gives rise to a new plant. There is a periodic fruiting pattern in Ulva, controlled by the lunar cycle. Gametes are released a few days before zoospores and fruiting occurs during a series of neap tides (e.g., U. pertusa) or during a series of spring tides (e.g., U. lobata).

Generation Change

There is an alternation of two morphologically similar generations (isomorphic alternation of generation). The haploid and diploid thalli are morphologically similar. The diploid thalli produce haploid quadriflagellate zoospores which give rise to haploid gametophytic plant body on germination. The haploid thalli produce gametes, which fuse to form a diploid zygote. The zygote gives rise to sporophytic plant body on germination.

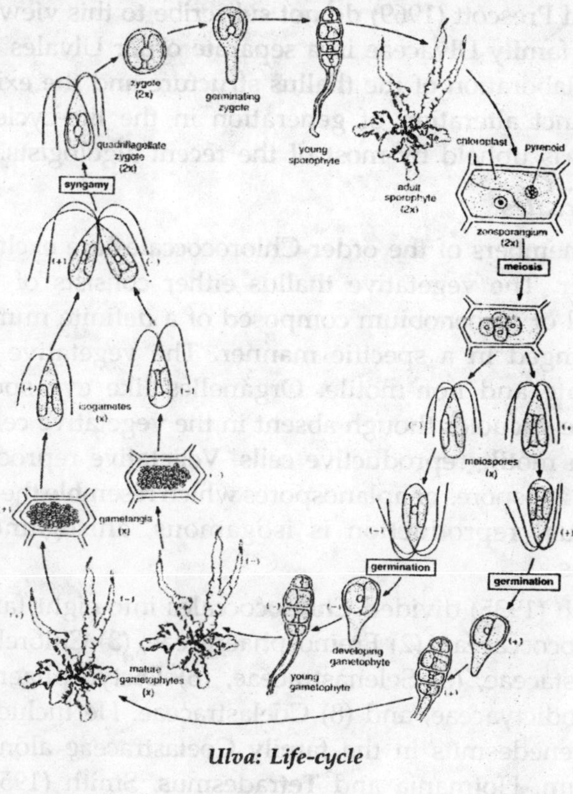

Ulva: Life-cycle

Taxonomic Position of Ulva

The taxonomic position of Ulva has been debated. The family Ulvaceae to which the genus Ulva was assigned was included by Fritsch (1935, 48) to the order Ulotrichales of the class Chlorophyceae. The inclusion of the family Ulvaceae in the order Ulotrichales is supported by the fact that young stages that of Ulva are uniseriate filaments closely resembling the simple filamentous plant body of Ulothrix. This treatment also receives support in the similarity of the cell structure, gametes and zoospores in Ulva and Ulothrix. Chapman and Chapman (1962, 73) and Lee (1980) even go as far as to state that Ulva is merely an elaborate form of Ulothrix.

On the other hand, Smith (1938, 55), Round (1965), Morris (1968) and Prescott (1969) do not subscribe to this view. They place the family Ulvaceae in a separate order Ulvales on the basis of elaboration of the thallus structure and the existence of a distinct alteration of generation in the life-cycle. This treatment is upheld by most of the recent algologists.

Chlorococcales

The members of the order Chlorococcales are exclusively freshwater. The vegetative thallus either consists of only a single cell or a coenobium composed of a definite number of cells arranged in a specific manner. The vegetative cell is uninucleate and non-motile. Organelles like eye spot and contractile vacuoles though absent in the vegetative cell, may appear in motile reproductive cells. Vegetative reproduction occurs by zoospores or aplanospores which resemble the parent cell. Sexual reproduction is isogamous, anisogamous or oogamous.

Fritsch (1935) divided Chlorococcales into eight families : (1) Chlorococcaceae, (2) Eremosphaeraceae, (3) Chlorellaceae, (4) Oocystaceae, (5) Selenastraceae, (6) Dictyosphaeriaceae, (7) Hydrodictyaceae, and (8) Coelastraceae. He included the genus Scenedesmus in the family Coelastraceae along with Coelastrum, Hofmania and Tetradesmus. Smith (1955) also recognised eight families in Chlorococcales but placed Scenedesmus in a separate family, Scenedesmaceae.

Family Chlorellaceae

The family Chlorellaceae includes non-motile, solitary (e.g.,

Chlorella) or colonial (e.g., Radiococcus) forms. The colonial forms show indefinite orientation of cells. The uninucleate cell has a single chloroplast, usually without pyrenoid. Reproduction takes place by autospores.

Chlorella

Systematic Position

Class	-	Chlorophyceae
Order	-	Chlorococcales
Family	-	Chlorellaceae

Occurrence: Species of Chlorella grow in a wide variety of habitats. They occur in fresh water, brackish water and also in terrestrial habitats. Most of the species are free living but some grow within the cells and tissues of invertebrates, forming symbiotic or parasitic association. Chlorella parasitica occurs as a parasite in Spongilla and Ophryodium; C. conductrix forms symbiotic association with invertebrates like Hydra, Stentor and Paramecium and C. lichina is an algal symbiotic partner of a lichen-Calicium chlorina.

The genus has about eight species. Chlorella ellipsoidea, C. parasitica and C. gonglomerata are common Indian species.

Structure: Chlorella is a unicellular non-motile alga. The cells are small, spherical to broadly ellipsoidal in shape and 2-12 μ m in diameter. The cell wall is made up of cellulose. The cell has a single cup-shaped parietal chloroplast, usually without pyrenoid. The chloroplast occupies major part of the cell. In the chloroplast, photosynthetic lamellae or thylakoids usually do not show grana-like organization, but in C. pyrenoidosa thylakoids are arranged in granal stacks. The cup-shaped cavity of the chloroplast is filled with colourless cytoplasm. The cytoplasm has a single nucleus and other cell organelles like endoplasmic reticulum, mitochondria, Golgi bodies and vacuoles. Changes in environmental conditions, like duration, intensity and quality of light, temperature and chemical composition of the growth medium greatly influence the micromorphology of the chloroplast, pyrenoid and cell wall.

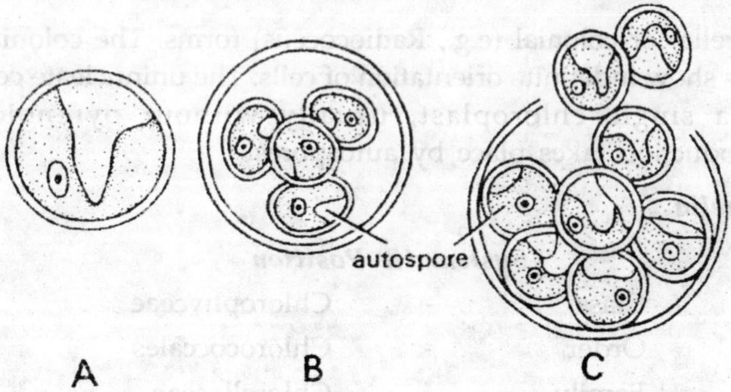

Chlorella: Formation of autospores; A. A vegetative cell, B-C. Cell with autospores

Reproduction: Chlorella reproduces only asexually by means of autospores (also called aplanospores). Motile spores or gametes are not formed.

During the formation of autospores, the cell grows in size and its contents divide into 2-16 daughter protoplasts. These divisions may occur both in light as well as in dark. Each daughter protoplast secretes a wall and is now exactly similar to the parent cell in its shape. The autospores are liberated by the rupture of the parent cell wall.

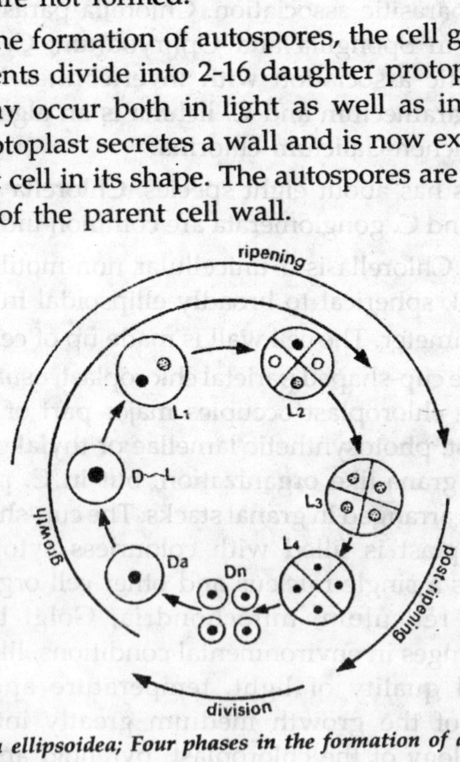

Chlorella ellipsoidea; Four phases in the formation of autospores

The formation of autospores and the factors governing the process have been studied in detail in C. ellipsoidea, and the following four phases have been identified in the life cycle of Chlorella.

1. *Growth Phase:* During this phase autospores grow in size by utilizing photosynthetic products.
2. *Early Ripening Phase:* In this phase growing cells prepare themselves for divisions.
3. *Post-ripening Phase:* In this phase the cell actually divides twice in dark or in light. Autospores formed in dark are known as dark nascent cells (Dn). These in turn give rise to photosynthetically active uninucleate cells, designated as dark active cells (Da). During transition from dark to light phase, cells grow in size.
4. *Division Phase:* During this phase the parent cell wall ruptures and autospores are liberated.

Economic Importance of Chlorella: Chlorella grows and multiplies fast in a variety of conditions and, therefore, it is a favourite material for the study of biochemical processes, particularly photosynthesis. It has high nutritive value. It contains proteins (50%), fats (20%) and carbohydrates (20%), besides minerals, vitamins and amino acids. It also contains an antibiotic, chlorellin, effective against many bacteria. Thus several developed countries like Japan, Germany and USA are working on the feasibility of growing Chlorella on commercial scale. But a very high cost of growing and processing has prohibited commercial exploitation of this alga so far.

Chlorella also acts as an aerating agent by removing carbon dioxide and restoring oxygen in photosynthetic process. Thus, growth of this alga in sewage treatment tanks helps in aerobic decomposition of sewage. It is also used as air purifier in nuclear submarines and space vehicles.

Family Hydrodictyaceae

The members of this family form non-motile coenobia which are more advanced than the coenobia of Coelastraceae. They reproduce asexually by zoospores. The zoospores become apposed to each other within the parent cell wall and as such a miniature colony is formed within the parent wall. The sexual reproduction is isogamous and the gametes are motile.

Hydrodictyon

Systematic Position

Class	-	Chlorophyceae
Order	-	Chlorococcales
Family	-	Hydrodictyaceae

Occurrence: Hydrodictyon or 'water net' is a free floating fresh water alga. It grows luxuriantly during late summer and spring season and sometimes almost completely covers the pond. The genus is represented by five species. Hydrodictyon indicum and H. reticulatum are common Indian species which occur in temporary or permanent water reservoirs.

Thallus Structure: Hydrodictyon is a macroscopic coenobial alga forming free floating 'water nets' in fresh water reservoirs. The mature coenobium is sausage-shaped and hollow in the centre, usually measuring 20-80 cm in length. The coenobium consists of few hundred to several thousand cells. The long cylindrical and pencil-shaped or ovoid cells are joined at the ends forming a design of hexagons and pentagons.

All the cells of a colony have the same general structure. Each cell is surrounded by a firm cell wall of cellulose. Inner to the cell wall, there is a thin layer of cytoplasm which encloses a large central vacuole. The young cell is uninucleate with a parietal chloroplast, but at maturity it becomes multinucleate and has a reticulate chloroplast. There are several pyrenoids in a mature cell.

The number of cells in a coenobium is fixed when it is young. Further growth of the coenobium is entirely due to the increase in cell size and not the number of cells.

Reproduction: Hydrodictyon reproduces mainly by asexual and sexual methods, but vegetative multiplication may also take place by fragmentation of the coenobium into small segments.

Asexual Reproduction: Asexual reproduction takes place by means of small, uninucleate and biflagellate zoospores (or swarmers). All the cells of a coenobium are capable of forming zoospores which eventually develop into new coenobia. The multiplication is very rapid. Before the formation of zoospores, pyrenoids disappear in the cell and starch grains are accumulated in the chloroplast. The multinucleate protoplast divides into 7,000-20,000 uninucleate segments (Fig. A, B). Each segment gets surrounded by a small amount of cytoplasm and a limiting membrane. It eventually develops into a biflagellate zoospore. Due to sudden contraction zoospores form a dense column in the centre of the mother cell.

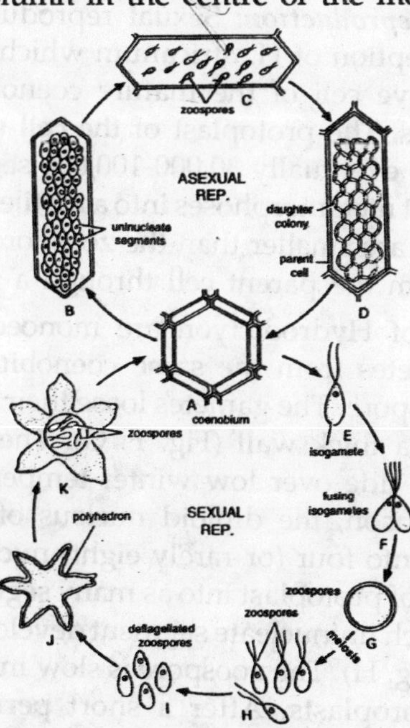

A-K. Hydrodictyon: A-D. Stages in asexual reproduction, E-K. Stages in sexual reproduction

The real swarming starts once the zoospores tear loose from the central column; then they swim freely and haphazardly inside the mother cell (Fig. C). The wall of the mother cell, however, does not rupture and the zoospores remain inside the mother cell. They buzz about rapidly within the mother cell. After a period of rapid movement, the zoospores loose their flagella and settle down on the inner side of the cell wall. They arrange themselves into characteristic hexagonal or pentagonal fashion to form a new net (Fig. D).

Although swarmers move quite haphazardly without any definite direction, the position of cells at the end of their movement decides the configuration of the future net.

The young coenobium is liberated by softening of the wall of the mother cell. In a young coenobium the cells grow mainly along their long axis.

Sexual Reproduction: Sexual reproduction is isogamous with the exception of H. africanum which shows anisogamy. Any vegetative cell of the mature coenobium is capable to form gametes. The protoplast of the cell undergoes repeated cleavage and eventually 30,000-100,000 segments are formed. Each segment metamorphoses into a motile biflagellate gamete. The gametes are smaller than the zoospores (Fig. E). They are liberated from the parent cell through a pore in its wall.

Species of Hydrodictyon are monoecious (homothallic). Two isogametes from the same coenobium fuse laterally to form a zygospore. The gametes loose their motility after fusion and secrete a thick wall (Fig. F, G). The zygospore has the capability to tide over low winter temperature. At the onset of spring season, the diploid nucleus of zygospore divides meiotically into four (or rarely eight) nuclei. This is followed by cleavage of protoplast into as many segments as the number of nuclei. Each uninucleate segment develops into a biflagellate zoospore (Fig. H). The zoospore is slow moving and has many discoid chloroplasts. After a short period of motility, the zoospore looses its flagella and comes to rest (Fig. I). It secretes a thick spiny wall and develops into a polyhedral cell, called polyhedron (Fig. J). The haploid nucleus of a polyhedron divides to form 200-300 zoospores (Fig. K). These zoospores eventually arrange themselves into a new daughter net within the polyhedron. The young net is liberated by the rupture of polyhedron wall.

Since the net of Hydrodictyon is formed by joining together of a large number of zoospores (each of which acts as an individual cell of the net), some investigators consider it as a collection of a number of individual plants.

Family Coelastraceae

These algae form two, four, or eight celled coenobia. The cells are uninucleate, variable in shape and specifically arranged with respect to one another. The members of the family reproduce only by means of autospores.

Scenedesmus

Systematic Position

Class	-	Chlorophyceae
Order	-	Chlorococcales
Family	-	Coelastraceae

Occurrence: Scenedesmus, a common freshwater alga has more than 100 species, abundant in ponds and lakes. It occurs as an almost pure culture in aquaria and in jars where water is left standing for sometime. Scenedesmus obliquus, S. dimorphus, S. quadricauda and S. denticvlatus are some common species of the genus.

Thallus Structure: The thallus of Scenedesmus is 4-,8- or sometime 16- celled coenobium. The cells are attached to each other by mucilage pads. The morphology of the coenobium can be modified considerably by varying the composition of the culture medium. In a medium with low phosphorus or low salt concentration, Scenedesmus can be induced to grow as unicells. The shape of the cells varies from fusiform (e.g., S.obliquus and S. dimorphus) to ellipsoidal (e.g., S. quadricauda and S. denticulatus, Fig. C, D).

The cell wall is smooth (e.g., S. bijuga,) or have processes that are elaborations of the cell envelope and are of taxonomic importance. For example, in S. quadricauda and S. armatus, the terminal cell bears long spines; in S. denticulatus and S. hystrix small spiny projections are present on the entire cell surface and in S. acuminatus, the cell has a tuft of bristles at each end. The presence of these processes is probably related to the planktonic mode of life and they help in floatation of the colony.

Each cell of the coenobium is uninucleate and possesses a single laminate chloroplast, containing one pyrenoid. The chloroplast occupies the entire cell cavity in a mature cell.

Reproduction: Reproduction in Scenedesmus is accomplished by cell division. Each cell of the coenobium is capable of giving rise to a new daughter colony, but all the cells of a colony rarely divide simultaneously. The protoplast undergoes successive transverse and vertical divisions forming four daughter protoplasts. Further divisions may also occur to give rise to eight or rarely sixteen daughter protoplasts. The protoplasts of the last generation give rise to autospores. Within the parent cell wall, the autospores undergo necessary adjustments according to the arrangement of cells in the coenobium. Fully developed coenobium is liberated by longitudinal rupture of the parent cell wall (Fig. A, F). Sexual reproduction is known to occur in S. obliquus, where it takes place by the formation of biflagellate gametes.

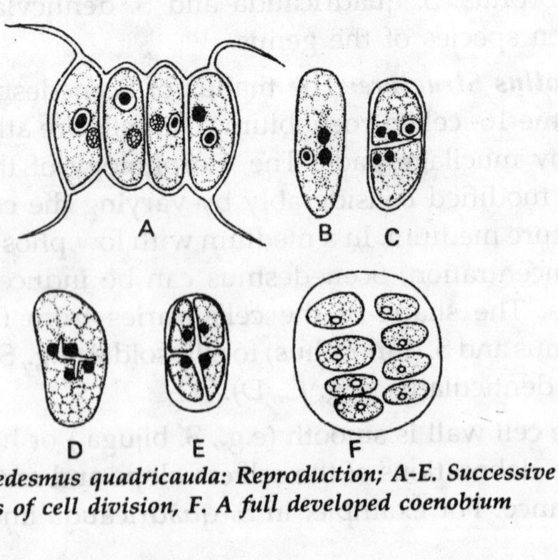

A-F. Scenedesmus quadricauda: Reproduction; A-E. Successive stages of cell division, F. A full developed coenobium

14

Functional Specialization

The Volvocales

The order Volvocales includes about 50 genera and 350 species. They mostly grow in freshwater and are abundant in waters rich in nitrogenous compounds. They are characterised by the presence of flagellated motile vegetative cells. There are usually two or sometimes four, flagella of equal length. The members are unicellular (e.g., Chlamydomonas, Dunaliella) or multicellular; in multicellular colonial forms the number of vegetative cells is in the multiple of two (e.g., 4-32 biflagellate cells in Pandorina, 4 or 16 in Gonium, 16-64 in Eudorina, 500-60,000 in Volvox). All the cells of a colony are surrounded by a common gelatinous sheath, but they are functionally independent. The cells are uninucleate and the nucleus divides only during reproduction. Each cell has a massive cup-shaped chloroplast with usually a single pyrenoid.

They reproduce both asexually and sexually. The asexual reproduction in the unicellular forms takes place by the division of protoplast into two or more segments, each forming a zoospore. In the colonial forms, cells undergo repeated divisions to form daughter colonies. The sexual reproduction ranges from isogamous to oogamous. In the isogamous and anisogamous forms both the fusing gametes swim freely in water, but in the oogamous forms the egg is located in the oogonium.

Fritsch (1935) divided this order into three sub-orders.

Order—Volvocales

Sub- orders

Chlamydomonadineae	Tetrasporineae	Chlorodendrineae
Families		
1. Chlamydomonadaceae	1. Tetrasporaceae	1. Chlorodendraceae
2. Sphaerellaceae	2. Palmellaceae	
3. Polyblepharidaceae		
4. Phacotaceae		

Family Chlamydomonadaceae: This family includes unicellular forms of the Volvocales. The cells are uninucleate with a definite cell wall. Each cell has two or four flagella of equal length and a cup-shaped chloroplast. The asexual reproduction is by zoospores and the sexual reproduction is from isogamous to oogamous.

Chlamydomonas, Chloromonas, Cymbomonas, Platymonas, and Carteria are common examples of this family. Chlamydomonas, a typical representative of the family, is described in details.

Chlamydomonas

Systematic Position

Class	-	Chlorophyceae
Order	-	Volvocales
Family	-	Chlamydomonadaceae

Occurrence: Chlamydomonas is one of the simplest unicellular fresh-water alga. It has about 500 species, occurring in fresh water ponds, lakes and tanks, which are rich in nitrogenous matter. It provides green turbidity to the ponds and lakes.

There are also some terrestrial species which are found along the banks of lakes and pools on the moist soil. Some species of Chlamydomonas occur in unusual habitats, e.g., C. yellowstonensis grows on the snow covered mountains of Yellow Stone National Park of the U.S.A., imparting them yellowish green colour, and C. nivalis, which is found on snow in alpine and arctic regions, imparts red colour to the snow due to the presence of a red coloured pigment (haematochrome). C. halophila and C. ehrenbergii occur in saline water.

In India, the genus is represented by about 18 species, of these C. moewusii, C. monadina, C. subcaudata, C grandistigma and C. iyengarii are common.

Thallus Structure: Chlamydomonas is a unicellular motile alga. The cell is biflagellate, spherical, ellipsoidal or pear-shaped; about 30 μ m in length and 20 μ m in diameter. The anterior end of the cell is usually papillate and the posterior relatively broader but in C. caudata the posterior end is pointed.

The cell is bounded by a thin but firm cellulose wall. Glycoproteins rich in hydroxyproline are the main structural component of the cell wall. In some species, the cellulose wall is surrounded by a distinct gelatinous sheath. The protoplast is surrounded by a thin semipermeable plasma membrane. Most of the species of Chlamydomonas have an massive cup-shaped parietal chloroplast.

It usually occupies almost entire posterior broader part of the cell. However, the chloroplast is 'H'-shaped in C. biciliata, discoid in C. alpina, reticulate in C. reticulata, ridged in C. stenji and axile in C. eradians. The chloroplast has a single pyrenoid, lying in its posterior part along the median line. But in C. kleinii, C. debaryanum and C. sphagnicola there are more than two pyrenoids in each chloroplast.

In C. reticulata the chloroplast is, however, without pyrenoid. The pyrenoid has a central protein core surrounded by starch plates, and it functions as the centre for starch synthesis.

The cup-shaped structure, formed by the chloroplast in the anterior part of the cell is filled with cytoplasm. The nucleus lies in the centre of the cytoplasm. In those species where chloroplast is other than cup-shaped, the nucleus is axial or eccentric. At the anterior pointed end of the cell there are two contractile vacuoles which are concerned with the osmoregulation of the cell. Each cell has an eye spot or stigma, which lies near the anterior end. The shape and position of the eye spot varies in different species. In C. pluristigma there are 2-3 eye spots in each cell. The eye spot has a colourless biconvex photosensitive lens and a curved pigmented plate. It is a photoreceptive organ and functions as a primitive eye.

The anterior papillate part of the cell bears two flagella which are equal in length. The flagella are of whiplash type and are associated with neuromotor apparatus of the cell. The neuromotor apparatus consists of three main parts: (i) blepharoplast, a small granular structure present at the base of each flagella; (ii) paradesmos, a transverse filament connecting the two blepharoplasts, and (iii) rhizoplast, a thin and soft descending thread connecting one of the blepharoplasts with the centrosome of the nucleus. The centrosome remains connected with the nucleolus by many fine fibrils.

Fine Structure of Chlamydomonas Cell: Ultrastructure of Chlamydomonas has been studied by Roberts et al. (1972) and Hill et al. (1973). Under electron microscope the cell wall appears multilayered, made up of cellulose fibrils. The plasma membrane, which lies inner to the cell wall, is differentiated into an outer and an inner proteinaceous layer and a middle lipid layer. There are many ribosomes, mitochondria and endoplasmic reticulum in the cytoplasm, present between the plasma membrane and chloroplast. The cytoplasm, which fills the cup-shaped cavity of the chloroplast, contains a nucleus, volutin granules and golgi apparatus, besides other organelles. The two layered nuclear membrane has many pores. Although a distinct nucleolus is present in the nucleus, chromosomes are not visible.

The cavity of the chloroplast, bounded by a double layered limiting membrane, is called stroma. It is filled with a homogeneous granular matrix. The stroma has many stacks of tubular photosynthetic lamellae. The lamellae are lipoproteinaceous in nature and are called thylakoids. Each stack has 2-6 thylakoids. The thylakoids are always oriented in the direction of the long axis of the cell. The pyrenoid is also a lamellate structure, and starch plates are synthesized on its periphery. The granular matrix of the chloroplast also contains many ribosomes, but they are smaller than those present in the cytoplasm. Ribosomes are absent in the pyrenoid.

Electron microscopic studies reveal that the eye spot consists of two or three parallel rows of fat globules (Fig. B). They are laden with carotenoids and lack limiting membrane.

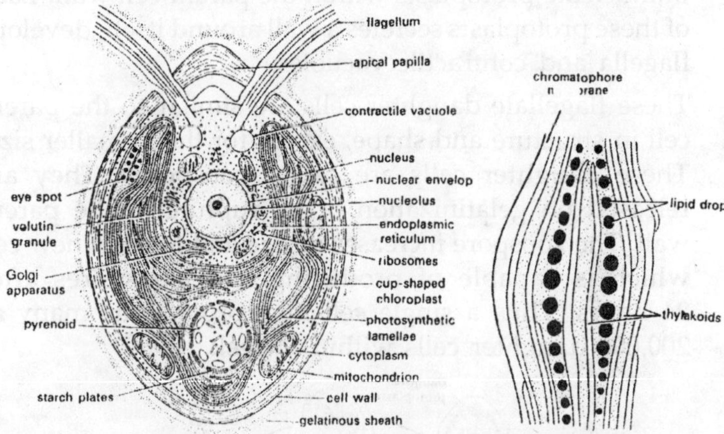

A-B. Chlamydomonas: a. Cell as seen under the electron microscope, B. Eye spot (note the presence of thylakoids)

Nutrition: Most species of Chlamydomonas are phototrophs, i.e., they synthesize their own food in the presence of chloroplast, light, water and carbon dioxide. A part of the sugar synthesized is utilized in the metabolic processes of the cell and the rest is stored around the pyrenoid in the form of starch grains. However, C. dysosmos is a facultative heterotroph and can grow in dark in the presence of acetate as a carbon source.

Reproduction: Chlamydomonas reproduces by asexual and sexual methods.

Asexual Reproduction: Asexual reproduction takes place mainly by zoospores. But some other types of asexual spores, such as aplanospores, hypnospores and palmella stage are also found.

1. *By Zoospores:* Zoospores are formed under favourable conditions, usually during the night. At the time of zoospore formation, the cell withdraws its flagella and becomes non-motile. The contractile vacuoles disappear, and the protoplast withdraws from the cell wall and divides longitudinally into two daughter protoplasts. This is followed by the successive longitudinal divisions at right angles to each other, forming 4, 8, 16 or more uninucleate protoplasts within the parent cell wall. Each of these protoplasts secretes a wall around itself, develops flagella and contractile vacuoles.

   These flagellate daughter cells are similar to the parent cell in structure and shape, except for their smaller size. These daughter cells are called zoospores. They are released on gelatinization or by rupture of the parent wall. The zoospore increases in size and forms a new cell which is capable of producing new zoospores after 24 hours. Thus a single cell may produce as many as 200,000 daughter cells within a week.

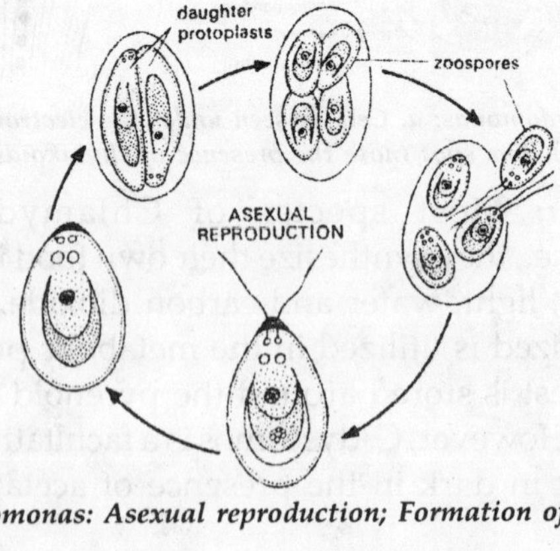

Chlamydomonas: Asexual reproduction; Formation of zoospores

Sometimes zoospores fail to escape from the gelatinized parent wall and this stage is known as palmella stage.

2. *By Aplanospores:* These spores are formed during unfavourable conditions. Under conditions of drought, the cells withdraw flagella and their protoplasts divide into 2-16 daughter protoplasts. Each protoplast secretes a thin wall but does not develop flagella. These non-motile cells are called aplanospores. In some species (e.g., C. nivalis), aplanospores secrete a thick wall under conditions of severe drought. These thick walled non-motile spores are called hypnospores. They germinate under favourable conditions and form new cells directly or their protoplasts divide to form zoospores.
3. *Palmella Stage:* Under adverse conditions (but not very dry) the protoplast of the parent cell divides to form 4-8 daughter cells. These cells, however, do not develop flagella and are non-motile and remain within the matrix formed by the gelatinization of the parent wall. This temporary colony of hundreds or thousands of cells, in a common gelatinous matrix, is called palmella stage. In C. nivalis, all the cells of the palmella stage secrete a thick wall and form hypnospores. The wall of the hypnospore is provided with haematochrome, a red. pigment, imparting red colour to the hypnospore.

The palmella stage is a non-motile and temporary reproductive phase. With the recurrence of favourable conditions, these cells form flagella and become motile. They come out of the gelatinous matrix and develop into large vegetative cells.

Sexual Reproduction: Sexual reproduction in Chlamydomonas takes place in the absence or deficiency of nitrogenous compounds in the medium. It varies from isogamous to primitive oogamous.

1. *Isogamy:* Most species of Chlamydomonas are isogamous. The sexual process starts with the division of the cell

protoplast into 8, 16, 32 or 64 biflagellate gametes. The gametes are usually' naked, without any cell wall. In many isogamous species the vegetative cells directly function as gametes without undergoing any division (hologamy). The isogamous species are characterised by structural similarities between the fusing gametes. In homothallic species (e.g., C. debaryanum, C. longistigma) the gametes produced in the same parent cell fuse to form the zygote (Fig.), whereas in heterothallic species (e.g., C. moewusii, C. reinhardii) gametes from the cells of two different strains fuse.

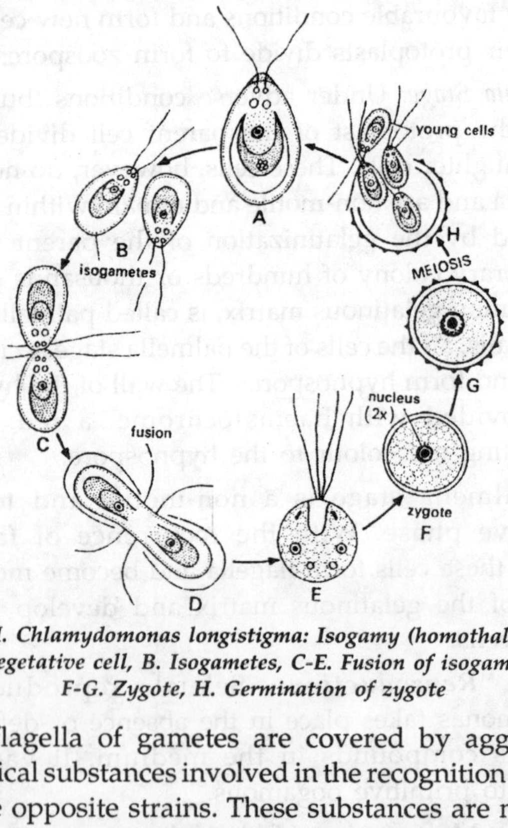

A-H. Chlamydomonas longistigma: Isogamy (homothallic); A. Vegetative cell, B. Isogametes, C-E. Fusion of isogametes, F-G. Zygote, H. Germination of zygote

The flagella of gametes are covered by agglutins, the chemical substances involved in the recognition of gametes of the opposite strains. These substances are not present in the flagella of the vegetative cells. In heterothallic species when gametes of 'plus' strain come in contact with those of 'minus' strain, the flagella of two gametes of opposite strains adhere because of agglutins. Initially, the gametes clump in groups of up to 50 with their flagella towards the center.

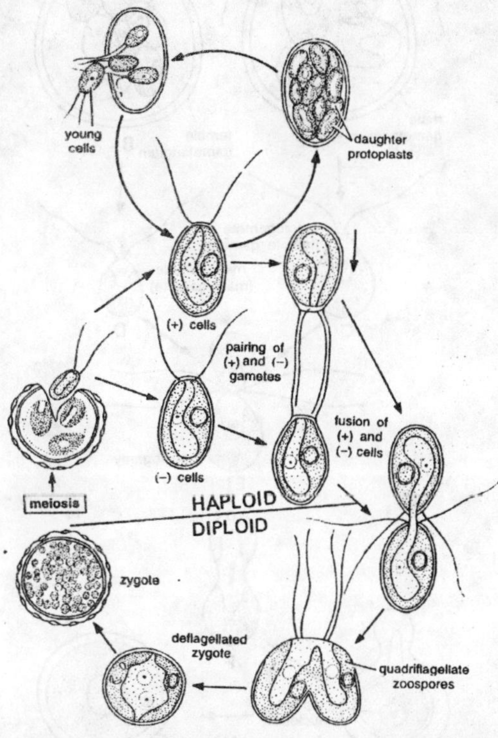

Chlamydomonas moewusii: Isogamy (heterothallic); life-cycle

In each clump the number of plus and minus gametes is variable. Eventually, gametes of opposite strains fuse in pairs at their anterior ends. The flagella become free and the paired gametes swim away from the clump.

2. *Anisogamy:* C. braunii and C. suboogama are the typical examples of anisogamy. Here the fusing gametes (male and female) are of different sizes. In the female gametangium only two or four macrogametes (female gametes) are formed which are similar to the mother cell in shape and size (Fig. A, C). In the male gametangium

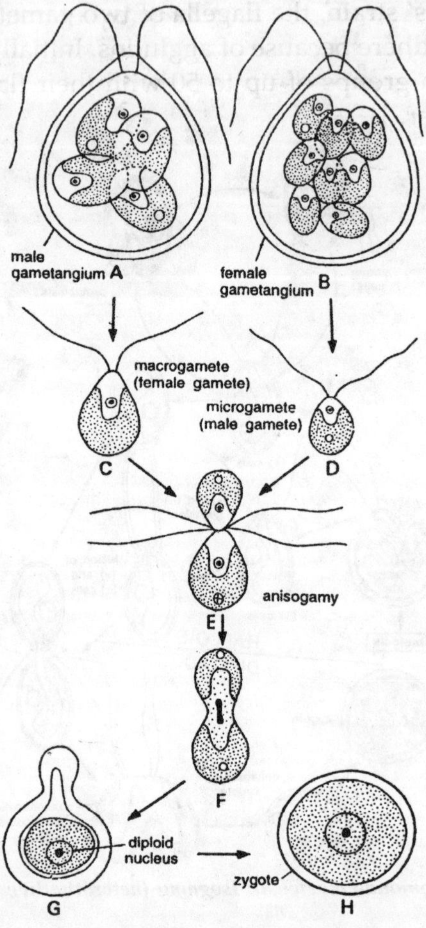

A-H. Chlamydomonas braunii: Anisogamy; A. Female gametangium, B. Male gametangium, C. female gamete, D. Male gamete, E-F. fusion of gametes, G-H. Zygote

8-16 microgametes (male gametes) are formed which are nearly half the size of the mother cell and are much narrower and elongated (Fig. B, D). The microgametes are more active than the macrogametes, and the latter come to rest sooner than microgametes. The active male gamete comes close to the female gamete and both fuse at their anterior ends to form a zygote (Fig.E-H).

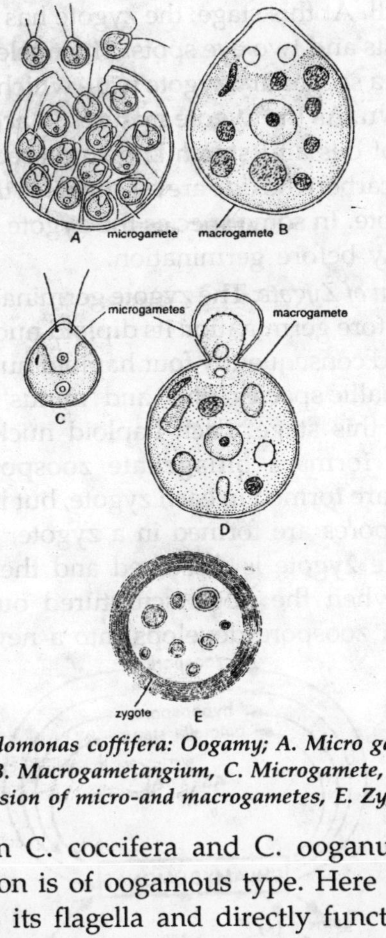

A-E. Chalmydomonas coffifera: Oogamy; A. Micro gametangium, B. Macrogametangium, C. Microgamete, D. Fusion of micro-and macrogametes, E. Zygote

3. *Oogamy:* In C. coccifera and C. ooganum, the sexual reproduction is of oogamous type. Here the female cell withdraws its flagella and directly functions as a non-motile macrogamete (it is equivalent to egg or ovum). The protoplast of the male cell divides repeatedly to form 16 segments (Fig. A). Each segment develops into a biflagellate micro or male gamete. The flagellated male gamete is active and comes close to non-motile female gamete. The two fuse at their anterior ends to form a zygote (Fig. C, H).
4. *Zygote:* The wall at the point of fusion between the two gametes soon dissolves. The quadriflagellate zygote swims for a while before it settles down. It secretes a primary

zygote wall. At this stage, the zygote has two nuclei, two chloroplasts and two eye spots. The nuclei fuse within 24 hours and a secondary zygote wall, which is ornamented, is laid down. As the zygote matures, it accumulates large amounts of oils and starch and turns reddish in colour. Light and carbon dioxide are essential for the development of the zygote. In some species the zygote enlarges in size enormously before germination.

5. *Germination of Zygote:* The zygote germinates inside water in dark. Before germination its diploid nucleus undergoes meiosis and consequently four haploid nuclei are formed. In heterothallic species 'plus' and 'minus' strains become distinct at this stage. Each haploid nucleus with some protoplast forms a biflagellate zoospore. Thus, four zoospores are formed in each zygote, but in C. reinhardtii eight zoospores are formed in a zygote. The inner wall layer of the zygote is dissolved and the zoospores are liberated when the outer sculptured outer wall splits open. Each zoospore develops into a new cell.

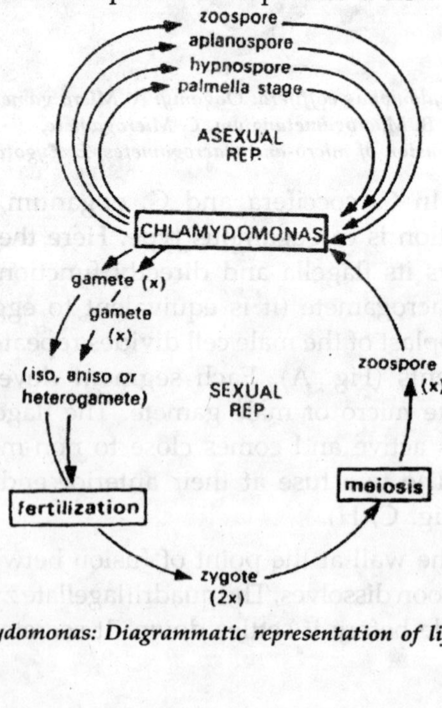

***Chlamydomonas*: Diagrammatic representation of life-cycle**

In Chlamydomonas sexuality is controlled by environmental and genetic factors. The amount of ammonium nitrogen in the growing medium is the most important. Its deficiency in the medium inhibits sexuality in Chlamydomonas and the gametes behave simply as vegetative cells.

Besides this, the intensity of light and temperature, and the presence of calcium in the medium also affects mating of gametes. In high light intensity larger number of zygotes are formed than in low light intensity.

The fusion occurs between the gametes of opposite strains (+ and -) in heterothallic species. It has been observed that non-flagellate mutant forms of Chlamydomonas are completely asexual. This suggests that flagella are important in mating processes. The presence of agglutin coating on the surface of flagella probably is essential for sexual processes.

Generation Change

The haploid phase is dominant in the life-cycle of Chlamydomonas. The vegetative cell is haploid and it produces haploid gametes. The diploid phase is attained in the zygote by gametic fusion. The diploid nucleus of the zygote divides meiotically before it germinates. Thus the diploid phase occurs for a short duration.

Origin and Differentiation of Sex: The various reproductive processes exhibited by different species of Chlamydomonas throw light on the evolution of sex in algae. When environmental conditions are favourable, Chlamydomonas produces motile, biflagellate zoospores, each of which is capable to develop into a new individual.

In this process, no fusion of cells occurs and hence it is a method of asexual multiplication. However, at the close of the growing season when environmental conditions become unfavourable Chlamydomonas resorts to sexual reproduction. It involves fusion of two sex cells resulting in the formation of a thick-walled resting structure, the zygote. The resting period of the zygote generally coincides with the duration of the unfavourable period for growth. The cells which fuse to form zygote are called gametes. The origin of gametes is as such the origin of sex. In isogamous species of Chlamydomonas like C. debaryanum, C. moewusii and C. longistigma, isogametes are formed in a similar manner as zoospores. Both the zoospores and gametes are remarkably similar in their form, structure, development and mode of liberation. They differ only in their size and subsequent behaviour. The gametes are invariably smaller in size and are incapable of growing into a new individual alone.

The marked resemblances between the gametes and zoospores suggest that isogametes have originated by the subdivision of swarmers (zoospores) into still smaller swarmers in response to unfavourable environmental conditions. These smaller swarmers (reduced zoospores) have become too small and weak to grow into new plants by themselves due to insufficiency of stored food and perhaps some other factors. Most of these perish in their wanderings, but a few meet by chance and fuse in pairs. This pairing increases their vitality and vigour to give rise to a new individual. This clearly illustrates that gametes are just reduced zoospores which are incapable to develop into new individuals. These reduced zoospores transformed into gametes and their union brings about sexual reproduction: The sex in Chlamydomonas is thus originated by accidental fusion of undersized zoospores in response to unfavourable environmental conditions. Since the process proved to be advantageous, it was maintained.

During further advancement gametes became of dissimilar size as in C. braunii. This was a step towards anisogamy. The next step in the evolution of sex is demonstrated by C. coccifera. Here, a vegetative cell instead of dividing enlarges and loses its flagella and motility. This cell acts as an ovum (egg cell), and fertilized by an active male gamete (antherozoid) formed in a manner as zoospores. Of these two gametes, one (the ovum) becomes sluggish and stores food material to support the new plant until it is established on its own, and the other (the male gamete or antherozoid) attains motility to reach the immotile ovum for fertilization. Thus, C. coccifera provides an example of the most advanced level of sexuality that can be traced in Chlamydomonas.

Family Volvocaceae

This family includes those flagellated forms of Volvocales which form colonies. The number of cells in a colony is in multiple of two, and each genus has a definite number of cells. Such colonies with definite number of cells are called coenobia (singular = coenobium). The shape of the coenobium and the arrangement of cells are specific for a genus. The members of Volvocaceae are fresh water forms. They reproduce asexually as well as sexually. The sexual reproduction is isogamous, anisogamous or oogamous.

Volvox

Systematic Position

Class	-	Chlorophyceae
Order	-	Volvocales
Family	-	Volvocaceae

Occurrence: Volvox is the most highly evolved coenobial alga, both in its morphology and reproduction. It has about 20 species which occur in fresh waters of temporary or permanent pools as minute spherical or ovoid balls of pin-head size. The colonies multiply so rapidly during rainy season that the entire pond is covered by them and looks greenish. The genus is represented in India by five species; of these V. globator, V. aureus, V. prolificus and V africanus are fairly common.

Thallus Structure: Volvox is a colonial alga, and the colonies are oval to spherical in shape, each with 500-60,000 cells. The cells are biflagellate and are arranged in a single layer within the periphery of the gelatinous colonial envelope (Fig. A). Each cell of the colony has its own gelatinous sheath and the sheaths of the individual cells may be distinct or confluent. In most of the species adjacent cells are connected by cytoplasmic strands which are formed during cell division. In some species (e.g., V. mononae, V. tertius) cytoplasmic strands are absent. The central hollow cavity of the colony is filled with a watery (e.g., V. globator) or gelatinous (e.g., V. aureus) substance.

The cells in a colony are usually pyriform in shape with broad posterior and narrow anterior end. They are eukaryotic and mostly show a typical chlamydomonad structure. But in V. globator and V. rousseletii it is of Sphaerella type. The proloplast is enclosed within a semipermeable plasma membrane, which forms the outer boundary of the cytoplasm. There is a single nucleus, a cup-shaped chloroplast, two contractile vacuoles, two flagella and an eye spot in each cell.

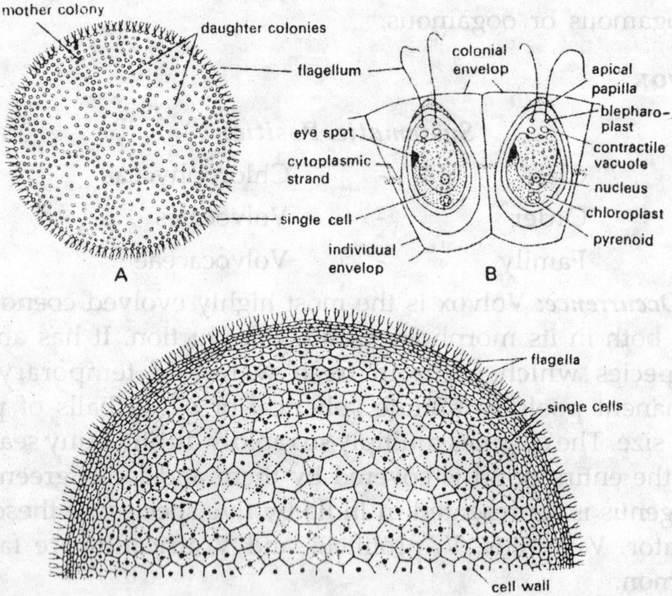

A-C. Volvax: A. A coenobium, B. A part of coenobium, C.A part of coenobium

The nucleus is connected with the flagella through the neuromotor apparatus of the cell. The eye spot is towards the external face of the cell. It has a colourless biconvex photosensitive lens and a curved pigmented plate. The massive cup-shaped parietal chloroplast occupies the posterior broader part of the cell. It has a single pyrenoid made up of a central protein core and peripheral starch plates.

Most cells of a colony are vegetative and only a few are reproductive. The reproductive cells do not have flagella and are relatively large than the vegetative cells. There are only a few or no reproductive cells at the anterior end of the colony, whereas the posterior end has many reproductive cells. The cells at the anterior end of the colony have relatively large eye spots than those at the posterior end. Thus, the colony of Volvox shows polarity.

Reproduction: Volvox reproduces by asexual and sexual methods. Asexual reproduction usually takes place during spring and summer months, whereas sexual reproduction starts towards the end of the summer season.

Asexual Reproduction: Asexual reproduction takes place by special type of cells situated at the posterior end of the colony. These cells, called gonidia or parthenogonidia, are about ten times larger than the vegetative cells. They are spherical and highly vacuolated. They develop numerous pyrenoids within their chloroplast. The nucleus is suspended in the center of the cell. The gonidia do not have flagella and eye spots. Usually two gonidia are separated by several vegetative cells. A gonidium is embedded in a hollow gelatinous cavity of the colony along with its globular gelatinous sac. The protoplast of the gonidium divides and forms daughter colonies.

Development of daughter colony. As mentioned above, the daughter colony develops from the gonidium. The enlarged and spherical gonidium undergoes three longitudinal divisions (with respect to the colony) forming eight cells. These cells arrange themselves in a curved plate and this stage is called plakea stage.

Each of these eight cells again divides longitudinally forming 16 cells. At this stage the cells form an almost complete sphere except for a small opening, the phialopore, directed towards the periphery of the colony. The phialopore eventually becomes the posterior pole of the daughter colony. These 16 cells divide repeatedly and may form as many as 60,000 cells.

However, such large number of divisions usually do not occur and divisions stop after a fixed number of cells have been formed. At this stage the cells are naked and are in close contact with each other. The anterior papillate part of each cell is directed towards the center of the hollow sphere. The normal orientation of the cells of a colony is achieved by inversion of cells of this sphere.

The inversion starts with the formation of a constriction at the posterior end of the sphere (i.e., opposite the phialopore) which pushes the cells of the posterior end towards the phialopore.

In the mean time the phialopore becomes broader and the cells of the posterior part are pushed through it. This process is known as inversion. At the end of inversion, the anterior papillate ends of the cells face towards the periphery of the colony (Fig.A, I).

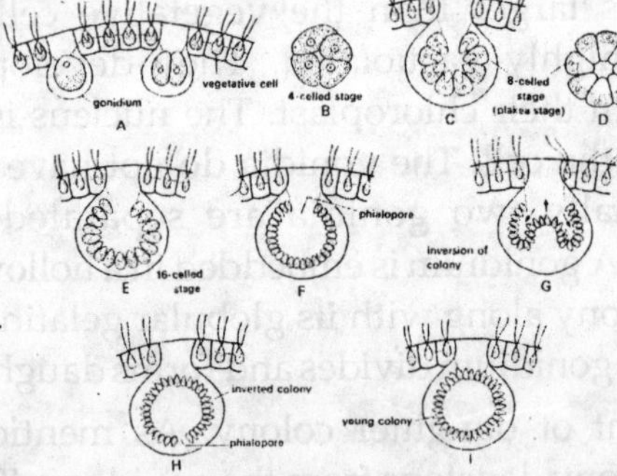

A-I. Volvox: Asexual reproduction (development of daughter colony); A-D. Formation of plakea stage, E. 16-celled stage, F-H. Stages of inversion, I. Daughter colony.

After inversion of the colony is completed the cells secrete gelatinous cell wall and develop flagella. In this way a daughter colony is formed inside each gonidium. Many daughter colonies may develop within a parent colony at the same time and they swim freely in the gelatinous matrix of the parent colony. The daughter colonies are released after the rupture or disintegration of the parent colony. The parent colony usually degenerates after the release of daughter colonies, but in V. africanus and V. carteri two to four generations of daughter colonies can be found within the parent colony.

Sexual Reproduction: Sexual reproduction in Volvox usually takes place towards the end of the growing season. It is of oogamous type. The species of Volvox are either monoecious (e.g., V. globator) or dioecious (e.g., V. aureus). Sometimes colonies of V. aureus have both antheridia and oogonia. The monoecious species are protandrous, i.e., antheridia mature before oogonia. Usually a colony involved in sexual reproduction does not have asexual daughter colonies. The cells developing into reproductive structures are located in the posterior part of the colony, and these cells are called gametangia. Usually the gametangia are in limited number. The male and female gametangia are called antheridia and oogonia respectively.

1. *Antheridium:* The gametangium developing into antheridium is larger than the vegetative cells of the colony. It withdraws its flagella, but remains connected with the other vegetative cells by cytoplasmic strands (Fig. A).

   The protoplast of the gametangium divides repeatedly to form 16-512 rod-shaped cells. Each cell develops into a naked, biflagellate and fusiform male reproductive cell or antherozoid (Fig. B-E).

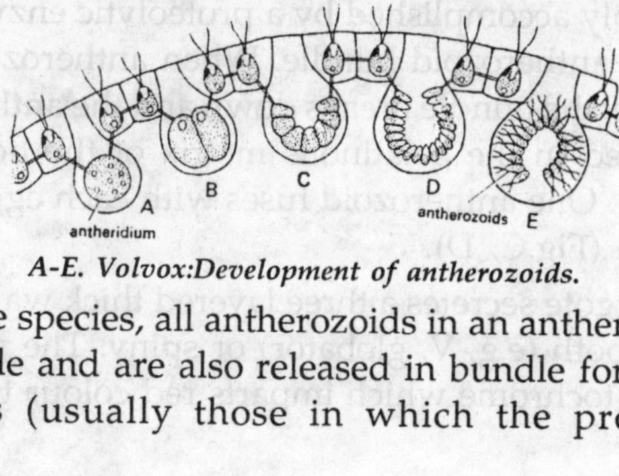

A-E. Volvox:Development of antherozoids.

In some species, all antherozoids in an antheridium form a bundle and are also released in bundle form. In other species (usually those in which the protoplast of antheridium divides to form more than 128 antherozoids), the antherozoids are arranged in a hollow sphere which shows inversion like asexual daughter colonies. Here the antherozoids are released individually.

2. *Oogonium:* The gametangium developing into oogonium is also larger than the other cells of the colony. It increases in size and eventually becomes flask-shaped. The entire protoplast of the oogonium finally forms a single uninucleate, non-flagellate and spherical egg.

   The egg has a parietal chloroplast and as such is green in colour. It is laden with reserve food material. The beak of the flask-shaped oogonium opens at the surface of the colony (Fig. C, D).

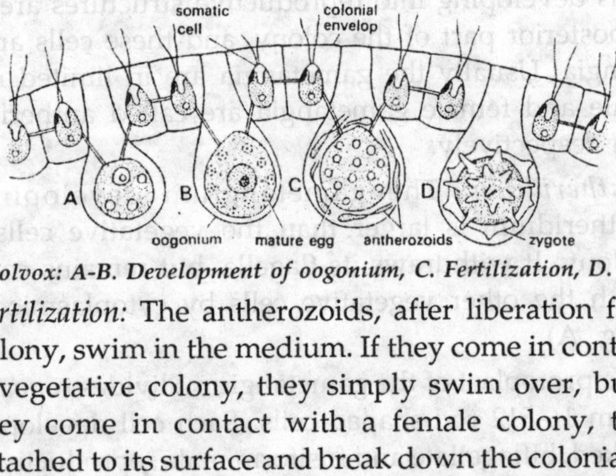

A-D. Volvox: A-B. Development of oogonium, C. Fertilization, D. Zygote.

3. *Fertilization:* The antherozoids, after liberation from the colony, swim in the medium. If they come in contact with a vegetative colony, they simply swim over, but when they come in contact with a female colony, they get attached to its surface and break down the colonial matrix of the colony in the area of their attachment. This task is probably accomplished by a proteolytic enzyme secreted by the antherozoid bundle. When antherozoids are in a bundle, the bundle breaks down and the antherozoids are liberated in the gelatinous matrix of the hollow female colony. One antherozoid fuses with each egg to form the zygote (Fig.C, D).

   The zygote secretes a three layered thick wall which may be smooth (e.g, V. globator) or spiny. The wall contains haematochrome which imparts red colour to the zygote.

The zygote may remain dormant for quite some time and is then liberated by disintegration of parent colony.

4. *Germination of Zygote:* The zygote usually does not germinate until the following spring. Before germination, the diploid nucleus of the zygote undergoes a meiotic division forming four haploid cells (Fig. A, C). These cells undergo further divisions. The outer layer of the zygote splits and the two inner layers protrude in the form of a vesicle. The cells migrate into this vesicle.

Species of Volvox vary considerably in further development of the zygote into a new colony. For example, in V. aureus and V. minor the protoplast of the zygote divides repeatedly and forms a new colony by the sequence of plakeal stages as is found in asexual reproduction. In V. campensis, the protoplast of the zygote forms many biflagellate zoospores, but only one of these survives and others degenerate. The surviving zoospore comes out of the vesicle and swims freely in the medium. It divides to form about 500 cells which arrange themselves into a hollow sphere. The hollow sphere undergoes inversion as in asexual daughter colony and develops into a new colony (Fig. E, F). In V. rousseletii, the entire protoplast of the zygote forms a single biflagellate zoospore which divides repeatedly and forms a new colony after inversion. The colonies developed by this method have relatively less number of cells.

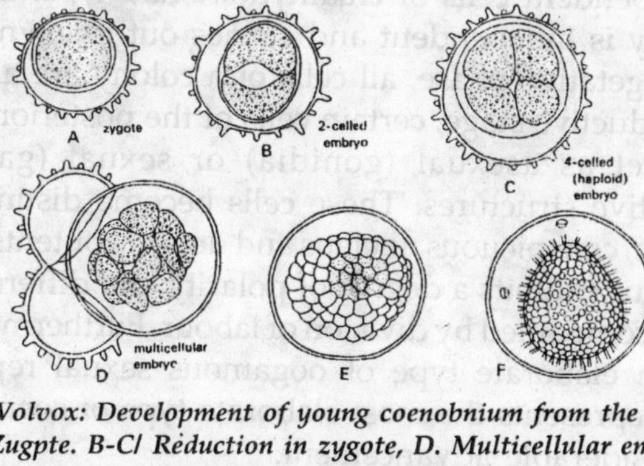

A-F. Volvox: Development of young coenobnium from the zygote; A. Zugpte. B-C/ Reduction in zygote, D. Multicellular embryo, E. Inversion, F. Young coenobium.

The life cycle of Volvox is of haplontic type; the zygote being the only diploid stage in the life-cycle. Its nucleus divides meiotically before the formation of zoospores which intimately develop into Volvox colony with haploid cells.

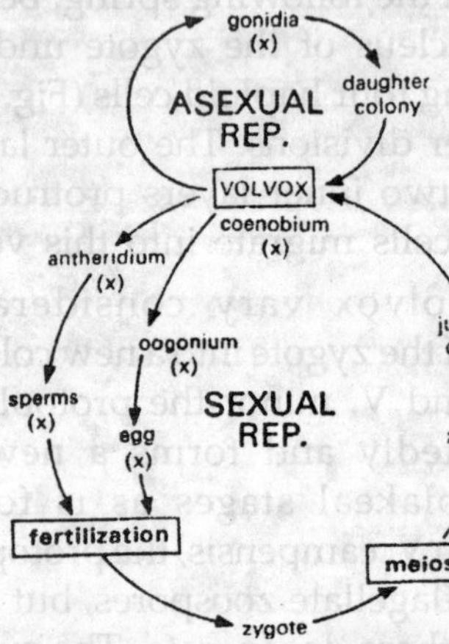

Volvox: Diagrammatic representation of life-cycle.

Thus, there is an alternation of a haploid phase with diploid zygote.

Specialization of Functions

(Division of Labour): Volvox shows several features of advancement over Chlamydomonas. The primitive unicellular motile Chlamydomonas cell performs all vital functions, both vegetative and reproductive. Volvox is an assemblage of similar and independent cells of chlamydomonad type. Each cell of the colony is independent and carries out its own functions. At the vegetative phase, all cells of a colony are alike. But at the reproductive stage, certain cells of the posterior end of the colony act as asexual (gonidia) or sexual (gametangia) reproductive structures. These cells become distinct by their larger size, conspicuous nucleus and denser contents. The adult colony thus exhibits a degree of polarity and differentiation of cells is accompanied by division of labour. Furthermore, Volvox shows an elaborate type of oogamous sexual reproduction. Thus, it represents the most elaborate type of coenobial form, with considerable advancement.

Chlorophyceae: The class Chlorophyceae is represented by 425 genera and 6500 species. Chlorophyll *a* and *b*, present in their plastids, impart them bright green colour, and as such they are called the green algae. Chlorophyceae are predominantly freshwater (about 90%), but there are some marine forms also (about 10%). The freshwater species are cosmopolitan in distribution, commonly inhabiting permanent or semipermanent pools, lakes, ponds, etc. The members of Oedogoniales and Conjugales are exclusively freshwater.

The marine species are mainly found in the tropical sea; members of Siphonales and Ulvaceae are predominantly marine. Certain forms of Volvocales, Cladophorales and Chaetophorales are found both in the fresh and marine waters. Species of Ulothrix, Vaucheria and Zygogonium are sub-aerial, and are usually found on damp soil in the form of sheets. Species of Protococcus and Trenteopohlia are epiphytic, found on tree trunks, damp woodwork and shady leaves. Many species of Cladophora and Characium are epizoic (growing on animal bodies), whereas a few unicellular forms (e.g., Chlorella) are endophytic. Cephaleuros and Rhodochytrium are parasitic, and they lack green colour. Cephaleuros causes red-rust disease of tea leaves.

Some species occur in unusual habitats; for example, Chlorella has some thermophilic species, found in hot springs, Chlamydomonas yellowstonensis and Haematococcus nivalis occur on the snow covered mountain tops and Chlamydomonas ehrenbergii inhabits in saline water. Some green algae in, association with certain fungi constitute lichens.

They following are important distinguishing features of the green algae:

(1) They exhibit a wide range in their thallus structure and morphology.

(i) Unicellular motile form e.g., Chlamydomonas, Sphaerella.

(ii) Unicellular non-motile form e.g., Chlorella, Chlorococcum.

(iii) Coenobium: e.g., Pandorina, Eudorina, Volvox, Pediastrum, Hydrodictyon.

(iv) Palmelloid: e.g., Tetraspora.

(v) Dendroid: e.g., Ecballocystis.

(vi) Filamentous unbranched e.g., Ulothrix, Spirogyra.

(vii) Filamentous branched e.g., Cladophora.

(viii) Heterotrichous e.g., Fritschiella, Coleochaete, Stigeoclonium, Draparnaldiopsis.

(ix) Siphonaceous e.g., Protosiphon, Vaucheria, Codium.

(x) Parenchymatous e.g., Ulva, Enteromorpha.

(2) The cell structure is eukaryotic.

(3) The cell wall generally has cellulose as the main structural polysaccharide except for siphonaceous forms, which have mannan or xylan. In Chara, the cell wall is impregnated with calcium carbonate.

(4) The protoplast is bounded by a thin and semipermeable plasma membrane. The cytoplasm possesses many small vacuoles or there is a large central contractile vacuole. The diastolic and systolic movements of the contractile vacuole possibly have osmoregulatory function. The large central vacuole in the mature cell pushes the protoplasm along the plasma membrane in the form of a thin layer, called primordial utricle.

(5) Like all other eukaryotes, the cells of the green algae have mitochondria, golgi bodies, ribosomes, endoplasmic reticulum, membrane bound plastids and other cell organelles.

(6) Their main pigments are chlorophyll *a* and *b*, but α— and β-carotenes and xanthophylls are also present. The pigments are found in the lamellae (thylakoids) of well organized chloroplast.

(7) The shape of the chloroplast shows much variation; it is cup shaped (e.g., Chlamydomonas), girdle shaped (e.g., Ulothrix), reticulate (e.g., Cladophora, Oedogonium),

parietal (e.g., Draparnaldiopsis), stellate (e.g., Zygnema), spiral (e.g., Spirogyra) or discoid (e.g., Vaucheria, Chara).

(8) The chloroplast normally contains pyrenoids surrounded by a sheath of starch. In Derbesia, pyrenoids disappear when the alga is grown in diffuse light.

(9) They have well organized nucleus. Usually there is a single nucleus in each cell, but the members of Siphonales and Cladophorales are coenocytic. The position, size and shape of nuclei are quite variable. The nucleoli and chromosomes remain suspended in the granular matrix of the nucleus. In general, there is a single nucleolus per nucleus but in Conjugales, there are several nucleoli in a nucleus.

(10) Most of the flagellate cells have a photosensitive red eye spot or stigma in the anterior portion, near the flagellar base.

(11) Their photosynthetic pathways are similar to those of higher plants. The reserve food is starch, composed of amylose and amylopectin.

(12) The motile vegetative or reproductive cells (zoospores or gametes) have two (rarely four) equal, smooth, whip-like flagella. However, in Oedogoniales the motile cells have a crown of flagella. In a few instances (e.g., Haematococcus), there are two pleuronematic flagella of equal length. Structurally, when viewed under electron microscope, flagella show a typical 9+2 fibrillar organization.

(13) They reproduce by vegetative, asexual and sexual methods. The vegetative propagation takes place by cell division and fragmentation. Akinetes, zoospores and aplanospores are the main asexual spores. Zoospores are often formed during night and are then liberated in morning.

(14) Sexual reproduction occurs in all members, except some Chlorococcales. It may be isogamous, anisogamous or oogamous.

(15) The zygote or oospore secretes a thick wall and undergoes a resting period. It germinates under favourable conditions and forms a new thallus.

Classification of Chlorophyceae: Fritsch (1935), on the basis of morphology and reproductive characters, recognised the following nine orders in Chlorophyceae:

(1) Volvocales (2) Chlorococcales
(3) Ulotrichales (4) Cladophorales
(5) Chaetophorales (6) Oedogoniales
(7) Conjugales (8) Siphonales
(9) Charales.

Smith (1955) classified the green algae into two classes-Chlorophyceae and Charophyceae. The class Chlorophyceae was divided into 12 orders, and only a single order was included in Charophyceae.

Class 1. Chlorophyceae

Order (1) Volvocales
(2) Tetrasporales
(3) Ulotrichales
(4) Ulvales
(5) Schizogoniales
(6) Cladophorales
(7) Oedogoniales
(8) Zygnematales
(9) Chlorococcales
(10) Siphonales
(11) Siphonocladales
(12) Dasycladales

Class 2. Charophyceae

Order (1) Charales

Mclean and Cook (1956) divided Chlorophyceae into the following eight orders:

(1) Volvocales
(2) Chlorococcales
(3) Chaetophorales
(4) Ulotrichales
(5) Oedogoniales
(6) Conjugales
(7) Siphonales
(8) Siphonocladales

The members of the order Charales were placed by them in a separate division Charophyta.

The Nostocales

This is the largest order of the class Cyanophyceae. The members of this order are non-heterotrichous filamentous forms without branching, but some members show false branching. Heterocysts are found in most of the members. Some genera show polarity with heterocysts and/or akinetes at the base and a hyaline hair at the apex of the trichome. Cyanophyceae reproduce by hormogonia, akinetes, exospores, endospores or hormospores.

Desikachary (1959) divided this order into five families: Oscillatoriaceae, Microchaetaceae, Nostocaceae, Scytonemataceae and Rivulariaceae.

Family Oscillatoriaceae

This is the largest family of the class Cyanophyceae. The members of this family are filamentous; the filaments are unbranched and uniseriate. All the cells of a filament are alike. In most of the members the trichomes have an envelope of homogeneous mucilagenous substance. All cells of a trichome are capable of division. They reproduce by hormogonia. Oscillatory movements of trichomes and the absence of heterocysts are the two main characteristic features of the family. The common examples of this family are: Oscillatoria, Lyngbya, Spirulina, Arthrospira, Phormidium and Trichodesmium.

Oscillatoria

Systematic Position

Class	-	Cyanophyceae
Order	-	Nostocales
Family	-	Oscillatoriaceae

Occurrence: Oscillatoria is found in a wide variety of freshwater (such as rain water pools, streams, ponds and lakes, and even in polluted water) and sub-aerial habitats. The alga forms extensive sheets along the muddy banks of rivers and ponds, rocky cliffs and on damp soil. It also occurs in the form of a thin layer at the bottom of shallow ponds. O. prolifica is a planktonic species and O. terebriformis is found in hot water springs. O. selina and O. limosa occur in saline waters. O. rubescens forms dense deep-water blooms in temperate lakes and ponds and its migration and colonization in deep water lakes is often associated with a high level of eutrophication. Some species of Oscillatoria have specific colours; for example, O. prolifica is red, O. formosa is blue, and O. angusta is colourless. In India this genus is represented by about 105 species, of these O. formosa, O. angusta, O. sancta, O. princeps, O. chlorina and O. decolorata are common.

Thallus Structure: Oscillatoria is an unbranched filamentous alga. The filaments (or trichomes, as they are called) occur singly or are matted together to form thin or thick sheets. Each trichome is made up of many cells, arranged in uniseriate fashion. There are no constrictions between adjacent cells. All cells of a filament are similar in shape and structure. Their breadth is usually greater than their height, but species with narrow trichomes have cylindrical cells (e.g., O. splendida). The trichome is slightly different at the anterior end and shows a definite polarity. The free end is mostly rounded or convex (e.g., O. annae, O. limosa;) but it is acuminate in O. acuminata and O. brevis and discoid in O. princeps. In some species (e.g., O. proboscidea;) the apical cell of the filament carries a thickened membrane, known as calyptra. The above mentioned differences are due to variations in turgor pressure of the terminal cell.

The trichomes are naked or are enveloped in a thin gelatinous sheath.

Cell Structure: All cells of a trichome have similar structure. The cell is surrounded by a thin and firm cell wall which lies external to plasmalemma. The wall is differentiated into four layers, designated as LI, LII, L III and LIV; L I being next to plasmalemma. Each layer is about 10 nm in thickness. In structure and composition the cell wall is similar to that of Gram-positive bacteria. It is composed of mucopeptide together with carbohydrates, amino acids and fatty acids. The L I and L III layers are electron transparent, whereas the L II and L IV layers are electron opaque. Contractions in the fibrils of the cell wall are responsible for gliding movements of Oscillatoria.

The cell exhibits a typical prokaryotic structure. The protoplast is differentiated into the peripheral pigmented chromoplasm and the central centroplasm (Figs.). The chromoplasm is principally composed of thylakoids and their associated structures, phycobilins. The thylakoids are lamellate structures with two membranes joined at the ends, and enclosing an intrathylakoid space of variable width. Photosynthetic pigments are found on the surface of thylakoids. Thus membrane bound plastids, like those of eukaryotic cells, are absent. The amount of c-phycoerythrin and c-phycocyanin is higher than any other pigment.

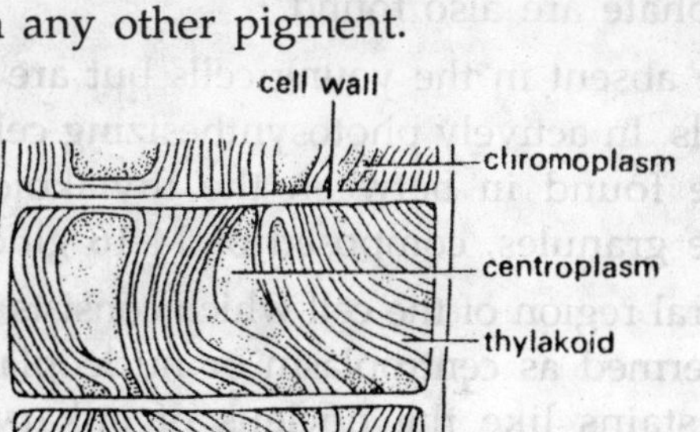

Oscillatoria: Cell structure as seen under light microscope.

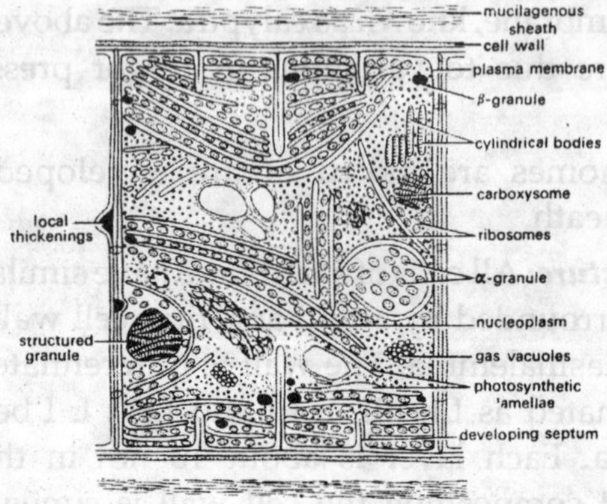

***Oscillatoria*: Cell structure as seen under electron microscope.**

In floating species, rounded or hollow cylindrical and tubular gas vesicles are found in the cytoplasm. The wall of the vesicle is exclusively composed of protein ribs or spirals. Since species with gas vacuoles are present near the surface of water, the vacuoles probably help in light shielding, and/or provide buoyancy.

The gas vacuoles are usually formed in low light intensity when filaments are near the bottom of the pond. Increase in buoyancy brings trichomes to the surface of water where they burst due to high light intensity, and again go deep into water.

Some angular polyhedral bodies are found near the centroplasm. They store carboxysomes which contain CO_2 fixing enzyme ribulose 1,5-bisphosphate carboxylase. In addition to these, some small spherical polyphosphate bodies containing stored phosphate are also found.

They are absent in the young cells but are prominent in the older cells. In actively photosynthesizing cells polyglucon granules are found in between the thylakoids. These are carbohydrate granules, composed of 14-16 glucose units.

The central region of the cell which consists of the nuclear material is termed as centroplasm or nucleoplasm. Although this region stains like the nucleus of eukaryotes, nuclear membrane and nucleoli are absent. It is not sharply differentiated, and commonly interfingers with the peripheral chromoplasm.

Gaidukov Phenomenon

Gaidukov (1902) observed that species of Oscillatoria develop different shades when exposed to lights of different wave lengths. For instance, a species of Oscillatoria can assume a green colour in red light, a reddish - blue in green light, and a bluish-green in yellow light.

Such a capacity to change colour complementary to that of the light is known as chromatic adaptation or Gaidukov phenomenon. This adaptation is of definite advantage since it enables the alga to maximally absorb the available light for photosynthesis.

Growth: Oscillatoria exhibits intercalary growth. All the cells of a trichome are capable of division and they divide at right angles to the long axis of the trichome. Cell division occurs by median constriction of the cell. The two inner wall layers, L I and L II, grow inward in a ring like fashion, pushing the thylakoids inwards, and gradually cutting across them as the new septum is formed.

When growth is rapid, most of the cells of the filament divide simultaneously. Usually the septum is not complete, and it leaves a fine interconnection between the two adjacent cells, called plasmodesmata.

This type of cell division is considered by some as primitive mitotic, while others prefer to call it amitotic.

Movement: Oscillatoria exhibits spontaneous movements. The rate of movement differs with the species. It is also affected by temperature, pH, viscosity of the medium and light intensity. The following two types of movements are recognised:

Gliding or Creeping Movement

When the trichome exhibits rhythmic movements on any solid surface without the help of any visible organ it is called gliding movement. This type of movement is characterised by the absence of a distinct change in the shape of the organism. It takes place within the mucilage sheath, the sheath sticking to the substrate and then being left behind by the advancing trichome. Gliding is always along the longitudinal axis of the trichome. In some species gliding movement is accompanied by rotation of the trichome.

Oscillatory Movement

The trichomes of Oscillatoria move to the left and right side of the axis just like the pendulum of a wall clock. Because of the presence of such oscillatory movements, the genus is named Oscillatoria.

Several views have been put forward to explain the movement in Oscillatoria. Some important ones are:

(1) The movement is due to invisible cilia present at the apex of the trichomes.

(2) The movement is accomplished by the secretion of mucilage (through the small pores present on the surface of the trichome) against substrate. The physical changes in the extruded mucilage also help in the movement.

(3) The undulating surface of the trichome is responsible for the movement.

(4) The movement results from rhythmic contraction waves in helical fibrils on the cell surface.

Nutrition

Main photosynthetic pigment is chlorophyll a, and the accessory pigments are c-phycocyanin and c-phycoerythrin. These pigments trap the solar energy and synthesize starch and glycogen, which are later converted into cyanophycin starch and glycoprotein.

Reproduction: Oscillatoria reproduces vegetatively by the following methods:

By Hormogonia: During vegetative growth of the trichome, some cells are filled with mucilage and become dead. These cells are called necredia (singular = necredium). They loose their turgidity, and due to the pressure of transverse walls of the adjacent cells become biconvex.

Such dead cells function as separation discs and they provide weak links in the trichome. The cells of the trichome between the two necredia form a hormogonium.

Soon after separation into hormogonia, the terminal cell of the broken end becomes spherical, conical or tapered, depending on the species. The hormogonia are capable of movement, and by repeated cell divisions they form new trichomes.

By Fragmentation: Sometimes the trichome divides into small pieces due to insect bite or mechanical injury. The fragments, so formed, have the capability of growing into long filaments by cell division, just like hormogonia.

Spirulina

Systematic Position

Class	-	Cyanophyceae
Order	-	Nostocales
Family	-	Oscillatoriaceae

Occurrence: Spirulina occurs in a wide variety of fresh water habitats, often found densely entangled with the Oscillatoria filaments. Some species are found in brackish water, and can tolerate very high salt concentrations.

For example, S. subsalsa can grow and multiply in solutions that are more concentrated than 3M sodium chloride. Some common species of Spirulina are S. albida, S. major, S. princeps, S. subsalsa and S. versicolor.

Species of Spirulina are rich in proteins, essential vitamins and unsaturated fatty acids. They are receiving considerable attention for their use as a supplement to diet.

Structure: It is an unicellular alga, the trichome consists of a single, cylindrical cell, wounded into a loose or close helix.

The cell exhibits a typical prokaryotic structure. The protoplast is differentiated into a peripheral pigmented chromoplasm and a central centroplasm. Cyanophycin granules are, however, absent.

Movement: Spirulina shows spontaneous movements. The motion is usually creeping or gliding in the direction of the longitudinal axis and may take place both backward and forward. In some species, gliding movement is accompanied by the rotation of the trichome, taking place in the direction of coiling.

Reproduction: Spirulina reproduces vegetatively by the process of fission.

Family Nostocaceae

The filamentous forms included in this family occur singly or in a common mucilagenous matrix. The trichomes are unbranched and uniseriate, and all the cells of the trichome are similar. They do not exhibit polarity, i.e., they are not differentiated into basal or apical ends.

Heterocysts are found in all members of the family; they may be terminal or intercalary in position. The reproduction takes place by akinetes and hormogonia.

Nostoc

Systematic Position

Class	-	Cyanophyceae
Order	-	Nostocales
Family	-	Nostocaceae

Occurrence: Species of Nostoc occur in freshwater as well as in terrestrial habitats. The fresh water species form macroscopic colonies on the surface of water which vary from few millimeters to 8 cm or more in diameter (Fig. A). Each colony is bounded externally by a pellicle like membrane. N. commune, a terrestrial species of alpine regions, forms leathery or rubbery sheet on the damp soil. Some species of Nostoc are endophytic and they occur in symbiotic association within other plants. For example, N. punctiforme is found in the coralloid roots of Cycas and Anthoceros thallus, and N. sphaericum and N. collema are the components of lichens and occur in symbiotic association with fungi.

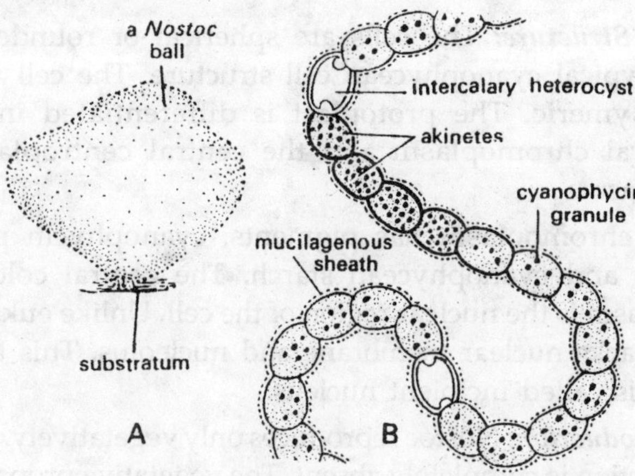

A-B. Nostoc: A. Gelationous colony, B. Trichome.

Some species of Nostoc are grown in rice fields. They have the capability to fix atmospheric nitrogen, thus helping in increasing fertility of the soil.

In India, Nostoc is represented by about 26 species, of which N. calcicola, N. endophyllum, N. ellipsoporum, N. muscorum and N. punctiforme are common.

Thallus Structure

The trichomes of Nostoc are uniseriate, usually contorted and twisted in various ways. An important feature is the presence of prominent constrictions between the adjacent cells which give a characteristic moniliform appearance to the trichome (Fig.B).

Each trichome is enveloped in a gelatinous sheath which is transparent, hyaline or coloured. Usually many trichomes aggregate together, and their gelatinous envelopes dissolve to form colonies of various shapes and sizes.

The trichomes of Nostoc are characterised by the presence of heterocysts which are intercalary or occasionally terminal, usually solitary and of much the same size and shape as the vegetative cell. The intercalary heterocysts have two polar nodules and the terminal ones have only one (basal) polar nodule.

Cell Structure: The cells are spherical or rounded and exhibit typical cyanophycean cell structure. The cell wall is mucopolymeric. The protoplast is differentiated into the peripheral chromoplasm and the central centroplasm or nucleoplasm.

The chromoplasm has pigments, cyanophycin protein granules and cyanophycean starch. The central colourless centroplasm is the nuclear region of the cell. Unlike eukaryotic cells, it lacks nuclear membrane and nucleolus. This type of nucleus is called incipient nucleus.

Reproduction: Nostoc reproduces only vegetatively. Sexual reproduction is completely absent. The vegetative propagation takes place by the following methods.

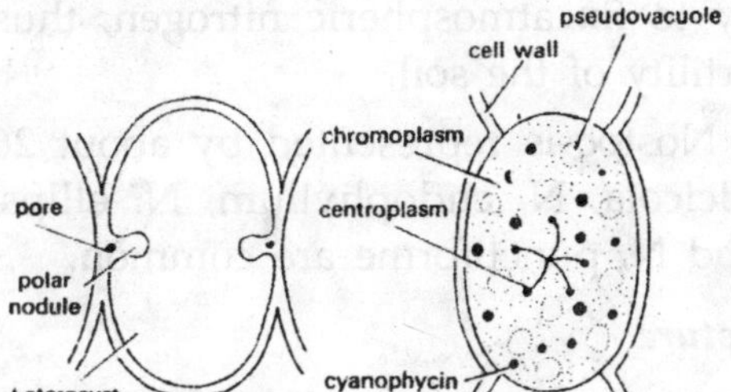

A-B. Nostoc: A. Heterocyst, B. Cell structure as seen under light microscope.

By Fragmentation: The colony may break into small fragments due to mechanical, physiological or other factors. Each fragment has the capability to develop into a new colony.

By Hormogonia: The trichome breaks into small segments due to the degeneration of intercalary vegetative cells or because of the presence of intercalary heterocysts. Multicellular fragments so formed are called hormogonia.

They come out of the gelatinous sheath of the colony, grow rapidly and form new colonies. Very often, the hormogones fail to come out of the parent colony and divide inside the gelatinous sheath of the parent colony. This results in a large number of trichomes inside the parent colony.

By Akinetes: During unfavourable period, some cells of the trichome are transformed into resting spores or akinetes. They have an additional three layered coat outside the normal cell wall. The outer layer is variously sculptured. The protoplasm is full of food reserves.

Development of akinetes normally takes place in a mature colony. It is always the vegetative cell adjacent to the heterocyst which is the first to metamorphose into akinete. Subsequently, many vegetative cells are transformed into akinetes. It is not unusual to find all the cells between the two successive heterocyst transformed into akinetes. Metamorphosis of the vegetative cell into akinete is probably stimulated by certain chemical substances secreted by heterocysts.

The akinetes have tremendous resistance for cold and drought. During favourable conditions, the protoplasm becomes active and breaks the thick outer wall and forms a new trichome.

By Heterocysts: In some species of Nostoc (e.g., N. commune) heterocysts act as resting spores. The protoplasm of the heterocyst of such species becomes functional and germinates to form a new trichome.

At the time of germination the protoplast divides by a transverse wall to form two cells. Both these cells divide again to form a 4-celled germling (Fig. A, C). The thick wall of the heterocyst ruptures at this stage and the germling comes out and develops into a new trichome.

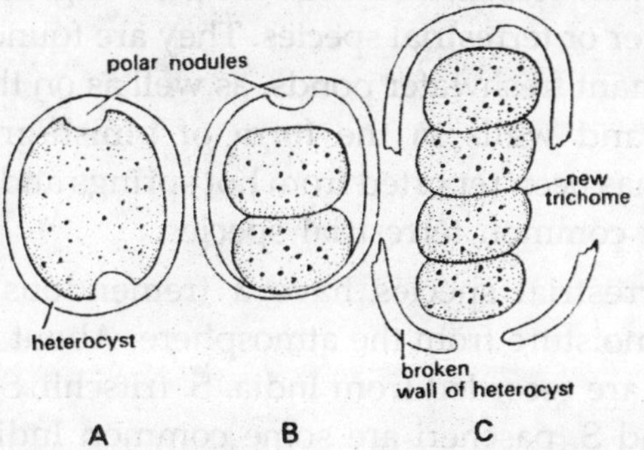

A.C. Nostoc: Stages of reproduction by heterocyst.

By Endospore: In some species of Nostoc (e.g., N. commune, N. microscopicum) the protoplasm of the heterocyst divides successively to form endospores.

Unlike akinetes, the endospores are thin walled and hence can not be regarded as resting spores. Disintegration of the capsule wall results in their liberation, and the rounded spores later germinate to form new trichomes.

Fusion of Filaments

Although true sexual reproduction and sex organs are absent in Nostoc as in other blue-greens; yet fusion and anastomosis of trichomes have been observed in N. muscorum. This phenomenon can be compared with somatogamy.

Family Scytonemataceae: The members of this family are filamentous forms with false branches. Except Plectonema, all members of the family possess spherical or oval heterocysts which are usually intercalary. They reproduce mostly by hormogonia. In India, the family is represented by members like Scytonema, Plectonema, Tolypothrix and Scytonematopsis.

Scytonema

Systematic Position

Class	-	Cyanophyceae
Order	-	Nostocales
Family	-	Scytonemataceae

Occurrence: This filamentous blue-green alga includes about 40 freshwater or terrestrial species. They are found in brackish water, stagnant freshwater ponds, as well as on the moist soil, tree-barks and walls in the form of bluish-grey scum. S. caldarium has been reported from hot springs and S. ocellatum is the most common terrestrial species.

The terrestrial species have a tremendous capacity of absorbing moisture from the atmosphere. About 33 species of Scytonema are recorded from India. S. fritschii, S. amplum, S. iyengari and S. pascheri are some common Indian species.

Thallus Structure: The plant body is filamentous, and the filaments intertwine to form a scum of bluish-green, blackish-green or yellowish-green colour. Each filament, made up of spherical, oval or squarish cells, is surrounded by an individual mucilagenous sheath. The sheath may be hyaline or coloured, smooth, homogeneous or lamellated. The terrestrial species, in particular, possess pigmented sheaths due to the presence of a pigment, scytonemin.

Cell Structure: The cell shows a typical cyanophycean structure. The cell wall is typically polymeric in nature, composed of mucopolypeptides and muramic acids. The protoplast is differentiated into the peripheral pigmented chromoplasm and the central centroplasm. The chromoplasm, in addition to chlorophyll- *a*, β-carotene and xanthophylls, also contains two important phycobilin pigments, c-phycoerythrin and c-phycocyanin. The centroplasm is relatively dense and it represents the (incipient) nucleus of the cell. It does not have the limiting nuclear membrane and nucleolus.

Branching: A characteristic feature of Scytonema is the presence of false branches, which are formed by the breaking of the filament. When a filament breaks, one or both the broken ends protrude out of the mucilagenous sheath and grow like branches. If both the broken ends come out of the sheath and grow like branches, it is called geminate branching, and if only one, then it is said to be false branching.

The false branches are formed by the following methods.

(a) *By Loop Formation:* Certain parts of the filament increase in length due to rapid cell division. Such localized growth results in loop formation that protrudes through the sheath. Eventually the loop breaks in the middle, cell division continues and two false branches are formed (Fig. A).

(b) *By Formation of a Separation Disc:* Sometimes one of the intercalary cells becomes dark coloured due to the diffusion of chromatophores in the protoplast, and its cell wall becomes thin. This cell forms a biconvex separation disc due to the pressure of the adjacent cells. The disc ultimately degenerates, resulting in the breaking of the trichome. One of the broken ends protrudes through the sheath as false branch (Fig. D).

(c) *By Rupture of Filament near the Intercalary Heterocysts:* In trichomes with intercalary heterocysts a zone of weak adhesion lies where the vegetative cell and the heterocyst abut each other.

If the trichome breaks on one side of the heterocyst, a single branch is formed and if it breaks on both the sides of the heterocyst, the branching is geminate.

(d) *By Degeneration of one or more Intercalary Cells:* This is a very common method of the formation of false branches. One or more intercalary cells degenerate and leave gaps in the trichome. One or both the broken ends of the trichome may protrude through the sheath and develop into branches.(Fig. B).

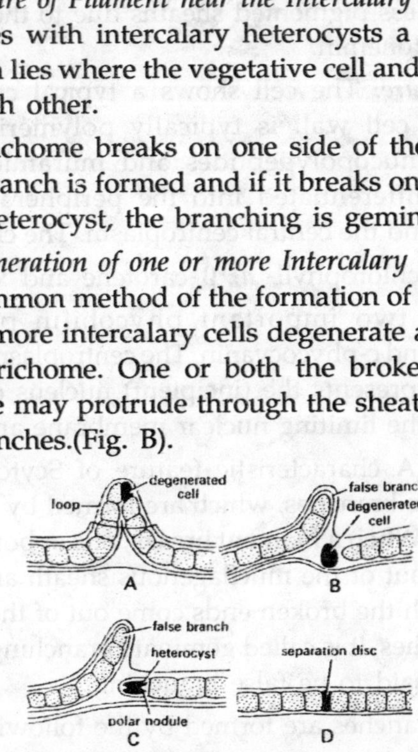

A-D. Scytonema: Modes of branching: A. Loop formation, B. Degeneration of intercalary cell, C. Breaking of filament near the intercalary heterocyst, D. Formation of separation disc.

Growth: In Scytonema growth occurs by repeated transverse divisions in the apical or sub-apical cell. In terrestrial species, the apical cell remains active for sometime and then degenerates. Thereafter the sub-apical cell functions as apical cell. Besides the apical and sub-apical cells, some intercalary cells also divide transversely and add to the growth of the trichome.

Reproduction: Multiplication of trichomes in Scytonema takes place by the formation of hormogonia or akinetes.

1. *By Hormogonia:* Many factors, such as the presence of separation discs and heterocysts, and disintegration of intercalary cells help in the fragmentation of the trichome. These fragments are called hormogonia, and each hormogonium is capable of growing into a new trichome.
2. *By Akinetes:* Some species of Scytonema (e.g., S. rivulare) form thick walled perennating cells, called akinetes. These resting cells, besides multiplication, also help in tiding over the unfavourable conditions. On return of favourable conditions, the akinetes germinate into a new trichome.

Family Rivulariaceae

This family includes freshwater forms with uniseriate tapering trichomes. The sheaths surrounding the trichomes are homogeneous or lamellated, and have a firm texture. The trichomes usually form spherical, hemispherical or irregular colonies. Except few non-heterocystous genera like Amphithrix and Calothrix, most of the members of this family are heterocystous. The false branches, which arise near the tapering ends, are formed by the breaking of the trichome just below an intercalary heterocyst.

Gloeotrichia

Systematic Position

Class	-	Cyanophyceae
Order	-	Noslocales
Family	-	Rivulariaceae

Occurrence: It is a fresh water alga, found floating in pools and ponds in the form of spherical mucilagenous colonies. A few species are attached to submerged plants. In India, the genus is represented by about 10 species, of these G. indica and G. ghosei are common.

Thallus Structure: The plant body of Gloeotrichia consists of whip-like uniseriate trichomes which are broader at the base and narrow towards the apex. The trichome terminates into an erect or curved multicellular hair. At the base of each trichome there is a spherical or cylindrical heterocyst with thick wall and homogeneous contents.

The heterocyst has only one polar nodule. Just above the heterocyst, there are one or more elongated resting spores, known as akinetes. When there are more than one akinetes, they usually occur in a chain. Sometimes two akinetes are separated by a vegetative cell.

The heterocyst, akinetes and some basal cells of the trichome remain covered with a gelatinous sheath. The gelatinous sheaths of several neighbouring trichomes fuse together to form small spherical colonies. In a colony the heterocysts of all the trichomes are pointed towards the centre and the narrow distal ends towards the periphery.The cell shows typical cyanophycean structure. The growth of the trichome takes place by active transverse divisions in the intercalary cells, located just below the distal hair. These intercalary cells cut segments both on the upper and the lower side. This type of growth is called trichothallic.

Reproduction: The reproduction in Gloeotrichia is brought about by hormogonia and akinetes.

1. *By Hormogonia:* The hormogonia are small segments of the trichome formed by fragmentation. Each segment is capable of growing into a new trichome.
2. *By Akinetes:* The akinetes occur singly or in chains, and are formed under unfavourable conditions. They secrete a thick wall which help them in tiding over the unfavourable conditions. Under the favourable conditions the akinetes germinate, each forming a new trichome.

Rivularia

Systematic Position

Class	-	Cyanophyceae
Order	-	Nostocales
Family	-	Rivulariaceae

Occurrence: Rivularia (L. rivulus = a small brook) is found in a wide variety of fresh water and terrestrial habitats. It grows on damp soil near river beds, moist rocks, or even epiphytically on some submerged plants. R. atra is found in salt marshes, and species like R. bullata, and R. polyotis occur in supra-littoral and upper intertidal zones of sea coasts.

Thallus Structure: The unbranched trichomes of Rivularia are arranged in a radial fashion to form a spherical, hemispherical or irregular colony (Fig. A). The radial arrangement of trichomes in a colony is due to repeated false branching in their basal portions. The evidence for false branching is found only in the young colony and is lost in a mature colony due to large displacement of trichomes. The trichomes in a colony remain partially or completely surrounded by a confluent sheath. The sheath is of very firm consistency, tough and very frequently heavily encrusted with lime.

The trichome is broad at the base and gradually tapers into a hair towards the apex. At the base of the trichome, a single heterocyst with one polar nodule is present (Fig.C).

Cell Structure: The cell exhibits a typical cyanophycean structure. The mucilagenous components of the cell wall contain arabinose and glucose. The protoplast is differentiated into a peripheral pigmented chromoplasm and a central centroplasm. The chromoplasm contains two important phycobilin pigments, c-phycoerythrin and c-phycocyanin: The centroplasm is relatively dense and represents the incipient nucleus of the cell.

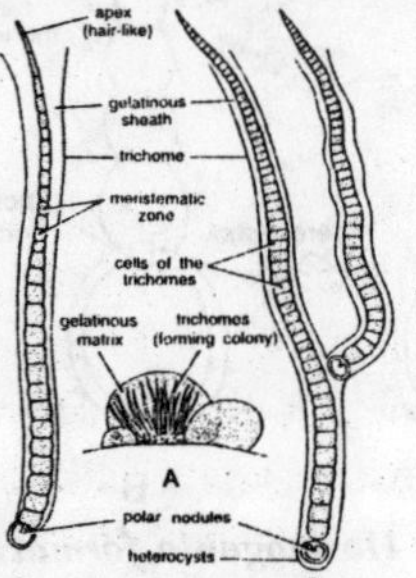

A-C. Rivularia: A. Colonies growing on a substratum, B. A colony showing trichome arrangement, C. Trichomes.

Growth: The growth of the trichome in Rivularia is trichothallic. It takes place by the activity of meristematic cells in the intercalary meristematic zone located at the base of the terminal hair. The meristematic cells are comparatively small in size and have denser granular contents.

Reproduction: Rivularia reproduces vegetatively by the formation of hormogonia and rarely by the heterocyst which gives rise to new trichome on germination. Sexual reproduction is not known. At the time of the formation of hormogonia, the trichome usually sheds its hair. The delimitation of hormogonia starts in the meristematic zone and gradually proceeds towards the basal portion of the trichome. Hormogonia are formed either singly or in series. In the latter case, they are separated by biconcave separation discs. At the time of germination, one of the two end cells of a hormogonium is modified into a basal heterocyst, and the other cell divides repeatedly in a single plane to form a long trichome (Fig. A, C).

Rivularia can be differentiated from Gloeotrichia in the absence of akinetes.

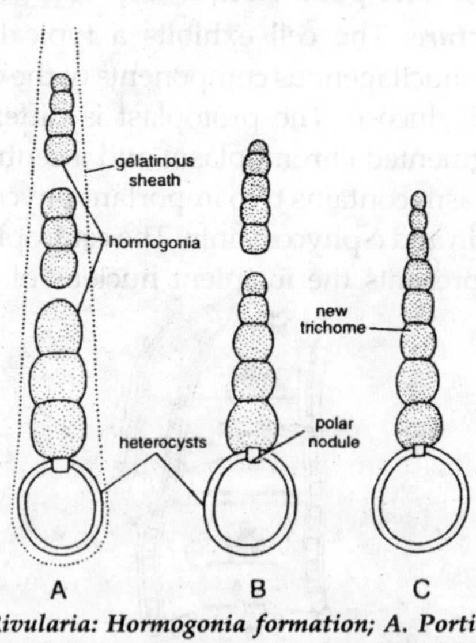

A-C. Rivularia: Hormogonia formation; A. Portion of a trichome with hormogonia in a series, B-C. Two stages in the germination of a hormogonium.

Chroococcales—Gloeocapsa: The order Chroococcales includes the most primitive members of the class Cyanophyceae. Almost all the members of the order are fresh water forms which are unicellular and are often organised into regular colonies; but they never show a trichome organisation. They reproduce chiefly by fission or by endospore formation. Heterocysts and hormogonia are absent.

The order comprises two families, Chroococcaceae and Entophysalidaceae.

Family Chroococcaceae

Members of the family are unicellular and free-living forms. They occur either singly or in colonies. The individual cells are usually spherical, ovoid, ellipsoidal or cylindrical. Multiplication takes place by fission or sometimes by nannocytes. The common examples of the family are Anacystis, Aphanocapsa, Chroococcus, Gloeocapsa, Microcystis and Synechococcus.

Gloeocapsa

Systematic Position

Class	-	Cyanophyceae
Order	-	Chroococcales
Family	-	Chroococcaceae

Occurrence: Gloeocapsa (Gr. gloia, glue; capsa, a box) is a terrestrial or sub-aerial alga, generally found on damp soils, wet rocks, banks, etc., as well as in the water reservoirs and lakes. Some species are marine, they occur as a black encrusting film on rocks at the upper limit of the high-.tide mark in the littoral zone.

Some species of the genus are endophytic. G. alpina constitutes algal component of lichens and occurs in symbiotic association with fungi, whereas G. gelatinosa is found on the bark of Hevea brasiliensis.

Some species of Gloeocapsa have the capability to fix atmospheric nitrogen and thus help in increasing fertility of the soil.

Thallus Structure: It is a colonial alga, and the colonies are spherical, irregular or expanded mass of mucilage. The number of cells in a colony ranges from 2-8. The cells are characteristically spherical, and each cell is surrounded by a conspicuous, variously coloured, homogeneous or stratified, concentrically lamellate gelatinous sheath. The sheath also surrounds the entire colony.

Cell structure. The cell exhibits typical cyanophycean structure. The protoplast is well differentiated into an outer pigmented chromoplasm and an inner centroplasm. The latter represents incipient nucleus; true nucleus is lacking. The chromoplasm possesses a special pigment said to be gloeocapsin.

In most species of Gloeocapsa photosynthetic thylakoids are arranged in rows at right angles to the plasma membrane, the outer thylakoids are directly connected with the plasma membrane.

Reproduction: Gloeocapsa reproduces only vegetatively by means of fission or by the formation of nannocytes.

Fission: The fission takes place in three planes at right angles to one another resulting in more or less spherical daughter cells. At the time of fission, individual cells are released from the colony by the disorganisation of the parent gelatinous sheath.

Each cell then grows into a new colony by simple cell divisions. The chromatin material in the centroplasm divides by a constriction into two tiny masses, followed by the division of the cytoplasm, resulting in two cells. The process is further repeated and as such a colony is formed.

Nannocytes: In some species such as G. granosa, G. crepidinum, nannocytes are formed. They are modified endospores, which are represented by small naked protoplasts formed by repeated divisions of the cell contents.

On liberation from the parent cell nannocytes germinate immediately without undergoing any resting stage.

Microcystis

Systematic Position

Class	-	Cyanophyceae
Order	-	Chroococcales
Family	-	Chroococcaceae

Occurrence: Microcystis (Gr. micro = small, kystis= bladder) is a fresh water colonial alga. Most species are euplanktonic and are aided by pseudovacuoles. In tropical countries, species of Microcystis multiply so prolifically that they impart a bluish-green colour to the pond water which appears like 'pea soup'. Some species of Microcystis, like M. aeruginosa, cause water blooms in lakes.

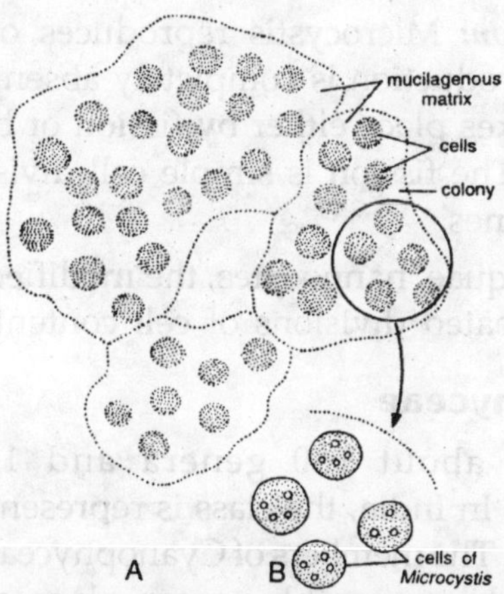

A-B Microcystic: A. A Colony, B. A part of the colony (magnified).

The prolific growth of some species of Microcystis (e.g., M. aeruginosa) in a water body often results in severe depletion of oxygen. This leads to mass mortality of fish and other aquatic animals due to suffocation. Certain species of Microcystis (e.g., M. aeruginosa) produce a toxin, called microcystin, which is highly toxic to fish and aquatic birds. The toxin is also fatal to domesticated animals that drink contaminated water. M. toxica is highly poisonous and is responsible for the death of thousands of sheep and cattles. In humans, some species of Microcystis cause gastric trouble, if they are ingested along with drinking water or during swimming. There are several species of Microcystis which inhibit the growth of bacteria (such as Staphylococcus, Clostridium), zooplanktons and several other algae.

Thallus Structure: The colonies of Microcystis may be round and definite in shape or irregular or sometimes clathrate, depending upon the species. There are hundreds or thousands of small, marble-like cells in each colony.

The spherical or elongated cells in a colony float freely in a colourless, homogeneous mucilaginous matrix. They lack individual sheath, and exhibit typical cyanophycean cell structure. Numerous pseudovacuoles are frequently present in a cell.

Reproduction: Microcystis reproduces only vegetatively and sexual reproduction is completely absent. The vegetative propagation takes place either by fission or by fragmentation of the colony. The fission is simple cell division which takes place in all planes.

In M. flos-aquae, nannocytes, the modified endospores are formed by repeated divisions of cell contents.

The Cyanophyceae

There are about 160 genera and 1500 species in Cyanophyceae. In India, the class is represented by 98 genera and 833 species. The members of Cyanophyceae (Myxophyceae) are commonly known as blue-green algae as their principal pigment is a bluish-green c-phycocyanin.

They also contain chlorophyll *a*, β-carotene and c-phycoerythrin. They differ from other algae: (i) in the absence of well organised cell organelles, (ii) the pigments are not restricted to definite chromatophores, but are distributed throughout the peripheral cytoplasm (chromoplasm), and (iii) the nucleus is of primitive type which lacks nuclear membrane and nucleolus. These characters show typical prokaryotic nature of the members of this group. These algae do not have flagella and their movement is brought about by gliding action.

Studies during the last three decades have clearly shown that the cell structure of blue-green algae is entirely different from the members of other algal groups. On the basis of their prokaryotic structure, many microbiologists consider blue-green algae as bacteria. Christensen (1962), on the basis of prokaryotic cell structure of blue-green algae and bacteria, placed both of them in a common phylum Prokaryota. Blue-green algae have also been named as Cyanobacteria. Although there are many similarities between bacteria ana blue-green algae, yet they show differences in their evolution.

Phycologists regard any organism with chlorophyll *a* and the thallus not differentiated into roots, stem and leaves to be an alga. Therefore, most of the algologists include Cyanophyceae in algae, rather than in bacteria.

The blue-green algae are found in a wide variety of habitats. Most of the species are fresh water, a few like Dermocarpa and Trichodesmium are marine. Species of Nostoc and Oscillatoria are found in terrestrial habitats. Some species of Nostoc and Anabaena grow as endophytes in roots of Cycas, leaves of Azolla and thalli of Anthoceros. Species of Chroococcus, Gloeocapsa, Nostoc, Scytonema and Stigonema are the main algal component (phycobiont) of lichens. The blue-green algae, growing in organically rich permanent waters, form planktons. The polluted water of lakes and ponds, exposed to intense sunlight and high temperature, provides a suitable habitat for the growth of planktonic blue-greens. They are also found on the bottom of ponds (benthos) and increase fertility of the underlying soil.

General Characters

Some important distinguishing features of blue-green algae are as follows:

(1) They are unicellular (e.g., Chroococcus), colonial (e.g., Gloeocapsa, Gloeotricha) or form chains of cells - trichomes (e.g., Oscillatoria, Scytonema). In colonial forms, all cells of a colony occur in a common gelatinous matrix. The cells of the colony may divide to form definite (e.g.,

Merismopedia) or indefinite and irregular (e.g., Aphanotheca) assemblage of cells. In filamentous forms (a trichome surrounded by a sheath is called a filament), each trichome has its own gelatinous sheath. The filaments are usually unbranched and uniseriate (e.g., Anabaena, Nostoc, Lyngbya) or may sometimes show pseudo-branching (e.g., Scytonema, Tolypothrix).

(2) Their cell structure is prokaryotic. They lack membrane bound organelles and the nucleus is of incipient type.

(3) The cell wall is composed of microfibrils held together in a mucilage sheath. The orientation of fibres may vary in different species, and may be reticulate (e.g., Nostoc) or long intertwining parallel (e.g., Lyngbya). It helps in sloughing off parasites and in retaining water, thus avoiding desiccation (particularly in those species which do not produce resting spores).

(4) The cell wall (external to plasmalemma) is differentiated into four layers. It is composed of mucopeptide, together with carbohydrates, amino acids and fatty acids.

(5) The principal pigments are chlorophyll *a*, c-phycocyanin and c-phycoerythrin, and they impart characteristic blue-green colour to these algae.

(6) They lack membrane bound plastids. Instead, pigments are found embedded within lamellae composed of two membranes joined at the ends. These structures are known as thylakoids.

(7) They show gliding or jerky movements; flagella being altogether absent.

(8) The reserve food material is a special type of starch (cyanophycean starch) and protein (cyanophycin granules).

(9) In some blue-green algae (e.g., Nostoc, Scytonema) specialised structures, known as heterocysts, are present. They may be terminal (e.g., Gloeotricha) or intercalary (e.g., Nostoc) in position. Definite function of heterocysts

is not known, but they are supposed to be the centre of N_2-fixation.

(10) Reproduction takes place by vegetative and asexual methods. Sex organs, gametes and flagellated zoospores are altogether absent. However, in some (e.g., Anabaena, Nostoc, Synechococcus, Cylindrospermum) genetic recombination has been reported.

(11) Vegetative reproduction takes place by cell division, fragmentation and hormogonia formation.

(12) Asexual reproduction takes place by akinetes, endospores, exospores, heterocysts and nannocytes.

Cell Structure

The blue-green algae have prokaryotic cell organisation. The internal membranes which separate DNA from the cytoplasm and isolate enzymatic machinery of photosynthesis and respiration in specific organelles are absent. The cell as such is devoid of organised nuclei, chromatophores and mitochondria. The cell wall contains a specific mucopeptide as its strengthening component.

Sheath and Cell Wall: The presence of the mucilagenous sheath is a constant feature of blue-green algae. The sheath consists of three layers of microfibrils reticulately arranged within an amorphous matrix. This gives a homogeneous appearance to the sheath. The mucilagenous sheath retains absorbed water which is useful during the periods of desiccation and for sloughing off parasites. A double-layered cell wall is present inside the sheath.

The inner layer of this wall contains mucopeptide and muramic acid. When viewed under electron microscope, the wall shows four layers, designated as LI, LII, LIII and LIV. Each layer is about 10 mm in thickness, and the LI lies next to plasmalemma. In structure and composition the cell wall of blue-green algae is similar to that of Gram-positive bacteria; the cell wall in both is composed of mucopeptide together with carbohydrates, amino acids and fatty acids. The LII layer has mainly peptidoglycan. The LI and LIII layers are electron transparent, whereas the LII and LIV layers are electron opaque.

A cytoplasmic membrane (plasmalemma) is present inner to the cell wall and this membrane consists of two electron opaque layers separated by a translucent layer. The cytoplasmic membrane invaginates inside the cell, and these invaginations are considered sites of biochemical functions normally associated with mitochondria, endoplasmic reticulum and Golgi bodies in eukaryotic cells.

Photosynthetic Lamellae: Electron microscopic studies have revealed the presence of a complex lamellar system (thylakoid) which is functionally analogous to the plastids of eukaryotic cells. The lamellae in cyanophycean cell are not separated from the remainder of the cytoplasm in a membrane-bounded organelle. They appear as elongated, flattened sacs consisting of two unit membranes, each about 75 Å thick. Adjacent thylakoids are separated from each other by a space of 50 nm, occupied by contiguous rows of discoidal phycobilisomes. These phycobilisomes contain photosynthetic pigments, c-phycocyanin and c-phycoerythrin.

Intracytoplasmic Inclusions: Various kinds of subcellular organelles have been observed in cyanophycean cells and these include cyanophycin granules, polyhedral bodies, ribosomes, polyglucoside bodies, polyphosphate bodies, α-granules, β-granules and gas vacuoles.

The significance and functions of these organelles are not fully understood. Ribosomes are involved in protein synthesis, cyanophycin granules are regarded as reserve food, and polyhedral bodies (now called carboxysomes) contain an important enzyme, ribulose diphosphate carboxylase. Several planktonic species of blue-green algae contain numerous small gas vacuoles (pseudovacuoles). These vacuoles are not tonoplast bound and contain metabolic gases. They are as such a provision for buoyancy regulation. Under electron microscope, the gas vacuoles appear to consist of packed arrays of cylindrical vesicles.

The gas vacuoles are produced in low light intensity and disappear in high light intensity. They refract light and as such cells appear purplish, brown or even black under microscope.

Nucleoplasm: The central region of a blue-green algal cell appears somewhat transparent and contains most of the genetic material. An organised nucleus with a nuclear membrane and nucleolus is absent. Since DNA is not associated with protein material (histones or protamines), organized chromosomes, characteristic of the eukaryotic cell, are not found in the cyanophycean cell. RNA is present in addition to DNA.

Heterocysts

These are specialized structures found in the members of Nostocales (except Oscillatoriaceae) and Stigonematales. They arise from somewhat enlarged vegetative cells, and are characterised by the presence of conspicuous pore at one or both poles. In terminal heterocysts, the pore is usually present only at one end, whereas intercalary heterocysts have pores at both ends. The position of heterocysts is a specific character; they are terminal in Gloeotricha, intercalary in Nostoc and Scytonema and rarely lateral as in Nostochopsis. They may be solitary or in pairs (e.g., Anabaenopsis). Intercalary heterocysts are usually spaced regularly along the filament.

Structure: Heterocysts are thick walled, pale yellow and barrel-shaped structures. The wall is differentiated into an outer fibrous, a middle homogeneous and an inner thin lamellar layer. It is characterised by the presence of pores either at one or both the poles. The wall layers become thicker in the polar region, and the pores are plugged with a refractive material, called polar granules (nodules).

Heterocysts contain dense and homogeneous cytoplasm. They contain mostly carotenoids, hence yellowish in colour. In the absence of phycobilins (which are the principal light absorbing pigments in oxygen-evolving photosynthetic reactions) and chlorophyll *a*, there is no CO_2 fixation or O_2 production in heterocysts.

The photosynthetic lamellae (thylakoids) are tightly packed, with reduced interlamellar spaces. They are concentrated along the periphery in a complex reticulate manner. These lamellae contain two lipids-glycolipids and acyl lipids, which are not found in vegetative cells.

Three types of granules (structural granules, crystalline granules, small granules) are associated with the peripheral lamellae, but they decrease in number and eventually disappear as the heterocyst matures. The mature heterocyst may remain attached to the filament for varying periods.

Development: Heterocysts develop from vegetative cells. One or both the daughter cells of a recently divided vegetative cell may behave as proheterocyst. A proheterocyst undergoes a number of changes to form heterocyst.

It enlarges in size and forms a multilayered envelop outside the cell wall. The wall becomes differentiated into an outer fibrous, a middle homogeneous and an inner lamellated layer. The outer fibrous layer is uniformly thick, whereas the middle and inner layers are thicker at the polar ends.

The pores appear at one or both poles, and through these pores cytoplasmic continuity is maintained with adjacent vegetative cells. There is gradual dissolution of storage granules and breakdown of photosynthetic thylakoids.

Sometimes the lamellae may form swollen vesicles and they become concentrated along the periphery and at the pores. In a mature heterocyst the polar pores become plugged with a refractive material (also known as polar nodules), due to which their cytoplasmic connections with adjacent cells are broken.

The frequency of heterocyst production is affected by the following external factors:

(1) The blue and green light inhibit heterocyst development, whereas the white and red light support it.
(2) The concentration of phosphate salts in the medium stimulates heterocyst development, whereas the absence

of Mg and Fe ions in the medium inhibits heterocyst development.

(3) The presence of combined nitrogen in the medium inhibits heterocyst development. The formation of heterocyst is inversely related to the amount of nitrogen in the medium.

(4) The cellular C: N ratio plays an important role in inducing and controlling the differentiation of vegetative cells into heterocysts.

Functions: Functions of heterocysts are not definitely known, and several views have been put forward from time to time. Early workers considered heterocysts as simply dead cells, thus having no functions.

But this view is no more valid as heterocysts have been shown to respire and germinate just like living cells. Brand (1903) considered them as sporangial or spore like structures, and Steinecke (1932) regarded them as secondary reproductive structures. There are evidences to suggest that heterocysts control sporulation.

If heterocysts are removed from the trichome, sporulation is inhibited. Besides this, the vegetative cells adjacent to heterocysts sporulate earlier than those away from it. Canabaeus (1929) found a sort of connection between the formation of gas vacuoles and heterocysts.

Fogg (1944) considered them as the sites of salt accumulation. Fritsch (1951) suggested that they secrete certain substances which stimulate growth and cell division.

Fay *et al.* (1968) considered heterocyst as N_2-fixing organ in blue-green algae. This view gets support from the fact that most of the nitrogen fixing blue-green algae possess heterocysts.

The factors inhibiting nitrogenase enzyme system also inhibit the formation of heterocyst, and experiments show strong reducing conditions within heterocysts.

Now it has been conclusively shown that heterocysts are the main sites of N_2-fixation in blue-greens, particularly under aerobic conditions. Under anaerobic conditions, the normal vegetative cells have the capability to fix nitrogen, but under aerobic conditions the nitrogenase enzyme becomes inactive due to the presence of oxygen.

Under such conditions, the thick-walled heterocysts provide anaerobic environment, necessary for nitrogenase activity. Stewart (1970) suggested that the main physiological roles of heterocysts are:

(i) control of sporulation,

(ii) involvement in growth and nitrogen metabolism, and

(iii) archaic reproductive units.

The various functions attributed to heterocysts clearly suggest that probably they have varied functions under diverse conditions.

Nitrogen Fixation by Blue-green Algae: An interesting aspect of blue-green algae is the capacity of several forms to fix atmospheric nitrogen. This ability of blue-greens makes them truly autotrophic. Beside blue-greens, the ability to fix atmospheric N_2 is found in few other prokaryotes like heterotrophic and photosynthetic bacteria.

The ability to fix nitrogen is restricted to about 22 filamentous species of Cyanophyceae. All these forms are characteristically heterocystous, e.g., species of Anabaena, Aulosira, Nostoc, Tolypothrix, Cylindrospermum, Mastigocladus, Anabaenopsis, Calothrix and Scytonema. Under anaerobic condition, certain non-heterocystous forms (e.g., Plectonema boryanum) also fix atmospheric nitrogen.

The blue-green algae, fixing nitrogen from the air, must possess an efficient mechanism to check oxygen inhibition of the enzyme nitrogenase. Thick-walled heterocysts provide a good locus for an efficient anaerobic functioning of nitrogenase, even in aerobic conditions.

It has been shown that in heterocystous blue-greens, under aerobic conditions, nitrogen fixation occurs only in heterocysts. But under anaerobic conditions, both heterocysts and vegetative cells possess active nitrogenase enzyme.

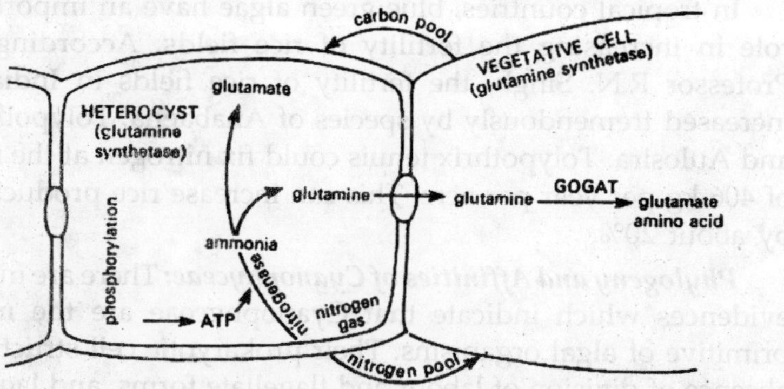

Schematic representation of nitrogen transfer from heterocyst to vegetative cell.

The mechanism of nitrogen fixation in blue-green algae has been studied by radioactive tracer technique. As mentioned above, heterocysts, which are nitrogen fixing sites, have nitrogen reducing enzyme nitrogenase. This enzyme reduces nitrogen gas into ammonia, which in turn is metabolized to glutamine by the enzyme glutamine synthetase (GS).

Glutamine is then transported from heterocyst to the adjacent vegetative cell, where it is converted into glutamate by the enzyme GOGAT. Both glutamine and glutamate, by various transamination reactions, form other amino acids.

The sequence of reactions in heterocysts and vegetative cells is shown in figure. The fixed nitrogen may be utilized in a number of ways: (i) it is assimilated by the blue-greens themselves, (ii) soluble nitrogenous compounds liberate from healthy cells into the culture medium, and (iii) after death and decay the nitrogenous compounds break into ammonia, which is eventually converted into nitrate by nitrifying bacteria, present in the soil.

In certain blue-green algae (e.g., Anabaena cylindrica) the nitrogenase enzyme is also present in vegetative cells, in addition to heterocysts. Besides this, in Gloeocapsa, although heterocysts are absent, it has the ability of fixing atmospheric nitrogen. Therefore, it shall be unfair to conclude that only hetcrocysts are the sites of nitrogen fixation.

In tropical countries, blue green algae have an important role in increasing the fertility of rice fields. According to Professor R.N. Singh, the fertility of rice fields in India is increased tremendously by species of Anabaena, Tolypothrix and Aulosira. Tolypothrix tenuis could fix nitrogen at the rate of 406 kg per year per acre. This can increase rice production by about 20%.

Phylogeny and Affinities of Cyanophyceae: There are many evidences which indicate that Cyanophyceae are the most primitive of algal organisms. Their prokaryotic cell structure, absence of division of labour and flagellate forms, and lack of sexuality are some of the characters which support their primitive nature. Fossil evidences indicate their origin in early Precambrian period. Their prokaryotic cell structure and absence of flagella lead to believe that they originated from unicellular forms devoid of flagella. Since most of the members of Cyanophyceae are terrestrial, many investigators believe that they originated from terrestrial ancestors rather than aquatic.

Similarity of blue-green algae with red algae in certain characters, such as the presence of similar phycobilins and complete absence of flagellated reproductive cells suggest some affinity in these two groups. However, some other important characters of blue greens, such as prokaryotic cell structure, and the presence of heterocysts, hormogonia and akinetes are not found in red algae. This indicates that there does not appear any close relationship between these two groups.

Since the recognition of blue-greens as a biological group in the early 19th century, they have been treated as a class or division of algae. But around 1960, when fundamental differences in cellular organisation between prokaryotes and eukaryotes were fully elucidated, it became evident that they are more closely related to the prokaryotic bacteria than to the eukaryotic algae. However, a detailed study of the nature of their photosynthetic pigments and capacity to perform oxygenic photosynthesis show that the only unifying character in Cyanophyceae and bacteria is prokaryotic structure of the cell.

The affinities of Cyanophyceae with Rhodophyceae and bacteria are summarised below:

Similarities with Red Algae

(1) Flagellated motile cells are absent in both groups.

(2) The blue (c-phycocyanin) and red (c-phycoerythrin) pigments present in Cyanophyceae are chemically similar to those found in Rhodophyceae (r-phycocyanin and r-phycoerythrin).

(3) Pit connections are present in the family Stigonemataceae of the class Cyanophyceae, and similar structures are also found in Rhodophyceae.

(4) The mechanism of fatty acid synthesis is similar in both groups.

Dissimilarities with Red Algae

(1) The blue-green algae show prokaryotic cell structure, whereas it is typical eukaryotic in the red algae.

(2) Cells of the blue-greens are covered with a mucilagenous sheath, which is absent in the red algae.

(3) Presence of hormogonia and heterocysts is a characteristic of the most blue-greens, but these structures are not found in the red algae.

(4) Sexual reproduction is absent in the blue-green algae, whereas the red algae show an advanced type of sexual reproduction.

Affinities with Bacteria

(1) Both, bacteria and blue-greens show typical prokaryotic structure.

(2) Like bacteria, the blue-green algae have an incipient nucleus (nuclear material without nuclear membrane), and lack membrane bound plastids.

(3) The mucilagenous sheath surrounding blue-green cells and the capsule present in many bacteria have a similar structure, as both are made up of extremely fine fibrils.

(4) Some bacteria (e.g., Thiothrix) are structurally similar to hormogonia, found in the blue-green algae.

(5) Beggiatoa, a sulphur bacterium, resembles with Oscillatoria in shape and movement.

(6) Like bacteria, the blue - green algae are also sensitive to antibiotics.

(7) Many metabolic processes (e.g., sulphur and nitrogen metabolism) are similar in both groups.

(8) Recombination of genetic material, similar to bacteria, has been reported in some blue-greens (e.g., Anacystis nidulans).

Classification of Cyanophyceae: Cyanophyceae has been divided into five orders (Fritsch, 1945). Papenfuss (1955) also recognised five orders of the blue green algae, but included them in the class Schizophyceae along with bacteria (class Schizomycetae) in the division Schizophyta. These five orders have been distinguished on the basis of morphological features, cell form, type of cell division and type of resulting colony, and some features of cell structure, such as presence of inclusions,, polar granules, gas vacuoles and specialized cells (akinetes and heterocysts).

1. *Order: Chroococcales:* Fresh water, marine or terrestrial, unicellular or colonial algae; colonies without polarity; reproduction by fission or endospores; heterocysts and hormogonia absent. Common examples: Aphanocapsa, Chroococcus, Gloeocapsa, Microcystis.
2. *Order: Chamaesiphonales:* Mostly marine, a few fresh water, unicellular or colonial algae; colonies with distinct polarity; reproduction by endospores or exospores; heterocysts and hormogonia absent. Example: Dermocarpa.
3. *Order: Pleurocapsales:* Fresh water or marine, filamentous algae; reproduction by endospores; heterocysts not formed. Example: Pleurocapsa.
4. *Order: Nostocales:* Mostly fresh water, a few marine, filamentous algae; filaments not differentiated into

prostrate and upright systems; multiplication by hormogonia; may or may not have heterocysts; spores and akinetes present. Common examples: Oscillatoria, Spirulina, Nostoc, Anabaena, Rivularia, Lyngbya, Cylindrospermum, Gloeotricha, Scytonema, Tolypothrix.

5. *Order: Stigonematales:* Freshwater, filamentous algae with distinct prostrate and upright filaments, showing true branching; reproduce by hormogonia or rarely by akinetes; heterocysts present. Example: Stigonema.

Bibliography

Abdelahad, N. and Bazzichelli G.: *Geitleria Calcarea Friedmann,* Cyanophycee Nouvelle, Italie, 1988.

Anderson R. and Adams M. S.: *Effect of Fire on Big Bluestem Production,* American Midland Naturalist, 1975.

Battarbee, R. W. and Gasse F.: *Past Climate Variability through Europe and Africa,* Springer, Netherlands, 2004.

Bebianno, M. J. and Company R. M.: *Antioxidant Enzymes and Lipid Peroxidation in Bathymodiolus Azoricus,* Aquatic Toxicology, London, 2005.

Berge, J. and Vader W.: *On the Taxonomic Status of the Antarctic Amphipod Genera Eclysis,* Organisms Diversity Evolution, 2005.

————————: *The Amphipod Genus Alexandrella,* Journal of Natural History, Washington, 2005.

Blanchette, C. A.: *The Effects of Biomechanical and Ecological Factors on Population and Community Structure of wave-exposed,* Oregon State University, 1994.

Bold, H. C. and M. J. Wynne: *Introduction to the Algae,* Englewood Cliffs, New Jersey, 1978.

Boudouresque, C. F. and Perret-Boudouresque M. M.: *A Checklist of the Benthic Marine Algae of Corsica,* GIS Posidonie Publishers, Marseille, 1987.

Bourrelly, P. and Dupuy P.: *Quelques Stations Francaises de Geitleria Calcarea, Schweizerische Zeitsch, Fur Hydrologie,* MacMillan, New York, 1973.

Boyer, C. S.: *The Biddulphoid Forms of North American Diatomaceae*, Natural Sciences of Philadelphia, Boyer, 1901.

Brodie, J. A. and Irvine L. M.: *Seaweeds of the British Isles*, British Museum, London, 2003.

Chapman, V. J.: *The Algae*, St. Martin's Press, New York, 1964.

Christiansen, J. S. and Gildberg A.: *Pagophilus Groenlandicus*, ICES Journal of Marine Science, 2004.

Claus, G.: *Algae and (their) Mode of Life (in) on the Baradla Cave of (at) Aggtelek*, Scientiarum Hungaricae, Budapest, 1955.

Claus, G.: *Data on the Ecology of the Algae of Peace Cave*, Nova Hedwigia Zeitsch, Kryptogamenkunde, 1962.

——————— : *The Reevaluation of the Genus Gomontiella*, Algol, 1960.

Coppejans, E.: *Iconographie d'Algues Mediterraneennes*, J. Cramer, Vaduz, 1983.

Delepine, R., Boudouresque C. F. and Frada-Orestano C.: *Algues et Autres Vegetaux Marins*, Mediterranee et Mer Noire, Roma, 1987.

Denny, M. W.: *Mechanical Limits to Size in Wave-swept Organisms*, Ecological Monographs, 1985.

——————— : *Predicting Physical Disturbance*, Ecological Monographs,1995.

Dixon, P. and Irvine L. M.: *Seaweeds of the British Isles*, British Museum, Natural History, London, 1977.

Evenset, A., Christensen G. N., Skotvold T, Fjeld E. and Gregor D.: *A Comparison of Organic Contaminants in two High Arctic Lake Ecosystems*, Science of the Total Environment, Norway, 2004.

Falk, Petersen S., Timofeev S., Pavlov V. and Sargent J. R.: *Climate Variability and the Effect on Arctic Food Chains*, Springer Verlag, Berlin, 2006.

Fletcher, R. L.: *Seaweeds of the British Isles*, British Museum, London, 1995.

Furevik, T., Gerdes R. and Berger W.: *The Nordic Seas*, American Geophysical Union, Washington, 2005.

Gantt, Elisabeth: *Handbook of Phycological Methods*, Cambridge University Press, England, 1980.

Gayral, P.: *Les Algues des Cotes Francaises*, Editions Doin, Paris,1966.

Guiry, M. D.: *A Concensus and Bibliography of Irish Seaweeds*, J. Cramer, Vaduz, 1978.

Gulliksen, B., Slagstad D. and Pamapanin D.: *Mytilus Thermophily*, Marine Ecology Progress Series, 2004.

Hakansson, H. and Stoermer E. F.: *Observations on the Type Material of Stephanodiscus Hantzschii Grunow in Cleve and Grunow*, Nova Hedwigia, 1984.

Hakansson, H.: *A Light and Electron Microscopical Investigation of the Type Species of Cyclotella*, Diatom Research, 1989.

Hanna, G. D.: *Pliocene Diatoms of Wallace County*, University of Kansas Science Bulletin, Kansas, 1932.

Haworth, E. Y. and M. A. Hurley.: *Comparison of the Stelligeroid Taxa of the Centric Diatom Genus Cyclotella*, O. Koeltz, Koenigstein, 1986.

Hellebust, J. A. and Craigie J. S.: *Handbook of Phycological Methods*, Cambridge University Press, Cambridge, 1978.

Hishock, S.: *A Field Key to the British Brown Seaweeds*, F.S.C. Publications, Shrewsbury, 1979.

Hoagland, K. D., Roemer S. C. and Rosowski J. R.: *Colonization and Community Structure of two Periphyton Assemblages*, American Journal of Botany, 1982.

Hollenberg G. J.: *Marine Algae of California*, Stanford University Press, California, 1976.

Hop, H. and Falk-Petersen S.: *Seasonal Lipid Dynamics and Life Strategy*, Marine Biology, 2005.

Irvine, L. M. and Chamberlain Y. M.: *Seaweeds of the British Isles*, British Museum, London, 1994.

———————— : *Seaweeds of the British Isles*, British Museum, London, 1983.

Kattner, K., Graeve M.: *The Arctic Pteropod Clione Limacina*, Marine Biology, 2005.

Kociolek, J. P. and Stoermer E. F.: *Ultrastructure of Cymbella Sinuata and its Allies*, Systematic Botany, 1987.

Kolosov, P. N.: *Upper Precambrian Paleoalgological Remains of the Siberian Platform*, Moscow,1982,

Krammer, K. and H. Lange Bertalot: *Bacillariaceae, Epithemiaceae, Surirellaceae*, Gustav Fischer Verlag, New York, 1988.

———————— : *Naviculaceae*, Gustav Fischer Verlag, New York, 1986.

Leake, D.V.: *The Algae of Crystal Lake*, Cleveland County, Oklahoma,1945.

Lebreton, B.: *World Bibliography on Subterranean Algae*, Bibliographie Mondiale des Algae Souterraines, 2000.

Lee, R. E.: *Phycology*, Cambridge University Press, Cambridge, 1980.

Littler, M. and Littler D.: *Handbook of Phycological Methods*, Cambridge University Press, Cambridge, 1985.

Lobban, Christopher S. and Harrison P. J.: *Seaweed Ecology and Physiology*, Cambridge University Press, Cambridge, 1994.

Makarikhin, V. V. and Kononova, G.: *Phytolites of the Karelian Lower Proterozoic*, Leningrad, Nauka, 1983,

Mann, D. G. and Stickle A. J.: *The Genus Craticula*, Diatom Research, 1991.

Mikhailova, N. S. and German, T. N.: *5th All Union Colloquium on Microfossils of the USSR Precambrian*, Izvestia Akadamie Nauk, 1987,

Nelson, E. N., Richardson J. K. and Bailey H. H.: *Aspects of the Occurrence of Aspidobothrid Parasites*, Proceedings of the Oklahoma Academy of Science, 1975.

Newman, G. A.: *Cowbird Parasitism and Nesting Success of Lark Sparrows in Southern Oklahoma*, Wilson Bull, 1970.

Newton, L.: *A Handbook of the British Seaweeds*, British Museum, London, 1931.

Oliver, G.: *Black Rat Snake Predation Upon Nesting Barn and Cliff Swallows*, Ornithological Society, Oklahoma,1970.

Osborn, R. S.: *Observations on the Spawning Ecology of Buffalos*, Oklahoma Academy of Science,1966.

Pampanin D. M., Camus L., Gomiero A., Marangon I., Volpato E. and Nasci C.: *Susceptibility to Oxidative Stress of Mussels*, Venice Lagoon, Italy, 2005.

Pamperl, M. A.: *Diatoms of Lake of the Ozarks*, Transactions, Missouri Academy of Science, Missouri, 1980.

Parks, J. M. and Barclay H. G.: *The Increasing Importance of Vines in Southern Oklahoma*, Oklahoma Academy of Science, 1966.

Parmelee, D. F.: *The Breeding Behavior of the Painted Bunting in Southern Oklahoma*, Bird-Banding, 1959.

Patten, B. C., Egloff D. A. and Richardson T. H.: *Total Ecosystem Model for a Cove in Lake Texoma*, Academic Press, 1975.

Patten, B. C.: *A Reservoir Cove Ecosystem Model*, Transactions of the American Fisheries Society, 1975.

———————: *Ecosystem Modeling and Reservoir Management*, Oklahoma Academy of Science, 1976.

Paulissen, M. A.: *A Technique for Marking Teiid Lizards for Field Observation*, Herpetology Review, 1985.

——————— : *Cnemidophorus Sexlineatus*, Journal of Herpetology, 1988.

Penfound, W. T., Shed J. S. and Jennisons M. C.: *A Plant Community Dominated by Vines*, Oklahoma Academy of Science, 1965.

Penfound, W. T.: *The Composition of a Black Willow Forest in South Central Oklahoma*, Oklahoma Academy of Science, 1961.

Pfiester, L.A. and Holt J.: *A Freshwater "Red Tide" in Texas*, Southwestern Naturalist, 1978.

Phillipi, A. and Tyrl R. J.: *Reproductive Biology of Proboscidea Louisianica (Martyniacea)*, Rhodora, 1979.

Pireh, W.: *Achillea Millefolium*, Oklahoma State Universi, Oklahoma, 1978.

Power, M. E. and Matthews W. J.: *Grazer Control of Algae in an Ozark Mountain Stream: Effects of Short-term Exclusion*, Ecology, 1988.

Proctor, V. W.: *Dispersal of Fresh-water Algae Water*, Science, 1959.

——————— : *The Growth of Basicladia on Turtles*, Ecology, 1958.

Prud, Homme: *A Taxonomic Revision of the European Sphacelariaceae*, Leiden University Press, Leiden, 1982

Purdue, J. R.: *The Phylogenetic Implications of Body Movements in Displaying Sceloporine Lizards*, Herpetology Review, 1970.

———————: *The Western Sandpiper in Oklahoma*, Oklahoma Ornithological Society, 1969.

Ratkova T. and Wassmann P.: *Sea Ice Algae in the White and Barents Sea*, Polar Research, 2005.

Reinke, D. C.: *New and Interesting Algae from Kansas*, State Biological Survey of Kansas, 1979.

Riedl, R.: *Fauna und Flora des Mitterlmeeres*, Verlag Paul Parey, Hamburg, 1983

Roemer, S. C., Hoagland K. D. and Rosowski J. R.: *Development of a Freshwater Periphyton Community as Influenced* Canadian Journal of Botany, 1984.

Rolbiecki L. and Walkusz W.: *Derogenes varicus*, Oceanological and Hydrological Studies, West Spitsbergen, 2005.

Round, F. E., Crawford R. M. and Mann D. G.: *The Diatoms: Biology and Morphology of the Genera*, Cambridge University Press, Cambridge, 1990.

——————— : *The Diatom Genus Stephanodiscus*, Archiv fur Protistenkunde, 1981.

Santi-Mazzini G.: *Aquarium*, Gribaudo, Cavallermaggiore, 2000

Setchell, W. A. and Gardner N. L.: *The Marine Algae of the Pacific Coast of North America,* University of California Press, California, 1925.

Shenfil, V. Y.: *Algae in the Riphean Deposits of the Yenesei Ridge,* Doklady Akad, Nauk, 1978,

South, G. Robin and A. Whittick.: *Introduction to Phycology,* Blackwell Scientific Publications, Oxford,1987.

Stegenga H., Bolton J. J., Anderson R. J.: *Seaweeds of the South African Coast,* University of Cape Town, 1997.

Stein R. and Macdonald R. M.: *The Organic Carbon Cycle in the Arctic Ocean,* Springer, New York, 2004.

Theriot, E., Stoermer E. and Hakansson H.: *Taxonomic Interpretation of the Rimoportula of Freshwater Genera in the Centric Diatom Family Thalassiosiraceae,* Diatom Research, 1987.

Troeger, W. W. and Menzel R. G.: *Benthic Diatoms in two Eutrophic Flood Detention Reservoirs in Oklahoma,* Journal of Freshwater Ecology, 1986.

Troeger, W. W.: *Colonization of Benthic Diatoms in a New Farm Pond and its Feeder Creek,* Southwestern Naturalist, 1983.

——————— : *Epiphytic Diatoms in Farm Ponds and Experimental Ponds in Bryan County,* Oklahoma Academy of Science, Oklahoma, 1978.

Usher, M. B., Callaghan T. V.: *Principles of Conserving the Arctic's Biodiversity,* Cambridge University Press, Cambridge, 2005.

Van, Landingham, S. L.: *Catalog of the Fossil and Recent Genera and Species of Diatoms and their Synonyms,* Cramer, Lehre, 1979.

Walcott, C. D.: *Middle Cambrian Algae,* Smithsonian Miscellaneous Collections, 1919.

Walter, M. R.: *Archean Stromatolites,* Princeton University Press, Princeton,1983,

Webber, H. J.: *Catalogue of the Flora of Nebraska,* Nebraska Board of Agriculture, Lincoln, 1890.

———: *The Fresh Water Algae of the Plains,* American Naturalist, 1889.

Wenke, T. L. and Eberle M. E.: *Diatoms from Streams in Ellis and Russell Counties, Kansas,* Transactions of the Kansas Academy of Science, 1986.

———: *Diatoms from Kansas,* Southwestern Naturalist, 1985.

Werner, D.: *The Biology of Diatoms,* University of California Press, Berkeley, California, 1977.

Wolle, F.: *Diatomaceae of North America,* Comenius Press, Bethlehem, Pennsylvania, 1894.

———: *Fourth Contribution to the Knowledge of Kansas Algae,* Bulletin of the Washburn College Laboratory of Natural History, 1889.

Wujek, D. E.: *Contribution to the Diatom Flora of Kansas,* Southwestern Naturalist, 1965.

Yankauskas, T. V.: *Upper Precambrian and Cambrian Plant Microfossils of the European USSR and their Stratigraphic Significance,* University of Moscow, Moscow, 1982,

Zmijewska M. I. and Bielecka L.: *Characteristics of the Arctic and Antarctic mesozooplankton in the neritic zone during summer,* Polish Polar Research, 2004.

Index

D

E

F

G

H

I

K

L

M

N

O

P

Q

R

S

T

U

V

❑❑❑

ENCYCLOPAEDIA OF ALGAE

ENCYCLOPAEDIA OF ALGAE

Vol. 4

Edited by

A.M. Bagulia

ANMOL PUBLICATIONS PVT. LTD.

NEW DELHI - 110 002 (INDIA)

ANMOL PUBLICATIONS PVT. LTD.
H.O.: 4374/4B, Ansari Road, Darya Ganj,
New Delhi-110 002 (India)
Ph.: 23278000, 23261597
B.O.: No. 1015, Ist Main Road, BSK IIIrd Stage
IIIrd Phase, IIIrd Block,
Bangalore - 560 085 (India)
Visit us at: www.anmolpublications.com

Encyclopaedia of Algae

First Published, 2008

ISBN 978-81-261-3528-8 (Set)

PRINTED IN INDIA

Printed at Mehra Offset Press, Delhi.

Contents

Preface

The first plants on earth were algae and these still thrive in a range of aquatic habitats. The land plants have evolved from the algae, more specifically green algae. Some 400 million years ago freshwater, green, filamentous algae invaded the earth. Algae are used as fertilizer and energy-source (biodiesel), etc.

Algae encompass several groups of living aquatic organism that get light-energy through photosynthesis, using it for transforming inorganic substances into organic substances. Algae make the photosynthetic organism found in most habitats. These are of single-celled to multicellular complex species, such as 'Giant Kelps' that grow to 65 metres in length. Conventionally, these have been regarded as simple plants, in the study of Botany and referred to as Eukaryota. Chromophyta have bacterial characteristics and some authorities consider them to be Protists.

Protists are traditionally considered more animal-like. The prokaryotic forms, referred to as blue-green algae are only half-algae with a mixture of bacterial characteristics. But they are quite different from the bacteria and are referred to as Cynochloronta. All other forms are known as true Eukaryota algae, in the study of Botany. They have a nucleus enclosed within a membrane. The protoctists are defined as eukaryotic microorganism with the exception of animals and plants but including fungi and algae.

All algae have photosynthetic machinery derived from cyanobacteria, and produce oxygen as a by-product of photosynthesis, unlike non-cyanobacterial photosynthetic bacteria. It is estimated that algae produce about 73 to 87 per cent of the net global production of oxygen, available for respiration.

Most of the simpler algae are unicellular flagellates or amoeboids, but colonial and non-motile forms have developed independently among several groups. Some more organisational levels, more than one of which may occur in the life cycle of a specie, are: Colonial, Capsoid, Coccoid, Palmelloid, Filamentous, and Parenchymatous.

This eminent work presents perfect and realistic features of the algae in an interesting manner. Expectedly, this study would generate a separate discipline in the study of algae.

Editor

1

Sporophyte

Family Chordaceae

Chorda filum, an annual seaweed of gregarious growth widely distributed in the Northern Hemisphere, though sometimes found between tidemarks, is commoner in the sublittoral region. The hollow, whip-like, unbranched thalli reach a length of 8 metres. Attachment is effected by a small discoid holdfast *(d)* and several thalli sometimes arise from the same base. In quiet localities the plants may form extensive submarine meadows, the upper flagelli-form portion floating on the surface. Spiral inrolling is not uncommon. In summer the thalli are often covered with hyaline hairs. C. *tomentosa* Lyngb., sometimes regarded as a variety of *C. filum,* is a smaller and rarer form, covered with an abundance of gelatinous hairs containing chromatophores.

According to Kylịn *C. filum* continues to lengthen throughout the summer with the help of a meristematic zone situated a short distance below the apex. Sporangia, which differentiate progressively from below upwards, cover a large part of the surface of the plant. The simple form of the thallus and the relatively unspecialised internal structure justify the reference of *Chorda* to a family of its own.

Family Laminariaceae

One of the simplest representatives of the true Laminariales is constituted by *Laminaria saccharina*, a widely distributed northern perennial, usually occurring at and to some slight depth below low-tide level. It favours quieter water. The smooth round stipe, which is sometimes as thick as a finger and upwards of a metre in length, terminates below in the richly branched holdfast characteristic of Laminariales generally, while above it expands into a flat, rather leathery lamina, reaching a length of 2 metres; old blades commonly show rows of alternate elevations and depressions within the ruffled margin.

As in all Laminariales, the principal formative region is situated in the transition zone between stipe and lamina. By its agency stalk and blade gradually increase in length during the growing season, while in the early months of the year a new blade *(nb)* is formed to replace that of the previous season *(ob)*. The new lamina *(nb)* first appears at the top of the stipe as a slight widening separated by a marked constriction from that of the preceding year *(ob)*, which by this time is usually much frayed and in course of destruction. As the new lamina enlarges, the old one becomes a mere appendage at its apex and ultimately breaks off completely. The unilocular sporangia appear towards the end of the season of active growth and constitute dark brown, irregularly circumscribed sori on both surfaces of the blade; they are usually produced only in older plants.

The embryonic stages are anchored by unicellular rhizoids, but relatively soon radial and tangential divisions in the cells at the bottom of the developing stipe result in the formation of a gradually broadening disc which is attached by rhizoids emerging from its lower cells. Already towards the end of the first season, however, a whorl of branched haptera arises from the disc and from the stipe above and further whorls are formed at successively higher levels on the latter as the plant grows older, usually at the time of the annual renewal of the lamina.

Among other species of *Laminaria* with an undivided lamina *(Simplices)*, *L. solidungula* J. Ag. is characterised by the persistence of the simple attaching disc throughout life. *L. longicruris* de la Pyl. attains to appreciable dimensions, with a hollow stalk reaching 5 metres and a broad blade and 4 metres long. Both species are Arctic forms, occurring also on the Atlantic coast of North America dpwri to Cape Cod. *L.japonica* Aresch., important on the coasts of Japan and Siberia, is another large form reaching a total length of 12 metres. The Japanese *L. gyrata* Kjellm is distinguished by the arrangement of the sori in small marginal, subcircular or transversely extended, groups. In *L. religiosa* Miyaoe the thickened middle part of the blade is depressed on one and elevated on the other surface, the former always facing the light; the sori first appear on the shaded side and are often restricted to it. This species therefore affords indications of dorsoventrally.

A number of species (e.g. the Mediterranean *L. Rodriguezii* Bornet; the Pacific L. *Sinclairii* (Harv.) Farl.; and the Arctic *L. longipes* Bory produce, from the region of the holdfast, horizontal stolons which sooner or later turn up at their tips and develop into new plants.

L. saccharina represents a simple type of thallus which is repeated with but little modification in the early stages of development of all Laminariales. A different habit is, however, exhibited by the mature plants of the two other species of *Laminaria* common on North European shores, viz. L. *digitata*, also found on the Atlantic shores of North America, and *L. Cloustoni (L. hyperborea* Foslie); the latter usually occupies a lower level than the former. In these species *(Digitatae)* the greater part of the blade is divided into numerous segments. *L. digitata* has a pliable stipe, while that of *L. Cloustoni* is firm and wrinkled, usually bearing numerous epiphytic Florideae. The sori of L. *digitata* form scattered patches on the blades, whilst in *L. Cloustoni* they appear as large irregular areas with but little free surface between them. The holdfasts resemble those of *L. saccharina*.

Young plants of both species have an undivided blade and are indistinguishable from those of L. *saccharina*. A new blade is formed in the same way during the winter and this is at first entire, but, as it enlarges, successive short longitudinal splits arise within the lower portion, the first ones appearing near the outer margin. Gradually the splits extend to the front edge of the new lamina, which consequently gets resolved into separate segments the middle ones commonly become completely free only after the old blade has worn away. In subsequent years the new blade likewise originates as a dilatation at the summit of the stipe and becomes segmented in the same way.

In some of the perennial digitate species such blade-renewal probably takes place repeatedly over a period of years, but this seems to vary with the latitude and depends on environmental conditions. Certain species are more long-lived than others. Sauvageau concludes that plants of *L. Cloustoni* persist for 10-20 years and Foslie records old specimens with stipes 5 metres long. According to Yendo, on the other hand, the *Laminarias* of Japan live for only two years and show only one period of blade-renewal. *L. ephemera* Setchell *(Renfrewia parvula* Griggs., an annual attached by a simple disc throughout life, exhibits relatively little splitting of the blade.

Several monotypic genera which are characteristic of the North Pacific, while retaining the simple habit of a *Laminaria saccharina*, exhibit diverse modifications in their fronds. In *Pleurophycus*, which is probably an annual, the middle of the blade shows a shallow fold simulating a broad midrib and contrasting markedly with the undulate wings. The blade of *Cymathere triplicata; Laminaria triplicata* Post. & Rupr. is provided with three narrow longitudinal folds *(r)* throughout its length; it may be as much as 4 metres long and 22 cm. broad, although the stipe does not usually exceed 5 cm. The sori are restricted to the base of the blade. *Cymathere* is fixed by a discoid holdfast and frequents relatively quiet water.

In *Costaria* the blade is traversed by five prominent

longitudinal ribs which project alternately on the two surfaces so that three are visible on the one and two on the other; on the surface opposite to each rib there is a corresponding depression, while between the ribs there are numerous puckerings *(b)*. The blade continues to widen after it has reached its full length and occasional perforations are formed in much the same way as in *Agarum*. In the young plant the midrib arises first and then, successively, the lateral ribs. The sori occupy the longitudinal depressions, as well as the hollows around the perforations.

A greater degree of specialisation is seen in *Agarum* and *Thalassiophyllum*. The former, represented by several perennial species in Northern Asia and America, has only a short stipe. The mature elongate blade is provided with a prominent midrib *(r)* and is pierced by numerous large perforations which are lacking in young plants. In *A. cribrosum* the lower margins of the lamina are distinctly inrolled. Most authorities state that the blade is renewed periodically as in a *Laminaria*, but according to Humphrey there is continuous growth at the base as the apex wears away. The sori are formed on either surface of the blade.

Young plants of *Thalassiophyllum* have a very short stipe bearing a broad cordate blade, the lower margins of which soon thicken; growth in width of the blade continues for a long time so that it soon becomes broadly reniform. At an early stage, as a result of unequal surface growth, the lower edges undergo inrolling *(s)* and their thickened margins come to lie in the centre of the developing scrolls. The degree of inrolling increases as the blade widens, while simultaneously the median part of the latter *(b)* undergoes progressive erosion.

This ultimately results in the complete separation of the two large scrolls, the basal parts of which have thickened to form two arms diverging from the apex of the original stipe. The keels constituted by the thickened basal margins of the lamin: are protected within their appropriate scrolls, but the exposed edge; of the two secondary blades undergo progressive erosion and so appear more or less ragged. The meristem that brings about the gradual enlargement of each secondary blade lies along the edge of the keel. As the blades spread out, they develop numerous perforations which are arranged in longitudinal series parallel with the keel; the edge of the perforations is at first revolute.

For a time the two secondary blades develop uniformly, but later one usually takes the lead and assumes a more or less erect position. As growth continues its basal portion elongates to form a false stipe, which usually becomes twisted. According to Rosenthal the branches found on mature plants are all adventitious, originating from the lower edge of certain perforations close to the thickened keel.

The adult plants, which grow in dense thickets just below low-water mark, reach to a height of about a metre. The thick twisted stipes bear numerous branches each terminating in a firm coriaceous blade inrolled at its base and pierced by numerous holes. Anchorage is effected by haptera *(h)* arising from the base of the original stipe. The sori form irregular dark brown areas on the blades.

A less complex form is attained by the allied genus *Hedophyllum,* the mature plants of which are attached by numerous haptera *(h)* arising from the thickened basal margins of the blade. The young plant has a short flattened stipe and at an early stage the base of the lamina becomes decumbent. In *H. sessile (Laminaria sessilis* Ag.; *L. apoda* Harv. the older blades split into two or more segments. In *H. subsessile,* however, the median part of the primary lamina wears away leaving two secondary blades *(sb)* at the ends of the thickened margins *(m)* of the former, which appear as bifurcations of the original stipe *(s);* the basal margins exhibit some inrolling, although this is more marked in *H. spirale* Yendo. There is much resemblance to *Thalassiophyllum,* but the inrolling is not pronounced and there are no perforations. *Hedophyllum sessile* appears to constitute a link between the digitate species of *Laminaria* and *H. subsessile* which in its turn, via *H. spirale,* approaches *Thalassiophyllum.* In *Arthrothamnus bifidus* (Gmel.) J. Ag. the primary lamina ultimately disappears completely except for its decumbent basal margins from which the secondary blades arise. The same process is repeated again and again and thus a dichotomising rhizome bearing numerous fronds is gradually produced.

A different structure is seen in *Saccorhiza bulbosa,* an annual found on the Atlantic shores of Europe and North Africa and sporadically in the Western Mediterranean; it is not infrequent at and below low-water mark in the British Isles. The mature plant has a digitate blade, sometimes exceeding 2 metres in length and borne at the end of a flattened stalk *(s)* which may be 60 cm. long. The stalk is Spirally twisted in its lowest portion and below the twist there is a large inverted bell-shaped outgrowth (6), which may reach 40 cm. in diameter and covers in the basal part of the stalk, as well as the original attaching disc. Numerous haptera *(h)* arise from the margin of the bell, while the outer surface often bears similar unattached structures.

The early stages are of the usual type. The bell develops from the transition zone and completely covers in the short stipe, while subsequently the same meristem gives rise to the flattened stalk which intervenes between bell and blade. This stalk ultimately develops undulating wings upon which the sori are usually formed, but they can also arise on the blade, as well as on the inner and outer surfaces of the bell. The latter survives throughout the winter after the rest of the plant has perished. Gardiner also records formation of sporangia on the haptera.

Another species, *S. dermatodea,* which is sometimes referred to a separate genus *Phyllaria* owing to the position of the sori at the base of the blade, is widely distributed in the Northern Hemisphere. Here the stalk is simple and the bell (Setcheirs rhizogen) is quite a small structure, although again serving for attachment. *Phyllaria reniformis* (Lamour.) Rostaf. occurs in the Mediterranean and the adjacent parts of the Atlantic.

Family Lesoniaceae

The splitting of the lamina, seen in diverse genera above considered, does not usually extend into the transition zone, so that the meristem remains a single entity; meristematic activity may, however, subsequently concentrate mainly in other regions, as for instance along the edge of the keel in *Thalassiophyllum*. In the forms, which Setchell groups as Lessoniaceae, on the other hand, the process of splitting extends also into the transition zone, so that the secondary blades are each provided with a part of the original 'meristem and can thus develop secondary stipes of their own. This is one of the characteristic features that underlies the construction of such genera as *Lessonia* and *Macrocystis*.

The first species to be mentioned is *Dictyoneurum californicum*, which is confined to Pacific North America. The young plant, so far as known, is of the usual pattern. Its short stipe soon becomes prostrate and develops into a flat and rather thin organ attached to the substratum by marginal haptera. The lamina segments by means of successive splits that originate in the transition zone and gradually extend outwards. The short stipes of the secondary blades become decumbent and are attached to the substratum in the same way as the primary one and, since this happens time after time, the front part of the attachment system gradually advances, while the older part behind dies away; in this way a single plant gives rise to a clump.

Older blades exhibit a network of coarse ribs. Broad irregular sori appear each year on both surfaces of the blades. *Dictyoneurum* is distinguished from other Lessoniaceae by the procumbent development of the stipe. In *Dictyoneuropsis* splitting takes place through the broad midrib of the young lamina, the secondary blades regenerating a new half along the exposed edges.

The chief centre of distribution of *Lessonia* is in the Southern Hemisphere. The young stages show the usual habit, but a median split soon arises near the base of the blade resulting in two segments provided with short pedicels *(p)* which appear as forks of the underlying stipe. The original meristem is thus segregated into two entities each contributing not only to the enlargement of the appropriate secondary blade, but also to the lengthening of the pedicel below. The secondary blades split in the same way, and this is repeated time and again. Commonly one fork develops more vigorously than the other and, since the pedicels of successive stronger forks become arranged in a straight line, a sympodial axis *(a)* is formed, upon which the weaker ones appear as laterals.

In some species (e.g. *L. laminariaeoides)* the primary stipe remains short, but in others it lengthens progressively to form a main trunk upon which the numerous secondary blades are borne. One of the most striking species is *L.flavicans,* the mature plants of which resemble small trees. The thick stipe divides again and again in its upper portion, the ultimate branches ending in long lanceolate blades. Harvey states that the plants may reach a height of 4 metres, while the main stipe may have the thickness of a human thigh. According to Skottsberg the stipe increases in thickness with each division of the lamina. The sori of *Lessonia* are formed on the basal parts of the blades.

Postelsia palmaeformis, an annual confined to the Pacific shores of North America, although of appreciably smaller dimensions, resembles the arboreal *Lessonias* in habit. One of the most striking of the Laminariales, it well deserves the popular name of Sea Palm.

The thalli have short thick upright stipes surmounted by a pendent mass of blades. Tuey grow gregariously in habitats exposed to the full violence of the waves, and it is a remarkable sight to see such a forest of *Postelsia* uncovered at low tide and unaffected by the battering of the waves to which it is subjected. The numerous blades arise by splitting in the transition zone, much as in *Lessonia.* The stipe, from 40 to 60 cm. high, is hollow, but the many short pedicels at its summit are solid. Each bears a falcate blade, traversed on both surfaces by deep longitudinal grooves within which the sori develop.

Nereocystis and *Macrocystis* are distinguished by greater elongation of the stipe and by the huge size attained by the mature thalli. The former, the Bull Kelp of the Americans, despite its dimensions, is probably in the main an annual seaweed lasting from February to the following November or December. The young plants are of the usual type, but soon exhibit rapid lengthening of the stipe, although at first it remains slender and very flexible. Near its summit and beneath the transition zone a spherical air-bladder *(v)* arises at an early stage and, soon after, the blade develops the first split *(s)*, which extends down to the bladder and divides the lamina into approximately equal halves. As the bladder enlarges, the short stalks *(p)* of the secondary blades become separated.

Successive splitting of the blades takes place, as the stipe elongates and the float enlarges, so that the older plant bears numerous lanceolate fronds, not uncommonly grouped in two or more bunches corresponding to the primary divisions of the lamina. The rather thick solid stipes of such plants are anchored by an elaborate holdfast, which is up to 30 cm. in diameter. The sori form elongate areas, reaching a metre in length, on both surfaces of the blades; after liberation of the swarmers the whole fruiting region drops out.

Frye states that the second and third splittings of the blade are often unequal, the larger segment being to the outside; he sees in this a resemblance to *Pelagophycus*. *Nereocystis* favours habitats where the tide is swift and usually grows in 5-8 fathoms of water; the length of the stipe depends on the depth. The plants often grow in dense masses and the largest are found in the vicinity of the Arctic.

The stipes were formerly employed as fish-lines by the natives, while the huge floats were used to siphon water from the boats. According to Macmillan the thallus reaches a length of 100 metres, with air-bladders 2-3 metres long and up to 15 cm. in diameter, while the blades are stated to become 15 metres long; most other authorities give considerably smaller dimensions and it remains doubtful whether the plants are ever more than 50 metres in length. The mature floats are more or less pear-shaped and separated by a constriction from the stipe below. They rest on the surface of the water, while the numerous blades are slightly submerged. Older stipes bear irregular outgrowths which have been regarded as homologous with the haptera.

Macrocystis, the largest of the perennial seaweeds, has its main centre in the Southern Hemisphere, although also well represented in the western America. It usually grows in 8-10 fathoms of water According to recent accounts the thalli of *M. pyrifera* attain a maximum length of 60 metres. Above the holdfast *(h)* there follows a long slender stalk *(st*, a centimetre or less in width), which often bears a number of scattered blades and trails out obliquely to the surface of the water, where the rest of the thallus floats horizontally.

This floating portion bears a dense double row of shortly stalked lanceolate blades. The blades have a corrugate surface and bear short spines along their margins, while at the base of each is a spherical to pyriform air-bladder *(b)*. On the older parts the blades usually disintegrate and at most only the air-bladders remain. The holdfast of M. *pyrifera* may be nearly a metre in diameter and is of the usual type, but in *M. integrifolia* Bory, found on the Pacific coasts of America, attachment is effected by a flattened, prostrate richly branched organ bearing haptera along its margins and dying away behind as it grows on in front (cf. *Dictyoneurum)*.

The development of the mature structure from the embryonic stage takes place in essentially the same way as in *Lessonia*. A median split divides the young blade into two equal halves, but before the split has extended to the apex of the lamina further splits *(s2)* are formed on either side.

As the tertiary blades become free, intensive marginal growth of the original lamina results in a rapid enlargement of the two outer blades, while the inner ones remain narrower; according to Skottsberg the next splits always arise in the outer segments.

In the subsequent course of events, after each process of splitting:

1. Larger dimensions are given by earlier authors. South African plants are small.
2. It has been suggested that this species, which occurs also in South Africa, is but a shallow-water form of *M. pyrifera*, but Smith regards it as distinct.
3. The drawing of Reinke, shows just the opposite. Skottsberg suggests that this specimen, drawn from herbarium material, had undergone displacement of its parts in mounting.
4. The later-formed haptera commonly take their origin from the branches of the first fork, so that several plants may appear to arise from the same holdfast or may even seem to originate adventitiously from the haptera.

Skottsberg was, however, unable to find any evidence for such an origin. It is always only the outer segment that undergoes further development. The splits soon commence to appear nearer the inner margin of this segment and as a result a relatively narrow piece *(bl)* gets detached from the large remaining part *(i)*. The latter, gradually assuming a falcate shape, comes to occupy a terminal position on the elongating sympodial stipe and constitutes a formative tract from which segments are detached unilaterally in a single series. The opposite margin of this terminal blade soon thickens to form a keel *(k)* which is continuous with the stipe below. Along its edge lies the meristematic zone, the region of maximum growth being located at the point of transition between the blade and the stipe. As successive segments become detached, the intervening parts of the keel elongate to form "inter-nodes" *(i)* whereby the individual blades become spatially separated. According to Brandt a frond lives for 4-7 months. Loose-lying forms are recorded in New Zealand.

The development of *Macrocystis,* like that of *Lessonia*> is therefore effected by successive splitting of the primary blade, but of the two segments thus formed in *Macrocystis* one remains arrested, while the other (the outer) continues to develop; the successive pedicels of the favoured segment combine to form a sympodial stipe which bears the other, narrower segments along its inner side. Before the segments of the functional blade becomes free, the basal portions enlarge to form the air-bladders *(b)*, while the short stalks of the blades *(p)* develop subsequently. Surface-growth of the blades continues for some time and leads to the characteristic corrugation, while the two-ranked arrangement is due to a subsequent readjustment. In robust plants other long shoots may arise by the further development of some of the first-formed inner segments.

The sori are usually formed on the blades of short forked shoots arising at the base of the plant and remaining deeply submerged. These fertile shoots do not exhibit the one-sided development that characterises the sterile ones, and in habit they are much like one of the smaller species of *Lessonia*. Sori, however, also occasionally occur on the ordinary blades at the water-surface; in either instance they lie mainly in the furrows.

It is not without interest to compare *Macrocystis* and *Nereocystis*. In the former the primary stipe remains quite short, while its branches unite to produce sympodial stipes of great length. In *Nereocystis*, on the other hand, it is the primary stipe that continues to grow, whilst the forks themselves remain quite short. In the former, flotation is secured by a bladder at the base of each blade, in the latter by a single bladder at the summit of the main stipe.

PC jgophycus; Nereocystis gigantea Aresch., confined to a small part of the coast of California, has a habit rather like that of *Nereocystis*. The stipe, usually 6-10 metres long, is surmounted by a large spherical air-bladder, beyond which it forks into two equal horizontal arms, each bearing at intervals 4-6 delicate secondary blades attached by slender stalks and reaching a length of 6 metres. The blades are formed by unilateral splitting of the two primary blades in much the same way as in *Macrocystis* and form the sori at their tips. Large individuals are stated to reach a length of 45 metres.

The monotypic and perennial *Lessoniopsis*. Is distinguished from other Lessoniaceae by the differentiation of fertile and sterile blades. The former are broad and oval, with a sharply demarcated stalk, and bear the sori *(so)* on both surfaces. The sterile blades *(v)*, on the other hand, are narrow and strap-shaped and merge gradually into the stalk. They are usually provided with an ill-defined midrib and segment by the formation of repeated splits *(si)* through the latter, whereas the fertile blades show no division. These sporophylls are outgrowths from the stalks of the sterile blades and develop in pairs from the transition region year after year. The short firm, often deeply furrowed, stipe bears at its summit numerous branches, produced by successive splitting, as in *Lessonia'*, older plants may bear as many as 800 blades. Like *Postelsia*, *Lessoniopsis* grows above low-tide limit in situations exposed to strong surf, although it usually occurs at a lower level than the former.

Family Alariaceae

The series of forms included in this family are, like *Lessoniopsis*, characterised by the restriction of the sori to special sporophylls, but the ordinary vegetative fronds do not here undergo splitting. The sporophylls arise as outgrowths, either from the stipe or from the margin of the blade. The young plants are again of the *saccharina* type.

The most familiar member of this family is *Alaria*, the main centre of distribution of which appears to be in the Arctic and the Northern Pacific, although a number of species are found in the North Atlantic. Of these, *A. esculenta*, popularly called Badderlocks or Henware, is not uncommon in exposed situations near low-water mark, though usually growing submerged. All species of *Alaria* are perennial. The relatively short stipe bears an elongate lamina, which often possesses a wavy margin and is decurrent on to the stipe; the blade is traversed by a stout midrib (r) which appears as an upward continuation of the stalk. During summer the latter produces in acropetal succession near its upper end two rows of thick tongue-shaped sporophylls which, after the maturation of their sori, are shed during autumn and winter, leaving small scars on the stipe. At this time of the year the blade becomes torn into segments perpendicular to the midrib and ultimately wears away more or less completely; fresh growth in the transition zone early in the following season leads to its replenishment. New sporophylls sprout out above those of the preceding season.

Alaria, Undarl

The mode of renewal of the lamina in *A. esculenta* and other species seems to take place somewhat differently in different localities. Thus, on Scandinavian shores the old blade wears away almost completely before the new one is formed and in the last months of the year the lamina is represented by the remains of the midrib. On French and British shores, on the other hand, as well as in other regions, renovation of the lamina takes place more as in *Laminaria,* since here the old blade persists until the new one has grown to a considerable size. Various authorities affirm that growth in the transition zone is continuous and that the blade is replenished basally in the same measure as it wears away apically. According to Yendo the Japanese *Alarias* last for only two years.

The midribs and sporophylls of *A. esculenta,* which have a sweet insipid taste, were occasionally consumed in the seventeenth century in parts of Scotland, while in Kamschatka the midribs of *A. fistulosa* are eaten. The blade of the former may reach a length of 6 metres, although the stipe does not exceed 30 cm.; the sporophylls may be 25 cm. long, with a width of about 2 cm. The largest species appears to be the North Pacific *A. fistulosa* Post. & Rupr., with blades up to 25 metres long and as much as 230 cm. wide; the midrib is in great part hollow, its successive chambers being separated by partitions composed largely of hyphae. The species of *Alaria* favour habitats exposed to surf and Macmillan describes forms of *A. nana* Schrader which, in their short sturdy habit, recall the *Postelsias* with which they are associated.

Some species (*A. marginata* Post. & Rupr.) have relatively large sporophylls and in this respect approach the monotypic *Pterygophora*. Here the long woody stipe (*st*), rounded at the base but gradually flattening above, bears sporophylls (*s*) and forming sori largely at their base. The terminal blade (*v*), which may likewise become fertile, is not much longer than the sporophylls. It is often in large part disintegrated, although, as in *Alaria*, it may subsequently regenerate from the base. New sporophylls are produced in each season above those of the preceding year.

In *Undaria*, the species of which are confined to Japan, the lower part of the lamina bears pinnately arranged outgrowths, whilst the flattened stalk is provided with an undulate wing (*w*) upon which the sori are produced. To this genus Okamura refers *Laminaria Peterseniana* Kjellm., as well as Yendo's genus *Hirome*. The latter is distinguished by the fact that sori (*so*) are also developed on the blade. Sauvageau suggests that it may be a hybrid. Yendo considers that the genera *Hirome*, *Undaria*, *Alaria*, and *Pterygophora* form a series in which the production of special fertile zones or sporophylls from the stalk becomes progressively more marked.

The systematic position of *Phyllogigas* in which the secondary blades that sprout out from the stipe are altogether like the terminal blade, is uncertain, since so far only sterile plants have been found.

In other Alariaceae the sporophylls arise as marginal pinnations of the lamina. In the Californian *Eisenia arborea* they appear already in the young plant and, whilst the apex of the primary blade gradually wears away, the pinnae themselves (*s*) enlarge and become as big or even bigger than the main blade. At this stage the lower margins of the latter become inrolled and, as this happens, the enclosed margins thicken.

Ultimately, when the primary blade has worn down to its very base, the involuted margins constitute two arms at the top of the stipe. The formation of pinnae is now transferred to secondary meristematic zones at the tips of these arms, but the original meristem persists at the top of the stipe which continues to grow in length. It becomes a stout and fairly rigid structure flattened near the top.

The sori form elongate patches on the basal parts of the sporophylls. Another species, *E. Cokeri* Howe, recorded from Peru, is only doubtfully distinct, but a further characteristic form is the Japanese *E. bicyclis* (Kjellm.) Setch.; *Ecklonia bicyclis* Kjellm., in which the sporophylls may themselves be pinnate.

The species of *ECklonia*, principally represented in the Southern Hemisphere, are well illustrated by *E. buccinalis*, the Bamboo Seaweed of South Africa, which somewhat resembles *Nereocystis* in habit, although the numerous blades are produced in quite a different manner. The hollow erect stalk, which may be 5 or more metres in length, enlarges at its summit (*ve*) but the primary lamina *(bl)* remains small and the greater part of it often wears away. The basal portion, however, persists and from it there grow out in aero-petal succession numerous long narrow sporophylls *(s)* bearing irregular sori. Some species propagate by stolons. Young plants show resemblance to those of *Eisenia*.

The greatest elaboration among Alariaceae is attained by *Egregia*, the species of which form extensive meadows at and below low-water mark on the western coasts of North America. Young plants of *E. Menziesii* are of the customary type, but at an early stage short flat ligulate segments sprout out from the lower margins of the blade *(b)* and from the flattening stipe *(st)*. The intercalary meristem contributes but little to the enlargement of the blade which ultimately wears away completely, whilst the flattened stalk exhibits continuous elongation, at the same time producing more and more lateral outgrowths.

The considerable branching is effected mainly from the basal parts of the thallus, where some of the outgrowths develop into long shoots. These likewise bear numerous marginal appendages and may grow to a length of more than 8 metres. Cylindrical at the base, they are markedly flattened throughout the greater part of their length and, in older plants, numerous perforations often appear in these band-shaped structures, while they are densely covered with blunt papillae *(p)*. The short laterals function as photosynthetic organs, and in *E. laevigata*, found in Lower California, vary from simple linear blades to pinnate structures with filiform segments. Here and there the base of one of the laterals swells up to form an air-bladder. Those that function as sporophylls are wrinkled pod-shaped structures bearing the sori between the ribs.

Except for the somewhat aberrant genus *Chorda*, the diverse seaweeds surveyed as post-embryonal modifications of a central type resembling such a form as *Laminaria saccharina*. Subsequent elaborations are due, on the one hand, to progressive splitting affecting either the lamina only and leaving the meristem intact (Laminariaceae) or extending also into the region of the latter with a resulting segregation of the same into numerous separate entities (Lessoniaceae), the latter process admitting of especially striking developments.

On the other hand, in the Alariaceae, elaboration is largely due to the development of outgrowths from the stipe or lamina, such outgrowths being frequently diverted to reproductive functions. The splitting of the lamina, as well as the perforations that occur in more than one evolutionary series, are no doubt adaptations admitting of the production of a large photosynthetic surface capable of resisting the rough seas to which the thalli are often exposed. The elaborate holdfasts, with production of increasing numbers of branched haptera as the thallus, enlarges, are well suited for anchorage in such a habitat, while the strong, though flexible, construction of the stipes admits of display of the blades to the best advantage.

Before proceeding to consider the internal structure, certain features of the growth of Laminariales may be noted. Elongation of the stipe and much of the enlargement of the lamina are due to the meristematic region in the transition zone, but, apart from annual forms with a long stipe *(Nereocystisdz)*, the growth of the latter is inconsiderable compared with the increase of the lamina. In a perennial *Laminaria, Alaria,* etc. the greater part of the stalk consists of mature tissue, except in so far as there may be secondary activity leading to increase in thickness. In the latter part of the season most of the lamina, too, is no longer enlarging, but during and soon after the time of blade-renewal an extensive basal region is in vigorous growth and further additions to the base of the blade may be made from the transition zone during the whole of the vegetative season. With the approach of winter this activity is arrested and it is then as a rule that the sori-are produced.

The growth of the stipe of *Nereocystis* is very rapid, about 2-5 cm. per, day during July and probably more than that in younger stages. According to Fallis the maximum growth of the blade in *Laminaria*$_y$ *Alaria, Costaria,* etc. takes place in the widest part near the base. The rate increases rapidly for a short distance above the base and beyond that again decreases. The rate of growth during the day is almost double that at night.

The average daily growth recorded by Fallis is 285 cm. in *Laminaria*$_y$ 223 cm. in *Alaria,* and 204 cm. in *Costaria,* these increments being largely due to the lengthening of the blades. According to Sauvageau *Chorda filurn* can reach a length of four metres in 4 months, while growth of the new blade of *Laminaria Cloustoni* is estimated at 0-5 cm. per day and plants of this species can reach a height of one metre in the first year. Lemoine found in young plants of L. *saccharina* a growth of only 24-*27cm and* in older ones of 2-52 cm. per month, but on the Welsh coast plants reach a length of 30 cm. in less than 3 months. The breaking strain of the stipes is considerable and plants are often detached complete during storms.

Anatomical Structure of the Sporophyte

Chorda: The zygotes of *Chorda,* like those of other Laminariales, divide transversely to produce a short thread, the basal cell of which lengthens into a rhizoid *(r).* A distinctive feature lies in the appearance of one or more apical hairs, which may develop while the embryo is still a simple filament.

The first longitudinal walls arise in planes at right angles to one another and divide the upper cells into quadrants; this is followed by transverse division and subsequently by the formation of radial and tangential longitudinal walls. Meanwhile a large number of the lower cells grow out into septate rhizoids.

The parenchymatous cylinder thus produced at first shows little differentiation, but soon a transverse meristematic zone is established near the top of the germling. By its activity new tissue is added both above and below throughout the summer, although that above the meristem dies away sooner or later.

Sections through the meristematic zone display a central group of cells formed from the original parenchymatous cylinder and a wide outer zone with radially arranged elements which no doubt originate by repeated tangential division of the surface-layer (meristoderm). Below the meristem the surface-cells undergo abundant anticlinal division, while the inner cells fail to divide and become more or less pronouncedly elongated. The marked surface-enlargement is probably directly responsible for the development of the central hollow.

In sections of the mature thallus the surface-layer *(s)* of isodiametric or palisade-like cells, with abundant chromatophores, is followed by one or two layers, the cells of which, though almost equally narrow, are markedly elongate. The greater part of the wall of the hollow thallus is, however, composed of wide elongate cells (*c*), which become narrower towards the interior. Here this cortex suddenly gives place to a system of very narrow elongate threads *(m)*, which exhibit the characters of "trumpet-hyphae".

A quantity of ordinary hyphae *(hy)* arise from the inner cells of the cortex and, growing mainly in the horizontal direction, extend into the hollow; they become aggregated at fairly even intervals to form bridging diaphragms. The hyphae are stated to contain chromatophores. At intervals the surface of the thallus bears tufts of colourless hairs *(h)*.

The central hollow contains mucilage and gases which may cause local inflations. The marked tensile strength of the thallus is ascribed by Pringsheim to the wide cortical cells, the walls of which exhibit two sets of striations, visible at a different focus and accompanied by slit-shaped pits. The inner striations give the inner margin of the wall a furrowed contour, while the outer ones are due to undulation of the surface of the inner layers. When a piece of the thallus is allowed to dry, it becomes markedly contorted, so much so that it may appear 8-shaped.

The Early Development of other Laminariales: The embryonic stages have been studied in numerous Laminariales and in general show considerable uniformity. The zygotes commonly germinate while still adhering to the empty oogonium, which is often perforated by the first rhizoid. The lengthening zygote undergoes transverse septation to produce a short erect thread, the cells of which divide by longitudinal walls chiefly orientated in the vertical direction; this results in a flat one-layered expanse which is usually broader above than below. The first rhizoid is soon supplemented by others emanating from the lower cells of the germling. The rhizoids may be septate or non-septate; in *Alaria* they are branched.

The polarity of the young plants of *Egregia* is determined by the direction of the incident light (us), while according to Williams the filamentous embryos are positively phototropic. Longitudinal division may set in at an early stage or a uniseriate filament of 6-8 cells may first be formed. According to Yendo the filamentous embryos of *Costaria* show apical growth.

Sauvageau describes a definite sequence of division in *Saccorhiza bulbosa*. The two uppermost cells of the four-celled thread first produced undergo longitudinal septation to form the one-layered apex of the young blade, while the two lower ones exhibit a further transverse division. The basal cell *(a)* develops into a rhizoid, the next above *(b)* furnishes the lower part of the future stipe, the uppermost *(d)* produces the basal part of the young blade, while the third from below *(c)* gives rise to the intercalary meristem of the transition zone. So clear a differentiation does not appear to occur in the embryos of other Laminariales.

Sooner or later longitudinal walls are formed also in other planes in the basal part of the embryo, whereby the cylindrical stipe is initiated; meanwhile abundant division is taking place in the region of the lamina which, however, for some little time usually remains monostromatic. Meristematic activity now begins to concentrate at the base of the lamina (transition zone) and from then onwards plays an important part in the further enlargement of blade and stipe. The meristematic zone is at first only a transverse row of cells, but it soon becomes more extensive. Over an increasing area of the central portion of the blade, extending outwards from the transition zone, the lamina becomes first two-, then three-, and then four-layered. Both marginal and apical regions, however, appear usually to remain one-layered, at least for a considerable length of time. Various authorities state that the young blade possesses a marginal meristem which contributes to its enlargement, but there is also appreciable surface-growth effected by crosswise anticlinal division of the cells of the two superficial layers, while the inner ones enlarge to keep pace with this growth.

The early development of the stipe has been followed by Sauvageau in *Saccorhiza bulbosa.* After the primary cells have divided into quadrants, periclinal walls cut off a peripheral layer *(m)* from a central group of cells *(ax),* which show no further division and constitute the first medullary elements. The peripheral layer, however, remains actively meristematic, segmenting by anticlinal, and especially by periclinal, walls and giving rise to a cortex *(co)* from which further additions are progressively made to the medulla; by degrees the inner cells develop thick mucilaginous walls. The surface-layer of actively dividing cells may appropriately be designated by the name of *meristoderm* proposed by Sauvageau.

The Structure of the Stipe: In transverse or longitudinal sections of an older stipe of a *Laminaria* or *Alaria,* taken at or near the transition zone, it is generally possible to distinguish four regions. At the surface is the actively dividing *meristoderm (m)* just considered, followed internally by one or more layers of similar small cells. The meristoderm is covered by a mucilage-layer *(mu)* occupying the position of a cuticle. Internal to this small-celled tissue follows an *outer cortex (oc)* of somewhat wider and paler cells, which are markedly elongate, but show no pronounced thickening of the wall; the cells commonly have pointed ends and exhibit frequent horizontal septa. Further to the interior lies the *inner cortex (ic)*, the cells of which are longer and have square ends, while the longitudinal walls show increasing gelatinisation; the cells of this region, too, exhibit numerous horizontal thread-like outgrowths (cross-connections and hyphae). The centre of the stipe is occupied by a mass of tangled threads, constituting the *medulla (me)*, which is commonly oval or elliptical in cross-section. It is not possible to draw a hard and fast limit between the different regions, since they grade over into one another.

When active growth is proceeding, the cells of the meristoderm at the level of the transition zone have a palisade-like shape and exhibit frequent tangential septation, with simultaneous radial divisions so that the cells remain small. They contain numerous small chromatophores and evidently constitute a photosynthetic layer. The cells cut off internally rapidly widen to form the outer cortex and lengthen, often assuming a more or less spindle-shaped form and, since there is no corresponding increase in chromatophores, the cells appear paler. The elongation of these cells is often accompanied by the appearance of a number of thin horizontal septa *(s)* so that each cell gives place to a vertical row.

Towards the interior of this tissue the longitudinal walls commence to thicken and there is a gradual passage to the inner cortex. The thickening is due to mucilage, which is stated to be formed from the middle lamella. The vertical files of cells constituting the inner cortex thus become separated by increasingly wide tracts of mucilage, except at certain points where there is no gelatinisation; these pits are often in the main situated on the tangential walls. As the cells separate more and more, the pits between them become horizontally extended and form short cross-connections which may become septate. This mode of formation of cross-connections has been described by several investigators.

According to Killian, however, such structures are only of a temporary nature and most of the numerous *cross-connections*, seen in the inner cortex and the medulla, arise by the fusion of papillose protrusions formed at corresponding levels on the longitudinal walls. These outgrowths originate from small cells cut off by an oblique septum at the upper end of the parent-cell, although later most of them come to lie opposite the horizontal walls between successive cells. These structures grow towards one another until they meet and thereupon elimination of the separating walls takes place just as in a process of conjugation, although there is no fusion of nuclei. The cross-connections thus formed elongate in the same measure as the vertical rows of cortical cells separate and, as this happens, the former undergo septation.

Meanwhile, as the inner cortex is supplemented from the outer, its innermost elements are added to the medulla, while further contributions are made to the outer cortex by the operation of the superficial meristem. Continuous formation of new cross-connections takes place in the inner cortex so that at its inner limit and in the region of the medulla a complete longitudinal network is produced. The mode of origin of cross-connections described by Killian is also reported by Sauvageau in *Saccorhiza*, but it remains to be established whether it is of general occurrence among Laminariales.

Somewhat later there begins the development of the structures known as *hyphae*, which differ essentially from the cross-connections only in the absence of fusion with a corresponding outgrowth. Hyphae can apparently grow out directly from the cells of the inner cortex, but, according to Killian, those of *Laminaria* arise, like the cross-connections, from small cells cut off in the middle of the cortical elements.

The hyphae develop into slender, often branched threads of considerable length, which grow mainly in the radial direction into the copious mucilage of the medulla. Here they commonly pursue a tortuous course so that it is difficult to trace them to their full extent.

The medulla thus includes structures of three different kinds, viz. *(a)* medullary cells derived from the innermost elements of the cortex and arranged in more or less evident longitudinal rows; *(b)* cross-connections *(co)*, primarily horizontal but later often pursuing an oblique course; and *(c)* the numerous hyphae *(hy)* which extend in all directions between the others, though prevalently radial. It is by no means always easy to distinguish these diverse elements from one another. Surface-enlargement and increase in width are taking place continuously in the transition zone and below it, so that there is a passive extension of the medullary cells and of their cross-connections, in both of which the capacity for growth is sooner or later lost.

Since they undergo no further septation, the component cells are drawn out, and this is accompanied by a reduction in width except at the septa, where the original dimensions are maintained. Many of the medullary elements consequently exhibit a characteristic widening at the septa which is responsible for the customary designation of these structures as *tnmpet-hyphae*\it appears, however, that the actual hyphae only rarely give rise to these elements. The surface-enlargement, leading to passive distention of the internal elements, often results in marked tensions between the outer and inner tissues; strips cut out of a fresh stipe become curved with the outer surface on the convex side.

The cells cut off internally by the meristoderm of the young stipe show the characteristics of outer cortical cells, but, as these lengthen in relation to the surface-growth and more elements are added by the meristoderm, the innermost exhibit thickening of the longitudinal walls and develop cross-connections; in this way the inner cortex is initiated. With further widening of the stipe and outgrowth of hyphae from the inner cortical cells, the medulla begins to differentiate, and thus the young stipe acquires the series of inter-grading tissue-regions that in later stages is met with especially in the transition zone. In the older plant increments to the various tissues appear to be made essentially only in this zone, since the primary regions above distinguished decrease in size as one passes for instance down the stipe of a *Laminaria*.

The medulla seems to undergo little or no increase below the level of the transition zone and in the maturer parts there is usually a sharp boundary between it and the cortex. Increase in width of the stipe is here due to some further additions to the outer cortex, but mainly to special secondary activity. It is not clear whether there is continued production of hyphae from the cells of the inner cortex in the older parts of the stipe, but this would appear to be so in certain instances.

The meristem of the transition zone is therefore of such a nature that it gives rise progressively to the various tissues of the stipe in a horizontal direction and, where such differentiation ceases, there is little further increase in length and the tissues are essentially mature. Fallis' data indicate that the region of elongation and differentiation is normally rather limited. The formative region of the stipe is, however, continuous with that of the lamina and there is no interruption of tissues in passing from the one to the other; the medulla in particular extends throughout the whole transition zone.

This is probably the explanation why, despite the existence of an active meristem in this region, there is no mechanical weakening at the level of the transition zone. The latter differs appreciably from the maturer parts above and below only in the more gradual transition, between cortex and medulla and in the marked evidence of elongation and horizontal septation in the outer cortical cells. In the more mature parts several layers of small cells with chromatophores commonly occur beneath the meristoderm.

In certain Laminariales; *Lessoniopsis* the meristoderm

appears to remain in operation throughout life, but in *Laminaria, Thalassiophyllum, Alaria,* and *Nereocystis,* for example, meristematic activity is subsequently transferred to a cortical layer, situated at a depth of about 4-8 cells beneath the surface. This secondary meristem brings about progressive increase in thickness of the outer cortex and the numerous cells it produces exhibit a radial arrangement. This secondary growth is responsible for the considerable thickness attained by the older stipes of *Lessonia* and of some *Laminarias.* The primary tissue towards the outside usually dies and is shed.

In the perennial forms the activity of the secondary meristem is periodic and the wide outer cortex that is gradually produced usually shows a concentric zoning recalling the annual rings of higher plants; although clearly apparent to the naked eye, the rings are often difficult to distinguish under the microscope. Their formation in *Laminaria, Pterygophora* and *Lessonia* appears to depend on periodic differences in the colour of the cell-contents and the size of the cells, whilst in *Thalassiophyllum* they are due to differences in the colour and thickness of the walls of the secondarily formed elements. It is still a matter of debate whether the successive rings represent annual increments. For the *Laminarias* of the Norwegian coast Printz relates them to the annual renewal of the blade and believes that each represents a year's growth. Others have brought them into relation with the formation of successive series of haptera.

In the mature stipe the horizontal walls of the inner cortical cells usually show a number of pits which are commonly arranged in a ring near the periphery; the pit-membranes are stated to be traversed by delicate cytoplasmic connections and similar plasmodesmae are reported in the septa of the hyphae and of the cross-connections. Church records numerous pits on the tangential walls in the secondary cortex. In the medulla and in the parts of the inner cortex immediately adjacent to it the cross-walls show *a,* fine sieve-like pitting which is often very evident on the septa of the young trumpet-hyphae; such structures are often referred to as sieve-tubes. According to

Sykes the young septum is traversed by numerous cytoplasmic threads, exhibiting a varied arrangement; in the older ones each thread becomes converted into a slime-string enclosed in a rod of callus.

Sieve-tubes of a more distinctive type are found in the inner cortex of *Macrocystis* and *Nereocystis*, and it might be better to restrict the name to these structures. In transverse sections they appear as large and rather thick-walled elements *(st)* which are arranged in more or less distinct radial rows and are surrounded by dense masses of hyphae *(hy)*; the septa are differentiated as typical sieve-plates *(sp)* perforated by coarse pores. These structures do not show the widening at the septa that characterises the trumpet-hyphae of the medulla. In general they pursue a longitudinal course, although neighbouring ones are linked by short cross-connections, the septa of which are likewise differentiated as sieve-plates. The considerable width and specialisation of the cortical sieve-tubes of these two genera is perhaps to be related to the great length attained by the slender stipes which necessitates conduction over a long distance.

According to Sykes "sieve-tubes", similarly disposed in radial rows, occur also in the inner cortex of *Laminaria saccharina* and probably of other species. These structures, however, exhibit neither the large size, nor the well-differentiated sieve-plates found in the sieve-tubes of *Macrocystis* and *Nereocystis*; in fact, except for the absence of marked inflation at the septa, they resemble the trumpet-hyphae of the medulla. There can be little doubt that all these structures are homologous with one another and that, in the older parts of the stipes where no further additions to the medulla are being made, "sieve-tubes", which have not been drawn out like the trumpet-hyphae, are differentiated in the inner cortex. It would seem, however, that this formation of extramedullary sieve-tubes is limited to the innermost region of the cortex.

In *Macrocystis* the young sieve-plate shows essentially the same features as the septa of the trumpet-hyphae. Later each primary cytoplasmic connection divides to form a group, each member of which forms its own callus-rod. Still later the individual strands of a group fuse to form a single slime-string, and in this way the coarse pores of the sieve-plates are produced.

Both in the cortical sieve-tubes and in the trumpet-hyphae of the medulla the older septa become obliterated by deposits of callus. This appears to take place centrifugally, since the medullary elements show such deposits at a time when the cortical ones exhibit few traces of them. In older trumpet-hyphae callus may be deposited almost throughout. The elements, which are thus obliterated, usually have appreciably thickened walls, the thickening consisting mainly of cellulose.

Both the trumpet-hyphae of Laminariales as a whole and the sieve-tubes of *Macrocystis* and *Nereocystis* are no doubt conducting elements, which according to Rigg are specially concerned with the transport of protein. Wille regards the cells of the cortex as storage and mechanical elements, and it is possible that in the young stipe they furnish the chief strengthening system. In the older plant, however, mechanical elements are principally located in the medulla and the adjacent parts of the inner cortex and it is no doubt these which are responsible for the marked tensile strength of the stipes.

The oval medulla of the mature stalk of *Alaria esculenta* consists exclusively of strongly thickened cells. In *Saccorkiza dermatodea* Setchell records, at the periphery of the medulla, a layer of fibre-like, thick-walled elements which can be removed from the other tissues by maceration and extends without interruption through the transition zone from stipe to blade. Analogous structures occur in *S. bulbosa* and *Pterygophora*, while a similar distribution of mechanical elements appears to obtain in *Lessoniopsis* and to a less marked extent in *Cymathere*.

These mechanical elements of the mature stipe are probably in most, if not in all, instances the trumpet-hyphae of younger stages which have become extremely attenuated, have undergone very extensive thickening of their walls, and have in part become filled with callus. Such changes are recorded in *Alaria, Nereocystis, Macrocystis* and *Laminaria*. The medulla of *Costaria* consists of nearly isodiametric cells.

Sauvageau's detailed description of the anatomy of *Saccorhiza bulbosa* at various stages of development differs in certain respects from the accounts given for other Laminariales, and it is not altogether clear whether the differences are peculiar to this species or are the outcome of differences in interpretation. According to him the medullary cells ("solenocysts") multiply, already in the young stipe, by the formation of outgrowths which extend both in the upward and downward directions and enlarge until they reach the width of the parent-cells; they may produce others in the same way.

These medullary elements, which may attain a great length and only rarely branch, are multinucleate and possess but few septa, although they show occasional constrictions *(t)*; they are regarded as conducting elements. Other outgrowths, which arise from the medullary cells, remain short and appear to correspond to the hyphae and cross-connections of other investigators. In the medulla of the blade Sauvageau distinguishes longitudinal elements called "cylindrostyles", composed of a number of elongate, thin-walled, multinucleate cells with few chromatophores; they are believed to arise from the "solenocysts". There can be no doubt that a comparative investigation of the histological structure of a number of Laminariales would furnish results of interest.

The Structure of the Lamina: The mature lamina shows the same basic structure as the stipe, but, owing to the great surface-enlargement, the elements of the medullary region are drawn out in all directions and do not show the prevalently longitudinal course which is characteristic of the stipe. Moreover, except where special ribs are developed, the inner elements exhibit a relatively loose aggregation.

The medulla occupies a rather narrow tangential zone, which is bounded on either surface by a cortex *(ic_y oc)* composed of a limited number of cell-layers; the inner cortex is often not clearly differentiated. The network of medullary cells and cross-connections found in the young blade soon loses its regularity as the latter enlarges. The rows of medullary cells are dragged out of their longitudinal course, while many of the cross-connections are so much displaced that they may even come to run longitudinally; in the older blade it is practically impossible to distinguish the two kinds of structures. In *Alaria* the medullary elements are mainly orientated at right angles to the midrib so that the blade tends to tear in this direction. Except in the midribs and other thickened parts, hyphae are not nearly as abundant as in the stipes; they exhibit a far more irregular course.

The early stages in the development of the blade have again been specially studied in *Saccorhiza bulbosa* by Sauvageau. According to him periclinal division of the surface-layer only takes place exceptionally. The meristematic region at the base of the blade—at first merely a single layer, later composed of two and then of four or more layers—furnishes by the anticlinal division of the cells the successive layers forming the cortex of the blade. The cells of the superficial layer divide crosswise by frequent anticlinal walls, whilst the subjacent layer or layers, which owe their origin to the meristem of the transition zone and not to division of the surface-layer, merely enlarge to keep pace with this surface-growth. Yendo reports essentially the same in *Costaria*.

The mode of origin of the medulla is scarcely clear. Sections of young blades, after the four-layered condition has been attained, show a central system of elongate elements which exhibit no direct relation to the peripheral layers and afford no satisfactory evidence of origin from the latter. Drew and Killian describe how the older germlings of *Laminaria* exhibit the gradual spread, from the transition zone upwards into the blade, of a dark area marking the appearance of the first medullary elements within the latter. This is in agreement with Sauvageau's investigations, according to which the medulla in the blade of *Saccorhiza* originates by the upward extension through the now elliptical transition zone of medullary cells ("solenocysts ") from the young stipe, their prolongations spreading out fan-wise within the mucilage between the cortical layers on the two surfaces of the blade. These observations furnish an explanation for the sharp contrast that is often evident, both in younger and older stages, between the cortex and medulla of the lamina. Sauvageau states that the blade is always thicker at the base than above and this fact, which is also true of *Laminaria,* agrees with a penetration from the stipe of tissue-elements, which at a higher level, is a result of the rapid widening of the blade, become spread out between the cortical layers. A repetition of Sauvageau's observations on *Saccorhiza* and other Laminariales is obviously indicated.

The medulla does, however, also receive additions from the cortex of the developing blade, although the extent of such contributions is at present unknown. In *Saccorhiza* Sauvageau describes the ingrowth of cortical cells into the medulla, the gap thus created in the cortex being filled by preclinical division of the surface-layer. The elements thus added to the medulla ("cellules multiclaves ") are multinucleate and become very long without developing septa they produce numerous short and narrow branches *(br)* by means of which they communicate with one another, as well as with other elements of the cortex and medulla.

Hairs are not as frequent as in most other orders of Phaeophyceae, but they occur on the blades of various genera. They are colourless and show the customary basal growth, usually forming tufts, which in later stages commonly occupy shallow pits owing to rapid growth of the surrounding cells. Dyes penetrate very rapidly into these hairs. In *Saccorhiza* the tufts often arise opposite the enlarged termination of a medullary cell ("solenocyst") which, according to Sauvageau, is continuous from the stipe into the blade. In *Alaria* and *Nereocystis* the hairs disappear in the mature plants.

In conclusion we may endeavour to form a general picture of the growth of a *Laminaria* or other similar, not markedly modified, member of Laminariales. The seat of active cell-division is doubtless in the vague region known as the transition zone which is usually approximately circular in section at the top rf the stipe, but more or less markedly elliptical at the base of the lamina. At periods of active growth the surface-layer in this region is undergoing copious division, both anticlinal and periclinal. In passing to the maturer parts of the stipe division of this meristoderm gradually ceases. In the blade, however, it continues to take place abundantly, although it is entirely or almost entirely anticlinal, so that great surface-enlargement occurs. The region of active cell-division in the transition zone is a hollow, and probably for the most part one-layered, cylinder which surrounds a core of mature or maturing tissue.

According to Lund the considerable thickness of the basal part of the older lamina is due to a great development of the cortical region. He also finds that the new blade of *Laminaria* is thicker than the old and that, as the former enlarges, it becomes progressively thinner. This is almost entirely due to a decrease in the thickness of the medulla.

Transition Zone: The medulla of the blade arises *part passu* with that of the stipe, as the first-formed medullary cells elongate in the upward and downward directions within the transition zone. It may well be that there is no direct extension of the medulla into the blade, but that, as additions are made to the four-layered base of the lamina, the latter acquires a medulla of elongate cells in the same measure as such a medulla differentiates in the transition zone. This picture endeavours to reconcile the diverse data to be found in the literature; how far it corresponds with the actual state of affairs only detailed observations on living plants can show.

Mucilage-canals

Mucilage-canals, which form an anastomosing system in the outer cortex, are met with in the stipes and blades of several Laininariales; in certain species of *Laminaria (L. saccharina, L. digitatd)* and in *Nereocystis* and *Cymathere* they occur only in the blades. As a general rule the canals of the stipe are arranged in a single layer in transverse section, but in *L. Lejolisii*, as well as in L. *Cloustoni*, they may form two rings, while in the older stipe of *Macrocystis* there are several concentric series. They are invariably situated on the outside of the secondary meristem above described. In the laminae the canals constitute separate systems on the two faces.

In *Laminaria* the mucilage-canals develop schizogenously at the upper and lower ends of the transition zone, and this is no doubt true also of other Laminariales. They first appear between the surface-cells as narrow slit-shaped spaces which are radially elongated and, as a result of the division of the meristoderm (*w*), gradually come to occupy a deeper position; as this happens, they widen and become spindle-shaped. A few millimetres beyond the transition zone the mucilage-cavities undergo tangential enlargement and, as a result, they fuse with one another, both in the transverse and longitudinal directions, to form a continuous network, which varies considerably in the shape and size of the meshes. Simultaneously they become protruded on their outer side and extend into the meristoderm, commonly penetrating to the surface, where they may even cause slight prominences; there is, however, no evidence of a direct connection with the external medium.

At the inner margin of the mucilage-cavities, not far beyond the place of their first formation, there appear small cells with a large nucleus and abundant granular cytoplasm. These multiply by division and gradually constitute an irregular layer over the inner surface of the space. In longitudinal sections, however, it is seen that these cells form isolated groups and, although they have the characteristics of secretory cells, they do not produce a connected epithelial layer. Oliver describes the formation of tylose-like structures from these cells.

According to Guignard the secretory cells develop secondarily in *Laminaria*. In *Macrocystis*, however, Skottsberg recognises them before the appearance of the mucilage-space, although, since the secretory cells do not keep pace with the radial elongation of the latter, they are subsequently found mainly at its inner margin. In the young lamina of *Nereocystis* the secretory cells form a complete investment to the canal.

The mucilage within the canals of *Laminaria* differs from that of the cell-membranes. The large elements recorded by Gain in the medulla of a *Lessonia,* and described as mucilage-canals, are of doubtful nature.

Glands of a different kind occur in *Undariaboi),* as well as in the young plants of certain species *of Laminaria* and *Alaria.* Those of *Undaria,* which appear as dark dots thickly scattered over both surfaces of the lamina, are single cells of large size situated in the hypodermal layer and formed by periclinal division of a surface-cell of the young blade. The overlying cell becomes flattened by the rapid enlargement of the gland and ultimately forms a thick hyaline roof to it. In a surface-view the rectangular areas overlying the glandular cells are readily seen. Yendo believed that these cells produced mucilage, but according to Sauvageau they are fucosan-receptacles. The fresh contents are colourless.

Origin of Splits and Perforations

As a general rule there would appear to be some factor causing local degeneration of the internal tissues during the development of splits and perforations, a process accompanied by the formation of furrows or depressions on the surface, but which of these changes is primary and which is secondary is not always evident from the published accounts. As the depressions gradually deepen with the progressive breakdown of the internal tissues, the meristoderm divides and provides a continuous covering. Ultimately, when the double layer thus formed extends through the whole thickness of the blade, complete rupture occurs, the freshly exposed edges being now entirely overgrown by the meristoderm. In most instances at least it seems that there is never more than a very temporary exposure of the inner tissues.

In *Laminaria Cloustoni* the splits are initiated by the appearance of longitudinal furrows which probably result from increased division of the cells of the meristoderm and are often seen sooner on one surface than on the other. As the furrows deepen, breaks appear in the internal tissues and the outer layer ruptures, although the exposed edges become rapidly overgrown. These data are based on a study of regions in front of already established splits, and they imply that, in the species in question, surface-changes are of primary importance, a conclusion apparently reached also by Grabendorfer in *Lessonia*. It is not altogether clear from

Killian's account whether the internal tissue are actually exposed. He is of the opinion that wave-action plays a considerable role in the extension of the splits towards the margin.

In *Macrocystis* internal changes are the first to be observed, the inner tissues undergoing gelatinisation which results in the formation of a cavity occupied by mucilage, whilst on either surface active division of the meristoderm causes the formation of furrows. This is in general agreement with the account given by Macmillan for *Nereocystis;* he describes how, as the opposite furrows meet, the separate meristoderrn layers unite without any destruction of cells. Wells, however, states that it is only the basal perforation that arises in this way, while the forward advance of the splits is due to excessive cortical division which causes rupture of the surf ace-layer and finally of the whole lamina.

This is also stated to be so in *Dictyoneurum*. In *Postelsia*, according to the same authority, the splits arise in regions in which division of the meristoderrn is practically at a standstill, so that the lamina here remains thin and gets worn by wave-action; much the same is recorded for *Lessoniopsis*.

Humphrey's account of the development of the perforations in *Agarum* again places stress on a primary growth of the meristoderrn. Rapid local multiplication of its cells results in a papillate protrusion on the one surface and of a corresponding depression on the other. The meristoderrn on this surface penetrates more and more deeply into the underlying tissues and, after reaching the medulla, continues to grow only along the sides of the depression, while that covering its base dies away. The sides thus grow inwards like a circular punch and cut out the part of the medulla still remaining in the centre of the perforation. In *A. ftmbriatum* Harv. only a few of the papillae become perforated.

The diverse observations agree that the edges of the splits are soon covered by rapid growth of the meristoderm and other superficial tissues. In this respect there is a marked contrast to the mode of healing of wounds. If the superficial layers of the lamina or stipe of a *Laminaria* are removed, the intact elements at the exposed surface divide to form perpendicular rows of cells which rapidly cover the wound. In his studies of regeneration in Laminariales Setchell found that it is always the inner cortex and medulla that grow out to form the new organs.

The same authority has drawn attention to certain special features connected with blade-renewal in the species of *Laminaria*. In *L. Sinclairii* the first sign of the inception of this process is the appearance of a slight dilation in the transition zone. This is followed by a transverse rupture at this level and the rapid separation of the two edges as the new blade *(nb)* develops from the more internal tissues; the ruptured surface layers form collars *(c)* at the top of the stipe and at the base of the old blade. Blade-renewal is thus effected solely by inner cortex and medulla; the outermost elements of the former divide actively at right angles to the surface and give rise to a new meristoderm by the division of which a new outer cortex is formed. Dr. Lund has observed similar events in *L. Cloustoni*, but it is not known whether comparable changes are met with in other Laminariales exhibiting blade-renewal. The shifting of meristematic activity to deeper layers is paralleled in the older stipe by the secondary growth which is initiated in the outer cortex.

Air-bladders

The walls of the air-bladders of *Nereocystis* and *Macrocystis* consist for the most part of cortex, although the inner part contains some sieve-tubes; in *Macrocystis* numerous hyphae project into the cavity. According to Macmillan a second meristematic zone is distinguishable in the inner part of the wall of the bladder of *Nereocystis.* The bladders of *Nereocystis* have a capacity of several litres. The contained gas is stated to include carbon monoxide (1-12 per cent), which is formed only in the presence of oxygen and is believed to be a product of respiration; Zeller and Neikirk, however, record only carbon dioxide in considerably larger quantities than in atmospheric air. The pressure within the bladder is often less than that of the atmosphere and varies at different times of the day; nevertheless the bladders can resist considerable external pressure without collapsing.

The Haptera

The succession of rhizoids that serves to attach the young plants is later replaced by the broadened base of the enlarging stipe, which forms a disc fastened by numerous rhizoids and showing no appreciable internal differentiation, although the inner cells are somewhat elongated. In older plants the disc is mostly replaced by special haptera which are arranged in approximate whorls and arise in succession from progressively higher levels on the stipe. They first appear as slight superficial puckerings and in *Laminaria Cloustoni* are formed at the time of most active growth, soon after the appearance of the new blade. The later ones often show considerable forked branching.

The haptera are produced by tangential division of the outermost cells of the stipe, although Church states that the older ones arise endogenously. Cell-division is localised in the apical region, where the cells are often arranged in well-defined rows. The inner part of the mature hapteron consists of elongate mechanical cells, with strongly thickened transverse walls which are often provided with numerous pits. There is no medulla, although a few hyphae may be present; mucilage-canals are sometimes found. When the substratum is reached, the tip of the hapteron spreads out and the surface-cells grow out into densely packed rhizoids. Growth in length may continue after attachment, so that the primary disc is lifted off the substratum.

In *Pterygophora* Macmillan describes growth-rings in the older haptera.

More specialised attaching organs are found in *Saccorhiza* and *Macrocystis*. In *S. bulbosa* most of the haptera arise from the bell and this is also so in *S. dermatodea*, where the latter is far less conspicuous. The development of the bell is largely due to expansion of the medulla, the cortex playing only a passive role In *Macrocystis pyrifera* Skottsberg distinguishes, apart from the ordinary haptera which become attached at their tips, others that curl round foreign objects and so help in fixation. The attaching organs of the Californian forms of *Macrocystis* have been fully studied by Setchell; those of *M. integrifolia*, in particular, are very elaborate, being prostrate and developing numerous floating shoots from their margins.

Asexual Reproduction

The extensive sori, the position of which in the different genera has already been described, invariably comprise only unilocuiar sporangia intermingled with abundant unicellular paraphyses. Their development takes place in essentially the same way in all Laminariales. The superficial cells grow out as a palisade-like layer, and each protuberance sooner or later divides tangentially into a small basal cell *(b)* and the future paraphysis *(p)* which continues to lengthen. As the thallus enlarges, the basal cells widen and the paraphyses separate; they now occupy only part of the surface of the basal cell. The apices of the paraphyses, however, usually broaden and remain in close contact with one another. During the outgrowth of the paraphyses the mucilaginous cuticle of the superficial cells is often merely stretched and constitutes a protective covering to the developing sorus; in *Saccorhiza bulbosa* this layer even undergoes further thickening.

The paraphyses, which are usually club-shaped, show a more or less marked gelatinous thickening of the membrane at their expanded outer ends. Here lie numerous chromatophores *(ch)*, whilst there are often abundant fucosan-vesicles in the interior. The oblong sporangia arise from the basal cells between the paraphyses and are thus protected during development. The efficiency of the paraphyses, however, no doubt varies. In *Laminaria* and other genera the thick outer membranes often flow together to form a continuous covering over considerable areas. On the other hand, in *Saccorhiza dermatodea* and in *Cymathere* the paraphyses have only a slightly thickened tip.

The number of zoospores produced in the sporangia seems to vary. 32 are recorded in *Laminaria, Alaria esculenta, Pterygophora,* and *Eisenia,* sometimes 64 in the last two. On the other hand, in *Chorda* there are 16, while in *Saccorhiza bulbosa* 128 is given as the probable number. Zoospore-development commences with successive nuclear division, accompanied by multiplication of the chromatophores (*c*), and takes place in essentially the same way as in Ectocarpales. Thus, in *Chorda* the 16 nuclei at first lie in the middle of the sporangium, while the 16 chromatophores *(c)* occupy a peripheral position; later the nuclei migrate outwards and become associated with the chromatophores, after which the rudiments are cut out by simultaneous cleavage. The sporangia of a sorus develop successively, although considerable numbers mature and liberate their contents at the same time. The wall at the apex of the mature sporangium usually becomes strongly thickened, and dehiscence occurs here. Sauvageau suggests that it is aided by the pressure of the gelatinous paraphyses on the sporangia. The whole contents, enveloped by a delicate membrane, are liberated with a jerk and gradually emerge from between the paraphyses.

The zoospores are of the usual type and generally possess a single plate-shaped chromatophore which often occupies a considerable part of the periphery; in *Chorda* there are a number of chromatophores. The swarmers commonly tend to move away from strong light, but the presence of an eyespot is variable.

Kylin showed in 1918 that the first nuclear divisions in the sporangium of *Chorda* are meiotic, and this has since been proved by McKay for *Pterygophora* and by Hollenberg for *Eisenia*. The haploid chromosomic-number is given as 13-15. There can be no doubt that all Laminariales normally show an alternation of cytological phases.

As a rule the sori differentiate only on mature parts, which have practically completed their growth. Their development usually coincides with the commencement of the cold season, and the fertile parts often persist through a considerable part of the winter. Precise data, which are rarely available, indicate some degree of variability. Harries showed that the three species of *Laminaria (digitata saccharina, Cloustoni)*, found on the Welsh coast, mature in the order named, and concludes that this is correlated with their position on the shore, whilst Sauvageau reports finding sori on adult plants of L. *digitata* and *L. saccharina* at all times of the year.

Gametophytes

Although the female prothalli are often more or less clearly pictured in some of the older figures depicting the development of the young plants, it was only in 1915 that Sauvageau reported the existence of heteromorphic alternation with two sexes of gametophytes. Prior to that it was generally held that the Laminariales showed no sexual reproduction; erroneous observations of Drew, on the sexuality of the swarmers produced by the sporophyte, were disproved by Lloyd Williams. In 1916 Sauvageau followed up his first discovery by the description of the gametophytes of *Laminaria digitata* and *L. saccharina* and later of those of *Alaria esculenta*. In the same year Kylin confirmed Sauvageau's observations on *L. digitata* and in 1918 described the gametophytes of *Chorda*. It was not, however, until 1921 that Williams observed the spermatozoids and the process of fertilization. Since then the gametophytes of numerous Laminariales have been described.

The swarmers of a sporangium give rise partly to male and partly to female gametophytes. This was established by

Sauvageau (HQ), who in *Saccorhiza bulbosa* observed the germination of the spores into the two kinds of prothalli while still enclosed within the sporangium; a similar observation was later made on *Laminaria Cloustoni*. By allowing the sporangia of a sorus of *Laminaria* to liberate their contents into solidifying gelatine, Schreiber isolated the individual groups of spores and by subsequent culture showed that male and female gametophytes are produced in equal numbers from a sporangium. Sex determination in this genus, and probably in all Laminariales, is therefore genotypic. The occurrence of swarmers of two sizes has occasionally been recorded, but such differences appear neither to be constant, nor to be related to sexual differentiation.

The early stages of germination of the two kinds of zoospores are identical and the differences between the sexes usually become apparent only after some time, although in *Chorda* they are evident before the gametophytes are a week old. The swarmer rounds off and secretes a thin membrane; the resulting embryospore puts out a slender outgrowth into which the bulk of the contents pass. Gradually the distal extremity of the tube becomes dilated and at this stage the germling often appears dumbbell-shaped.

Finally the contents collect within the swelling, which becomes separated by a septum, the orientation of which is independent of the direction of the incident light. It is this cell that divides to form the actual gametophyte and, even in late stages, the empty spore-membrane with its tubular outgrowth may remain attached to the prothallus. In *Alaria esculenta*, where the spore occasionally retains part of its contents, it may subsequently give rise to a second gametophyte, but this is exceptional.

The filamentous gametophytes are commonly well branched and possess rather elongate cells containing a number of approximately discoid chromatophores; their habit is heterotrichous and growth is usually slow. The natural substrata are probably rocks and other Algae. The male gametophytes are nearly always more copiously branched than the female, although the reverse condition has been reported in *Costaria*. On the other hand, the cells of the male gametophyte, though usually more numerous, are invariably smaller than those of the female and sometimes also paler in colour. In *Chorda* the gametophytes are exceptionally robust, richly branched, while in *Egregia* according to Myers both kinds consist of only a few cells.

The data on the size of the gametophytes are, however, based on observations made in cultures, where their development is likely to be retarded. Both form and size do in fact vary appreciably with the temperature, the light-intensity, and the kind of nutrient salts available. According to Schreiber gametophytes, as well as sporophytes *of Laminaria* exhibit good growth even at higher temperatures, although they remain sterile; when the temperature is lowered gametes are produced, whilst when it is raised their formation is arrested. This perhaps accounts for the practical restriction of Laminariales to colder seas. Regeneration of the gametophytes from fragments or isolated cells has been reported.

The *antheridia* arise singly or in groups at the tips of the erect threads or as lateral outgrowths from the upper side of the creeping threads; intercalary antheridia, formed by transverse division of the cells, are not uncommon. The antheridia are small, colourless or almost colourless structures, which at best contain only one or two much reduced chromatophores. The entire contents are utilised to form a single spermatozoid. The wall at the apex becomes protruded and markedly swollen, and here an aperture is formed through which the male cell (s) is liberated. The ellipsoidal or pyriform spermatozoids, about 4 *p*, long, possess the usual two laterally attached flagella; an eyespot is reported in *Pterygophora*, but appears usually to be lacking. The male gametophytes commonly die after liberation of the sperms. Sauvageau comments on the marked protandry exhibited by the species studied by him.

In the female prothalli any or all of the cells may enlarge to form oogonia (0). The latter are either pear-shaped with the upper extremity narrowed'or dilated, or they may be tubular or even lobed. The wall at the distal end of the oogomum usually becomes greatly thickened and in *Laminaria(i9S)* three distinct layers can be distinguished. The numerous chromatophores become crowded together at the apex of the mature organ with their long axes parallel to it, which probably indicates a marked internal pressure playing a role in dehiscence. The entire contents form a single ovum, which is normally liberated prior to fertilization through a narrow apical aperture. After the female cell has emerged, the thick elastic edges of the wall again meet and form a platform to which the ovum usually remains attached for some considerable time after fertilization. In *Chorda* the mature ovum merely lengthens out of the oogonium, whilst in *Macrocystis* it is stated to be retained until after fertilization; this is not confirmed by Papenfuss.

Nuclear fusion has been observed in *Laminaria (19 per cent)*, *Pterygophora*, *Macrocystis*, and *Eisenia*. In *Pterygophora* the fusion nuclei are in early prophase, while in *Eisenia* both gametes are stated to enlarge appreciably during fusion. The zygote secretes a thin membrane and develops into the sporophyte without any resting period. Parthenogenesis has been reported by Schreiber, but its occurrence in nature is not established; the germlings from the parthenogenetic ova show irregularities in division. The possibility for natural hybridisation must be considerable, but Schreiber's attempts to produce hybrids artificially were unsuccessful.

In *Saccorhiza bulbosa* the female zoospore commonly gives rise directly to an oogonium, without any vegetative phase. This is the simplest type of gametophyte known among Brown Algae, and its discovery has played a considerable role in the interpretation of the life-cycle of Fucales. Such extreme reduction seems to be frequent in this species, although occasionally much more extensive gametophytes are formed.

A tendency to reduce the female gametophyte to a single cell has been observed in cultures in other Laminariales; thus, it has been reported in *Laminaria*, *Macrocystis*, and *Arthrothamnus*. The one-celled condition seems to be frequent in *Laminaria Cloustoni* and in *Arthrothamnus*, but it remains to be determined how far such simple prothalli are the result of cultural conditions. According to Sauvageau the female prothalli *of Phyllaria reniformis* Rostaf are "parasitic" in *Lithothamnion lichenoides*.

Status of Laminariales

In cultures of *Laminaria saccharina* Pascher reports a precocious development of sporangia in embryos in which surface-development was just beginning. Other embryos, still in the filamentous (2-8-celled) stage, produced 4-8 swarmers from certain of their cells, and instances were even observed in which the fertilized ovum, without further segmentation, divided to form a number of reproductive elements. The last condition does not differ from that which obtains in an isogamous or oogamous Green Algae. These observations suggest that, under certain circumstances, production of zoospores might take place without the development of an elaborate sporophyte, and they lend support to the view of an ultimate isomorphic origin for the sharply distinct generations of Laminariales.

Moreover, both they and the early stages of development of the sporophyte point to simple filamentous ectocarpoid types for the derivation of these elaborate forms. It is noteworthy that, while the gametophytes are heterotrichous and in great part prostrate, the sporophytes, even in the reduced stages just described, are erect from the first. The Laminariales may in fact be regarded as exhibiting to an extreme the elaboration of the sporophyte so evident among Ectocarpales, while the filamentous heterotrichous habit has been retained by the gametophyte. The absence of a prostrate system in the sporophyte finds its parallel in other highly differentiated series.

The Laminariales would thus exhibit the ultimate development of the tendency towards a parenchymatous construction evident among Ectocarpales and realised in such forms as *Litosiphon* and *Stictyosiphon.* The relatively simple genus *Chorda* is relevant in this connection, for although it is quite like other Laminariales in reproduction and life-history, it is less specialised morphologically and anatomically. It may well owe its simplicity to the non-differentiation of an expanded lamina, since this probably necessitates a more elaborate mechanical construction, such as is supplied by the medulla of the stipe of most Laminariales. The marked activity of the meristoderm is foreshadowed in diverse parenchymatous Ectocarpales. The specialised sori of Laminariales are likewise already indicated in some of the more advanced representatives of the latter Order.

Both Laminariales and Fucales show the development of a far more elaborate plant-body than is usually found among Algae. Alike in their morphology and their anatomy they exhibit features, which find a certain degree of parallel among higher land-plants and are probably an expression of general tendencies in somatic evolution. Since the early stages in the origin of land-plants are likely to remain unknown, these elaborate Brown Algae may perhaps afford some data as to the possible lines of development that were followed. It must be emphasised, however, that the system of intergrading tissues met with in these forms is without parallel among higher plants.

Classification and Geographical Distribution

The classification—largely based on Setchell—that has been adopted in the preceding matter is as follows:

1. *Chordaceae:* Chorda.
2. *Laminariaceae:*
   (a) *Laminarieae:* Laminaria, Phyllaria, Saccorhiza.
   (b) *Cymathereae:* Cymathere, Pleurophycus.
   (c) *Agareae:* Agarum, Costaria, Thalassiophyllum.
   (d) *Hedophylleae:* Arthrothamnus, Hedophyllum.

3. *Lessoniaceae:*
   - (a) *Lessonieae:* Dictyoneuropsis, Dictyoneurum, Lessonia, Nereocystis, Postelsia.
   - (b) *Macrocysteae:* Macrocystis, Pelagophycus.
   - (c) *Lessoniopseae:* Lessoniopsis.
4. *Alariaceae:*
   - (a) *Alarieae:* Alaria, Hirome, Phyllogigas, Pterygophora, Undaria.
   - (b) *Ecklonieae:* Ecklonia, Eisenia.
   - (c) *Egregieae:* Egregia.

The Laminariales are characteristic of colder seas, the chief centres of distribution being circumpolar in the Arctic and Antarctic regions. Here they occur wherever there are suitable substrata, free from permanent ice. From the polar regions they extend into the temperate zones and, where cold ocean currents sweep along the coasts, as for instance along the shores of Lower California and Peru or the west coast of South Africa, certain genera penetrate far into subtropical or even tropical regions. The genera occurring in the Northern and Southern Hemispheres are in large part distinct.

Typical of the Northern Hemisphere are the Chordaceae, Laminarieae, Alarieae, and Agareae, although the last-named are restricted to North America, Greenland, and adjacent parts of Asia. The *Laminarias* are essentially northern in their distribution, although L. *Lejolisii* flourishes on parts of the Atlantic coast of Morocco and a few species are frequent in the Southern Hemisphere (e.g. *L. pallida* (Grev.) J. Ag. at the Cape. *Saccorhiza dermatodea* and *Alaria pylaii* (Bory) J. Ag. occur over a large area of the North Atlantic. South of the Cape Cod Peninsula Laminariales only play a small role on the eastern shores of North America. In the Mediterranean *Phyllaria reniformis* (Lamour.) Rostaf., *Saccorhiza bulbosa*, and *Laminaria Rodriguezii* have alone been recorded; the last is stated to occur mainly at depths of 100-150 metres. Berthold (H) lists no member of the order from the Gulf of Naples.

The Pacific shows a considerably greater diversity of forms than does the Atlantic, and both on the Asiatic and on the American shores members of the order extend far towards the south, although the two floras are in part distinct. Characteristic of the North Pacific, both on the American and Asiatic shores, are the Cymathereae, *Thalassiophyllum, the* Hedophylleae *(Hedophyllum, Arthrothamnus),* and *Costaria. Undaria* appears to be confined to Asia, while along the American shores and in part extending down to Lower California are found a number of characteristic Lessonieae *(Dictyoneurum, Postelsia),* as well as the genera *Costaria, Pelagophycus, Nereocystis, Les-soniopsis, Pterygophora, Eisenia,* and *Egregia*; of these *Pelagophycus* is restricted to the southern part of this region. As compared with the Arctic regions the Californian coast has fewer Alarieae and is much poorer in Laminarieae (here *L. Sinclairii)* and Agareae (only *Costaria).*

The characteristic Laminariales of the Southern Hemisphere are *Lessonia*> *Ecklonia,* and *Macrocystis,* while the imperfectly known genera *Phaeoglossum* and *Phyllogigas* are at present only recorded from the Antarctic. *Macrocystis* is probably most abundant on the shores of Patagonia, extending from here down to the area of permanent ice, but it is also widespread in Australia and New Zealand, and South Africa; in the last-named area it is practically confined to the east coast with its colder water. *Ecklonias* are well represented in Australia, as well as in South Africa, which harbours the largest species of the genus *(E. buccinalis).* Most of the striking *Lessonias* are confined to the South American region around Cape Horn, although *L. flavicans* is recorded from New Zealand. *L. laminarioides* is a characteristic component of the flora of Japan and of the adjacent shores of Siberia.

Macrocystis pyrifera has the widest distribution of all species of Laminariales. Its area extends over the whole of the Southern Subantarctic zone whence it has been distributed by Antarctie currents to many of the outlying islands (Kerguelen, Tristan d'Acunha, St Paul Islaad). It follows the Humboldt current up the western shores of South America (here accompanied by

Eisenia Cokeri and *Lessonia nigrescens)* as far as the equator, and, passing to the cold waters flowing in from the North Pacific, extends along the entire western coast of North America up to British Columbia.

Sphacelariales

The five orders considered in, all of which exhibit intercalary growth, are no doubt special developments of types of construction represented among Ectocarpales, although the members of the latter order do not attain to the same degree of differentiation or of reproductive specialisation. The three remaining orders of Phaeophyceae are characterised by possessing well-marked apical growth and comprise forms with a parenchymatous construction only; they include no parallel for the pseudoparenchymatous type exemplified by Cutleriales and Sporochnales. Sphacelariales and Dictyotates also differ in the nature of their life-cycle from the orders since alternation in both is almost certainly isomorphic; in this respect they differ markedly from the polystichous Ectocarpales.

The origin of Sphacelariales is probably to be sought among simple filamentous Ectocarpales with isomorphic alternation. The members of this order are usually copiously branched, although they mostly attain no great size. They afford marked evidence of heterotrichy and most are probably perennials, persisting with the help of the prostrate and the basal parts of the erect systems. The diverse branches of the latter are capped by a conspicuous apical cell cutting off a single row of segments, which for the most part undergo very regular septation so that the younger parts at least exhibit a definite tier-like structure. The individual branches are nearly always radial in symmetry. The prostrate system is usually well developed and in the more specialised types commonly takes the form of a many-layered crust.

Hairs, which are invariably provided with a sheath, occur in many species. The cells contain numerous, rather small chromatophores which are lenticular or granular and devoid of pyrenoids. Eau de Javelle causes the cell-walls of all

Sphacelariales to assume a black colouration which disappears after some time. The reaction is confined to the older layers of the wall and, when the latter is thin, affects only the middle lamella. According to Sauvageau there is also widespread occurrence of tannin, which is either diffused through all the cells *(Battersid)* or localised in certain peripheral or central elements of the erect system, as in some species of *Sphacelaria.*

In the early part of the century Sauvageau published a series of memoirs dealing with this order; these, which have been largely followed here, considerably amplified and in part corrected the earlier investigations. The morphology and anatomy, of the European species at least, are well known, although many details of reproduction and life-history still remain obscure. The few genera are distinguished mainly by the mode of origin of the lateral branches, the behaviour of the segments cut off from the apical cell, and the disposition of the sporangia. Only two of them *(Sphacelaria, Halopteris)* comprise any considerable number of species.

Habit and Geographical Distribution

Many Sphacelariales form short densely branched tufts, varying from a few millimetres to one or two centimetres in height. Several species of *Halopteris,* as well as *Cladostephus, Phloeocaulon,* and *Ptilopogon,* attain to larger dimensions, although probably never exceeding about 30 cm. in length; one of the largest of the *Sphacelarias* is *S. plumigera.* In diverse species of *Sphacelaria,* as well as in *Sphacella,* the basal system is endophytic and then usually consists of separate threads. Such endophytic forms often exhibit little or no branching of the upright threads.

In certain *Sphacelarias* (e.g. *S. olivacea, S. radicans*, *S. racemosa)* the few laterals are irregularly arranged and differ in no appreciable respect from the parent axes. As a rule, however, there is differentiation into branches of unlimited and limited growth and very commonly at least the majority of the latter exhibit a distichous arrangement resulting in a pinnate habit, although this may become obscured in the older plant by the outgrowth of dormant initials or other irregularities in branching. In the pinnately branched forms the short laterals are either opposite or alternate (species of *Halopteris* and *Phloeocaulon*). *Cladostephus* is distinguished by the whorled arrangement of these laterals.

The densely tufted, fastigiate habit assumed by *Halopteris scoparia (Stypocaulon scoparium* Kutz.) at certain times of the year is, according to Sauvageau, due to the abundant outgrowth of dormant branch-initials exhibiting a tetra-stichous arrangement, as well as to the sprouting of adventitious branches from the persistent stumps of the short laterals of the previous season. Similar features are no doubt found in other members of the order.

The Sphacelariales commonly inhabit the intertidal region, especially its lower zones, and often penetrate for some distance below low-tide level *(Halopteris scoparia, Cladostephus spongiosus).* *Chaetopteris plumosa* is usually found in deeper water, while *Sphacelaria olivacea* may occur in shaded situations near high-water mark. Reinke records a free-floating form of *Halopteris scoparia* in the Baltic. In the same region *Sphacelaria cirrhosa* at times assumes an aegagropilous habit, forming hollow balls reaching the size of a human head. Balls of uncertain origin consisting of numerous plants of *Halopteris funicularis* have been recorded from Tasmania.

The Sphacelariales are met with in all seas, although many are inhabitants of temperate and even polar waters. One group of species of *Sphacelaria* is apparently restricted to the Northern (e.g. *S. racemosa, S. hystrix, S. radicans)* and another to the Southern Hemisphere (e.g. *S. Borneti* Hariot, *S. pulvinata* Harv., *S. divaricata* Mont.). Sauvageau concludes that the Australasian seas are probably the richest in species of this genus. Some *Sphacelarias* are common to both hemispheres (e.g. *S. cirrhosa* (Roth) Ag., *S.furcigera),* while *S. tribuloides* Menegh. and *S. furcigera* are the species most frequently recorded from warmer seas. In the Mediterranean the genus is poorly represented. *Chaetopteris* is mainly confined to the North Atlantic.

Halopteris scoparia has been recorded from all parts of the world, while *H. filicina* is widely distributed in the warmer seas of the Northern Hemisphere, although its occurrence south of the equator remains uncertain. Most species *of Halopteris* are, however, to be found in the Australasian region, and *H. funicularis* (Kutz.) Sauv, is known from many parts of the Southern Hemisphere. *Phloeocaulon* and *Ptilopogon* are almost confined to Australasia, although *Phloeocaulon squamulosum* Geyler is recorded from South Africa. *Cladostephus verticillatus* seems to enjoy a worldwide distribution, while the occurrence of C. *spongiosus* south of the equator is still doubtful. Sauvageau's *Alethocladus* is known only from the Antarctic.

Apical Cell and Segmentation the Erect System

The Apical Cell and its Mode of Segmentation: The prominent elongate apical cells commonly contain abundant fucosan-vesicles, as well as a quantity of brown tanniniferous matter which often causes the tips to appear dark brown or black to the naked eye, and it is to this feature that *Sphacelaria* owes its name. This substance is regarded as a reserve, which is utilised during growth; a similar substance commonly occurs in dormant initials and in other parts of the mature plant. The cytoplasm of the apical cells and of the subjacent segments is markedly alveolar, while the large nucleus of the apical cell is provided with a centrosome.

The apical cell cuts off segments parallel to its base. When it has reached a certain length, a transverse septum separates a *primary segment (i)* from the apical cell proper *(a)*. The former again divides transversely into two *secondary segments—superior (s)* and *inferior (i)*—which are usually of equal size, and these divide further to produce the mature tissue. In many Sphacelariales the secondary segments, though undergoing progressive septation into smaller and smaller cells, exhibit no further evident enlargement. In other words, except in so far as there is production of a special cortical investment in the older parts, the diameter of the different branches is essentially that of their apical cell which effects the entire growth in length of the branch. These are what Sauvageau calls *leptocaulous* forms.

On the other hand, in the long axes of *Cladostephus*, *Phloeocaulon*, and *Ptilopogon* the secondary segments not only undergo abundant division, but also enlarge more or less appreciably in width and in length. The axis therefore increases in girth from the apex backwards, and the laterals tend to become more widely spaced in the older parts. These are the *auxocaulous* forms of Sauvageau. In *Phloeocaulon foecundum* the primary segments may enlarge to thirty times their original height, while *(Cladostephus)* the tenth segment is already more than three times the length of the first. This is due to continuous elongation of the cells formed by longitudinal division of the secondary segments, a process accompanied by repeated transverse septation. Increase in width is less pronounced, although there may be subsequent broadening as a result of secondary meristematic activity. The laterals of limited growth, even in the auxocaulous forms, show no enlargement of the segments.

The sequence of division behind the apical cell can often be traced without great difficulty, since the primary septa long remain recognisable. The two secondary segments soon divide by longitudinal walls. In *Sphacella* and a few species of *Sphacelaria (S. pulvinata, S. bracteatd)* such longitudinal division is quite restricted and may fail to occur in certain segments, but in most Sphacdariales it takes place plentifully so that, at a short distance behind the apex, the diverse branches are multiseriate and consist of a peripheral layer of cells surrounding a central group. The central cells commonly develop a slight thickening at the corners, whilst the outer walls of the peripheral cells always become markedly thickened.

In many species of *Sphacelaria (S. plumula, S. pulvinata, S. furcigera*, etc.) the secondary segments divide by longitudinal walls only, but in others *(S. radicans; S. olivacea, S. racemosd)* transverse septation also takes place sooner or later, and this occurs in all other Sphacelariales *(Chaetopteris, Halopteris*, etc.)

and is especially pronounced in the auxocaulous forms. The secondary segments may form only one series of transverse septa, as usually in *Sphacelaria radicans* and *S. racemosa*, but more commonly there are several series (e.g. *S. olivacea*, *S. plumigera*, *Halopteris*, etc.). Since the transverse septa all arise at about the same horizontal level the tier-like structure is retained.

In *Sphacelaria cirrhosa* and its allies all the longitudinal walls are radial, but in other species their orientation is more varied, and there is probably not the degree of regularity indicated by Geyler. Not uncommonly the first two walls are perpendicular to one another and divide the segment into quadrants, whilst the subsequent ones tend to be periclinal and to delimit a superficial layer, the cells of which multiply further by anticlinal division. Quite frequently, however, there are only one or two large central cells surrounded by an irregular peripheral series, the longitudinal septa having been largely periclinal from the first (e.g. *£. radicans S. plumula)*.

In the more specialised genera longitudinal septation is more regular. In *Cladostephus verticillatus* the formation of quadrants and the subsequent periclinal division is followed by abundant and rather irregular longitudinal septation of the inner cells, though crosswise division is frequent. In the larger branches of *Halopteris scoparia* the perpendicular septa of the quadrant division, the first (i) of which lies approximately in the plane of branching, are followed by further septa parallel to the previous ones; each quadrant thus undergoes regular crosswise division so that there result four central and twelve peripheral cells. Four of the latter", which appear triangular in the transverse section, constitute delayed branch-initials or rhizogenous cells (pericysts). Further septa, perpendicular to the third and fourth, result in a crosswise division of each central cell, as well as of each of the eight non-triangular peripheral cells. The 32 inner cells thus produced usually undergo no further longitudinal division, although the peripheral ones divide by a further radial wall.

The branches of limited growth of *H. scoparia* show a modified segmentation and no pericysts are formed. After the quadrant division there is periclinal segmentation leading to the delimitation of a peripheral layer. In the finer branches periclinal division may set in from the first so that the centre is occupied by a single large cell. This recalls the segmentation of *Sphacelaria,* and is probably found in the smaller branches of most of the more specialised members in which the major axes possess a complex structure similar to that of *Halopteris scoparia.*

Cortex-formation

The only means of secondary enlargement in leptocaulous forms is by cortication. In auxocaulous Sphacelariales, however, the peripheral layer sooner or later becomes meristematic and. gives rise to a *secondary cortex,* which is responsible for the later increase in girth. This secondary activity as a general rule appears to set in only after primary division of the segments has ceased. It is little marked in *Halopteris hordacea,* but much more evident in *Phloeocaulon, Ptilopogon,* and *Cladostephus,* where it follows upon the marked growth in length above referred to. The peripheral layer then becomes a meristoderm, the cells of which divide by successive periclinal walls to form a radially arranged secondary cortex. Meristematic activity continues for a variable period and is finally terminated by abundant anticlinal division of the surface-cells.

In many *Sphacelarias* (e.g. *S. olivacea)* the tiers remain recognisable throughout the plant, but in others they are obscured in the older parts by the production from the surface-cells of closely apposed, downwardly growing cortical threads. The extent of their development evidently varies not only with the species, but also in different individuals. Certain species (*S. plumula*) produce only few of these threads which form but an imperfect investment, but in S. *plumigera,* for example, they are so numerous and copiously branched that they ultimately form a dense several-layered investment which envelopes the parent axis *(a)* and the lower parts of the laterals and may be as thick as the parts covered. In this species the cortical threads originate from definite peripheral cells which lie in the plane of branching and are situated in the upper half of the inferior secondary segments *(i)*. Many of these cortical threads extend to the basal system and, spreading over it, contribute to its efficiency as an attaching organ.

The thick cortex of *Chaetopteris*, which encases the bases of the laterals, is composed of closely approximated threads growing obliquely downwards and again in part functioning in attachment. They arise without regularity from peripheral cells of both the superior and inferior segments, thin circular areas *(o)* in the thick outer walls of these cells becoming protruded to form the threads. The peripheral cells of the dense pseudoparenchyma formed by the richly branched threads subsequently give rise to the fructifying branches. *dadostephus* possesses a similar cortex.

The four triangular peripheral cells formed during segmentation in *Haloptens scoparia* usually constitute, with or without further division, initials either of late-formed branches or of rhizoids. These cells, called *pericysts* by Sauvageau, possess denser cytoplasmic contents and larger nuclei than the adjacent cells and usually occupy the complete height of a secondary segment. In the lower part of the plant a large proportion of these cells grow out into rhizoids, although the latter sometimes originate from the base of a branch. The well-branched rhizoids of *H. scoparia* acquire thick walls and undergo transverse and longitudinal septation. They gradually cover the older parts with a loose spongy envelope which extends to the substratum. Similar features are met with in other species of *Halopteris*, e.g. *H. hordacea*.

The secondary increase in thickness of the auxocaulous forms described above does not usually take place in the basal parts of the main axes where mechanical strength is attained by the formation of an investment similar to that of the leptocaulous types. In *Phloeocaulon foecundum*, for instance, a number of the peripheral cells grow out into branching filaments which appear to be fused from the first; the radial files in this extensive pseudoparenchyma are slightly deflected in the longitudinal direction. Except for the apparent fusion of the threads there is much resemblance to the cortex of *Chaetopteris*. *A* similar, but more extensive, envelope is formed in *Cladostephus verticillatus*, but here it arises also from the peripheral layer of the secondary cortex. Both here and in *Ptilopogon* Sauvageau finds evidence of an intermittent formation of the cortex; this is most pronounced in the latter genus, where the consecutive strata are delimited by marked thickening of the membrane of the outermost cells.

Branching

The Method of Branching: In *Sphacelaria* and *Chaetopteris* the laterals, whether of definite or indefinite growth, almost invariably arise from a cell occupying the whole height of a superior segment, or in other words half the length of a primary segment. Such branching is described as *hemiblastic* by Sauvageau. The insertion of the branch on the parent axis is limited above by a primary and below by a secondary septum. The branch-initials are probably for the most part peripheral cells, which cease to segment at an early stage, but they may consist of entire sectors. Pringsheim concluded that the initials were nearly always entire quadrants, but according to Sauvageau this is rarely so. In species with pinnate branching practically every superior segment produces a pair of functional branch-initials which usually grow out at an early stage, but when branching is less plentiful, a large number of the initials may remain dormant for a longer or shorter period.

These dormant initials are often readily recognisable in *Sphacelaria* especially in species in which the secondary segments undergo abundant septation, as larger cells with denser contents which do not exhibit transverse division; they frequently contain the same brown contents as the apical cells. In *S. radicans* and *S. olivacea* such dormant initials are present in each superior segment; in the former they probably grow out only in the reproductive phase.

The apical cell of the branch, formed by protrusion of the initial, segments in the customary manner. In the laterals of limited growth the apical cells ultimately cease to function, either persisting as pointed cells or dividing by variously orientated walls *(S. radicans)*. In *Sphacella* the branches arise as outgrowths from the upper ends of the cells.

In certain species of *Sphacelaria* (e.g. *S. hystrix, S. plumula, S. tribuloideSj*), as well as in *Chaetopteris*, the branches of limited growth bear colourless sheathed hairs *(h)* which are usually inserted opposite the transverse septa. Their peculiar mode of development shows much similarity with the method of branch-formation found in the more specialised Sphacelariales. The hair-initials are cut off as small lenticular cells a little to one side of the tip of the apical cell *(a)*. As the latter elongates, the initial comes to occupy a lateral position so that, when the apical cell divides, the septum *(t)* meets the convex wall of the initial approximately at right angles. Hence the position of the mature hairs opposite the septa. Sometimes the initial divides in the plane of branching (*S. radicans*) and forms two hairs, or division may go further resulting in a tuft.

If the apical cell ceases growth after the formation of the initial, the hair remains in an apical position and may be actually terminal (*S. plumula, S. furcigera*). Iji *Alethocladus*, which lacks the hairs found in most of the more specialised Sphacelariales, the lateral branches develop in identically the same way as these hairs of *Sphacelaria*. The laterals consequently He opposite the primary septa, their bases later extending onto the inferior segment above and the superior segment below.

The process of hair-formation in *Sphacelaria*, as well as the method of branching found in the more specialised Sphacelariales, is interpreted by Magnus and Sauvageau as sympodial in nature. The lenticular hair- or branch-initial is regarded as the actual apical cell, while the large remaining portion *(a)* of the original apical cell, which segments to form the continuation of the axis, is looked upon as a branch. Since this develops from the entire segment of the apical cell, this method of branching is styled *holoblastic* by Sauvageau; it differs from the hemiblastic type also in the oblique position of the septum. The main axes of *Alethocladus* are, on this interpretation, sympodia, comprising as many sections as there are laterals, each of the latter representing the actual termination of the underlying segment of the sympodium. Similarly, the above-described hairs of *Sphacelaria* are regarded as terminal and the axis bearing them as a sympodium, the successive joints of which are terminated by the hairs. The transverse septum, which abuts on the hair-initial of a *Sphacelaria* or the branch-initial of *Alethocladus* divides the original apical cell into a lower primary segment belonging to the joint terminated by the hair or branch, and an upper segment *(a)* constituting the apical cell of a new joint of the sympodium.

As evidence in support of this interpretation Sauvageau cites those instances in which such initials develop into terminal hairs. The hair then occupies its normal position at the apex of the axis. Further support for the hypothesis is to be found in the details of propagule-formation in *Sphacelaria*. The correctness of this interpretation is disputed by Reinke and Pringsheim.

The holoblastic branching of *Halopteris, Phloeocaulon* and *Ptilopogon* shows greater complexity. Here the branch-initial *(in)* at an early stage divides horizontally and cuts off a small cell *(ax)* on its upper side. According to Sauvageau's interpretation the latter *(ax)* is the true apical cell, while the lower larger cell represents its only segment which elongates and divides to form the branch; in other words this branch is actually a lateral of a diminutive axis capped by the cell *ax*. The septum *(t)*, arising in the original apical cell *(a)* of the axis, again abuts on the lenticular branch-initial, but always on its lower larger segment. As this lengthens and divides to form the branch, the overlying cell *(ax)* remains in its axil, and in the further course of events either remains arrested or develops in diverse ways into hairs or reproductive organs. It is the presence of such axillary growths that distinguishes *Halopteris*, etc. from *Alethocladus.* The laterals of limited and unlimited growth arise in the same way. Branching in the holoblastic types is necessarily alternate and, if hairs are present, they are always axillary in position. The successive apical cells either cut off several ordinary segments between the intervals of branch-formation or only one segment is formed. This determines whether the laterals are separated by marked internodes or not.

In most species of *Halopteris* (e.g. *H. scoparia)* the axillary cell divides and gives rise to a tuft of hairs. *H. filicina* is, however, distinguished by the usual absence of hairs and by the fact that, although the numerous laterals of a given axis are for the most part arranged in two alternating rows, as in other species of the genus, the first two are usually both situated on the adaxial side, the lowest not uncommonly appearing in a pseudoaxillary position. This is because, in *H. filicina,* the axillary cell mostly develops into a branch *(b')* resembling the next one *(b),* which is as usual produced on the adaxial side. In occasional individuals, however, the axillary branch may be replaced by hairs.

In *Phloeocaulon* the axillary initial gives risf to a group of cells which soon disorganise, while in *Ptilopogon* it produces a cushion of cells which remain dormant until the time o fruiting.

Branching in these holoblastic forms may also be hemiblastic, although laterals formed by the latter method often appear only in older plants. They originate, as in a *Sphacelaria,* from dormant initials (pericysts) formed at an early stage in the segments of the apical cell. Such branching has not been observed in *Halopteris filicina*, but is frequent in *H. scoparia, H. funicularis* and *Ptilopogon;* in the last some of the fruiting branches are probably produced in this way. Most authorities speak of these branches as adventitious, but one must agreed with Oltmanns in questioning the validity of such a term; despite their tardy outgrowth, they are primary in origin.

Whatever attitude may be taken towards the interpretations

of Magnus and Sauvageau, it is manifest that the origin of the branch-initials in *Halopteris, Phloeocaulon,* and *Ptilopogon,* from small segments of the apical cell itself, results in these genera in a distinct type of morphological construction. The hemiblastic mode of branch-formation is probably to be regarded as primary, while the holoblastic method was perhaps initiated in connection with hair-development.

A subsequent modification resulted in the cutting off from the hair-initial of a segment which could grow out into a lateral, giving the mode of branch-formation found in most species of *Halopteris* where, however, it is still accompanied by the development of hemiblastic branches. *H. filicina,* which apparently no longer has the capacity for hemiblastic branch-formation, would thus represent the most specialised member of its genus, and it is significant in this connection that the axillary hairs of the other species are here usually replaced by a branch. *Alethocladus* would represent the culmination of this line of development, since here the tendency of the initial to give rise to a hair is completely lost.

In *Cladostephus verticillatus* the first whorls of laterals of limited growth arise a little way behind the apex by the protrusion of peripheral cells of the superior secondary segments the members of a whorl do not all necessarily develop at the same time. The first segment or segments, cut off from the apical cells thus constituted, become part of the axis and undergo copious division to form a cushion of tissue *(b).* In transverse sections through older whorls the interspaces between the individual members *(br)* are occupied by a tissue *(c)* which by degrees completely envelopes their bases and is formed by radial elongation and periclinal division of the peripheral cells situated just above a young whorl, the rows of cells thus produced being slightly deflected in the downward direction.

As a result of the abundant septation and elongation taking place behind the apex, the older whorls appear to arise only from the summits of the superior segments, although originally hemiblastic in origin. The peripheral tissue occupying the lower portions of these segments—actually representing the bases of the primary branches—sooner or later gives rise to other whorls of laterals. Since these originate only from a part of the secondary segment, they are spoken of as *meriblastic* by Sauvageau. Other secondary whorls, in part often incomplete, subsequently arise from the peripheral cells of the secondary inferior segments *(m')*. Despite its whorled branching and other specialised features, *Cladostephus* approaches closely to *Sphacelana* in the mode of formation of the primary laterals and probably represents a separate evolutionary line.

The long branches of *Cladostephus*, which commonly form a dichotomising system, are hemiblastic in origin. The initials, constituted by one of the first-formed peripheral cells of a superior segment, do not undergo the anticlinal divisions seen in the adjacent cells; they thus correspond to a number of the initials forming the laterals of limited growth. Pringsheim concluded that branching of the long axes was dichotomous, but, according to Sauvageau, the segmentation he describes occurs only if the apical cell is damaged.

The laterals of limited growth do not exhibit the secondary increase in girth shown by the long axes. The members of the first and last whorls formed in any given season are simple, while the intermediate ones are more or less strongly branched, the branches being commonly situated on the abaxial side. This branching is always holoblastic, with formation of axillary hairs, as in *Halopteris*; similar shoots of limited growth may arise direct from the basal disc. The axillary hairs usually occur in two pairs, the one in front of the other, the two members of a pair lying in a plane perpendicular to that of the branching of the lateral. Each pair of hairs originates from a separate lenticular cell, the two cells being cut off successively from the branch-initial Sauvageau describes numerous modifications of the normal plan.

Dichotomous branching is characteristic of the *monotypic Disphacella* which is imperfectly known. Branching is initiated by the apical cell becoming bilobed; a longitudinal wall then separates the lobes which constitute two apical cells. The structure of the axes resembles that of the simpler *Sphacelarias* and pericysts are differentiated which rather exceptionally grow out into branches. Oltmanns suggests that this may be an abnormal form.

The diverse methods of branching typical of the mature plant of Sphacelariales undergo some modification as a result of damage to one of the major axes or when the latter die back in the autumn. In *S. radicans* Sauvageau records an apparently natural truncation of all the erect shoots which is followed by the outgrowth of dormant branch-initials into laterals bearing the reproductive organs. When the upper part of an erect shoot is removed, one or more of the youngest laterals carry on the growth or a new shoot arises from the adjacent inferior segment. If the remaining part bears mature laterals only, the intact cells at the exposed surface can grow out into new shoots. Such regeneration from the stumps of shed laterals is partly responsible for the fastigiate habit sometimes shown by *Halopteris scoparia*. Production of adventitious growths from the remains of the fruiting axes of the previous season is also seen in *Cladostephus verticillatus*, where the medullary cells at the exposed surface grow out into branches.

Damaged apical cells of *Sphacelaria* are regenerated from the underlying segment. If this has undergone transverse division only, the upper half constitutes a new apical cell, while if longitudinal walls have already arisen one of the resulting cells grows out laterally into a new apex. Inferior segments, if they have already divided longitudinally, cannot produce a new apex, which is then formed from the underlying superior segment. According to Sauvageau damage to the apical cell of *Cladostephus* often leads to the regeneration of two new apicals, resulting in the dichotomous branching described by Pringsheim as a normal phenomenon. If the apical cell of *Sphacelaria* is exposed to a low light-intensity, it grows out into a rhizoid (inversion of polarity).

Prostrate System

A well-developed discoid prostrate system is probably typical for the order. Good examples are seen in many species of *Sphacelaria*, as well as in *Phloeocaulon* and *Chaetopteris*. In all of these the erect shoots arise from a basal crust which consists of several layers and exhibits marginal growth. Viewed from below the basal layer appears as a system of radiating coalescent filaments, which give rise to approximately vertical files of cells forming the crust; there is considerable resemblance to a *Ralfsia* or *Lthoderma*. As already noted these crusts are strengthened by the downgrowth of rhizoids from the erect axes. Not uncommonly several crusts grow over one another. The superjacent strata often belong to distinct individuals, but they may also be due to the outgrowth of surface-cells from older crusts which, spreading over the latter, produce a new superposed stratum. Very elaborate crusts of this kind are formed in *Cladostephus verticillatus*. Foreign Algae are sometimes enclosed between the successive strata.

The thick crusts just described are probably always perennial, and the species of *Sphacelaria* possessing them rarely form propagules; all of them show transverse septation of the secondary segments. The bases of the erect axes are invariably buried amid the vertical rows of the crust so that the upright shoots represent a further development of certain of the latter.

Basal discs of the type just described are rarely met with in *Halopteris*, in most species of which the corticating threads envelop the bases of the axes with a spongy mass of considerable width, a condition well seen in *H. scoparia*. The individual threads spread over the substratum. A less highly differentiated prostrate system is also found in diverse epiphytic *Sphacelarias*. In *S. cirrhosa*, for instance, the disc consists of a single layer of cells, whilst in *S. plumula* the short irregular filaments of which it is composed are only partly coalescent. In the endophytic species the prostrate system generally consists of loosely branched threads *(S. pulvinata)*, which sometimes form bundles *(S. Reinkei)* penetrating to some depth into the substratum. Other endophytes possess an epiphytic disc, bearing branches which penetrate into the host (e.g. *Si Novae Caledoniae). S. caespitula* Lyngb. forms a several-layered crust of uneven thickness within the tissues of *Laminaria Cloustoni* and *Saccorhiza bulbosa* the little-branched erect threads serve mainly to bear the reproductive organs. In *Sphacella,* inhabiting *Carpomitra costata,* the endophytic system varies.

Reduction of the erect system is commonly encountered among the endophytic *Sphacelarias,* but may occur also in epiphytic species like *S. olivacea.* Here the plant is at times represented solely by the basal crust—a stage described by Kuckuck as a distinct genus *Sphaceloderma.* In such instances the unilocular sporangia are borne directly on the basal stratum and are the only parts of the erect system present.

The rare lithophyte *Battersia,* first discovered at Berwick-on-Tweed, may be a similar reduced state of some *Sphacelaria,* although at present there is no evidence to prove that it is not an independent genus; it is certainly a reduced form. The thick, many-layered crusts (up to 4 cm. in diameter) consist of a number of separate strata while the little-branched erect threads *(e)* exhibit apical growth and occasional longitudinal division of their cells. They bear unilocular sporangia *(u)* and are aggregated to form scattered groups.

In diverse *Sphacelarias* the basal disc plays a role in the spread of the individual over the substratum. Thus, in *S. radicans* certain peripheral cells grow out into simple or branched horizontal stolons possessing the same structure as the upright threads. The tips develop into new discs which either produce erect threads or give rise to further stolons. Much the same occurs in *S. Novae Caledoniae.* In *S. cirrhosa* some of the lower laterals may become prostrate and form new discs *(d)* at their tips. The stolons of *S. bipinnata(4)* and *S. hystrix* have a different origin. Here the bases of the erect shoots, which arise from the bundle of endophytic threads, become covered with a weft of rhizoids, some of which *(st)* spread horizontally and give rise at their tips to other plants.

The elaborate crusts of *Cladostephus verticillatus*, which are composed of several irregularly superposed strata delimited from one another *(t)* by strong thickening of the outer walls of the surface-cells, show a marked capacity for the formation of stolons from the marginal and superficial cells. These produce new crusts at their tips or along their length while some of them grow on to the parent crust and contribute a further stratum. According to Sauvageau the erect axes of this species always arise from stolons.

Vegetative Reproduction

The stolons described above serve as a means of local vegetative reproduction, but many species of *Sphacelaria* possess propagules of characteristic form, which constitute an efficient method of dispersal. In several, propagules are produced far more frequently than sporangia, and such species evidently rely largely on this method of reproduction. Propagules and sporangia may be formed in different seasons *(S. hystrix)*, while in other species they occur simultaneously, but on different individuals or on different branches (e.g. *S. biradiata)*. Not uncommonly, however, the two types of structures are intermingled *(S. cirrhosa, S. bipinnata)*.

The propagules invariably arise from dormant branch-initials which are nearly always situated in the superior segments. In *S. plumula* and other species the initial of the propagule becomes protruded as an apical cell which first cuts off a number of transverse segments, during which it commonly broadens. The first septum, which arises beyond the surface of the parent axis, marks the point of subsequent detachment of the propagule so that, after the latter is shed, a short stalk (sterigma of Janczewski) remains from which a further propagule can develop. The subsequent segments (2 or 3 in number) divide into two secondary ones (s, *i)* in the usual way. After this the apical cell cuts off at its summit a small lenticular cell which either undergoes no further development *(S. plumui, S. fusca)* or sooner or later grows out into a colourless hair *(S. biradiata*, sometimes in *and. hystrix)*. The large remaining portion of the apical cell, which corresponds to a last primary segment, soon spreads out laterally to form the arms of the propagule and usually divides by a median longitudinal wall. Either before or after this, a curved septum arises in each arm and cuts out an apical cell which produces two or more segments. The further subdivision of these segments and of the cells composing the body of the propagule varies according to its final shape.

According to Janczewski detachment of the propagule is due to pressure exerted by its basal cell upon the sterigma; this causes rupture of the external membrane which forms a collar around the persisting sterigma. Each time a new propagule develops, a further sterigma is formed at its base so that, when repeated innovation has occurred, the point of insertion exhibits a series of collars.

The propagule may have the shape of a broad *(S. tribuloides)* or narrow *(S. plumula)* wedge, or the two arms may be produced and more or less cylindrical *(S. furcigera)* or fusiform (S. *biradiatd).* In certain species *(S. hystrix; S. fusca\S. cirrhosa)* the body of the propagule is prolonged into three arms lying in different planes and the number of arms is then not uncommonly variable; there may be 2, or 4 or 5. Since the lower part (foot) of the propagule constitutes an additional arm, the two-armed types after detachment, appear triradiate, the three-armed ones quadriradiate. In *S. furcigera* one or both of the arms may be branched, and this is more marked in *S. divaricata* Mont., where the propagules commonly show repeated forking in one plane, each fork being surmounted by an arrested lenticular cell.

The propagules are clearly modified branches. Particularly when the arms are developed as cylindrical rays, they afford a clear example of "dichasial" or "trichasial" branching which, in *S. divaricata*, may moreover take place repeatedly. As Oltmanns points out, the mode of development of the propagules, especially in those instances in which the apical cell grows out into a hair, lends considerable support to the view that the holoblastic branching is sympodial in character.

The rayed types of propagules probably float easily and will readily become anchored among surrounding algal growth, while the cuneate types being more massive perhaps harbour more reserve food. Any one of the arms (including the foot) can develop into a new plant if it comes into contact with a suitable substratum- It gives rise to a short filament which expands into a disc from which the erect axes *(e)* arise. The arms, not in contact with the substratum, usually produce a tuft of hairs (A), one of which is formed from the apical initial.

No other member of the order possesses the specialised propagules met with in *Sphacelaria,* but many can probably propagate vegetatively with the help of detached branches, as has been observed in *S. plumula,* as well as in *Halopteris scoparia.* In either instance the peripheral cells of the lower segments grow out into plentiful rhizoids, and in *H. scoparia* these are stated to be capable of producing new erect axes.

Arrangement of the Sporangia

Both uni- and plurilocular sporangia, not uncommonly found on distinct individuals, are known in a considerable number of Sphacelariales. The two kinds of organs usually occupy a similar position, but there is much diversity in their mode of arrangement.

In *Sphacelaria* the sporangia are generally borne terminally on simple or branched laterals, which are often short and may be reduced to a single cell. In *S. radicans* the paired sporangia are sessile. The fertile branches mostly, if not invariably, arise from dormant branch-initials. In species with a simple structure (e.g. *S. olivacea)* there is little distinction between them and the vegetative branches, but in those showing abundant septation *(S. plumigera, S. racemosa,* etc.) the fertile laterals remain comparatively simple and are not uncommonly uniseriate. Special instances are afforded by *S. bracteata,* where the fertile branches are pseudoaxillary, which is due to the fact that the first abaxial lateral *(st)* remains sterile; and by *S. Borneti* and *S. Reinkei,* where the fertile branches are sympodial in structure and usually bear all the sporangia on the adaxial side. In

Battersia and *Sphacella* the sporangia are terminal or lateral on the ordinary erect threads.

In *Chaetopteris* formation of fertile branches is relegated to the peripheral cells of the rhizoidal cortex (cf. *Desmarestia*). These branches show quite an irregular distribution and possess a simple multiseriate structure; the sporangia are borne at the ends of few-celled laterals, but there are no other branches.

The sporangia of *Halopteris* are found in the axils of laterals of higher orders, which constitute the "bracts" of Reinke and Sauvageau; the sporangia develop from the axillary initials so that on Sauvageau's interpretation they are actually terminal in position. In *H.filicina* they usually occur singly on short stalks in the axils of branchlets (bracts) and replace the "axillary" laterals present on the larger axes; sometimes, however, the initial divides so that two or even four sporangia occur in the same axil. In other species (e.g. *H.funicularis*) the axillary initial undergoes more plentiful division and forms a cushion of cells, each of which can grow out into a branched uniseriate (rarely biseriate) thread bearing a number of sporangia. A more advanced condition is seen in *H. scoparia*, in which such sori *(sp)* with their subtending "bracts" *(br)* are united into spikelets situated at the ends of the branches. The lower laterals *(st)* of these spikelets are sterile and distichous, whilst the upper fertile ones *(br)* are tetrastichous. Much the same occurs in *H. hordacea*, but here the sessile sporangia are crowded together on the axillary cushions.

The fertile systems are likewise arranged in spikelets in *Phloeocaulon*, but here the cushions produced from the axillary initials give rise, on either side of the sorus, to a short branch which occupies the position of a "bracteole" with respect to the subtending "bract". In *P.foecundum* additional sporangia arise adventitiously from the peripheral cells of the axis of the spikelet.

In *Ptilopogon* most of the fertile branches are again axillary, arising from large-celled cushions *(ax)*, which are produced in the axils of the laterals of limited growth; these remain dormant for a considerable time and as a result become deeply embedded in the secondary cortex. When the reproductive period sets in, the cushions develop into branches which remain apposed to the subtending lateral and at or near the surface of the cortex give rise to a tuft of threads bearing the sporangia. Similar fertile branches also arise from other points on the axis, their exact mode of origin being not yet clear.

The fertile branches of *Cladostephus verticillatus*, lastly, are formed successively during winter from the basal parts of the older axes, at a time when the upper portions have often largely perished. The fertile laterals ultimately constitute dense cushions, which more or less completely envelop the perennating parts. They arise from surface-cells of the secondary cortex, or of the rhizoidal cortex, and continued growth of these tissues often results in the bases of the fertile laterals becoming embedded. The fertile branches are in general simpler than the vegetative ones, but they vary considerably in the amount of septation and the number of sporangia, as well as in the presence or absence of lateral hairs and branches, both of which are holoblastic in origin; the sporangia are borne on the superior secondary segments. The mode of origin of the fertile branches of *Cladostephus* resembles that of *Chaetopteris*.

The unilocular sporangia are generally globular or oblong in form. At maturity the contents enclosed in a common mucilage-envelope escape through an apical aperture, the individual swarmers slowly separating from one another. The plurilocular sporangia are commonly spherical or oblong and never markedly elongate. They generally possess only two series of compartments, often surrounding a central cavity. The swarmers are either liberated individually from each loculus *(Sphacelaria; Cladostephus)* or the septa disappear at maturity and all the contents are set free through an apical aperture *(Halopteris, Sphacelaria olivacea)*.

Proliferation of the unilocular sporangia is frequent. The stalk-cell may grow into a vegetative branch, but more usually another sporangium is formed. In *Cladostephus verticillatus* the underlying cell bulges into the sporangium before it has discharged its contents. Except where heterogamy is suspected, there is no essential difference between the swarmers from the two types of sporangia. They possess one or more chromatophores *(c)*, one of which commonly bears an eyespot *(s)*. According to Kuckuck both kinds of sporangia are usually formed during the cold season.

The Life-cycle: The unilocular sporangia are no doubt normally organs of asexual reproduction, reduction during spore-formation having been demonstrated in *Sphacelaria bipinnata, Halopteris scoparia*. Since plurilocular sporangia commonly occur on distinct individuals and their swarmers have been shown to behave as gametes in *Cladostephus spongiosus* and *Sphacelaria bipinnata*, such plants must be haploid, although this has not up to the present been proved cytologically. Plurilocular sporangia may, however, also occur on the same individuals as the unilocular ones and, in *S. bipinnata*, their swarmers have been shown to be diploid, as in many Ectocarpales. The two types of sporangia are found on the same plant in several other species of *Sphacelaria*, although in most of these individuals bearing plurilocular sporangia alone are also recorded, suggesting the probable existence of haploid plants. About 90 per cent of the individuals of *S. bipinnata* examined by Papenfuss bore unilocular sporangia only.

It is significant that the species of *Halopteris*, apart from *H. filicina*, never bear the two types of sporangia on the same plant. This segregation is also found in *Phloeocaulon, Ptilopogon, Chaetopteris* and *Cladostephus*. This implies that, in the more specialised Sphacelariales, the occurrence of plurilocular sporangia (presumably always diploid) on individuals bearing the unilocular type has, as in Ectocarpales, either become exceptional or is altogether suppressed.

Fusion of the swarmers from the plurilocular sporangia has been demonstrated in *Cladostephus spongiosus*. Schreiber observed abundant copulation and established that there are positive and negative strains. The two types of gametes are indistinguishable, but the female round off at an early stage and, losing their flagella, are sought out by the more active male gametes *(m)*. Schreiber concludes that, in C. *spongiosus*, there is an isomorphic alternation between a sporophyte with unilocular and a gametophyte with plurilocular sporangia (gametangia). Both types of gametes can also develop apogamously. The same authority failed to obtain fusion of the swarmers from the plurilocular sporangia of *Chaetopteris plumosa* and therefore regards them as diploid, but it is equally possible that some factor in the cultures prevented the occurrence of sexual fusion.

In *Sphacelaria bipinnata* Clint records the fusion of the haploid swarmers from the unilocular sporangia, although the data are not convincing and give the impression of abnormal occurrences. Papenfuss did not observe any such fusion, although the germlings from these swarmers did not develop beyond the one-celled stage; this is perhaps related to the endophytic habit of the species in question. The swarmers from the plurilocular sporangia of diploid plants likewise failed to fuse, but in plants bearing such sporangia only they function as gametangia. This species is monoecious, and both gametes are motile when they copulate; the two corresponding flagella of the zygote are usually wound round one another and the product remains motile for several hours.

Swarmers from the unilocular sporangia have also been stated to copulate in *Halopteris filicina* and *H. scoparia*, although Higgins failed to corroborate this in the latter species. The possibility of such fusion cannot be denied in view of its occurrence in Ectocarpales. Sauvageau(si), however, grew practically mature plants of *H. scoparia* from the swarmers of the unilocular sporangia without observing fusion, a fact which is strongly in favour of their being asexual. On the other hand, the swarmers from the plurilocular sporangia of *Cladostephus verticillatus* developed directly into vigorous plants, which either implies that they were diploid or that apogamous development readily occurs in this species.

The gametangia of *Cladostephus spongiosus* and *Sphacelaria hipinnata* are all of one type, and this appears to be true of several other *Sphacelarias* (e.g. *S. olivacea*). Several species *of Sphacelaria* and *Halopteris,* however, produce plurilocular sporangia of two kinds and in all of these the unilocular sporangia seem to occur on distinct individuals. In *Sphacelaria hystrix* the two types of sporangia differ in the size of their compartments; the swarmers of the one are relatively large and possess several chromatophores *(c),* whilst those of the other type are small and colourless. Fusion has not been observed, nor has it been possible to secure germination of either kind of swarmer. It is probable that these swarmers are gametes and that the plants producing them are gametophytic. Both they and the individuals bearing unilocular sporangia produce propagules, which are, however, formed at a later stage on special threads arising from dormant initials. Two types of plurilocular sporangia are also known in *S. furcigera,* but here both have relatively large compartments (megasporangia and mciosporangia).

Sauvageau likewise records two types of plurilocular sporangia in *Halopteris filicina, Phloeocaulon,* and *Ptilopogon.* They differ in colour and in the size of their compartments, those of the supposed female gametangia being twice as large as those of the supposed male *(m).* The same authority has published data that indicate a possible further differentiation in certain species of *Halopteris.* In the spikelets of a dried specimen of *H. scoparia,* he observed reproductive organs of two types, viz. *(a)* "antheridia", with a single apical aperture and containing numerous small cytoplasmic masses arranged as in a plurilocular sporangium, although septa were not recognisable, and *(b)* somewhat pyriform "oogonia", containing a single strongly contracted mass of cytoplasm. Similar structures are also believed to occur in *H. hordacea.*

The observations relating to *Halopteris* and *Phloeocaulonvttrz* made on herbarium specimens in which the reproductive organs were no doubt poorly preserved, but they cannot for this reason be dismissed as of no value, coming as they do from so reputable an investigator. They indicate with a considerable degree of probability that in many, if not in all, the species of these two genera there are two types of plurilocular sporangia. The implied existence of heterogamy is quite in conformity with the high degree of anatomical and morphological differentiation exhibited by the genera in question. Greater doubts may be felt as to the existence of oogonia with a single ovum in certain species of *Halopteris*, although this cannot be regarded as altogether improbable. Those who are favourably situated for the study of these species should attempt to elucidate these important problems.

Early Stages of Development

The first stages of germination of the spores resemble those of other Phaeophyceae already considered. In *Sphacelarias*, provided with a discoid basal system, the latter probably develops first. In *Halopteris hordacea* and *H. scoparia* a compact parenchymatous disc is formed before the erect shoots develop.

In *H. scoparia* this disc generally gives rise to a single upright thread having a simple multiseriate structure and bearing a number of hairs (A), though otherwise unbranched. After some little time its growth terminates, but before this happens it produces from one of it's lower segments a single lateral which develops into a thread of the second generation. This is rather more robust and bears a number of distichous branches, as well as hairs. This second axis, in its turn, ultimately ceases to grow and produces from near its base a still more vigorous lateral of the third generation. Ultimately a fourth arises in the same way, and this develops into the mature plant. The production of each successive generation from a basal segment of the previous one recalls the late formation of laterals from dormant branch-initials in the mature individuals of *Halopteris*.

Such a development probably occurs also in other species of *Halopteris*. In *H. filicina*, although the first-formed erect thread may develop into an adult shoot, its growth is commonly limited, the functional mature shoot then arising as a lateral branch from near its base. The first erect thread of this species soon branches and develops rhizoids from its lower cells. These envelop the part of the axis below and spread out over the substratum, where they expand into small discs which may fuse to form larger ones and which give rise to other erect axes; these may exhibit a similar development.

The small disc produced from the swarmers of *Cladostephus verticillatus* always gives rise in the first place to a long hair (A), after which a simple erect thread of limited growth is formed. Subsequently other more robust axes develop from surface-cells of the enlarging disc, but it is only after a number of these have been formed that there arises a shoot of unlimited growth which develops the characteristic whorled branching and other features of the mature axis. The successive primary shoots are all of limited growth and show increasing complexity. In the first many of the lower secondary segments do not undergo longitudinal division, but in the later ones such segmentation occurs more regularly. The first-formed shoot often bears solitary hairs only, while the subsequent ones produce geminate hairs and show occasional holoblastic branching with axillary hairs.

Further complication is introduced by the production, from certain cells of the disc, of curved stolons which are homologous with the upright shoots. They spread in all directions over the substratum, developing occasional coralloid haptera, especially near the bases of the upright axes that originate from the stolons. Other stolons, as already mentioned, arise as prolongations of some of the radial files of cells composing the disc.

Genus *Choristocarpus*

Choristocarpus tenellus (*Ectocarpus tenellus* Kiitz.), found in deep water in the Mediterranean, is a small form with uniseriate filaments exhibiting well-marked apical growth and alternate branching. The segments of the apical cell undergo no further appreciable elongation. The thin-walled cells contain numerous rounded or oblong chromatophores which are without pyrenoids; they are densely packed in the apical cells, but less closely in the others so that the elongate cells present a rather characteristic transparent appearance.

The approximation to one of the simpler Sphacelariales is increased by the abundant production of propagules, although these do not possess the characteristic shape of those of *Sphacelaria*. They are oblong structures borne on a unicellular stalk and usually composed of two cells, although sometimes undivided or three-celled. Each cell is uninucleate and has alveolar cytoplasm; in the nature of the cell-contents there is much resemblance to the monospores of Tilopteridales. After detachment of the propagule the stalk can produce a second.

Plants bearing such propagules also produce sessile unilocular sporangia occupying the same position as the former and producing about 16 swarmers delimited by mucilage-septa. The swarmers, which are set free by an apical rupture of the sporangium, are large and move slowly; they possess numerous chromatophores, with an eyespot next to the point of attachment of the flagella. Mature plants bearing propagules have been grown from these swarmers. Other individuals, apparently devoid of propagules, bear plurilocular sporangia which are likewise sessile and spherical or oblong in shape. The fate of the contents is unknown.

Affinities

Kjellman placed the rare algae *Discosporangium (10)*, only once found in the Bay of Naples, together with *Choristocarpus* in a family Choristocarpaceae. Falkenberg had already emphasised the general resemblances in vegetative characters, although the branches in *Discosporangium* arise from the middle of the parent-cell. The sole reproductive organs, so far observed, are plurilocular sporangia which form a one-layered plate around the middle of the elongate vegetative cells; in some ways they recall those of *Zosterocarpus*.

The systematic position of *Discosporangium* remains altogether obscure. The apical growth and other features suggest an affinity with *Choristocarpus*, but the reproductive organs scarcely lend support for such a relationship. Schmidt contributes nothing.

Affinities and Status of the Sphacelariales

In their pronounced apical growth and characteristic segmentation the Sphacelariales form a clearly defined group which is almost certainly a natural one. The numerous modifications that occur are all traceable to a common plan and warrant the assumption of an ancestry among simple filamentous types with apical growth. Such a type is furnished by *Choristocarpus*, which Kylin refers to Sphacelariales, a point of view with which one may well concur. Apart from the resemblance in mode of growth and cell-structure, the occurrence of propagules in *Choristocarpus* serves to emphasise the affinity. Kuckuck drew attention to the marked similarity between the young propagules of *Sphacelaria tribuloides* and those of *Choristocarpus*. The greater degree of elaboration of the mature propagules of *Sphacelaria* has perhaps gone hand in hand with the differentiation of the vegetative system. There is much similarity, too, between *Choristocarpus* and *Sphacella*. In the latter the simple structure is possibly a result of reduction and, without further investigation, it is impossible to say whether this may not also be true of *Choristocarpus*. Be that as it may, both genera exemplify little-differentiated forms which, except in their apical growth, are not readily separated from the simpler Ectocarpales.

It has been pointed out above that in many, if not in all, Sphacelariales the life-cycle probably involves an isomorphic alternation between asexual and sexual individuals; in this respect the Sphacelariales contrast markedly with the polystichous Ectocarpales. The type of alternation also directs attention to forms like the simpler isomorphic Ectocarpaceae for the origin of the members of this order. They probably represent one of the numerous lines of evolution arising from simple filamentous Brown Algae, *Choristocarpus* being possibly an offshoot from this line.

An affinity between Sphacelariales and Tilopteridales has also been suggested, the monospores of the latter being regarded as equivalent to the propagules of the former. Despite certain resemblances, the importance of which it is difficult to evaluate in the present state of our knowledge, the Tilopteridales appear to be widely removed from the Sphacelariales by their intercalary (trichothallic) growth and by the fact that the monospores, whatever be their function, are of the nature of spores, since they are formed within definite sporangia.

Oltmanns, largely on the basis of Sauvageau's investigations, subdivides the order into:

1. *Sphacelariaceae:* Battersia, Chaetopteris, Disphacella, Sphacella, Sphacelaria.
2. *Stypocaulaceae:* Alethocladus, Halopteris (incl. Stypocaulon), Phloeocaulon, Ptilopogon.
3. *Cladostephaceae:* Cladostephus.

To these may be added:

4. *Choristocarpaceae:* Choristocarpus.

Dictyotales

The genera classed in this order form a natural group, distinguished not only by the apical growth, the usual dichotomous branching in one plane and the lack of anatomical differentiation of the foliose thalli, but also by the oogamous sexual reproduction and the probable occurrence of isomorphic alternation in all the members. The asexual phase bears sporangia which produce large motionless spores, commonly in fours though other numbers sometimes occur. The antheridia and oogonia of the sexual phase are commonly grouped in well-defined sori, and the single ovum is, as usual in Phaeophyceae, liberated prior to fertilization. The antheridia resemble the pturilocular sporangia of otter produce numerous minute colourless spermatozoids. There are no evident affinities with other specialised Phaeophyceae exhibiting apical growth, and the nearest relatives must be sought among the less highly differentiated forms. The Dictyotales, like the Sphacelariales, have probably originated from simple filamentous types with an isomorphic life-cycle, although evolution has here followed a different course. Most of the non-European genera are imperfectly known.

The Dictyotales are for the most part inhabitants of warmer seas. The three largest genera —*Dictyota, Padina,* and *Dictyopteris (Neurocarpus, Halyseris)*—are practically worldwide in their distribution, but, except for the first with two, each is represented only by a single species on North European shores. The northern part of North America, both on the Atlantic and Pacific coasts, is practically destitute of representatives of this order. The considerable genus *Zonaria* is widely distributed in warmer regions, a common tropical species being Z. *variegata* several species are recorded from New Zealand, although none occur in North Temperate regions. *Taonia atomaria* is found over a large part of Northern 'Europe and is known from the Canary Isles and the Mediterranean. The Dictyotales, especially species of *Zonaria,* are well represented in the Southern Hemisphere, and the remarkable genus *Lobospira* is so far only recorded from Australia.

Most Dictyotales grow in permanently submerged situations and the normal habitat is probably in water of some depth. Certain species *Dictyota dichotoma, Padina pavonid)* are, however, commonly found in rock-pools between tide-levels and it is evident that various Dictyotales can exist in a variety of situations.

Vegetative Structure

Dictyota dichotomy, the commonest British representative, is a widely distributed annual. The familiar forked fronds, all the branches of which are normally, situated in the same plane, arise from a cylindrical rhizome attached to the substratum by tufts of sometimes branched rhizoids. These can also develop from cells of the flat thallus, which is sometimes partly prostrate and attached. The lower part of the erect fronds is likewise

Cylindrical and commonly gives rise to adventitious laterals and some of which may develop as horizontal stolons.

Both the lower cylindrical parts and the flattened fronds, into which they gradually expand, grow by means of a single lenticular apical cell which, when viewed from above, appears circular in the cylindrical and elliptical in the flattened portions. On the broader fronds the apical cell is often slightly sunk, but when the thallus is more finely divided (var. *implexa)* the segments gradually narrow to the apical cell.

The latter cuts off a single series of segments which broaden considerably behind the apex and undergo regular septation. In the flat thallus the first curved wall runs approximately parallel to the plane of flattening and divides the segment into two unequal cells. The second wall arises in the same plane in the larger cell and separates a central *(c)* from two peripheral cells, which divide by consecutive longitudinal walls perpendicular to the surface; this is accompanied by some vertical segmentation of the central cell.

The mature thallus thus comes to consist of three layers, a middle one *(c)* of large cells with few or no chromatophores bounded by layers *(p)* of small cells, densely packed with chromatophores. The central cells contain, apart from fucosan-vesicles, conspicuous groups of large refractive globules *(g)* which are suspended by cytoplasmic strands; they appear to constitute some kind of food-reserve (fat according to Hansen) and are stated to be responsible for the iridescence commonly exhibited by this algae. The rather thick walls separating the cells possess conspicuous pits. Tufts of colourless hairs, with the usual basal meristem, are scattered over both surfaces of the thallus, but are shed during the reproductive phase. In the cylindrical thallus there are from 3 to 6 layers of internal cells.

Dichotomy is effected by longitudinal division of the apical cell into two equal halves, the classical instance of such branching. Adventitious laterals, which develop from single marginal cells and expand in the same plane as the rest of the thallus, are not uncommon; they tend to arise especially at points of injury and, after death of the parent, become independent and constitute a prolific source of vegetative reproduction. According to Schreiber pieces of the thallus provided with an apical cell readily develop into new plants.

In other species *of Dictyota* the dichotomy is commonly not so regular and, as a result of more vigorous development of the one limb, a pinnate habit (e.g. *D. Binghamiae J.* Ag.) may be realised. In certain species the fronds are toothed (e.g. *D. ciliata J.* Ag.). The flat' thalli sometimes possess two or more layers of central cells and species exhibiting this feature have been referred to a separate genus *Dilophus*, but the maintenance of such a distinction presents considerable difficulties.

Padinapavonia is a rare perennial found near mid-tide level on the southern shores of England, but it is abundant in the Mediterranean and other warmer seas, where the majority of the other species occur. The numerous stalked fan-shaped fronds, 5-12 cm. high, the larger ones often loosely rolled on their longitudinal axis like a cornet, are distinctive of all the species of the genus. The stalk of each frond is the upward continuation of a branch of the prostrate perennial rhizome which is richly branched and attached to the substratum by tufts of rhizoids *(r)*. The stalks of the fans bear a considerable number of laterals, which are terete at the base but flatten somewhat at their summit. Many of them do not develop beyond this stage, but some may widen out and form secondary fans, the stalks of which produce other fronds in the same manner. In this way extensive branch-systems may arise. Except for the distinctive shape of the fronds, the morphology of *Padina* recalls that of *Dictyota*.

The concentric zones on the fans are due to hairs, which are arranged in 4-8 rows. The zones of hairs occur alternately on the two surfaces, but are more strongly developed on the upper and are stated to be specially prominent in plants exposed to strong light. The reproductive organs are formed on either side of these zones, although during their development the hairs are shed. There is often a thin incrustation of carbonate of lime, especially on the upper surface; as a result the fronds appear whitish, although olive-green or reddish hues are also frequent.

The cylindrical shoots possess a single apical cell like that of *Dictyota;* the segments divide into a central and four peripheral cells, all of which undergo further segmentation. The young flabellate fronds also grow by means of a single apical cell, but the mature fans possess a marginal meristem.

When a young frond is about to develop into a fan, the segments of the apical cell undergo copious longitudinal division so that there is rapid broadening behind the apex; as a result the angle formed by the two edges of the thallus becomes more and more obtuse until it approaches 180°. The apical cell (a), flanked by later-formed segments, now appears in the middle of the broad front margin, while the outer cells produced by longitudinal division of the older segments, are situated

Collins described a sterile algae *(Dictyerpa)*, which was found washed up on the shores of Jamaica and appeared like "rolled and twisted strings, up to 20 cm. long" and 1-3 mm. thick. The terete strands showed di-or trichotomous branching, with occasional irregularly disposed laterals, each branch being capped by a hemispherical apical cell and traversed internally by a single row of cylindrical cells. This may prove to be the rhizome of some member of Dictyotales, although Svedelius suggests that it may be an unattached form of *Dictyota*, comparable to those known among Fucales and like them sterile.

All the cells at the front edge acquire dense cytoplasmic contents and become meristematic; they lengthen perpendicular to the margin and undergo repeated periclinal division, accompanied by some anticlinal division, whereby the number of files of cells is increased. The original apical cell either divides longitudinally or perishes, its place being taken by cells of the underlying segment. The semi-circular edge of the enlarging fan is now occupied by an extensive series of marginal initials which have the shape of bricks placed edgewise, with their greatest width perpendicular to the surface of the thallus.

The segments cut off from the cells of the marginal meristem soon divide longitudinally into two somewhat unequal parts, the smaller segment lying towards the morphologically upper side. As a consequence of more rapid elongation of these upper segments, the free edge of the fan becomes circinnately inrolled towards the lower side, a feature which is specially obvious in longitudinal sections; illumination of the fans from below causes disappearance of the inrolling, which may even set in the reverse direction. The lower layer of small peripheral cells is generally cut off Only at a considerable distance behind the apical meristem. In *P. Sanctae Crucis*, in fact, the two-layered condition persists throughout the plant. Both in *P. pavonia* and other species the middle layer of cells may later undergo further division parallel to the surface.

The hairs commence to develop in the inrolled part near the apex and, according to Carter, arise from cells distinguished by their dense cytoplasm and large nuclei. These cells cut off the small hair-initial by an oblique wall, usually on their distal side; the initial then divides transversely to form the hair, which later develops a basal meristem. The group of hairs is at first covered by the common cuticle *(cu)*, which ruptures sooner or later. Bitter deals with response of the thalli to various types of wounding.

In general habit *Taonia atomaria (Dictyota atomaria* Grev. resembles a robust *Dictyota*, the wedge-shaped segments having irregularly toothed margins. As in *Padina*, there are transverse zones of hairs forming zigzag bands (A), alternating on the two surfaces of the thallus. The fronds, which may reach a length of 30 cm., show repeated di- or poly-chotomous branching in one plane. They arise singly or in tufts from a dense felt of septate rhizoids which are produced in large numbers from the lower cells and spread out to form a false attaching disc. According to Robinson new fronds can develop from these rhizoids.

The young frond possesses a single apical cell, but this rapidly gives place to a marginal meristem occupying the almost straight front edge; its cells, as in *Padina* and *Zonaria,* are widest at right angles to the surface. Branching is due to cessation of division in one of the initials. According to Reinke the marginal teeth are formed by the segmentation of special wedge-shaped initials. The mature thallus is six-layered, the superficial cells being rather larger than in the genera previously considered. The occasional adventitious branches probably arise from surface-cells.

Closely related to the two genera just considered is *Zonaria.* The branched, usually erect fronds show transverse zones of hairs (A), which are generally more abundant on one surface than on the other. The method of attachment is as in *Taonia,* the basal weft of rhizoids often attaining considerable thickness and in diverse species forming a kind of pseudoparenchymatous stalk supporting the rest of the plant. In *Z. flava* (Clem.) Ag. new fronds can develop from the rhizoids. *Zonaria* and *Taonia* thus do not possess the rhizomatous attachment-system found in *Dictyota* and *Padina.*

A more aberrant form is constituted by *Zonaria variegata* Lamour., the young thallus of which is often completely prostrate, forming an expanse, 1-10 cm. wide, resembling an *Aglaozonia* and attached by numerous rhizoids. A number of thalli commonly grow over one another and become fused. The older thalli are erect and it appears that growth may be erect from the first.

The growth and branching of the *Zonaria-thallus* takes place in the same way as in *Taonia.* The segments of the diverse initials *(i)* divide into superficial and central cells *(c)*; the latter give rise to a number of layers, while the former divide by vertical walls so that the cells composing the thallus lie in regular vertical rows. In *Z. Farlomi* the surface-cells along the median line of the thallus produce adventitious branches which develop rhizoids at their base and readily become detached. *Z. flava* is stated to show iridescence.

Padina, Taonia, and *Zonaria* show many points of similarity and no doubt constitute a closely allied group. A more distinctive type is furnished by *Dictyopteris (Halyseris). D. membranacea* (Stackh.) Batt. *(D.polypodioides* Lamour.), a rare sublittoral perennial on British shores, is widely distributed over the surface of the earth, but the chief centre of the genus is in warmer seas. The habit is that of a small and rather delicate *Fucus,* a resemblance emphasised by the midrib traversing the thallus-segments; the stalk of older plants is formed by the persisting midrib. The base of the thallus is covered by a dense felt of rhizoids from which, as in the preceding genera, a number of plants usually arise. Tufts of hairs are scattered over the whole surface; Boergesen's figure of *D. delicatula* indicates that they are formed in the same way as in *Padina.*

The apical initials have the same shape as in *Padina.* Four or five of the median ones (i), which are longer and filled with denser cytoplasm than the rest, give rise to the midrib; they often occupy a slight depression. These particular initials exhibit abundant division, but the activity of those on either side gradually diminishes towards the margin, and Oltmanns is perhaps justified in regarding the median initials as the true meristem. The dichotomous branching commences with a multiplication of these cells, which is followed by the differentiation of two separate groups. Adventitious branches may arise in four longitudinal rows from single surface-cells situated along the edges of the midrib. As in *Zonaria* they develop rhizoids at their base and no doubt serve for vegetative propagation, which is also stated to take place readily from pieces of the thallus.

The wings consist of two layers at the margins and of 4-6 layers near the midrib. The latter shows some differentiation into elongate medullary and more or less isodiametric cortical cells *(c).* The midrib increases in thickness by tangential division of the surface-layer and, as the wings wear away in the older parts, the exposed edges of the midrib become covered with a similar meristematic layer. In certain species (*D. delicatula*) a special group of small elongate cells, with oblique end-walls and thick cellulose membranes traversed by pit-canals, are found within the margins of the wings.

The remaining genera have not been so fully studied. In *Spatoglossum* the broad thallus-segments are branched in a pinnate or subpalmate manner as a result of unequal development of the dichotomies; growth is by means of a marginal meristem, The monotypic *Lobospira* shows pronounced pinnate branching in a single plane. The older axes, as a result of growth in thickness, gradually become cylindrical, while the numerous short flattened, and not uncommonly incurved, laterals become orientated in diverse directions owing to twisting of the axes.

Asexual Reproduction

Asexual Reproductive Organs

These, the so-called- tetrasporangia, constitute one of the characteristic features of Dictyptales. They are nearly always found on distinct asexual individuals, although Holden has recorded them both on male and female plants of *Dictyota*. The sporangia themselves display considerable uniformity, but their diverse arrangement in the different genera affords features of taxonomic importance. They are spherical *(Dictyota)* or pear-shaped, superficial outgrowths, which are usually separated from the parent-cell by a transverse septum although this is lacking in some species of *Zonaria*. The small basal cell sometimes *(Dictyopteris)* undergoes a second transverse division. It is usually embedded in the peripheral layer, although in *Dictyota* it is stated to become elevated into the sporangium like a columella as a result of accumulation of mucilage between wall and protoplast and to project markedly above the surface after dehiscence; it is suggested that this protrusion of the basal cell may play an active part in spore-discharge. The sporangia generally project well above the surface, but in *Taonia* they are partially sunk.

During the enlargement of the sporangium the nucleus increases greatly in size, often becoming elliptical with the long axis parallel to the surface of the thallus and, when this is so, it twists through a right angle before division takes place. Centrosomes have usually been recorded. The first nuclear division, after the basal cell has been cut off, is meiotic, and is normally followed by only one further division. After this the cytoplasm undergoes cleavage about the four nuclei to form the naked tetraspores, which are usually arranged crosswise when viewed from above, although a tetrahedral grouping is not uncommon. In some species of *Zonaria* there is a third division in the sporangium resulting in the formation of eight spores, which in Z. *Farlowii* are separated by delicate septa. The mature spores generally round off before being set free through the gelatinised apex of the sporangium. Soon after liberation the large motionless spores secrete a cellulose membrane and can then develop directly into a new plant; in fact germination sometimes occurs within the sporangium.

In *Dictyota* the tetrasporangia occur singly or in small groups on both surfaces of the flat fronds and show no relation to the tufts of hairs. In most of the other genera, however, their distribution is connected with that of the hairs. Thus, in *Dictyopteris* the sporangia are grouped around the tufts. In *Padinapavonia* they develop in several rows on either side of the zones of hairs situated on the lower (sometimes also on the upper surface of the fan. Since the hairs usually die during the development of the sori, the latter form characteristic double, brown-coloured, bands upon the surface. In weak plants the sorus develops only along the anterior margin of each series of hairs, and this is the normal condition in *P. Sanctae Cruets*, where, moreover, a sorus is associated only with every second band.

Another variation is seen in *P. gymnospora,* where the sori are situated between the bands of hairs.

The sori of *Taonia* exhibit essentially the same arrangement. Both here and in *Padina* scattered sporangia are also found, and this compares with the condition in *Zonaria* where the sori form clearly circumscribed areas between the bands of hairs when these are present; in *Z. variegata* they occur on the upper surface of the prostrate thallus. The sori of *Zonaria* often show a centrifugal development and are distinguished by the presence among the sporangia of colourless multicellular paraphyses of characteristic form which give the sori a speckled or even whitish appearance. In the little-known genus *Glossophora* all the reproductive organs are borne on small ligulate outgrowths which arise in large numbers from the upper parts of the forked band-shaped thalli.

The developing sporangia of a sorus gradually elevate the "cuticle" which, although it may persist for some time as a protective covering, is ultimately burst. In *Zonaria variegata* the marked gelatinous thickenings of the outer walls of the sterile cells intermingled with the sporangia aid in raising the cuticle, a role perhaps fulfilled by the paraphyses of other *Zonarias.*

In *Taonia* the germlings develop into mature thalli in about 30 days and during this period there are formed some 60 zones of tetrasporangia so that a definite correlation between the daily tidal rhythm and the initiation of the zones is suggested. A similar periodicity in soral development is likewise probable in *Padina* and other instances, where the sori arise successively from a marginal meristem. In *Dictyota,* on the other hand, there is no evidence of any such relation; Williams in fact states that sporangia of all ages are to be found throughout the season on the thalli of all British Dictyotales. Discharge of the spores in *Dictyota* and *Dictyopteris* is strongly favoured by light.

Sexual Reproductive Organs

The oogonia and antheridia are usually borne on distinct individuals, although a few species (e.g. *Padina pavonia)* are monoecious. The sex organs are for the most part arranged in definite projecting sori. Although the oogonia of *Spatoglossum* and of *Dictyopteris membranacea* are scattered singly or in small groups. In *Dictyota dichotoma* both kinds of sori are distributed over the greater part of the thallus, appearing elliptical when viewed from above and fan-shaped in sections. The outgrowing surface-cells remain in close apposition and gradually lift the "cuticle" *(cu)* which envelops the sorus until maturity; a small basal cell is cut off at an early stage. Each oogonium produces a single ovum which is liberated through the platinised usually advancing centrifugally; there are 25-50 oogonia in a sorus. Occasionally there are functionless oogonia" at the margin. The deep-brown colour of the female sori contrasts with that of the male which appear as white glistening spots upon the thalli. The future male sori become recognisable at an early stage by the disintegration of the chromatophores in an elliptical group of surface cells which grow out to form the antheridia. The cells at the margin of the group elongate like the others, but in them the chromatophores persist and retain their pigment. These cells remain undivided and constitute three or more rows of involucral cells at the edge of the sorus; they are usually.

Sexual Reproduction

Interpreted as sterile antheridia. The actual male organs *(a)* first segment by a vertical wall, perpendicular to the long axis of the thallus; this is followed by further vertical walls in other planes, and during this there is also extensive horizontal septation. In the mature, altogether colourless, antheridium 32-64 compartments separated by delicate septa are visible in the transverse and 20-24 tiers in the longitudinal section. There are 100-200 antheridia in a sorus and each forms about 1500 sperms.

Like the sporangia, the sex organs of *Padina pavonia* are produced in relation to the hairs which are shed during their development. The oogonia form double concentric zones on either side of the bands of hairs on the lower surface. The basal cell usually divides transversely and the upper half may give rise to a second oogonium after the first has dehisced. The zones of oogonia are interrupted at variable intervals by radial rows of antheridia, the differentiation of which commences in the middle of each row and advances centrifugally towards apex and base of the thallus. The mature antheridia show about 32 chambers in transverse section.

The imperfectly known oogonial sori of *Taonia* form short double transverse rows which are scattered over the thallus, while the milk-white male sori are situated on either side of and between the zones of hairs. In section they resemble those of *Dictyota*, although the antheridia do not all ripen simultaneously; the sori usually exhibit sterile marginal cells with yellowish chromatophores.

In *Zonaria* the sex organs are independent of the hairs and the sori lack the paraphyses found in the asexual ones. In *Z. flava* they form continuous or interrupted zigzags, while those of *Z. Farlowii* constitute irregular areas between the bands of hairs, the white male sori *(a)* in both species being larger than the female (o). In the latter only some of the cells grow out into oogonia, while in the male sori all the surface-cells produce antheridia and involucral cells are usually absent, although occasionally found in *Z. flava* and *Z. lobata*. In the mature antheridium of *Z. Farlowii* there are 20-40 tiers of cells, with 8 or 16 cells in each tier.

The scattered male sori of *Dictyopteris* are of very diverse size, containing between 3 and 100 antheridia.

The discharge of eggs and sperms takes place rapidly at about the time of daybreak. Both the septa and the walls of the ripe antheridia are dissolved so that the numerous minute male cells lie in a mass of mucilage occupying the space bounded by the involucral cells, when these are present. At this stage the male cells, formed singly from each compartment, appear rounded. Lloyd Williams first showed that they possess flagella, although Thuret seems to have recognised something of the kind. According to the former, swarming of the sperms in *Dictyota* and *Taorda* depends on exposure to a sufficiently bright light and on the realisation of a certain temperature. The pear-shaped spermatozoids have a large posterior nucleus and an eyespot situated at the anterior end, while the single long, forwardly directed, flagellum arises near the eyespot; there is reason to suspect the presence of a second, very short, backwardly directed flagellum, although this has not so far been demonstrated. Williams, who observed fertilization, states that freshly discharged ova exert a great attraction on the sperms, although, if the eggs are not fertilized within an hour, they become invested by a membrane and germinate parthenogenetically. The male nucleolus can be recognised within the fusion-nucleus after fertilization; the single centrosome later divides into two.

In British waters *Dictyota dichotoma* shows a regular fortnightly production of sexual sori from July onwards, corresponding to the interval between two spring tides. The sori are initiated during a period of neap tides, and liberation of the sexual cells takes place at a variable number of tides after the highest spring tide (3-5 in the Menai Straits; 7-12 at Plymouth). At this time the rudiments of the next crop of sori begin to appear. A similar rhythm is observed in plants kept in the laboratory, and according to Hoyt it is also shown by the new thallus-branches that arise in cultures and have never been exposed to the influence of the tides.

The same rhythm is, moreover, maintained when the light-conditions are experimentally reversed or when the tides are modified by wind. Williams concludes that the periodicity in question is hereditary, but that on British shores the time of development of the sexual sori is regulated by the greater illumination available during the low water of spring tides. There is some evidence that a similar periodical production of oogonia takes place in *Dictyopteris* but other Dictyotales do not appear to exhibit this rhythm. Comparison should be made with similar features in *Nemoderma*.

The behaviour of *Dictyota* has also been investigated in other localities. A similar, though less marked, fortnightly production of sexual sori takes place at Naples where D. *dichotoma* occurs at some depth and the tidal differences are slight; here, as in the laboratory cultures, illumination can scarcely be a factor. On the other hand, in Jamaica, where the tides are irregular, periodicity is scarcely evident.

On the shores of North Carolina the successive crops of sexual sori are produced at monthly intervals (spring tides of full moon).

Early Development of the Thallus

The germination of the tetraspores shows certain differences in the individual genera. The first wall is stated to be perpendicular to the direction of the incident light in *Dictyota* and *Dictyopteris*, although Reinke concluded, from the observation of spores germinating within the sporangia of *Dictyota*, that it is the end of the spore directed away from the parent-thallus that becomes the apex of the new plant. In *Dictyota* the upper segment becomes the apical cell of the embryo, while the lower produces a rhizoid. Reinke records diverse abnormalities. The segments of the apical cell divide into an inner and four outer cells, the latter dividing anticlinally to form the surface-layer of the cylindrical embryo. Later colourless multicellular rhizoids grow out from the superficial cells. The summit of the embryo soon broadens to form the band-shaped thallus, while its basal portion gives rise to more or less numerous lateral branches, some of which become prostrate and form the rhizome.

Iujnpst Dictyotales, however, the germinating spores first produce a multicellular oval body, well seen in *Padina*. One of the superficial cells of this *nodule* becomes protruded to form the apical cell of a cylindrical shoot which later flattens at its apex and slowly passes over into the adult fan. Other cells of the nodule subsequently grow out in the same manner, although some of the resulting shoots usually remain arrested; other surface-cells give rise to rhizoids *(r)*. In *Taonia* one of the cells formed by the first transverse division of the spore grows out into a rhizoid and the formation of the nodule takes place subsequently. One of its superficial cells, determined by the direction of the incident light, enlarges to form an apical cell which, however, only cuts off a few segments before dividing longitudinally to form a number of initials.

Much the same occurs in *Dictyopteris,* although here the cells of the nodule sometimes grow out into branched uniseriate threads *(p)* attached to the substratum by rhizoids and growing by means of an apical cell; from this protonema a number of fronds *(t)* arise as lateral outgrowths. More usually, however, the outgrowing cell of the nodule develops direct into a mature plant. The young fronds may remain one-layered for a long time. The occasional liberation and germination of the entire sporangial contents in individuals that do not produce any tetraspores. Robinson observed similar phenomena in *Taonia,* although here such abnormal stages are encountered side by side with tetraspores. In *Padina* they show no differences in germination, but in *Taonia* they exhibit greater powers of resistance and a greater capacity for germination than the tetraspores, while the resulting plants are larger and display more vigorous growth.

The germination of zygotes and tetraspores appears to take place in the same way, although that of the former has not often been studied.

Life-cycle

The full life-cycle has only been investigated in *Dictyota dichotoma, Padina pavonia,* and *Zonaria Farlowii.* The isomorphic alternation, indicated by the distribution of the reproductive organs, is fully confirmed by the cytological data and has also been demonstrated experimentally in *Dictyota* and *Padina variegata.* Hoyt and Wolfe used the faculty of tetraspores and zygotes to germinate on oyster-shells placed in cultures of asexual and sexual plants respectively, the shells bearing the germlings being subsequently submerged in their natural habitats; all adjacent plants of the species experimented with were removed. It was thus established that, as a general rule, tetraspores give rise to sexual and zygotes to asexual plants.

Data were also obtained that indicated a production of equal numbers of male and female plants, suggesting segregation of the two sexes at the reduction divisions. Schreiber, using methods similar to those employed for *Laminaria,* definitely proved that in *Dictyota* two of the four spores of a sporangium develop into male and two into female plants. It can scarcely be doubted that this will be found to apply to Dictyotales generally and that all members of the order possess an isomorphic alternation with genotypic sex determination. Although the two generations often occur simultaneously, there is evidence of some seasonal succession.

Parthenogenesis does not appear to play an important role among Dictyotales, although germination of unfertilized ovate not uncommon. Lloyd Williams found that such ova exhibit irregularities in nuclear division and that segmentation soon ceases; this is followed by the death of the germlings.

Many investigators comment on the great preponderance of asexual over sexual plants and in some regions only tetrasporangiate individuals have been found. No adequate explanation of these facts is at present forthcoming, but the repeatedly recorded germination of the entire contents of a sporangium—presumably without the occurrence of reduction—and the occasional greater vigour of such germlings as compared with those formed by tetraspores, may indicate a method of multiplication of the diploid phase without the intervention of a sexual one. Abundant vegetative propagation may, moreover, help in the rapid multiplication of the asexual generation.

On the other hand, although Carter only met with tetrasporic plants of *Padina pavonia* in British waters, he nevertheless found that they formed tetraspores in the normal manner with accompanying reduction. This shows that the foregoing explanations will not account for all the facts. Since it is hardly likely that all the haploid spores perish, one is forced to assume either that there is a microscopic sexual generation or perhaps more probably that the latter inhabits deep water. In this connection it may be noted that, according to Lloyd Williams, the spores and zygotes of *Dictyota* germinate in the summer, but the germlings remain small until the following June.

Affinities of the Dictyotales

Owing largely to the possession of tetraspores and of male cells, which were long thought to be motionless, the Dictyotales have often been regarded as related to the Red Algae and have indeed occasionally been directly referred to that class. A matter of historical interest it may be mentioned that both Thuret and Cohn regarded the oogonial sori as equivalent to the cystocarps of the Rhodophyceae. There is, however, no sound basis for envisaging any affinity with Red Algae. The pigmentation of the chromatophores, the metabolic products, the type of swarmer as exemplified by the spermatozoid, and the plurilocular nature of the antheridia all indicate a close affinity with other Phaeophyceae.

Even the tetrasporangia no doubt merely represent a specialisation of the unilocular sporangium, and in this connection the frequent formation of eight spores in *Zonaria* is significant. It should, moreover, be recalled that in diverse Ectocarpales and Tilopteridales the unilocular sporangia produce only few spores, which are then occasionally developed as aplanospores or liberated as motionless cells. The exact relation of the Dictyotales to other Brown Algae is obscure and they appear as a specialised line that originated from simple isomorphic forms and has departed rather widely from the normal type. Sauvageau supports an affinity with Cutleriales, a view which is largely based on the close resemblance in habit and structure between *Zonaria variegata* and the *Aglaozonia*-stages. Particular stress is laid on *Aglaozonia canariensis*, for which at present no corresponding *Cutleria*-stage is known. Here the pyriform sporangia in the small and scattered sori are not in lateral contact and are not cut off from the cell which produces them so that there is considerable resemblance with the sori of *Zonaria variegata*.

Genus Dictyotopsis

It is difficult to assign a position to this remarkable genus which in various respects recalls one of the simpler Jungermanmales *(Metzgerid).* The cells, however, possess brown-coloured chromatophores devoid of starch and include probable fucosan-vesicles. The thallus is one-layered and devoid of a midrib; it shows dichotomous branching, with one fork usually growing more strongly than the other, while the recurved margins bear multicellular, little-branched rhizoids *(r).* The large apical cell is two-sided. The only method of reproduction so far observed is by ovoid, one-layered propagules borne on two-celled stalks along the margins of the thalli. After cutting off a certain number of segments the apical cell of the propagule remains dormant until detachment has occurred; the propagules are thus arrested laterals.

The algae is a member of the Eastern Mangrove-community and in part inhabits mud that is inundated only at the time of spring tides; at lower levels it constitutes a thick covering on the projecting respiratory roots of *Sonneratia.*

The type of apical cell and its mode of segmentation are unique among Brown Algae. The abundant formation of propagules recalls *Sphacelaria,* with which, however, there is practically no other resemblance, and, until other types of reproductive organs become known, the systematic position must remain quite doubtful.

Fucales

The numerous seaweeds belonging to this order are, like the Laminariales, diploid and characterised by marked morphological differentiation and anatomical complexity, but in many respects the two orders diverge widely from one another. No member of Fucales attains the huge dimensions reached by some Laminariales. The broad discoid holdfasts are very distinctive, while all the typical Fucales exhibit apical growth by means of an apical cell of characteristic shape. Apart from the copious sexual reproduction, multiplication is effected solely by vegetative means. The sex organs are located within the conceptacles which constitute one of the chief characteristics of the order. Reduction occurs during gametogenesis, and no distinct haploid generation is organised so that there is no alternation. The peripheral cells harbour numerous oval chromatophores, and centrosomes are usually recognisable during nuclear division.

Despite great diversity in outward habit, the germlings developed from the zygotes show considerable uniformity and this, no less than other distinctive features (apical growth, conceptacles) speaks for a monophyletic origin. There are, however, a number of aberrant forms (Hormosiraceae, Durvilleaceae) whose exact relation to the main series is not at present clear.

In the Northern Hemisphere the Fucales are largely surfioving forms, many of them inhabiting the littoral region where they often display a marked zonation in relation to a varying capacity for drought-resistance. Several genera *(Halidrys, Cystoseira)*, however, occur in the sublittoral region or in permanent pools near low water mark, where they are only exposed occasionally during spring tides. The warm-water forms *(Turbinaria, Sargassum)* practically always grow permanently submerged, while the bulk of the New Zealand Fucales grow at and below low tide level.

The Fucales attain their greatest development in the colder seas. Species of *Fucus* are abundant in North Temperate waters and with them are associated *Pelvetia, Ascophyllum, Himanthalia* and *Halidrys.* The large genus *Cystoseira* is also largely confined to the Northern Hemisphere, though more abundantly represented in the warmer seas (Mediterranean, Indian and Pacific Oceans); some species are known from South Africa. The Fucates also show an extensive development in the Southern Hemisphere, and diverse distinctive genera are recorded from. Australia and New Zealand *(Carpoglossum, Phyllospora,, Scaberia, Cystophora, Marginariella,, Seirococcus, Landsburgia, Myriodesma, Xiphophora, Notheia, Hormosira))* the last three are endemic, whilst *Landsburgia* and *Marginariella* are confined to New Zealand. The wealth of Fucales in the Australasian region is in striking contrast to their paucity in South Africa. The Australasian region seems also to be one of the chief centres for *Sargassum,* although the latter has representatives in all warmer seas and is widely distributed in tropical waters. *Turbinaria* seems to have its main centre in the Eastern Tropics, though a few species extend south to the Cape and Australia and northwards to China. The remarkable *Durvillea* is subantarctic in distribution (New Zealand, Falkland Isles, Cape Horn, Kerguelen). *Coccophora* is so far only known from Japan, where too most species of *Cystophyllum* are found.

External Morphology

Fucus and its Allies (Fucaceae)

The many species of *Fucus* are essentially characteristic of North Temperate and Arctic regions. The repeatedly forked band-shaped thalli, narrowed below into a short stalk *(s)* terminating in the discoid attaching organ, are familiar to every student of Botany. The branching is dichotomous and in one plane, although one or other limb is commonly favoured. The flattened segments are traversed by a prominent midrib *(r)* which thickens in the older parts where it passes over into the stalk; the latter is, in fact, largely composed of the midrib of the first-formed part of the thallus, the wings of which have been worn away by wave-action. The tip of every growing branch is occupied by a narrow groove extended in the plane of flattening and filled with mucilage; at the base of this the apical cell is situated. The wings usually bear irregularly scattered sterile conceptacles or *cryptoblasts,* while at certain times of the year the tips of the branches are more or less markedly swollen and harbour the large fertile conceptacles, within which the sexual organs are produced. In these fertile tracts or *receptacles* the midrib is lacking; their inflation is due to abundant production of mucilage.

The young plants, which to begin with are nearly spherical

as in all Fucales, soon become cylindrical and then club-shaped as a result of broadening at the top; at this stage an apical depression appears, from which a number of hairs *(K)* arise. By degrees most of the embryo flattens, although the basal part remains cylindrical and constitutes the beginnings of the stalk. As forking sets in, the mature thallus-form is assumed. Adventitious laterals occasionally develop from the stalk of the young plant. In *F. dichotomus* Sauv., in-which the fronds last for one year only, these laterals regenerate the plant in the next season.

The British species are in general readily distinguished and comprise: P. *serratus*, with a serrate margin and flattened receptacles; *F. vesiculosus*, with a smooth margin, conspicuous and commonly paired air-bladders *(v*\absent in var. *evesiculosus)*, and swollen receptacles; *F. spiralis (F. platycarpus* Thur.), smaller than either of the preceding, devoid of air-bladders, and with dilated irregularly swollen fertile tips which have a sterile border *(b)*; and *F. ceranoides* L., especially common in brackish water, with a delicate thallus, showing regular dichotomy and bearing slender fusiform receptacles. Soft irregular swellings, due to internal accumulation of gases, are not uncommon in the first three species, especially in brackish water. *F. spiralis* is hermaphrodite (similarly *F. dichotomus; F. vesiculosus* and F.

Compared with *Fucus, Pelvetia* appears as a specialised form suited to an extreme environment. *P. canaliculata* is widely distributed in the North Atlantic, where it forms a characteristic girdle at the uppermost limit reached by spray. The thallus shows adaptations to the prolonged exposure to desiccation during intertidal periods. It never reaches any considerable dimensions and dwarfing is the more marked, the higher the plants grow, the uppermost ones rarely exceeding 3-5 cm. in height. The narrow segments of the repeatedly forked thallus are channelled, as implied by the specific name; during low water moisture is retained for some time within the channel, whilst the somewhat edgewise position of the wings reduces the degree of exposure to the midday sun. There is no midrib.

The depression harbouring the apical cell is readily seen. There are no cryptoblasts, whilst the irregularly swollen receptacles render fertile plants very conspicuous. The thallus is stated invariably to be occupied by an ascomycetous fungus (*Mycosphaerella pelvetiae*. P. *fastigiata* (J. Ag.) De Toni, abundant in somewhat less extreme habitats on North Pacific coasts, though larger, has essentially the same habit. The receptacles are not so sharply defined and there are occasional cryptoblasts. Another similar form found in the same region is *Pelvetiopsis*.

A different habit is seen in *Ascophyllum nodosum*, abundant on somewhat protected shores where it occurs together with *Fucus vesiculosus* in the upper part of the littoral region. The mature thallus consists of dichotomously branched strap-shaped axes *(a)* bearing numerous short laterals *(la)*, which often arise in tufts from narrow marginal slits and are commonly subopposite in arrangement. *Ascophyllum* thus combines dichotomous and monopodial branching. The fronds may be 1£ metres long; they are devoid of a midrib and are usually dilated at intervals by conspicuous vesicles *(b)* so that at high water the thallus spreads out near the surface. In adult plants the tips of the upper laterals become swollen and constitute receptacles. According to Tobler the greater part of the long shoots perishes at the end of each season, but this is probably not so in all areas.

The young plants soon become elongate and show but little broadening above the basal stalk; the apex is truncate and devoid of hairs. In somewhat older germlings a number of laterals *(bs)* arise near the base, but these at first remain short, whilst the main axis grows rapidly and soon forms marginal branches of limited growth at a later stage the basal laterals *(bs)* also lengthen, those originating close to the base subsequently appearing to arise directly from the attaching disc. If the axis is broken or damaged near its base, one of the basal shoots takes its place, whereas injury to the upper part of an older plant leads to the outgrowth of a number of the laterals into long shoots and results in a bushy habit. There is altogether much diversity in the appearance of this seaweed.

A greater resemblance to *Fucus* is shown by the Australasian *Xiphophora* where the flat ribless thallus appears monopodially branched below, although the ultimate fertile parts are dichotomous. There are no air-bladders. *Axillaria constricta* (*Carpoglossum constrictum* Kutz.; *Ozothallia* (*Ascophyllum*) *constricta* Ag.), a sublittoral South African seaweed, approaches nearer to *Ascophyllum.* The characteristic notches on the longer fronds subtend vegetative laterals *(la)* or short pear-shaped receptacles (*r*); both emerge from marginal slits extended in the plane of the thallus. The essential difference from *Ascophyllum* lies in the absence of dichotomous branching, in which respect *Axillaria* agrees with several allied genera. In these, however, the principal fronds bear leafy laterals which are probably homologous with the notches of *Axillaria.*

Such forms are well illustrated by the Australian *Seirococcus axillaris; Scytothalia axillaris* Kutz. The flat fronds here bear large leafy outgrowths arising in two alternating rows. The apex *(g)* is occupied by a rounded depression covered in by the rudiments of these laterals and with the slit harbouring the apical cell (*i*) at its base. The leafy laterals are stated not to grow by means of an apical cell, but are formed by the division of groups of surface-cells. Branches repeating the structure of the main axis arise at rare intervals. The small and crowded moniliform receptacles (*r*) for the most part occupy the adaxial margins of the laterals; they are simple or branched and each swelling harbours a single conceptacle, the male usually at the base, the female nearer the top.

A number of genera resemble *Seirococcus* in the possession of leafy laterals and like it are no doubt derivatives of *Fucus-like* types. In *Scytothalia dorycarpa* (Turn.) Grev. the prominent receptacles mainly occupy the margins of the axis between the "leaves". This is also so in *Cystosphaera,* which is distinguished by dichotomous branching and the possession of axillary vesicles. *C. Jacquinotii* is a characteristic sublittoral Antarctic species. Similar stalked air-bladders occur in *Marginariella* (*Marginaria* and, like the crowded spindle-shaped receptacles, arise from the edges of the thallus-branches. In *Phyllospora* the numerous toothed laterals, borne on the flattened leathery axes are narrowed basally so that there is a greater resemblance to foliar organs. The smaller receptacular branches *(r)* arising between them are often provided with a large basal air-bladder *(v)*. The laterals are again devoid of an apical cell and develop without order. P. *comosa* is one of the largest Australian Fucoids growing in exposed situations near low-tide level. A more aberrant form is the Australian *Scaberia* in which the laterals are verrucose pseudopeltate structures harbouring the conceptacles in their dilated ends and in part replaced by large globular air-bladders covered with warts.

The genera grouped as Fucaceae all exhibit more or less marked flattening of the greater part of the thallus, although a basal stalk is often present in older plants. The least specialised member is *Fucus*, with its essentially dichotomous branching; *Pelvetia* readily falls into line with it. In *Ascophyllum* and its allies *(Axillaria, Seirococcus,* etc.) there is a progressive replacement of dichotomous by monopodial branching, combined with the differentiation of more or less clearly marked fertile laterals. The affinity between the diverse members of the *Ascophyllum-alliance* is evident from a comparison of the structure of the growing apices and the details of branch-development. Moreover, the mature plants of all Fucaceae so far investigated possess a four-sided apical cell, which is met with in no other representatives of the order.

Himanthalia

The Genus: This striking seaweed is abundant near low-tide level on moderately exposed shores, often growing attached to steep cliff-faces. The thallus is sharply differentiated into sterile and fertile portions. The former is represented by a stalked *Peziza-like* structure, which may reach a diameter of several centimetres. The stalk *(s)*, attached by the customary disc *(d)* and between 1 and 2 cm. in length, is somewhat flexible, so that the cup can yield to wave-action. The upper surface of the cup is concave, and in relatively young stages a central dot marks the position of the apical pit; in older plants 2-4 such pits can be distinguished. During the second or third year of its life from 1-4 (generally 2) long, band-shaped, dichotomously branched receptacles, bearing conceptacles throughout most of their length, grow out from the concave surface and, after liberation of the sexual cells, the whole plant usually dies away. The receptacles, though only 1-2 cm. broad, can reach a length of 2 metres; the apical slit is here orientated at right angles to the surface. A small basal part, rather narrower than the rest, is sterile and is stated to be especially flexible. The young receptacles are negatively geotropic. Young stages of *Himanthalia* are pear-shaped and much like those of *Fucus* occasional abnormal individuals retain this form permanently. Most soon broaden and become concave at the top, whilst the lower part differentiates as the stalk. The receptacles represent one fork of a dichotomy, the other limb of which remains arrested.

This rather isolated type appears to be most nearly allied to *Fucus*. The vegetative part of the plant somewhat resembles a young stage of the latter, though modified in shape and with branching suppressed until the time of production of the receptacles. On the other hand, in the restricted development of the primary axis, there is some resemblance to *Cystoseira* and other specialised forms.

Halidrys, Cystoseira, and their Allies (Cystoseiraceae)

The series of genera now to be considered are for the most part monopodially branched and, since branching is usually copious, they often present a bushy habit. Few, if any, appear to be able to withstand much desiccation, and they are found either in deep rock-pools or in the sublittoral region. They are best represented in warmer seas. In certain genera the bulk of the branching takes place in one plane so that the thallus, as in Fucaceae, is bilaterally organised, and, since these are probably the more primitive types, they may be described first.

In *Halidrys siliquosa* the numerous, usually slightly flattened, branches are for the most part disposed in one plane and arise alternately from the margins of their respective parent axes. Most of the lower laterals develop into long shoots, some of which may reach the same length as the main axis. At higher levels the few long branches are scattered quite irregularly among the many short ones, some of which appear as mere teeth (2-3 mm. long). Certain laterals develop into characteristic, chambered and mucronate air-bladders resembling a septate siliqua, while some of the upper ones differentiate into stalked receptacles *(r)* which are often grouped in a racemose manner. According to Reinke many of the short branches are shed at the end of the season. In *H. dioica* flattening of the thallus is more marked.

Bifurcaria (Pycnophycus) tuberculata, frequent on the Atlantic shores of Europe and Africa, is exceptional among Fucales in possessing a richly branched, perennial rhizome attached to the substratum by small adhesive discs. A similar rhizome is found in the South African *B. sisymbrioides* Kiitz. That of *B. tuberculata* is a gnarled structure with numerous short, irregularly swollen branches, many of which remain arrested; these show the apical depression *(a)* very clearly. Other branches *(e)* bend up and give rise to forked, terete shoots which grow to a height of 30-45 cm. and die down in the late autumn; constrictions occur at the points of branching and where the erect axes arise from the rhizome. Despite the apparent forking the branching is, according to Gruber, actually monopodial and bilateral.

The rhizome is probably a lateral outgrowth from the axis of the young plant, which is no doubt erect and radial as in other Fucales. The tuberculate receptacles, which are often exposed at low water, arise at the tips of the upright shoots, the branches bearing them sometimes developing air-bladders.

The South African *B. laevigata* Kiitz., which has a discoid holdfast and differs in other respects, has recently been referred to a distinct genus, *Bifurcariopsis*.

Brief mention may be made of certain distinctive Australasian genera. *Carpoglossum,* in the restricted sense adopted by Gruber, comprises only C. *confluens,* which is clearly less specialised than *Halidrys.* The mode of branching of the broad elongate fronds is similar, but no special receptacles are differentiated, the conceptacles merely occupying the ultimate laterals *(r).*

Greater differentiation is evident in *Platythalia.* In P. *quercifolia; Carpoglossum quercifolium* J. Ag. the laterals of limited growth are foliar with serrate or almost pinnate margins and a narrowed base; some are sterile, others harbour the fertile conceptacles. The branching is bilateral, but by twisting of the axis the laterals assume a somewhat radial disposition. In the otherwise similar *Platylobium mertensii* Kiitz. special fertile shoots with marginal conceptacles arise from the edges of the leafy laterals, while large spherical air-bladders sometimes replace the receptacles or even an entire lateral.

In other Cystoseiraceae most of the branches are laid down radially from the first. This is so in *Cystoseira* several species of which occur in Britain near and below low-tide level, where they often form conspicuous submarine bushes, generally densely occupied by epiphytes. The genus is, however, more characteristic of warmer temperate and tropical seas; its species form a prominent belt below low-water mark in many parts of the Mediterranean. The Pacific C. *osmundacea* (Menz.) Ag. may grow to a length of 6 metres, whilst C. *canariensis* Sauv. reaches a height of only 14 cm. Several species show a striking iridescence, and the majority favour well-illuminated habitats, although C. *Montagnei* J. Ag. penetrates to depths of 50 metres in the Mediterranean. Species of *Cystoseira* are occasionally attacked by a Pyrenomycetous fungus.

There is usually a distinct main axis, but this mostly reaches no appreciable height; it is rounded or angular in cross-section and attached by a basal disc *(d)* which is often somewhat conical. In a few species (e.g., C. *Abies-marina* C. Ag.) the main axis is prostrate, whilst in C. *fibrosa* C. Ag. it is elongate and flattened and bears distichous branches. As a general rule, however, the primary laterals are disposed in a 2/5 spiral and grow to a much greater length than the main axis, which may be altogether hidden among them (e.g. C. *granulata*). Its height is given as 1-2 cm. in C. *amentacea*, 1-10 cm. in C. *abrotanifolia*, 10-30 cm. in C. *crinita*.

The laterals always show extensive monopodial branching. In some species the ultimate ones are very short so that the habit is ericoid (C. *ericoides* Ag.). In others (C. *abrotanifolia*, C. *osmundacea*, etc.) the basal parts of the primary laterals, and sometimes of those of higher orders, are flat and more or less markedly lobed, with all the branches arising in one plane. The young plants then present a characteristic appearance, altogether different from that of the mature stage, in which the primary laterals and their branches have lengthened, become terete, and acquired the normal radial branching. In most species, however, all the branches are terete and radial from the first. Air-bladders are commonly present and usually occupy the lower parts of the ultimate laterals, often occurring in short series. The receptacles are frequently spindle-shaped.

All *Cystoseiras* are perennial, although a few are sometimes monocarpic (C. *abrotanifolid).* During the annual lengthening of the main axis it often undergoes considerable increase in thickness. On the approach of the dormant season a more or less considerable part of the branch-system is usually shed, and the plant may assume a strikingly different appearance. The laterals of the first order usually develop continuously and finally become detached as a whole, leaving a stump or a mere scar on the main axis. In certain species, however, which are distinguished by the dilated bases ("tofuli" of Valiante of the primary laterals, their development takes place in two stages, the swollen bases arising towards the end of the period of active growth and remaining dormant until the following season when they grow out into branches. The basal dilation persists after the upper part of the branch is shed. In some species (C. *Montagnei* J. Ag., C. *spinosa* Sauv.) all the primary laterals develop in this way, and the surface of the main axis becomes covered with an uninterrupted series of tubercles, but in others (*C. granulata*, *C. platyclada* Sauv.) some of the first-formed primary laterals of each season lack the basal dilation. The axis then bears alternating series of tubercles and scars by means of which its age can be determined.

Shedding of the branches occurs both in littoral and sublittoral species, but does not take place in young plants. In the new season branches in the tuberculate forms arise not only from the apical tubercles, but also adventitiously from the swollen bases of older laterals. In certain species the long laterals, after they are shed, are replaced by numerous short branches (C. *mediterranea* (Ag.) Sauv.) or, as in C. *foeniculacea*, by flattened shoots with distichous laterals, devoid of air-bladders.

The tubercles are probably storage-organs. They consist of thin-walled cells, are devoid of hyphae, and are traversed by a central strand of elongate cells. They are commonly covered with spiny outgrowths which also often occur abundantly on other parts of the thallus.

The young plants resemble those of *Fucus*, especially in species like *C. abrotanifolia* in which the upper part of the embryo flattens at an early stage, but so far as present evidence goes, the primary axis of the embryo does not develop beyond a certain stage. When it has reached a length of 5 or 6 mm., an apical cell arises near its base *(d)* and soon becomes sunk in the usual depression. The active division of this apical results in the production of a secondary shoot which becomes the definitive axis of the plant and, as the latter becomes established, the primary axis above its place of origin is gradually turned to one side. The basal part of the axis is therefore a sympodium. All the later laterals, however, develop monopodially and in acropetal succession from the definitive axis which persists throughout the life of the plant. The first two laterals arise on opposite sides of the axis, and it is only after this that a 2/5 divergence is established, although subsequent displacement causes the first two laterals to fall into the 2/5 spiral. The early development has only been examined in a few species, and it remains to be seen whether the sequence just outlined is always followed.

In certain species (C. *foeniculacea*, C. *abrotanifolia)* several fronds arise from the basal disc in the older plant. This is due to the development of adventitious branches from the lower part of the axis, which gradually become incorporated in the enlarging holdfast.

The large genus *Cystophyllum* widely distributed in Japanese waters, has the habit of a *Cystoseira* and the two genera are no doubt nearly allied. It is not easy to draw a sharp distinction between the mature plants of the two genera. According to Yendo the only general difference lies in the restriction of air-bladders to the ultimate branchlets in *Cystophyllum*. Another imperfectly known genus of this affinity is the Australasian *Cystophoral*, which is chiefly distinguished by its stalked air-bladders. In some species the entire vegetative system is bilateral, the fertile branches alone showing a radial arrangement.

A rather different type is seen in *Landsburgia*, where the almost terete main axes bear radially disposed laterals, some of which resemble the axis, while others develop as lobed leafy structures; smaller laterals of the latter type *(r)* harbour the conceptacles. There is some resemblance to *Platythalia*. The leafy laterals of *Landsburgia* recall those of the young plants of such *Cystoseiras* as *C. abrotanifolia,* and there is also similarity to the flat basal parts of the primary laterals of certain *Cystoseiras*. Oltmanns may well be right in looking upon *Landsburgia* as a relatively primitive representative of the radial Cystoseiraceae, although Gruber regarded it as a specialised form.

The radial organisation of *Cystoseira* and its allies is probably secondary, and there is reason to believe that bilateral forms like *Halidrys* represent the more primitive condition within the family. Their derivation from *Fucus-like* types presents no difficulty. It may be recalled that some *Cystoseiras* retain a bilateral organisation in the early branching of their primary laterals. All Cystoseiraceae appear to possess a three-sided apical cell.

Sargassum and its Allies (Sargassaceae)

The genera included in the Sargassaceae, which are in part rather vaguely defined, exhibit the most complex morphology among Fucales. Like the Cystoseiraceae some show bilateral, others radial organisation, and the former are no doubt more primitive. The branching is always monopodial. All the members of the family are distinguished by the special features of the lateral branch-systems. These bear at their base one or two leaf-like organs, in the "axil" of one of which the whole system often appears to arise. The three-sided apical cell is lodged at the base of a deep funnel-shaped depression. Species of *Sargassum* are not native to Britain, although *S. natans (S. bacciferum* Ag. is sometimes cast up after storms. A few (S. *Hornschuchii* Ag., *S. linifolium)* occur in the Mediterranean, but the main centre of distribution is in the Southern Hemisphere and especially in the warmer seas.

With Oltmanns the South African *Sargassum (Anthophycus) longifolium* may be taken first as representing a relatively unspecialised type. The lateral branch-systems here show a distichous arrangement throughout the plant. On the lower parts of the long axes they are represented merely by leafy outgrowths flattened in the plane of branching, but above they constitute fertile systems comprising a number of distinct structures. Near the base these bear two alternate' leaves "*(I)* which are again expanded in the plane of branching; the upper adaxial one usually surmounts a prominent air-bladder *(v)* which is occasionally alone present. Beyond this the axis of the lateral bears a series of repeatedly branched receptacles (*r*). In several of the distichous species both the axes and the laterals are flattened, and a midrib *(m)* may be present throughout. Agardh regarded such species as primitive.

From forms like these there are many gradations to the typical *Sargassums* (Sect. *Eusargasswri)* which are radially organised and present a bushy habit like that of a *Cystoseira.*

As in that genus, too, the main axis is commonly only a few centimetres high, and it is the richly branched primary laterals that constitute the greater part of the vegetative system. The *Eusargassums* comprise about two-thirds of the known species and are essentially confined to tropical seas. In them the terete or angular branches of the first and higher orders are spirally arranged and bear numerous laterals of limited growth, in which the basal "leaf" is usually the most conspicuous. These foliar organs, which are sometimes very narrow (e.g. *S. patens* Ag.), commonly possess a serrate margin and are usually provided with an obvious midrib. In certain species the long axes bear wart-like outgrowths giving them a spiny or verrucose appearance. Parts of the branch-system may be shed periodically and later replaced by new growth.

Both the primary and the subsidiary branch-systems develop a basal" leaf" which more or less clearly subtends the remainder of the branch-system. The "leaves" are, as a general rule, flattened in the vertical plane, although commonly expanded horizontally in the lower parts and sometimes over a considerable part of the plant, for the most part probably as a result of twisting of their stalks. On the lower parts of the long shoots some of the laterals comprise little but this leaf, although there are sometimes axillary rudiments, but in the upper regions of mature plants many of the "leaves" subtend axillary branch-systems, which appear condensed as compared with those of *Anthophycus*. The lowest member or members of these branch-systems usually develop as stalked air-bladders *(v)* which may or may not be crowned by a leafy termination. Not uncommonly, however, the lowest member may develop as a leaf, while conversely the subtending leaves are sometimes replaced by air-bladders; the latter are no doubt homologous with the "leaves". The subsequent members of the axillary branch-systems, which are more or less richly branched, are cylindrical or flattened and constitute the receptacles. While it is convenient to speak of "leaves" and "axillary branches ", the whole lateral no doubt represents a single branch-system, the first member of which appears as the subtending leaf.

This brief account is inadequate to afford a true conception of the range of form which is presented by the species of *Sargassum* and which will be evident from a perusal of the works cited in the bibliography. The lateral branch-systems exhibit a varied differentiation, both in different individuals and in different species. In *Acystis*, recorded from the Red Sea, the erect shoots arise from a prostrate and flattened rhizome and there are no air-bladders.

Certain bilateral flattened forms, devoid of air-bladders on the axillary systems and characterised by the possession of an elongate main axis, are usually referred to a separate genus *Carpophyllum*, well represented in the Australasian region. Judging by Gruber's figure the subtending leaf is sometimes suppressed, the branched receptacles *(r)* then arising direct from the margins of the flattened shoots. These forms possibly constitute a distinct evolutionary line.

In *Turbinaria* the radially organised plants are compact and cone-like. The laterals take the form of firm, more or less peltate structures bearing the short well-branched receptacles *(r)* in clusters on the upper sides of their stalks. Certain species possess a conspicuous air-bladder below the peltate enlargement, while in *T. trialata*, for example, the stalk is winged. Attachment is effected by a system of spreading branches at the base of the main axis.

The monotypic *Coccophora* has a perennial basal disc from which long shoots arise in each season. At the beginning and towards the end of their growth they produce dichotomously branched filiform laterals, but over most of the older part they bear triangular scale-like appendages. Axillary systems that bear secondary filiform laterals appear sooner or later, but on the lower parts of the axes these laterals remain small, while on the upper many of them lengthen and branch extensively so that the shoots assume a densely tufted appearance. More than a year after the appearance of the long shoots, the upper axillary systems undergo marked elongation, producing scales *(s)* similar to those borne on the main axis and, after some time, developing at their tips a number of stalked rounded receptacles (r); soon after this the whole shoot dies. No clear information is available as to the relation between the axillary systems and the filiform laterals from which they arise, but *Coccophora* would appear to be one of the most highly specialised members of the order.

Hormosiraceae

All the Fucales hitherto noticed possess a single apical cell, but the Australasian genera *Hormosira* and *Notheia*, although they resemble the others in many respects, differ in the presence of several apical cells. Oltmanns classes them as *Anomalae*, but they are perhaps better considered as constituting a separate family.

Hormosira is a plant of characteristic habit, the dichotomously branched thalli consisting of much enlarged hollow segments *(v)* separated by short solid stalks *(s)*. Sterile and fertile conceptacles are irregularly scattered over the vesicular segments. In transverse sections through base or apex of the latter the central hollow is produced into four (rarely three) bays, but towards the middle of the segment these disappear and there is a wide cavity bounded by a large-celled tissue penetrated by hyphae.

Notheia is a small epiphyte about 12 cm. high, so far only found on *Hormosira* and *Xiphophora*. The terete thallus is branched monopodially and bears conceptacles throughout. The branching is peculiar in the fact that the laterals arise from the bases of the conceptacles which at the same time also harbour oogonia or antheridia. According to Barton the zygote gives rise to a septate filament which penetrates between the cells of the host and branches profusely; in the older plant the endophytic portion consists of a mass of elongate colourless cells. In segments of *Hormosira* occupied by *Notheia* the central hollow is filled by proliferation of the medulla.

The relation of these two genera to other Fucales is not at present clear.

Durvilleaceae

Durvillea, a striking member of the vegetation of the Subantarctic seas, presents many resemblances to one of the larger *Laminarias.* In *D. antarctica (D. utilis* Bory) the circular attaching disc *(a)* may be more than half a metre in diameter and bears several fronds which may reach a length of 10 metres. The relatively short stipe *(st),* which increases in thickness year by year and may become 7-5 cm. thick, is flattened at its upper end, where it passes over gradually into the leathery blade. The latter divides into numerous narrow whip-like segments, in which the central tissue breaks down with the formation of a number of chambers which become inflated with gases. *D. antarctica* is an inhabitant of the littoral zone, but favours localities where there is strong wave-action so that it is always covered with spray. Where the growth is luxuriant, the segments of the blades become densely tangled. *D. Harveyi* is considerably smaller and the undivided portion of the blade is more extensive.

Durvillea possesses neither an apical, nor an intercalary meristem and growth is diffuse. The segments of the blade are stated to arise as local outgrowths Despite the *Laminaria-like* habit, fertile conceptacles differing only in minor particulars from those of other Fucales are scattered over the whole thallus. On the whole, therefore, the resemblances to Fucales are greater than those to Laminariales.

Skottsberg's *Ascoseira,* referred to a distinct family, has a repeatedly forked stipe, each branch of which terminates in a long and rather narrow, leathery blade, bearing conceptacles on both surfaces except in the basal portion. The conceptacles contain chains of reproductive organs, each producing eight reproductive cells of unknown nature. The relation to other Fucales remains quite obscure.

Early Development, Apical Growth, and Branching

The early development of the embryo, which is known in many genera, commences with rapid division of the zygote to form a spherical or pear-shaped mass of cells attached to the substratum by more or less numerous unicellular rhizoids arising from the basal cells. Periclinal division sets in at an early stage. The first wall, dividing the zygote into an upper and a lower cell, is usually followed by a further horizontal division in the latter, whereby a small rhizoid-cell *(rh)* is cut out at the base. At the same time or soon afterwards vertical division into quadrants takes place in the upper (i) and subsequently in the underlying cell, while at an early stage the upper cell usually divides transversely into two tiers. The base of the zygote sometimes lengthens into a rhizoid before the rhizoid-cell is cut off, while in many Fucales the latter divides vertically into several cells before the outgrowth of rhizoids occurs; the number of cells thus produced is often considerable among Sargassaceae and varies in different species.

The orientation of the first wall is, however, also influenced by the presence of adjacent germinating zygotes (group orientation) and, where the latter are near together, light cannot overcome this stimulus. Du Buy and Olson attribute this to the presence of growth regulators which are specially abundant in the ova.

The rhizoids always grow out on the side away from the light. The rhizoids appearing on the less strongly illuminated side of the zygote, and their position is not determined by the actual direction of the incident light. If zygotes are successively illuminated from two directions perpendicular to one another, the rhizoids originate between the two and lie the nearer to the first direction of illumination, the longer the light from that source has acted. Zygotes *of Cystoseira* and *Pelvetia fastigiata* subjected to centrifugal force develop rhizoids centrifugally. On the other hand Rosenvinge could recognise no effect of gravity or contact. Whitaker stresses the importance of gradients of hydrogen-ion concentration and suggests that this may afford an explanation of group-orientation.

The early development of *Fucus* differs in certain respects from that of other genera. The side of the zygote away from the light grows out, usually prior to the appearance of any septa, into a rhizoid which subsequently divides to form a narrow multicellular strand. According to Nienburg the first wall in the zygote is followed by two other horizontal ones giving a row of four cells; the rhizoid is largely formed from the lowest cell, although the one above contributes something. The uppermost segment divides by perpendicular walls, followed by horizontal segmentation into two tiers, and then by anticlinal and periclinal walls leading to the formation of peripheral and central cells. At the same time the underlying primary segment has subdivided and, together with the products of the uppermost primary segment, gives rise to the body of the young plant. The prominent primary rhizoid is later supplemented by others growing out from the lower cells. A multicellular rhizoid has also been recorded in certain Japanese Fucales.

The subsequent development of the embryo probably always takes place in essentially the same way. Sooner or later there is abundant anticlinal division in the peripheral cells, whilst the inner ones enlarge without much segmentation. The embryo is now assuming an elongate form. Occasional periclinal divisions, whereby the number of central cells is increased, also take place in the peripheral layer *(me)* in the upper part of the embryo.

As a result, the latter widens, but the lower part remains narrow and constitutes a short stalk.

At this stage, amid the group of cells occupying the apex of the embryo of *Fucus*, one develops a specially conspicuous nucleus; this cell grows out into a first hair. As it lengthens, a meristem *(mh)* appears above the initial cell *(i)* which persists unaltered during the subsequent course of events. As the hair differentiates, the surrounding cells separate from it and initiate a funnel-shaped apical depression which rapidly widens, in part no doubt owing to active division of the neighbouring cells. By degrees other adjoining cells successively grow out into hairs giving the apical tuft mentioned. The first-formed hair soon disorganises, except for its basal cell *(b)* and from this, according to Nienburg, the apical cell of the young embryo is constituted. The origin of the apical cell below a trichothallic meristem recalls the condition met with in certain Ectocarpales. It is not at present known whether such an origin is traceable also in other Fucales. Hairs have not been observed at the apex of the young embryos of *Pelvetia* and *Ascophyllum* inter alia.

Whatever be its mode of origin in other Fucales, all who have studied the young plants record the appearance of a three-sided meristematic cell in the apical depression at a certain stage; in longitudinal section it appears more or less biconvex. Except in Fucaceae, this apical cell functions throughout the life of the plant. The segments cut off from its three faces undergo periclinal division, the outer cells becoming part of the peripheral layer, the inner ones dividing further to form the internal tissue of the embryo. For a time hairs with a basal meristem are produced from the peripheral cells in *Fucus* but, when they have been carried on to the edge of the apical depression as a result of the apical activity, they are usually shed. Production of hairs ceases before the apical cell has been long established.

Soon after the formation of the apical depression the greater part of the embryo of *Fucus* begins to flatten, only the basal third retaining a cylindrical form. In the flattened part a midrib soon appears, while cryptoblasts develop along the margins. At the same time the depression becomes a slit extended in the plane of flattening, the sides of the slit adjacent to the thallus-surfaces sloping steeply, whilst the marginal parts do so more gradually. It is at this stage that, in Fucaceae, the three-sided is replaced by a four-sided apical cell.

The latter has the shape of a truncated pyramid, the rectangular base of which is directed inwards and the somewhat convex apex outwards; the lateral faces are slightly arched and of unequal size, those parallel to the plane of flattening being narrower than those at right angles to it. The distinctive shape is specially well seen in *Pelvetia* and *Ascophyllum*. It is probable that this definitive apical cell arises from the three-sided one of the embryo by longitudinal division. It is well protected within the apical slit which is filled with mucilage.

In *Fucus* segments are cut off successively parallel to the base *(bs)*, to the two narrower faces *(ss* and then to one (sometimes two) of the other faces; this last segment *(ms)* is often as big as the remainder of the apical cell. In *Pelvetia* and *Ascophyllum* the segmentation is stated to be less regular. The basal segments divide to form cells of the medulla, while the others divide into inner and outer cells as before.

Dichotomous branching generally takes place in *Fucus* soon after flattening has become apparent. It is initiated by a segment (usually representing half the initial cell) from one of the lateral faces assuming the form and functions of an apical cell. According to Oltmanns the two apicals thus constituted cut off a considerable number of large segments, which for a time undergo little division so that the initials are at first difficult to distinguish, although later they stand out clearly. Each apical cell then proceeds to form one limb of the dichotomy, while the intervening segments produce a mound of tissue which divides the apical slit into two. The dichotomous branching of *Pelvetia* and *Ascophyllum* takes place in the same way.

The abundant monopodial branching of *Ascophyllum* is initiated close to the apex. Longitudinal sections through the apical slit parallel to the surface of the thallus disclose, on either side of the main apical cell *(a)*, a number of initials distinguished by denser cytoplasmic contents; they are derived from segments of the apical cell which have already undergone some further subdivision. Each such initial becomes sunk within a depression and, as further segments are cut off from the main apical cell, the initial is gradually carried over the margin of the apical slit on to one of the edges of the thallus. Here the depression assumes the form of a narrow slit drawn out parallel to the long axis of the thallus; in longitudinal sections it is often distinctly saccate at the base. Meanwhile, as the marginal slit lengthens, the initials multiply, and it commonly becomes divided into several bays, with an apical cell at the bottom of each. At the base of the germling the primary lateral initial often grows out into a branch without forming others, but on the more mature parts twelve or more initials may arise within a single slit. At first only one of these grows out, but others soon follow suit giving the familiar tufts of branchlets. The presence of numerous apical cells within the marginal slits admits of almost unlimited regeneration of laterals.

Seirococcus shows similar features. The slit harbouring the four-sided apical cell *(a)* lies at the base of a rounded depression, covered in by the rudiments of the leafy laterals. The segments of the apical cell, situated between the initials, here divide to form mounds from which the leafy laterals are produced. As in *Ascophyllum,* the initials situated in the intervening bays increase in number and in part come to occupy the adaxial edge of the lateral; these secondary initials lie in depressions separated by small mounds which are clearly equivalent to the leafy laterals. Only few of these initials develop into branches of unlimited growth; most give rise to the fertile branches *(r)* which are of different ages.

The leaf-like laterals, which constitute the chief difference between *Seirococcus* and *Ascophyllum,* are in the latter represented by the inconspicuous mounds separating the initials; were these to grow out in the same way, the resemblance to *Seirococcus* would be very marked. In the latter, neither the leafy laterals, nor the small outgrowths between the fertile branches, possess an apical cell.

The change from the three- to the four-sided apical cell in Fucaceae is evidently associated with the prevalent flattening and bilateral symmetry. In Cystoseiraceae and Sargassaceae, with a marked tendency towards radial organisation, the three-sided cell of the germling persists throughout life. In most of them the apical depression is a deep funnel filled with mucilage, but in *Halidrys siliquosa* it takes the form of a deep slit, perpendicular to the plane of flattening, and this is also essentially so in *Himanthalia*.

The lateral initials are produced in the same way as in *Ascophyllum*. In *Halidrys* they arise alternately and with some regularity from certain segments, which undergo little division until the definitive three-sided apical cell is cut out within them by oblique walls. When the initials have passed out of the apical pit, each becomes lodged in a separate depression and gives rise either to a long or a short branch; when the latter has completed its growth, the initial divides into a group of cells and the apical depression disappears. Except for the differences in the early development, matters are much the same in *Cystoseira*, once the apical cell of the axis is established; in the radially branched species, however, the laterals are laid down on a 2/5 spiral.

The formation of the lateral branch-systems of *Sargassum* commences in the usual way. In *S. linifolium* the apical depression becomes drawn out in the direction of the future branch and, when several branches arise at about the same level, it may appear star-shaped when seen from above. Soon after the initial has passed out of the apical pit, it cuts off a secondary initial on its distal side; the depression harbouring the two initials assumes the form of a radial slit, and they become separated by an intervening mound of tissue *(d)*. After a short time the distal initial divides more rapidly than the other and develops into the subtending "leaf" whilst the proximal one *(bi)* comes to lie more or less definitely in its axil. The "leaf" *(b2)* is clearly a lateral which has pushed aside the parent axis *(bi)*. The apical cell of the latter soon produces further initials, which are arranged on a 2/5 spiral and give rise to the other branches of the system, although the "leaf" does not form part of this spiral, in this respect behaving like a prophyll. It would be well if the development of the lateral branch-systems were studied in other species of *Sargassum*.

The segments of the three-sided apical cell of *Himanthalia* divide to form the tissue of the cup, whilst the receptacles represent one branch of a dichotomy. The cup can perhaps be compared to the main axes of a *Cystoseira* and the receptacles to the successively formed laterals.

The four (rarely three) apical cells, characteristic of *Hormosira,* are readily recognisable in a transverse section through the apex. They are separated by four thicker walls which are approximately perpendicular to one another and can be traced both in the peripheral direction and backwards into the substance of the thallus, where they terminate in the four bays of the central hollow. The apicals are situated at the base of a shallow groove. Preparatory to branching the number of apical cells is increased by longitudinal division. In *Notheia* transverse sections of the apex show three apical cells which project slightly; for the details of branching. Little is known of the early development of these two genera.

Anatomical Structure

Although the anatomy of the mature thallus has been examined in a considerable number of Fucales, certain details still remain obscure. Investigation of fresh material and of material fixed and stained by new methods would probably shed light on obscure points. There are appreciable resemblances to the Laminariales, although the Fucales have not attained to the degree of anatomical specialisation that is evident in some of the former. The cells normally contain but a single nucleus which shows some differences in structure in the different families.

As in Laminariales, the thallus is built up from two sources, viz. by the primary meristem (here usually an apical cell) and by the activity of the surface-layer of cells (meristoderm). The more definite character of the former renders the elucidation of its role in the formation of the thallus easier than in Laminariales. The first medullary elements arise from the inner cells of the embryo before the establishment of an apical cell. Subsequently the medulla is increased by contributions from the inner halves of the segments of the latter; also from the basal segments in the Fucaceae, while the outer halves of these segments add progressively to the meristoderm (*me*). The early separation of the medullary cells (*m*), as a result of gelatinisation of the middle lamellae of the longitudinal walls, is even more marked in the elements contributed from the apical cell. Here and there, however, the walls remain thin and constitute pits *(p)*. Little mucilage appears to be formed in the medulla in Cystoseiraceae.

Soon after the differentiation of the apical cell, the inner parts of its segments divide a number of times without gelatinisation of the walls, and thus there arises a cortex of relatively large, but closely arranged cells. Many authorities report tangential division of the surface-layer of cells with resulting additions to the cortex, and the gradual transformation of the innermost cells of the latter into medullary elements, but the degree of importance of these processes as compared with the contribution from the apical meristem is not clear. In the stipe of *Durvillea* there is evidently active division of the meristoderm. The general resemblance to the mode of origin of the different tissues of a *Laminaria* is obvious, and another point of similarity is that, despite the extensive gelatinisation of longitudinal walls, especially in the inner tissues, the transverse walls remain unaffected.

The cells of the peripheral layer at first multiply by anticlinal (often crosswise) division. This active surface growth, as in Laminariales, leads to passive stretching of the inner elements which, particularly in the flattened parts of the thalli, tend to be dragged out of their longitudinal course so that their arrangement often becomes very irregular, nowhere more evident than in the receptacles of *Fucus*, etc. Tissue-tensions, similar to those of Laminariales, are reported. In rapidly lengthening structures (e.g. the receptacles of *Himanthaliadwtytiit)* elongation of the medullary cells may be accompanied by frequent horizontal division.

Fucales

The surface is covered by a mucilaginous "cuticle" which is sometimes thick and stratified, as at the apices of the fronds of *Halidrys*. The peripheral cells, especially in the younger parts, have a palisade-like shape, and the hypodermal layer may exhibit a similar form. The surface-cells are usually crowded with chromatophores and abundant fucosan-vesicles, although according to Wille there are few chromatophores in the outermost layer of the mature receptacle of *Himanthalia* which he suggests acts as a light-screen. In surface-view the cells of the peripheral layer are usually separated by thin walls, which sometimes show undulation although this may be an artefact. The cells are connected with the internal elements by pits.

As in Laminariales, the cortical cells undergo little further division and gradually assume an elongate form; towards the interior the longitudinal walls usually gelatinise so that there is commonly an imperceptible gradation to the medulla. There is evidence of the formation of cross-connections by extension of the pits left in these longitudinal walls, but the special mode of formation of such connections, described for Laminariales, has not been observed in Fucales. That cross-connections develop in some quantity is shown by the network of medullary elements found in the wings of the thallus of a *Fucus*, for instance.

The cortical cells often contain much fucosan and have been regarded as fulfilling a storage function, while their frequent strong thickening indicates also a mechanical rdle; Henckel regards the thickening as subserving water-storage in *Pelvetia canaliculata*. Simons describes the central tissue of the axis of *Sargassum filipendula* as consisting of narrow elongate cells, the inner with thin, the outer with thick walls; in the leaves only the former are present.

The extent of production *of* hyphae varies greatly in different Fucales. In *Bifurcaria*, *Pelvetia*, and *Notheia* they are confined to the basal parts of the thallus, while a considerable number of the published figures of other genera afford no evidence of their presence. Except near the attaching organs, they seem to be less prominently developed in Cystoseiraceae and Sargassaceae than in Fucaceae. As in Laminariales, the hyphae usually arise from the lower ends of the cells at the periphery of the medulla.

In *Fucus* the numerous hyphae of the midrib are produced only at a considerable distance behind the apex; as the midrib thickens, more and more develop so that, in sections of older parts, the original medullary cells *(m)* are embedded amid copious hyphae *(h)* which in the main run longitudinally. There are far fewer hyphae in the wings and these follow an irregular course, which adds to the confused picture presented by the medulla. These structures develop later in *Halidrys* than in *Fucus,* but they become very abundant at the periphery of the medulla in the older parts. The *Ascophyllum*-thallus possesses only scanty hyphae, although they are formed in considerable numbers near the slits from which the laterals emerge and also appear abundantly in the base of the lateral itself; these hyphae penetrate into the axis and intertwine with those of the latter, whereby a firm attachment is secured. The hyphae in the receptacle of *Himanthalia* run prevalently in a radial direction.

Although probably similar in origin, two kinds of hyphae have to be distinguished, viz. those with thin and those with strongly thickened walls; there are numerous transitions. Those of the former type, which may contain chromatophores and possibly play a conducting role, are widely distributed in the flattened parts while those of the latter type, which are no doubt essentially mechanical (Wille's "Verstarkungshyphen"), are to be met with in midribs and other more or less cylindrical structures, especially in the older parts of the plant; they also occur in the narrow basal part of the receptacle of *Himanthalia*. Where such strengthening hyphae are lacking, the cells of the inner cortex evidently satisfy the mechanical requirements. The mechanical system may thus either be located towards the periphery or occupy a more or less central position, the latter arrangement evidently occurring in the main in the more permanent organs or in parts exposed to special strains (base of receptacle of *Himanthalia).*

Hypha-development is most marked in the neighbourhood of the *attaching discs,* which finally consist of a mass of branching and impenetrably interwoven hyphae. The discs of young plants are formed partly by the outgrowth of peripheral cells and partly by the development, from the internal cells of the stalk of the embryo, of hyphae which penetrate to the outside. Further hyphae continually grow over the surface and extend the area of the disc, whilst increase in thickness is brought about by the production of new hyphae in the interior. The stalk of *Fucus* also gradually becomes overgrown with a layer of hyphae, but its marked thickness in older plants, as also that of the axes of *Cystoseira* for example, is only in part due to this superficial felt and to the internal production of hyphae. In the older regions there is marked *secondary meristematic activity* on the part of one or more layers of the cortex, the original surface-layer dying and being gradually shed. There is at present no information whether such secondary activity occurs also in other genera.

Sieve-like perforations traversed by cytoplasmic connections have been recorded in the pit-membranes, on the thin horizontal walls of the cortical and medullary cells, as well as on the septa of their cross-connections and of the conducting hyphae, by Wille and others *(ds)*. Other investigators have failed to find such perforations, although the contracted protoplasts commonly show a tendency to adhere to the pit-membranes. Callus is rarely reported. Certain workers have described continuity of the cytoplasm through large openings in the walls of the medullary and cortical cells, but this requires verification. According to Hansteen the medullary cells in *Sargassum* show scalariform thickenings on their walls.

The *air-bladders* originate not far from the apex as a result of surface growth, accompanied by increase in thickness of the cortex; this leads to rupture of the medulla, remnants of which are commonly found around the edge of the central hollow'. In *Fucus vesiculosus* and *Ascophyllum (289)* the inner surface of the bladder is occupied by a felt of short unbranched threads containing chromatophores, while in *Cystoseira* the cavity is traversed by one or more longitudinal strands of medullary tissue. In *Cystoseira* and in other Fucales the walls of the vesicles show varying degrees of tissue-tension, while they are stated to be thicker in individuals of *Ascophyllum nodosum* growing in deeper water. The diaphragms in *Halidrys* consist of isodiametric cells and are connected by longitudinal strands of fibre-like cells apposed to the wall of the bladder. Structures simulating air-bladders, found occasionally on the fronds of certain species of *Fucus*, have been studied by Richard.

It has been stated that there is no carbon dioxide within the bladders (us), but according to Colla they contain the same gases as the atmosphere, carbon dioxide being at a maximum and oxygen at a minimum during the night. Those of *Ascophyllum nodosum* are stated to contain a higher proportion of oxygen, the amount varying seasonally, as well as from day to day. Wille and Colla believe that the bladders have a respiratory significance and that the oxygen is that formed in photosynthesis, while others see in them merely a mechanism for decreasing specific gravity.

Wound-healing and Adventitious Shoots

The surfaces exposed as a result of the wearing away of the wings of *Fucus* become covered with a secondary cortex, the formation of which commences at the edge of the wound. On the other hand, when the inner tissues are laid bare by the breaking off or devouring of part of a branch, as happens especially in the Fucales of the littoral zone, the exposed medullary cells alone participate in the healing of the wound. The outermost divide actively by radial and tangential longitudinal walls to produce a new layer that in certain respects recalls the callus of higher plants. At first this layer of dividing medullary cells is discontinuous, consisting of small mounds separated by undivided cells, and in *Pelvetia* complete closure never seems to result. In *Fucus vesiculosus*, however, medullary cells below the wound also become active and, pushing between the surface-cells, divide like the latter so that the surface becomes completely covered. Wounding is nearly always followed by the formation of *adventitious branches* which are often produced in some numbers from the small mounds formed by the medullary cells, a phenomenon which is particularly conspicuous in *Fucus vesiculosus* and *Pelvetia canaliculata*.

In *F. vesiculosus*, the formation of adventitious growths is mostly confined to regions containing hyphae (basal disc, stalk, midrib), although, except in the basal disc, these do not actively cooperate in their development. The shoots arising from the basal disc are formed by division of internal hyphal cells distinguished by denser cytoplasm, and the product breaks through the overlying tissue so that the development is endogenous. Numerous adventitious shoots commonly originate on the stalk from the regions where the wings have worn away. These diverse growths altogether resemble the young germlings, but few show any considerable development. They are not so common in *F. serratus*, but are frequently seen in *Pelvetia*, although here apparently not formed on the attaching organ. The adventitious shoots of *Cystoseira* have already been noticed. Herriott records them on all parts of the stipe and frond of *Durvillea*, where they apparently develop endogenously.

Tobler states that the bladders of *Ascophyllum* frequently contain Mytilidae (and Turbellarians) which enter as larvae through holes caused by Crustaceans. The interior of such bladders is lined by a compact cortex produced by division of the cells on the outside of the threads occupying the inner surface of the normal vesicle.

Development of the Conceptacles

The differentiation of conceptacles commences within the apical depression in the immediate vicinity of the apical cell, although they are carried out of the depression before development is complete. Bower first showed that each conceptacle originates from a single initial cell; further details were elucidated by Oltmanns and especially by Nienburg to whom our present knowledge is largely due. Only a limited number of species have so far been studied. The considerable variations in the details of development fall into line with the classification of Fucales here adopted. Sterile and fertile conceptacles show the same early development.

The conceptacle nearly always originates from a superficial cell, which is distinguished by its larger size and a slower rate of division. In *Cystoseira, Sargassum* and *Carpophyllum* the initial is flask-shaped and provided with a large nucleus. This divides unequally and the products become separated by a markedly curved septum; the characteristically shaped upper cell *(t)* may with Simons be styled the *tongue-cell*. The lower cell *(basal cell, b)* then segments by two perpendicular walls. Meanwhile division of the surrounding cells leads to a gradual sinking of the products of the initial and to separation of the tongue-cell from the adjoining tissue. In *Sargassum* and *Carpophyllum* the tongue-cell merely elongates or sometimes divides to form a row of a few cells, whilst in *Cystoseira* it develops into a multicellular hair with a basal meristem. The inner lining of the conceptacle is almost entirely formed by vertical division of the lower cell *(b)* of the rudiment, which results in a gradual widening of the cavity; only the uppermost part is produced from the adjacent segments. Development of the conceptacles takes place in essentially the same way in *Halidrys*.

In *Fucus* and its allies the initial soon becomes lodged at the base of a deep and narrow cavity owing to rapid growth and division of the surrounding cells. The septum between tongue and basal cells is straight and, after segmentation of the latter by vertical walls, the resulting cells divide transversely to form two layers; by this time the rudiment is already deeply sunk. The further division of the products of the basal cell provides only the tissue occupying the floor of the conceptacle, the greater part of the lining being derived from the adjacent cells. The tongue-cell commonly exhibits no further development and sooner or later collapses, but in *F. spiralis* it sometimes divides to form a hair. Heine records much the same series of events in *Xiphophora*.

A somewhat different development is found in *Pelvetia fastigiata*, for here the initial first segments longitudinally and this is followed by transverse division of the resulting cells. The two basal cells thus produced segment rather irregularly to form the floor of the conceptacle, although the greater part of the wall again arises from the adjacent cells; the two upper cells *(t)* develop into short hairs. According to Nienburg the first division is not always vertical, and this irregular segmentation is the rule in *Ascophyllum* where, despite occasional vertical division, the initial usually gives rise to a mound of cells occupying the floor of the conceptacle; in other respects, however, the development is like that of other Fucaceae.

The rather scanty data indicate that, in Cystoseiraceae and Sargassaceae, a considerable part of the wall of the conceptacle is formed from the basal segment of the initial cell, whilst in Fucaceae this gives rise only to the floor. The characteristic tongue-cell, with its rather marked degree of persistence, is moreover found only in the former, i.e. in the forms with a three-sided apical cell. An intermediate type is furnished by *Bifurcaria*, where the tongue-cell soon collapses and a relatively large part of the conceptacle-wall is formed by the adjacent cells. In both respects there is resemblance to *Fucus serratus*.

In *Himanthalia* the surface-cells within the depression at the summit of the receptacle soon grow out into hairs with a basal meristem. As this adds to the hair, the tip of the latter progressively disorganises and, when the hair has been carried out of the depression, only the sunken basal cell *(b)* still persists. Certain of these basal cells function as conceptacular initials, dividing first by a vertical and then by transverse walls. The central cells thus produced do not grow as actively as the marginal ones, which results in the formation of the cavity of the conceptacle, the entire lining of which arises from the products of the initial cell. The conceptacles of *Bifurcariopsis* perhaps develop in much the same way.

The development of the conceptacle of *Himanthalia* shows that the tongue-cells of Cystoseiraceae and Sargassaceae, which are no doubt homologous with the upper cell of the rudiment in the Fucaceae, represent vestigial hairs which are suppressed to a varying degree. This means that the conceptacles take their origin to a more or less appreciable extent from the basal cell of a hair, and it is profitable in this connection to recall the perfectly similar derivation of the apical cell in the young embryo of *Fucus*. The initial cells of the conceptacles may, in fact, be regarded as the equivalents of branch-initials which, like the latter, become lodged in a depression (the conceptacle), although diverted to a special purpose.

From this point of view the development of the conceptacle of *Notheia* is of special interest. After transverse division of the initial the upper cell develops into a hair *(k)*, while the basal cell *(b)* gradually separates from the adjoining ones, its products apparently forming a large part of the lining of the conceptacle. As the latter enlarges, further hairs arise from the wall and form a projecting tuft. Later the upper parts of the hairs are shed, although their basal cells *(be)*, which have rich cyto-plasmic contents, persist. There is some uncertainty as to the exact mode of development of the laterals which arise from the conceptacles; Gruber believed that they were formed from the basal cells of the hairs. Several laterals may develop, although only one lengthens into a branch. A similar development of (adventitious) branches from cryptoblasts has been recorded in a salt-marsh form of *Fucus ceranoides*.

The association of reproductive organs with groups of hairs is frequent in the more specialised Phaeophyceae. In the Encoeliaceae these hairs commonly occupy the base of a depression, and it has been suggested that this may show the mode of origin of the conceptacles of Fucales. The resemblance is, however, remote, and the little that is known as to the origin of these structures in Encoeliaceae implies an altogether different mode of development. In view of the probable absorptive functions of the hairs, their frequent association with reproductive organs requiring a localised food-supply may be but the outcome of a physiological need. On the other hand, the presence in young plants of many Browa Algae of an apical hair, at the base of which lies the meristem or at the base of which it subsequently develops, shows that in Phaeophyceae a hair frequently precedes or is associated with meristematic activity, and this leads to the same conclusion as that arrived at above, viz. that the conceptacle is to be regarded as an arrested and specially modified branch or branch-system.

Structure of the Mature Conceptacles

All Fucales produce their sex organs within conceptacles and in many *(Pelvetia canaliculata, Ascophyllum, Halidrys, Bifurcaria, Himanthalid)* fertile conceptacles are alone present. In *Fucus, Sargassum turbinaria,* and others, cryptoblasts also occur scattered over the thallus. In *Sargasswn filipendula* and various species of *Cystoseira* they even occur on the receptacles among the fertile conceptacles.

The lining of the conceptacles consists of a number of layers of flat cells formed by tangential division of the primary cells and containing plentiful chromatophores. This wall separates the cavity from the loose medullary tissue. In the cryptoblasts the middle of the floor is occupied by an often extensive group of colourless unbranched hairs with a basal meristem *(m)*, similar to those found in other Phaeophyceae; they project in a tuft from the aperture, and in older cryptoblasts only the basal portions usually persist. Intermingled with them, and appearing in increasing numbers in later stages, are other shorter pigmented hairs with diffuse growth; at first one-celled and more or less clavate, they later become multicellular and produce more or less numerous, usually unicellular branches. Their cells have dense granular contents and numerous small chromatophores. According to Nordhausen they are largely responsible for producing the mucilage that fills the conceptacles.

The development of cryptoblasts is to some extent influenced by habitat conditions. They are specially abundant in those species of *Fucus* which are capable of withstanding prolonged drought and are particularly prominent in the salt-marsh forms of *F. vesiculosus,* where they tend to occupy a marginal position.

These have been specially studied in species of *Fucus,* but are no doubt of wide distribution.

According to Gardner *F.furcatus* is distinguished by having altogether closed cryptoblasts (caecostomata) devoid of hairs. The conceptacles of *Cystophyllum sisymbrioides* are likewise without an aperture, the long hairs pushing their way through the superficial layer. In certain *Cystoseiras* (e.g. *C. Myricd)* the narrow cryptoblasts occur singly within small spinous branchlets, occupying their entire length. Henckel describes a *Cystoseira* lacking cryptoblasts and with hairs arising directly from the surface.

In the fertile conceptacles of *Fucus,* and probably also of other genera, the colourless hairs are restricted to a small area near the aperture, from which they project markedly in *F. spiralis.* The greater part of the wall bears branched hairs *(paraphyses* H, in which the septa are often oblique and the component cells barrel-shaped with scanty chromatophores. In the male conceptacles these structures bear the antheridia and are usually much more richly branched than in the female conceptacles, where indeed they sometimes remain unbranched. Between the paraphyses and the colourless hairs are others with abundant chromatophores, which correspond to the branched hairs of the cryptoblasts and exhibit transitions to the paraphyses. In *Himanthalia* structures of the third type are not recorded. Only the hairs near the aperture take up stains readily.

The apparent identity in the early development of cryptoblasts and fertile conceptacles, quite apart from other similarities indicated by the above data, implies that the two are homologous. In *Sargassum* antheridia in various stages of abortion have been found in the cryptoblasts. The young conceptacles of *Notheia* possess the basal tuft of hairs characteristic of a cryptoblast, although these later disappear. In various species of *Cystoseira* (*C. abrotanifolia*, *C. foeniculacea*), also, long hairs resembling those of a cryptoblast occupy the centre of-the fertile conceptacles, the sex organs arising between them and the aperture. It may, therefore, be concluded that in forms possessing cryptoblasts the whole thallus once bore fertile conceptacles and that the present-day relegation of the latter to special receptacles is a secondary feature. The Australasian *Myriodesma*, with a habit altogether like that of *Fucus*, in fact bears fertile conceptacles over the whole thallus.

Sexual Reproduction

The distribution of sex organs varies considerably in different genera and even in one and the same species. Several are dioecious (many species of *Fucus* and *Sargassum; Ascophyllum; Himanthalia; Halidrys dioica\Coccophora; Hormosira\ Durvillea*, although the monoecious condition is perhaps the more usual. In many monoecious Fucales *(Pelvetia, Halidrys siliquosa, Bifurcaria tuberculata, Bifurcariopsis, Carpoglossum, Platylobium* the two sexes are found in the same conceptacle, but in others (species of *Sargassum, Phyllospora, Seirococcus, Scytothalia)* they occur in distinct conceptacles although on the same receptacle.

In the *Cystoseiras* above referred to, in which the conceptacles possess a basal tuft of hairs, the same receptacle harbours male, female, and hermaphrodite conceptacles. A similar condition occurs in some species of *Sargassum*, although many (e.g. *S. Horneri)* are dioecious. In this genus differences are sometimes apparent between the plants of the two sexes or more commonly between the two types of receptacles, the male being smooth, the female spinous, etc.; in *S. Horneri* the male receptacles are longer and more slender. Differences between male and female plants are also recorded in *Halidrys dioica*. A few Fucales vary between monoecism and dioecism.

In hermaphrodite conceptacles the oogonia and antheridial hairs are generally intermingled, but in *Bifurcaria tuberculata* and in many species of *Cystoseira* the oogonia occupy the base and the antheridial hairs the sides of the conceptacles; in the former the oogonia develop later than the antheridia so that the young conceptacle appears to be male.

In *Cystoseira montagnei* and C. *opuntioides* Sauvageau records two periods of fertility, conceptacles first appearing on the young branches and later, after these have grown to their full length, in terminal receptacles.

The oogonia are nearly always borne directly on the wall of the conceptacle and are usually provided with a definite stalk-cell which is either embedded in the wall (*Cystoseira*) or more usually stands above it a stalk-cell is lacking in *Pelvetia canaliculata.* In *Sargassum* and *Carpophyllum* the oogonia (0) are themselves to a large extent embedded. A different arrangement obtains in *Durvillea,* where the female organs are usually borne on branched hairs, although sometimes also arising directly from the wall. The occurrence on hairs is of interest, since it furnishes a parallel to the usual arrangement of the antheridia in Fucales.

The young oogonium, after the formation of the stalk-cell, is uninucleate, devoid of vacuoles. During its ralluration there are probably always three successive nuclear divisions. The first two follow rapidly upon one another and, at their conclusion, the four nuclei show a tetrahedral grouping; after this, there is normally a period of rest before the third division takes place. Strasburger, as well as Farmer and Williams, proved that reduction in chromosome-number occurs during the first nuclear divisions, while Yamanouchi subsequently showed, both for the oogonium and antheridium of *Fucus,* that these divisions are meiotic; this has since been established also in diverse other genera. Centrosomes have usually been observed during the nuclear divisions in the oogonia and antheridia.

It is only in *Fucus* and *Notheia* that the cytoplasm undergoes cleavage into eight parts and that all the eight nuclei are used in the production of ova, which in *Fucus* exhibit some diversity in arrangement. In the remaining Fucales only some of the nuclei are employed in the organisation of ova, the supernumerary ones being extruded, sometimes together with a small quantity of cytoplasm. In *Ascophyllum, Xiphophora, Bifurcariopsis, Hormosira,* and *Durvillea* there are four, in *Pelvetia* two, and in all other Fucales only one functional ovum. It is clear, however, that the number of ova is not altogether rigid and diverse variants have been reported. In *Phyllospora* the single ovum is so large as to be visible to the naked eye.

In the mature oogonia of *Fucus, Xiphophora* and *Hormosira* the ova are separated by delicate septa, connected with the innermost layer of the oogonium-wall, although neither Oltmanns nor Yamanouchi mention this feature. The presence of such septa would imply that the oogonium is plurilocular in nature. The two ova of *Pelvetia canaliculata i* re separated by a thick gelatinous septum, although this is not so in *P.fastigiata,* where the ova show a collateral or oblique arrangement.

The production of four or eight ova is largely confined to the less specialised members of Fucales, and an oogonium producing eight female cells no doubt represents the primitive condition. It is of interest that the aberrant genera *Notheia* and *Hormosira* exhibit relatively primitive features in this respect. The supernumerary nuclei, representing aborted ova, sometimes show a considerable degree of persistence. In *Cystoseira* they may attract spermatozoids, and actual fusion between a spermatozoid and one of these nuclei has been observed, although the fate of the product could not be followed.

The unused nuclei take up diverse positions. In *Pelvetia* they lie in the equatorial region between the two ova, while in *Ascophyllum* they form a central group. Those of *Cystoseira* float in the colourless liquid that Occupies the space between the female cell and the oogonium-wall. In *Sargassum* the ova contain eight peripheral nuclei at the time of discharge, seven undergoing disintegration during fertilization; much the same is true of *Coccophora.* The single functional nucleus in *Hesperophycus* and *Pelvetiopsis* undergoes considerable enlargement, while the remaining seven are cut off at the base of the oogonium, together with a small amount of cytoplasm which is liberated together with the ovum; the further fate of this cytoplasmic mass is not known.

Application of pressure or subjection to higher temperatures may cause a number of the ova of *Fucus* to fuse; the resulting mulfinucleate eggs are apparently capable of fertilization and sometimes develop into (polyploid) germlings. By addition of potassium chloride to the sea-water, Tahara caused persistence of all eight nuclei in *Sargassum.* If such ova are replaced in normal sea-water, all the nuclei are utilised in segmentation, though some of the germlings are abnormal.

The septum between stalk and oogoniur$_{n}$ remains thin, but the rest of the wall of the latter thickens and becomes differentiated into several layers of which there are usually three. Of these the innermost (endochi te, *en)* and outermost (exochite, *ex)* are thin, whilst the intervening (one (mesochite, *m)* is thick and separated from the exochite by a space *(g).* When the oogonium of *Fucus* is mature, the exochite ruptures and the contents, still enveloped by the two inner layers, are set. The packets of ova are extruded from the aperture of the conceptacle in a quantity of mucilage; five layers are not stated to be distinguishable.

In the sea the inner part of the mesochite gelatinises and, as it swells, it ruptures apically and, slipping backwards exposes the ova within the endochite; rapid solution of the latter and of the intervening septa leads to the liberation of the naked female cells. The ovum of *Himanthalia* escapes through a small lateral rupture in the enveloping membranes.

In diverse Fucales the ova remain enclosed in part of the oogonial wall for a more or less prolonged period. A familiar example is furnished by *Pelvetia canaliculata,* where the two inner layers persist around the eggs until relatively late stages of embryo-development. The mesochite j& here very thick and mucilaginous, consisting of an structureless layer and an inner stratified one interrupted at the equator; in this region the mucilage seems to be specially soft so it allows of the penetration of the sperms. The thick persistent envelopes are generally regarded as an adaptation to the extreme exposure to which this species of *Pelvetia* is subjected. Amongs$_{t}$ the more specialised

Fucales the liberated ovum often remains enclosed in the mucilaginous oogonial membranes, which form a kind of stalk adhering to the receptacle; the female cell is thus retained during fertilization and the early stages of embryology.

In certain *Cystoseiras* the ova are liberated in a naked condition and are discharged successively and with considerable force from the conceptacle; the receptacles commonly mature from below upwards. In others the ova remain enclosed in the meso- and endochite and the former swells to form a long tube by means of which the ova are held on the outer surface of the receptacle. The same appears to occur in *Coccophora*.

The oogonia of *Cystophyllum sisymbrioides*, and perhaps also of other species of the genus, are caught in the tangle formed by the projecting mucilaginous hairs of the conceptacle. In several of the investigated species of *Sargassum* and *Cystophyllwn*, as well as in *Bifurcaria brassiciformis*, the extruded oogonia remain for some days firmly attached to the inside of the conceptacle by means of long gelatinous stalks which, in *Sargassum Horneri* and *Bifurcaria* are formed by the thickened apex of the oogonial wall.

The lengthening distal end of the latter curves back and becomes fixed to the inside of the conceptacle so that, as the stalk straightens, the oogonium is pushed out of the aperture in an inverted position. Much the same is reported for *Carpophyllum*. In *Marginariella* the swelling mesochite forms a long tube, through which the contents of the oogonium pass before the three nuclear divisions occur and to the distal end of which the ovum remains attached. In *Bifurcariopsis* the four ova are extruded separately on mucilage stalks formed from the mesochite.

The *antheridia* usually occur in large numbers on the richly branched paraphyses of the male conceptacles, very commonly occupying more especially the lower branches. A study of the development of these hairs, however, shows that the male organs actually occupy much the same position as the oogdnia. The first antheridium arises from a cell of the wall, the resulting papilla later dividing into a stalk-cell and the antheridium proper *(a)*. The former usually produces a branch, which pushes the antheridium to one side and itself gives rise to a second antheridium. Since this series of events is generally repeated indefinitely, richly branched paraphyses like those of *Fucus* or *Pelvetia* result. In the final stages a number of sterile branches are produced.

Such richly branched structures are not, however, formed in all Fucales. Thus, in *Pelvetia fastigiata* and *Notheia* the antheridia usually arise directly from the wall of the conceptacle, and this is sometimes also so in *Cystoseira, Phyllospora* and *Durvillea*; in *Turbinaria*, too, the hairs are stated rarely to bear more than two or three antheridia.

The antheridial initial of *Fucus* has a large nucleus with a prominent nucleolus, a number of yellow-green chromatophores, and alveolar cytoplasm. The first nuclear divisions, bringing about reduction, are succeeded by four further simultaneous mitoses, accompanied by multiplication of chromatophores which, however, no longer contain any pigment at the 8-nucleate stage. When 32 nuclei have been formed, the cytoplasm becomes divided by finely granular lamellae, similar to those observed in the maturing unilocular sporangia of Ectocarpales.

The last nuclear division results in the formation of 64 male cells. At this stage, according to Kylin, the somewhat enlarged chromatophores, of which there is ordinarily one to each nucleus, again develop an orange-yellow or yellow pigment, which is stated to appear on one side of the plastid. Numerous fucosan-vesicles and oil-drops are present.

In the Fucaceae, and probably also in certain other Fucales, the wall of the mature antheridium is two-layered. The male cells, still enclosed in the inner layer, are extruded from the opening of the conceptacle as oblong packets embedded in mucilage. In the sea the enveloping membrane gelatinises at one, or rarely at both, ends to liberate the minute spermatozoids. Detachment of the entire antheridium takes place in *Halidrys*, *Cystoseira*, *Phyllospora*, and others, the spermatozoids being liberated within the conceptacle or in the sea; even when the former obtains, however, they probably pass out in a group.

The spermatozoids are generally pear-shaped, with the anterior end pointed and with the longer flagellum directed posteriorly. Their detailed structure has been much disputed. Behrens and Strasburger distinguished a large nucleus with a thin envelope of cytoplasm, while Guignard concluded that the bulk of the body consisted of cytoplasm with a relatively small posterior nucleus. More recent investigations by Kylin and others have, on the whole, supported Strasburger's view that the greater part of the body consists of nucleus.

The eyespots of the spermatozoids are formed from the chromatophores, which reacquire pigment during the final stages of antheridial development; they are stated to contain only carotene, possibly with some xanthophyll. In *Pelvetia* and certain species of *Cystoseira* an eyespot is lacking. Retzius first observed other inclusions in the cytoplasm of the mature sperm, which he suggested were mistaken by Guignard for the nucleus.

They commonly occur singly, although sometimes there are several, and are believed to be comparable to the plastomeres frequently found in animal spermatozoa. The flagella arise from a blepharoplast which Meves regards as homologous with a centrosome. The longer posterior flagellum extends backwards within the body as far as the eyespot, where it passes to the outside. According to Retzius the flagella taper at their ends.

The mechanism of liberation of the sexual products is still not clear. Their escape can be observed if ripe receptacles are placed in a damp chamber. In the littoral Fucaceae mucilage, containing packets of ova and spermatozoids, is often seen at the apertures of the conceptacles during low water; in male plants these masses are commonly yellow in colour. With the advent of the next tide the gametes are freed and fertilization can take place. It is tempting to suppose that in such forms the contraction of the thallus during low water plays some part in squeezing the packets out of the conceptacles, but this is in all probability not so. Extrusion frequently takes place during continuous submergence and Schreiber observed the most marked liberation of sexual packets in *Fucus serratus* after the thallus had been under water. Oltmanns states that it ceases after some time in drying plants, but recommences vigorously when they are returned to sea-water. Baker concluded that the species growing at higher levels on the shore require a longer period of exposure for a maximum liberation of sexual products.

Oltmanns suggests that the most important factors causing liberation are increased turgor of the cells of the wall and the pressure of the mucilage within the conceptacle. This is in agreement with the fact that more marked extrusion is observable in exposed plants after showers of rain, a phenomenon which can be reproduced artificially in the laboratory. Increased turgor in the cells of the wall might result after inundation owing to concentration of the sap during exposure. In the many Fucales growing permanently submerged *(Cystoseira, Halidrys,* etc.) such external factors cannot come into play, although the rise and fall of the tide will involve alterations in light-intensity and in hydrostatic pressure that may affect turgor. Extrusion in *Cystoseira* has in fact been found to be accelerated by illumination and in nature takes place during the early morning hours.

Among submerged Fucales a periodic liberation of the sexual products, not necessarily at regular intervals, has frequently been reported; Moser records a regular fortnightly liberation in all the Fucales he examined. In such instances an intermittent production of mucilage within the conceptacle or periodic swelling of the mucilage is perhaps part of the liberation-mechanism. Escape of the sexual cells no doubt often follows the tidal rhythm, and this may remain so for some time during constant submergence.

As in other oogamous Phaeophyceae, fertilization in Fucales takes place after liberation of the ova. If sexual packets from a ripe receptacle of *Fucus* are washed into a watch-glass, the liberation of the gametes and the process of sexual fusion are readily observed, although fertilization has been followed only in a few instances. Large numbers of sperms surround the ova. to which they become attached by the anterior flagellum, while the posterior one continues to move vigorously; in *Fucus* and *Ascophyllum* this is sufficient to set the egg in rotation. As soon as a spermatozoid has penetrated, the attraction ceases; Whitaker describes the ensuing physical changes.

During the rapid passage of the minute male nucleus to the female, a centrosome surrounded by faint radiations becomes recognisable, whilst, after the male has penetrated the female nucleus, a second centrosome *(c')* appears at the spot where the sperm *(m)* has entered. Occasionally several sperm-nuclei pass into the female nucleus and, when this is so, a corresponding number of centrosomes appears, which proves that the second centrosome observed in normal fertilization is associated with the penetration of the sperm. This centrosome is possibly derived from the blepharoplast of the spermatozoid. During the first hours succeeding fertilization there is a great increase in oxygen-consumption. Abe failed to observe centrosomes in *Sargassum*.

Fertilization, Hybridisation

In *Halidrys*, at the time of fertilization, the surface of the female cell is covered with conical projections bearing a number of fine threads, although these disappear after 3-5 minutes. Similar features have been recorded in *Cystoseira* and in *Fucus*-eggs treated with hypotonic solutions. Winkler observed penetration of spermatozoids into non-nucleated fragments of *Cystoseira*-eggs, some of which afforded germlings of small dimensions; Moser failed to obtain their germination.

In view of the frequent dioecism fertilization can only be achieved, if there is a simultaneous production of large numbers of gametes which become intermingled by water-movements; there must, however, be enormous wastage. The heavy ova tend to sink, although in the sublittoral Fucales the devices keep the female cells *in situ*. According to Kotte the spermatozoids of *Fucus serratus* are negatively phototactic and aerotactic, as well as positively geotactic, so that they will tend to follow the sinking ova. Chemotactic attraction is stated to be due to a substance excreted from the latter and the response is a phobic one. Bordet, however, maintains that the sperms are mainly sensitive to contact.

Parthenogenesis, which has been induced experimentally in *Fucus*, has not been observed in nature and germination of unfertilized eggs is only rarely recorded. In suggesting possible parthenogenesis in *Notheia*, Williams does not refer to Barton's record of antheridia.

In view of the mixed populations of Fucales found on many shores, it is likely that hybridisation will occur, although there are few satisfactory records of the presence of hybrids in nature. When spermatozoids of one species are brought together with ova of another species of the same or of a different genus, there is active attraction, although fusion only ensues in certain instances.

Thuret first recorded fertilization of the ova of *Fucus vesiculosus* by the spermatozoids of *F. serratus* and the putative hybrid has been described by Sauvageau. Thuret was, however, unsuccessful with other Fucaceae and failed to obtain the reciprocal cross between the species mentioned, although this was subsequently achieved by Kniep. Card describes a supposed hybrid between *F. spiralis* and *F. ceranoides*. Williams obtained some positive results in crossing *Ascophyllum* with species of *Fucus* and records a plant intermediate between the two genera.

Kniep achieved a large number of crosses between F. *spiralis* and *F. vesiculosus,* with either parent as the male. Ova of the former were obtained by incapacitating the sperms by brief immersion in a 1 per cent solution of chloral hydrate in sea-water or in 10 per cent alcohol; mature oogonia were also dissected out of the conceptacles. The sperms are readily separated from the rapidly sinking ova. Sauvageau has described probable hybrids between the two species; these resemble *F. vesiculosus* in the presence of bladders and in being dioecious, while the receptacles are more like those of *F. spiralis.* It still remains doubtful, whether these and other similar transitional types, found at the limits of the zones occupied by the various species of *Fucus,* are actual hybrids or habitat-forms. In commenting on this point, Kniep suggests that the apparent scarcity of hybrids in nature may be due to unsuitability of the habitat or to inability to withstand competition with the parent-forms.

2

Ectocarpales

As mentioned at the outset the delimitation of Ectocarpales here adopted agrees essentially with that of Oltmanns. The order, as thus circumscribed, is characterised primarily by its heterotrichy as well as by the comparatively simple organisation of the thallus. There can of course be no serious objection to the elevation of some of the families or groups of families here recognised to ordinal rank, but the differences between them, save in vegetative construction, are slight and scarcely of the nature usually accepted for distinguishing major systematic units. With the discovery of heteromorphic alternation in *Dictyosiphon*, a new taxonomic feature became available and the separation of the Dictyosiphonaceae from other Ectocarpales was advocated. Sauvageau, with more reason, later established the order Sporochnales.

Recently Kylin has laid special stress on the features of the life-cycle in the classification of the Phaeophyceae as a whole, distinguishing a series Isogeneratae, with isomorphic, from another series Heterogeneratae, with heteromorphic alternation; to the former he refers Ectocarpales, Sphacelariales, Cutleriales, Tilopteridales, and Dictyotales, to the latter Chordariales, Sporochnales, Desmarestiles,

Punctariales, Dictyosiphonales, and Laminariales. The Ectocarpales are restricted to those forms that show isomorphic alternation (Ectocarpaceae, some Myrionemataceae) and, since they include both simple filamentous types and genera, exhibit a considerable vegetative range. Cfiordariales, Punctariales, and Dictyosiphonales are distinguished only on vegetative grounds.

In the preceding section the view has been advanced that forms with heteromorphic alternation are to be regarded. If this view be accepted, a classification based on the life-cycle is scarcely feasible among the simpler Brown Algae.. Moreover the Punctariaceae and their allies, for example, are closely linked by way of such forms as *Phloeospora* with the simple filamentous types, and it scarcely appears advisable to class them in a separate order.

All the diverse kinds of thallus-development among Ectocarpales are, in fact, derivable from simple filamentous, types, such as are illustrated by the Ectocarpacieae. On the other hand, both Sporochnales and Desmarestiales not only snow a very marked degree of vegetative specialisation in different directions, but exhibit true oogamy which, so far as present evidence goes, is never realised in the Ectocarpales. Both the former are indeed as sharply characterised as the Cutleriales, which have long been classed as a distinct order.

It is scarcely feasible to trace in further detail the lines of evolution that have resulted in the diverse types of structure met with among Ectocarpales. Their interrelation depends largely on their derivation from a common filamentous type. Two significant evolutionary advances were the inception of longitudinal septation, leading to the parenchymatous habit, and the substitution of apical for trichothallics seen in both the big series of Ectocarpaceae (Spermatochnaceae, Dictyosiphonaceae). It is also noteworthy that, among the polystichous forms, there is commonly a tendency for the aggregation of the sporangia into sqri, accompanied by the development of associated paraphyses which probably often fulfil a protective function.

In adding an epitome of the classification here adopted, it may be emphasised that it is essentially one of convenience and that a larger number of families should no doubt be distinguished; the exact relationships of a considerable number of the genera are, moreover, at present scarcely clear.

Simple and in part probably primitive forms:

1. *Ectocarpaceae:* Ascocyclus, Bodanella, Ectocarpus, Geminocarpus, Mikrosyphar, Phaeostroma, Pleurocladia, Pylaiella, Sorocarpus, Streblonema, Streblonemopsis, Zosterocarpus.

Haplostichous forms:

2. *Myrionemataceae:* Chilionema, Compsonema, Hecatonema, Heribaudiella, Lithoderma, Mesospora, Microspongium, Myrionema, Nemoderrna, Petroderma, Ralfsia, Symphyocarpus.
3. *Elachistaceae:* Elachista, Giraudya, Halothrix, Leptonema, Symphoricoccus.
4. *Leathesiaceae:* Corynophlaea, Cylindrocarpus, Leathesia, Myriactula.
5. *Mesogloeaceae:* Analipus, Bactrophora, Caepidium, Castagnea (incl. Eudesme), Chordaria, Gontrania, Haplogloea, Hapterophycus, Heterochordaria, Mesogloea, Myelophycus, Myriockdia, Myriogloea, Sphaerotrichia, Strepsithalia.
6. *Acrotrichaceae:* Acrothrix.
7. *Spermatochnaceae:* Chordariopsis, Halorhiza, Nemacystus, Spermatochnus, Stilophora, Stilopsis.
8. *Splachnidiaceae:* Splachnidium.

Polystichous forms:

9. *Punctariaceae:* Corycus, Desmotrichum, Litosiphon, Omphalophyllum, Phaeosaccion, Phloeospora, Punctaria, Stictyosiphon, Xanthosiphonia.

10. *Asperococcaceae:* Asperococcus, Isthmoploea, Myriotrichia, Petalonia (Phyllitis, Ilea), Protasperococcus, Scytosiphon, Striaria.
11. *Encoeliaceae:* Chnoospora, Cladochroa, Colpomenia, Hydroclathrus, Iyengaria, Soranthera.
12. *Dictyosiphonaceae*: Coilodesme, Delamarea, Dictyosiphon, Scytothamnus.

Tilopteridales

The few filamentous Algae included in this order are usually found in the sublittoral region, often at some depth, and so far have been recorded almost only from the Atlantic shores of Europe and North America. Although widely distributed, they are sporadic in their occurrence and are mostly found only for a brief period during spring. As a consequence our knowledge of Tilopteridales, and especially of the life-cycle, is imperfect. Certain peculiar features, in particular the large motionless monospores formed singly within a sporangium, have been given diverse interpretations.

The order was created in 1855 by Thuret who, recognising that the *Ectocarpus Mertensii* of Harvey possessed motionless monospores, assigned it to a new genus *Tilopteris.* Subsequently Kjellman added a second genus *Haplospora,* later studied more fully by Reinke(ig). To this genus Bornet referred *Ectocarpus Vidovichii* Kiitz. *(E. geminatus* Menegh.), although Kuckuck subsequently placed this species 'in a distinct genus *Heterospora.* Bornet also established for *Ectocarpus pusillus* Griff. A genus *Acinetospora,* distinguished by the production of large scarcely mobile spores from the plurilocular sporangia. Later Sauvageau showed that *A. pusilla* possessed rnonospores, as well as unilocular sporangia, and referred it to the Tilopteridales. He also concluded that there were no grounds for the generic separation of *Heterospora* and *Acinetospora.* Since the latter name has priority, *Heterospora vidovichii* becomes *A. vidovichii,* and this has been followed here. Sauvageau draws attention to the close resemblance between *Ectocarpus crinitus* Carm. and

Acinetospora pusilla. Newton lists it as a variety of the latter species.

Vegetative Features

As the preceding paragraph shows, nearly all Tilopteridales have in the past been placed in the genus *Ectocarpus*, which is due to the similar habit and arrangement of the reproductive organs. The well-branched plants, which possess a heterotrichous habit, form tufts attached to diverse substrata, those of *Tilopteris* often occurring on mud-covered rocks. The prostrate system is incompletely known, but Sauvageau describes a compact basal disc in *Tilopteris* and Reinke's data for *Haplospora* indicate a similar condition. The erect filaments of *Acinetospora* (are uniseriate throughout, but in *Tilopteris* and *Haplospora* the lower parts become multiseriate and in their regular septation recall a *Sphacelaria*. Growth in the Tilopteridales is, however, always intercalary, the diverse branches usually terminating in hairs *(h)* with a basal meristem. In addition to this well-marked trichothallic growth, the longer threads usually show other regions of active division, as in some species of *Ectocarpus*.

The branches generally arise from the top of the parent cell, but in *Acinetospora* both they and the reproductive organs commonly grow out from its middle. In *A. pusilla* the numerous short and spine-like branches serve to link the filaments with one another, as well as with those of other algae. Rhizoids, which sometimes show longitudinal division, often arise in some numbers from the lower cells. The cells contain a considerable number of chromatophores; pyrenoids are only recorded in *Acinetospora bidovichii*.

Reproductive Organs

Characteristic of the order are the globose or oblong *monosporangia*, which are generally situated at the ends of short, often one-celled, laterals, although commonly sessile in *Acinetospora pusilla*. In *Tilopteris*, where they are usually paired, they occupy an intercalary position. As a general rule the entire contents are liberated as a single monospore, which is usually provided with a firm but delicate membrane. The relatively wide aperture of the sporangium is in *Tilopteris* lateral in position. The cytoplasm of the spore is markedly alveolar. Division of the nucleus into four or more (up to twelve) parts has in certain instances been reported before the liberation of the spore. Germination has frequently been observed and not uncommonly occurs within the sporangium.

In *Tilopteris* formation of a rhizoid is followed by the outgrowth of a thread *(e)* from the opposite end; subsequently the body of the spore may become divided by septa in diverse directions. A similar type of germination is recorded in *Haplospora,* although in cultures division of the quadrinucleate spore into four parts, prior to the outgrowth of either rhizoids or threads, was observed; Reinke regarded this as the normal method of germination, but this is not supported by Sauvageau's observations. In *Acinetospora pusilla, A. Vidovichii,* and *Tilopteris* a new generation producing monospores has been grown from the monospore.

Plurilocular sporangia are known in all genera and are always accompanied by monosporangia. In *Tilopteris* and *Haplospora* they are usually situated at or near the bases of the branches, although in *Haplospora* they sometimes constitute the entire lateral. The numerous small compartments of these sporangia contain only one or two chromatophores and form a single series surrounding a central hollow (/); the individual compartments open separately to the exterior. The swarmers in *Tilopteris* are hyaline with a red eyespot *(s)* and are stated to resemble the spermatozoids of *Fucus* \ Reinke's incomplete data for *Haplospora* indicate similar features. The plurilocular sporangia of these two genera in fact show most, if not all, the characteristics of antheridia and, since Thuret's time, have generally been interpreted in this sense.

The plurilocular sporangia of *Acinetospora* on the other hand, are of a different nature. They are always terminal on short laterals, possess relatively few and large compartments, and are without a pronounced central hollow. The swarmers, which commonly behave as aplanospores, are large and provided with several chromatophores. They germinate readily and grow into plants again bearing plurilocular sporangia. Recently Dammann records at the base of the "antheridia" of *Tihpteris* large, dark brown compartments, with abundant chromatophores, the contents of which are liberated as amoeboid spores; some of these grow out into short filaments, although the majority fail to germinate. It is possible that these large compartments are to be compared with those of the plurilocular sporangia of *Acinetospora,* although they occur in the same organ as produces the supposed spermatozoids.

Unilocular sporangia, producing a number of large swarmers similar to those of the plurilocular sporangia, are recorded only in *Acinetospora;* the swarmers are devoid of an eyespot. In *A. pusilla* they are found on the same plants as bear the phirilocular sporangia, whilst in *A. vidovichii* they apparently occur on distinct individuals; in the former species the zoospores have been observed to germinate within the sporangium.

Vegetative reproduction takes place readily from fragments of the threads, which easily form rhizoids. Sauvageau also records the formation of akinetes from the prostrate system and the lower polysiphonous portions of the erect threads in *Tilopteris* and suggests that these structures constitute a means of perennation.

Problems Presented by the Life-cycle

In *Haplospora* there are two kinds of individuals, the one producing monosporangia only, the other with monosporangia and the above described "antheridia". The second was first recorded as a separate genus *Scaphospora,* but since Reinke's memoir *Scaphospora speciosa* Kjellm. (incl. *S. arcticd)* has generally been regarded as a phase of *Haplospora globosa.* There appear to be no vegetative differences between the two plants. Reinke established that individuals liberate a quadrinucleate monospore provided with a definite membrane, whereas the monosporangia of the *Scaphosporaphnts* set free a naked uninucleate spore. In *Haplospora* the monosporangia are generally borne terminally on short lateral branches, whilst in *Scaphospora* they are partly immersed in the threads and generally arise after a longitudinal division from half the parent-cell; both halves may, however, develop into monosporangia or the latter may be produced without division.

These facts led Reinke to conclude that there was an isomorphic alternation between an asexual *Haplospora-phase* producing quadrinucleate monospores, homologous with the tetraspores of Dictyotales, and a sexual *Scaphospora-phase* bearing oogonia and antheridia, the monosporangia liberating naked uninucleate spores on these individuals being interpreted as female organs. Nienburg described synezesis and an indistinct diakinesis stage during monospore-formation in *Haplospora globosa*, thus affording some evidence for the occurrence of a reduction division, although the data for halving of the chromosome-number are scarcely satisfactory. All those who have studied this genus report a great scarcity of *Scaphospora-mdividuals*; on the shores of Heligoland, where *Haplospora* is frequent, *Scaphospora* has never been found.

In *Tilopteris mertensii* Reinke observed two, four, or sometimes more, nuclei in monospores provided with a membrane and derived from plants bearing these structures only. He therefore concluded that the individuals in question represented the asexual phase of this species, while the plants previously recorded by Thuret, which formed both monosporangia and "antheridia ", were regarded as constituting the sexual phase. Reinke thus assumed that *Tilopteris* showed the same isomorphic alternation as *Haplospora*, although in the former genus there is no difference either in position or form between the monosporangia of the asexual plant and the supposed oogonia of the sexual individual; no cytological data are available. Kuckuck found but few individuals of *Tilopteris* with antheridia, while the monospores in his material were always naked and uninucleate.

Reinke's hypothesis as to the nature of the life-cycle has been accepted by various authorities, although there is no evidence of fertilization, nor of any attraction of the spermatozoids by the naked uninucleate monospore. There is thus at present nothing to show how the diploid state is re-established; yet, if the *Haplospora*- and TYoptem-plants bearing monospores only are asexual and the observations as to the occurrence of reduction are correct, then there must be a nuclear fusion to restore the diploid state. A further objection lies in the absence of any parallel for the complete similarity between monosporangia and female organs as is postulated by Reinke's theory. Sauvageau also emphasises the great range in size of the uninucleate monospore, a feature which is very unusual in a sexual cell.

Acinetospora, in which "antheridia" are unknown and the mono-spores are always uninucleate and provided with a membrane before liberation, does not fit into the theory at all. Kylin, in fact, regards this genus as having no affinity with the others. It must be recognised, however, that *Acinetospora* shows considerable similarities, both in structure and other respects, with the other two genera; much emphasis cannot be laid on the absence of longitudinal division.

A somewhat different complexion` has been put on the problem by the demonstration that *Tilopteris* liberates two kinds of reproductive cells from its plurilocular sporangia, viz. "spermatozoids" and others which resemble the units produced from the plurilocular sporangia recorded for the two species of *Acinetospora.* The latter may represent female gametes, and further investigation may disclose the presence of antheridia in *Acinetospora.* The Tilopteridales would then show anisogamy comparable to that of Cutleriales. On this view the monosporangia, of the sexual individuals at least, would be relegated to the position of accessory asexual reproductive organs, an interpretation which Sauvageau has maintained since 1899. The fact that some individuals possess these structures only is paralleled by what is known of the distribution of reproductive organs in other series of the lower Brown Algae.

Sauvageau regards the monospores as organs of vegetative propagation comparable to the propagules of Sphacelariales, a view based largely on the structural resemblances between them and the propagules of *Choristocarpus*. The monospores of Tilopteridales are, however, clearly spores and the organs which produce them sporangia. They are better interpreted as modified unilocular sporangia which, so far as present evidence goes, have altogether replaced the usual type in *Tilopteris* and *Haplospora*. A cytological study can alone shed light on these matters and settle the role of the unilocular sporangia, which are known in *Acinetospora* only.

In this genus they may prove to be the seat of the reduction divisions, whilst in the other two, on present evidence, these would appear to take place in the monosporangium of the asexual individuals. *Acinetospora* seems to be the most primitive member of the order.

In their trichothallic growth and simple filamentous habit the Tilopteridales appear as an offshoot of the Ectocarpales. They show specialisation, however, in diverse respects, viz. in the monospores, in the differentiated "antheridia ", in the tendency towards a multiseriate construction, and in the specialised unilocular sporangia of *Acinetospora*. On the other hand the affinity with Sphacelariales, which some have supported, does not appear to be based on any strong evidence, and such resemblances as there are likely to be due to parallel development.

The correspondence between the propagules of *Choristocarpus* and the monosporangia of Tilopteridales is, however, certainly striking. The distinctive features are sufficiently marked to warrant regarding the Tilopteridales as an order allied to, but distinct from, the main body of the Ectocarpales.

The genus *Masonophycus*, established by Setchell arid Gardner and doubtfully referred to this order, is distinguished by the possession of tetrasporangia producing four cruciate spores. At present so little is known about it, that its reference to a separate family appears premature.

Cutleriales

Cutleria and *Zanardinia*, the only known members of this order, would hardly merit separation from Ectocarpales on the basis of their vegetative structure. The anisogamous sexual reproduction, however, falls little short of the oogamy of the more specialised Phaeophyceae, whilst the alternating generations of *Cutleria* imply an advance from isomorphic to heteromorphic alternation which is probably significant as an indication of specialisation.

The life-cycle of *Cutleria* presents many complications, and the normal sequence will first be considered. The species of *Cutleria* are annuals, occurring in the sublittoral region for a few months during the favourable period of the year. *C. multifida* and *C. adspersa*, the latter mainly confined to the Mediterranean, the former known also in northern latitudes, are found in relatively shallow water, whilst the Mediterranean C. *monbica* occurs at depths of 20-80 metres. *C. multifida* has a repeatedly branched band-shaped thallus, up to 20 cm. in length; all the branches usually lie in one plane and terminate in tufts of assimilatory hairs *(a)*. *C. monoica* is similar but smaller, whilst *C. adspersa* has a lobed fan-shaped thallus bearing numerous assimilatory hairs along its margin.

These plants are gametophytes bearing gametangia of two kinds, usually on distinct individuals which may show certain inconstant differences. *C. monoica*, however, is monoecious, the two kinds of gametangia sometimes even occurring in the same sorus. The zygote develops into a flat encrusting thallus of considerable size, known as the *Aglaozonia*-stage. These crusts are apparently perennial and represent the resistant phase in the life-cycle; they grow permanently submerged like the gametophytes and can evidently flourish at appreciable depths. They were long regarded as species of a distinct genus, *Aglaozonia*. It is now known, however, that *A. parvula (A. reptans)* is the sporophyte of *C. multifida*, *A. melanoidea* that of *C. adspersa*, and *A. chilosa* that of *C. monotca*. The *Aglaozonia-stage* produces on its upper surface dense sori of unilocular sporangia, the swarmers of which reproduce the gametophyte.

'*Zanardinia collaris* Crouan, which is probably perennial, is likewise a deep-water form known mainly from the Mediterranean, although it is reported from the Azores and occurs as far north as the Channel Islands. The thallus, which takes the form of a rounded disc, leathery in texture and 4-20 cm. in diameter, somewhat resembles a shallow *Peziza*\ older plants are often irregular in shape with parts of the margin worn away. The discs sometimes possess a slight stalk and their edge is fringed with assimilatory hairs. The plants are either sexual or asexual, the two generations being alike; the sexual phase is monoecious. In both genera reduction occurs at the first division in the unilocular sporangium.

Sexual Phase

Although so different in outward habit, the gametophytes of both genera show essentially the same trichothallic mode of growth and the same ultimate structure. The peripheral assimilatory hairs, the cells of which contain abundant chromatophores, usually form two superposed series. At the base of each hair lies an intercalary meristem, giving off segments both to the outside and to the inside. Those cut off externally contribute to the lengthening of the hairs, whilst those formed below the meristem undergo further transverse and longitudinal septation and, with some tangential enlargement, fuse to form the compact tissue of the thallus a short distance beneath the individual meristems.

Uniseriate branches, which arise, as in the trichothallic species *of Ectocarpus* or as in *Elachista*, from the segments below the meristem, add to the number of assimilatory hairs and lead to a progressive broadening of the thallus. The branching of *C. multifida* is due to periodic failure of the filaments to fuse. The gameto-phytic thalli of Cutleriales are thus produced by congenital fusion of numerous distinct threads which appear as separate entities in and beyond the region of the meristem.

In radial longitudinal or transverse sections through the growing margin of the disc of *Zanardinia* the products of the two superposed sets of threads can be traced back for long distances into the compact thallus. The several layers composing the mature plant originate from them by division parallel to the surface. The inner segments gradually enlarge, while the outer undergo plentiful anticlinal division. In transverse sections through the margin of the disc the cells are seen to multiply largely by crosswise division, as in many polystichous Ectocarpales.

The structure and mode of formation of the thallus of *Cutleria adspersa* is almost identical. C. *monoica* differs in so far as the assimilatory threads exhibit considerable branching also above the meristem; as a result the apical tufts appear relatively bulky. Attachment to the substratum is in both genera effected by a mass of little branched rhizoids growing out from the lower cells. Adventitious branches, recorded in both genera originate from surface-cells which grow out into a tuft of assimilatory hairs; these soon acquire intercalary meristems and development proceeds in the way above described.

A cross-section of the mature thallus shows several layers of cells which only differ in the larger size of the internal ones. In *Zanardinia* the upper surface of the disc consists of vertical rows of small cells.

The gametangia of *Cutleria* are borne laterally on uniseriate threads which arise in tufts on both surfaces of the thallus and appear as irregularly scattered, dark-coloured excrescences to the naked eye. The threads bearing the female gametangia are simple or little branched, whilst those bearing the numerous male organs are plentifully branched, although occasional male gametangia may arise directly from the superficial cells. In *Zanardinia* the gametangia are situated terminally on simple threads that sprout out from the upper surface and form ill-defined sori.

Both types of gametangia are multiseriate and set free the gametes separately from each compartment. The dark-brown female gametangia have relatively few and large compartments, each of which produces a swarmer with numerous chromatophores; an eyespot *(s)* is attached to one of the anterior chromatophores and at this point the two flagella are inserted. The bright yellow male gametangia possess numerous small chambers, each liberating a minute spermatozoid, with an eyespot *(s)* next to the point of insertion of the flagella. Kuckuck states that chromatophores are lacking, but according to Yamanouchi a few small ones are present.

The female cells soon withdraw their flagella and round off, and in this condition they are fertilized by a male cell *(sp)* which penetrates at the colourless end of the female. At the time of fusion the nucleus of the female is in the resting condition, while that of the male shows 22 separate chromosomes. The secretion of a membrane follows and germination ensues almost immediately.

Asexual Phase

As above mentioned, the zygote of *Zanardinia* gives rise to a diploid asexual plant which is in other respects indistinguishable from the haploid phase and develops in exactly the same way. In *Cutleria*, on the other hand, the zygote gives rise to the *Aglaozonia*-stage, as was first established by Falkenberg, although the existence of alternation was suspected by Reinke. The zygote produces a short erect thread, attached to the substratum by rhizoids *(r)* and later undergoing longitudinal division to form a club-shaped upgrowth, which not uncommonly bears an apical tuft of hairs. This *column* (foot-embryo of Church, however, remains small and meristematic activity soon becomes restricted to its base and results in the formation of a parenchymatous crust. This is the *Aglaozonia*-stage, which continues to grow by means of a large-celled marginal meristem till it may reach the size of a human hand. Attachment is effected by a number of septate, little-branched rhizoids.

In *A. parvula* the two surfaces are bounded by small cells which enclose 2-3 layers of larger ones. The crusts of *A. melanoidea* are appreciably thicker and there are several rows of small cells on either side. The upper surface bears tufts of hairs which possess the usual basal meristem and are often somewhat sunk. Those of *A. parvula* are stated to form linear groups parallel to the direction of growth of the thallus. The dark colour of *A. melanoidea* is, according to Sauvageau, due to tannin in the surface-cells.

Both in *Zanardinia* and *Cutleria* the unilocular sporangia are formed by tangential division of surface-cells and appear in dense sori on the upper surface. The segment, which grows out to form the sporangium of *Cutleria*, cuts off a basal cell so that the sorus is separated by two layers of cells from the large inner ones; in *Zanardinia* there are 3-5 of these layers. The nucleus of the sporangium divides successively, the first divisions being meiotic. Cleavage ensues only after the full number of nuclei have been formed, and at this stage the numerous chromatophores are grouped about the individual nuclei. In *Cutleria multifida* usually 8 swarmers are produced, whilst in *Zanardinia*.

The zoospores are liberated through an apical aperture formed by gelatinisation of the wall and closely resemble the female gametes. Both in zoospores and gametes Yamanouchi records a row of deeply staining granules connecting nucleus and plasma-membrane, the outermost granule lying at the point of insertion of the flagella.

In *Cutleria multifida* they give rise to an erect thread attached by one or more branched rhizoids *(r)*. The early growth of this thread is probably diffuse, but an intercalary meristem soon arises near its base which cuts off numerous segments on its lower side. These produce successive branches which exhibit a similar mode of growth, and, as they multiply, become closely apposed to one another. Later, the parts of the threads below the individual meristems coalesce to form a compact cylinder which bears the free threads as an apical tuft. The cylinder soon branches and so the mature thallus is gradually established. In C. *adspersa* Sauvageau describes a similar sequence, but here the cylinder widens into a funnel which bears a double row of threads along its margin, a condition very nearly identical with that found in *Zanardinia*. Later the funnel splits longitudinally to form the mature fan-shaped thallus.

The Problems Presented by the Life-cycle of Cutleria: The life-cycle of *Cutleria* has mainly been investigated at Plymouth, Naples, Heligoland, and in the Adriatic. Part of the population no doubt follows the cycle described above, although the time of development and of fertility of the two generations of C. *multifida* varies with the geographical location. Thus, at Plymouth, Church found the gametophyte at its maximum in summer and no longer represented in October, whilst at Naples it occurs chiefly from December to April. The sporophyte is most abundant at Plymouth in October and November and becomes fertile in March and April, whereas at Naples it fructifies in the late autumn. At Heligoland, where the gametophyte is very rare, the *Aglaozonia-stege* is fertile between June and August. In the north of England the sporophytic stage is frequent, although the gametophyte is unknown, while on the shores of Scandinavia it is much more abundant and more widely distributed than the latter.

At Plymouth and Naples there is, thus, a certain seasonal sequence between the two phases, the sexual one being represented only during the favourable period of the year, whilst the perennial asexual phase becomes fertile a few months before the sexual one reaches its maximum; these facts tend to indicate a normal alternation. Schlosser has recently, in cultures undertaken at Naples, grown fertile *Aglaozonia-plante* from the zygotes, while Sauvageau has obtained recognisable sexual plants from the zoospores of the *Aglaozonia-stage*. Parthenogenetic development of the female cells has frequently been recorded. Germination is rather slow and the products are diverse; sometimes they resemble young *Cutlerias*, sometimes they are more like *Aglaozonias*, and sometimes they are of an intermediate or rudimentary character. A similar diversity has also occasionally been encountered in the products of germination of *Aglaozonia*-zoospores.

In *Cutleria multifida* Reinke and Falkenberg observed sexual fusion only and unfertilized ova failed to exhibit any further development. On the other hand, Thuret and Church record parthenogenetic development only, and both state that in the habitats studied the female plants were in the great majority. The germination-stages obtained by these two investigators were, however, fundamentally different, those of Thuret (from the English Channel) apparently resembling the young filamentous stages of *Cutleria*, whilst those of Church were young *Aglao-zonias*. Parthenogenetic development is also recorded by Crouan from Brest and by Derbs and Solier from Marseilles. It has likewise been observed in *Zanardinia*, although only early stages of germination were obtained.

Comparable data are available for *Cutleria adspersa*. The male and female gametophytes are widely distributed and equally common in the Mediterranean, although the corresponding *Aglaozonia-stage* has been found only in one or two localities. Janczewski found that the female cells germinated only after fertilization and that they gave rise to young *Aglaozonta-stages*. Sauvageau, on the other hand, both in the Mediterranean and in the Bay of Biscay, where male plants are more abundant than female, failed to obtain sexual fusion; the ova in his cultures, however, likewise developed into recognisable *Aglaozonia-stages*. The numerous germlings found on the mature thalli of the gametophyte in nature were partly *Aglaozonias* and partly young *Cutlerias*, while a few were of the filamentous type to be described below ("forma Church")- Sauvageau concludes that all these had probably developed without sexual fusion, although the possibility of some being derived from fertilized ova is discussed.

The faculty of the female cells to develop apogamously clearly varies. Schlosser was able to stimulate unfertilized ova to further development by treatment with various chemicals and by centrifuging; the resulting plants produced oogonia. If strong centrifugal force was applied during the first division of such ova, sterile *Aglaozpnia-stages* resulted; these are suspected of being diploid, but for this there is at present no cytological proof.

Church found that many of the zoospores of the *Aglaozonia*-stages of *C. multifida* developed into branched upright filaments, longitudinally divided in their lower part and bearing antheridia, a form already encountered by Kuckuck and described by him as var. *confervoides.* It is strange, in view of the scarcity of male plants at Plymouth, that Church obtained only antheridia in these cultural forms. This may indicate a dimorphism of the two sexes under certain conditions in nature. In some of Church's filamentous stages the lower part broadened into a small *Aglaozonia (ag)* and such germlings are usually described as the "forma Church". The zoospores of *A. melanoidea* afford in cultures mainly filamentous stages, but also a few *Aglaozonia*-forms, while those of *A. chilosa* have produced confervoid stages with oogonia, as well as the "forma Church".

The considerable diversity in the products of germination of parthenogenetic ova and zoospores is not easily elucidated. The filamentous forms obtained by Church and others from zoospores may well be cultural stages of the normal dioecious gametophyte; they recall the precociously fertile sporophytic stages met with in diverse Ectocarpales under conditions of culture. On the other hand parthenogenetic ova produce not only fresh gametophytes, but evidently rather more frequently *Aglaozonia-stages* which must presumably be haploid. Zoospores can seemingly also produce such stages, which then arise as a vegetative outgrowth from the base of the primary filament; this likewise suggests the existence of haploid individuals. The probable occurrence of many haploid *Aglaozonias* in nature, presumably originating in the first place by apogamous development of female cells, has been strengthened by

Sauvageau's observations on *Cutleria monoica*. An apogamously produced plant of *Aglaozonia chilosa* was observed to give rise by vegetative proliferation.

A survey of the facts indicates that, side by side with a normal alternation, there is in many regions extensive apogamous development of the female cells, which results for the most part in the production of haploid *Aglaozonia-individuals*. In diverse localities, in fact, it would appear as though the majority of these stages may be haploid and that these possibly propagate themselves indefinitely by haploid zoospores. At the same time, however, the female cells and the swarmers of the haploid *Aglaozonias* can evidently also give rise to sexual plants. The actual state of affairs can probably only be settled by an extensive cytological study of *Aglaozonias* in different regions.

Affinities of Cutleriales

The mode of development of the thallus in Cutleriales shows analogies with that seen in Elachistaceae. In view, however, of the probable simple filamentous nature of the gametophyte in the latter, the resemblance in vegetative features is unlikely to imply any close affinity. So far as present evidence goes, the Cutleriales must be regarded as a direct development from the primitive heterotrichous ectocarpoid stock with its trichothallic growth, isomorphic alternation, and marked tendency towards oogamy. Sauvageau, laying special stress on the *Aglaozoma-stsige* which presents many points of resemblance to certain *Zonarias*, believes in an affinity with Dictyotales.

There can be little doubt that alternation in this order is primarily isomorphic. The appearance of an *Aglaozonta-crust* as a basal outgrowth from a gametophytic filament in the "forma Church", usually after the latter has undergone longitudinal septation in its lower part, suggests that the column of the normal *Aglaozonia* is an arrested and sterile structure, homologous with the *Cutleria*-thallus. The typical sporophyte shows a clear derivation from a heterotrichous ancestry, the column representing the erect, the crust the prostrate system. The fact that the latter is here formed secondarily does not appear to be of significance. The gametophyte, on the other hand, shows no trace of heterotrichy. It is possibly this dominance of the erect system that leads also in *Aglaozonia* to the primary formation of an upright growth. It is to be noted that the two generations of *Zanardinia* and the *Cutleria*-gametophyte exhibit the trichothallic growth characteristic of many of the specialised members of Ectocarpales (cf. also Sporochnales), whereas the *Aglaozonia-crusts* show the marginal (apical) growth which is invariably met with in the prostrate system.

The instances discovered by Sauvageau, in which new *Cutleria*-plants arise vegetatively from the *Aglaozonia-stage* of *C. monoica*, illustrate the presence of both parts of the heterotrichous system in what is to all intents and purposes a single individual. They possibly afford a picture of the ancestral condition when the alternation in *Cutleria* was an isomorphic one between two heterotrichous individuals, neither system of which necessarily showed the specialisation seen in the present-day forms. In the course of further evolution the sexual individual lost the prostrate and the asexual one the upright system, except for the relic preserved in the shape of the column.

Such an interpretation explains *inter alia* the frequent association of gametophytic threads bearing sex organs with a prostrate *Aglaozonia-crust*. This condition is evidently not confined only to the precociously fertile gametophytic stages, since, according to Gran, *Aglaozonia-stages* can arise from young *Cutleries* which already show definite coalescence of threads. The view that the heteromorphic alternation of *Cutleria* is derived from an isomorphic one accords with the occurrence of strict isomorphic alternation in *Zanardinia*. Young plants of the latter show many resemblances to the early stages of development of *C. adspersa*.

Oltmanns looks upon the *Aglaozonia-stage* as a new structure originating from the base of the *Cutleria-plant* and to which the production of unilocular sporangia was gradually relegated. This necessitates the assumption that at one time asexual organs were borne upon the *Cutleria-plant* itself; since they occur on identical individuals in *Zanardinia*, this is quite plausible. The Cutleriales are, however, clearly related to the Ectocarpales with their marked heterotrichy, and it seems far more likely that the *Aglaozonia-stage* represents a further development of an ancestral prostrate system than that it is a new structure. An elaborate development of the prostrate system at the expense of the upright one is well shown also in other series of Phaeophyceae.

Sporochnales

The members of this order 0 are distinguished by the possession of highly organised sporophytes which bear unilocular sporangia only. Their zoospores give rise to minute ectocarpoid gametophytes bearing antheridia and oogonia. Although sexual fusion has not yet been observed, its occurrence in nature can scarcely be doubted, especially in regions favourable to the development of these plants. The Sporochnales thus show heteromorphic alternation combined with oogamy and this, no less than the peculiar characteristics of the sporophyte, warrant their being regarded as a distinct order. The few genera, included in the single family Sporochnaceae, apparently have their chief centre of distribution in the Southern Hemisphere, although a number of species are widely distributed in warmer seas; *Sporochnus pedunculatus* and *Carpomilra costata*. Batt. are rare British seaweeds.

Bellotia, Encyothalia, and *Perithalia* are known only from Australia and the adjacent islands, and this is true also of a considerable number of the species of the other genera. A few species of *Nereia* occur in the Mediterranean and on the west coast of Africa, while those of *Carpomitra* are found on the western shores of Europe, on the Pacific coast of North America, in South Africa, and Australasia. *Sporochnus* is represented in the Australian seas and on the west coasts of Europe, *S. pedunculatus* occurring as far north as Scandinavia. Certain species found in the Northern Hemisphere possibly owe their presence there to transport on ships' bottoms. All Sporochnaceae tend to grow below low-tide level, often in deeper water.

Sporophyte

The members of this order are moderately large seaweeds, characterised by the fact that the apices of all the branches of a growing plant are crowned by a tuft of free, unbranched assimilatory hairs. Each of these lengthens with the help of a basal or suprabasal meristem, but these meristems take no part in the growth of the actual thallus below. *Sporochnuspedunculatus* is a short-lived annual, with well-branched cylindrical axes, up to 45 cm. in length and bearing numerous branchlets of limited growth, each of which terminates in such a tuft of hairs.

Encyothalia has a similar habit. *N. filiformis* is less richly branched; the laterals of limited growth are Rere not always so clearly defined, while the apical tufts of hairs are particularly strongly developed. In *Bellotia eriophorum,* recorded from New Holland and Tasmania, the elongate laterals of limited growth arise in whorls. A rather different habit is seen in *Carpomitra costata,* in which all the branches of the irregularly dichotomous thallus are flattened and provided with a broad midrib *(m);* there are no laterals of limited growth.

The various genera are specially distinguished by the positions of the sori. In *Nereia* these show no very definite location and are not always conspicuous, although they tend to develop on or near the shorter branches. The sori of *Bellotia* and *Encyothalia* constitute swollen regions in the middle of the laterals. In *Sporochnus* they likewise develop on the laterals, occupying a longer or shorter stretch just beneath the tuft of assimilatory hairs so that such fertile laterals appear as stalked pear-shaped structures. The sori of *Carpomitra,* lastly, are differentiated at the tips of the branches beneath the apical hairs, from which they are separated by an annular swelling.

Longitudinal sections of the mature thalli appear to show considerable uniformity. They exhibit a number of longitudinal rows of cells which in *Carpomitra* are sometimes slightly twisted; the central cells are elongate and colourless, while the peripheral ones *(c)* are shorter and packed with discoid chromatophores. In *Nereia* the surface-layer bears densely aggregated 2-3-celled paraphyses, but in the other genera similar outgrowths are only produced in the fertile regions. The mature thallus is formed from a horizontal meristem which lies beneath the tuft of assimilatory hairs. Its mode of origin will be best understood from a description of the early stages in development which are known only for the European genera.

The young plants, produced from the gametophytes, at first consist of a simple erect thread with diffuse growth, the *proembryo* of Sauvageau this is anchored to the substratum by rhizoids. The proembryo soon becomes differentiated into three regions. The lower part, comprising a row of often narrower cells, constitutes a pedicel which in *Carpomitra* is of some length, although relatively short in *Nereia*; this region plays no further part in the development of the mature thallus. The upper part differentiates a basal meristem and constitutes the first hair of the future tuft. The single flat cell, situated between these two regions, is destined to give rise to the meristem from which the mature thallus is produced.

The cell in question undergoes longitudinal division in various planes to form a progressively broadening meristematic plate. Both before and during the septation of the primary meristematic cell, successive protuberances arise on the upper side and gradually develop into further assimilatory hairs; the later ones originate from the individual units of the meristem as it broadens. Each cell of the meristem also gives rise on its lower side to an outgrowth which lengthens into a downwardly growing multicellular thread. The individual threads thus produced coalesce to form a multicellular pseudoparenchymatous body, which is the beginning of the future thallus and is intercalated between the pedicel and the apical tuft. Church quite unwarrantably speaks of a parenchymatous organisation.

It is not altogether clear how long the addition of further threads from the margin of the meristem continues. According to Sauvageau only a limited number are produced in *Carpomitra*, and further widening of the thallus takes place by periclinal divisions in the threads, accompanied by a corresponding division in the cells of the meristem. Continued lengthening of the rows of cells thus produced is effected by horizontal division of the meristematic cells. The pedicel undergoes no further change and merely serves to raise the embryo above the substratum. Sooner or later the cells at the base of the young thallus grow out into bifurcating rhizoids which gradually hide the pedicel and spread out upon the substratum.

According to Sauvageau only the midrib of *Carpomitra* arises in this way, the wings *(w)* being formed by periclinal division of the segments of an initial cell (i), recognisable just below the meristem and regarded as a reflexed marginal cell of the latter. Another distinctive feature of this genus lies in the development, from the margin of the meristem, of a horizontal tuft of coalesced threads with enlarged distal cells that ultimately gives rise to the marked annular swelling situated just below the apical tuft.

In the richly branched genera the numerous groups of assimilatory hairs must constitute an important photosynthetic system, and it is significant that, in *Carpomitra* and *Sporochnus*, the sori are located in their immediate neighbourhood. These tufts of threads, so similar to those found in *Elachista* and in Cutleriales, here play no part at all in the growth of the thallus, which is produced entirely from the separate meristem arising from an intermediate cell of the proembryo. It is indeed difficult to find a parallel among other Brown Algae for the peculiar mode of thallus-development obtaining in the Sporochnales.

The position of the sori has already been mentioned. In the

small sporangia arise either from the peripheral cells that bear the paraphyses or from the basal cells of the latter; the dilated end-cells of the paraphyses form a protective covering. In the remaining genera the sporangia are borne on special fertile threads that grow out from the surface-cells only in the region of the sori. The fertile threads of *Sporochnus* and *Carpomitra* are branched, with enlarged terminal cells, and bear the several sporangia laterally. Those of *Bellotia* are elongate and often unbranched. The sporangia open by a broad apical rupture. The zoospores appear to be of the usual pattern; Sauvageau describes those of *Carpomitra* as having a single posterior chromatophore provided with an eyespot.

Gametophyte

The gametophytes of *Carpomitra* were discovered in 1926 by Sauvageau; a year later he described those of *Nereia filiformis*, and subsequently also those of *Sporochnus pedunculatus*, although of the last only sterile prothalli have so far been found. In *Carpomitra* and *Sporochnus* the cells of the resulting prothalli each possess only a single parietal much dissected chromatophore, which contrasts with the numerous discoid chromatophores found in those of the sporophyte.

The prothalli are always ectocarpoid in form; those of *Nereia* are usually dioecious, while those of *Carpomitra* are monoecious. In the latter the antheridia *(an)* are one-celled and generally form terminal clusters on the ultimate branches. The well-branched male prothalli of *Nereia* bear the numerous small, conical antheridia *(an)* in a lateral or terminal position upon the upper cells; the antheridia are here two-celled and each compartment opens separately to the exterior. In both genera the male organs are colourless and composed of uninucleate cells. There can be little doubt that each cell forms a single spermatozoid, although these have not so far been observed.

The female organs, which give rise to the proembryos by transverse division, are in *Carpomitra* enlarged terminal cells of elongate laterals; in *Nereia* they occupy the apices of the unbranched female prothalli. So far only apogamous development of the cells, without liberation of an ovum, has been observed. In view of the production of numerous antheridia from which the contents are liberated, it is probable that fertilization sometimes occurs in nature. On the other hand, as Sauvageau points out, it is possible that the inflated terminal cells are not actually oogonia, but merely cells of the gametophyte giving rise vegetatively to a new sporophyte. The numerous antheridia, which are evidently functional, would be difficult to explain if this were so. The single chromatophore of *Carpomitra* gradually breaks up into numerous discs as the proembryo develops.

Affinities of the Sporochnales

The Sporochnales constitute a highly specialised group of uncertain affinity. As regards the characters of the sporophyte they stand altogether isolated, while the gametophytes show resemblances to those of Desmarestiales and Laminariales. The unicellular antheridia of *Carpomitra*, with their colourless contents, are remarkably like those of the latter order, and there is also similarity in the position of the supposed oogonia. The two-celled antheridia of *Nereia* possibly indicate a stage in reduction from the plurilocular type. It is significant of the specialisation of the sporophyte that all indications of heterotrichy have disappeared.

Berthold assumed a relationship with Mesogloeaceae, whilst Oltmanns refers especially to *Myriogloea*. It is difficult to detect any real affinity with either, since the mode of formation of the thallus in Sporochnales is quite unique. A certain intercalary cell of the erect filamentous proembryo divides to form the horizontal meristem from which the mature thallus arises as a system of downwardly growing threads. This to some extent recalls the condition in *Desmarestia*, where the bulk of the thallus is produced from downwardly growing cortical threads, although these here arise at diverse points along the uniseriate main axis. The analogy is heightened in certain instances observed by Sauvageau in *Carpomitra*, in which more than one transverse meristem arises in the course of a proembryonal thread.

Desmarestiales

The three genera included in this order possess sporophytes showing considerable anatomical specialisation. *Desmarestia* is a widely distributed lithophyte in the colder seas of both hemispheres. The common British species, *D. viridis* (Mull.) Lamour. and ZX *aculeata* (L.) Lamour., occur on the Atlantic shores of Europe and North America, extending north to Greenland, although they are absent from the Mediterranean. The genus is well represented on the western shores of North America (from Behring Island to Southern California), and Setchell and Gardner distinguish a considerable number of species, of which only *D. ligulata* is also recorded from the Atlantic. Another group of species, in part with large thalli, occur in the Antarctic, while the monotypic *Phaeurus* is known only from the latter region. *Arthrocladia,* with the single species *A. villosa*, is found on most European shores (incl. the Mediterranean), as well as on the Atlantic coasts of North America. The Desmarestiales, though sometimes found near low-tide level, usually inhabit the sublittoral region, often extending to some depth; in the Antarctic they play a dominant role in the sublittoral vegetation, the most conspicuous species being *D. compressa* and D. *anceps* Mont.

Arthrocladia$_y$ as well as certain *Desmarestias (D. viridis, D. ligulata)*, are annuals, but other species of the latter survive for more than one year. Thus, in the European *D. aculeata* the older portions of the thalli persist through the winter and give rise to numerous fresh shoots in the early part of the next season. The plants always bear unilocular sporangia only, and the zoospores of *D. aculeata* have been shown to give rise to minute filamentous prothalli producing oogonia and antheridia and closely resembling those of Laminariales. The zoospores of *Arthrocladia)* also develop into ectocarpoid growths, but these have so far exhibited only an apogamous production of sporophytes.

Structure of the Sporophyte

The species of *Desmarestia* are usually richly branched plants of appreciable dimensions. Thus, *D. aculeata* may reach a length of 2 metres, although the annual *D. viridis* is smaller. The Antarctic *D. compressa* Reinsch may grow to 3 metres, *D. anceps* to 4 metres with an attaching disc up to 20 cm. in diameter, while the Pacific *D. herbacea* reaches 2 J metres. *Arthrocladia* is smaller, though it may be nearly a metre in length.

In a number of species of *Desmarestia* the diverse branches of the thallus are narrow and subcylindrical, although the principal axes appear always to show some flattening. In the so-called ligulate species this flattening is more pronounced, and most of the branches are developed as leafy expanses with dentate margins and often with more or less clearly marked midribs; there is, however, always a short cylindrical portion at the base of the whole plant and at the base of each branch.

All the branching generally takes place in one plane and the longer laterals are commonly arranged in opposite pairs, however, shows alternate branching. A characteristic feature of the growing plants, both in *Desmarestia* and *Arthrocladia,* is the presence of numerous, richly branched laterals of limited growth, appearing as short feathery tufts to the naked eye. These are dichotomously arranged in *Desmarestia* and whorled in *Arthrocladia*. In *Phaeurus* these laterals are coarse, unbranched structures forming a dense and relatively persistent covering to the longer axes. The thalli are usually attached by a broad discoid holdfast.

The general plan of construction of the thallus is similar in *Arthrocladia* and *Desmarestia*. Transverse or longitudinal sections through any part of the plant show an axial series of large and much elongated cells *(a)* which, in the older regions of a *Desmarestia* have thick walls. In the mature parts these axial cells are surrounded by a broad cortical envelope *(co)* exhibiting differentiation into a few layers of small peripheral cells, packed with lenticular chromatophores, and a wider inner region; its large cells contain few or no chromatophores and are elongate with blunt, tapering ends. Interspersed in the cortex of *Desmarestia* are more or less numerous smaller cells, which belong to hyphae produced from the inner cortical elements of several layers of small cells with chromatophores. *Arthrocladia* lacks these complications.

The origin of the mature structure can be traced in the brushlike tufts crowning the tips of the diverse branches in young growing plants. Essentially the same features are shown by the numerous laterals of limited growth referred to above; these are far more conspicuous in ZX *aculeata* than in *D. ligulata*. The cells composing these tufts are densely filled with chromatophores, and they no doubt constitute an important photosynthetic system in the growing thallus. Examination of one of the terminal tufts shows that, towards the base of the evident axial row (a), there is an intercalary meristem *(m)* composed of flat cells and cutting off elements both above and below. The former replenish the terminal hair (A), the uppermost cells of which are constantly wearing away, and each of these segments usually bears one or two unbranched laterals, which lengthen with the help of a basal meristem.

The numerous segments cut off below the meristem grow out at their upper ends into uniseriate laterals which are paired in *Desmarestia* and formed in whorls of four in *Arthrocladia*; at first these lengthen by means of a basal meristem. These primary laterals later produce delicate branches in opposite pairs (sometimes whorled in *Arthrocladia* and thus arise the numerous short pinnate tufts that are arranged along the sides of the long axes. Subsequently the segments bearing the primary laterals exhibit further transverse divisions—mostly two in *Desmarestia*—so that in the older parts only every fourth cell of the series usually bears laterals; since the axial cells gradually undergo marked elongation, 'the laterals ultimately become widely spaced.

From the basal cells of the primary laterals, just below the intercalary meristem, the first constituents of the cortical envelope are formed. According to recent accounts the basal cells cut off successively a number of small elements, which multiply by anticlinal division and gradually form a complete one-layered envelope around the elongating axial cell below the lateral. The cells of this envelope soon undergo periclinal division and, in the further course of events, the outer layer thus produced behaves as a meristem cutting off on its inner side successive elements which enlarge and lengthen to form the cortex enveloping the axial cells. As a result the basal parts of the laterals are gradually buried.

According to Jdnsson the cells cut off from the superficial meristematic layer enlarge, but show no further division, the meristem accommodating itself by frequent anticlinal division to the progressive widening and elongation of the axis. Its cells at first exhibit a palisade-like shape and, both in this respect and in the faculty of adding new elements to the internal tissue, this layer resembles the meristoderm of Laminariales. In older parts of the thalli the small-celled surface-layers constitute the main photosynthetic system and contrast sharply with the large and often nearly colourless cells of the cortex. It is not clear how long the meristoderm remains active or whether, as seems probable, it divides afresh in each season in the older perennial parts of *D. aculeata*.

As the cortical envelope arises, the meristerns of the primary laterals take up a position above their base and commence to divide like that at the tip of the main axis; the lower (inner) segments lengthen and divide in the same measure as the lateral becomes buried, so that its meristem remains at the edge of the expanding axis. Such growth of the laterals is specially marked in the ligulate (leafy) species of *Desmarestia*, where the superficial meristoderm undergoes very plentiful anticlinal division so that the cortical envelope rapidly broadens out to form a flat expanse.

In *Desmarestia* the primary growth above described is succeeded by considerable secondary activity. The cells of the inner cortex grow out into narrow, septate, branching hyphae which push their way between the elements of the cortex and grow mainly in the downward direction at first their cells contain chromatophores. These hyphae constitute the small elements seen interspersed between the larger ones in transverse sections. They are abundant in the older thalli and are specially numerous in the midribs of the ligulate species; in D. *aculeata* they increase in number with each season. Those in the basal parts have thick walls and, as a result of extensive branching, form the bulk of the tissue. These structures no doubt fulfil a mechanical function and in this connection it may be noted that the breaking strain of D. *aculeata* may amount to 12 Ib.

In the interior of the attaching discs hyphae vastly exceed in number the few primary cells. The upper surface of these discs is, however, composed of vertical rows of rectangular cells, which show a sharp demarcation between the growth of successive seasons owing to the greater length of the cells formed in the early part of each year.

A further secondary development, that sets in at an early stage, leads to the formation of an internal system of cells with chromatophores. This originates from richly branched hyphae produced from the cortical cells, immediately surrounding the axial thread. The structures in question grow both in the upward and downward directions and become compacted to form a 3-4-layered envelope of small cells with abundant chromatophores which directly invests the axial cylinder. It has been suggested that this utilises the carbon dioxide of respiration which is unable to diffuse into the surrounding water. Occasional larger elements surrounded by a similar envelope of small cells with chromatophores may be observed in the inner cortex of older parts. The large elements in question arise as downgrowing branches of the axial cells or of their primary laterals. According to Pease such structures are abundantly developed in the ligulate species of *Desmarestia,* where they constitute a branching system of large-celled threads connected directly or indirectly with those of the axial row. They may also occur amid the hyphae of the attaching disc. In *D. aculeata* Jonsson describes a further system of cells with chromatophores, produced by outgrowth of those surrounding the axial elements and constituting a weft of hypha-like threads within the walls of the latter.

The elongate axial cells traversing all parts of the thalli are thus provided with more or less numerous branches, and the whole probably constitutes a conducting system. The horizontal septa of the component elements show fine pits, which are at first irregularly scattered, although later aggregated into groups, usually of 4 or 5; there is some evidence that these pits are traversed by cytoplasmic threads. Jonsson also records pits in the transverse septa of the hyphae and, on the walls between the axial cells and the enveloping secondary system, numerous semi-lunar areas provided with a fine pitting.

The growth of a main axis of a *Desmarestia-thallus* therefore involves (i) the production from the intercalary meristem of new axial segments bearing laterals; (ii) the formation of the cortical envelope and its gradual increase in thickness by division of the surface-layer; and (iii) the secondary activity due to the production of hyphae from the cortical cells and of downgrowing branches from the main axial series. In *Arthrocladia* only the first two stages of this growth are seen. Only a small proportion of the primary laterals develop into branches of unlimited growth. The majority cease to grow when they have reached a certain length, and at the end of the active season both the lateral and its meristem are shed; according to Pease such abscission takes place at a definite predetermined point. Simultaneously the cessation of meristematic activity in the long axes is marked by the shedding of the terminal tuft. In the ligulate forms the basal parts of the branches are entirely buried in the flat expanse of cortical tissue, within which they sometimes appear as lateral veins, while in *D. aculeata* they persist for a time as the thorn-like outgrowths, which are a familiar feature of this species when active growth is not taking place. In *Arthrocladia* Sauvageau also records the origin of laterals of limited growth from the superficial layer of the cortical envelope.

In *Desmarestia aculeata* the new growth in spring originates from the axils of the laterals of previous seasons, or more rarely from the axils of some of the thorns. At these points there are groups of cells derived from the axial series and forming small swellings; according to Jonsson such rudiments may remain dormant for one or more seasons or be permanently arrested. They give rise to two shoots, only one of which usually develops into a long branch; this may grow to a length of upwards of 30 cm. in a single season.

Reproductive Processes

The sporangia of *Desmarestia* formed by tangential division of a Surface-cell, lie embedded in small groups in the superficial layer of the cortical envelope. In *D. aculeata* they are produced in winter after the tufts have been shed. The sporangia are of small size and give rise only to a few zoospores; Abe (0 records reduction in the first divisions in *D. viridis*. The zoospores of *D. aculeata* possess a single chromatophore *(c)* with the usual apposed eyespot *(s)*; it has not been possible to trace the origin of the long flagellum to this point.

In *Arthrocladia* the sporangia form moniliform chains on the adaxial side of the branches of the first or second order of the whorled laterals of limited growth; each sporangium produces about 16 swarmers. Dehiscence by a lateral pore is practically simultaneous in all the sporangia of a chain. Even young plants of *Arthrocladia* may already bear sporangia.

The zoospores of *Desmarestia* germinate according to the usual pattern. The two types of prothalli become distinguishable at an early stage, since the female have larger cells than the male. The tubular oogonia dehisce apically to liberate the single female cell *(o)* which, as in Laminariales, usually remains adhering to the gelatinised aperture. The male gametophytes are generally more richly branched and comprise more numerous cells which contain fewer chromatophores. The numerous oblong antheridia *(a)* occur singly or in small groups at the ends of the branches, their highly refractive contents including only a single pale chromatophore. The solitary spermatozoid is liberated through a narrow apical aperture. Actual fertilization has not been observed.

Abe describes the fusion of swarmers from unilocular sporangia of the same individual; the germlings of these zygotes resemble those formed from the zoospores, although they remain sterile.

The zoospores of *Arthrocladia* give rise to extensively branched heterotrichous growths which appear to be all of one kind. The erect threads bear at their tips and along their sides large rounded cells *(r)* of uncertain function. In Sauvageau's cultures young sporophytes arose from the terminal cells of ordinary branches and he interpreted them as oogonia developing apogamously. He also recognised doubtful antheridia, exhibiting occasional dehiscence, borne at the summit of narrower branches of the same prothalli. Germination-stages found entangled amid the latter suggest by their dimensions an origin from the escaped contents of the putative antheridia; the further fate of these germlings was not ascertained. It may be doubted whether these prothalli were normal.

Development of the Sporophyte

The zygotes of *Desmarestia* undergo successive transverse division to form an unbranched erect thread *(e)*. If the zygote becomes detached from the oogonium before germination, which is likely to be the rule in nature, the formation of branched rhizoids *(r)* sets in at an early stage. In the older embryos plentiful and mainly opposite branching of the uniseriate primary axis takes place. Schreiber draws attention to the similarity between such stages and one of the lateral hair-tufts on the adult thallus.

When the embryo is 3 to 4 weeks old, corticating threads grow out from the lower cells of the main axis in close apposition to the cells below. As they undergo septation and develop branches, the lower part of the axis becomes covered by a continuous cortical investment. At the base of the embryo the threads diverge to form a group of rhizoids (r), but farther up their cells divide to form a several-layered cortex. At this stage the axis of the embryo develops, a little way below its apex, a definite intercalary meristem *(m)*. The early development of *Arthrocladia* is unknown.

Status and Affinities of Desmarestiales

The general structure and mode of growth of the sporophytes of *Arthrocladia* and *Desmarestia* present so many points of similarity that, despite the difference in the mode of arrangement of the sporangia, a real affinity is evident. It is therefore hardly justifiable to refer *Arthrocladia* to a separate order, although its inclusion in a distinct family, as suggested by Kylin, is plausible on the basis of our present knowledge. Much will depend on the nature of the sexual generation of *Arthrocladia*. In its sporophyte it is clearly less specialised than *Desmarestia*.

The marked degree of correspondence between the prothalli of *Desmarestia* and those of Laminariales induced Schreiber actually to include the Desmarestiaceae in the latter order. This is, however, altogether unwarranted, in view of the fundamental differences between the structure of the sporophyte in the two orders. The Desmarestiales are essentially uniaxial forms acquiring a pseudoparenchymatous construction secondarily as a result of the development of a cortical investment. This cortical envelope is fundamentally homologous with the corticating threads of a *Spermatochnus* or *Sphacelaria*. In the Laminariales, on the other hand, the parenchymatous construction results from the segmentation of a primary filament; in other words the members of this order are highly specialised polystichous types. The Desmarestiales must have their origin among uniaxial forms such as occur in diverse Ectocarpales.

Despite the different mode of growth, there are considerable analogies between the structure of a *Spermatochnus* and a *Desmarestia*, as Jonsson has particularly emphasised.

Like other specialised forms the Desmarestiales have lost all traces of heterotrichy in the sporophyte, the growth of the embryo being erect from the first. This, as well as the high degree of elaboration in vegetative structure and the specialisation of the gametophytes, justifies their being grouped in a separate order.

Laminariales

The sporophytes of Laminariales include some of the largest forms known among Algae, while they surpass all other Phaeophyceae in morphological complexity and anatomical specialisation. The gametophytes, by contrast, are minute filamentous growths, bearing either oogonia or antheridia. The sporophyte is usually differentiated into a cylindrical stalk or *stipe* and one or many blades or *laminae*.

The main region of cell-division lies in the transition zone between stipe and lamina, and this intercalary growth constitutes an important distinction from Fucales. More significant differences are, however, to be found in the processes of reproduction and in the nature of the life-cycle. The sporangia of Laminariales usually form extensive sori on the blades.

The members of this order are for the most part lithophytes attached to rocks by elaborate holdfasts, although they are sometimes epiphytes or found on movable substrata. They usually occur in the sublittoral region and not uncommonly form a dense belt not far below low-tide limit where the greater part of the thallus is deeply submerged; they sometimes constitute natural breakwaters.

On North European shores they generally follow immediately below the characteristic Fucaceous girdle of the littoral region. Many of the larger Pacific forms, such as *Macrocystis* and *Nereocystis*, are anchored in relatively deep water, the fronds alone floating at the surface by virtue of the great elongation of the stipes and the presence of air-bladders at the bases of the blades.

Life-cycle in Ectocarpales

Of the two main types of reproductive organs met with in Ectocarpales, the unilocular sporangia appear, almost without exception, to constitute asexual organs which produce zoospores germinating without fusion to form a potential gametophyte. The first nuclear divisions in the unilocular sporangium have been shown to bring about reduction in diverse Ectocarpales a fact first demonstrated by Knight and confirmed by Dammann. While there are probably a number of instances in which meiosis is suppressed, there is little evidence that reduction can occur at any other stage in the life-cycle. Unless there is proof to the contrary, therefore, the swarmers formed from unilocular sporangia must be regarded as haploid, and the plants bearing such sporangia as diploid and sporophytic.

On the other hand the motile elements produced in the plurilocular sporangia probably always have the same chromosome-number as the parent plant. Where such sporangia occur on plants also bearing the unilocular type, the resulting swarmers are no doubt normally diploid and constitute merely a means of accessory propagation of the sporophyte. When plurilocular sporangia occur on what is known to be a haploid phase, they are invariably unaccompanied by unilocular sporangia and have in diverse instances been shown to liberate gametes which fuse to form a zygote producing a new diploid phase. The frequent absence of observed fusion cannot be taken as definite proof that the swarmers in question lack sexuality, since the artificial environment of a laboratory is only too likely to create conditions inimical to the occurrence of sexual fusion; this is moreover clearly dependent on outside factors, as shown by various data in the literature. The usual close resemblance between the accessory plurilocular sporangia on diploid individuals and those which function as gametangia on the haploid phases, is one of the reasons why the life-cycle of Ectocarpales so long remained obscure.

The investigations of the last two decades have made it

clear that the complete life-cycle in this order involves an alternation between *(a)* a diploid individual bearing asexual (unilocular) sporangia (often accompanied by accessory plurilocular sporangia) and *(b)* a haploid individual bearing plurilocular sporangia which function as gametangia. In the simple filamentous Ectocarpaceae the two generations are essentially alike and the alternation is isomorphic.

The elaboration, which the thallus has undergone in the more advanced members of the Ectocarpales, has in a few instances affected both sporophyte and gametophyte, so that the alternation is likewise isomorphic. More usually, however, only the sporophyte has been elaborated, whilst the gametophyte remains a small ectocarpoid or streblonemoid structure so that alternation has become heteromorphic, Fundamentally, however, wherever alternation occurs in Ectocarpales it is derived from a state with two identical generations, i.e., using the term in its old sense, alternation throughout the Ectocarpales is originally homologous.

In considering this interpretation it must be borne in mind that the complex sporophyte of the Ectocarpales practically always originates from an erect thread or threads of an heterotrichous filament, essentially—though not in detail—resembling the simple filamentous gametophyte. The sporophyte does not ordinarily become fertile until the mature structure has been attained, whereas the gametophyte, in forms with heteromorphic alternation, forms reproductive organs while still in a filamentous condition. Precocious fertility on the part of the sporophyte is, however, frequent.

Forms with Isomorphic Alternation

Two practically identical generations are probably not uncommon in *Ectocarpus* and its allies. In *E. siliculosus* Berthold showed that the plants found at Naples during spring reproduce sexually by means of gametes liberated from the plurilocular sporangia, the only kind of reproductive organs present. The fate of the zygote was not clearly established, although some evidence was produced that it developed into a small plant bearing uni- and plurilocular sporangia. Sexual reproduction in *E. siliculosus* at Naples has since been studied by diverse investigators and Knight showed that the plants producing gametes are haploid. Berthold's conclusions as regards the sporophyte have also been confirmed; at Naples it is a winter-form, considerably smaller than the gametophyte and bearing, apart from unilocular, accessory plurilocular sporangia.

A similar alternation has been established in *E. siliculosus* by Foyn on the west coast of Norway and by Papenfuss at Woods Hole in North America. The diploid plants at Woods Hole, however, are more robust than the haploid ones, possessing bigger cells and considerably larger plurilocular sporangia. It would thus seem that in northern waters, by contrast to the condition obtaining at Naples, the diploid plant of the species under discussion is more vigorous than the haploid one. This is in harmony with the fact that Knight in the Isle of Man found only diploid plants, while Kylin, on the west coast of Sweden, records plants (assumed to be diploid) with mainly plurilocular sporangia that liberate asexual swarmers. The occurrence of reduction in the unilocular sporangium of *E. siliculosus* first established by Knight, has since been confirmed on material from diverse regions.

While, therefore, *Ectocarpus siliculosus* in various localities exhibits a distinct alternation with varying preponderance of sporophyte or gametophyte, there is some evidence that yet another type of life-cycle may occur. Knight, in her British material, records fusion of the swarmers from the unilocular sporangia, and a similar state of affairs has been reported from the Adriatic. Knight's data are perhaps not altogether convincing, but analogous observations since published strengthen the indications that such a behaviour on the part of these normally asexual swarmers is possible.

All who have studied the sexual reproduction of *E. siliculosus* have reported that a certain percentage of the gametes fail to fuse. Such gametes (apparently mainly the female ones, however can develop apogamously with the production of further plants bearing plurilocular sporangia only, and in this way accessory reproduction of the haploid phase takes place. In the normal life-cycle of the species under discussion, therefore, either phase can propagate itself indefinitely, the diploid phase with the help of the accessory swarmers from the plurilocular sporangia, the haploid phase with the help of apogamous gametes.

The successive individuals, which may thus for long periods perpetuate the diploid or haploid phase as the case may be, do not in *Ectocarpus* ordinarily appear to display any essential differences from the normal adult thallus. Instances are, however, on record in which both the accessory sporophytic phases derived from the swarmers of the plurilocular sporangia and the gametophytic phases derived from the swarmers of the unilocular sporangia, show simplification as compared with the adult thallus. Thus, in *E. tomentosus* such stages are described as simple ectocarpoid structures, showing nothing of the fasciculate character typical of the adult plant; they are also distinguished by the hairs which they bear.

There are indications that other species of *Ectocarpus* exhibit an analogous life-cycle. Thus, according to Sauvageau, *E. Hincksiae* possesses two types of individuals, the one bearing unilocular, the other plurilocular sporangia; in either type plurilocular sporangia may also occur on the prostrate system. The individuals with plurilocular sporangia only are invariably epiphytic on *Saccorhiza*. The lack of viability of the swarmers from the plurilocular sporangia on the erect filaments is regarded as a possible indication of their sexual nature, although fusion has not been observed. The scarcity of plants of *E. Hincksiae* with unilocular sporangia may be due to their being winter-forms or to their growth in deeper water. The reproduction of *E. Mitchellae* also furnishes evidence of isomorphic alternation.

Such a life-cycle evidently also occurs in *Pylaiella*. The individuals of the true *P. littoralis* on the coasts of Scandinavia for the most part bear either uni- or plurilocular sporangia, the diploid ones being in general larger than the haploid and showing certain other differences. In Britain Knight records plurilocular sporangia also on the diploid individuals, while haploid plants are very rare (cf. *Ectocarpus siliculosus*). Kylin finds diploid plants all the year round on *Ascophyllum* (more rarely on *Fucus*), while the haploid ones, occurring on *Sertularia* or on *Ceramium rubrum* epiphytic on *Ascophyllum*, appear in spring and disappear by midsummer. In the Isle of Man the haploid phases are found in spring and, like the diploid ones during early summer, occur on *Ascophyllum*, while later in the season the diploid phase spreads to species of *Fucus*. A seasonal alternation between a diploid spring phase and a probable autumnal haploid one is also implied by the observations of Johnson and York.

Fusion of the swarmers from the gametangia on the haploid phases has been observed, although apogamy is also recorded. Both Knight and Kylin have in cultures grown well-developed gametophytes bearing plurilocular sporangia from the zoospores of the unilocular sporangia of the diploid individuals. The former also records fusion of zoospores, although this was not observed by either Dammann or Kylin. In view of the scarcity of haploid plants in the Isle of Man, *Pylaiella littoralis* may display the same tendency as *Ectocarpus siliculosus* towards an elimination of the gametophyte.

Dammann's record of unilocular sporangia on the haploid individuals of *Pylaiella*, the only instance of the kind so far reported, is believed by Kylin to be due to confusion between the two species of *Pylaiella*. In *P. rupincola* only individuals with unilocular (and occasional plurilocular) sporangia are known, and the swarmers of the former produce new individuals with unilocular sporangia. This can only be explained by the supposition that there is here no reduction in the unilocular sporangium.

Among the more specialised Ectocarpales isomorphic alternation appears to be rare, although it probably occurs in *Nemoderma*, *Lithoderma*, and possibly other Myrionemataceae.

It must be noted, however, that neither in *Lithoderma fatiscens* Kuck., nor in *Nemoderma,* has a mature plant with unilocular sporangia been grown from the zygote, nor are there at present any cytological data. Apogamy has been observed in both genera.

Another probable instance of isomorphic alternation is furnished by *Heterochordaria abietina* in which uni- and plurilocular sporangia occur on distinct individuals. Since heteromorphic alternation obtains in diverse Mesogloeaceae, the isomorphic life-cycle of *Heterochordaria* would remove it from that family. Reduction in the unilocular sporangium has been established and, according to Abe, the swarmers of the plurilocular sporangia are gametes; there is strict dioecism, the female swarmers being slightly larger and alone capable of apogamous development. Abe also describes a fusion of swarmers from the unilocular sporangia, usually from those of different individuals.

Forms with Heteromorphic Alternation

Heteromorphic alternation is definitely established for members of six families of Ectocarpales. These are: Mesogloeaceae (*Castagnea virescens; Mesogloea vermiculata*); *Sphaerotrichia divaricata* Spermatochnaceae (*Spermatochnus paradoxus; Stilophora rhizodes* Punctariaceae (*Phloeospora brachiata*); Asperococcaceae (*Asperococcus bullosus*); *A. fistulosus;* Encoeliaceae (*Soranthera ulvoidea);* and Dictyosiphonaceae (*Dictyosiphon foeniculaceus*). In all instances the gametophyte is a minute filamentous plant), with the general characteristics of a small *Ectocarpus* or similar form. The occurrence of reduction in the unilocular sporangium has been established in *Castagnea vtrescens, Mesogloea vermiculata, Myriogloea Sciurus, Spermatwhnus paradoxus, Phloeospora brachiata,* and *Asperococcus bullosus.*

Plurilocular sporangia do not occur frequently on the diploid individuals of the species listed above, being recorded only in *Castagnea virescens, Spermatochnus paradoxus Phloeospora brachiata*, and *Asperococcus bullosus*; even in these, moreover, individuals with unilocular sporangia only are often reported. Among the more advanced Ectocarpales there is thus a tendency to eliminate, either entirely or for the most part, the accessory reproduction by diploid swarmers from plurilocular sporangia which is so frequent among the less specialised members. In this respect the Ectocarpales show a definite evolutionary trend in the direction of the higher Phaeophyceae where such accessory reproduction of the diploid phase does not occur.

Among the species cited above the full life-cycle has only been established in *Asperococcus bullosus* and *Phloeospora brachiata*, but the available data suffice to indicate that the general outline is the same in the other species. The lacunae concern particularly the observation of sexual fusion and the development of a new sporophyte from the zygote. The type of alternation under consideration was first demonstrated in *Dictyosiphon* by Sauvageau.

Diverse records in the literature indicate the possibility of occasional fusion of the swarmers from the unilocular sporangia among more advanced Ectocarpales. This was first reported by Areschoug in *Dictyosiphon hippuroides*, although it remains doubtful whether this was a true instance of fusion records such fusion in *Asperococcus fistulosus*, *Myriotrichia clavaeformis*, and *Punctaria plantaginea*, while Hygen assumes its occurrence in *Sphaerotrichia divaricata*. There are, therefore, data supporting the existence of a tendency towards the elimination of the gametophyte in these advanced types, similar to that noted in Ectocarpaceae.

Evidence of the occurrence of heteromorphic alternation is available for diverse Ectocarpales in addition to those listed above. In many of these the macroscopic plant bears both uni- and plurilocular sporangia and, although one or other type usually predominates at a certain stage of the life of the individual, there is generally a longer or shorter period of overlap. The unilocular sporangia may be formed before the plurilocular ones (*Leathesia difformis*, but more usually the sequence is reversed (e.g. *Stilophora adriatica; Punctaria latifolia; Myrionema strangulans; Litosiphon pusillus; Myriotrichia repens; Protasperococcus myriotrichiformis*. In certain instances the two kinds of sporangia tend to occur on distinct individuals which, in *Asperococcus* for example, may show differences in size. In general the swarmers from the two types of organs are alike, but in *Litosiphon pustllus* and *Spermatochnus* those from the plurilocular sporangia are described as being markedly larger and exhibiting a less rapid movement.

There can be little doubt that in all the Ectocarpales now under consideration the ordinary plant is diploid and that the swarmers produced in its plurilocular sporangia are accessory and bring about multiplication of the sporophyte before the unilocular sporangia appear. With few exceptions these macroscopic thalli are non-perennial structures, only found for a few months, although the time of their appearance varies in different geographical regions.

The restricted period of occurrence is particularly evident in those genera which possess a large thallus, where presence or absence is easily established. The manner in which such forms survived during the period of absence of the macroscopic plant was long a mystery, since no definite resting stages have ever been observed in the life-cycle of Ectocarpales or for the matter of that in any seaweed. Sauvageau was the first to elucidate this point.

He provided evidence that, during the period of absence, these forms persist as minute and largely prostrate filamentous stages, arising from the swarmers either of the uni- or plurilocular sporangia of the macroscopic plant. These germlings are sometimes sterile, but more usually they bear plurilocular sporangia, the swarmers of which can produce several successive generations of such plants.

Sauvageau calls the macroscopic stage the delophycean and the microscopic ephemeral one the adelophycean. The latter sometimes shows a curious dimorphism, some of the plants being discoid or myrionemoid in form, whilst others are streblonemoid, a phenomenon referred to by Sauvageau as heteroblasty. Kylin is of the opinion that the difference results from the degree of rapidity with which the swarmer settles on the substratum. According to Knight the alternation between the macro- and microscopic phases may be related to the availability of nitrates and phosphates.

Sauvageau distinguishes three different categories among these filamentous stages. When the swarmers which they produce exhibit sexual fusion, he speaks of the parent-structures as *prothalli.* When there is no evidence of sexuality on the part of the swarmers, the filamentous stages—whether derived from pluri- or unilocular sporangia—are designated *plethysmothalli.* Sometimes such stages directly produce a new macroscopic plant and are then termed *protonemata;* they may at the same time bear plurilocular sporangia.

All filamentous stages derived from swarmers of unilocular sporangia must, until evidence to the contrary is forthcoming, be regarded as haploid; they are potential gametophytes and the reproductive organs they bear are gametangia. The frequent apogamous development of their swarmers may or may not be a natural condition. It is important to note that, wherever the fate of the swarmers from unilocular sporangia has been studied among the more advanced Ectocarpales, the resulting prothalli bear *plurilocular sporangia only;* there is, moreover, no good evidence that they ever give rise directly to the adult thallus.

The swarmers from the plurilocular sporangia of the prothalli have often failed in cultures to afford any evidence of sexuality, but have germinated direct to give similar prothalli bearing similar sporangia, and in certain instances several successive generations of such stages have been obtained. These are some of the plethysmothalli of Sauvageau. They are, however, certainly different in nature from the plethysmothalli formed from the swarmers of the plurilocular sporangia of the sporophyte. This is in agreement with the opinions expressed by Kylin.

It is possible that extensive apogamous development of the gametes may take place in nature as in cultures, and that a succession of gametophytes is formed, of which the ultimate one produces gametes capable of sexual fusion and thus re-establishes the diploid phase; alternatively, the first-formed gametophyte may give rise to zygotes which develop a succession of similar plethysmothalli. Perhaps this is one means of perpetuation during the unfavourable period of the year. The fact, however, that sexual fusion has so rarely been observed in cultures may not be taken as a definite indication of its scarcity or absence in nature, since the conditions afforded by laboratory cultures are only too likely to be somewhat abnormal. In this connection attention may be drawn to the fact that in *Asperococcus bullosus* Sauvageau failed to observe sexual fusion, whereas this has now been established; the same is true of *Mesogloea vermiculata, Spermatochnus paradoxus,* and *Castagnea virescens.*

The diploid swarmers from the plurilocular sporangia of the macroscopic plant give rise to filamentous stages quite similar to the gametophytes, but in no instance among the many studied has a sexual fusion of the swarmers they produce been observed. It is for these minute diploid stages that, in my opinion, the term *plethysmothallus* should be reserved. Mostly they bear plurilocular sporangia, from which a succession of such (presumably diploid) plantlets can be produced; sometimes, however, they form both kinds of sporangia, as in *Myriotrichia repens* and *M. adriatica* and *Litosiphonpusillus,* although this appears to be unusual.

A feature, which speaks strongly in favour of the diploid nature of the plethysmothalli, is that sooner or later they commonly give rise directly to new plants (probable sporophytes;). In other words their erect threads are capable of developing into the macroscopic plant as soon as conditions render this possible. In cultures of the diploid swarmers from the plurilocular sporangia of the sporophyte, therefore, a succession of plethysmothalli may be obtained or a new sporophyte may be produced at an early stage. Which of the two occurs, no doubt depends not only on the season of the year in which the cultures are set up, but also, and perhaps more generally, on conditions in the cultures being suitable for the development of the macroscopic plant.

The plethysmothallus may give rise to the sporophyte as a lateral outgrowth, without itself producing any reproductive organs and, when this is so, Sauvageau speaks of a *protonema*. In such instances, however, as always among Ectocarpales, the adult thallus is merely an erect branch or branches of the primary heterotrichous stage constituting the so-called protonema. Its mode of development does not differ from that found in those plethysmothalli which simultaneously bear plurilocular (sometimes also unilocular) sporangia.

Such differences appear to be fortuitous, the plethysmothallus being sterile or fertile according to circumstances, and I am of the opinion that no importance is to be attached to such variations. The diploid plethysmothallus must be regarded as a sporophyte, which remains for long periods arrested at the ectocarpoid stage, during which it can multiply by swarmers from plurilocular sporangia; similarly, it may continue to form such swarmers on the primary filaments when some of its erect branches are developing into the macroscopic sporophyte.

The succession of plants formed from the diploid swarmers of the plurilocular sporangia of an *Ectocarpus siliculosus* or other simple form are just as much plethysmothalli; but here, owing to the non-elaboration of the sporophyte, they are in general identical in their morphological features with the latter. A large proportion of the advanced Ectocarpales thus evidently possess a life-cycle with a heteromorphic alternation. The diploid phase is constituted by the elaborate thallus, which bears either unilocular sporangia only or plurilocular ones as well, the latter then being most commonly produced at an earlier stage than the former. The haploid phase is a small filamentous plant bearing gametangia only. Either phase can probably propagate for several generations by means of small ectocarpoid stages, the diploid one by the swarmers from the plurilocular sporangia which give rise to plethysmothalli, the haploid one by means of apogamously developing gametes which give rise to accessory gametophytes. The plethysmothalli are to be regarded as arrested (juvenile) stages of the sporophyte. In how far this life-cycle can be short-circuited by the fusion of swarmers from the unilocular sporangia remains to be seen.

Where both plethysmothalli and gametophytes are known, they do not in general appear to differ much in their vegetative structure, and sometimes there is almost complete identity. This is what one would expect on the view above propounded that the alternation is derived from an isomorphic one in which the one generation (the sporophyte) has been elaborated, while the other has remained at the stage of the simple filament. Certain differences in the early stages of germination and in the later growth are noted by Kylin in *Litosiphon pusillus*, as well as by Sauvageau in *Punctaria latifolia* and *Spermatochnus*, but in all such instances the differences are not fundamental. Sauvageau records the persistence of sterile gametophytes over long periods.

Modified Types of Life-cycle

It is evident that some Ectocarpales show more or less marked departures from the type of life-cycle indicated above. In *Striaria attenuata* and *Dictyosiphon Chordaria*, in which only unilocular sporangia have so far been found, the swarmers in cultures give rise to filamentous growths, which, after some time, produce erect branches that divide to form the adult thallus regards these forms as diploid ones in which reduction in the unilocular sporangium has been suppressed and on the available evidence this would appear to be the only valid interpretation. In *Hapterophycus* Hollenberg found evidence of failure of meiosis in certain of the unilocular sporangia, and this possibly indicates how the condition in the species under consideration has arisen. It is to be noted that some of their immediate allies show a normal alternation.

In *Dictyosiphon chordaria,* Fsyn succeeded in growing four successive generations without seeing any plurilocular sporangia. A similar condition seems to exist in *Asperococcus compressus,* where, however, rare plurilocular sporangia have also been recorded. Sauvageau has grown new plants from the swarmers of the unilocular sporangia.

Other instances are afforded by *Pylaiella rupincola* and *Elachista fucicola.* In *E. stellaris,* where the macroscopic plant also bears plurilocular sporangia, the swarmers from the unilocular ones can reproduce the plant directly. For all three Kylin advances the same explanation as above.

It has long been known that the thalli of *Scytosiphon, Petalonia,* and *Colpomenia* bear plurilocular sporangia only. Fusion of the swarmers has repeatedly been reported in *Scytosiphon lomentaria;* Kuckuckd also recorded such fusion in material of *Petalonia zosterifolia* from the Adriatic. Others find that the swarmers of *Scytosiphon* give rise, without fusion, to a system of creeping filaments which may perhaps reproduce by plurilocular sporangia, but more usually unite to form a disc and regenerate a new plant. For *Petalonia debilis* and *P. Fascia* it has likewise been shown that the swarmers can give rise to new plants direct. According to Sauvageau a similar development occurs in *Colpomenia.*

It is difficult to reconcile the data on direct development of a new thallus with the records of sexual fusion between the swarmers from the plurilocular sporangia. They suggest that there may be several types of the latter, and this is supported by the recent observations on *Colpomenia.* Kylin, who regards the three genera under discussion as diploid forms which have lost the unilocular sporangia, considers the fusions that have been described as misinterpretations. In this connection attention may be drawn to Reinke's record of the occurrence of double zoospores *Petalonia,* although he found no evidence of copulation. The possibility, however, that reduction might take place during the development of certain of the plurilocular sporangia cannot be altogether ruled out.

Much the same position exists in *Giraudya*, where Sauvageau records a succession of plethysmothalli before the adult thalli reappear. Several species of *Stictyosiphon* too are only known to possess plurilocular sporangia, although unilocular ones are found in *S. adriaticus*. Kuckuck records fusion of the swarmers in *S. tortilis,* whereas Sauvageau observed a direct development into new plants in *S. Corbierei.*

Mention must also be made of *Ascocyclus orbicularts,* in which direct development of new individuals from the product of germination of the swarmers of the plurilocular sporangia appears to be the rule, although Sauvageau believes he has on one occasion obtained some evidence of sexual fusion. The recent discovery of unilocular sporangia supports the view that *Ascocyclus* is diploid and renders improbable the opinion expressed by Parke that this algae is a stage in the life-cycle of *Castagnea Zosterae*.

3

Phaeophyceae

The numerous seaweeds have long been recognised as a coherent entity, though with increased knowledge of the processes of reproduction and of the course of the life-cycle diverse views as to their taxonomic arrangement have been formulated. The delimitation here adopted corresponds in many respects with the Ectocarpales of Oltmanns, although he included in the latter the Sporochnales and Desmarestiales, which are now regarded as constituting separate orders.

General Characteristics

The Ectocarpales as thus delimited include the least specialised members of the Brown Algae and present a wide diversity of structure, ranging from the simple filamentous types of the Ectocarpaceae to relatively elaborate thalli. The latter are produced either by aggregations of filaments and of their branches (multiaxial as in *Leathesia*, and *Castagnea*, uniaxial as in *Spermatochnus*) or are formed by the repeated subdivision of the cells of primary erect filaments (parenchymatous types like *Punctaria*,and *Scytosiphon*). Despite the considerable range in outward form and internal structure of the mature sporophytes, the younger stages of nearly all Ectocarpales show a clear derivation from the heterotrichous filament seen in *Ectocarpus* and other Ectocarpaceae. Development commences with the production of a system of prostrate filaments from which a number of upright branches sooner or later arise; one or more of the latter develop further to produce the mature plant. In other words the thallus, in all advanced Ectocarpales, is an elaboration of the erect threads of a primary heterotrichous stage. The growth of the thallus is diffuse or often accomplished by well-defined intercalary meristems, but apical growth is met with in certain advanced forms.

Many Ectocarpales bear colourless multicellular hairs which commonly grow with the help of a basal meristem. A production of hairs seems to be a general characteristic of the primitive heterotrichous forms, being seen also in the Chaetophorales and in the Nemalionales among Red Algae, although the type of hair is variable. In the Ectocarpaceae the hairs usually terminate the branches of the filaments, whilst in the more elaborate Ectocarpales they either occur scattered or united into characteristic groups, which are sometimes sunk in cavities. In some genera, especially among Myrionemataceae, the hairs are provided with a basal sheath. Very similar to the colourless hairs are the structures (assimilatory hairs) found in *Elachista* and its immediate allies, in which the cells contain chromatophores. It can hardly be doubted that the two are homologous, although it is at present not clear which is the more primitive.

Berthold was of the opinion that the colourless hairs serve as a light-screen, whilst others have regarded them as structures serving mainly for the absorption of mineral nutriment and for gaseous exchange. The degree of development of the hairs is very variable, but little is known of the conditions that determine this; Kylin found that germlings of *Stilophora rhizodes* and *Asperococcus bullosus* produced hairs only in those cultures in which the solution was changed daily.

The typical life-cycle in the Ectocarpales comprises an alternation between diploid and haploid phases, either of which seems to be capable of indefinite propagation, the diploid by accessory swarmers, the haploid by apogamous gametes. Although diverse departures from this normal life-cycle are known to occur, an alternation of phases is probably of wide occurrence. In the Ectocarpaceae this alternation is essentially isomorphic, but in most of the advanced Ectocarpales it is heteromorphic, the elaboration of the upright system of the primary heterotrichous stage affecting only the diploid I phase (sporophyte), while the haploid one (gametophyte) remains a mere filament. In other words, the sporophyte has evolved, whilst the gametophyte remains at the simple level of the Ectocarpaceae. In a few specialised Ectocarpales, however, the alternation is seemingly isomorphic, both phases having undergone elaboration.

Apart from the underlying heterotrichy, the Ectocarpales are characterised by asexual reproduction by zoospores and by a process of sexual reproduction which is for the most part isogamous, though there is usually a physiological distinction between the fusing gametes. This physiological anisogamy, which also commonly betrays itself in a different behaviour of the two gametes, is, in a small number of Ectocarpales, accompanied by marked morphological differentiation, in certain instances leading to a state of anisogamy which is not far removed from true oogamy. So far, however, no example of sexual reproduction has become known among Ecto- carpales in which the female cell is non-motile.

As in other Phaeophyceae, the reproductive cells are invariably produced in specially differentiated organs, the unilocular and plurilocular sporangia. The former are always confined to the diploid phase and undergo meiosis during the formation I of zoospores, so that the products of these zoospores are normally\haploid and gametophytic. Plurilocular sporangia occur both on the diploid and haploid phases, being the only organs of reproduction found on the latter, where they constitute gametangia. The plurilocular sporangia of the diploid plant exhibit no reduction division in the production of their swarmers, which are therefore diploid and serve to propagate the phase upon which they occur. The two phases of the life-cycle thus contrast also in their reproductive organs, the Haploid bearing only plurilocular, the diploid usually both pluri- and unilocular sporangia.

Geographical Distribution and Occurrence

The Ectocarpales attain their chief development, both as regards abundance of individuals and numbers of genera and species, on temperate and polar (us) shores. Many of the genera and species that occur in the colder seas of the Northern Hemisphere are, however, also recorded from the Mediterranean (40), (64), (98) and other warmer seas (33). In the actual Tropics, on the other hand, representatives of this order are relatively rare, although members of the Encoeliaceae *(Colpomenia, Hydroclathrus, Chnoospord)* are widespread and the genus *Ectocarpus* seems to be practically ubiquitous. Certain genera (e.g. *Myriogloea, Splachnidium, Scytothamnus)* appear to belong essentially to the Southern Hemisphere, although penetrating some way towards the Tropics, but information as to the distribution of the southern Ectocarpales is still rather fragmentary. *Coilodesme* is characteristic of the Pacific coast of North America, though a few species also occur in Japan; another widespread North Pacific type is *Heterochordaria*. Several genera *(Geminocarpus, Caepidium)* appear to be confined to the Antarctic.

The majority of the Ectocarpales are annuals, the diploid phases often being represented only for a few months in the year, while at other times they are lacking. During the period of absence they persist as minute filamentous sporo- or gametophytic stages, although their mode of occurrence in nature is at present hardly known. The sporophytes in large part inhabit the lower part of the intertidal zone or the sublittoral region, but *Ralfsia*, some species of *Pylaiella, Elachista,* and *Punctaria,* as well as *Chordariaflagelliformis, Dictyosiphonfoeniculaceus,* and *Scytosiphon Lomeniaria* are commonly met with in the upper part of the littoral.

General Features of Vegetative Organisation

Ectocarpus, with numerous species found especially in the colder seas, appears from every point of view to be the most primitive of living Brown Algae. The prostrate system of the heterotrichous filament is well marked in diverse species, although it is uncertain whether it occurs in all; as in other heterotrichous types, it evidently varies in its degree of development. The erect system not uncommonly, and perhaps typically, consists of but few erect, copiously branched filaments, but there would seem to be all gradations between this condition and that in which the prostrate base bears a more or less considerable number of little branched erect threads and itself constitutes an essential part of the vegetative system; such forms grade over to the reduced types considered below and to the Myrionemataceae.

Species with well-branched erect threads form the brown tufts which are familiar objects attached to the rocky sides of tidal pools and to other larger Algae, both in the littoral and sublittoral regions. There is commonly a marked difference in size between the main axes and the later branches, which usually terminate either in a point or in a series of elongate tapering cells, with few or no chromatophores, forming a hair *(E. siliculosus)*. The branches arise from just beneath the septa and frequently retain their lateral position, although detachment may occur, resulting in apparent dichotomy.

Attachment is commonly aided by the outgrowth of rhizoids from the lower cells of the main axis, and these may arise in such numbers as to form a basal cortication. In several species the erect threads tend to coil around one another or to become matted together. The former condition is commonly seen in *E. fasciculatus* Harv. and *E. siliculosus* while in *E. breviarticulatus* and *E. tomentosus*—the latter a frequent epiphyte on *Fucus*—the main filaments become twisted to form tufts, from 1-3 mm. thick, the individual strands of which are bound together by means of numerous short bent branches or by rhizoid-like structures. This condition recalls that found in *Spongomorpha*. Regeneration of wounded threads has been studied by Prowazek.

Growth of the prostrate system is apical, but that of the erect threads shows considerable diversity. Apical growth seems to be exceptional, although it is recorded in *E. lucifugus* Kuck. and *E. chantransioides* S. & G. In most species (e.g. *E. siliculosus)* the erect threads show diffuse growth, although in some instances there may be localisation of division to certain places. Definite intercalary meristems are found only in a limited number of species. *(E. paradoxus* Mont., *E. Lebehi* Crouan, *E. irregularis* Kiitz., etc.) in which the entire growth of the individual branches is restricted to certain zones, each situated at the base of a hair. The cells of the meristem cut off segments both above and below, the former, which are few, adding to the length of the terminal hair while the latter give rise to ordinary photosynthetic cells from which also branches are produced. This is the *trichothallic growth* characteristic of diverse Ectocarpales, as well as of other orders of Brown Algae. Some species exhibiting this type of growth (e.g. *E. irregularis*) also show divisions in the main axes.

Hairs are by no means confined to the species possessing trichothallic growth, occurring also in some with diffuse growth *(E. siliculosus)*. In certain of these (e.g. *E. criniger*) the hairs are in part narrower than the underlying thread and originate from a well-defined basal meristem.

When the erect threads are well developed, the sporangia are terminal on short lateral branches or not uncommonly are sessile on the larger ones; in a few species *(E. irregularis; E. coniferus* Boerg. They may appear in a pseudoaxillary position. The plurilocular sporangia are not infrequently surmounted by a few hair-cells. *Pylaielld* has the same structure as *Ectocarpus,* except for a tendency to longitudinal division of the cells. It is characterised by the usual intercalary position of the sporangia, the unilocular ones *(u)* occurring in short series. In the Pacific *P. Postelsiae,* in which only plurilocular sporangia are known, entire branches become converted into sporangia.

According to Kylin the common *P. littoralis* includes two different species, the one an epiphyte for which this name is retained, the other essentially a lithophyte, which is vegetatively distinguished by the coiling of the lower threads around one another; there are also differences in reproduction. The rare *P. fulvescens* Bornet is distinguished by the loose felting together of the erect threads by means of numerous rhizoid-like branches.

Several genera with well-developed erect threads show essentially the same vegetative features as *Ectocarpus*. *Pleurocladia lacustris*, a rather rare freshwater form, is distinguished by the abundant deposition of carbonate of lime between the upright threads, which gives the minute, yellowish-brown cushions a certain rigidity. The numerous branches are short and tend to be unilateral; most of them ultimately bear unilocular *(u)* or plurilocular sporangia. *Sorocarpus uvaeformis* Pringsh. a rare marine epiphyte, is characterised by bearing the small (plurilocular) sporangia in grape-like clusters on short condensed branch-systems, usually situated at the base of a hair.

In the Mediterranean *Zosterocarpus*; incl. *Prototilopteris* the plurilocular sporangia form a peripheral girdle around a central cell from which they have been cut off by longitudinal walls. The genus is also distinguished by the position of the laterals opposite the septa and the presence of enlarged fucosan-cells with greyish contents (brown after death;). Its affinities are obscure, but the longitudinal septation that occurs in the formation of the sporangia recalls the condition found in the polystichous Ectocarpales, and *Zosterocarpus* is possibly a reduced member of that series.

In the species of *Ectocarpus* with short erect threads the sporangia may arise terminally upon them *(E. faeroensis* Boerg.), but in others they are borne directly on the prostrate base *(E. elachistaeformis*, JR *Battersii*). The latter is then usually well developed and may branch so copiously as to become.

It is scarcely possible to draw a hard and fast line between the reduced forms of *Ectocarpus* and *Streblonema*; Levring, in fact, advocates the removal of *E. parasiticus* to this genus. In the typical species of *Streblonema* the plant consists almost wholly of the endophytic prostrate system, which consists of a loosely branched thread. The upright system is represented only by hairs and by the sporangia, which are often borne directly on the creeping threads and may not project beyond the surface of the host. A freshwater species epiphytic on *Compsopogon* has been described. More marked reduction is evident in *Bodanella*, a lithophyte found in the continental lakes; here there is only the creeping system with unilocular sporangia borne on its upper surface.

It is not altogether easy to demarcate *Phaeostroma* from *Streblonema. S. aequale* Oltm. regarded by Kuckuck as belonging to the former genus, is nowadays usually retained in *Streblonema*.

One of the simplest of the prostrate and probably reduced Ectocarpaceae is *Mikrosyphar*, an endophyte in *Zostera* and in Red Algae *M. Polysiphoniae* occupies a subcuticular position within its host, whereas *M. Porphyrae* develops numerous penetrating threads traversing the walls and extending from face to face of the *Porphyrathus*; there are also some projecting hairs (A). Swarmers are liberated singly from the cells of short 2-4-celled laterals, which Kuckuck regards as simple plurilocular sporangia.

A more markedly parasitic type is constituted by *Streblonemopsis*, the branched threads of which, anastomosing to form a network, occur within the surface of small richly lobed galls on Mediterranean *Cystoseiras;* it is probable that the galls are caused by the presence of the endophyte. There are few upright threads. De Toni would refer this genus to Reinsch's *Entonema*.

Ascocyclus occurs as minute epiphytic discs, superficially resembling those of a *Coleochaete* and usually circular in outline. They bear uniseriate plurilocular sporangia, hairs with a basal meristem *(h)*, and long ascus-like cells *(a)* to which the genus owes its name. These ascocysts, when young, are filled with numerous fucosan-vesicles, while in older plants they are empty. Feldmann has recently recorded unilocular sporangia. Many authorities place this genus among Myrionemataceae.

Haplostichous Ectocarpales

The further evolution of the filamentous type illustrated by the Ectocarpaceae appears to have taken place in the two directions already referred to above, viz. either by the aggregation of threads and their branches affording multiaxial and uniaxial thalli or by the longitudinal septation of the primary erect threads. The two types of construction have been designated *haplostichous* and *polystichous* respectively by Kuckuck, and they are considered separately as a matter of convenience. The two series do not, however, necessarily represent divergent lines, and some of the polystichous genera are perhaps related to certain haplostichous forms.

Crust- and Cushion-forming Types

Those Ectocarpaceae, in which the prostrate base bears several little-developed erect threads, lead over to the forms included in Myrionemataceae. The species of *Myrionema* are for the most part epiphytes forming minute, rounded or irregular, olive-brown discs; one of the commonest is *M. strangulans* Grev. (*M. vulgare* Thur.), found on Ulvaceae and other Algae. *M. aecidioides* Sauv. is an endophyte in *Laminaria*. In *M. strangulans* and other species there is a compact one-layered basal stratum, which represents the first-formed part of the plant; it exhibits marginal growth and consists of clearly distinguishable radiating threads which show pseudodichotomous branching. In older individuals only the peripheral parts are directly attached to the substratum, while the central cells may produce rhizoids of diverse length. In *M. aecidioides* the prostrate system creeps beneath the surface layer of the host, and a compact disc is only formed when the plant becomes fertile.

Except in the marginal growing region nearly every cell of the basal stratum produces an erect thread. The majority of these upgrowths take the form of short simple (or rarely once-branched) photosynthetic threads which, though compactly aggregated and sometimes enclosed in common mucilage, are not laterally fused. As a general rule these erect filaments develop progressively from the centre outwards and attain to a uniform height; their growth is largely apical. In *M. siliquosum* they are comparatively elongate and moniliform towards the summit. More or less plentiful colourless hairs provided with a basal sheath and elongating by means of the usual basal meristem, project from among the photosynthetic threads. They usually originate directly from the cells of the prostrate system and develop earlier than the surrounding filaments.

The large unilocular sporangia are generally borne laterally on the basal cells of the photosynthetic threads, although in some individuals many arise directly from the basal stratum; such are sessile or supported on a one-celled stalk. The uniseriate plurilocular sporangia of *M. strangulans* usually show the latter disposition, although sometimes lateral on the erect threads. In some species (e.g. *M. patagonicum* Skottsb.) the basal system bears hairs and sporangia only.

Other genera of Myrionemataceae show much the same structure. *Ulonema rhizophorum* Foslie is very similar to *Myrionema strangulans* and is indeed included in that species by Levring. In *Hecatonema* the photosynthetic threads are of unequal height and in part of considerable length; moreover many of the cells in the outer part of the basal stratum produce no erect growths. Kylin is no doubt right in including in this genus *Ectocarpus terminate* Kiitz. which, in its sheathed hairs and other features, closely resembles the species of *Hecatonema*. In *Chilionema* the basal stratum bears dense tufts separated by areas in which few or no erect growths are produced. Both in *Hecatonema* and *Chilionema*, which are closely related, the cells of the basal layer undergo horizontal division. Only plurilocular sporangia are known. *Compsonema*, a lithophyte recorded from the Adriatic and the Canary Isles, would seem to belong to this affinity, despite the extensive development of the erect threads.

More distinctive types are constituted *by Petroderma* and *Nemoderma*. In the former, which has been recorded also from Clare Island, the elongate erect threads are held together only by mucilage. Both types of sporangia are terminal. *Nemoderma* forms small brownish gelatinous crusts in the littoral and upper sublittoral regions of warmer seas. It was first described from near Tangier in Morocco, but has since been found also in the Mediterranean and in the Canary Isles. The one- to three-layered basal stratum bears elongate upright threads *(e)*, which are held together only by the gelatinisation of their walls; in the older crusts the lower cells contain few or no chromatophores and appear to serve for storage only.

The abundant hairs are sunk in groups in the crusts, their basal meristem in later stages producing ordinary thallus-cells below. Unlike the forms the unilocular sporangia develop from intercalary cells in the upper parts of the threads, while the sex organs are borne laterally at about the same level but on distinct individuals. The greater specialisation, evident in the structure of the thallus, as well as in the sexual process, is also shown by the restriction of the reproductive organs to a special zone near the surface of the crust.

A different development of the type of structure under consideration is found in *Ralfsia*; incl. *Stragularia*, where the erect threads are coalescent. The species are perennials and form dark brown, almost black, often leathery crusts, frequent on rocks in the littoral region. The crusts are circular, although older ones often possess a lobed margin; they frequently show concentric zonation and may attain to 15 cm. or more in diameter. Older ones often exhibit considerable differences in thickness and have a rather uneven surface. Adjacent crusts commonly overgrow one another so that a complex structure may appear in transverse section.

The basal layer is either attached directly to the rock or by means of septate rhizoids, and every cell bears an upright thread. These threads show apical growth and frequently fork a little way above the base in some species they are arched.

Long colourless hairs with a basal meristem, borne on the ends of short threads or arising directly from the basal system, appear embedded among the fused threads; they either occur singly or appear in tufts which sometimes arise from funnel-shaped pits. The single chromatophore is usually apposed to the upper wall of the cell.

In the reproductive phase, which sets in late autumn, the erect threads grow out over localised areas as free filaments and form a number of sori appearing as flat warts on the crusts. The unilocular sporangia are borne laterally on the free threads, while the plurilocular ones are terminal in position. After liberation of the swarmers the fertile threads are shed, while the parts below continue to grow. Vegetative enlargement, involving both increase in thickness and marginal growth, takes place during spring and summer. *Mesospora* combines the lateral position of the unilocular sporangia seen in *Ralfsia* with a lack of fusion between the erect threads as in *Nemoderma*.

The similar crusts formed by *Lithoderma* are at first light-coloured; the species are characteristic of colder seas and are found more particularly in winter in the sublittoral region. The coalescence of the erect threads, even in fertile plants, is very marked. There is only a single sorus composed of uni- or plurilocular sporangia. The cells contain several chromatophores, although in those 01 the free threads there is generally only one; in the northern forms hairs are completely lacking.

The principal species, *L. fatiscens* Aresch. (L. *extensum* (Crouan) Hamel), was described as bearing the plurilocular sporangia laterally on projecting threads. Kuckuck subsequently found an otherwise similar form with terminal plurilocular sporangia and suggested that Areschoug's plant was wrongly ascribed to this species. Most authorities have followed Kuckuck in regarding the sporangia as terminal, while some even describe their position as variable. Svedelius, however, regarded the two forms as distinct and referred

KuHmck's algae to a separate genus *Pseudolithoderma.* Recent investigations of Lund lend considerable support for this view, and the *Lithoderma* of Kuckuck with terminal uni- and plurilocular sporangia, normally borne on distinct individuals, must be regarded as different from that of Areschoug with lateral plurilocular sporangia. Whether generic separation is justified remains to be seen. Lund advances considerable evidence in favour of the view that *Ralfsia ovata* Rosenv., which is known only with lateral unilocular sporangia, may be the asexual phase of Areschoug's *Lithoderma fatiscens.*

Certain freshwater forms found in rapid-flowing streams, formerly included in *Lithoderma,* have recently been again referred to Gomont's genus *Heri-baudiella.* The distinctive feature lies in the plurilocular sporangia, which arc constituted by five or six of the uppermost cells of an erect thread, each of which without further septation forms a number of swarmers. The large unilocular sporangia are also terminal. The vegetative cells contain a number of discoid or lobed chromatophores. Klebahn's *Lithodorabz)* is probably a *Heribaudiella,* intimately admixed with some other algae.

In *Symphyocarpus,* an epiphyte in the sublittoral region, the erect filaments are short and bear the plurilocular sporangia terminally. There are occasional hairs, while at certain points the upright threads are replaced by large cells full of fucosan-vesicles; these were regarded by Kuckuck as homologous with the paraphyses of *Scytosiphon.*

It is not improbable that some of the genera are reduced types, although in the present state of knowledge it is difficult to get adequate evidence for this point of view.

A more elaborate structure is found in the Elachistaceae, well exemplified by *Elachista.* Here and in the families subsequently to be considered a greater degree of anatomical specialisation exists than in the Myrionemataceae. The thalli are differentiated into a colourless, basal or central, medullary region and a peripheral photosynthetic cortex. The latter produces three kinds of structures which in the subsequent matter are distinguished as follows: *(a)* long threads of more or less indefinite growth with cells containing chromatophores, the *assimilatory hairs; (b)* shorter and often copiously branched threads of limited growth, the *assimilators* (paraphyses of many writers); *(c)* long threads with colourless cells, the *colourless hairs.*

Elachista comprises a number of widely distributed epiphytes, which are specially common on Fucales; thus, *E. scutulata* is found on the receptacles of *Himanthalia,* the swarmers germinating in the conceptacles from which the cushions of the epiphyte later project, while the very common *E. fucicola* Fries is chiefly found on species of *Fucus.* The mature plants form small, firm, often hemispherical cushions which are densely pilose. The lower part of the adult thallus is constituted by a more or less extensive medulla composed of dense vertical rows of almost colourless cells *(m),* which in some species at least are firmly joined together. The upper part (cortex, r) is likewise often compact, although the component threads are not laterally united.

They comprise more or less numerous assimilators (*a*), which are commonly curved, monaliform, and frequently clavate in shape, and, projecting from amid them and often preponderating, the unbranched assimilatory hairs *(h).* The cells of the latter are densely filled with chromatophores, except for some of the uppermost which tend to decay early. These structures, which sometimes reach a length of a centimetre, in many species probably constitute the chief photosynthetic system. There are no colourless hairs.

Young stages of *Elachista* exhibit the usual discoid basal system of a single layer of cells. From the latter arise numerous unbranched assimilatory hairs, the basal cell of which is sometimes much narrowed. These erect threads at first exhibit uniform division of their cells, but subsequently differentiate a little above the base an intercalary meristem which cuts off series of photosynthetic cells to the outside. A cell below the meristem sooner or later puts forth a lateral branch which develops into a secondary assimilatory hair. This in its turn acquires a meristem and branches in the same manner, and so there appear upon the basal system a number of primary tufts which gradually close together to form a cushion. The medulla arises through enlargement of the cells that bear branches, in some species perhaps supplemented by an active formation of new cells from the overlying meristem. The assimilators, like the secondary assimilatory hairs, originate from the cells below the meristem and as a rule probably only commence to develop at a later stage; ultimately they are often produced in considerable numbers *(E. fucicold)*, although scanty in some species *(E. intermedia)*, and it is not impossible that there may be some degree of variability in this respect in one and the same form.

The assimilators are no doubt homologous with the assimilatory hairs, but growth becomes arrested before the intercalary meristem is established. The meristems of the diverse assimilatory hairs tend to be located at approximately the same horizontal level and lie just above the point at which the cortex commences. Characteristic of Elachistaceae are the large, pear-shaped, usually sessile unilocular sporangia, which are borne laterally on the basal parts of the assimilators and often occupy a zone a little way above the medulla.

The plurilocular sporangia either form intercalary sori on the upper parts of the hairs *(E. lubrica)* or occur terminally on the assimilators, spring and dying away during the winter, although sometimes fresh assimilatory hairs can apparently arise in spring from persisting parts of the thalli. Formation of reproductive organs normally begins in autumn and may continue through most of the winter; in *E. fucicola* Rosenvinge records the production of unilocular sporangia in quite young plants and this is clearly not uncommon. *Symphoricotcus* was established by Reinke for an algae with the structure of an *Elachista* but bearing uni- and plurilocular sporangia laterally, both on the bases of the assimilators and on the upper parts of the assimilatory hairs. To this genus Kuckuck referred

E. stellaris Aresch., which shows the same features. Many authorities hold, however, that the establishment of a distinct genus is not warranted.

Other members of Elachistaceae form minute epiphytic growths, which possess but an ill-defined medulla. It is so in *Leptonema fasciculatum,* which Hamel refers to Ectocarpaceae; the affinity with *Elachista* is, however, shown not only by the long unbranched assimilatory hairs, but by the large pyriform unilocular sporangia that arise at their base. The plurilocular sporangia are formed in the upper parts of the hairs. Another species of this genus, L. *lucifugum* Kuck., which appears to be not uncommonly associated with *Rhodochorton Rothii* in caves, shows considerable branching of the upright threads. Whether this is really one of the Elachistaceae remains to be seen. *Halothrix* is distinguished by the fact that the sori of plurilocular sporangia (the only ones known) form dense girdles around the assimilatory hairs.

The somewhat divergent genus *Giraudya,* a rather rare sublittoral European form, is included in the Elachistaceae by most authorities. The upright threads, which form minute tufts, here develop into fusiform parenchymatous structures, 8-10 mm. long; these bear terminal and lateral sheathed hairs, with a basal growing point, which are lacking in other Elachistaceae. The assimilatory threads are widest in the region of the intercalary meristem, which is situated towards their base. According to Sauvageau the ill-defined medulla is formed by interlacing rhizoids originating from the lower cells of the erect threads; these rhizoids can produce new assimilatory filaments, although the latter also originate from the basal parts of the primary threads. The medulla of *Giraudya* would thus appear to differ in origin from that of *Elachista,* The apical parts of the successive assimilatory threads gradually wear away or are used in the production of sporangia. *Giraudya* though a polystichous form, has much in common with the Elachistaceae in its method of branching.

Only plurilocular sporangia have been observed and of

these Sauvageau distinguishes three types. One kind, arising by division of the peripheral cells in the upper parts of the assimilatory threads, forms dense sori partly encircling the latter. Others *(e)*, occurring in a similar position, form small excrescences on the surface, while those of the third type, which are not uncommonly branched, are clustered around the bases of the threads beneath the intercalary meristem.

Gelatinous Cushion-forms

A series of forms, conveniently grouped as Leathesiaceae; Corynophlaeaceae of Kuckuck and Oltmanns, appear closely related to Elachistaceae. They are characterised by a considerably greater production of mucilage. As in *Elachista*, it is possible to distinguish a large-celled medulla composed of compacted threads and a peripheral photosynthetic cortex *(a)* from which numerous colourless hairs *(h)* project. In *Myriactula*, which is commonly placed among Elachistaceae and most of the species of which were originally assigned to *Elachista*, the chief distinctive feature is indeed constituted by these colourless hairs which replace the assimilatory hairs of the latter genus. The facts suggest that the two types of structures are homologous.

Myriactula pulvinata, not uncommonly epiphytic on Fucales forms small gelatinous cushions, the lower part of which is completely embedded in the conceptacles. The cortex consists of numerous assimilators *(a)* which taper at each end and bear the hairs *(h)* laterally. The early development appears to be much as in *Elachista*. In *M. stellulata (Elachuta stellulata* Griff., a rare endophyte in *Dictyota dichotoma*, creeping threads *(c)* extend horizontally beneath the superficial cells of the host and function as stolons from which a series of cushions can arise. The medullary threads penetrate the surface-layer of the host, which is always wanting at the place where a cushion occurs. In *Myriactula* both kinds of sporangia are borne as lateral outgrowths at the base of the assimilators.

In this genus, as in *Elachista*, medulla and cortex are as a general rule of approximately equal size, but in *Leathesia* and related genera there is an exceptionally large medulla. *L. difformis* is an abundant annual on north temperate-coasts, forming brown, rounded or lobed, gelatinous growths, up to 5 cm. in diameter, conspicuous between tide levels during summer on all kinds of substrata (rocks, other Algae). *Corynophlaea,* with a number of rarer species, exhibits only trifling differences (mainly the wider cortex), and it is often merged in *Leathesia.*

The prostrate base in these genera is a more or less continuous expanse and is as usual older than the erect-growing portion. In the mature thallus the branching threads composing the cushion consist for the most part of large cells with few or no chromatophores and widely separated by mucilage *(g).* This extensive medulla is covered by a narrow cortex *(c)* composed of short assimilators, the cells of which are packed with chromatophores; in *Leathesia* proper each thread consists of only 2-4 cells, the end one being enlarged. From amid the cortical threads hairs *(h)* project in considerable numbers.

The development of the common *Leathesia* has not been fully studied, but from what is known of it and allied species it would appear that the prostrate system first gives rise to upright threads composed of small cells these grow by division of the basal cells and multiply by branching. The medulla is formed secondarily by progressive enlargement of the lower cells of the threads and is gradually intercalated between those first formed and the basal stratum. Between medulla and cortex there is an intermediate region in which the cells are still relatively small. In *L. difformis* the older cells of the medulla assume an irregular stellate form and exhibit a considerable number of anastomoses, so that the medulla constitutes a framework of considerable stability, the wide meshes of which are occupied by mucilage. The older thalli are commonly hollow, a condition which is more rarely realised in *Corynophlaea.* The two kinds of sporangia arise from the basal cells of the assimilators or from the perimedullary cells.

Another member of this family is *Cylindrocarpus* (*Petrospongium* Naeg.) in which the medullary threads are bound together by hyphae. The plurilocular sporangia have been described by Hanna in C. *Berkleyi* (Grev.) Crouan (*Petrospongium Berkleyi* Naeg.; *Leathesia Berkleyi* Harv. C. *microscopicus* Crouan (*Ectocarpus investiens* Hauck), which is sometimes endophytic in *Gracilaria* has been fully studied by Kuckuck.

Myrionemataceae, Elachistaceae, and Leathesiaceae constitute a progressive series (though not necessarily an evolutionary one) with increasing anatomical differentiation, in all the members of which the thallus is composed of numerous equivalent upgrowing filaments emerging from the basal system. The Elachistaceae show the commencement and the Leathesiaceae the further elaboration of a proximal medulla and a distal photosynthetic cortex, the latter consisting of the ultimate branches of the threads forming the medulla. These types are therefore multiaxial. *Leathesia* can in some ways be compared with *Codium Bursa* among Siphonales.

Cylindrical Uni- and Multiaxial Types

In another series of gelatinous forms with essentially the same construction, the thallus takes the form of branched cylindrical strands of a varying degree of firmness. The principal members may be grouped as Mesogloeaceae, which comprise such common genera as *Castagnea, Mesogloea,* and *Chordaria.* Kuckuck regards the Mesogloeaceae as the direct descendants of the Leathesiaceae, while Kjellman's Chordariaceae included not only both of these, but also the Myrionemataceae.

The radially symmetrical thallus of the Mesogloeaceae commonly exhibits differentiation into a central large-celled medulla, a peripheral small-celled cortex *(c),* and an intervening subcortex with cells of intermediate size. Longitudinal sections of the mature thalli nearly always display a multiaxial structure. The axial threads form the more or less compact medulla, while their numerous off standing branches, which are not united, give rise in their inner parts to the subcortex and in their outer parts to the cortex; the latter consists largely of assimilators, between which there usually arise hairs (A) of the customary type.

The axial threads originate from the first-formed prostrate system and, as a rule, several threads arising from the latter form the basis of the medulla. In *Mesogloea* and *Myriocladia,* however, there is only one such axial filament, the primary laterals of which for a time run parallel to the parent-thread and thus, together with it, constitute the axial strand composing the medulla; the axial thread is usually clearly recognisable in the apices of the thallus branches. The contrast between the basic construction of\ *Castagnea* and *Mesogloea,* which in other respects show considerable resemblances, is found also in other heterotrichous groups, where likewise in related types the erect fronds may in the one be derived from a single thread, in the other from a number of upgrowing threads.

Mesogloea vermiculata Le Jol. (M. *vermicularis* Ag.) is a frequent annual near low-water mark, often attached to rocks with a thin covering of sand, though sometimes epiphytic. The more cartilaginous *M, Leveillei,* sometimes placed in a distinct genus *Liebmannia,* is commoner in warmer seas. The single axial thread is crowned by an assimilator *(a)* of some length, at the base of which lies a well-marked meristem *(g).* Below the latter the cells rapidly enlarge and produce primary laterals, which grow apposed to the axial thread for a considerable distance before bending outwards to terminate in a curved assimilator *(a)* like that at the apex numerous hairs. Another species, C. *Zosterae,* mostly epiphytic on *Zostera,* is smaller and less branched. Both of these are usually placed in *Eudesme.* The species of *Castagnea* proper\are rarer in northern waters, although not uncommon in warrnet seas like the Mediterranean. C. *contorta* is known from British and French shores.

Sauvageau has given a detailed account of the early development of C. *Zosterae.* Many of the filaments, which arise from the centre of the basal disc, are at first moniliform and devoid of terminal hairs so that they closely resemble the cortical assimilators of the mature thallus; subsequently the cells divide copiously and form an elongate medullary thread. The basal disc, however, also bears occasional sheathed hairs *(sh)* with a basal meristem, the lowest cell of which later likewise divides to form a medullary thread *(me)* which then bears a terminal hair *(ht)* devoid of a sheath; the sheaths at the base may persist or vanish.

The lower cells of the upgrowing threads elongate and remain narrow, but produce no laterals. Intercalary division, on the other hand, takes place in most of the upper cells and is accompanied by the outgrowth of numerous assimilators and of occasional sheathed lateral hairs *(hi)*. The rapid growth of these hairs often results in displacement of the distal portion of the primary filament so that the hair appears terminal. Growth is then carried on by division of the cell beneath the hair and leads to the production of further laterals, one or other of which may assume the terminal position, thus leading to a sympodial construction. Diverse other investigators have arrived at the conclusion that the development of the primary filaments in *Castagnea* takes place sympodially, the axial threads being built up of the successive basal portions of laterals, the apical parts of which are deflected to form the subcortex and cortex. This is in contrast to most Mesogloeaceae, in which the axial threads branch monopodially,

The cells of the first-formed axial filaments of *Castagnea* thus come to bear numerous well-branched laterals, but this primary condition is complicated by considerable secondary activity leading to a more compact structure. Supplementary medullary threads, bearing numerous secondary laterals, are produced by the development of downwardly growing threads from the basal cells of some of the primary laterals. In C. *Zosterae* and *C.fistulosa* they constitute a large-celled perimedullary envelope and can scarcely be distinguished from the original axial threads except by their mode of origin. Their numerous laterals *(a)* push in between the primary ones and contribute a further system of assimilators to the cortex; the basal cells of these secondary assimilators sometimes repeat the same process. According to Parke such downgrowing threads are not produced in C. *virescens.*

Perimedullary threads also appear to originate by the erection of certain primary laterals which grow parallel to the original axial threads for a considerable distance (cf. *Mesogloea)* and contribute other secondary assimilators to the cortex. According to Sauvageau this is due to intercalary division of the basal cells of the primary lateral. Branching hyphae, which are probably largely mechanical in function, may grow out from any cell. In *C. fistulosa* they form an investment to the medulla.

In the mature thalli of *Castagnea* the medullary threads often separate, leaving a cavity occupied by gas. The inner cells of the laterals forming the cortex and subcortex commonly elongate appreciably, while the smaller outer ones bear copious branches, but in many species (e.g. C. *virescens)* there is no very sharp boundary between the two regions. On the other hand, in *C. crassa (Mesogloea crassa* Suring., which Kylin makes the type of a separate genus *(Tinocladia],* cortex and subcortex are sharply differentiated. The ultimate laterals, the assimilators, are usually moniliform and not uncommonly curved. Branching of the thallus is effected by the outgrowth of series of medullary filaments.

Castagnea and *Mesogloea,* which thus differ appreciably in their basic construction, probably represent divergent developments from a common source. In both the primary threads and their laterals normally terminate in assimilators, and the development of the medulla in *Mesogloea* by erection of laterals is repeated in *Castagnea* on the part of each individual primary filament. The early development of the thallus in C. *Zosterae* shows similarities to that of *Leathesia*. In both assimilators are first produced, while the medulla originates secondarily.

The cartilaginous thalli of *Chordarta* show a more compact structure. In C. *flagelliformis,* a frequent annual between tidemarks, the elongate main axis bears whiplike branches which exhibit little further branching; the entire surface is clothed with long hairs with very gelatinous membranes. Transverse sections of the thallus show a broad, compact medulla and a narrow cortex *(c)*; the cells are smaller than in other members of this family. The medulla consists of larger (primary) cells *(mf)* intermingled with numerous narrow hyphae *(hy)*; the latter are outgrowths from the medullary cells and grow vertically upwards and downwards. There is no subcortex, the comparatively short and well-branched assimilators arising directly from the medullary threads; the branches of the assimilators are produced abaxially and combine to form a dense layer from amid which the hairs *(h)* project.

The end-cells of the assimilators and of their branches are often enlarged and provided with a strongly thickened outer wall; in the Australian C. *Chordaria* the assimilators consist of only two cells, of which the outer is elongate. In the growing tips there are a number of axial threads, each with an ill-defined meristematic region at the base of the apical assimilator; their laterals, terminating in branched assimilators, apparently insert themselves continuously between those already present, giving the apices a fanlike appearance. The branching of the axial threads is monopodial, as in *Mesogloea*.

Numerous variants on the types of construction above considered are found among Mesogloeaceae. A number of genera lack the colourless hairs found in other members of the family. Thus, in the highly mucilaginous' *Myriogloea Sciurus; Myrioladia Sciurus* Harv., which is not uncommon in the Southern Hemisphere and has much the habit of a *Castagnea*, the multiaxial medulla is encased in a dense cortex composed of long unbranched assimilatory hairs. Each of the medullary threads possesses a sharply defined intercalary meristem, surmounted by one of these hairs and, as the axial threads increase in number by branching *(br)* below the meristem, the outermost ones diverge in a fanlike manner. There is resemblance to *Elachista* in the presence of assimilatory hairs and in the relation of the different parts to one another. The medullary threads give rise to branched hyphae which run horizontally between the former. According to Parke the sporangia arise laterally at the level of the intercalary meristem and the two kinds occur on distinct plants. Kylin, however, states that *M. Sciurus* lacks plurilocular sporangia, while those of other species are formed in the upper parts of the assimilatory hairs. The southern types, with plurilocular sporangia originating near the bases of the assimilatory hairs, are placed in the new genus *Levringia.*

The American *Mesogloea Andersonii* Farl., which was included by Kuckuck in *Myriogloea,* is primarily uniaxial and possesses numerous short curved assimilators amid which the sporangia arise. Levring refers it to a distinct genus *Haplogloea.* Another uniaxial form, commonly listed as *Mesogloea divaricata* (Ag.) Kiitz. or *Chordaria divaricata* Ag., which is common on the Atlantic coasts of the United States, is according to Kylin distinguished by the fact that cell-division is confined to the intercalary meristem situated beneath the 2-4-celled assimilator *(a)* that caps the axial thread. Kuckuck referred this algae to *Nemacystus,* because he believed it grew by an apical cell. Kylin makes it the type of a distinct genus *Sphaerotrichia.*

Certain other, rather imperfectly known, genera show resemblances to *Chordaria. Heterochordaria abietina* (Rupr.) Setch. & Gardn. (*Chordaria abietina* Rupr.), a small Pacific form with numerous short branches giving the plant a fir-like habit, differs considerably from the true species of *Chordaria* and seems to possess an isomorphic life-cycle; it forms extensive communities below low-tide level. Both in the general habit and in the possession of a well-developed, probably perennial, basal stratum there is resemblance to the Antarctic *Caepidium.* The erect multiaxial fronds of the latter, which here arise from cuplike depressions in the lobed basal crust *and* attain a height of 16 cm., have a cortex composed of usually 3-celled assimilators, with unilocular sporangia *(u)* embedded between them. The striking feature is the production from the basal crust at all times of the year of vesicular growths of diverse size, resembling those of *Leathesia*. At first these develop as small solid warts showing the same structure as the ordinary erect fronds, but later they enlarge and become hollow; they bear single hairs lodged in a deep pit.

These vesicles bear the plurilocular sporangia, which are formed in large irregular sori and sometimes cover nearly the whole surface; interspersed between them are large cells *(s)* filled with dark-brown contents (fucosan). Cotton suggested that two distinct Algae might be involved, but Skottsberg's reinvestigation leaves little doubt that the vesicles belong to the same algae as the erect fronds. In view of the probable close affinity between Mesogloeaceae and Leathesiaceae there is nothing very surprising in the erect growths from the basal stratum repeating the structural characteristics of the two families. The Pacific *Hapterophycus*, in which the perennial crusts produce upright branches bearing unilocular sporangia, may also belong to this affinity. Another debatable genus is *Analipus* recorded only from the Behring Sea.

The well-developed basal system is several-layered and shows some anatomical differentiation. It bears a number of cylindrical unbranched erect fronds reaching a height of 6 cm. and having much the same structure as in *Chordaria*; the unilocular sporangia arise at the base of the few-celled cortical assimilators *(a)*. *Acrothrix gracilis*, which is referred to a distinct family by Kylin, is an infrequent sublittoral northern type. The uniaxial thallus is rather delicate, though well-branched, and attains a length of 30 cm.

The main axial filament is terminated by a hair, with a sharply defined basal meristem *(g)*, the segments of which *(af)* enlarge *without further division*. They produce whorls of primary laterals of limited growth, the basal cells *(b)* of which undergo marked increase in size and divide to form short large-celled perimedullary threads *(p)* constituting a continuous one-layered envelope around the axial filament. The second cell of each lateral usually develops an adaxial hair *(h)* whilst the others constitute a terminal assimilator *(a)*. Other unbranched assimilators *(a)* arise from the basal cells of the laterals, as well as from the perimedullary cells, and in their entirety they form a narrow but complete envelope to the axial system. Later the perimedullary threads separate from the axial one, which either undergoes great attenuation or disorganises so that the thallus becomes fistular. Branching is effected by the occasional development of a primary lateral into a new axial thread. Except for its intercalary growth and the presence of a terminal hair, *Acrothrix* shows much resemblance to the Spermatochnaceae considered below.

This is specially obvious in *A. novae-angliae* Taylor.

The series of genera constitute a plexus of interrelated forms showing points of contact, on the one hand with Elachistaceae, on the other hand with Spermatochnaceae. They exhibit a progressive restriction of division to the subterminal meristem, as is shown by a comparison of *Mesogloea* with *Sphaerotrichia* and *Acrothrix*, all uniaxial types with monopodial branching. Among multiaxial forms *Myriogloea* and *Castagnea* show the same method of growth as *Mesogloea*, although *Castagnea* stands out by the sympodial build of its threads; *Chordaria*, on the other hand, seems to approach *Sphaerotrichia* in the restriction of division to the region of the meristem.

The often large unilocular sporangia are always borne laterally on or near the bases of the assimilators, while the plurilocular sporangia are usually formed from the distal cells of the latter; in *Heterochordaria*, however, they are developed from the inner cells of the assimilators. Protection of the sporangia is most marked in *Chordaria*, where the often enlarged and thickened end-cells of the assimilators unite to form a protective cap, which is often supplemented by a layer of mucilage.

Strepsithalia, an endophyte in various gelatinous Algae (*Mesogloea, Helminthodnadin*), shows a loosely branched basal system bearing at intervals compact mucilaginous tufts composed of numerous branched assimilators. Hairs *(h)* are formed both on the creeping threads and on the lower cells of the assimilators. The latter closely resemble those of certain species of *Castagnea*, a resemblance which is heightened by the large unilocular sporangia borne at their base. The plurilocular sporangia usually arise in the same position, but both kinds can also originate directly from the creeping threads. *Strepsithalia* is clearly a reduced form, which was referred by Kuckuck to the Leathesiaceae, but a closer affinity with the Mesogloeaceae is perhaps to be suspected.

Spermatochnaceae

The members of this family, while showing the same general structure as many Mesogloeaceae, are specially distinguished by apical growth of the axial filaments, the segments cut off from the hemispherical apical cell undergoing no further division. The uniaxial *Nemacystus*, in particular, shows much resemblance to *Acrothrix*, and it can hardly be doubted that here, as in other groups of Phaeophyceae, the trichothallic meristem has been replaced by a definite apical cell. According to Kuckuck the apical hair is shed in older plants of *A. gracilis* and, subsequent to this, the cells of the axial thread exhibit enlargement only. If, however, the meristem remained active, we should have reached what is tantamount to apical growth and this apparently obtains in *A. novae-angliae* Taylor, a species which comes very close to *Nemacystus*.

This genus is probably the most primitive of the uniaxial Spermatochnaceae. In *N. flexuosus (N. ramulosus* Derb. et Sol.), a Mediterranean species with a branched filiform thallus, the basal cells of the whorled laterals keep pace with the elongation of the segments of the axial thread, forming a large-celled envelope around it, much as in *Acrothrix*; later the thallus becomes hollow and the axial thread may disappear. The second cell of each lateral, as in other Spermatochnaceae, produces an adaxial hair, while the remaining cells constitute a primary assimilator. The large-celled perimedullary envelope is augmented by the outgrowth of short threads *(d)* which bear numerous secondary assimilators *(sa)*. The two kinds of sporangia arise laterally from the lower cells of the assimilators. In the Japanese *N. decipiens (Mesogloea decipiens* Suring. *Cladosiphon decipiens* Okam. the thallus remains solid.

Spermatochnus paradoxus (Stilophora Lyngbaei xlarv., a deep-water annual growing on sand or mud, has a firm dichotomously branched thallus in the older parts. The axial thread, arising from the usual prostrate base, bears alternating whorls of generally four clavate assimilators at the upper end of each segment. The basal cell *(b)* of each such assimilator at an early stage grows out on its under side into a branched cortical thread which keeps pace with the increasing elongation of the axial segment so that the group of threads from each whorl completely envelops the part of the segment below.

This is followed by the outgrowth from the epibasal cells of similar threads which cover those first formed. Subsequent periclinal division of the cells of the outer threads results in a firm, several-layered envelope around the axial thread. Later, owing to the extensive peripheral growth, the axial thread separates from the surrounding envelope, the resulting hollow *(c)* being occupied by aqueous mucilage. The persisting axial thread remains connected to the cortical envelope by the elongate basal cells *(b)* of the primary assimilators. Branching is effected by one of the primary laterals developing into a new axial thread. According to Jonsson the cells enveloping the axial thread may give rise to hyphae which penetrate the walls and later the cell-cavities of the former; their cells may contain chromatophores.

When the reproductive phase sets in, the projecting portions of the assimilators are shed and only the basal *(b)* and epibasal *(e)* cells remain. From the latter hairs *(h)* and secondary assimilators *(so)* are produced, the uni- and plurilocular sporangia usually arising as lateral outgrowths from the lower cells of these assimilators. The sori thus formed occupy the positions of the former primary assimilators, but similar sori can subsequently originate at any point on the cortex.

Spermatochnus departs appreciably from the forms hitherto considered in the fact that the assimilators are of quite subordinate importance until the time of reproduction, the photosynthetic region of the vegetative plant being constituted by a cortex formed by the close juxtaposition of cortical threads produced from the lower cells of the whorled laterals. In this respect it resembles the Desmarestiales. The abundant production, in the reproductive phase, of assimilators, which, as in Mesogloeaceae, act both as a protection for the developing sporangia and as a local photosynthetic apparatus, is again a feature seen in many advanced members of the Brown Algae.

Stilophora and *Halorhiza* are multiaxial, although it is uncertain whether this structure is primary or secondary. The peripheral cells of the axis bear curved assimilators which overtop the summit, although more or less widely separated in the older parts as a result of the marked elongation of the axial cells. The cortical envelope is formed in the same way as in *Spermatochnus,* its peripheral cells finally dividing to form a small-celled superficial layer bearing the simple or branched secondary assimilators *(so);* these are either uniformly distributed *(Halorhiza)* or appear in localised areas *(Stilophora).* They bear the two kinds of sporangia on their lower cells. In the older parts the medullary threads separate, the central hollow being occupied by hyphae which are outgrowths from the medullary cells.

Stilophora rhizodes is a not uncommon annual on diverse substrata near low-water mark and favours situations where there is an influx of fresh water into the sea. The thallus, up to 65 cm. long, shows subdichotomous branching, the rounded wartlike sori being scattered over the whole surface. *Halorhiza vaga* Kiitz. is so far known only from the Baltic.

Kylin establishes the genus *Chordariopsis* for the South African *Chordaria capensis* Aresch., a uniaxial form distinguished by its firm parenchymatous structure. Assimilators are completely lacking on the vegetative parts, while on the fertile tracts they appear as clavate, usually one-celled, structures *(a)* forming a protection for the unilocular sporangia *(u)* which arise from the peripheral cells.

Genus Splachnidium

There is every reason to believe that this anomalous monotypic genus is a specialised member of the haplostichous Ectocarpales. *S. rugosum*, with a gelatinous, monopodially branched, hollow thallus attached by a basal disc, is a perennial algae of the Southern Hemisphere (Cape of Good Hope, Australasia). In the past it has usually been referred to Fucales owing to the occurrence of the unilocular sporangia in conceptacles, although the considerable differences were already clearly pointed out by Mitchell and Whitting. The "apical cell" recognised by earlier investigators is an endophytic *Codiolum* which appears to be invariably present.

The apices of the thallus-branches are occupied by monopodially branched threads, capped by numerous short, clavate assimilators *(a)*. The axial threads grow by a subapical meristem *(me)*, but separate widely a short distance behind the apex; as in *Leathesia*, however, they remain connected by anastomoses bridging the large mucilaginous interspaces which are also traversed by numerous branching hyphae *(hy)*. In the older parts the assimilators are shed and the surface is formed by a compact cortex *(co)* of small cells.

The conceptacles appear to originate in the immediate neighbourhood of the apex by localised cell-division leading to overarching of certain parts of the surface. The centres around which the conceptacles arise are usually occupied by a *Codtolu individual*. This may imply that the presence of the endophyte exerts a stimulus or the association may be due to a local alteration of the surface facilitating colonisation. Hairs are restricted to the inner surface of the conceptacles, although they project from the aperture, much as in Fucales. The large unilocular sporangia *(u)* arise successively from the cells forming the inner lining *(w)*. Laing has observed the escape of zoospores.

Splachnidium constitutes a striking example of parallel development, the conceptacles though so similar to those of Fucales being borne on a thallus with an altogether different structure and mode of growth.

Polystichous Ectocarpales

The formation of a thallus by longitudinal septation of a primary filament is infrequent, although found in Ulvaceae and Prasiolaceae among Chlorophyceae and in Bangiales among Rhodophyceae. In the Phaeophyceae, on the other hand, it has resulted in the evolution of the highly specialised Laminariales and Fucales. On these grounds it might appear justifiable to treat the polystichous Ectocarpales as a separate order. Their simplest representatives are, however, but little removed from Ectocarpaceae, both in structure and reproduction, so that a separation of the polystichous forms from other Ectocarpales would be artificial. The early stages of development, moreover, are ectocarpoid and evidently heterotrichous, one or usually several of the erect threads giving rise to mature thalli.

Occasional longitudinal division is found in diverse Ectocarpaceae, and this tendency, though little apparent in the vegetative parts of the haplostichous forms, is shown by the frequent multiseriate character of the plurilocular sporangia. The diverse genera to be considered form a closely related assemblage and their segregation into families presents considerable difficulties. The grouping adopted below, based on morphological characters, is largely one of convenience.

Punctariaceae

One of the simplest polystichous forms is *Phloeospora*. The erect filaments of P. *brachiata* Born. (*Ectocarpus brachiatus* Harv.), a widely distributed but rather rare summer epiphyte on

Rhodymenia palmata, arise from a system of endophytic threads and are at first.uniseriate. Longitudinal walls appear at rather regular intervals and, at these points, branches *(br)* and, at a later stage, reproductive organs arise; the branches are commonly opposite. Longitudinal septation later extends also to other parts, but the apices *(a)* of the filaments remain permanently uniseriate. Division of the primary cells is sometimes plentiful and leads to a multiseriate, though slender, thallus, occasionally with a slight differentiation in size between central and peripheral cells. Rhizoids, which often develop in considerable numbers from the lower cells, serve to strengthen the attachment to the substratum.

Unilocular sporangia develop as embedded structures directly from the peripheral cells at and near the points of branching, while plurilocular sporangia are rare. Sporangia commonly develop in individuals which are still for the most part uniseriate, a feature encountered in most of the genera.

Phloeospora agrees with the more typical Punctariaceae in the lack of any marked differentiation among the cells and the immersed position of the sporangia. Several genera exhibit a surface-development and betray a considerable degree of parallel with the Ulvaceae.

The leafy thalli of *Punctaria*, which usually grow in tufts, are almost exact replicas of *Ulvas*. In *P. latifolia*, a frequent annual lithophyte, the olive-brown fronds, which may reach a length of 40 cm., taper into a short slender stalk terminating in a basal attaching disc. The mature thallus consists of up to seven layers of cells, of which the inner are often longer than the outer. At certain points the superficial cells grow out into colourless sheathed hairs, forming tufts which often appear slightly immersed. The embedded sporangia form sori, the unilocular ones (M) differentiating when the thallus is only two-layered and becoming immersed by further division of the surrounding cells, while the plurilocular ones develop later and project slightly above the surface. The sori of plurilocular sporangia, in particular, appear as dots to the naked eye. The young stages are simple unbranched uniseriate filaments arising from a creeping base. Longitudinal division leading to surface expansion soon commences, although apex and base remain uniseriate. The basal part soon becomes overgrown by rhizoids produced from the lower cells. Marginal hairs appear before the superficial ones.

Forms having the same essential structure but possessing a ribbon-shaped thallus are often placed in a separate genus *Desmotrichum*; inch *Rhabdtno-cladia*. The young stages of *D. undulatum*, a rare epiphyte found near low-tide level, show evident heterotrichy, the prostrate base producing a few hairs *(h)* and a number of erect uniseriate threads *(t)* bearing one or more apical hairs. In cultures these threads may form plurilocular sporangia, and such stages recall the algae described as D. *balticum* Kiitz. in which the filaments, though showing considerable longitudinal septation in later stages, do not exceed a centimetre in height and bear plurilocular sporangia either in an intercalary position or as lateral outgrowths. This is possibly but a precociously fertile juvenile stage of *D. undulatum*. Lund records structures like the ascocysts *of Ascocyclus* on certain individuals of *D. balticum*.

The subantarctic *Corycm* has a club-shaped thallus which is solid below and hollow above. In other respects it resembles *Punctaria*.

The parallel between Punctariaceae and Ulvaceae is also illustrated by *Phaeosaccion*, which has a tubular thallus simulating that of *Enteromorpha*. The wall of the tube is composed of a single layer of cells, each with a lobed chromatophore. The plurilocular sporangia are commonly formed only by a single division of the vegetative cell. In *Omphalophyllum* the thallus is probably at first a hollow sack, which later tears open irregularly (cf. *Monostroma)*.

The early stages of *Litosiphon ptistllus* resemble an unbranched *Phloeospora*, but there is a greater degree of ultimate differentiation, since the central cells are much larger than the peripheral ones which contain the bulk of the chromatophores.

Hairs occur scattered over the surface of the thallus. The immersed sporangia cover considerable areas, the plurilocular ones developing before the unilocular. The basal parts continue to elongate after the upper have become fertile. The unbranched filaments of *L. ptisillus* are frequent on *Chorda filum,* often covering the latter in late summer.

In *L.filiformis* (Reinke) Batt. (*Pogotrichum filiforme* Reinke, an epiphyte on *Laminaria,* the central cells are exceptionally large. Kuckuck found plurilocular sporangia arising directly from the prostrate system, individuals exhibiting this phenomenon commonly failing to produce erect threads.

Stictyosiphon (incl. *Cladothele* Hook. & Harv. differs principally in the extensive ramification of its thallus; the numerous branches end in colourless hairs. *S. tortilis* Reinke is found on diverse substrata in the littoral and sublittoral regions. Rosenvinge describes the basal disc as parenchymatous and composed of large quadrangular cells arranged in vertical rows. The structure of the erect thallus and the arrangement of the reproductive organs does not differ in any appreciable respect from that of *Litosiphon filiformis.* In *S. soriferus* (Reinke) Rosenv.; *Kjellmania sorifera* Reinke the sporangia are either embedded or form projecting sori. There can be little doubt that the genera *Litosiphon* and *Stictyosiphon* are closely allied, although Kjellman placed them in distinct families on the grounds that in the former intercalary division persists longest at the base, whilst in the latter it persists longest below the apex.

Asperococcaceae

Petalonia (Phyllitis) and *Scytosiphon* while showing essentially the same thallus-structure, differ from the genera in the manner in which the reproductive organs are borne. *Petalonia Fascia (Phyllitis Fascia (Mull.)* Ktitz.; *P. caespitosa* Le Jol.; *Laminaria Fascia* has undulate, greenish or olive-brown thalli which may reach a length of 30 cm. and are narrowed basally into a short flattened stalk; it is usually found near low-water mark. Sections show a central medulla of large and somewhat elongate, colourless cells, interspersed with hyphae *(hy)*, and a cortex *(c)* of small cells. The early stages are of the usual heterotrichous type; those described by Yendo are probably pathological.

Scytosiphon Lomentaria (Chorda Lomentaria Lyngb., common in rock-pools between tide-levels, has narrow unbranched fistular thalli, up to 50 cm. long, with occasional constrictions and gradually tapering at base and apex; they generally grow in tufts, in which smaller individuals showing little or no constriction are intermingled with the larger ones. The mature thalli originate from the primary heterotrichous stages by longitudinal division of the erect threads, each of which terminates in a hair. Later similar hairs arise in a lateral position; in the mature thallus they may occupy pits like those of *Colpomenia* (p. in). As the fronds enlarge, the central cells undergo marked increase in size and produce a long-celled medulla *(m)*, while the peripheral ones divide abundantly to form the 1-2-layered photosynthetic cortex *(c)*. The subsequent separation of the medullary cells fails to occur at fairly regular intervals, where the central hollow is interrupted.

Both *Petalonia* and *Scytosiphon* produce only plurilocular sporangia, which are formed by outgrowth of the superficial cells over considerable areas. In *Petalonia* the densely packed sporangia *(p)* possess 4-6 rows of compartments, while the sori of *Scytosiphon* are interrupted by occasional large, ovoid or pyriform cells (paraphyses, *a)* which overtop the sporangia *(p)*. *Striaria attenuata* is a sublittoral epiphyte with cylindrical thalli reaching a length of 50 cm. The commonly opposite branches (sometimes in whorls of three) taper to either extremity and, in the fruiting condition, usually show a transverse banding due to the sori of unilocular sporangia; each branch terminates in a hair. Longitudinal septation of the primary filament sets in at an early stage, but intercalary growth continues for a long time in the subapical region *(in)*. The large central cavity in the mature thallus is usually surrounded by only two layers of cells. The sessile unilocular sporangia are intermingled with unicellular paraphyses *(a)* and hairs *(h)*.

In the projecting sori of certain allied genera multicellular assimilators are associated with the sporangia. One of the simplest of these is *Myriotrichia*, some forms of which barely depart from the filamentous condition. Its species are minute epiphytes, which never exceed a few centimetres in length and form dark, olive-brown, rather soft tufts. *M. clavaeformis*, not uncommon on *Scytosiphon Lomentaria* and other Algae, has a somewhat club-shaped axis, which is capped by a hair and is uniseriate below, where intercalary division continues for some time. Short, sometimes whorled, assimilators (*a*), intermingled with occasional colourless hairs *(h)*, occur in large numbers; in rare instances one of the assimilators develops into a branch. The uni- and plurilocular sporangia, generally found on distinct individuals, arise directly from the superficial cells, being either scattered or arranged in ill-defined annular zones. According to Kuckuck assimilators may be altogether lacking in plants bearing plurilocular sporangia. Both in *M. clavaeformis* and *M. repens* unilocular sporangia are sometimes borne directly on the prostrate system.

The erect threads of *M. repens (Dichosporangium repens* Hauck show little or no longitudinal septation, although there is differentiation into short nodal cells (*w*), bearing hairs *(h)* and sporangia, and long internodal cells *(i)*. Longitudinal division, when it occurs, is usually confined to the nodes.

Sauvageau regards the Mediterranean M. *Protasperococcus* Berth., an unbranched form with sporangia in well-defined annular zones, as the type of a distinct genus, *Protasperococcus*. The little-known *Buffhamia* will perhaps find a place among this series of forms.

Isthmoploea sphaerophora; Ectocarpus sphaerophorus Carm., a small epiphyte with opposite or whorled branches found on diverse Red Algae between tide-levels, is usually classed with the forms under consideration. Longitudinal septation is sometimes confined to the basal parts and varies in extent. The sessile unilocular sporangia generally project in opposite pairs, while the plurilocular sporangia are immersed among the superficial cells; there are no paraphyses. *Isthmoploea* shows points of contact with certain Punctariaceae.

The tendency to form a hollow thallus reaches its extreme in *Asperococcus*, in which the plants are usually sack- or bladder-like, with occasional irregular constrictions. The wall consists of only a few layers of cells, of which the outermost are the smallest. The sori, appearing to the naked eye as dark dots irregularly scattered over the surface, include numerous 2-3-celled assimilators, intermingled with long hairs *(h)* of the customary type. The two kinds of sporangia are sometimes found in the same sorus.

The two commonest species are *A. fistulosus* Hook. *(A. echinatus* Grev. and *A. bullosus; A. Turneri* Hook., both of which frequently attain to considerable dimensions; the former occurs in tide-pools, while the latter is found in the sublittoral region. The minute *A. scaber* lacks the usual central hollow, the axis being occupied by four large medullary cells. In this species (plurilocular sporangia have also been found arising directly from the compact prostrate system. The germlings are of the usual type. Impoverished forms, which scarcely get beyond the primary filamentous stage, are not rare.

Encoeliaceae (sensu Oltmanns) Soranthera ulvoidea, an epiphyte on Florideae on the western shores of North America, shows many resemblances to *Asperococcus*. The mature plants are hollow, sometimes irregularly lobed, vesicles, reaching a diameter of 7 cm. and with a wall of several layers of cells; the young plants are solid. Attachment is effected by a system of threads, some of which *(e)* are endophytic; however. The scattered sori of unilocular sporangia, *u)* are interspersed with numerous clavate assimilators *(a)*. The sporangia are grouped around a central tuft of hyaline hairs *(h)*, commonly situated at the base of a depression.

The irregularly lobed and often sinuous vesicles of *Colpomenia sinuosa* are not unlike a *Leathesia* in appearance;

older thalli often tear open irregularly. This seaweed is widely distributed in warmer seas, but early in the present century another form, of more delicate texture and less sinuous, became naturalised on the Atlantic shores of Europe, where the earlier literature is summarised). According to Sauvageau this constitutes a distinct variety (var. *peregrina),* agreeing in many respects with specimens described from the Pacific coast of North America, whence he believes the European form to have been introduced. Its rapid multiplication has caused considerable damage to the French oyster-beds; when filled with gas as a result of photosynthesis, the often large thalli lift the young oysters and carry them out to sea.

The mature structure has been described by Mitchell, while the accounts of the early development are somewhat conflicting. According to Sauvageau certain cells of the primary creeping filaments divide to form clusters, from which, evidently much as in *Soranthera,* the young globular thalli originate. Killian and Werner, on the other hand, state that it is cells of the erect threads that divide by radial walls to form the mature thallus.

The large central hollow arises as a result of extensive surface-growth effected by repeated crosswise division of the superficial cells. *Colpomenia* possesses only plurilocular sporangia which, in the warm-water form, usually constitute dense sori, intermingled with unicellular paraphyses functioning as fucosan-receptacles; the centre of the sorus is occupied by a tuft of hairs *(h)* arising from an often ill-defined depression. As the sorus develops, the surface-pellicle (cuticle, *cu)* is lifted and ultimately burst. In var. *peregrina* the association of sporangia with the depressions is for the most part fortuitous, most of the sori arising quite independently of them.

According to Mitchell the hairs are superficial outgrowths which become sunk owing to abundant division of the surrounding cells. In the French form (var. *peregrina),* however, they originate from cortical cells surrounding a small intercellular space, which at first communicates with the exterior only by a narrow gap between the superficial cells. This space arises prior to the development of hairs and enlarges by division of the surrounding cells.

Hydroclathrus cancellatus Bory, widely distributed in tropical and subtropical seas, is distinguished by the reticulate perforation of the older thalli. According to Boergesen interlocking rhizoids may grow out from the cells bounding some of the narrower perforations. Sporangia seem to occur principally on the younger unperforated individuals, where they cover a large part of the surface.

Another member of this series is *lyengaria stellata* Boerg.; *Rosenvingea stellata* Boerg., forming extensive communities above low-water mark in tropical India; the thallus is provided with projections which give it a somewhat stellate appearance.

A different habit is shown by *Chnoospora*, which is apparently confined to warmer seas (especially the Pacific). Kuckuck evidently intended to refer it to a separate family. The small, dichotomously branched thalli bear numerous prominent tufts of hairs which emerge from often barely indicated depressions. The thalli are composed of a number of layers of cells and grow by means of a subapical meristem. Plurilocular sporangia form dense sori, often aggregated around the tufts of hairs *(h)*. Skottsberg's *Cladochroa*, recorded from the Falkland Islands, has a similar habit and structure, but produces unilocular sporangia. The North Pacific *Myelophycus* is at present imperfectly known.

Dictyosiphonaceae

The polystichous Ectocarpales hitherto considered for the most part exhibit diffuse growth, although in older plants cell-division is sometimes largely confined to a suprabasal *(Scytosiphon,; Litosiphori)* or subapical position *(Stictyosiphon chnoospora)*. Diverse Dictyosiphonaceae, on the other hand, possess apical growth, effected by a dome-shaped apical cell the segments of which, as they enlarge, divide longitudinally; Kuckuck and others have, however, emphasised that some transverse division occurs in the older parts.

There is evidence that, as in Spermatochnaceae, the apical growth is secondary and derived from a trichothallic mode of growth.

Thus, young plants of *Dictyosiphon foeniculaceus* bear an apical hair with the usual basal meristem. According to Sauvageau this hair arises at an early stage, but it plays no part in the growth of the underlying portion, which results from transverse division of the hemispherical cell *(ac)* at its summit. Sooner or later the hair is shed, and the apical cell occupies the actual tip. The apical hair suggests affinity with certain Punctariaceae, with which there are also other points of resemblance. The thalli in this family bear only unilocular sporangia.

The species of *Dictyosiphon* are littoral epiphytes or lithophytes. ZX *foeniculaceus,* a slender form attaining a length of half a metre, is richly branched, while *D. Chordaria* Aresch. shows little ramification. The structure of the mature thallus recalls that of *Litosiphon* or *Stictyosiphon,* although showing rather more differentiation. In the younger parts the axis is occupied by four elongate cells which are surrounded by layers of progressively smaller ones (co), the outermost containing discoid chromatophores; some of these grow out into hairs *(h)* of the usual type. Most of the mature thallus is hollow, owing to separation of the axial cells which ultimately disorganise, but the upper parts of the branches and the extreme base of the plant remain solid. The peripheral· layers originate by tangential division of the surface-cells. The inner cells have large pits on the transverse and longitudinal walls, Hyphae arise chiefly from the peripheral cells and mainly in the older parts, often appearing in the basal regions of the larger branches and penetrating from there into the main axes.

The numerous large sporangia usually develop singly from superficial cells, but subsequently become embedded by division of the surrounding cells. The algae known as *Gobia baltica; Cladosiphon balticus* Gobi, with a little-branched thallus, appears to be a depauperate form of *D. Chordaria.*

In *Scytothamnus*, the species of which occur in the Southern Hemisphere, the apical growth is less well defined. In *S. australis* the tip of the thallus is occupied by a group of apical cells, while in *S. hirsutus* there appears to be only one. The elements of the central tissue are widely separated by mucilage, which is traversed by numerous hyphae. The large sporangia occupy the same position as in *Dictyosiphon.*

Delamarea attenuata (Kjellm.) Rosenv.; *Physematoplec attenuata* Kjellm., first described from the Arctic but latei found at Heligoland, is probably a related form. The sporangia are embedded amid numerous unicellular assimilators which arise from the cortical cells.

Coilodesme is commonly referred to Dictyosiphonaceae on account of the mode of arrangement of the sporangia, although Setchell and Gardner place it in a family by itself. The early stages afford no evidence of a distinct apical cell and the growth appears to be largely, if not entirely, diffuse. The usually unbranched thalli, which often reach a considerable size, include a large cavity which is surrounded by a few layers of cells; the basal stalk remains solid. The sporangia are completely immersed, only a narrow aperture remaining above each. This is due to the upgrowth from the peripheral layer of usually 3-celled assimilators the outermost cells of which meet to form an almost complete canopy.

In *C. californtca* the sporangia are very irregular in shape and vary much in size. This species is widely distributed on the Pacific coast of North America, whilst other species (e.g. *C. bulligera)* are subarctic in their distribution.

General Features of Reproduction

The situation of the uni- and plurilocular sporangia has been fully described; they commonly arise in the same position. The details of their development are known only in a few forms. The young *unilocular sporangium* has a large and conspicuous nucleus and the cytoplasm increases in amount until it completely fills the rudiment; a number of parietal chromatophores are usually present. The first nuclear divisions are no doubt as a rule meiotic, although there are certain probable exceptions. For a time repeated nuclear division takes place, accompanied by multiplication of the chromatophores; in *Pylaiella* Knight records a long pause after the quadrinucleate condition has been reached.

During the earlier stages of division nuclei *(n)* and chromatophores *(c)* are distributed throughout the cytoplasm, but subsequently they assume a peripheral arrangement, with the chromatophores *(c)* commonly in a profile position perpendicular to the wall; at this stage the chromatophores are often elongate and sometimes more or less wrapped round the nuclei. As the sporangium matures, the contents once again acquire a more even distribution, and this is followed by cleavage of the cytoplasm into uninucleate portions, each including one or more chromatophores; according to Kuckuck the initials are sometimes separated by mucilaginous septa. At this stage a brownish red eyespot (tf), in intimate contact with a chromatophore (r), is recognisable in each initial. Chadefaud states that the chromatophores in the initials of *Mesogloea* become constricted, the stigma *(s)* arising on the outer surface of one half. According to him the eyespot in Ectocarpales consists of an oblong colourless highly refractive mass enveloped by pigment. Knight, in *Ectocarpus siliculosus,* records a conspicuous granule, regarded as a blepharoplast, during all stages of sporangium-development.

The contents of the sporangium are usually liberated through an apical aperture (lateral in *Pylaiella,*); prior to dehiscence, the wall in this region develops a mucilaginous middle layer, the swelling of which causes rupture. Most of the contents are usually extruded in a single group enclosed within a thin mucilage-envelope but release of the individual swarmers *(b)* takes place almost at once. Extrusion is often violent and affords evidence of a considerable internal pressure. Occasionally small groups of swarmers fail to separate from one another. The cell below the sporangium may grow into the empty cavity and form either a new one or continue as a vegetative thread.

The swarmers are haploid and as a general rule behave as asexual zoospores; they show the characteristic features of the motile cells of Phaeophyceae and contain one or several chromatophores. They are often markedly phototactic. Certain species (e.g. *Ectocarpus tomentosus; E. granulosus; Pylaiella fulvescens* produce zoospores of specially large size which often show little motility, apart from amoeboid movement.

After swarming for some time, the zoospores, in some species of *Ectocarpus,* become attached with the help of a sucker-like enlargement formed at the tip of the anterior flagellum. By the gradual contraction of the latter the body of the swarmer is drawn on to the substratum; this is followed by the withdrawal of the other flagellum, rounding off, and secretion of a membrane, after which immediate germination to form a filament ensues. It is not known whether this method of attachment is frequent, and it seems that the zoospores commonly settle directly on a substratum. In the more specialised Ectocarpales the rounded cell formed from the zoospore *(embryospore,)* puts out a tubular, prolongation into which most or all of the contents pass, after which the prolongation is cut off by a septum; it is this cell which divides to produce the prostrate system of the gametophyte.

The *plurilocular sporangia* are generally more elongate than the unilocular ones; those of *Ectocarpus siltculosus* become progressively shorter as the season advances. As a general rule the rudiment undergoes successive divisions by transverse septa, and the row of flat cells thus produced may constitute the mature sporangium (e.g. *Ascocyclus,*; many Myrionemataceae,; *Myriactula,*). Commonly, however, longitudinal walls appear at a fairly early stage leading to division of the primary compartments into quadrants, and this may be followed by further segmentation by vertical septa so that as many as 24 chambers may be present in the cross-section of the sporangium; the degree of segmentation probably often varies within fairly wide limits in one and the same species. As a result of this regular septation the plurilocular sporangia usually possess cubical compartments of approximately even size.

In most Ectocarpales septation of the developing plurilocular sporangium seemingly ensues after each nuclear division, in conjunction with division of the chromatophores. In *Lithoderma*, however, nuclear division is completed before any formation of septa occurs, while in *Stictyostphon tortilis*, as *Phloeospora subarticulata* Aresch.) several divisions of the chromatophores and of the nucleus may occur before the first septa are laid down. As Oltmanns points out, such sporangia in their development show appreciable approximations to the unilocular type. According to Knight the young plurilocular sporangium of *Ectocarpus siliculosus* possesses dense cytoplasmic contents and reduced chromatophores; a probable blepharoplast-granule is recognisable during the later divisions.

Dehiscence of the plurilocular sporangium is commonly at least effected by a single apical aperture; the septa between the individual compartments disappear completely *(Stictyosiphon tortilis)* or only the central region is dissolved. The swarmers, produced singly from each compartment, pass out in a slow stream or in a series of bursts through the apical aperture, a method of liberation contrasting with that observed in the unilocular sporangium. In *Zosterocarpus, Castagnea, Lithoderma* and certain species of *Ectocarpusy* however, each compartment of the sporangium opens separately to the outside. No septa are formed in conjunction with the later divisions of the plurilocular sporangia of *Heribaudiella,* so that each of the primary compartments furnishes a number of swarmers; the latter are stated to lack chromatophores in the mature condition.

As already mentioned, the plurilocular sporangia of Ectocarpales include structures of diverse nature. Those found upon diploid individuals, also bearing unilocular sporangia, produce diploid swarmers which constitute an accessory method of multiplication of the sporophyte. Such diploid accessory swarmers either reproduce the sporophyte direct or give rise for a period to dwarf ectocarpoid stages reproducing by diploid plurilocular sporangia. Occasionally they germinate directly within the sporangium. The haploid swarmers from the unilocular sporangia produce gametophytes, which are either identical in form with the sporophytes or more usually are represented by minute filamentous stages. The plurilocular sporangia, borne upon the haploid gametophytic stages, are gametangia which form haploid gametes. The zygotes directly or indirectly produce the new diploid phase. Although there are diverse exceptions and the full life-cycle has only been followed in a few species, the preceding statement probably applies to many Ectocarpales.

Most members of the order appear to be morphologically isogamous and monoecious, although *Ectocarpus siliculosus* is usually dioecious with distinct male and female strains. Goebel reported an altogether isogamous sexual fusion in *Ectocarpus globifer* Kiitz. (as *E. pusillus* Kiitz.), but this has never been confirmed, and it is probable that physiological anisogamy is the rule. The sexual process of *E. siliculosus*, first described by Berthold and often since corroborated, clearly demonstrates the sharp degree of differentiation.

Despite the morphological identity of the two kinds of gametes, the female move less actively and for a shorter period; they come to rest on the illuminated side of the containing vessel, attaching themselves, like the zoospores, by their anterior flagellum. Each such passive female swarmer becomes surrounded by numerous gametes *(m)* of the opposite sex which, still showing vibrating movements, in part at least fix themselves to the former by their forward flagellum. Soon the anterior flagellum of one of the male gametes contracts so that its body is brought into contact with that of the female, after which the two progressively fuse; the remaining male gametes swarm away. This type of clump-formation in which each group gives rise to a single zygote is, so far, recorded only in *Ectocarpus* although not always found there.

A fusion of morphologically isogamous gametes is also reported *inter alia* for *Ascocyclus, Lithoderma, Mesogloea venniculata, Phloeospora brachiata, Asperococcus bullosus, Punctaria, Sorocarpus,* and *Heterochordaria,* although few details are available. In *Sphaerotrichia divaricata (Nemacystus divaricatus* Hygen larger female and smaller male gametes are recorded, and the former are stated also to develop apogamously. *Soranthera* is described as having large, dark-coloured, sluggish female gametes and small, transparent, active male gametes, as well as swarmers of an intermediate type; both of the latter can apparently fuse with the female gametes. In neither instance do the published figures afford clear evidence of the existence of more than one type of plurilocular sporangium on the gametophytes, and further investigation is needed to establish the true state of affairs.

According to Hartmann the two types of gametes in *Ectocarpus silictdosus* can be distinguished by *intra vitam* staining and contain specific sexual substances. It is not altogether easy to harmonise this with the occurrence of relative sexuality, i.e. the fusion between two different groups of gametes belonging to the same strain. Hartmann concludes that sex determination in this species is phenotypic.

Several species of *Ectocarpus* possess more than one kind of plurilocular sporangium, the compartments being of different sizes and different types of swarmers being produced. The one kind (meiosporangium) seemingly serves to produce accessory asexual swarmers, while the others (megasporangia and microsporangia) constitute gametangia. Kniep, however, regarded the megasporangia as asexual and the meiosporangia as female organs. In view of the marked anisogamy shown by some of these species, they were referred by Batters to a distinct genus *Giffordta*, but such separation is not warranted, since the heterogamous species of *Ectocarpus* are in part closely allied to isogamous ones.

In the rare *E. secundus*, in which meiosporangia appear only towards the end of the growing season, the female gametangia have large compartments containing a number of deep brown chromatophores, while the male *(mi)* have small, pale-coloured, compartments, each with a single minute yellow chromatophore. Sexual fusion occurs between the two kinds of gametes after the manner above described for *E. siliculosus*, the female soon coming to rest and being sought out by the males. Apogamous development of the megagametes is not uncommon in the later part of the season when the male gametangia are lacking and, in *E. Padinae* Sauv. with similar gametangia, this is always so, so far as present observations go, and it is not known whether the micro-gametes are ever functional. In *E. Padinae* meiosporangia, with compartments intermediate in size between those of the micro- and megasporangia, are found throughout the season; the germlings of the meiospores show narrower dimensions than those formed from megaspores.

In the two species just discussed no individuals with unilocular sporangia have been discovered, and the plants so far studied must be presumed to be gametophytes which can propagate apogamously, as well as by means of the swarmers from the meiosporangia. The sexual fusion in *E. secundus*, however, implies the existence of diploid thalli that will probably be found to bear unilocular sporangia, and the same may be true of *E. Padinae*.

In another species, *E. Mitchellae* Harv. (*E. virescens* Thur.), in which megasporangia and meiosporangia are borne on distinct individuals, unilocular sporangia have recently been found on plants bearing also meiosporangia. Cultures of the swarmers from the unilocular sporangia afford plants bearing megasporangia and microsporangia. The swarmers produced by the former show attendency to loss of movement, while those of the latter are colourless, save for the eyespot. No fusion of the two kinds of swarmers has been observed, but the megagametes can give rise to further generations bearing the same two types of reproductive organs. It is probable, however, that sexual fusion occurs in nature and that the zygotes develop into the plants bearing the unilocular sporangia which are otherwise inexplicable. Svedelius has shown that plants bearing megasporangia or meiosporangia both possess the same number of chromosomes, which suggests a formation of meiosporangia also on the gametophyte (cf. *E. secundus* and *E. Padinae).* Those on the plant with unilocular sporangia can hardly be other than diploid.

Two kinds of plurilocular sporangia have also been recorded in *E. Lebelii* Crouan, the one with very small compartments producing swarmers like the male gametes of *secundus,* the other with deep brown compartments. No fusion of the swarmers has been observed, and it remains doubtful whether those of the second kind represent meiosporangia or megasporangia. Askenasy also describes two types of plurilocular sporangia in *E. indicus* Sond. The large-chambered plurilocular sporangia recorded in *E. confervoides* are, according to Sauvageau, the result of the attack of a parasite.

Plurilocular sporangia of different sizes are not, however, confined to *Ectocarpus.* Howe records two kinds in *Streblonema Cokeri.* In *Myrionema strangulans* Sauvageau describes some with large compartments regarded as megasporangia and other rarer ones with compartments of about half the size, interpreted as meiosporangia; Kylin, who observed only those of the former type, regards them as comparable to the meiosporangia of *Ectocarpus Mitchellae.* Karsakoff also recorded in *Myriotrichid filiformis* two kinds of sporangia, the swarmers of which were stated to copulate. This has never been confirmed and, in view of what is known of the sexual reproduction of allied forms, the reported fusion of swarmers is improbable. Several kinds of plurilocular sporangia have also recently been reported in *Colpomenia sinuosa*. In spring the plants are stated to bear male and female gametangia, the male gametes being considerably smaller than the female and devoid of a chromatophore; fusion is described.

In *Giraudya spkacelarioides* Berthold observed megasporangia forming sori on the assimilatory threads and liberating swarmers with several chromatophores, while recent workers have seen only meiosporangia with swarmers having a single chromatophore and occurring in the three types of sori. The only record of sexuality in this algae is that of Goebel, who describes conjugation of the swarmers from the megasporangia. Recent workers have generally regarded this as a misinterpretation.

Incipient oogamy, similar to that of *Ectocarpus secundus*, is also seen in *Nemoderma*, where the male and female gametangia are borne laterally on the upper parts of the erect threads, while the intercalary unilocular sporangia occur on distinct individuals. The gametes are liberated singly from the compartments of the gametangia. The large female swarmers, which become attached by the anterior flagellum, are sought out by the small male swarmers *(m)* which lack a chromatophore, although possessed of a distinct eyespot *(s)*. The female gametes can also develop apogamously. The sexual cells ripen fortnightly and are liberated in the interval between two successive spring tides.

Although some Ectocarpales thus attain 10 an advanced state of anisogamy, true oogamy does not appear to have evolved in this order.

4

Seaweeds

The ensuing account of geographical distribution is intended as an introduction to the topic and makes no claim to be complete. The local variations in the flora are on the whole insignificant when compared with the marked diversities which are encountered in different parts of the world. Thus, the algal populations of the Old and New Worlds and of the Northern and Southern Hemispheres present numerous points of contrast, while that of the Atlantic is in large part distinct from that of the Indian and Pacific Oceans.

A certain number of marine Algae enjoy a wide geographical distribution (e.g. *Digenea simplex* in all warmer seas), while some would appear to be almost cosmopolitan (*Ulva Lactuca, Colpomenia sinuosa, Ectocarpus siliculosus, Sphacelaria furcigera, Melobesia farinosa, Catenella Opuntia, Gracilaria confervoides, Hypnea musciformis*), but the vast majority are restricted to definite, though often relatively wide, areas. For most seaweeds the warm water of the equatorial belt constitutes an impenetrable barrier, and *Scytosiphon Lomentaria*, for instance, although recorded from the colder regions of both the Northern and Southern Hemispheres, is lacking in the Tropics.

Temperature is no doubt one of the principal factors determining the latitudinal distribution of marine Algae, a view specially advanced by Setchell. He showed that the distribution of the genera of Laminariales is markedly related to the mean maxima for the hottest months and that it coincides with the isotheres of 10°, 15°, 20° and 25°C. Later Setchell draws attention to the importance of the mean minima of the surface-waters and suggests that the temperature-interval between minima and maxima (as a rule probably about 10°C.) may determine whether reproduction can take place and therefore whether survival is possible. He distinguishes nine climatic zones (Boreal, Upper Boreal, North Temperate, North Subtropical, Tropical, South Subtropical, South Temperate, Upper Austral, Austral), determined by temperature-intervals (surface-isotheres) of 5°C., and concludes that, when a species occurs in two or more zones, its normal habitat is only in one of them. Thus, *Ascophyllum nodosum,* an Upper Boreal species, in the southern portion of its range fruits only in winter. The southern limit of *Fucus virsoides* in the Adriatic is determined by temperature, while Isaac emphasises the importance of this factor in South African waters.

The effect of temperature is, however, not easily separated from that of illumination which, variable both in duration and intensity in different latitudes, probably also plays a role in determining the quality of their algal vegetation; one can, however, scarcely admit Tilden's theories, which are quite unsupported by fact. Temperature, owing to its influence on the rate of respiration, will affect the capacity of a sublittoral algae to thrive below a certain depth. Lami draws attention to the possible importance of mists and prevalent cloud in relation to the distribution of Algae. The effects of diminished salinity have already been considered.

Zones of algal distribution do not necessarily correspond to degrees of latitude, since ocean currents traversing the coasts often influence the temperature of the water and cause apparent anomalies in distribution. Reference need only be made to the marked differences at equivalent latitudes between the algal flora on the eastern shores of North America and the western shores of Europe due to the influence of the Gulf Stream, and to the pronounced contrast between the seaweed vegetation on the eastern and western shores of the Cape Peninsula, due to the Antarctic current skirting the western shores. The flora of the latter comprises *Laminaria pallida, Ecklonia* and *Macrocystis pyrifera*, which are practically wanting on the opposite coast.

Boergesen's and Jonsson's (H) analysis of the distribution of seaweeds in northern latitudes, with special reference to the flora of the Faeroes, also essentially recognises the temperature-factor. They classify the northern seaweeds in five groups, viz. *(a)* an arctic element (not represented in the Faeroes); a subarctic element with species represented in the Arctic Sea and in part extending southwards to England *(Chaetopteris plumosa, Agarum cribrosum, Alaria pylaii, Phyllophora brodiaei, Halosaccion ramentaceum*, etc.); *(c)* a boreal-arctic element extending from the Arctic as far south as the Atlantic coast of North Africa *(Ascophyllum nodosum, Bangia fuscopurpurea, Ahnfeltia plicata*, etc.); *(d)* a cold-boreal element, extending northwards to Iceland, the Faeroes and North Norway and southwards to the Mediterranean and North Africa *(Elachista scutulata, Castagnea virescers, Alaria esculenta, Laminaria sacchartna, L. Cloustoni, Gigartina stellata, Cystoclonium purpureum, Furcellaria fastigiata*, etc.); and *(e)* a warm-boreal element *(Leathesia difformis, Cutleria multifida, Halidrys siliquosa, Arthrocladiavillosa, Sporochnus peduncu-latus Taonia atomarta, Chondrus crispus, Plocamium coccineum, Laurencia pinnatifida, Callithamnion corymbosum, Rhodophyllis bifida, Gracilaria confervoides, Sphaerococcus coronopifolius*, etc.). This grouping has been adopted by many subsequent worker. The marine flora of Holland, though poor in species, is essentially like that of the adjacent coasts of Britain and France.

On the Atlantic coast of North America, Harvey first pointed out that the Cape Cod Peninsula sharply segregates the northern from the southern seaweeds, and this corresponds to the isothere of 20°C. The northern area, which probably extends to Greenland, includes many species found on the Scandinavian coast, while the larger *Fuci* and other Brown Algae occur only in colder spots south of the peninsula. A similar dividing line is recognisable on the coasts of Spain and in the region of the southern Japanese sea. Gail, on this basis, distinguishes the entire seaweed population of the Northern Hemisphere into a northerly *Laminaria* region (with Ectocarpales, Bangiales and Gigartinales) and a southerly *Sargassum* region (with Siphonales, Dictyotales, and Ceramiales). On the coast of North Carolina the transition between northern and southern floras is indicated by the simultaneous presence of *Fucus vesiculosus* and of numerous Florideae with a marked southerly range. According to Collins and Hervey the flora of Bermuda has much in common with that of Florida and the West Indies.

The Pacific shores of North America afford an analogous picture, although the cold current that skirts a considerable part of the coast results in a pronounced southerly displacement of the Laminariales. The southern shores of the Japanese islands harbour a considerable number of seaweeds belonging to the Indo-Pacific region, which extend southwards to Australia and westwards to the Red Sea, while comparatively few of these appear to occur on the American Pacific shores, where the similarities with the Japanese flora refer more to the subarctic element comprising several striking members of Laminariales. Okamuram has made a detailed analysis of the distribution of marine Algae in Pacific waters.

The tropical seaweed flora is distinguished especially by the abundance and variety of Siphonales and the dominance of Florideae (especially Ceramiales); the most important of the Phaeophyceae are Dictyotales and species of *Sargassum* and *Turbinaria*, while Laminariales are completely lacking. It would seem that the algal flora of the Tropics shows a greater measure of resemblance to that of the Southern than to that of the Northern Hemisphere. A considerable number of seaweeds (*inter alia* diverse Sporochnales and Sphacelariales, many

Rhodomelaceae and Delesseriaceae) are restricted to the Southern Hemisphere. The marine flora of New Zealand has something in common with that of Australia, but also includes an evident Antarctic element, some of the representatives of which are also found in South Africa; there are a considerable number of endemic species. *Carpophyllum* occurs throughout New Zealand, while *Durvillea* (associated with *Macrocystis* and *Marginariella)* is dominant in the south..The flora of the Chatham Islands shows a general resemblance to that of New Zealand.

The Subantarctic and Antarctic floras include several species with a circumpolar distribution, as examples of which *Scytothamnus australis, Halopteris funicularis, Durvillea antarctica, Macrocystis pyrifera, Iridaea cordata, Ballia cattitricha* and *Gigartina radula* may be mentioned. In Antarctic seas species of *Desmarestia* play a very prominent part in the sublittoral region, accompanied by *Lessonia* and *Scytothalia,* while *Durvillea* and *Macrocystis* are more particularly characteristic of the Subantarctic zone.

There is often an appreciable degree of correspondence between the flora of coasts separated by thousands of miles of ocean or by intervening land-barriers. These similarities can in general only be explained by taking into consideration the changes that have affected the distribution of oceans and of landmasses within recent geological epochs. During the Tertiary period, when there was a land-connection between Europe and America by way of the Faeroes and Iceland, the arctic flora to the north of it probably differed from that to the south; when the connection was broken, a mingling of the two floras ensued. During the glacial epochs this vegetation was forced southwards into the Atlantic and Pacific Oceans and a flora, which was arctic in character, probably occurred as far south as England and Northern France. With the advent of a warmer era many of the arctic species were again driven northwards, while others capable of adapting themselves to a wider range of temperature remained to form the present-day element common to the North Pacific and Atlantic Oceans. Kjellman established that nearly 60 per cent of these species also occur in the Arctic Ocean.

There are, however, many seaweeds common to the Atlantic and Pacific which are lacking in the Arctic (e.g. *Chorda filum, Fucus vesiculosus, Saccorhiza dermatodea)* and these are probably more southern types which migrated northwards during the warmer post-glacial epochs, at a time when the habitat was still continuous. Simmons regards the greater measure of similarity between the seaweed flora of the Polar Sea and the North Atlantic, as contrasted with the North Pacific, as due to the existence of greater opportunities for migration afforded by the more numerous coastal paths. In the White Sea the Atlantic element has only penetrated into the northern part.

Rosenvinge concludes that the algal flora of the Danish seas migrated into it from the North Sea after the glacial epoch, while Levring regards that of the Baltic as a remnant of the North Atlantic flora which became established during the *Littorina*-period. There is also a considerable subarctic element.

It has long been assumed that certain genera of sea-grasses *(Posidonia, Cymadocea)*, which are found both in the Mediterranean and in the Indian Ocean, owe their present distribution to migration from the latter into the former during Tertiary times when there was an open connection between the two. A similar relation is probable for *Acanthophora Delilei*, for the species of *Vidalia*, for those of the *Bursae* section of the genus *Codtum*, and for several other genera. Apart from this Indo-Pacific element, the Mediterranean flora also includes a number of genera and certain species (e.g. *Halimeda Tuna, Hydroclathrus cancellatus, Hypnea musciformis, Digenea simplex, Taenioma perpusillum)* that are widespread in all warmer seas, but the bulk of the flora is composed of an element that has arrived from the Atlantic. Part of this consists of species common to the Mediterranean and the tropical and subtropical Atlantic (e.g. *Caulerpa prolifera, Acetabularia medtterranea, Nereiafiliformis, Gracilaria armatd)*, while another part comprises species widely distributed in the North Atlantic, many of which reach the southern limit of their distribution in the Mediterranean (e.g. *Stilophora rhizodes, Taonia atomaria, Sphaerococcus coronopifolius, Chondria dasyphylld)* this second group probably includes many species which were forced southwards during the glacial epochs. The Guernsey flora includes a number of Mediterranean species.

Many species and varieties found in the Mediterranean are closely related to forms found in the adjacent Atlantic, a fact which has been specially emphasised by Sauvageau in relation to the species of *Cystoseira*. He points out that those inhabiting the eastern Mediterranean differ markedly from the oceanic species, although the latter show much agreement with the *Cystoseiras* of the western Mediterranean. Feldmann also instances the Mediterranean *Phyllophora nervosa* (DC) Grev., which is closely related to the North Atlantic P. *epiphylla*. The marine flora of the Azores includes many species common to the Mediterranean and Northwest Europe, but there is also a tropical element. That of the Red Sea has been analysed by Nasrds.

Boergesen has shown that several seaweeds of the Indian and Pacific Oceans (e.g. *Anadyomene stellata, Dictyosphaeria favulosa, Neomeris annulata* and other Siphonales, *Liagora valida* Harv.) also occur in the West Indies, though lacking at the southern extremities of Africa and America. Murray, who first drew attention to this fact, explained the discontinuity by assuming a warmer climate at the Cape in earlier epochs. During the Pliocene and Oligocene, however, the Caribbean Sea was connected to the Pacific Ocean, the isthmus of Panama being of comparatively recent date, and Svedelius ascribes the above-mentioned similarities to this interoceanic connection. This view is based primarily on the present-day distribution of the tropical sea-grasses and of the species of *Caulerpa*, but stress is also laid on the occurrence of pairs of closely related (vicarious) species of diverse genera *(Udotea, Padina, Scinaia, Galaxaura, Martensia*, etc.) in the Indo-Pacific region and the West Indies respectively. The present distribution indicates that the Indo-Pacific Ocean was the centre from which most of the new types migrated into the Atlantic, although the species of *Dasycladus*, for example, are assumed to have originated in the Atlantic. Since species showing a discontinuous distribution must have existed prior to the severance of the sea-connection between the two oceans, the bulk of tropical Phaeophyceae and many of the Rhodophyceae must be of younger date than the marine Chlorophyceae.

Many species, which are common to the West Indies and to the Canaries, seem to reach the eastern limit of their distribution on the shores of the latter or possibly on the adjacent coast of Africa, which is yet scarcely explored. Half the species recorded for the Canaries, moreover, also occur on the Atlantic coasts of America, and the existence of this common element lends some support to the theory of a former land-connection between America and Africa. Taylor concludes that the marine flora of Brazil presents much similarity to that of the West Indies.

Many striking instances of discontinuous distribution, for which there is at present no satisfactory explanation, are afforded by the presence of species, found at the Cape and in West and South Australia, also to the north of the Tropic of Cancer (Japan, Arabian Sea, although they are apparently lacking in the intervening tropical zone. This is paralleled by the occurrence, both in the Arctic and Antarctic, of a number of species which are not known to occur in the equatorial belt, although no species of Laminariales or Fucales is common to the two poles. The marked contrast between the algal flora of the Japan sea and that of the Pacific coast-line of Japan is taken by Okamura (no) to indicate the comparatively late origin of the former which until that time is supposed to have formed, together with the Japanese Islands, part of the continent of Asia.

Our knowledge of the geographical distribution of marine Algae is still too limited to provide much positive evidence as to the existence of endemic genera and species. Attention may, however, be drawn to the numerous Florideae (especially Rhodomelaceae) at present only known from the Australasian region and to the genus *Coccophora* apparently confined to the Japan sea. Feldmann recognises *Laminaria Rodriguezii* and *Rissoella verruculosa* as endemic species in the Mediterranean, the latter characteristic of wave-swept rocks in the littoral region.

Certain marine Algae have doubtless been introduced into new geographical areas, probably for the most part conveyed on ships' bottoms. Comparatively recent arrivals in Europe are *Colpomenia* and the species of *Asparagopsis*, which have been brought from Australia and Japan. After storms considerable masses of seaweeds are often cast up on the shore, many of them bearing viable reproductive organs. It is known, moreover, that many Florideae in particular can regenerate from quite small fragments of their thalli.

Ocean currents will likewise serve to distribute detached portions of seaweeds, while birds may play a part. The fact that seaweeds become established on volcanic and newly arising coral islands indicates that agencies such as those mentioned must be effective. It remains doubtful, however, whether any very considerable number of marine Algae possess the capacity to become naturalised on shores to which they are transported by these means, and it would appear that the spread of most seaweeds takes place rather along coastal stretches than over intervening areas of sea.

General Features of Vegetative Construction

The heterotrichous filamentous habit exhibited by the Ectocarpaceae, seemingly the most primitive members of Phaeophyceae, shows a marked degree of parallel with that found in Chaetophorales among Chlorophyceae. As there, the Ectocarpaceae include an assemblage of reduced types in which essentially only the prostrate system is present. The parallel may not, however, be pressed too far. The heterotrichous habit thus displayed in the simplest members of the class is recognisable, at least in early stages of development of the sporophyte. The persistence of heterotrichy among many of the less specialised Phaeophyceae and its disappearance (i.e. elimination of the prostrate system) in the more advanced forms is altogether paralleled among Florideae. In those Brown Algae, which exhibit heteromorphic alternation, the gametophyte (except in *Cutleria)* always shows traces of the heterotrichous habit.

The bulk of the seaweeds belonging to Phaeophyceae are compact multicellular types with a more or less evident differentiation of tissues. This is specially marked among Laminariales and Fucales, where it is combined with a marked degree of external differentiation and, in some members of the former, with dimensions that vie with those of the larger land-plants. The elaborated sporophytes of Ectocarpales are produced in one of two ways, viz. either *(a)* from a single uniseriate filament or from aggregates of such filaments by dense juxtaposition of their numerous branches (uni- and multiaxial construction) giving a pseudoparenchymatous structure, or *(b)* by division in diverse planes of the cells of a primary filament giving a parenchymatous structure.

The filaments by the elaboration of which the nature body is thus produced belong to the erect system of the primary heterotrichous stage. The details are given under the individual forms, and it must suffice to indicate here that in *Leathesia* and *Castagnea*, for instance, the construction is multiaxial (cf. *Codium* among Siphonales), in *Spermatochnus* uniaxial (cf. *Dasycladus* among Siphonales), whilst parenchymatous members of Ectocarpales are exemplified by such types as *Punctaria, Scytosiphon*, and *Asperococcus*. In uniaxial and parenchymatous types several erect-growing sporophytes commonly originate from the same prostrate base, and this is no doubt sometimes also true of multiaxial forms. While special types of uni- or multiaxial thalli are met with in Desmarestiales, Sporochnales and Cutleriales, most Phaeophyceae are true parenchymatous forms.

The structures produced in the diverse ways indicated above undergo strengthening and elaboration by two methods which are probably interrelated, since in both there is an outgrowth from already mature cells of threads. In many uniaxial *(SpermatocKntis, Desmarestia)* and parenchymatous (e.g. Sphacelariales) types such filamentous outgrowths arise superficially and, growing intimately apposed to the surface and to one another, become compacted to form an enveloping cortex which may increase in thickness by cell-division. Such superficial threads usually contain are most suitably designated *cortical threads.* Quite similar structures often arise in the basal parts of the thallus and, spreading over the primary attachment system, fulfil a subsidiary role to anchorage, so that in these instances they could with equal justification be called *rhizgids.* In many Brown Algae *(Chordaria,* Laminariales, Fucales), however, the outgrowing threads originate from internal cells and ramify in diverse directions through the tissues. Such structures are usually devoid of chromatophores and may be called *hyphae.* For the most part they probably fulfil conducting and mechanical functions. There is no sharp distinction between cortical threads, hyphae, and rhizoids, all of which are merely diverse expressions of the capacity of the algal cell to form thread-like outgrowths. The latter commonly originate from the lower ends of the parent-cells, affording evidence of polarity.

In many parenchymatous types, after the primary longitudinal septation demarcating a peripheral from a central series of cells, further cell-division tends to be restricted to the superficial layer. The cells of this meristem divide periclinally, the segments cut off internally being added to the central tissue, while occasional anticlinal division in the surface-layer enables it to keep pace with the increasing girth of the whole. This superficial meristematic activity, which is little marked among the parenchymatous Ectocarpales, becomes a pronounced feature in Laminariales and Fucales, where it is responsible for the marked increase in thickness attained by the older parts. It is also met with in some of the pseudoparenchymatous forms.

The often abundant branching of the laterals borne on the axial threads of the pseudoparenchymatous forms, which leads to the formation of a compact cortical region, and the superficial growth of parenchymatous forms just discussed, often result in marked tensions between the external and internal tissues. Such tissue-tensions are due to a state of compression in the outermost tissues, while the internal elements are more or less markedly stretched.

As a consequence longitudinal strips of the thalli of *Dictyosiphon Chordaria, Desmarestia, Chorda*, etc., especially when placed in water, curve in the outer tissues on the convex side; not uncommonly such strips roll up into a complete circle. Similar tensions are realised in the stipes of Laminariales and in all instances they probably contribute to the flexibility of the structures involved. The extensibility of such thalli varies (so), being slight in *Fucus,* which has Considerable tensile strength and more considerable in *Halidrys, Ascophyllum*, and *Laminaria*. The massive tissues formed by many Brown Algae are for the most part completely devoid of intercellular spaces.

The Phaeophyceae afford some of the best examples of intercalary growth among lower plants. The intercalary meristems are typically located at the base of a hair, to the increase in length of which they contribute to some extent, although the bulk of the segments are cut off below the meristem. This is found in many Ectocarpales, Desmarestiales and Cutleriales. Intercalary growth is also met with in Laminariales, where the seat of the meristem is the transition zone between stipe and blade. Such well-defined intercalary meristems are not, however, represented in all Ectocarpates.

Intercalary growth has, moreover, been replaced by morphological growth in a number of evolutionary lines among Ectocarpales as well as in Sphacelariales, Dictyotales, and Fucales. There is considerable evidence to show that, in certain Ectocarpales and Fucales, the apical cell of the mature plant is at first differentiated beneath a hair with a basal meristem. Although no such sequence is observable in other apical growth, the instances cited and the prevalence of intercalary growth among the less specialised members.

The Phaeophyceae exhibit not only marked internal differentiation, but also display a great range especially in the more advanced orders. This is fully illustrated in the subsequent detailed consideration. In many Ectocarpales the thallus consists Q£ branched cylindrical or flattened strands presenting little outward differentiation, but the Punctariacgae exhibit a surface-development presenting many analogies with that of the Ulvaceae among Green and of the Bangiales among Red Algae. Marked morphological elaboration, accompanied by the development of pseudoaxillary branch-formation, is met with in some Sphacelariales.

A foliose differentiation of certain parts of the thallus is specially characteristic of Laminariales and Fucales, the leafy expanses of which are either differentiated parts of the thallus *(Laminarid)* or represent special outgrowths (Alariaceae, *Sargassum)*. At the same time the Laminariales exhibit a method of thallus-elaboration by progressive splitting of the primary foliar expanses that is unique among Thallophyta.

Although many Brown Algae *(Ralfsia, Lithoderma,* many Laminariales and Fucales) are lithophytes, a considerable number (Ectocarpales, Sphacelariales, etc.) are epiphytes. In diverse of these, some of the filaments penetrate into the foreign tissue to an appreciable depth and the non-endophytic part of such forms is commonly reduced to a more or less marked extent.

As a general rule the endophytic filaments possess chromatophores and only traverse the gelatinous walls of the host, although diverse instances are on record of actual penetration of the cell-cavities. It is scarcely possible to speak of parasitism since, although the contents of the host's cells adjacent to the endophytic filaments are commonly disorganised or destroyed, there is no evidence that the endophyte draws any nourishment from them.

Cell wall

The cell walls of Phaeophyceae are commonly gelatinous and appear to consist of pectic substances with a layer giving the reactions of cellulose adjacent to the protoplast. The latter, especially abundant *Laminaria*, though found also in *Fucus* and in very small quantities in *Ascophyllum*.

Protoplast

The cytoplasm of Phaeophyceae does not exhibit the remarked viscosity seen in many Red Algae. Although complete rounding off of the protoplast is rarely obtained, there is no marked adhesion to the cell-walls. The capacity for deplasmolysis varies. The mitochondria, appearing as rounded or filiform bodies, have been studied especially by the French school. They are not affected by *intra vitam* stains that colour the vacuolar contents, but as a general rule are stained by iron haematoxylin after fixation (10 per cent formalin in sea-water).

The cells commonly possess numerous small vacuoles separated by delicate tonoplast in particular, the numerous vacuoles give an alveolar structure to the cytoplasm. A single central vacuole is, however, reported in *Dictyota*, species of *Fucus, Sporochnus*, etc. According to Chadefaud the sap includes colloidal substances (metachromatin), very abundant in *Desmafestia*, which are responsible for the *intra vitam* staining with neutral red or cresyl blue and are sometimes slowly precipitated by the latter as granules or crystalline clusters; some of these may be due to iodides.

In diverse species of *Desmarestia* the extract is stated to have a pH of about one, although there is difference of opinion as to the nature of the acid present. The bluish colour assumed by certain species of this genus after death is due to the action of the acid sap on the fucoxanthin of the chromatophores. The osmotic pressure, determined cryoscopically, lies between 39-2 and 31-8 atmospheres in diverse Fucales, while Hurd by a less precise method determined it as 227 atmospheres in *Nereocystis*. Kylin gives much higher values and points out that the to be more widely distributed; that of Laminariales is stated to be a polymer of mannuis. It has many economy uses, being employed the preparation of adhesives, artificial silk, etc. According to Miwa, very similar to algin and likewise contains sulphur. Callus is widely distributed in the sieve-tubes of Laminariales.

The abundant mucilage of the larger Brown Algae is only in certain Laminariales located in definite mucilage-canals. It can hardly be doubted that it plays a significant role in the littoral Fucales. Younger cells often have a somewhat higher osmotic pressure than the older ones.

Chromatophores and Pigments

The chromatophores in Phaeophyceae are recorded only in *Pylaiella*. Where they are stellate structures something like those of a *Zygnema* and present to the number of one or two in each cell. In the vast majority of Brown Algae the photosynthetic cells contain numerous small discoid chromatophores.

Among Ectocarpales there is, however, some variety of form. In diverse genera the cells contain a single chromatophore. In others the number of chromatophores is limited. Thus, in *Lithoderma* there are 8 and in *Hecatonema terminate* 2-3 curved plate-shaped chromatophores in the cells, often fitting closely together; much the same obtains in *Dictyosiphon*. Ribbon-shaped chromatoghores occur in some species of *Ectocarpus* and in *Phloeospora*, where they are branched; in *E. confervoides* the ribbon is often narrow and spirally twisted. In the older cells of these species of *Ectocarpus*, however, one often finds a considerable number of small discs. The occurrence of single chromatophores in diverse Ectocarpales is in conformity with the general low level of differentiation exhibited by this order. It is probably correct to assume that, as in other classes of Algae, this marks the primitive sedition and that the occurrence of discoid chromatophores represents a more advanced state. Dissected chromatophore shows contrast to the numerous discoid chromatophores of the sporophyte.

The small chromatophores within the apical cells of *Fucus* and *Cladostephus* undergo appreciable enlargement in the outer, although they remain small in the inner cells of the mature thalli. A similar enlargement and flattening of the "chromatophores is recorded by Senn in diverse Dictyotales. The stigma in the motile elements of Phaeophyceae is, according to Chadefaua, invariably differentiated as part of a plastid.

Many Brown Algae show movements of chromatophores in response to changes in the intensity and Direction of illumination. When the light is of medium intensity the chromatophores take up a position against the external walls, while in bright light there is movement to the side walls. The fucosan-vesicles remain stationary during such changes. Hypertonic solutions cause chromatophores of Dictyotales to take Tip a surface position, whilst in hypotonic solutions they tend to shift to the side walls. According to Senn the chromatophores of diverse Phaeophyceae (Dictyotales, *Zanardinia, Asperococcus)* assume a spherical or ellipsoidal shape in weak light or darkness, while contraction of those of *Saccorhiza* in bright light. The single band-shape chromatophore of *Petalonia* becomes inrolled-in weak light. Some Phaeophyceae *(Cystoseira, Dictyota)* exhibit bluish or greenish iridescence in strong light; in *Cystoseira*, as in Red Algae, this is due to the aggregation of special protein-masses beneath the external walls.

Bodies of doubtful nature, but since Kuckuck's investigations usually described as *pyrenoid* (Phaeophycean starch of Schmitz, occur in diverse Ectocarpales, as well as in *Haplospora, Nereia,* and *Arthrocladia*; they appear to be lacking in the higher orders. They are commonly more or less pyrenoid and attached by their narrower ends to the surface of the latter.

In other instances they appear more or less globose. According to Knight pyrenoids of *Ectocarpus siliculosus* are either closely attached to the chromatophore by strand of cytoplasm the latter condition being regarded as secondary; some evidence is produced for the origin of these bodies from the chromatophores. They are present also during zoospore-formation.

Treatment with a o-1 per cent solution of picric acid in sea-water causes bursting of the fucosin vesicles, while leaving the pyrenoids intact and their true character and function still remain obscure. They may represent a metabolic product of a nature different from that of the pyrenoids of other Algae, a view which finds some support in the fact that they can apparently become detached from the chromatophores.

The characteristic colouration of Phaeophyceae, varying from olive-yellow due to an accessory carotenoid pigment, *fucoxanthin* (also called phycoxanthin)), which was first isolated by Willstatter and Page, although its presence had been recognised long before. It is accompanied by chlorophyll with only small traces of chlorophyll *b* \ according to Seybokf and Egle the latter is altogether wanting. Howard finds evidence of a chlorophyll in *Nereocystis.* Kylin has recently cast doubts on these conclusions. Lubimenko also states that the amount of chlorophyll in Brown Algae is only 30 per cent of that in *Zoster* whilst Seybold and Egle find that the Brown Algae investigated by them contain just as much chlorophyll and carotenoid pigments per unit of surface as the Chlorophyceae and that the molecular ratio of green and yellow pigments is the same Jguboth. All Phaeophyceae so far examined contain and flavins are widely represented.

An impure solution of fucoxanthin is readily extracted from the dead algae with water, leaving the chromatophores green. The formula is $C_{40}H_{66}O_6$ according to Willstatter and Page. The aqueous solution is orange yellow, whilst the crystals are red-brown with a bluish sheen. There is nothing very distinctive about the absorption spectrum, save for the strong absorption of the blue rays. The amount of fucoxanthin varies; *Laminaria, Dictyota, Petalonia Fascia, Pylaiella littorcdis,* and species of *Ectocarpus,* for instance, are rich, while the species of *Fugus* are poor in it/ According to Montfoirt the last, like most of the littoral Brown Algae, are rich in xanthophyll and fucoxanthin. Of the two modifications of fucoxanthin previously recognised by Kylin one has proved to be a postmortem oxidation-product of the other.

Until recently it has generally been doubted whether the fucoxanthin participates in any way in the process of photosynthesis. Montfort has, however, shown that Brown Algae rich in fucoxanthin exhibit much higher rate of photosynthesis which are poor in it; elimination or blue rays effects marked reductions in the photo many of the properties and is usually regarded as a derivative. Hansteen's conclusion that the contents were — a carbohydrate and represented a first product of photosynthesis is due to confusion with laminarin.

The latter originate plasm, independently of the chromatophores. Mangenot believed that they arose primarily in the vacuoles. The evidence for Chadefaud's view that fucosan-vesicles develop from mitochondria is inadequate and for the present they may be regarded as vacuoles of a special kind, the mode of origin of which is not clear. According to Chadefaud they are formed only in the apical cells in Sphacelariales.

The vesicles not uncommonly show movements which may be accompanied by change of shape and are sometimes rapid; it is not clear whether these movements result cytopiasmic streaming or from physical causes. Chadefaud provides data indicating a possible excretion of fugosan.

Comments are frequently made on the absence of fucosan-vesicles from the cells adjacent to the endophytic filaments of "parasitic" Algae, as for instance in *Ascophyllum* attacked by *Polysiphonia fastigiata*; this is evidence of a metabolic disturbance, but has not imply that the vesicles are of the nature of a food reserve. There does not appear to be any good evidence that they accumulate in fact in *Ectocarpus siliculosus* an increase during prolonged dąrkness is recorded. Knight reports diminution during nuclear division. Fucosan-vesicles are often found in the young cells of hairs.

Products of Photosynthesis

Of frequent occurrence are dextrm-like polysaccharides grouped as *laminarin*. They appear to constitute food-reserves and are believed by Kylin to arise from the simple sugars of photosynthesis. After extraction laminarin appears as a white, tasteless powder soluble in water, the solution being laevorotary. It is coloured by iodine and is transformed into dextrose on hydrolysis with sulphuric acid, as well as after treatment with diastase. Laminarin is particularly abundant in *Laminaria*, whilst *Fucus vesiculosus* and *Ascophyllum nodosum* contain relatively little; in *Laminaria* it occurs copiously during the summer. It appears to "be Jacking in diverse other Phaeophyceae that have been investigated (e.g. *Halidrys siliquosa. Chorda filum, Chordaria flagelliformis*, etc.;. Enzymes capable of hydrolysing carbohydrates have been demonstrated in *Lamtnana* and *Nereocystis*.

Mannitol appears to be more widely distributed. Haas and Hill indeed suggest that it may be several. According to Kylin mannitol occurs as a frequent food reserve. Gin Laminariales, Fucales, etc., but he failed to find it for instance in *Desmarestia viridis* and *Stilophora rhizodes;* in *Laminaria* the amount varies in different seasons. The mannitol is sometimes present as an anhydride (mannitan). Qxalates are recorded in diverse Phaeophyceae commonly present being recorded in Fucales (especially *Ascophyllum),* in the young sporangia of *Chorda filum,* in the photosynthetic tissues of *Laminaria,* is a greater amount of fat in exposed forms *(Pelvetid)* than in those which are mainly submerged *(Laminarid)*; moreover the fats in the former are more highly, saturated. There is also some seasonal variation.

The facts indicate that Brown Algae possess a carbohydrates metabolism which shows many distinctive features, although clearly following a course parallel with that of other holophytic plants. Little is known of the nitrogen metabolism. Nitrates are sometimes recognisable in the growing apices, although absent from older parts. In *Laminaria* they are found but there is little in the blade or in the transition zone; according to Wille the newly developing blades are richer in nitrogen and phosphorus than the old blades or the stipes. Although this was lacking in the forms inhabiting the lower zones. It is suggested that this is due to interruption of the normal metabolism of the former owing to the long periods of emersion. In artificial cultures growth is usually markedly increased by the addition of nitrates to the sweater. Phosphates are easily demonstrated in diverse Phaeophyceae, while *Desmarestia* is rich in sulphates. The larger Brown Algae have been an important source. Diverse Laminariales and Fucales contain considerable quantities of *vitamins*.

Iodine

The larger Brown Algae were long one of the sources of commercial iodine. Whereas sea-water only contains this element largely. Kylin gives the following percentages of KI in the fresh weight: *L. saccharina*, stipe, 0-19; transition zone, 0-30; *L. Cloustont*, stipe, 0-30; other data for Laminariales are: *Nereocystis*, 0-08-0-22 per cent; *Eisenia bicyclis* (*irj*) 0-15-0-35 per cent.

Himanthalid) although Scruti gives for *Sargassum linifolium* cr 1 2 per cent or less. Kylin also quotes low values for many other Brown Algae. In *Laminarias* there is moreiodine in younger parts than in the older parts, and thę amount is greatest in July and decreases markedly in autumn. Opinions differ as to the form in which the iodine occurs according to some most of it is present as an organic compound soluble in water.

On the other hand, the latter are distinguished "by the liberation of iodine from the living algae under natural circumstances. It is very marked in certain seasons in the European species of *Laminaria*, as well as in *Ascophyllum nodosum* and *Pelvetia*, less obvious in *Alaria* and in the species of *Fucus* it takes place from the surface-layer and is readily demonstrated by "the blue colour assumed by starch-paste. Kylin showed that the escape of iodine depends on the presence of iodide-oxidases. In the presence of oxygen and of an acid reaction in the membrane, these enzymes act upon the iodides that have diffused into the walls. Liberation of iodine is in fact increased by immersion and exposure, by variations in temperature, wave-action, etc.

Nucleus

The nuclei of Phaeophyceae are usually vary their size being often closely related to that of the cell. Those of apical cells are commonly especially conspicuous and this is particular marked in Sphacelariales, the nuclei of which have, as a consequence, been frequently investigated. A multinucleate condition exceptional, although recorded in some in the older cells of *Halopteris*. The nuclei possess a large and readily stain vacuoles, and a delicate network, usually with little chromatic material, although obvious chromatin masses are a few Fucales; some of the latter have nuclei with two nucleoli.

No other class of Algae affords such clear examples of which, however, seem to be most readily demonstrated in the higher orders and in certain instances have only been recognised in the reproductive cells. They are very conspicuous in the apical cells of Sphacelariales and in Fucales.

At the commencement of mitosis the nucleus usually enlarges and two centrosomes with accompanying radiations become visible. Enuring prophase the chromatin in the outer nucleus increases in amount and gradually aggregates to form the small chromosomes; simultaneously the nucleolus stains less deeply and becomes more evidently vacuolate, although as a general rule persisting up to metaphase. Diverse earlier workers concluded that the nucleolus contributed material for the development of the chromosomes, but this is unlikely. The spindle always appears to be intranuclear; the nuclear membrane is in fact very persistent, except at the poles, where the fibres of the spindle converge on the centrosomes. When aggregated on the equatorial plate, the chromosomes are rounded or slightly elongated. There is often a considerable contraction of the nucleus at metaphase. The cell-plate is formed by arrangement in one plane of the cytoplasmic lamellae between the alveoli. There is never any relation to the nuclear spindle.

The reduction division, which except Fucales is always located in the unilocular sporangium, presents no special features. A distinct spireme stage is mostly recorded, followed by often long-continued synezesis and diakinesis. Centrosomes have usually Seen observed and the spindles are intranuclear. The haploid number of chromosomes is given as: 8-10 in Ectocarpales, although that number is 20 in *Heterochordaria*; usually' 16 in Sphacelariales and Dictyotales; 13-15 in Languaariales; 24 in Cutleriales. In Fucales Yamanouchi and most recent workers report 32 chromosomes, and the lower numbers given for *Sargassum* by Kunieda (us), as well as by earlier investigators, are probably erroneous. The data suggest that there may be a greater measure of uniformity than is at present apparent.

General Facts of Reproduction

Except for the Dictyotales and Fucales, all Brown Algae reproduce by zoospores produced in definite sporangia, which are borne on the diploid phase. Among Ectocarpales and Sphacelariales there are commonly two kinds of sporangia, spoken of as *unilocular* and *plurilocular* respectively, but in all other zoosporic Phaeophyceae only the unilocular sporangium is represented on the diploid plant. This sporangium, as the name implies, is not partitioned, and the more or less numerous motile elements produced within it are formed simultaneously, after completed nuclear division, by separation of the cytoplasm into as many uninucleate portions. The sporophytes of Dictyotales possess a special type of unilocular sporangium producing usually four motionless spores. Meiosis in all Brown Algae, except Fucales, is associated with the first nuclear divisions in the unilocular sporangium so that the spores derived from it are normally haploid.

The plurilocular sporangium, on the other hand, is divided by transverse, and often also by longitudinal, septa into numerous small compartments, each forming a single swarmer. No reduction occurs during the differentiation of the plurilocular sporangium and the zoospores it produces are therefore diploid. In the more advanced Ectocarpales and Sphacelariales such sporangia are commonly rare or lacking on the diploid phase, an indication of the evolutionary trend that has resulted in their complete disappearance in the sporophytes of the other zoosporic orders (Cutleriales, Desmarestiales, Laminariales).

The zoospores from the unilocular sporangia produce haploid gametophytic stages, while those from the plurilocular sporangia constitute an accessory means of reproduction of the diploid phase. The gametophytic stages in the less specialised Phaeophyceae (Ectocarpales, Cutleriales, Sphacelariales) bear reproductive organs, in general character altogether identical with the plurilocular sporangia of the diploid phase, although unilocular sporangia are never present. The plurilocular sporangia of the haploid phase are, however, gametangia, and their swarmer, identical in form with those of the diploid phase, behave as gametes, although frequent apogamy occurs, at least in artificial cultures. The complete similarity of the gametangia of the haploid, and of the plurilocular sporangia of the diploid, phases long encumbered the elucidation of the life-cycle of Ectocarpales.

In most Ectocarpales, and probably also in many Sphacelariales, the fusing gametes are morphologically identical, but pronounced anisogamy is found in a limited number of Ectocarpales, as well as in Cutleriales. All other Phaeophyceae exhibit oogamy, usually with the production of a single ovum in the oogonium, although a few Fucales produce several. The ova are almost invariably extruded prior to fertilization. The male cells, always motile, are commonly produced singly in the antheridia, although in Dictyotales the male organ has the form of a plurilocular sporangium and in Fucales considerable numbers of spermatozoids are produced in the antheridium. The zygote invariably germinates without a resting period.

The motile cells, whether zoospores, gametes or spermatozoids, are of a uniform type. The two flagella are attached laterally, commonly rather nearer to the posterior end of the pear-shaped body; they are of unequal length, the longer usually directed forwards, and the shorter backwards during movement. The eyespot is usually adjacent to the point of attachment of the flagella, and is directly apposed to the often single chromatophore or (in the spermatozoids of Fucales) represents the chromatophore itself; when several chromatophores are present, one bearing the eyespot lies adjacent to the place of origin of the flagella. The tip of the longer flagellum often acts as a sucker when the swarmer settles on a substratum, while it commonly serves to attach the male to the female cell. As already indicated the uniformity of the motile element in Phaeophyceae implies a common origin for the members of this class from unicellular organisms presenting these particular characteristics.

So far as the evidence goes, the life-cycle normally involves an alternation between two phases, with reduction in the unilocular sporangium. In Ectogjrpaceae and a few other Ectocarpales the two phases are alike and the alternation is isomorphic; this is also so in Dictyotales, *Zanardinia,* and probably all Sphacelariales. On the other hand the majority of Ectocarpales, as well as Sporochnales, Desmarestiales, and Laminariales, exhibit a marked heteromorphic alternation, with minute filamentous gametophytic stages. It is evident that in the Ectocarpales the heteromorphic alternation is derived from an isomorphic one, and this is likewise probable in *Cutleria.* There is, moreover, every reason to suppose that in Desmarestiales and Laminariales also the marked contrast between the two generations is secondary. In other words isomorphic alternation probably represents the primitive condition in this class. The Fucales afford no evidence of an alternation of phases, the diploid plant bearing sex organs as the only organs of reproduction. The life-cycle is here probably a highly derived one, the elucidation of which presents many difficulties.

Classification and Status of Phaeophyceae

Most of the orders of Phaeophyceae have long been recognised and are so clearly defined that there is no difference of opinion about their delimitation, Since a better knowledge of the life-cycle of the simpler Brown Algae has become available, rather diverse views have, however, been expressed as to their classification and still another grouping is adopted here. The reasons for this attitude are given. The old subdivisions—Phaeosporeae, Acinetae and Cyclosporeae—can now scarcely be maintained.

Desmarestiales: Uniaxial forms with complex cortication and intercalary growth; asexual reproduction by zoospores formed in unilocular sporangia; sexual reproduction oogamous.

Sphaceluriules: Relatively small filamentous forms showing evident heterotrichy; apical growth; asexual reproduction by zoospores formed in unilocular sporangia; sexual reproduction isogamous (and oogamous); isomorphic alternation probable. Little differentiated parenchymatous forms, often with dichotomous branching; apical growth; asexual reproduction by motionless tetraspores formed in a unilocular sporangium; sexual reproduction.

Fucales: Parenchymatous forms with complex morphological and anatomical differentiation; apical growth; no asexual reproduction; sex organs borne in conceptacles on the diploid phase and exhibiting meiosis during the formation of the sexual cells; no evident alternation.

The Phaeophyceae present no obvious affinities with any other class and are indeed in most respects so sharply circumscribed that little opportunity is afforded for speculations on their relationships.

Neither argument has, however, stood the test, since there is evidence to show that the Dictyotaceous tetrasporangium is a special form of the ubiquitous unilocular sporangium.

On present evidence this class must be regarded as an altogether distinct evolutionary line (Phaeophyta). Its many approximations in morphological and anatomical respects (foliar differentiation, axillary branching, conducting system with sieve-tubes, secondary-growth) to higher plants are no doubt but an expression of the general evolutionary trend within the plant-kingdom. As such and as examples of parallelism they are of the greatest possible interest, but they are undoubtedly without any direct phylogenetic significance. Attention may be drawn to the fact that similar specialisations are encountered also in Florideae with their altogether different basic construction and highly distinctive reproduction.

Among the few fossil types that have been ascribed to Phaeophyceae the most striking is *Prntnfaxtps (Nematophycus, Nematophyton* from the Silurian and Devonian. The remains, some of which reach a diameter of 1 metre, consist of loosely aggregated forlcedthreads, sometimes interwoven with much narrower ones (hyphae). Periodic differences in the thickness of the walls of the threads occasionally give the appearance of growth-rings. In certain specimens threads running perpendicular to the medullary region have been recognised. This form appears to have a multiaxial construction and might represent a large member of Ectocarpales.

5

Kelp Forests

Occurring worldwide throughout temperate and polar coastal oceans, Kelp Forests are recognised as one of the most productive and dynamic ecosystems on Earth. Physically formed by brown macroalgae of the order Laminariales, kelp forests provide a unique three-dimensional habitat for marine organisms and are a source for understanding many ecological processes. Over the last century, they have been the focus of extensive research, particularly in trophic ecology, and continue to provoke important ideas that are relevant beyond this unique ecosystem. For example, kelp forests can influence coastal oceanographic patterns and provide many ecosystem services to humankind. However, the influence of humans has often contributed to kelp forest degradation. Of particular concern are the effects of overfishing nearshore ecosystems, which can release herbivores from their normal population regulation and result in the over-grazing of kelp and other algae. This can rapidly result in transitions to barren landscapes where relatively few species persist. The implementation of marine protected areas (MPAs) is one management strategy useful for addressing such issues since it may limit the impacts of fishing and buffer the ecosystem from additive effects of other environmental stressors.

Kelp

The term 'kelp' refers to marine algae belonging to the taxonomic order Laminariales (Phylum: Heterokontophyta). Though not considered a taxonomically diverse order, kelps are highly diverse structurally and functionally. The most widely recognised species are the giant kelps (Macrocystis *spp.*), although there are numerous other genera such as Laminaria, Ecklonia, Lessonia, Alaria and Eisenia.

Frequently considered an ecosystem engineer, kelp provides the physical substrate and habitat about which kelp forest communities are built. In algae (Kingdom: Protista), the body of an individual organism is known as a thallus rather than as a plant (Kingdom: Plantae). The morphological structure of a kelp thallus is defined by three basic structural units:

- The holdfast is a rootlike mass that anchors the thallus to the sea floor, though unlike true roots it is not responsible for absorbing and delivering nutrients to the rest of the thallus;
- The stipe is analogous to a plant stalk, extending vertically from the holdfast and providing a support framework for other morphological features;
- The fronds are leaf- or bladelike attachments extending from the stipe, sometimes along its full length, and are the sites of nutrient uptake and photosynthetic activity.

In addition, many kelp species have pneumatocysts, or gas-filled bladders, usually located at the base of fronds near the stipe. These structures provide the necessary buoyancy for kelp to maintain an upright position in the water column.

The environmental factors necessary for kelp to survive include hard substrate (usually rock), high nutrients (e.g., nitrogen, phosphorus), and light (minimum annual irradiance dose > 50 E m^{-2}). Especially productive kelp forests tend to be associated with areas of significant oceanographic upwelling, a process which delivers cool nutrient-rich water from depth to the ocean's mixed surface layer. Water flow and turbulence facilitate nutrient assimilation across kelp fronds throughout the water column. Water clarity affects the depth to which sufficient light can be transmitted. In ideal conditions, giant kelp (*Macrocystis spp.*) can grow as much as 30-60 centimetres vertically per day. Some species such as Nereocystis are annual while others like Eisenia are perennial, living for more than 20 years. In perennial kelp forests, maximum growth rates occur during upwelling months (typically spring and summer) and die-backs correspond to reduced nutrient availability, shorter photoperiods and increased storm frequency.

Kelps are primarily associated with temperate and arctic waters worldwide. Of the more dominant species, *Laminaria* is mainly associated with both sides of the Atlantic Ocean and the coasts of China and Japan; *Ecklonia* is found in Australia, New Zealand, and South Africa; and *Macrocystis* occurs throughout the northeastern and southeastern Pacific Ocean, Southern Ocean archipelagos, and in patches around Australia, New Zealand and South Africa.

The region with the greatest diversity of kelps (>20 species) is the northeastern Pacific, from north of San Francisco, California, to the Aleutian Islands, Alaska. Although kelp forests are unknown in tropical surface waters, a few species of *Laminaria* have been known to occur exclusively in tropical deep waters. This general absence of kelp from the tropics is believed to be mostly due to insufficient nutrient levels associated with warm, oligotrophic waters.

Ecosystem Architecture

The architecture of a kelp forest ecosystem is based on its physical structure, which influences the associated species that define its community structure. Structurally, the ecosystem includes three guilds of kelp and two guilds occupied by other algae:

- *Canopy* kelps include the largest species and often constitute floating canopies that extend to the ocean surface (e.g., *Macrocystis* and *Alaria*);

- *Stipitate* kelps generally extend a few metres above the sea floor and can grow in dense aggregations (e.g., *Eisenia* and *Ecklonia*);
- The benthic *assemblage* is composed of other algal species (e.g., filamentous and foliose functional groups, articulated corallines) and sessile organisms along the ocean bottom;
- *Encrusting* coralline algae directly and often extensively cover geologic substrate.

Multiple kelp species often coexists within a forest; the term *understory canopy* refers to the stipitate and prostrate kelps. For example, a *Macrocystis* canopy may extend many metres above the seafloor towards the ocean surface, while an understory of the kelps *Eisenia* and *Pterygophora* reaches upward only a few metres. Beneath these kelps, there may be a benthic assemblage of foliose red algae.

The dense vertical infrastructure with overlying canopy forms a system of microenvironments similar to those observed in a terrestrial forest, with a sunny canopy region, a partially shaded middle, and darkened seafloor. Each guild has associated organisms, which vary in their levels of dependence on the habitat, and the assemblage of these organisms can vary with kelp morphologies.

For example, in California *Macrocystis pyrifera* forests, the nudibranch *Melibe leonina* and skeleton shrimp *Caprella californica* are closely associated with surface canopies; the kelp perch *Brachyistius frenatus*, rockfish Sebastes *spp.* and many other fishes are found within the stipitate understory; brittle stars and turban snails *Tegula spp.* are closely associated with the kelp holdfast, while various herbivores such as sea urchins and abalone live under the prostrate canopy; many seastars, hydroids and benthic fishes live among the benthic assemblages; solitary corals, various gastropods and echinoderms live over the encrusting coralline algae. In addition, pelagic fishes and marine mammals are loosely associated with kelp forests, usually interacting near the edges as they visit to feed on resident organisms.

Trophic Ecology

Classic studies in kelp forest ecology have largely focused on trophic interactions (the relationships between organisms and their food webs), particularly the understanding of bottom-up and top-down trophic processes. Bottom-up processes are generally driven by the abiotic conditions required for primary producers to grow, such as availability of light and nutrients, and the subsequent transfer of energy to consumers at higher trophic levels. For example, the occurrence of kelp is frequently correlated with oceanographic upwelling zones, which provide unusually high concentrations of nutrients to the local environment.

This allows kelp to grow and subsequently support herbivores, which in turn support consumers at higher trophic levels. By contrast, top-down processes are those in which predators limit the biomass of species at lower trophic levels through consumption; in the absence of predation, these lower level species would flourish because the resources that support their energetic requirements are non-limiting. In a well-studied example from Alaskan kelp forests, sea otters (*Enhydra lutris*) control populations of herbivorous sea urchins through predation.

When sea otters are removed from the ecosystem (for example, by human exploitation), urchin populations are released from predatory control and grow dramatically. This leads to increased herbivore pressure on local kelp stands. Deterioration of the kelp itself results in the loss of physical ecosystem structure and subsequently the loss of other species associated with this habitat. In Alaskan kelp forest ecosystems, sea otters are the keystone species that mediates this trophic cascade. In Southern California, kelp forests persist without sea otters and the control of herbivorous urchins is instead mediated by a suite of predators including lobsters and large fishes. The effect of removing one predatory species in this system differs from Alaska because there is redundancy in the trophic levels and other predatory species can continue to regulate urchins. However, the removal of multiple predators can effectively release urchins from predator pressure and allow the system to follow trajectories towards kelp forest degradation. Similar examples exist in Nova Scotia, South Africa, Australia and Chile. The relative importance of top-down versus bottom-up control in kelp forest ecosystems and the strengths of trophic interactions continue to be the subject of considerable scientific investigation.

The transition from macroalgal (i.e. kelp forest) to denuded landscapes dominated by sea urchins (or 'urchin barrens') is a widespread phenomenon, often resulting from trophic cascades like those described above; the two phases are regarded as alternative stable states of the ecosystem. The recovery of kelp forests from barren states has been documented following dramatic perturbations, such as urchin disease or large shifts in thermal conditions. Recovery from intermediate states of deterioration is less predictable and depends on a combination of abiotic factors and biotic interactions in each case.

Though urchins are usually the dominant herbivore, others with significant interaction strengths include seastars, isopods, kelp crabs, and herbivorous fishes. In many cases, these organisms feed on kelp that has been dislodged from substrate and drifts near the ocean floor rather than expend energy searching for intact thalli to feed on. When there is sufficient drift kelp, herbivorous grazers do not exert pressure on attached plants; when drift subsidies are unavailable, grazers directly impact the physical structure of the ecosystem. Many studies in Southern California have demonstrated that the availability of drift kelp specifically influences the foraging behaviour of sea urchins. Drift kelp and kelp-derived particulate matter have also been important in subsidising adjacent habitats, such as sandy beaches and the rocky intertidal.

Patch Dynamics

Another major area of kelp forest research has been directed at understanding the spatial-temporal patterns of kelp patches. Not only do such dynamics affect the physical landscape, but they also affect species that associate with kelp for refuge or foraging activities. Large-scale environmental disturbances have offered important insights concerning mechanisms and ecosystem resilience. Examples of environmental disturbances include the following:

- Acute and chronic pollution events have been shown to impact southern California kelp forests, though the intensity of the impact seems to depend on both the nature of the contaminants and duration of exposure. Pollution can include sediment deposition and eutrophication from sewage, industrial by-products and contaminants like PCBs and heavy metals (for example, copper, zinc), runoff of organophosphates from agricultural areas, anti-fouling chemicals used in harbours and marinas (for example, TBT and creosote) and land-based pathogens like faecal coliform bacteria.
- Catastrophic storms can remove surface kelp canopies through wave activity but usually leave understory kelps intact; they can also remove urchins when little spatial refuge is available. Interspersed canopy clearings create a seascape mosaic where sunlight penetrates deeper into the kelp forest and species that are normally light-limited in the understory can flourish. Similarly, substrate cleared of kelp holdfasts can provide space for other sessile species to establish themselves and occupy the seafloor, sometimes directly competing with juvenile kelp plants and even inhibiting their settlement.
- El Nino-Southern Oscillation (ENSO) events involve the depression of oceanographic thermoclines, severe reductions of nutrient input, and changes in storm patterns. Stress due to warm water and nutrient depletion can increase the susceptibility of kelp to storm damage and herbivorous grazing, sometimes even prompting phase shifts to urchin-dominated landscapes. In general, oceanographic conditions (that is, water temperature, currents) influence the recruitment success of kelp and its

competitors, which clearly affect subsequent species interactions and kelp forest dynamics.

- Overfishing higher trophic levels that naturally regulate herbivore populations is also recognised as an important stressor in kelp forests. The drivers and outcomes of trophic cascades are important for understanding spatial-temporal patterns of kelp forests.

In addition to ecological monitoring of kelp forests before, during, and after such disturbances, scientists try to tease apart the intricacies of kelp forest dynamics using experimental manipulations. By working on smaller spatial-temporal scales, they can control for the presence or absence of specific biotic and abiotic factors to discover the operative mechanisms. For example, in southern Australia, manipulations of kelp canopy types demonstrated that the relative amount of *Ecklonia radiata* in a canopy could be used to predict understory species assemblages; consequently, the proportion of *E. radiata* can be used as an indicator of other species occurring in the environment.

Human Use

As exemplified by the Aleut hunting of sea otters, kelp forests have been important to human existence for thousands of years. Modern economies are based on fisheries of kelp-associated species like lobster and rockfish. Humans also harvest kelp directly to feed aquaculture species like abalone and to extract the compound alginic acid, which is used in products like toothpaste and antacids. Kelp forests are valued for recreational activities such as Scuba diving and kayaking; the industries that support these sports represent one benefit related to the ecosystem and the enjoyment derived from these activities represents another. All of these are examples of ecosystem services provided specifically by kelp forests.

Threats and Management

Given the complexity of kelp forests – their variable structure, geography and interactions – they pose a considerable challenge to environmental managers. It is difficult to extrapolate even well-studied trends to the future because interactions within the ecosystem will change under variable conditions, not ali relationships in the ecosystem are understood, and there are can be non-linear thresholds to transitions that are not yet recognised. With respect to kelp forests, major issues of concern include marine pollution and water quality, kelp harvesting and fisheries, invasive species and climate change. It has been argued that the most pressing threat to kelp forest preservation is the overfishing of coastal ecosystems, which by removing higher trophic levels facilitates their shift to deprecate urchin barrens. The maintenance of biodiversity is recognised as a way of generally stabilising ecosystems and their services through mechanisms such as functional compensation and reduced susceptibility to foreign species invasions.

In many places, managers have opted to regulate the harvest of kelp and/or the taking of kelp forest species by fisheries. While these may be effective in one sense, they do not necessarily protect the entirety of the ecosystem. Marine protected areas (MPAs) offer a unique solution that encompasses not only target species for harvesting but also the interactions surrounding them and the local environment as a whole. Direct benefits of MPAs to fisheries (for example, spillover effects) have been well-documented around the world. Indirect benefits have also been shown for several cases among species such as abalone and fishes in Central California. Most importantly, studies have demonstrated that MPAs can be effective at protecting existing kelp forest ecosystems and may also allow for the regeneration of those which have been impacted.

Dominant Species

The rocky nearshore environment of the Monterey Bay National Marine Sanctuary (MBNMS), like the rest of central California, is characterised by dense forests of kelp growing at depths from 2 m to more than 30 m (Foster and Schiel 1985).

The giant kelp *Macrocystis pyrifera* and the bull kelp *Nereocystis luetkeana* are the dominant canopy-forming kelps in this region, and make up the major forests within the MBNMS. Giant kelp forms dense beds in the sanctuary from Cambria to Año Nuevo except in the area between Monterey and Santa Cruz where the sandy substrate is unsuitable for kelp attachment (NOAA 1992). The rocky coastline south of Monterey Bay is characterised by hard granite substrates with only a few sandy beaches which are limited to small coves (McLean 1962). North of Santa Cruz, the bull kelp, which occurs from Point Conception to Unimak Island in the eastern Aleutians (Abbott and Hollenberg 1976, Miller and Estes 1989), becomes the dominant canopy-forming kelp (Foster 1982, Foster and Schiel 1985).

Along the central California coast where the distributions of these two species overlap, giant kelp outcompetes bull kelp for light. Giant kelp dominates areas of relatively low water motion, and is dominant in years with relatively calm sea conditions. Bull kelp is more tolerant of high water motion and dominates more exposed areas (McLean 1962, Foster 1982, Harrold et al. 1988). The shallow areas inshore of these kelp forests are often characterised by canopies of the feather boa kelp *Egregia menziesii*, the intertidal giant kelp *Macrocystis integrifolia*, and the Fucalean algae *Cystoseira osmundacea* (Foster and Schiel 1985).

Foster (1982) observed that at Greyhound Rock, 24 km North of Santa Cruz, kelp canopies become sparse due to the combined effects of unsuitable substrate and increased water motion. This may help explain the lack of continuous kelp forests extending from Año Nuevo to the northern end of the MBNMS at Rocky Point.

Seasonal Patterns and Kelp Life Histories

The seasonal patterns in the kelp forests of central California are very different from those observed in southern California. In central California, giant kelp and bull kelp exhibit their greatest recruitment in the spring and maximum canopies in early fall (Foster 1982) while kelp canopies in southern California reach their maximum in the winter. Recruitment of giant kelp sporophytes in southern California is greatest during periods of low temperatures and high nutrients, called "recruitment windows" (Deysher and Dean 1986). These conditions are nearly continuous in central California but are particularly evident during spring upwelling (McLean 1962; Breaker and Broenkow 1994) when light is also high because of canopy thinning by winter storms. These new kelp sporophytes grow from the substrate to the surface where they are supported by gas-filled bladders called pneumatocysts, and may form very dense surface canopies.

Although individual giant kelp plants can live up to seven years (Rosenthal et al. 1974), plants in central California may be shorter lived because they are removed during periods of high water motion associated with winter storms (McLean 1962; Foster 1982; Reed and Foster 1984; Harrold et al. 1988). During these periods, giant kelp survivorship is positively correlated with the hardness of the substrate to which the plants are attached (Foster and Schiel 1985). Another major source of kelp mortality along central California has been grazing by sea urchins (*Strongylocentrotus* spp.) (Pearse and Hines 1979, Watanabe and Harrold 1991). The effect of urchins however, has decreased due to predation by the sea otter *Enhydra lutris* (McLean 1962, Kenner 1992, Watanabe and Harrold 1991).

Although much is known about growth and survivorship of adult kelp sporophytes, relatively little is known about the ecology of their microscopic stages. These stages are probably highly vulnerable to grazing (Leonard 1994) and sedimentation (Devinny and Volse 1978, Deysher and Dean 1986) such as that from sewage discharges, which may thus be important in determining the distribution of kelp forests within the sanctuary. Studies on microscopic stages of giant kelp suggest that they are also sensitive to the toxins associated with municipal and industrial waste discharges. For example, growth and fertility are reduced when gametophytes are exposed to elevated copper concentrations in discharge effluent (Anderson et al. 1990).

Algal Assemblages Associated with Kelp Forests

Within the MBNMS, there are rich algal assemblages associated with the kelp forests. Subsurface canopies of the stipitate kelps *Pterygophora californica*, *Eisenia arborea*, and several species of *Laminaria* occur beneath the surface canopies (McLean 1962, Foster and Schiel 1985, Harrold et al. 1988). Although they occur throughout the MBNMS, these understory kelps are more characteristic of exposed areas (Harrold et al. 1988). Other algae, such as fleshy red species, can form dense algal turfs under the canopies (Table 1; Breda and Foster 1985, Harrold et al. 1988) and are often distributed along a depth gradient (Harrold et al. 1988) with the more robust species occurring shallower and the more delicate species occurring deeper (McLean 1962, Devinny and Kirkwood 1974).

Kelp canopies alone or in combination with one another can reduce the amount of light reaching the substrate to less than 1 per cent of surface irradiance (McLean 1962, Reed and Foster 1984). During the winter months along the central California coast, increased water motion from winter storms removes kelp canopies thereby increasing the amount of light reaching the substrate, which in turn can have dramatic effects on the algal assemblages beneath them (Foster 1982, Reed and Foster 1984, Breda and Foster 1985). One common phenomenon occurring in areas where surface canopies have been removed is the recruitment of the brown algae *Desmarestia ligulata* (Foster 1982, Reed and Foster 1984). This species forms a dense subsurface canopy which can inhibit recruitment of other algal species including giant kelp (Dayton et al. 1992).

Nongeniculate or encrusting coralline algae e.g., *Lithothamnium* spp. and *Lithophyllum* spp. and upright articulated or geniculate coralline algae e.g., *Bossiella* spp and *Calliarthron* spp. occur throughout the kelp forests and are generally more tolerant of increased water motion and thus abundant in exposed sites (Harrold et al. 1988). They also are apparently tolerant of low light and can dominate the substrate under multiple kelp canopies. In exposed areas like those at Point Santa Cruz, water motion and sand abrasion associated with storms cause an overall decrease in fleshy red algae in the winter, which then increases in the summer. This leads to an overall increase in species diversity as compared to more protected sites like those at Hopkins Marine Station in Monterey Bay (Breda and Foster 1985).

Invertebrate and Vertebrate Assemblages Associated with Kelp Forests

The kelp forests of the MBNMS provide habitat for a large variety of invertebrates, fishes, birds and mammals which are distributed among three different regions of the forests; the surface canopies, the midwater and the substrate.

- Invertebrates

The holdfasts of giant kelp and mats of geniculate coralline algae provide microhabitats for an abundant and species rich association of invertebrates (Andrews 1945, Foster and Schiel 1985, Dearn 1987). Andrews (1945) found approximately 23,000 individuals from nine invertebrate phyla residing in five giant kelp holdfasts collected from the Monterey and Carmel Bays, the most common of which were polychaetes, amphipods, decapods, gastropods and ophiuroids.

Outside the holdfasts, sponges, tunicates, anemones, cup corals and bryozoans are probably the most commonly occurring sessile animals within kelp forests (Foster and Schiel 1985). McLean (1962) observed 204 species of invertebrates living in a bull kelp forest along an exposed coast south of Carmel during 30 Scuba dives. In addition to these bottom dwelling species, a large number of invertebrates, such as the isopod *Idotea resecata* and the bryozoan *Membranipora tuberculata* occur within the canopies, while diverse assemblage of planktonic species such as jellyfish, crustaceans and fish larvae live in the water column of the kelp forests.

A wide variety of motile grazers, the majority of which do not remove entire kelp plants but graze upon their tissue and other associated algae (Foster and Schiel 1985), also occur in the forests. Some species, such as sea urchins (*Strongylocentrotus franciscanus* and *S. purpuratus*), may completely remove entire kelp plants by grazing through their holdfasts (Pearse and Hines 1979, Foster and Schiel 1985, Kenner 1992). However, these species usually do not graze on attached plants under normal circumstances, but rather feed on drift algae (Ebeling et al. 1985, Foster and Schiel 1985, Watanabe and Harrold 1991). Other species, like the gastropods *Tegula* spp., graze along the entire thallus of the kelp plants from the substrate to the surface (Watanabe 1984). In addition to herbivores, several species of predatory sea stars, snails and crabs inhabit the kelp forests, but comparatively little is known about the dynamics of these organisms (Foster and Schiel 1985).

Fishes

Although they are not limited to kelp forests, a variety of fish occur within them, including many of economic importance. The commercial fisheries associated with kelp forest fishes are reviewed in Leet, Dewees and Haugen (1992). In addition to the commercial fisheries, a large sport fishery is associated with kelp forests, particularly with rockfish, *Sebastes* spp. (Karpov et al. 1995).

The heterogeneous environment of the forest provides an important source of food and shelter for many species (Foster and Schiel 1985, Bodkin 1988). These can be categorised according to where they reside in the forest (Foster and Schiel 1985). Midwater species of the kelp canopy, such as the senorita (*Oxyjulius californica*) and the surfperch (*Brachyistius frenatus*) browse on the small crustaceans associated with both the kelp fronds and canopies. Other midwater predatory fishes including the common plankton-feeding blue rockfish (*S. mystinus*), the blacksmith (*Chromus punctipinnus*) and juveniles of the predatory kelp rockfish (*S. atrovirens*), olive rockfish

(*S. serranoides*) and black rockfish (*S. melanops*). Compared to southern California kelp forests, MBNMS kelp forests have relatively few tropically-derived fish species and even fewer families, but generally host more species per family.

Along the central California coast, fish diversity and abundance decrease in areas where the kelp canopies have been removed (Bodkin 1988). In addition, mass mortality of kelp forest fishes, particularly *Sebastes* spp. have been observed as a result of large oceanographic waves at the southern end of the MBNMS at San Simeon (Bodkin et al. 1987). Variations in fish abundance may have significant impacts on other communities. For example, juvenile rockfishes associated with kelp forests in Monterey Bay can reduce the amount of barnacle larvae reaching the intertidal to 2 per cent of the level found in the absence of fish (Gaines and Roughgarden 1986).

California Mammals and Birds

Harbour seals (*Phoca vitulina*) and California sea lions (*Zalophus californianus*) are common in and around MBNMS kelp forests. Harbour seals feed on shallow-dwelling kelp forest fishes, while California sea lions, which feed mainly on pelagic fishes, probably limit their use of the forests to transitory feeding (Foster and Schiel 1985). Gray whales (*Eschrichtius robustus*) have been observed entering kelp forests to escape predation from killer whales (*Orcinus orca,* Baldridge 1972) and also to feed on invertebrates such as midwater crustacean swarms (Nerini 1984).

One of the most recognisable marine mammals associated with the kelp forests in the MBNMS is the sea otter (*Enhydra lutris*). The current California sea otter population occupies a linear coastal range of about 280 km in central California. Most of the population occurs within the boundaries of the MBNMS. Otters feed on invertebrates associated with kelp forests (Foster and Schiel 1985; 1988, Estes et al. 1986), and must consume 25 per cent of their body weight per day to meet their energy needs (Costa 1978). Otters also use kelp forests as a refuge from predation by white sharks and winter storms, and as nursery areas for females with pups (Foster and Schiel 1985).

Otters can have significant impacts on sea urchin populations, which in turn may affect the kelp forests themselves (Foster and Schiel 1988). In this respect, otters have been called a "keystone species," suggesting that they occur high in the food web and by controlling their prey species (sea urchins) they greatly alter the community as a whole (Estes and Palmisano 1974, Estes and Duggins 1995), although this interaction may not be as important in California kelp forests (Foster and Schiel 1988, Foster 1990). Ostfeld (1978) observed otters colonising a new kelp forest off Santa Cruz ate mostly red sea urchins (*Strongylocentrotus franciscanus*) at first, but as urchin densities declined, they switched to kelp crabs and clams.

A wide variety of birds use MBNMS kelp forests although their relationship with the forests is poorly known. Foster and Schiel (1985) report that kelp provides three distinct habitats used by birds: the kelp forest made up of living attached plants associated with rocky substrata, drift kelp floating in the open sea, and the kelp wrack, i.e. detached kelp deposited on the beach by water motion. Kelp forests provide a large potential source of invertebrate and fish prey as well as a refuge from storms. Birds commonly observed in this habitat are gulls, terns, snowy egrets, great blue herons and cormorants (Foster and Schiel 1985). Birds associated with drift kelp, like phalaropes, feed on the associated plankton and larvae. The kelp wrack provides an important food source and habitat for kelp flies, maggots and small crustaceans on which several species of shore birds, starlings, common crows, black phoebes and warblers feed (Davis and Baldridge 1980).

Human Uses and Impacts

Aside from the fisheries which kelp forests support, the kelp forests of California are themselves an important economic and recreational resource. Giant kelp is harvested commercially in both southern and central California, and in the mid-1980's kelp harvesting supported an industry worth more than $40 million a year (Tarpley 1992). Kelp was originally harvested as a source of potash for making gunpowder during World War I (Frey 1971, Tarpley 1992) but currently the emphasis is on the production of algin, which serves as an emulsifying and binding agent in food and pharmaceutical products (Frey 1971) and food for use in abalone farms. Currently between 100,000 and 170,000 wet tons are harvested from California waters each year (Foster and Schiel 1985; Tarpley 1992).

In 1993, more than 4,700 wet tons were harvested from within the MBNMS boundaries, but due to severe thinning of the kelp forests by strong winter storms the following year, only approximately 553 wet tons were harvested in 1994 (P. Reiley pers comm.). At present, Kelco Company of San Diego is the major harvester of kelp in California waters, although the Pacific Kelp Company harvests kelp off central California (Tarpley 1992). Interest in commercial use and concern over kelp harvesting has stimulated considerable research on kelp forests since the late 1950's. In a review, North and Hubbs (1968) reported no adverse effects were observed as a result of kelp harvesting (reviewed in Foster and Schiel 1985). Likewise, Kimura and Foster (1984) found no adverse effects of a single kelp harvest on a kelp forest in Carmel Bay.

In addition to harvesting, the Sanctuary's kelp forests provide an important source of recreational activities, which range from hook-and-line and spear fishing to sport diving and underwater photography (Foster and Schiel 1985). To regulate these activities, a variety of special areas along the coast have been established, some because of their proximity to kelp forests. Marine reserves such as the one at Point Lobos, which was established in 1959 as the nation's first ever marine reserve (Mott 1970) and Hopkins Marine Life Refuge in Monterey Bay, regulate access and restrict fishing and collecting of organisms (Foster and Schiel 1985). These reserves may serve as valuable natural baselines to measure impacts of human activities, such as those described in I.B., in less protected MBNMS kelp forests.

We probably know more about kelp forests in the Monterey Bay area than anywhere else in the world except southern California. However, knowledge is lacking on the kelp forests south of Carmel Bay and north of Santa Cruz, and many processes are still poorly understood, e.g. the effects of local fisheries on kelp forest fish populations. Considerable research currently in progress at several institutions within the MBNMS addresses issues like these; the results will greatly enhance our understanding of the structure and dynamics of the kelp forests of the Monterey Bay National Marine Sanctuary.

6

Technical Aspects

Harmful Algal Blooms

Harmful algae are microscopic, single-celled plants that live in the sea. Most species of algae or phytoplankton are not harmful and serve as the energy producers at the base of the food web, without which higher life on this planet would not exist.

Occasionally, the algae grow very fast or "bloom" and accumulate into dense, visible patches near the surface of the water. "Red Tide" is a common name for such a phenomenon where certain phytoplankton species contain pigments and "bloom" such that the human eye perceives the water to be discoloured. Blooms can appear greenish, brown, and even reddish orange depending upon the type of organism, the type of water, and the concentration of the organisms. The term "red tide" is thus a misnomer because blooms are not always red, they are not associated with tides, they are usually not harmful, and some species can be harmful or dangerous at low cell concentrations that do not discolour the water.

Unfortunately, a small number of species produce potent neurotoxins that can be transferred through the food web where they affect and even kill the higher forms of life such as zooplankton, shellfish, fish, birds, marine mammals, and even humans that feed either directly or indirectly on them.

Scientists now prefer the term, HAB, to refer to bloom phenomenon that contain toxins or that cause negative impacts.

Species Responsible for Harmful Algal Blooms

Only a few dozen of the many thousands of species of microscopic and macroscopic algae are repeatedly associated with toxic or harmful blooms. Some species, such as the dinoflagellate *Alexandrium tamarense* and the diatom *Pseudonitzschia australis* produce potent toxins which are liberated when the algae are eaten.

Other species kill without toxins, like this *Chaetoceros* species (left) which has spines with serrated edges which can lodge in fish gill tissues, causing irritation, over production of mucous, and eventual death. Each of these species, and many others, need careful study at the organismal level if we are to understand the population dynamics and trophic impacts of HAB's throughout the US Photos by L. Fritz, J. Rines, and R. Horner.

Human Illness Associated with Harmful Algae

Man is exposed principally to the naturally-occurring toxins produced by harmful algae through the consumption of contaminated seafood products. The most significant public health problems caused by harmful algae are:

- Amnesic Shellfish Poisoning (ASP)
- Ciguatera Fish Poisoning (CFP)
- Diarrhetic Shellfish Poisoning (DSP)
- Neurotoxic Shellfish Poisoning (NSP)
- Paralytic Shellfish Poisoning (PSP)

Each of these syndromes are caused by different species of toxic algae which occur in various coastal waters of the US and the world. With the increase in interstate and international transport of seafood, as well as international travel by seafood consumers, there are virtually no human populations that are free of risk. Since 1978, illnesses in the US due to natural algal toxins have included PSP, NSP, CFP, and ASP. No incidents of DSP have yet been verified in this country. Although records are incomplete because reporting to the Centres for Disease Control (CDC) is voluntary, evidence indicates that ciguatera was responsible for about half of all seafood intoxications. A growing body of evidence indicates that incidents of ASP are on the increase and that DSP may shortly make its debut in the United States, since the causative organisms occur throughout the temperate coastal waters of the US.

Amnesic Shellfish Poisoning (ASP)

Causative Organisms: Pseudonitzschia SP

Toxin Produced: Domoic Acid—ASP can be a life-threatening syndrome. It is characterised by both gastrointestinal and neurological disorders (Bates et al., 1989). Gastroenteritis usually develops within 24 hours of the consumption of toxic shellfish; symptoms include nausea, vomiting, abdominal cramps, and diarrhoea. In severe cases, neurological symptoms also appear, usually within 48 hours of toxic shellfish consumption. These symptoms include dizziness, headache, seizures, disorientation, short-term memory loss, respiratory difficulty, and coma. In 1987, four victims died after consuming toxic mussels from Prince Edward Island, Canada. Since that time, Canadian authorities have monitored both the water column for the presence of the causative diatom, and shellfish for the presence of the toxin, domoic acid. Shellfish beds are closed to harvesting when the domoic acid concentration reaches 20 μg/g shellfish meat. Fish and crab viscera can also contain domoic acid, so the risk to human consumers and animals in the marine food chain is more significant than previously believed.

A very diverse groups of freshwater algae are the Chlorophytes Or Green Algae. Based on the compounds of the photosynthetic pigments and several other characteristics they seem closest related to plants.

A common green algae is *Hydrodictyon,* the water net. It is a related to *Pediastrum* (top image) But it forms a bag-shaped colony. Like *Pediastrum* each individual cell can develop into a new colony. You can imagine that since the colony contains thousands of cells *Hydrodictyon* can reproduce very rapidly. And unlike *Pediastrum, Hydrodictyon* can grow large, almost 30 cm. in length. Blooms of *Hydrodictyon* can be a real problem for water treatment plants.

The image shows a part of a small colony (left) and three individual cells of a big colony. Inside each of these cells a new colony can be formed.

Many species are flagellated and motile, other are immobile but often have a flagellated stage at some time in their life cycle. Certainly one of the most spectacular flagellated green algae is *Volvox.* It forms a spherical colony. All the small cells of the colony possess two flagella and a small eyespot. With this, the colony is able to swim towards the light. *Volvox* has interesting ways of reproduction.

Volvox is a Chlorophyte, or green algae. It exists as a grand spherical colony. Each little algae within the colony bears two flagella, whiplike hairs. The individual algae are connected to each other by thin strands of cytoplasm that enable the whole colony to swim in a coordinated fashion. The individual algae also have small red eye spots.

The colonies even have what we could call a front and rear end. Or, since Volvox resembles a little planet, a 'north and south pole'. In the northern region, the eyespots are more developed. This helps the colony to swim towards the light. This differentiation of cells make *Volvox* quite unique. It is a colony that comes really close to being a multi-celled organism.

Watching groups of colonies under the light microscope is a breathtaking sight. If you leave enough space under the cover slip (with the aid of spots of Vaseline under the corners) the spheres will swim slowly towards the light of the microscope, (use dark field illumination!).

Volvox can be found relatively easy. They need quite clean nutrient rich water (Eutrophic) and it has to be warm so summer is the time of the year for a good *Volvox* catch. The best thing to do is to use a plankton net but if you are lucky you can also find them by squeezing pond scum. In a jar of pond water, they will swim towards the light so they are often easy to find near the surface.

The two common species in my country are *Volvox aureus* and *Volvox globator*. *Volvox aureus* is usually smaller and has less individual cells. *Volvox globator* may reach a size of 2 millimetres so they can be easily seen with the naked eye. If you are short sighted the larger *Volvox aureus* colonies can also be visible with a size of half a millimetre as small green dots.

Although the colony is bigger, the individual cells of *Volvox globator* are slightly smaller (4 micron) than the cells of *Volvox aureus* (5 to 8 micron). The cells of *Volvox aureus* are also a bit more egg-shaped. An easier way to determine the species is to look at the sexual reproduction of *Volvox*.

Reproduction

It is not only their 'planetary movements' that make *Volvox* a sure candidate for one of the seven wonders of the micro world. It is also their interesting methods of reproduction that are really spectacular. They have asexual as well as sexual reproduction.

One of the first things that you notice on *Volvox* is that most colonies have spheres inside. These are 'daughter' colonies, called gonads. It is a means of asexual reproduction. The gonads grow from cells around the equator of the colony. These cells enlarge and undergo a series of cell divisions until they form a small sphere. There is only one complication, the flagella will be on the inside of the new sphere. It has to turn itself inside out so the flagella will be situated towards the outside of the colony.

Volvox also has a sexual reproduction. As in the asexual reproduction special cells are formed around the equator. These cells develop into special germ cells. Male and female colonies form different germ cells. Sperm cells are formed by division. The female germ cells do not divide. They enlarge to form an ova.

Most species have male and female colonies. In some species the colonies are hermaphroditic. They are able to form sperm as well as ova. But not at the same time so self fertilization is prevented.

This images shows packages of sperm cells. They will soon swarm out in search of ova in other *Volvox*.

These are fertilized egg cells, the zygotes, of *Volvox globator*. After fertilization the egg cells form a hard protective layer. Zygotes are able to withstand harsh conditions and will survive a winter.

These zygotes make it easy to distinguish *Volvox globator* from *Volvox aureus*. *Volvox aureus* has zygotes with a smooth surface. The zygotes of *Volvox globator* are starlike.

Sometimes, we can find inhabitants within the colony. There is a small rotifer, called *Proales parasita* that lives inside *Volvox* and feeds on the cells.

The rotifer is able to detect when the colony becomes immobile. When we observe *Volvox* under the coverslip it may be trapped by the pressure of the coverslip. *Proales* will quickly make a hole and escape.

A small rather inconspicuous green algae without undulipodia is *Chlorella*. It can be found as an endosymbiont inside ciliates, hydra and other animals. They raise *Chlorella* as if they grow crops in a greenhouse.

Many green algae form long filaments. The cells stay attached after they divide. Some genera, like *Spirogyra*, *Mougeotia* and *Zygnema* can become so numerous that they form dense mats of growth in surfaces of ponds, so-called pond scum. This pond scum is interesting for a study under the microscope. If you squeeze it in a jar you will collect many interesting organisms. And the filamentous algae are at least as interesting. The chloroplasts have distinct shapes. In *Spirogyra,* the chloroplast runs through the cell like a helix. The image also shows the nucleus hanging on fine threads.

Diatoms

Diatoms are a major group of eukaryotic algae, and are one of the most common types of phytoplankton. Most diatoms are unicellular, although some form chains or simple colonies. A characteristic feature of diatom cells is that they are encased within a unique cell wall made of silica (hydrated silicon dioxide) called a frustule. These frustules show a wide diversity in form, some quite beautiful and ornate, but usually consist of two asymmetrical sides with a split between them, hence the group name. Fossil evidence suggests that they originated during, or before, the early Jurassic Period.

General Biology

There are more than 200 genera of living diatoms, and it is estimated that there are approximately 100,000 extant species. Diatoms are a widespread group and can be found in the oceans, in freshwater, in soils and on damp surfaces. Most live pelagically in open water, although some live as surface films at the water-sediment interface (benthic), or even under damp atmospheric conditions. They are especially important in oceans, where they are estimated to contribute up to 45 per cent of the total oceanic primary production. Although usually microscopic, some species of diatoms can reach up to 2 millimetres in length.

Diatoms belong to a large group called the heterokonts, including both autotrophs (e.g. golden algae, kelp) and heterotrophs (e.g. water moulds). Their yellowish-brown chloroplasts are typical of heterokonts, with four membranes and containing pigments such as fucoxanthin. Individuals usually lack flagella, but they are present in gametes and have the usual heterokont structure, except they lack the hairs (mastigonemes) characteristic in other groups.

Most diatom species are non-motile but some are capable of an oozing motion. As their relatively dense cell walls cause them to readily sink, planktonic forms in open water usually rely on turbulent mixing of the upper layers by the wind to keep them suspended in sunlit surface waters. Some species actively regulate their buoyancy with intracellular lipids to counter sinking

Diatoms cells are contained within a unique silicate (silicic acid) cell wall comprised of two separate valves (or shells). The biogenic silica that the cell wall is composed of is synthesised intracellularly by the polymerisation of silicic acid monomers. This material is then extruded to the cell exterior and added to the wall. Diatom cell walls are also called Frustules or Tests, and their two valves typically overlap one other like the two halves of a petri dish. In most species, when a diatom divides to produce two daughter cells, each cell keeps one of the two valves and grows a smaller valve within it. As a result, after each division cycle the average size of diatom cells in the population gets smaller. Once such cells reach a certain minimum size, rather than simply divide vegetatively, they reverse this decline by forming an auxospore. This expands in size to give rise to a much larger cell, which then returns to size-diminishing divisions. Auxospore production is almost always linked to meiosis and sexual reproduction.

Decomposition and decay of diatoms leads to organic and inorganic (in the form of silicates) sediment, the inorganic component of which can lead to a method of analysing past marine environments by corings of ocean floors or bay muds, since the inorganic matter is embedded in deposition of clays and silts and forms a permanent geological record of such marine strata.

Classification: The classification of heterokonts is still unsettled, and they may be treated as a division (or phylum), kingdom, or something in-between. Accordingly, groups like the diatoms may be ranked anywhere from class (usually called Bacillariophyceae) to division (usually called

Bacillariophyta), with corresponding changes in the ranks of their subgroups. The diatoms are also sometimes referred to as Class Diatomophyceae.

Diatoms are traditionally divided into two orders: Centric diatoms (Centrales), which are radially symmetric, and Pennate diatoms (Pennales), which are bilaterally symmetric. The former are paraphyletic to the latter. A more recent classification divides the diatoms into three classes: centric diatoms (Coscinodiscophyceae), pennate diatoms without a raphe (Fragilariophyceae), and pennate diatoms with a raphe (Bacillariophyceae). It is probable there will be further revisions as understanding of their relationships increases.

Round & Crawford (1990) and Hoek *et al.* (1995) provide more comprehensive coverage of diatom taxonomy.

Ecology: Planktonic forms in freshwater and marine environments typically exhibit a "boom and bust" (or "Bloom and bust") lifestyle. When conditions in the upper mixed layer (nutrients and light) are favourable (*e.g.* at the start of spring) their competitive edge allows them to quickly dominate phytoplankton communities ("boom" or "bloom"). As such they are often classed as opportunistic r-strategists (*i.e.* those organisms whose ecology is defined by a high growth rate, *r*).

When conditions turn unfavourable, usually upon depletion of nutrients, diatom cells typically increase in sinking rate and exit the upper mixed layer ("bust"). This sinking is induced by either a loss of buoyancy control, the synthesis of mucilage that sticks diatoms cells together, or the production of heavy Resting Spores. Sinking out of the upper mixed layer removes diatoms from conditions inimical to growth, including grazer populations and higher temperatures (which would otherwise increase cell metabolism). Cells reaching deeper water or the shallow seafloor can then rest until conditions become more favourable again. In the open ocean, many sinking cells are lost to the deep, but refuge populations can persist near the thermocline.

Ultimately, diatom cells in these resting populations re-enter the upper mixed layer when vertical mixing entrains them. In most circumstances, this mixing also replenishes nutrients in the upper mixed layer, setting the scene for the next round of diatom blooms. In the open ocean (away from areas of continuous upwelling), this cycle of bloom, bust, then return to pre-bloom conditions typically occurs over an annual cycle, with diatoms only being prevalent during the spring and early summer. In some locations, however, an autumn bloom may occur, caused by the breakdown of summer stratification and the entrainment of nutrients while light levels are still sufficient for growth. Since vertical mixing is increasing, and light levels are falling as winter approaches, these blooms are smaller and shorter-lived than their spring equivalents.

In the open ocean, the condition that typically causes diatom (spring) blooms to end is a lack of silicon. Unlike other nutrients, this is only a major requirement of diatoms so it is not regenerated in the plankton ecosystem as efficiently as, for instance, nitrogen or phosphorus nutrients. This can be seen in maps of surface nutrient concentrations - as nutrients decline along gradients, silicon is usually the first to be exhausted (followed normally by nitrogen then phosphorus).

Because of this bloom-and-bust lifestyle, diatoms are believed to play a disproportionately important role in the export of carbon from oceanic surface waters. Significantly, they also play a key role in the regulation of the biogeochemical cycle of silicon in the modern ocean.

The use of silicon by diatoms is believed by many researchers to be the key to their ecological success. In a now classic study, Egge & Aksnes (1992) found that diatom dominance of mesocosm communities was directly related to the availability of silicate. When silicon content approaches a concentration of 2 mmol m^{-3}, diatoms typically represent more than 70 per cent of the phytoplankton community. Raven (1983) noted that, relative to organic cell walls, silica frustules require less energy to synthesize (approximately 8 per cent), potentially a significant saving on the overall cell energy budget.

Other researchers have suggested that the biogenic silica in diatom cell walls acts as an effective pH buffering agent, facilitating the conversion of bicarbonate to dissolved CO_2 (which is more readily assimilated). Notwithstanding the possible advantages conferred by silicon, diatoms typically have higher growth rates than other algae of a corresponding size.

Evolutionary History

Heterokont chloroplasts appear to be derived from those of red algae, rather than directly from prokaryotes as occurs in plants. This suggests they had a more recent origin than many other algae. However, fossil evidence is scant, and it is really only with the evolution of the diatoms themselves that the heterokonts make a serious impression on the fossil record.

The earliest known fossil diatoms date from the early Jurassic (~185 Ma), although recent molecular clock and sedimentary evidence suggests an earlier origin. It has been suggested that their origin may be related to the end-Permian mass extinction (~250 Ma), after which many marine niches were opened. The gap between this event and the time that fossil diatoms first appear may indicate a period when diatoms were unsilicified and their evolution was cryptic. Since the advent of silicification, diatoms have made a significant impression on the fossil record, with major deposits of fossil diatoms found as far back as the early Cretaceous, and some rocks (diatomaceous earth, diatomite, kieselguhr) being composed almost entirely of them.

Although the diatoms may have existed since the Triassic, the timing of their ascendancy and "takeover" of the silicon cycle is more recent. Prior to the Phanerozoic (before 544 Ma), it is believed that microbial or inorganic processes weakly regulated the ocean's silicon cycle. Subsequently, the cycle appears dominated (and more strongly regulated) by the radiolarians and siliceous sponges, the former as zooplankton, the latter as sedentary filter feeders primarily on the continental shelves. Within the last 100 year, it is thought that the silicon cycle has come under even tighter control, and that this derives from the ecological ascendancy of the diatoms.

However, the precise timing of the "takeover" is unclear, and different authors have conflicting interpretations of the fossil record. Some evidence, such as the eviction of siliceous sponges from the shelves, suggests that this takeover began in the Cretaceous (146 Ma to 65 Ma), while evidence from radiolarians suggests "takeover" did not begin until the Cenozoic (65 Ma to present). Nevertheless, regardless of the details of the "takeover" timing, it is clear that this most recent revolution has installed much tighter biological control over the biogeochemical cycle of silicon.

Collection: Living diatoms are often found clinging in great numbers to filamentous algae, or forming gelatinous masses on various submerged plants. Cladophora is frequently covered with Cocconeis, an elliptically shaped diatom; Vaucheria is often covered with small forms. Diatoms frequently present as a brown, slippery coating on submerged stones and sticks, and may be seen to "stream" with river current.

The surface mud of a pond, ditch, or lagoon will almost always yield some diatoms. They can be made to emerge by filling a jar with water and mud, wrapping it in black paper and letting direct sunlight fall on the surface of the water. Within a day, the diatoms will come to the top in a scum and can be isolated.

Since diatoms form an important part of the food of molluscs, tunicates, and fishes, the alimentary tracts of these animals often yield forms that are not easily secured in other ways. Marine diatoms can be collected by direct water sampling, though benthic forms can be secured by scraping barnacles, oyster shells, and other shells.

The silicious shells of diatoms are among the most beautiful objects which can be examined with the microscope. To obtain perfectly clean mounts requires considerable time and patience, but once the material is cleaned, preparations may be made at any time with very little trouble.

Genome Sequencing

The entire genome of the centric diatom, Thalassiosira pseudonana, has been sequenced, and the sequencing of a second diatom genome from the pennate diatom Phaeodactylum tricornutum is in progress. The first insights into the genome properties of the *P. tricornutum* gene repertoire was described using 1,000 ESTs. Subsequently, the number of ESTs was extended to 12,000 and the Diatom EST Database was constructed for functional analyses. These sequences have been used to make a comparative analysis between *P. tricornutum* and the putative complete proteomes from the green algae Chlamydomonas reinhardtii, the red algae Cyanidioschyzon merolae, and the centric diatom *T. pseudonana*.

Asterionella

The cells are joined at the apex by mucilage pads to form stellate or spiral colonies. Some populations are particularly susceptible to chytrid infestations.

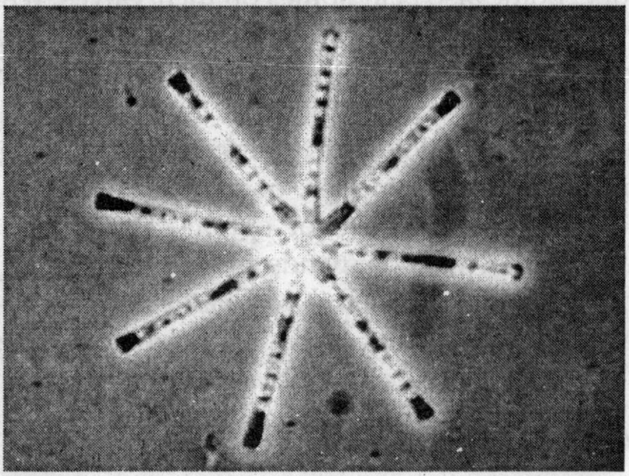

Asterionella is a genera of freshwater diatoms found worldwide.

Auxospore

In phycology, Auxospores are special cells produced by diatoms that expand cell size back to that which is normal for vegetative cells. They are necessary because successive mitotic cell divisions in diatoms leads to a diminuation in cell size for one of the two daughter cells. This occurs because each daughter cell produced by cell division inherits one of the two valves that make up the silica cell wall, and then grows a smaller valve within it. Consequently, each division cycle decreases the average size of diatom cells in a population. As well as restoring cell size, auxospores are also often zygotes (the product of sexual reproduction) or resting stages (used to survive unfavourable periods such as winter).

Centrales

Order Centrales is a traditional subdivision of the heterokont algae known as diatoms. The order is named for the shape of the cell walls (or valves or frustules) of Centric diatoms, which are circular or ellipsoid in valve view. The valves often bear radially symmetrical ornamental patterns that can appear as dots when viewed with an optical microscope. Some also bear spines on their valves, which may either increase cell surface area and reduce sinking, or act as a deterrant to zooplankton grazers. Unlike pennate diatoms, centric diatoms never have a raphe. In terms of cell cycle, vegetative cells are diploid and undergo mitosis during normal cell division. In sexual species, oogamous meiosis produces haploid gametes, either ova or sperm cells. These fuse to produce a zygote which expands in size to develop into an auxospore from which full-size vegetative cells are produced. In some taxonomic schemes, the centric diatoms are known instead as Order Coscinodiscophyceae.

Clepsydra (Genus)

Clepsydra is a recently described genus of diatoms, including the species Clepsydra truganiniae. It was found in Tasmania.

Didymosphenia Geminata

Didymosphenia geminata, commonly known as Didymo or Rock Snot, is a species of diatom that grows in warm and shallow water. If it overgrows, it can form large mats on the bottom of lakes, rivers and streams. It is not considered a significant human health risk, but it can affect stream habitats and sources of food for fish and make recreational activities unpleasant. It is considered a nuisance organism. The microscopic algae can be spread in a single drop of water.

In New Zealand

In 2004, didymo was discovered in New Zealand, the first time it was found in the southern hemisphere. To restrict its spread, the whole of the South Island of New Zealand was declared a controlled area in December 2005. All items, such as boats, fishing gear, clothing, and vehicles, that have been in a stream, river or lake, must be cleaned before they enter another waterway. *Biosecurity New Zealand* working with *Environment Southland*, *AgriQuality* and *Fish and Game* launched an extensive public awareness campaign to encourage river users to clean their equipment after use in affected waterways. This campaign was highly successful, with up to a 90 per cent decrease in use of affected rivers.

As of October 2005, further populations of didymo were discovered in the Hawea, Buller, Oreti and Upper Clutha rivers. Delimiting testing is being undertaken in North Island waterways and other South Island river systems to find out how far didymo has spread in New Zealand.

In the United States

Didymo has been confirmed in the tailwaters of Norris, Cherokee, Wilbur and South Holston dams — hydroelectric facilities of the Tennessee Valley Authority.

Preventing the Spread of Didymo

The following methods have been recommended to prevent the spread of didymo:

Check: Before leaving the river, remove all obvious clumps of algae and look for hidden clumps. Leave them at the site. If you find clumps later don't wash them down the drain, treat them with the approved methods below, dry them and put them in a rubbish bin.

Clean: Soak and scrub all items for at least one minute in either hot (60°C) water, a 2 per cent solution of household bleach or a 5 per cent solution of salt, nappy cleaner, antiseptic hand cleaner or dishwashing detergent.

Dry: If cleaning is not practical (e.g. livestock, pets), after the item is completely dry wait an additional 48 hours before contact or use in any other waterway.

Pennales

Order Pennales is a traditional subdivision of the heterokont algae known as diatoms. The order is named for the shape of the cell walls (or valves or frustules) of Pennate diatoms, which are elongated in valve view. The valves may be linear or oval in shape, and usually bear bilaterally symmetrical ornamental patterns. These patterns are composed of a series of transverse lines (known as striae) that can appear as rows of dots when viewed with an optical microscope. Some pennate diatoms also exhibit a fissure along their longitudinal axis. This is known as a raphe, and is involved in gliding movements made by diatom cells.

In terms of cell cycle, vegetative cells are diploid and undergo mitosis during normal cell division. Periodically, meiosis produces morphologically-identical haploid gametes (isogametes), which fuse to produce a (sometimes binucleate) zygote that develops into an auxospore (from which full-size vegetative cells are produced).

In some taxonomic schemes, the pennate diatoms are divided into two groups: pennate diatoms without a raphe (Order Fragilariophyceae), and pennate diatoms with a raphe (Order Bacillariophyceae).

Pinnularia

Pinnularia is a fresh water algae and it is a type of Bacilliophyta (diatoms).

Occurrence

It is a fresh Water algae. It is usually found in Ponds and moist soil.

General Structure

Cell Wall and Valves: It is Unicellular. The Cell is elongated and elliptical. Cell wall is chiefly composed of pectic substances. Silica is impregnated in it, therefore, their wall becomes very hard. Their wall is composed of two halves called valves. These halves overlap like petri dish. The margins of the two valves are covered by a connecting band called cingulum. The two valves with their inner protoplasts are called frustule. The frustule has two views.

The surface view is called valve view and band view is called girdle view. The outer larger valve is called epithcea. The smaller inner valve is called hyptheca. The cell is covered by mucilaginous layer. The surface of valve has minute pores or pits. It produces characteristic markings on the surface of valves. These markings are bilaterally symmetrical. The marking free area is called axial field. The markings are arranged in linear rows.

The axial field contains a longitudinal slit raphe. The raphe is not continuous throughout the length of the valve. It has circular thickening called central nodules in the centre. Similar Thickening also present at the outer ends of raphe. These are called polar nodules.

Cell Structure

Cytoplasm is arranged in parietal layer. It is called primordial Uticle. A large central vacuole is present inside the cell. Single nucleus Is suspended in the centre of Vavuole by transverse cytoplasmin bridge. Two chloroplasts are present along the sides of the cells. They contain diatomin pigment. One or two pyrenoids are present in the chloroplast. But they do not synthesise starch. Their storage compound is fat.

Reproduction

Vegetative Reproduction: Vegetative reproduction in diatoms occurs by simple cell division. Nuclear division occurs by mitosis and cell divides into two parts. Each half receives on half of the parent cell. It synthesise new valve. New valve fits into the parent valve. Thus new valves are always smaller than the parent halves. Thus one generation gradually become smaller in size. It reaches to minimum size. Then its size is restored by auxospore formation. But second generation remains of same size.

Auxospore Formation

The protoplast of cell secretes a large amount of mucilage, therefore, two valves are separated from each other. It separates the inner protoplast. The protoplast grows to its maximum size. It secrets silica rich pectic membrane. This membrane is called prizonium. New valves are secreted inside the prizonium. Nuclear division does not occur. Therefore, auxospores are formed by parthenogenisis.

Heterokont

The Heterokonts or Stramenopiles are a major line of eukaryotes presently containing about 10,500 known species. Most are algae, ranging from the giant multicellular kelp to the unicellular diatoms, which are a primary component of plankton. Other notable members of the Stramenopila include the (generally parasitic) oomycetes, including Phytophthora of Irish potato famine fame and Pythium which causes seed rot and damping off.

Chloroplasts

Heterokont algae are chromists with chloroplasts surrounded by four membranes, which are counted from the outermost to the innermost membrane. The first membrane is continuous with the host's chloroplast endoplasmic reticulum, or cER. The second membrane presents a barrier between the lumen of the endoplasmic reticulum and the primary endosymbiont or chloroplast, which represents the next two membranes, within which the thylakoid membranes are found. This arrangement of membranes suggest that heterokont chloroplasts were obtained from the reduction of a symbiotic red algal eukaryote, which had arisen by evolutionary divergence from the monophyletic primary endosymbiotic ancestor that is thought to have given rise to all eukaryotic photoautotrophs. The chloroplasts characteristically contain chlorophyll a and chlorophyll c, and usually the accessory pigment fucoxanthin, giving them a golden-brown or brownish-green colour.

Most basal heterokonts are colourless, suggesting they diverged before acquisition of chloroplasts within the group. However, fucoxanthin-containing chloroplasts are also found among the haptophytes, and evidence suggests that the two groups have a common ancestry, as well as possible a common phylogenetic history with cryptomonads. In this case, the ancestral heterokont was an algae, and all colourless groups arose through loss of the secondary endosymbiont and hence its chloroplast

Motile Cells

Many heterokonts are unicellular flagellates, and most others produce flagellate cells at some point in their life-cycle, for instance as gametes or zoospores. The name heterokont refers to the characteristic form of these cells, which typically have two unequal flagella. The anterior or *tinsel* flagellum is covered with lateral bristles or *mastigonemes*, while the other flagellum is whiplash, smooth and usually shorter, or sometimes reduced to a basal body. The flagella are inserted subapically or laterally, and are usually supported by four microtubule roots in a distinctive pattern.

Mastigonemes are manufactured from glycoproteins in the cell's endoplasmic reticulum before being transported to its surface. When the tinsel flagellum moves, these create a backwards current, pulling the cell through the water or bringing in food. The mastigonemes have a peculiar tripartite structure, which may be taken as the defining characteristic of the group, thereby including a few protists that do not produce cells with the typical heterokont form. They have been lost in a few lines, most notably the diatoms.

Classification

As noted above, classification varies considerably. Originally the heterokont algae were treated as two divisions, first within the kingdom Plantae and later the Protista:

- Division Chrysophyta
- Class Chrysophyceae (golden algae)
- Class Bacillariophyceae (diatoms)
- Division Phaeophyta (brown algae)

In this scheme, however, the Chrysophyceae are paraphyletic to both other groups. As a result, various members have been given their own classes and often divisions. Recent systems often treat these as classes within a single division, called the Heterokontophyta, Chromophyta or Ochrophyta. This is not universal, however - for instance Round *et al.* treat the diatoms as a division.

The discovery that oomycetes and hypochytrids are related to these algae, rather than fungi as previously thought, has led many authors to include them among the heterokonts. Should it turn out that they evolved from coloured ancestors, the group would be paraphyletic in their absence. Once again, however, usage varies. David J. Patterson named this extended group the stramenopiles, characterised by the presence of tripartite mastigonemes, mitochondria with tubular cristae, and open mitosis. He used the stramenopiles as a prototype for a classification without Linnaean ranks. Their composition has been essentially stable, but their use within ranked systems varies.

The origin of the name stramenopile is explicited in the following quote from Adl et al. (2005):

"Regarding the spelling of stramenopile, it was originally spelled Stramenopile. The Latin word for "straw" is stramine-us, -a, -um, adj. [stramen], made of straw—thus, it should have been spelled Straminopile. However, Patterson (1989) clearly stated that this is a common name (hence, lower case, not capitalised) and as a common name, it can be spelled as Patterson chooses. If he had stipulated that the name was a formal name, governed by rules of nomenclature, then his spelling would have been an orthogonal mutation and one would simply correct the spelling in subsequent publications (e.g. Straminopiles). But, it was not Patterson's desire to use the term in a formal sense. Thus, if we use it in a formal sense, it must be formally described (and in addition, in Latin, if it is to be used botanically). However, and here is the strange part of this, many people liked the name, but wanted it to be used formally. So they capitalised the first letter, and made it Stramenopiles; others corrected the Latin spelling to Straminopiles."

Thomas Cavalier-Smith treats the heterokonts as identical in composition with the stramenopiles; this is the definition followed here. He has proposed placing them in a separate kingdom Chromista, together with the haptophytes and cryptomonads. This is one of the most common revisions to the five-kingdom system, but has not been generally adopted, partly because some biologists doubt their monophyly. A few treat the Chromista as identical in composition with the heterokonts, or list them as a kingdom Stramenopila.

Dinocyst

Dinocysts are produced by around 10 per cent of living Dinoflagellates as a dormant zygotic stage of their life cycle. The next-generation Dinoflagellate exits the cyst (excysts) via the archaeopyle, a hole which may take the form of the removal of the apex of the cyst (apical archaeopyle), or the removal of one of the tabulae (Precingular archaepyle).

Dinocysts are composed of Dinosporin, an organic compound similar to sporopollenin, but unique to Dinoflagellates.

The tabulation of the Dinoflagellate is mirrored in the 'paratabulation' of the Dinocyst, allowing species to be deduced from the cyst (Fensome and Taylor, 1933). Dinocysts are essentially the only evidence in the fossil record of Dinoflagellates; none the less they have been discovered as early as the Triassic.

Dinophyceae

The Dinophyceae are the main class of dinoflagellates. They include all species where the nucleus remains a dinokaryon throughout the entire cell cycle, which is typically dominated by the haploid stage. All the "typical" dinoflagellates, such as *Peridinium* and *Gymnodinium*, belong here. Others are more unusual, including some that are colonial, amoeboid, or parasitic. Dinoflagellates are classified by morphology. Species with a theca are divided into four orders, based on the arrangement of the armour plates:

- Peridiniales - e.g. *Peridinium*
- Gonyaulacales - e.g. *Ceratium, Gonyaulax*
- Dinophysiales - e.g. *Dinophysis*
- Prorocentrales - e.g. *Prorocentrum*

The Peridiniales are probably paraphyletic to the others, and on rRNA trees they are mixed with the species that lack thecae. The other three orders are probably monophyletic, with the Dinophysiales and Prorocentrales as close relatives, united by the presence of a sagittal suture dividing the theca in two. However, on rRNA trees the Prorocentrales are split up.

The groups of dinoflagellates without theca are understood to be artificial, and are mostly polyphyletic. Many of the genera, such as *Gymnodinium* and *Amphidinium*, are also polyphyletic. However, some may approximate monophyletic groups, such as the Suessiales, and some have not been studied phylogenetically:

- Gymnodiniales - e.g. *Gymnodinium, Amphidinium*
- Ptychodiscales
- Suessiales - e.g. Symbiodinium

- Desmocapsales
- Phytodiniales - e.g. *Dinamoeba,* Pfiesteria (includes Dinococcales
- Thoracosphaerales

There is also a group of parasitic dinoflagellates, the Blastodiniales, that do not have dinokarya during their trophic stage. Because of this, they have been treated as a separate class Blastodiniphyceae, but some or all may actually have developed within the Dinophyceae.

Gymnodinium

Gymnodinium is a genus of dinoflagellates. It is one of the few naked dinoflagellates lacking armour (cellulosic plates).

Karenia brevis

Karenia brevis (formerly known as *Gymnodinium breve* and *Ptychodiscus brevis*) is a marine dinoflagellate common in Gulf of Mexico waters, and is the organism responsible for Florida red tide. *K. brevis* is a microscopic, single-celled, photosynthetic organism that can "bloom" frequently along Florida coastal waters. Each cell has two flagella that allow it to move through the water in a spinning motion. *K. brevis* naturally produces a suite of potent neurotoxins collectively called brevetoxins, which are responsible for large die-offs of marine organisms and seabirds.

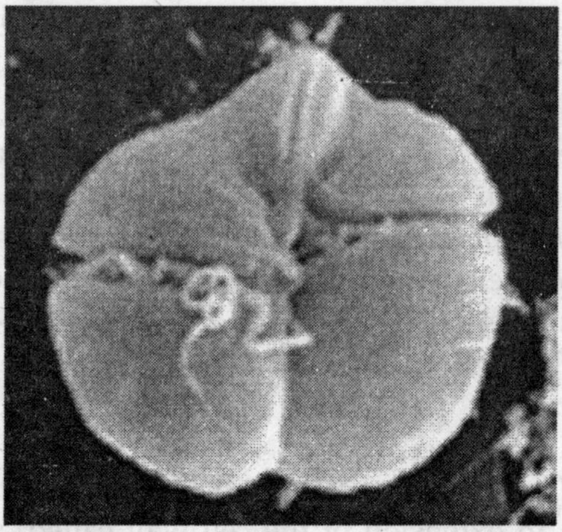

Noctiluca Scintillans

Noctiluca Scintillans, also published as *Noctiluca miliaris*, is a free-living non-parasitic marine-dwelling species of dinoflagellate that exhibits bioluminescence. The bioluminescent characteristic of *N. scintillans* is produced by a luciferin-luciferase system located in thousands of spherically shaped organelles, or "microsources", located throughout the cytoplasm of this single-celled protist. Non-luminescent populations within the genus Noctiluca lack these microsources.

Diet

N. scintillans is a heterotrophic (non-photosynthetic) organism that engulfs its food (phagotrophic) which primarily consists of plankton, including diatoms and other dinoflagellates, as well as fish eggs and bacteria. Diatoms are often found in the vacuoles (internal membrane-bound storage compartments) within these single-celled creatures. The diatom *Thalassiosira sp.* has been noted in the literature as a favoured food source of these organisms.

Distribution

N. scintillans can be found widely distributed throughout the world, often along the coast, in estuary, and shallow areas of the continental shelf that receive plenty of light which promotes the growth of the phytoplankton that make up a large portion of *N. scintillans's* diet.

Size: The size of the single-celled *N. scintillans* ranges from 200 to 2,000 cm in diameter, assuming the generally spherical shape. *N. scintillans* lacks the armour plates possessed by other types of dinoflagellates. And, unlike many other dinoflagellates, the chromosomes of the *Noctiluca* are not clearly-visible and condensed throughout its life cycle.

Anatomy

N. scintillans has a ventral groove within which is located a flagellum, an extension of the cell wall called a tooth, and a striated tentacle involved in ingestion that projects posteriorly. The flagellum does not move the organism and therefore, the non-motile *N. scintillans* depends upon regulation of its buoyancy within the water column – perhaps by controlling its cellular concentration of ions and ammonia.

At least one study has shown that a string of mucus is produced by *N. scintillans* extending from the tip of the tentacle which then adheres to plankton as it ascends rapidly through concentrations of its prey in the water column.

"Blooms"

High concentrations of their plankton food source that likely result from environmental conditions such as well-mixed nutrient-rich waters and seasonal circulation factors are implicated in population blooms of *N. scintillans*, known as "red tides". Runoff from agricultural pollution may contribute to the severity of these blooms, but is not required to cause such events of explosive population growth.

N. scintillans itself does not appear to be toxic, but as they feed voraciously on phytoplankton high levels of ammonia accumulate in these organisms which is then excreted by *N. scintillans* into the surrounding area which may add to the neurotoxic chemicals being produced by other dinoflagellates, such *Alexandrium spp.* or *Gonyaulax spp.*, that do result in the death of other aquatic life in the area.

Not all blooms associated with *N. scintillans* are red. The colour of *N. scintillans* is in part derived from the pigments of organisms inside the vacuoles of *N. scintillans*. For instance, green tides result from *N. scintillans* populations that contain green-pigmented prasinophytes (green plant algae, Subphylum Chlorophyta) that are living in their vacuoles.

Evolutionary Development

At least one study, based on small subunit (SSU) RNA data, found that the dinoflagellates are a monophyletic group and that within this group *N. scintillans* was among the earliest to diverge.

Another study, based on large sub-unit rRNA, supports the claim that heterotrophic species of dinoflagellates such as *N. scintillans* preceded the emergence of photosynthetic species. Photosynthetic species would have evolved by acquiring photosynthetic capability through secondary and even tertiary endosymbiosis.

N. scintillans is also placed within a classification scheme that has a class Diniferea, or Dinophyceae, which includes non-parasitic dinoflagellates that lack armour plating.

Noctilucales

The Noctilucales are a peculiar order of marine dinoflagellates. They differ from most others in that the mature cell is diploid and its nucleus does not show a dinokaryotic organisation. These cells are very large, from 1 to 2 millimetres in diameter, and are filled with large buoyant vacuoles. Some may contain symbiotic green algae, but there are no chloroplasts. Instead, they feed on other plankton, and there is usually a special tentacle involved in ingestion.

Noctilucales reproduce mainly by fission, but sexual reproduction also occurs. Each cell produces numerous gametes, which resemble more typical athecate dinoflagellates and have the dinokaryotic nuclei. Evidence suggests that they diverged from most other dinoflagellates early on, and they are generally placed in their own class.

The most common species is Noctiluca scintillans, also called *N. miliaris*. It is usually bioluminescent when disturbed, as are various other dinoflagellates, and large blooms can sometimes be seen as flickering lights on the ocean.

Oodinium

Oodinium (also known as *Piscinoodinium*) is a genus of microscopic parasitic dinoflagellates. They live off salt and fresh water fish, causing a type of fish velvet disease (also called gold dust disease).

The host typically develops small yellow or gold dust

scattered on its head, fins and body. At this stage, the infestation is already severe. The attack usually starts at the gills at which stage it is difficult to notice. The host is irritated and often swims in fuzziness while rubbing itself against rocks. The yellowish spots are more vivid under sunlight or flashlight. It is very similar to Ichthyophthirius, though the oodinium spots are yellowish and smaller.

The life cycle of oodinium starts as a dinospore that swims in the water to look for a suitable host. As it attaches itself onto the host skin, it forms a hard shell protecting itself against the outside environment while it is eating the fish skin cells. This is the cyst stage seen as dust covering the fish skin. After few days, the cyst sinks to the bottom, freeing new generation of dinospores. And the cycle repeats. The dinospore must find a host within 48 hours, otherwise the dinospore would die.

Treatment is possible by diluting the commercially available copper salts. Free swimming dinospore is extremely vulnerable to copper. Bringing the water temperature to 30°C helps to release the dinospore from cyst.

Pfiesteria

Pfiesteria is a genus of dinoflagellates, named for the late Dr. Lois Pfiester, a biologist who did much of the early research on the genus.

Pfiesteria (*Pfiesteria piscicida*) is a dinoflagellate which some researchers claimed was toxic and in large blooms was responsible for fish kills in the 1980s and 1990s on the coast of North Carolina and in tributaries of the Chesapeake Bay. While some experts say the algae is not toxic, peer-reviewed evidence exists showing that it is.

While some peer-reviewed research by government and university dinoflagellate experts has suggested that its living cycle is extremely complex (these scientists have found at least 24 different stages, spanning from cyst to several amoeba-like forms), other peer-reviewed research by government and university dinoflagellate experts has found only a simple life cycle with no toxic amoebic stage. The original research claimed that the organism apparently moves through many different stages as environmental conditions require. However, research from 2002 claims the cycle is much simpler than previously thought, and that the true Pfiesteria is non-toxic. Some peer-reviewed research based on advanced DNA analyses shows that the organism lacks certain genetic structure to be capable of making the type of toxic proteins associated with typical dinoflagellates.

However, researchers from the NOAA National Centres for Coastal Ocean Science, the National Institute of Standards and Technology, the Medical University of South Carolina, and the College of Charleston (S.C.) have formally isolated and characterised the toxin in the estuarine dinoflagellete Pfiesteria piscicida and also have identified how the organism transforms from a non-toxic to toxic state. The findings were published in the peer-reviewed science journal, Environmental Science and Technology, on January 11, 2007.

Pfiesteria Piscicida

Pfiesteria Piscicida is a dinoflagellate that some researchers claim is responsible for many blooms in the 1980s and 1990s on the coast of North Carolina. Pfiesteria is named after Lois Ann Pfiester (1936-1992), who researched dinoflagellates, and its species name means "fish-killer." The algae was discovered by JoAnn Burkholder at North Carolina State University. An in-depth story of her and her discovery can be found in *And the Waters Turned to Blood* by Rodney Barker.

Life Cycle

While some peer-reviewed research by government and university dinoflagellate experts has shown that its living cycle is extremely complex (these scientists have found at least 24 different stages, spanning from cyst to several amoeba-like forms), other peer-reviewed research by government and university dinoflagellate experts has found only a simple life cycle with no toxic amoebic stage. The original research claimed that the organism apparently moves through many different stages as environmental conditions require. However, the new research from 2002 claims the cycle is much simpler than previously thought, and that the true Pfisteria is non-toxic.

Toxicity

Some peer-reviewed research based on advanced DNA-analyses shows that the organism lacks certain genetic structure to be capable of making the type of toxic proteins associated with typical dinoflagellates. Researchers from the NOAA National Centres for Coastal Ocean Science, the National Institute of Standards and Technology, the Medical University of South Carolina, and the College of Charleston have formally isolated and characterised the toxin in the estuarine dinoflagellete Pfiesteria Piscicida and also have identified how the organism transforms from a non-toxic to toxic state. The findings were published in the peer-reviewed science journal, Environmental Science and Technology, on 11 January 2007.

Human Health

Very little research on the human health effects of Pfiesteria toxins has been conducted. At a multi-state workshop at the Centres for Disease Control and Prevention (CDC) in Atlanta, Georgia, at the end of September 1997, attendees agreed on clinical symptoms that characterise the adverse health consequences of exposure to Pfiesteria toxins. These clinical features include:

- memory loss
- confusion
- acute skin burning (on direct contact with water); or
- three or more of an additional set of conditions (headaches, skin rash, eye irritation, upper respiratory irritation, muscle cramps, and gastrointestinal complaints (i.e., nausea, vomiting, diarrhoea, and or abdominal cramps).

With these criteria and environmental qualifiers (e.g., 20 per cent of a 50-fish sample, all of the same species, have lesions caused by a toxin), it is likely that Pfiesteria-related surveillance data can better track potential illnesses.

Pfiesteria toxins have been blamed for causing adverse health effects in people who have come in close contact with waters where this organism is abundant. Since June 1997, the Maryland Department of Health and Hygiene has been collecting data from Maryland physicians through a state-wide surveillance system on illnesses suspected of being caused by Pfiesteria toxin. As of late October 1997, illness was reported by 146 persons who had been exposed to diseased fish or to waters that were the site of suspected Pfiesteria activity. Many of these persons are watermen and commercial fishermen.

The strongest evidence of adverse human health effects so far comes from case studies of two research scientists who were both overcome in their North Carolina laboratory in 1993. They still complain of adverse effects on their cognitive abilities, particularly after exercising. Duke University Medical Centre researchers conducted experiments on rats, which showed that the toxin appeared to slow learning but did not affect memory.

Pfiesteria shumwayae

Pfiesteria shumwayae is a dinoflagellate that was recently found in Taskinas Creek, a tributary of the York River in James City County, Virginia. The toxicity of *P. shumwayae* is unknown, although it is considered potentially dangerous. A recent Nature article suggests that *P. shumwayae* kills fish not by secreting an exotoxin but by micropredation. One implication is that the fears about possible human health risks may be overblown.

Prorocentrales

The Prorocentrales are a small order of dinoflagellates. They are distinguished by having their two flagella inserted apically, rather than ventrally as in other groups. One flagellum extends forward and the other circles its base, and there are no flagellar grooves. This arrangement is called desmokont, in contrast to the dinokont arrangement found in other groups.

Accordingly, the Prorocentrales may be called desmoflagellates, and in some classifications were treated as a separate class Desmophyceae.

All members have chloroplasts and a theca, which is composed of two large plates joined by a sagittal suture. This structure is shared with the Dinophysiales, and they are probably sister groups. This does not show up on rRNA trees, which also show two separate groups of *Prorocentrum*, but they leave their relationships mostly unresolved.

Syndiniales

The Syndiniales are an order of dinoflagellates, found exclusively as endosymbionts of marine animals and protozoa. The trophic form is often multinucleate, and ultimately divides to form motile spores, which have two flagella in typical dinoflagellate arrangement. They lack a theca and chloroplasts, and unlike all other orders, the nucleus is never a dinokaryon. A well-studied example is *Amoebophrya*, which is a parasite of other dinoflagellates and may play a part in ending red tides.

Zooxanthella

Zooxanthellae are golden-brown intracellular endosymbionts of various marine animals and protozoa, especially anthozoans. They are members of the phylum Dinoflagellata and are typically dinoflagellate algae, although algae such as diatoms can also be zooxanthellae. They may be acquired by direct ingestion, and subsequently reproduce by splitting apart; a process known as budding. In other cases, zooxanthellae may be transmitted by the coral eggs and planulae. Most are autotrophs and provide the host with energy in the form of translocated reduced carbon compounds derived from photosynthesis. Zooxanthellae can provide up to 90 per cent of a coral's energy requirements. In return, the coral provides the zooxanthellae with protection, shelter, and a constant supply of the carbon dioxide required for photsynthesis. Their population in the host tissue is limited by available nutrients and incident light, and by expulsion of excess cells. However, zooxanthellae do not appear to be digested by their hosts.

Hermatypic (reef-building) corals have zooxanthellae and are largely dependent on them, limiting their growth to the photic zone. The symbiotic relationship is probably responsible for the phenomenal success of corals as reef-building organisms in tropical waters. However, when corals are subjected to high environmental stress, they can lose their zooxanthellae by either expulsion or digestion and die. The process known as coral bleaching occurs when the zooxanthellae densities within the coral tissue become low or the concentration of photosynthetic pigments within each zooxanthella decline. Colour loss is also attributed to the loss or lowering of concentrations of Green Fluorescent Proteins (GFP) from the cellular pigments of the cnidarian itself. The result is a ghostly white calcareous skeleton, absent of zooxanthellae, with the inevitable death of the coral unless conditions improve, allowing for the zooxanthellae to return.

Coral are under constant disturbance, which is ultimately felt by the zooxanthellae living within their tissues. Exposure to air during extremely low tides or damage from intensifying solar radiation in shallow water environments are some of the ecological stressors zooxanthellae face. Temperature changes have provided the most stress to the zooxanthellae-coral relationship. A rise in temperature of 1-2 degrees Celsius for 5-10 weeks or a decline in temperature of 3-5 degrees Celsius for 5-10 days has resulted in a coral bleaching event. Strong temperature changes shock the zooxanthellae and cause them to suffer cell adhesion dysfunction which sees the detachment of the cnidarian endodermal cells from the zooxanthellae.

Other organisms which may have zooxanthellae include jellyfish, clams, sea slugs, and radiolaria. There are several different species of zooxanthellae, typically grouped together as the genus *Symbiodinium*, which appears to be monophyletic.

Bibliography

Abdelahad, N. and Bazzichelli G.: *Geitleria Calcarea Friedmann,* Cyanophycee Nouvelle, Italie, 1988.

Anderson R. and Adams M. S.: *Effect of Fire on Big Bluestem Production,* American Midland Naturalist, 1975.

Battarbee, R. W. and Gasse F.: *Past Climate Variability through Europe and Africa,* Springer, Netherlands, 2004.

Bebianno, M. J. and Company R. M.: *Antioxidant Enzymes and Lipid Peroxidation in Bathymodiolus Azoricus,* Aquatic Toxicology, London, 2005.

Berge, J. and Vader W.: *On the Taxonomic Status of the Antarctic Amphipod Genera Eclysis,* Organisms Diversity Evolution, 2005.

—————— : *The Amphipod Genus Alexandrella,* Journal of Natural History, Washington, 2005.

Blanchette, C. A.: *The Effects of Biomechanical and Ecological Factors on Population and Community Structure of wave-exposed,* Oregon State University, 1994.

Bold, H. C. and M. J. Wynne: *Introduction to the Algae,* Englewood Cliffs, New Jersey, 1978.

Boudouresque, C. F. and Perret-Boudouresque M. M.: *A Checklist of the Benthic Marine Algae of Corsica,* GIS Posidonie Publishers, Marseille, 1987.

Bourrelly, P. and Dupuy P.: *Quelques Stations Francaises de Geitleria Calcarea, Schweizerische Zeitsch, Fur Hydrologie,* MacMillan, New York, 1973.

Boyer, C. S.: *The Biddulphoid Forms of North American Diatomaceae,* Natural Sciences of Philadelphia, Boyer, 1901.

Brodie, J. A. and Irvine L. M.: *Seaweeds of the British Isles,* British Museum, London, 2003.

Chapman, V. J.: *The Algae,* St. Martin's Press, New York, 1964.

Christiansen, J. S. and Gildberg A.: *Pagophilus Groenlandicus,* ICES Journal of Marine Science, 2004.

Claus, G.: *Algae and (their) Mode of Life (in) on the Baradla Cave of (at) Aggtelek,* Scientiarum Hungaricae, Budapest, 1955.

Claus, G.: *Data on the Ecology of the Algae of Peace Cave,* Nova Hedwigia Zeitsch, Kryptogamenkunde, 1962.

—————— : *The Reevaluation of the Genus Gomontiella,* Algol, 1960.

Coppejans, E.: *Iconographie d'Algues Mediterraneennes,* J. Cramer, Vaduz, 1983.

Delepine, R., Boudouresque C. F. and Frada-Orestano C.: *Algues et Autres Vegetaux Marins,* Mediterranee et Mer Noire, Roma, 1987.

Denny, M. W.: *Mechanical Limits to Size in Wave-swept Organisms,* Ecological Monographs, 1985.

—————— : *Predicting Physical Disturbance,* Ecological Monographs,1995.

Dixon, P. and Irvine L. M.: *Seaweeds of the British Isles,* British Museum, Natural History, London, 1977.

Evenset, A., Christensen G. N., Skotvold T, Fjeld E. and Gregor D.: *A Comparison of Organic Contaminants in two High Arctic Lake Ecosystems,* Science of the Total Environment, Norway, 2004.

Falk, Petersen S., Timofeev S., Pavlov V. and Sargent J. R.: *Climate Variability and the Effect on Arctic Food Chains,* Springer Verlag, Berlin, 2006.

Fletcher, R. L.: *Seaweeds of the British Isles,* British Museum, London, 1995.

Furevik, T., Gerdes R. and Berger W.: *The Nordic Seas*, American Geophysical Union, Washington, 2005.

Gantt, Elisabeth: *Handbook of Phycological Methods*, Cambridge University Press, England, 1980.

Gayral, P.: *Les Algues des Cotes Francaises*, Editions Doin, Paris,1966.

Guiry, M. D.: *A Concensus and Bibliography of Irish Seaweeds*, J. Cramer, Vaduz, 1978.

Gulliksen, B., Slagstad D. and Pamapanin D.: *Mytilus Thermophily*, Marine Ecology Progress Series, 2004.

Hakansson, H. and Stoermer E. F.: *Observations on the Type Material of Stephanodiscus Hantzschii Grunow in Cleve and Grunow*, Nova Hedwigia, 1984.

Hakansson, H.: *A Light and Elcctron Microscopical Investigation of the Type Species of Cyclotella*, Diatom Research, 1989.

Hanna, G. D.: *Pliocene Diatoms of Wallace County*, University of Kansas Science Bulletin, Kansas, 1932.

Haworth, E. Y. and M. A. Hurley.: *Comparison of the Stelligeroid Taxa of the Centric Diatom Genus Cyclotella*, O. Koeltz, Koenigstein, 1986.

Hellebust, J. A. and Craigie J. S.: *Handbook of Phycological Methods*, Cambridge University Press, Cambridge, 1978.

Hishock, S.: *A Field Key to the British Brown Seaweeds*, F.S.C. Publications, Shrewsbury, 1979.

Hoagland, K. D., Roemer S. C. and Rosowski J. R.: *Colonization and Community Structure of two Periphyton Assemblages*, American Journal of Botany, 1982.

Hollenberg G. J.: *Marine Algae of California*, Stanford University Press, California, 1976.

Hop, H. and Falk-Petersen S.: *Seasonal Lipid Dynamics and Life Strategy*, Marine Biology, 2005.

Irvine, L. M. and Chamberlain Y. M.: *Seaweeds of the British Isles*, British Museum, London, 1994.

————————: *Seaweeds of the British Isles*, British Museum, London, 1983.

Kattner, K., Graeve M.: *The Arctic Pteropod Clione Limacina*, Marine Biology, 2005.

Kociolek, J. P. and Stoermer E. F.: *Ultrastructure of Cymbella Sinuata and its Allies*, Systematic Botany, 1987.

Kolosov, P. N.: *Upper Precambrian Paleoalgological Remains of the Siberian Platform*, Moscow,1982,

Krammer, K. and H. Lange Bertalot: *Bacillariaceae, Epithemiaceae, Surirellaceae*, Gustav Fischer Verlag, New York, 1988.

———————— : *Naviculaceae*, Gustav Fischer Verlag, New York, 1986.

Leake, D.V.: *The Algae of Crystal Lake*, Cleveland County, Oklahoma,1945.

Lebreton, B.: *World Bibliography on Subterranean Algae*, Bibliographie Mondiale des Algae Souterraines, 2000.

Lee, R. E.: *Phycology*, Cambridge University Press, Cambridge, 1980.

Littler, M. and Littler D.: *Handbook of Phycological Methods*, Cambridge University Press, Cambridge, 1985.

Lobban, Christopher S. and Harrison P. J.: *Seaweed Ecology and Physiology*, Cambridge University Press, Cambridge, 1994.

Makarikhin, V. V. and Kononova, G.: *Phytolites of the Karelian Lower Proterozoic*, Leningrad, Nauka, 1983,

Mann, D. G. and Stickle A. J.: *The Genus Craticula*, Diatom Research, 1991.

Mikhailova, N. S. and German, T. N.: *5th All Union Colloquium on Microfossils of the USSR Precambrian*, Izvestia Akadamie Nauk, 1987,

Nelson, E. N., Richardson J. K. and Bailey H. H.: *Aspects of the Occurrence of Aspidobothrid Parasites*, Proceedings of the Oklahoma Academy of Science, 1975.

Newman, G. A.: *Cowbird Parasitism and Nesting Success of Lark Sparrows in Southern Oklahoma*, Wilson Bull, 1970.

Newton, L.: *A Handbook of the British Seaweeds,* British Museum, London, 1931.

Oliver, G.: *Black Rat Snake Predation Upon Nesting Barn and Cliff Swallows,* Ornithological Society, Oklahoma,1970.

Osborn, R. S.: *Observations on the Spawning Ecology of Buffalos,* Oklahoma Academy of Science,1966.

Pampanin D. M., Camus L., Gomiero A., Marangon I., Volpato E. and Nasci C.: *Susceptibility to Oxidative Stress of Mussels,* Venice Lagoon, Italy, 2005.

Pamperl, M. A.: *Diatoms of Lake of the Ozarks,* Transactions, Missouri Academy of Science, Missouri, 1980.

Parks, J. M. and Barclay H. G.: *The Increasing Importance of Vines in Southern Oklahoma,* Oklahoma Academy of Science, 1966.

Parmelee, D. F.: *The Breeding Behavior of the Painted Bunting in Southern Oklahoma,* Bird-Banding, 1959.

Patten, B. C., Egloff D. A. and Richardson T. H.: *Total Ecosystem Model for a Cove in Lake Texoma,* Academic Press, 1975.

Patten, B. C.: *A Reservoir Cove Ecosystem Model,* Transactions of the American Fisheries Society, 1975.

————————: *Ecosystem Modeling and Reservoir Management,* Oklahoma Academy of Science, 1976.

Paulissen, M. A.: *A Technique for Marking Teiid Lizards for Field Observation,* Herpetology Review, 1985.

———————— : *Cnemidophorus Sexlineatus,* Journal of Herpetology, 1988.

Penfound, W. T., Shed J. S. and Jennisons M. C.: *A Plant Community Dominated by Vines,* Oklahoma Academy of Science, 1965.

Penfound, W. T.: *The Composition of a Black Willow Forest in South Central Oklahoma,* Oklahoma Academy of Science, 1961.

Pfiester, L.A. and Holt J.: *A Freshwater "Red Tide" in Texas,* Southwestern Naturalist, 1978.

Phillipi, A. and Tyrl R. J.: *Reproductive Biology of Proboscidea Louisianica (Martyniacea),* Rhodora, 1979.

Pireh, W.: *Achillea Millefolium,* Oklahoma State Universi, Oklahoma, 1978.

Power, M. E. and Matthews W. J.: *Grazer Control of Algae in an Ozark Mountain Stream: Effects of Short-term Exclusion,* Ecology, 1988.

Proctor, V. W.: *Dispersal of Fresh-water Algae Water,* Science, 1959.

————— : *The Growth of Basicladia on Turtles,* Ecology, 1958.

Prud, Homme: *A Taxonomic Revision of the European Sphacelariaceae,* Leiden University Press, Leiden, 1982

Purdue, J. R.: *The Phylogenetic Implications of Body Movements in Displaying Sceloporine Lizards,* Herpetology Review, 1970.

————— : *The Western Sandpiper in Oklahoma,* Oklahoma Ornithological Society, 1969.

Ratkova T. and Wassmann P.: *Sea Ice Algae in the White and Barents Sea,* Polar Research, 2005.

Reinke, D. C.: *New and Interesting Algae from Kansas,* State Biological Survey of Kansas, 1979.

Riedl, R.: *Fauna und Flora des Mitterlmeeres,* Verlag Paul Parey, Hamburg, 1983

Roemer, S. C., Hoagland K. D. and Rosowski J. R.: *Development of a Freshwater Periphyton Community as Influenced* Canadian Journal of Botany, 1984.

Rolbiecki L. and Walkusz W.: *Derogenes varicus,* Oceanological and Hydrological Studies, West Spitsbergen, 2005.

Round, F. E., Crawford R. M. and Mann D. G.: *The Diatoms: Biology and Morphology of the Genera,* Cambridge University Press, Cambridge, 1990.

————— : *The Diatom Genus Stephanodiscus,* Archiv fur Protistenkunde, 1981.

Santi-Mazzini G.: *Aquarium,* Gribaudo, Cavallermaggiore, 2000

Setchell, W. A. and Gardner N. L.: *The Marine Algae of the Pacific Coast of North America*, University of California Press, California, 1925.

Shenfil, V. Y.: *Algae in the Riphean Deposits of the Yenesei Ridge*, Doklady Akad, Nauk, 1978,

South, G. Robin and A. Whittick.: *Introduction to Phycology*, Blackwell Scientific Publications, Oxford,1987.

Stegenga H., Bolton J. J., Anderson R. J.: *Seaweeds of the South African Coast,* University of Cape Town, 1997.

Stein R. and Macdonald R. M.: *The Organic Carbon Cycle in the Arctic Ocean*, Springer, New York, 2004.

Theriot, E., Stoermer E. and Hakansson H.: *Taxonomic Interpretation of the Rimoportula of Freshwater Genera in the Centric Diatom Family Thalassiosiraceae*, Diatom Research, 1987.

Troeger, W. W. and Menzel R. G.: *Benthic Diatoms in two Eutrophic Flood Detention Reservoirs in Oklahoma,* Journal of Freshwater Ecology, 1986.

Troeger, W. W.: *Colonization of Benthic Diatoms in a New Farm Pond and its Feeder Creek*, Southwestern Naturalist, 1983.

————————: *Epiphytic Diatoms in Farm Ponds and Experimental Ponds in Bryan County*, Oklahoma Academy of Science, Oklahoma, 1978.

Usher, M. B., Callaghan T. V.: *Principles of Conserving the Arctic's Biodiversity*, Cambridge University Press, Cambridge, 2005.

Van, Landingham, S. L.: *Catalog of the Fossil and Recent Genera and Species of Diatoms and their Synonyms*, Cramer, Lehre, 1979.

Walcott, C. D.: *Middle Cambrian Algae*, Smithsonian Miscellaneous Collections, 1919.

Walter, M. R.: *Archean Stromatolites*, Princeton University Press, Princeton,1983,

Webber, H. J.: *Catalogue of the Flora of Nebraska,* Nebraska Board of Agriculture, Lincoln, 1890.

———————: *The Fresh Water Algae of the Plains*, American Naturalist, 1889.

Wenke, T. L. and Eberle M. E.: *Diatoms from Streams in Ellis and Russell Counties, Kansas,* Transactions of the Kansas Academy of Science, 1986.

———————: *Diatoms from Kansas,* Southwestern Naturalist, 1985.

Werner, D.: *The Biology of Diatoms,* University of California Press, Berkeley, California, 1977.

Wolle, F.: *Diatomaceae of North America,* Comenius Press, Bethlehem, Pennsylvania, 1894.

———————: *Fourth Contribution to the Knowledge of Kansas Algae,* Bulletin of the Washburn College Laboratory of Natural History, 1889.

Wujek, D. E.: *Contribution to the Diatom Flora of Kansas,* Southwestern Naturalist, 1965.

Yankauskas, T. V.: *Upper Precambrian and Cambrian Plant Microfossils of the European USSR and their Stratigraphic Significance,* University of Moscow, Moscow, 1982,

Zmijewska M. I. and Bielecka L.: *Characteristics of the Arctic and Antarctic mesozooplankton in the neritic zone during summer,* Polish Polar Research, 2004.

Index

A

B

Index

A

B

G

H

Q

R

S

T

U

V

W

□□□